EMBRY-RIDDLE AERONAUTICAL UNIVERSITY
3 1932 00032 4688

D0760890

Semiconductor Physical Electronics

MICRODEVICES
Physics and Fabrication Technologies

Series Editors: Ivor Brodie and Julius J. Muray†

SRI International
Menlo Park, California

ELECTRON AND ION OPTICS
Miklos Szilagyi

GaAs DEVICES AND CIRCUITS
Michael Shur

ORIENTED CRYSTALLIZATION ON AMORPHOUS SUBSTRATES
E. I. Givargizov

THE PHYSICS OF MICRO/NANO-FABRICATION
Ivor Brodie and Julius J. Muray

PHYSICS OF SUBMICRON DEVICES
David K. Ferry and Robert O. Grondin

THE PHYSICS OF SUBMICRON LITHOGRAPHY
Kamil A. Valiev

SEMICONDUCTOR LITHOGRAPHY
Principles, Practices, and Materials
Wayne M. Moreau

SEMICONDUCTOR PHYSICAL ELECTRONICS
Sheng S. Li

† *Deceased.*

A Continuation Order Plan is available for this series. A continuation order will bring delivery of each new volume immediately upon publication. Volumes are billed only upon actual shipment. For further information please contact the publisher.

Semiconductor Physical Electronics

Sheng S. Li

Department of Electrical Engineering
University of Florida
Gainesville, Florida

Plenum Press • New York and London

Library of Congress Cataloging-in-Publication Data

Li, Sheng S., 1938-
Semiconductor physical electronics / Sheng S. Li.
p. cm. -- (Microdevices)
Includes bibliographical references and index.
ISBN 0-306-44157-8
1. Semiconductors. 2. Solid state physics. I. Title.
II. Series.
TK7871.85.L495 1993
621.3815'2--dc20
92-26807
CIP

ISBN 0-306-44157-8

A Division of Plenum Publishing Corporation
233 Spring Street, New York, N.Y. 10013

Printed in the United States of America

Preface

The purpose of this book is to provide the reader with a self-contained treatment of fundamental solid state and semiconductor device physics. The material presented in the text is based upon the lecture notes of a one-year graduate course sequence taught by this author for many years in the Department of Electrical Engineering of the University of Florida. It is intended as an introductory textbook for graduate students in electrical engineering. However, many students from other disciplines and backgrounds such as chemical engineering, materials science, and physics have also taken this course sequence, and will be interested in the material presented herein. This book may also serve as a general reference for device engineers in the semiconductor industry.

The present volume covers a wide variety of topics on basic solid state physics and physical principles of various semiconductor devices. The main subjects covered include crystal structures, lattice dynamics, semiconductor statistics, energy band theory, excess carrier phenomena and recombination mechanisms, carrier transport and scattering mechanisms, optical properties, photoelectric effects, metal–semiconductor devices, the p–n junction diode, bipolar junction transistor, MOS devices, photonic devices, quantum effect devices, and high-speed III–V semiconductor devices. The text presents a unified and balanced treatment of the physics of semiconductor materials and devices. It is intended to provide physicists and materials scientists with more device backgrounds, and device engineers with a broader knowledge of fundamental solid state physics.

The contents of the text is divided into two parts. In Part I (Chapters 1–9), we cover subjects on fundamental solid state and semiconductor physics that are essential for device applications. In Part II (Chapters 10–15), we deal with basic device physics and structure, operation principles, general characteristics, and the applications of various semiconductor devices.

Chapter 1 presents the classification of solids, crystal structures, the concept of reciprocal lattice and Brillouin zone, the definition of Miller indices, chemical bondings, and crystal defects. Chapter 2 deals with the thermal properties and lattice dynamics of crystalline solids. The lattice specific heat, the dispersion relation of lattice vibrations, and the concept of phonons are also described. Chapter 3 is concerned with the derivation of the three semiconductor statistics, namely, the Maxwell–Boltzmann (M–B), Bose–Einstein (B–E), and Fermi–Dirac (F–D) distribution functions. Chapter 4 covers energy band theory, the concept of effective mass, and the density-of-states function for the bulk semiconductor and the superlattice. Chapter 5 describes the equilibrium properties of both the intrinsic and extrinsic semiconductors. Derivation of electron and hole densities, and a discussion of the properties of shallow- and deep-level impurities are also covered. Chapter 6 presents the excess carrier phenomenon and recombination mechanisms in a semiconductor. The six basic equations which govern the transport of excess electrons and holes in a semiconductor are described. Chapter 7 is concerned with the derivation of transport coefficients from the Boltzmann equation, and low-field galvanomagnetic effects in a semiconductor. Chapter 8 deals with various scattering mechanisms in a semiconductor. The relaxation time and mobility expressions for ionized and neutral impurity scattering, acoustic and optical phonon scattering are derived. Chapter 9 is concerned with the optical properties and photoelectric effects in a semiconductor. The fundamental optical absorption and free carrier absorption processes, the photoconductive, photovoltaic, and photomagnetoelectric effects in a semiconductor are also discussed. Chapter 10 covers the basic physical principles and properties of metal–semiconductor contacts. Both the Schottky barrier and ohmic contacts on a semiconductor are discussed. Carrier transport in a Schottky barrier diode, methods of determining and modifying the barrier height, and the formation of ohmic contacts are presented. Chapter 11 deals with the basic device theory and properties of a p–n junction diode and a junction-field effect transistor (JFET). Chapter 12 is concerned with the device physics, the structure and characteristics of various photonic devices such as photodetectors, solar cells, light emitting diodes (LEDs) and diode lasers. Chapter 13 describes the bipolar junction transistor (BJTs) and p–n–p–n four-layer devices (e.g., SCRs or thyristers). Chapter 14 presents the silicon-based metal–oxide–semiconductor (MOS) devices. The device physics and characteristics for both the MOS field-effect transistors (MOSFETs) and charge-coupled devices (CCDs) are also described. Finally, in Chapter 15, we cover some novel high-speed devices using GaAs and other III–V compound semiconductors. These include GaAs-based metal–semiconductor field effect transistors (MESFETs), high electron mobility transistors (HEMTs), heterojunction bipolar transistors (HBTs), and transferred electron devices (TEDs).

Throughout the text, the author stresses the importance of relating fundamental solid state physics to the properties and performance of various semiconductor devices. Without a good grasp of the physical concepts and an understanding of the underlying device physics, it would be difficult to tackle the problems encountered in material growth and device fabrication. The information presented in this book should provide a solid background for understanding the fundamental physical limitations of semiconductor materials and devices. This book is especially useful for those who are interested in strengthening and broadening their basic knowledge of solid state and semiconductor device physics.

The author would like to acknowledge his former students for the many useful discussions and comments on the text. In particular, he is grateful to Drs. D. H. Lee and C. S. Yeh, and Robert Huang, for their assistance in the preparation of illustrations and the solutions manual, and the proofreading of the manuscript. He is indebted to his son, Jim, for setting up the MicroTex and TEXtures programs with which this book was prepared, and to his wife, Jean, and daughters Grace and Jeanette for their love and support during the course of preparing this book.

Sheng S. Li

Gainesville, Florida

Contents

CHAPTER 4. Energy Band Theory

CHAPTER 5. Equilibrium Properties of Semiconductors

CHAPTER 6. Excess Carrier Phenomenon in Semiconductors

CHAPTER 7. Transport Properties of Semiconductors

CHAPTER 8. Scattering Mechanisms and Carrier Mobilities in Semiconductors

CHAPTER 9. Optical Properties and Photoelectric Effects

CHAPTER 10. Metal–Semiconductor Contacts

CHAPTER 11. p–n Junction Diodes

CHAPTER 12. Photonic Devices

CHAPTER 13. Bipolar Junction Transistor

CHAPTER 14. Metal–Oxide–Semiconductor Field-Effect Transistors

CHAPTER 15. High-Speed III–V Semiconductor Devices

Semiconductor Physical Electronics

1

Classification of Solids and Crystal Structure

1.1. INTRODUCTION

Classification of solids can be based upon atomic arrangement, binding energy, physical and chemical properties or the geometrical aspects of the crystalline structure. In one class, the atoms in a crystalline solid are set in an irregular manner, without any short- or long-range order in their atomic arrangement. This class of solids is commonly known as noncrystalline or amorphous materials. In another class, the atoms or group of atoms of the solid are arranged in a regular order. These solids are usually referred to as crystalline solids. The crystalline solids can be further divided into two categories: the single crystalline and the polycrystalline solids. In a single crystalline solid, the regular order extends over the entire crystal. In a polycrystalline solid, however, the regular order only exists over a small region of the crystal, ranging from a few hundred angstroms to a few centimeters. A polycrystalline solid contains many of these small single crystalline regions surrounded by the grain boundaries. Distinction between these two classes of solids—amorphous and crystalline—can be made through the use of X-ray or electron diffraction techniques.

Classification of solids can also be made according to their electrical conductivity. For example, while the electrical conductivity of an insulator is usually less than 10^{-8} mho/cm, the electrical conductivity of a metal is on the order of 10^{6} mho/cm at room temperature. As for a semiconductor, the room-temperature electrical conductivity may vary from 10^{-4} to 10^{4} mho/cm, depending on the doping impurity density in the semiconductor. Furthermore, the temperature behavior of a metal and a semiconductor may be quite different. For example, the electrical conductivity of a metal is nearly independent of the temperature (except at very low temperatures), while the electrical conductivity of a semiconductor is a strong function of the temperature.

In this chapter, we are concerned with the classification of crystalline solids based on their geometrical aspects and binding energies. Section 1.2 presents the seven crystal systems and fourteen Bravais lattices. The crystal structure, the concept of reciprocal lattice and Brillouin zone, and the definition of Miller indices are described separately in Section 1.3 through Section 1.5. Section 1.6 discusses the classification of solids according to the binding energy of their crystalline structure. Of particular interest is the fact that many important physical properties of a solid can be understood, at least qualitatively, in terms of its binding energy. Finally, defects in a semiconductor including vacancies, interstitials, impurities, dislocations, and grain boundaries are depicted in Section 1.7. It should be noted that these defects play an important role in controlling the physical and electrical properties of a semiconductor.

1.2. THE BRAVAIS LATTICE

In a crystalline solid, the atoms or group of atoms are arranged in an orderly or periodic pattern. It can be distinguished from all other aggregates of atoms by the three-dimensional periodicity of the atomic arrangement. Thus, by properly choosing a small polyhedron as a basic building block, it is possible to construct the entire crystal by repeatedly displacing this basic building block along the three noncoplanar directions of the crystal lattice by translational operation. The suitable geometrical shapes for the basic building blocks are a regular cubic dodecahedron, a truncated octahedron, and any arbitrary parallelepiped. The basic building block of a crystal is called the unit cell. Although a variety of unit cells may be chosen for a particular crystalline structure, there is generally one which is both the most convenient and the most descriptive of the structure. If the shape of the unit cell is specified, and the arrangement of all the atoms within the unit cell is known, then one has a complete geometrical description of the crystal. This is due to the fact that there is only one way in which the unit cells can be stacked together to fill the entire space of the crystal.

The various possible arrangements of unit cells in a crystalline solid can be readily achieved by means of the space lattice, a concept introduced by Bravais. The space lattice is an arrangement of lattice points in space such that the placement of points at any given point in the space is the same for all points of the lattice. In general, the periodic translational symmetry of a space lattice may be described in terms of three primitive, noncoplanar basis vectors, $\mathbf{b}_1, \mathbf{b}_2, \mathbf{b}_3$, defined in such a way that any lattice point, $r(n_1, n_2, n_3)$, can be generated from any other lattice point, $r(0, 0, 0)$, in the space by translational operation

$$r(n_1, n_2, n_3) = r(0, 0, 0) + \mathbf{R} \tag{1.1}$$

where

$$\mathbf{R} = n_1\mathbf{b}_1 + n_2\mathbf{b}_2 + n_3\mathbf{b}_3 \tag{1.2}$$

is the translational vector, and n_1, n_2, n_3 are arbitrary integers. A lattice generated by such a translational operation is called the simple Bravais lattice, and the parallelepiped spanned by the three basis vectors, $\mathbf{b}_1, \mathbf{b}_2$, and $\mathbf{b}_3$ is called the unit cell of the Bravais lattice. Figure 1.1 shows a parallelepiped unit cell defined by the length of three basis vectors, $\mathbf{b}_1, \mathbf{b}_2, \mathbf{b}_3$ and the three angles, α, β, and γ. The smallest unit cell (in volume) in a Bravais lattice, which is formed by the three noncoplanar primitive basis vectors with lattice points located only at the vertices of the parallelepiped, is called the primitive cell. The number of atoms per primitive cell is given by the number of atoms in the basis.

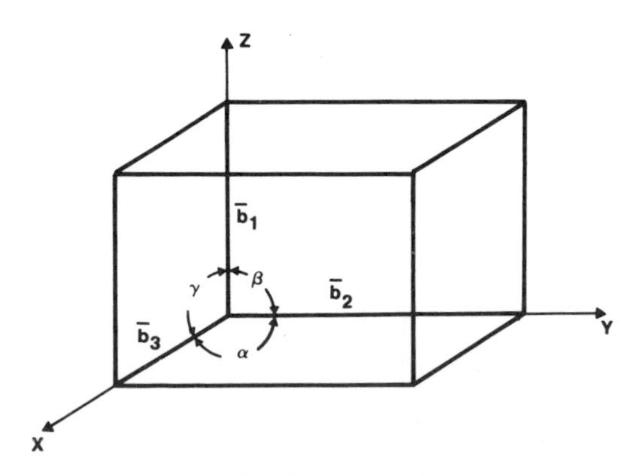

FIGURE 1.1. A parallelepiped unit cell for a Bravais lattice.

Depending on the lengths of three noncoplanar basis vectors, the angles between them, and the number of lattice points in a unit cell, the space lattice may be divided into seven lattice systems and fourteen Bravais lattices. These lattice systems and Bravais lattices, which are generated by using Eq. (1.1), are shown in Table 1.1 and Fig. 1.2, respectively. The fourteen Bravais lattices, which are all that are known to exist in nature, are formed by using all the possible arrangements of lattice points in each unit cell. Each of the Bravais lattices is unique in that it cannot be generated by any of the other thirteen Bravais lattices. Rather, they are generated through any combination of a simple lattice, a base-centered lattice, a face-centered lattice, and a body-centered lattice. Each of the Bravais lattices has different symmetry properties. A simple Bravais lattice contains lattice points only at the vertices of a parallelepiped. A base-centered Bravais lattice has lattice points located at the center of the top and bottom faces as well as at the vertices of the unit cell. In a face-centered Bravais lattice, in addition to the vertex lattice points, it has lattice points located at the center of all six faces. Finally, a body-centered Bravais lattice has an extra lattice point located at the volume center of the unit cell.

Symmetry is a very important consideration in the crystalline solids because many of the physical, electrical, magnetic, elastic, and thermal properties of the solids are strongly dependent on the symmetry properties of their crystal lattice. For example, the electrical conductivity of a cubic crystal is isotropic and independent of its crystalline orientations, while the electrical conductivity of a trigonal crystal can be highly anisotropic along different crystalline axes. The symmetry of a real crystal is determined by the symmetry of its basis and of the Bravais lattice to which the crystal belongs. In addition to the translational symmetry, each Bravais lattice may have different degrees of rotational, reflection, and inversion symmetry. The rotational symmetry of a crystal lattice is obtained when rotation about a certain crystal axis through an angle of $2\pi/n$ radians leaves the lattice invariant. The lattice is said to have an n-fold rotation axis. Due to the requirements of translational symmetry, the possible values of n are limited to 1, 2, 3, 4, and 6. There is no fivefold rotational symmetry in a Bravais lattice. Examples of rotation axes can be seen in the (100) axes of a cubic crystal which has a fourfold rotational symmetry, and in the body diagonal (111) axis which has a threefold rotational symmetry. Another type of symmetry, known as the reflection symmetry, is possessed by a crystal lattice when it is invariant under reflection in a plane through the lattice. For example, the six faces of a cubic lattice are the reflection planes for that lattice. Finally, it is noted that all Bravais lattices possess the inversion symmetry. A crystal lattice with an inversion symmetry will remain invariant despite the replacement of every lattice point at the coordinate r with the point at $-r$. Although all monatomic crystals do have a center of

TABLE 1.1. Seven Lattice Systems and Fourteen Bravais Lattices

Lattice systems	Angles and basis vectors	Bravais lattices
Triclinic	$b_1 \neq b_2 \neq b_3$ $\alpha \neq \beta \neq \gamma$	Simple
Monoclinic	$b_1 \neq b_2 \neq b_3$ $\alpha = \beta = 90° \neq \gamma$	Simple, base-centered
Orthorhombic	$b_1 \neq b_2 \neq b_3$	Simple, based-centered body-centered, face-centered
Tetragonal	$b_1 = b_2 \neq b_3$ $\alpha = \beta = \gamma = 90°$	Simple, body-centered
Trigonal	$b_1 = b_2 = b_3$ $\alpha = \beta = \gamma \neq 90°$	Simple
Hexagonal	$b_1 = b_2 \neq b_3$ $\alpha = \beta = 90°, \gamma = 120°$	Simple
Cubic	$b_1 = b_2 = b_3$ $\alpha = \beta = \gamma = 90°$	Simple, body-centered, face-centered.

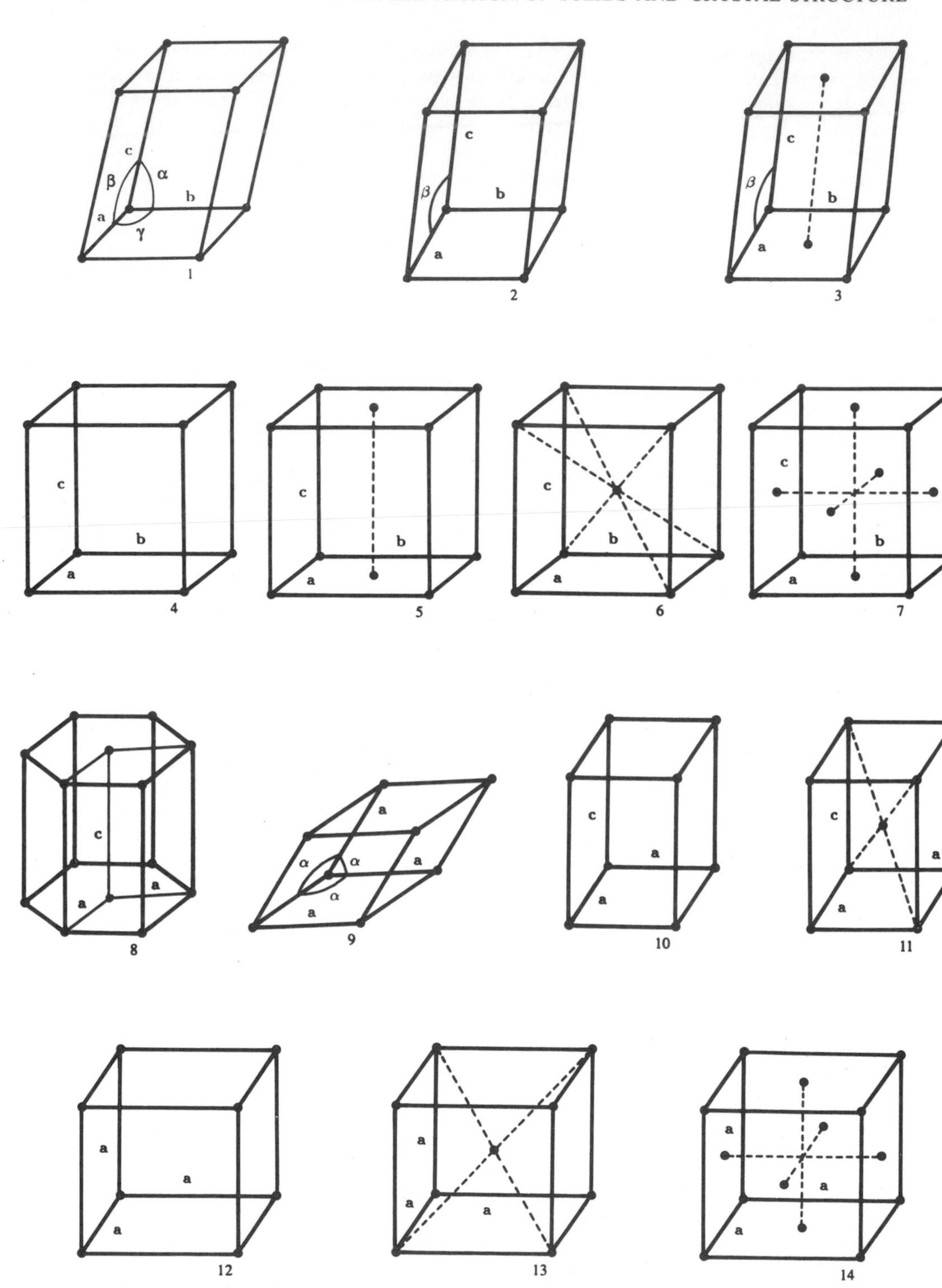

FIGURE 1.2. The fourteen Bravais lattices: (1) triclinic, simple; (2) monoclinic, simple (3) monoclinic, base-centered; (4) orthorhombic, simple; (5) orthorhombic, base-centered; (6) orthorhomic, body-centered; (7) orthorhombic, face-centered; (8) hexagonal; (9) rhombohedral; (10) tetragonal, simple; (11) tetragonal, body-centered; (12) cubic, simple; (13) cubic, body-centered; (14) cubic, face-centered.

inversion, this type of symmetry is not a general property of all crystals. The different types of symmetries that a crystalline solid possessed can be identified through the use of the X-ray diffraction technique.

1.3. THE CRYSTAL STRUCTURE

The Bravais lattice discussed in the preceding section is a mathematical abstraction which describes the periodic arrangement of lattice points in space. In general, a real crystal is not a perfect replica of a Bravais lattice with identical atoms at every lattice point. In fact, there is generally a set of atoms, whose internal symmetry is restricted only by the requirement of the translational periodicity, which must be associated with each lattice point of the corresponding Bravais lattice. This set of atoms is known as the basis, and each basis of a particular crystal is identical in composition, arrangement, and orientation. A crystalline structure is formed when a basis of atoms is attached to each lattice point in the Bravais lattice. Figure 1.3a and b show the distinction between a space lattice and a crystal structure. Many metals and semiconductors have a simple crystal structure with high degrees of symmetry. For example, alkali metals such as lithium, sodium, and potassium have the face-centered cubic (FCC) structure, while elemental and compound semiconductors have either the diamond, zinc blende, or wurtzite structure. Figure 1.4 shows the four most commonly observed crystal structures in a semiconductor. The diamond structure shown in Fig. 1.4a is actually formed by two interpenetrating face-centered cubic lattice with the vertex atom of one FCC sublattice located at (0, 0, 0) and the vertex atom of another FCC sublattice located at $(a/4\ a/4, a/4)$, where a is the lattice constant. In the diamond lattice structure, the primitive basis of two identical atoms located at (0, 0, 0) and $(a/4, a/4, a/4)$ is associated with each lattice point of the FCC lattice. Elemental semiconductors such as silicon and germanium belong to this crystal structure. The zinc blende structure shown in Fig. 1.4b is similar to the diamond structure except that the two FCC sublattices are occupied alternatively by two different kinds

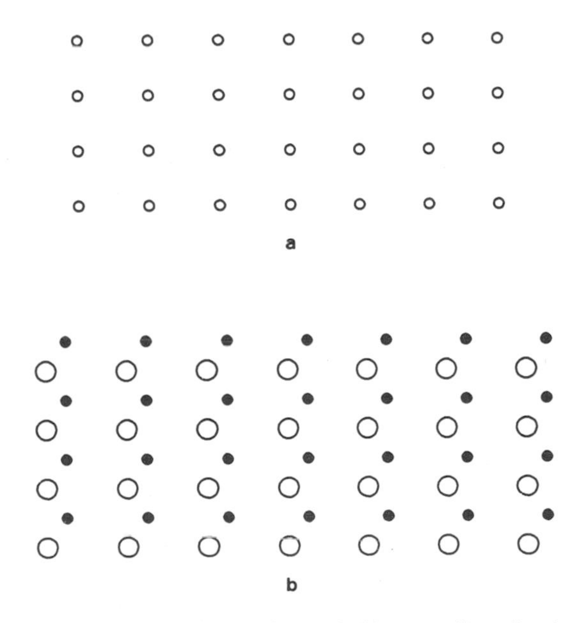

FIGURE 1.3. (a) A two-dimensional space lattice, and (b) a two-dimensional crystal structure with basis of atoms attached to each lattice point.

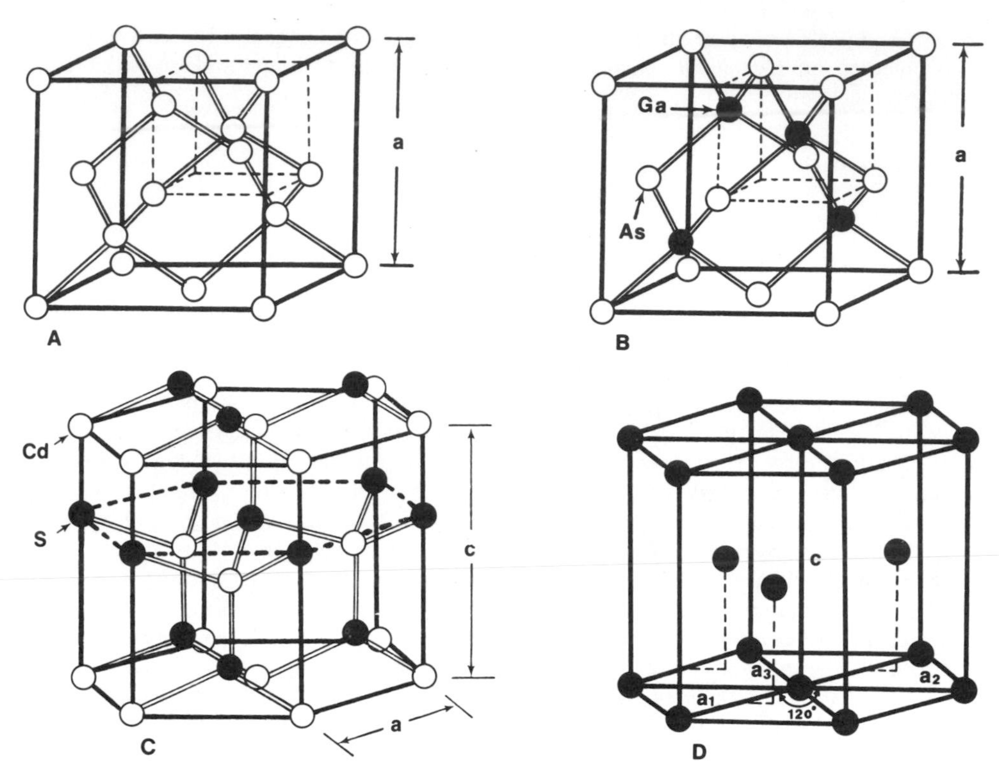

FIGURE 1.4. Four important crystal structures in semiconductors: (A) diamond structure, (B) zinc blende structure, (C) wurtzite structure, and (D) hexagonal closed-packed structure.

of atoms (e.g., Ga and As in a GaAs crystal). III-V compound semiconductors such as GaAs, InP, and InSb have zinc blende structure. The wurtzite structure shown in Fig. 1.4c is formed by two interpenetrating hexagonal closed-packed structures occupied alternatively by two different kinds of atoms. II–VI compound semiconductors such as CdS, CdTe, ZnS, and ZnSe have this type of crystalline structure. Both the diamond and zinc blende structures belong to the tetrahedral phase with each atom surrounded by four equidistant nearest-neighbor atoms at the vertices of a tetrahedron. Figure 1.4d shows a hexagonal closed-packed structure. It should be noted that some of the III–V and II–VI compound semiconductors including GaP, ZnS, and CdSe may be crystallized either in a zinc–blende or a wurtzite structure. Table 1.2 lists the crystal structures and the lattice constants for some important semiconductors.

1.4. MILLER INDICES AND THE UNIT CELL

The orientation of a crystal plane can be determined by three numbers, h, k, and l, known as the Miller indices. They are related to the orientations of a crystal plane in the following manner: If h', k', and l' represent the intercepts of a particular crystal plane on the three crystal axes (i.e., x, y, z) in units of lattice constant, a, then the three smallest integers h, k, and l which satisfy the relation

$$hh' = kk' = ll' \tag{1.3}$$

TABLE 1.2. Crystal Structures and Lattice Constants for Some Elemental and Compound Semiconductors

Semiconductors	Elements	Lattice structure	Lattice constant (Å)
Elemental semiconductors	Ge	Diamond	5.62
	Si	Diamond	5.43
IV–IV Semiconductor	SiC	Zinc blende	4.36
III–V Compound semiconductors	GaAs	Zinc blende	5.65
	GaP	Zinc blende	5.45
	GaP	Wurtzite	$a = 5.18$, $c = 5.17$
	InP	Zinc blende	5.87
	InAs	Zinc blende	6.06
	InSb	Zinc blende	6.48
II–VI Compound semiconductors	CdS	Zinc blende	5.83
		Wurtzite	$a = 4.16$, $c = 6.75$
	CdSe	Zinc blende	6.05
		Wurtzite	$a = 4.30$, $c = 7.01$
	CdTe	Zinc blende	6.48
	ZnSe	Zinc blende	5.88
	ZnS	Zinc blende	5.42
		Wurtzite	$a = 3.82$, $c = 6.26$
IV–VI Compound semiconductors	PbS	Cubic	5.93
	PbTe	Cubic	6.46

are called the Miller indices. As an example, Fig. 1.5 shows an arbitrary plane which intercepts the three crystal axes at $h' = 2a$, $k' = a$, and $l' = a$, where a is the lattice constant. In this case, the smallest integers which satisfy Eq. (1.3) are $h = 1$, $k = 2$. and $l = 2$. These three integers are called the Miller indices, and the plane defined by them is called the (122) plane. If a plane is parallel to one of the crystal axes with no interception, then the corresponding Miller index for that axis is zero (i.e., $k' \to \infty$, and $k = 0$). For example, a plane set parallel to the y–z plane and intercepted at the x-axis is called the (100) plane. Furthermore, a set of equivalent planes can be represented collectively by enclosing the Miller indices with a brace. For example, the $\{100\}$ planes represent a set of family planes consisting of the (100), (010), (001), $(\bar{1}00)$, $(0\bar{1}0)$, and $(00\bar{1})$ planes. The bar on the top of a particular Miller index represents a plane

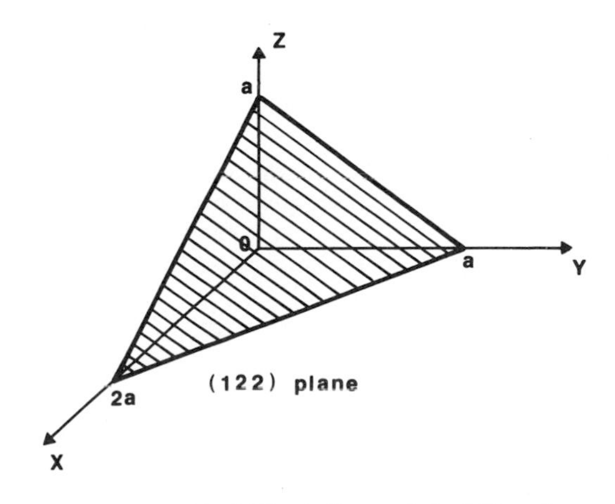

FIGURE 1.5. The Miller indices and the lattice plane.

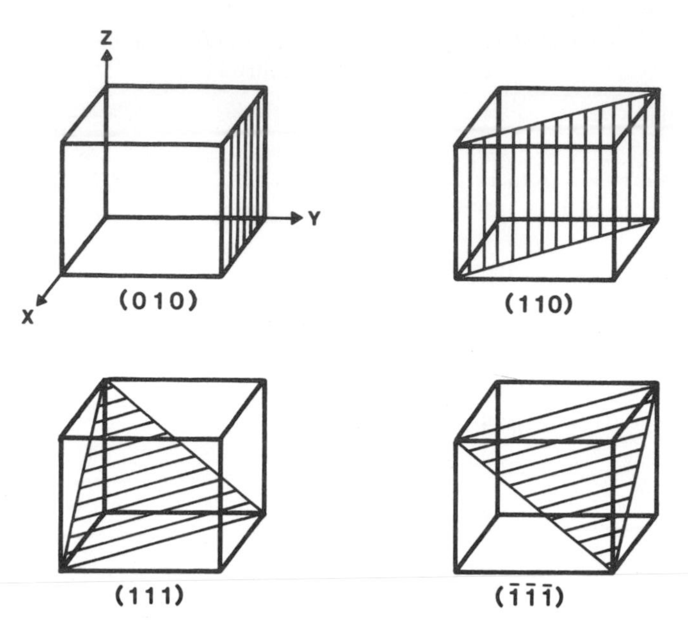

FIGURE 1.6. Some lattice planes in a cubic crystal.

which is intercepted at a negative crystal axis. Figure 1.6 shows some crystal planes for a cubic crystal.

1.5. THE RECIPROCAL LATTICE AND BRILLOUIN ZONE

The Bravais lattice described in Section 1.2 is a space lattice, which has the translational symmetry in the real space. Since the motion of electrons in a crystal is usually described in both the real space and the momentum space (or k-space), it is useful to introduce here the concept of reciprocal space and reciprocal lattice. Analogous to any periodic time-varying function which is described in terms of the sum of Fourier components in the frequency domain, the spatial properties of a periodic crystal can be described by the sum of the components in Fourier space, or the reciprocal space. For a perfect crystal, the reciprocal lattice in the reciprocal space consists of an infinite periodic three-dimensional array of points whose spacing is inversely proportional to the distance between lattice planes of a Bravais lattice.

The reciprocal lattice is a geometrical construction which allows one to relate the crystal geometry directly to the electronic states and the symmetry properties of a crystal in the reciprocal space. Many important physical, electrical, and optical properties of semiconductors and metals can be understood by using the concept of reciprocal lattice. The unit cell in the reciprocal lattice is also known as the Brillouin zone or the Wigner–Seitz cell. The importance of the Brillouin zone in a solid will become clear when we discuss the lattice dynamics and the energy band theories in Chapter 2 and Chapter 4, respectively.

While the basis vector of a direct lattice has the dimension of length, the basis vector of a reciprocal lattice has the dimension of reciprocal length. The translational basis vector of a direct lattice is defined by Eq. (1.2). In a reciprocal lattice a set of reciprocal basis vectors, $\mathbf{b}_1^*$, $\mathbf{b}_2^*$, and $\mathbf{b}_3^*$, can be defined in terms of the basis vectors, $\mathbf{b}_1$, $\mathbf{b}_2$, and $\mathbf{b}_3$ of a direct lattice. This is given by

$$\mathbf{b}_1^* = \frac{2\pi(\mathbf{b}_2 \times \mathbf{b}_3)}{|\mathbf{b}_1 \cdot \mathbf{b}_2 \times \mathbf{b}_3|}, \qquad \mathbf{b}_2^* = \frac{2\pi(\mathbf{b}_3 \times \mathbf{b}_1)}{|\mathbf{b}_1 \cdot \mathbf{b}_2 \times \mathbf{b}_3|}, \qquad \mathbf{b}_3^* = \frac{2\pi(\mathbf{b}_1 \times \mathbf{b}_2)}{|\mathbf{b}_1 \cdot \mathbf{b}_2 \times \mathbf{b}_3|} \tag{1.4}$$

The reciprocal lattice vector can be defined in terms of the reciprocal basis vectors and Miller indices by

$$\mathbf{K} = h\mathbf{b}_1^* + k\mathbf{b}_2^* + l\mathbf{b}_3^* \tag{1.5}$$

where $\mathbf{b}_1^*$, $\mathbf{b}_2^*$, and $\mathbf{b}_3^*$ are given by Eq. (1.4), and h, k, l are the Miller indices. The reciprocal lattice vector defined by Eq. (1.5) may be used to generate all the reciprocal lattice points in the entire reciprocal space with its unit cell spanned by the reciprocal basis vectors defined by Eq. (1.4). Some important properties of the reciprocal lattice are summarized as follows:

1. Each reciprocal lattice vector in the reciprocal lattice is perpendicular to a set of lattice planes in the direct lattice. This is illustrated in Fig. 1.7. From Eqs. (1.2) and (1.5) we obtain

$$\mathbf{R} \cdot \mathbf{K} = 2\pi(n_1 h + n_2 k + n_3 l) = 2\pi N \tag{1.6}$$

or

$$\exp(i\mathbf{K} \cdot \mathbf{R}) = 1 \tag{1.7}$$

where N, n_1, n_2, and n_3 are integers. Equation (1.6) shows that the projection of the translational vector $\mathbf{R}$ in the direction of $\mathbf{K}$ has a length given by

$$d_{hkl} = 2\pi N/|K| \tag{1.8}$$

where d is the spacing between two nearby planes of a direct lattice.

2. The volume of a unit cell in the reciprocal lattice is inversely proportional to the volume of a unit cell in the direct lattice. The denominator of Eq. (1.4) represents the volume of the unit cell of a direct lattice, which is given by

$$V_d = |\mathbf{b}_1 \cdot \mathbf{b}_2 \times \mathbf{b}_3| \tag{1.9}$$

The volume of the unit cell of a reciprocal lattice is defined by the three reciprocal basis vectors, which can be expressed by

$$V_r = |\mathbf{b}_1^* \cdot \mathbf{b}_2^* \times \mathbf{b}_3^*| = \frac{8\pi^3}{V_d} \tag{1.10}$$

The factor $8\pi^3$ given in Eq. (1.10) is included so that the reciprocal lattice is defined in such a way that the dimension of the reciprocal lattice vector is the same as the wave vector of phonons or electrons in the reciprocal space or the momentum space. This will be discussed further in Chapter 2 and Chapter 4.

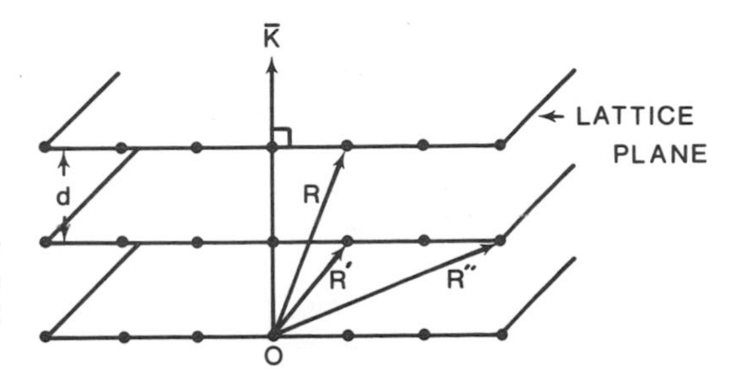

FIGURE 1.7. The reciprocal lattice vector and the lattice planes of a direct lattice, where **K** denotes the reciprocal lattice vector and **R** is the translational lattice vector.

3. A direct lattice is the reciprocal of its own reciprocal lattice. This can be shown by using Eq. (1.10).
4. The unit cell of a reciprocal lattice need not be a parallelepiped. In fact, one always deals with the Wigner–Seitz cell of the reciprocal lattice, which is also known as the first Brillouin zone in the reciprocal space, as is shown in Fig. 1.8.

We shall next discuss the construction of a Brillouin zone in the reciprocal lattice. The first Brillouin zone is the unit cell of the reciprocal lattice. It is the basic building block with the smallest volume in the reciprocal space, centered at one reciprocal lattice point, and bounded by a set of planes which bisect the reciprocal lattice vectors connecting this reciprocal lattice point to all its neighboring reciprocal lattice points. As an example, Fig. 1.8a shows the construction of the first Brillouin zone for a two-dimensional reciprocal lattice. It is obtained by first drawing a number of reciprocal lattice vectors from the center reciprocal lattice point, say (0, 0), to all its nearest-neighbor reciprocal lattice points, and then drawing the bisecting lines perpendicular to these reciprocal lattice vectors. The smallest area enclosed by these bisecting lines is called the first Brillouin zone, or the unit cell. The first Brillouin zone for a three-dimensional crystal lattice can be constructed in a similar way as that of a two-dimensional reciprocal lattice described above. This is done by first drawing the reciprocal lattice vectors from a chosen reciprocal lattice point to all its nearest-neighboring points and then drawing the bisecting planes perpendicular to these reciprocal lattice vectors. The smallest volume enclosed by these bisecting planes normally will form a polyhedron about the central point, and this polyhedron is called the first Brillouin zone or the Wigner–Seitz cell of the reciprocal lattice. Figures 1.8b and c show the first Brillouin zones for a face-centered cubic lattice and a body-centered cubic lattice, respectively. It is noted that the first Brillouin zone for a diamond lattice and a zinc blende lattice structure is identical to that of a face-centered cubic lattice shown in Fig. 1.8b.

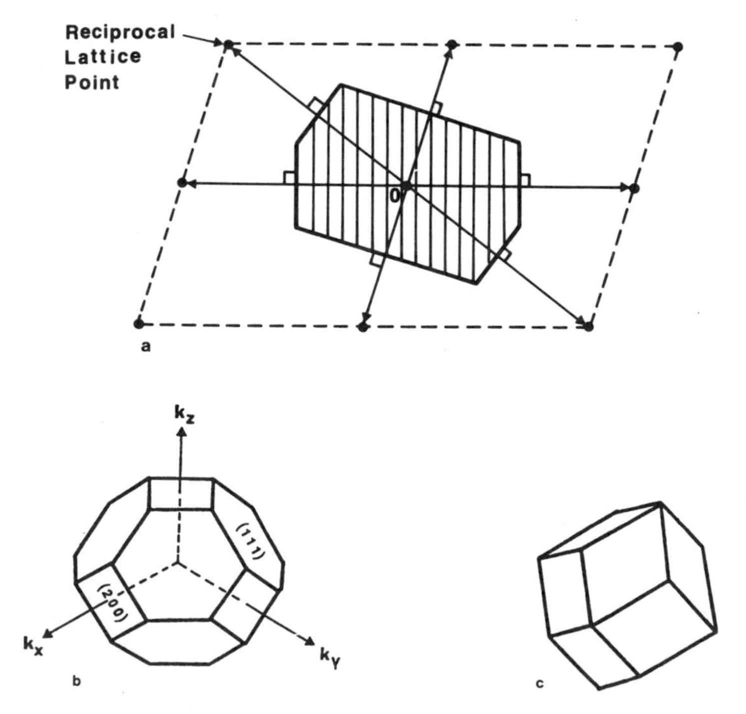

FIGURE 1.8. Construction of the first Brillouin zone for (a) a two-dimensional reciprocal lattice, (b) first Brillouin zone of a face-centered cubic lattice, and (c) of a body-centered cubic lattice.

The importance of the first Brillouin zone can be best illustrated by considering the wave function of an electron wave packet in a crystalline solid, which is described by the wave vector k in the momentum (or the reciprocal lattice) space. In a periodic crystal, it can be shown by using translational operation that for any given wave vector k' of the electron wave packet in the reciprocal space, there is a corresponding wave vector k inside the first Brillouin zone, which is related to k' by

$$k' = k + K \tag{1.11}$$

where K is the reciprocal lattice vector defined by Eq. (1.5). Therefore, for a given reciprocal lattice point in the reciprocal space, there is a corresponding reciprocal lattice point in the first Brillouin zone which can be obtained through the translational operation by substituting Eq. (1.7) in Eq. (1.11). In fact, we can show that except for a phase factor difference, the wave function of an electron at any given reciprocal lattice point in the reciprocal space is identical to the wave function of a corresponding reciprocal lattice point in the reciprocal space is identical to the wave function of a corresponding reciprocal lattice point in the first Brillouin zone obtained via translational operation of Eq. (1.7). This is important since it allows one to describe the physical properties of electrons or phonons in the first Brillouin zone of the reciprocal space using the reduced zone scheme. In fact, the phonon dispersion relation and the electronic states (or the energy bands) in a solid can be described by using the concept of reciprocal lattice and the first Brillouin zone depicted in this section. This will be discussed further in Chapter 2 and Chapter 4.

1.6. TYPES OF CRYSTAL BINDINGS

In Section 1.2, we described the classification of solids based on the geometrical aspects of the crystal lattice. In this section, we shall classify solids according to their bindng energy (i.e., the energy responsible for holding the atoms of a solid together). Based on the types of chemical binding energies which hold the atoms together in a solid, we can divide the crystalline solids into four categories. These are discussed next.

Ionic Crystals. In an ionic crystal, the electrostatic bonding normally comes from the transfer of electrons from alkali atoms to halogen atoms, resulting in the bonding of positively and negatively charged ions by the Coulomb attractive force. Typical examples are alkali metals such as sodium and potassium, in which each of these atoms has one extra valence electron to transfer to the atoms of halogens such as chlorine and bromine to form an alkali-halide salt (e.g., NaCl, KCl, NaBr, etc.). The II–VI (e.g., CdS, ZnSe, and CdTe) and III–V (e.g., Ga As, InP, and InSb) compound semiconductors also show certain ionic crystal properties. The ionic crystal usually has high binding energy due to the strong Coulombic force between the positive and negative ions. Ionic crystals formed from group I and VII elements in the periodic table belong to this category. Although they are good electric insulators at room temperature due to their large binding energy, these ions may become mobile at very high temperatures and diffuse through the crystal which results in an increase of electrical conductivity. The electrical conductivity of an ionic crystal is usually many orders of magnitude smaller than the electrical conductivity of a metal since the mass of the ion is about 10^4 times larger than the electron mass in a metal. The conductivity of an ionic crystal at elevated temperatures is related to the diffusion constant D of the mobile ion by

$$\sigma_i = \frac{Nq^2D}{k_BT} \tag{1.12}$$

where σ_i is the electrical conductivity of the ionic crystal, N is the density of mobile ions, q is the electronic charge, and k_B is the Boltzmann constant.

One important feature of the alkali-halide crystals is that they are transparent to visible and infrared (IR) optical radiation, and hence are widely used as optical window materials in the visible to IR spectral ranges. For example, NaCl crystal, which is transparent to the optical radiation from 0.4 to 16 μm, is often used as the prism material for a grating monochromator in this spectral range.

Covalent Crystals. In a covalent crystal, the binding energy comes from the reciprocal sharing of valence electrons among the nearest-neighboring atoms rather than from the transfer of valence elecrons as in the case of ionic crystal. Elemental semiconductors such as silicon and germanium are the typical covalent crystals.

The structure of a covalent crystal depends strongly on the nature of bonding itself. Covalent crystals such as germanium, silicon, and carbon have four valence elecrons per atom which are shared reciprocally with the nearest-neighboring atoms, contributing to the bonding of the crystal. Figure 1.9 illustrates the charge distribution of a silicon crystal lattice. Each silicon atom has four valence electrons which are shared reciprocally by its neighboring atoms and form a tetrahedral bonding. The diamond structure of a silicon crystal is a structure in which each atom is at the center of a tetrahedron, symmetrically surrounded by the four nearest-neighbor atoms located at the vertices. The tetrahedral bonding in silicon and germanium crystals can be explained by a linear combination of the s- and p-like atomic orbitals, which is called the sp^3 hybrids.

High-purity covalent crystals can have very high electrical resistivity and behave like insulators at room temperature. However, the binding force holding the valence electrons in an orbit is not as strong as that of an ionic crystal. For example, while the energy required to break an ionic bond in most ionic crystals may be as high as 10 eV, the energy necessary to break a covalent bond is much smaller, having values ranging from 0·1 to around 2·7 eV. Therefore, at room temperature or higher, the thermal energy may be sufficient to break the covalent bonds and freeing the valence electrons for electrical conduction in a covalent crystal. Furthermore, the broken bonds left behind by the valence electrons may be treated as free holes in the covalent crystal which in turn can also contribute to the electrical conductivity in the valence band. In fact, both electrons and holes can contribute to the electrical conduction in an intrinsic semiconductor, as will be discussed further in Chapter 5.

The fact that III–V compound semiconductors such as GaAs, InP, and InAs crystallized in the zinc blende structure implies that covalent bonding occurs in these crystals. However, in order to form covalent (homopolar) bonding in III–V semiconductors, the sp^3 orbitals surrounding each group-III and group-V atom would need four valence electrons per atom. This means that a transfer of one electronic charge from the group-V atom to the group-III

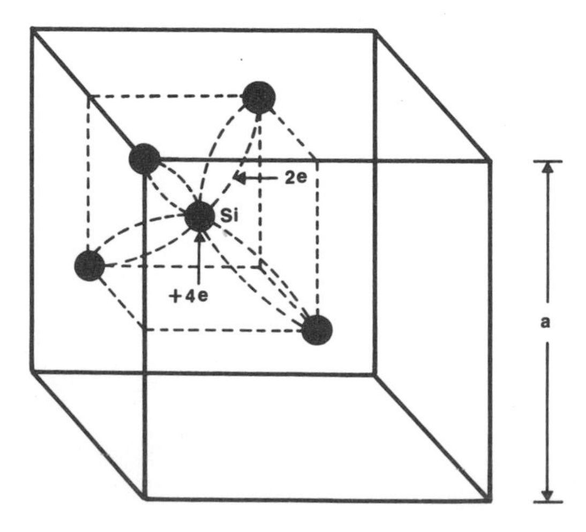

FIGURE 1.9. Charge distribution in a diamond lattice structure, showing the tetrahedral covalent bonds of a silicon lattice.

atom will occur in the III–V compound semiconductor crystals. This would result in group-III atoms becoming negatively charged (III^{1-}) and group-V atoms becoming positively charged (V^{1+}). The (III^{1-}) and (V^{1+}) is a nonneutral situation involving Coulombic interaction and hence ionic bonding. This partially ionic bonding characteristic is responsible for some striking differences between the III–V compound semiconductors and silicon.

Metallic Crystals. One of the most striking features of a metal is its high electrical conductivity. The binding energy of a metal comes mainly from the average kinetic energy of its valence electrons, and there is no tendency for these electrons to be localized within any given portion of the metal. For example, in a monatomic metal such as sodium or potassium, there are some 10^{23} cm^{-3} valence electrons which can participate in the electrical conduction process. In the classical theory of metals, valence electrons are treated as free electrons which can move freely inside the metal. The valence electrons form an electron sea in which the positive ions are embedded. Typical examples are the 2s electrons in a lithium crystal and the 3s electrons in a sodium crystal, which are responsible for the binding force of these metallic crystals.

In general, the binding energy of a monatomic metal is mainly due to the average kinetic energy of the valence electrons, which is usually much smaller than that of the ionic and covalent crystals. However, for transition metals, the binding energy, which is due to the covalent bonds of the d-shell electrons, can be much higher than that of monatomic metals.

Molecular Crystals. Argon, neon, and helium are solids which exhibit properties of molecular binding. These substances generally have a very small binding energy, and consequently have low melting and boiling temperatures. The binding force which holds the saturated molecules together in the solid phase comes primarily from the van der Waals force. This force is found to vary as r^{-6}, where r is the distance between the two molecules. To explain the origin of this force, it is noted that molecules in such a substance carry neither net electric charge nor permanent electric dipole moment. The instantaneous dipole moment on one molecule will give rise to an electric field which induces dipole moments on the neighboring molecules. It is the interaction of these instantaneous dipole moments which produces the cohesive energy of a molecular crystal. Since the individual molecules of a molecular crystal are electrically neutral and interact only weakly with one another, they are good electrical insulators, showing neither electronic nor ionic conductivity.

1.7. DEFECTS IN A CRYSTALLINE SOLID

It is generally known that a perfect crystal lattice is only mathematically possible, and in fact does not exist in real crystals. Defects or imperfections are always found in all the crystalline solids. The existence of defects usually has a profound effect on the physical properties of a crystal, which is particularly true for semiconductor materials. Therefore, it is important to discuss various types of defects which are commonly observed in a crystalline solid.

In general, defects may be divided into two broad categories: One class, which is called the dynamic defects, includes phonons, electrons, and holes. Another class, which is known as the stationary defects, consists of point defects (e.g., vacancies, interstitials, antisite defects, and foreign impurities), line defects (e.g., dislocations), and surface defects (e.g., grain boundaries). The physical properties and formation of these stationary defects are discussed as follows:

1.7.1. Vacancies and Interstitials

Both vacancies and interstitials are defects of atomic dimensions, which can only be observed through the use of the modern field-ion microscopy or infrared microscopy technique.

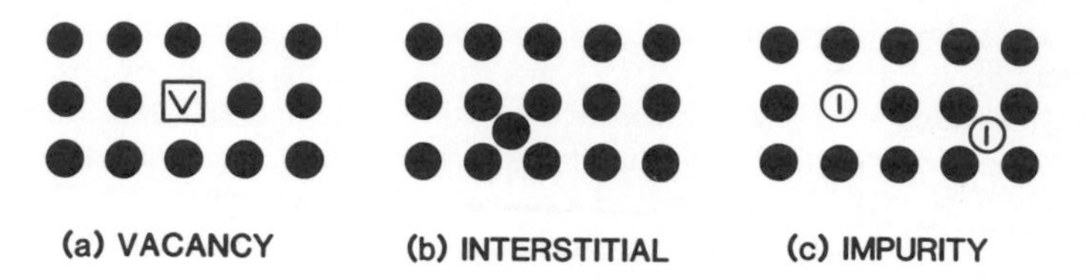

FIGURE 1.10. Vacancy, interstitial, and impurity point defects in a crystal lattice.

Vacancies are always present in the crystal, and the density of vacancies depends strongly on temperature. In fact, thermal fluctuation can cause a constant creation and annihilation of vacancies in a crystal. Figure 1.10 shows the formation of vacancy, interstitial, and impurity defects in a crystal lattice.

A vacancy is created when an atom migrates out of its regular lattice site to an interstitial position or to the surface of the crystal. The energy required to remove an atom out of its regular lattice site is defined as the activation energy of the vacancy. Two types of defects are usually associated with the creation of vacancies, namely, the Frenkel and Schottky defects. A Frenkel defect is created when an atom is moved from its regular lattice site to an interstitial site, while a Schottky defect is formed when the atom is moved from its regular lattice site to the surface of the crystal. Figure 1.11 shows both the Frenkel and Schottky defects. Another type of point defect which is often found in a semiconductor is created by the introduction of foreign impurities into the crystal, either intentionally (by thermal diffusion or ion implantation), or unintentionally (due to metallic or chemical contaminations). Figure 1.10c shows a point defect introduced by a foreign impurity.

The density of vacancies in a crystal can be calculated by using classical statistics and thermodynamic principles. In thermal equilibrium, the entropy of a crystal is increased by the presence of disorders, and thus a certain number of vacancies are always present in the crystal. According to thermodynamics, the equilibrium condition of a system at a finite temperature is established when the free energy of the system is at a minimum. If there are n vacancies distributing randomly among N lattice sites, then the increase of entropy and free energy in the crystal can be calculated as follows: Let E_v be the activation energy of a vacancy, and the total incremental internal energy U of the crystal due to the creation of n vacancies is equal to nE_v, where n is the number of vacancies at temperature T. The total number of ways of arranging n vacancies among N lattice sites is given by

$$P = \frac{N!}{(N-n)!n!} \tag{1.13}$$

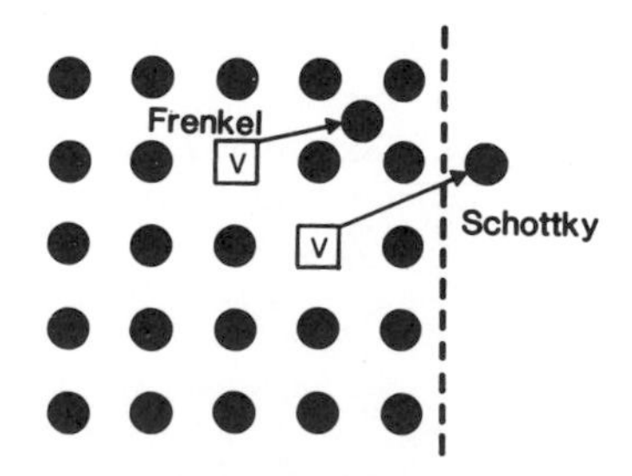

FIGURE 1.11. Formation of Frenkel and Schottky defects in a crystal lattice.

The increase of entropy due to the creation of n vacancies in a crystal can be expressed by

$$S = k_B \ln P = k_B \ln [N!/(N-n)!n!] \tag{1.14}$$

where S is the entropy, and k_B is the Boltzmann constant. Thus, the total change in the free energy of the system is given by

$$F = U - TS = nE_v - k_B T \ln [N!/(N = n)!n!] \tag{1.15}$$

In thermal equilibrium, the incremental free energy, F, must be at its minimum with respect to n. The factorials given in Eq. (1.15) can be simplified by using Stirling's approximation when n is very large (i.e., $\ln n! \simeq n \ln n - n$, for $n \gg 1$). Thus, for $n \gg 1$, Eq. (1.14) is reduced to

$$S \simeq k_B[N \ln N - (N-n),\ln (N-n) - n \ln n] \tag{1.16}$$

From Eqs. (1.15) and (1.16), the minimum free energy can be obtained by differentiating F with respect to n in Eq. (1.15) and setting the result equals to zero, which yields

$$n = (N-n) \exp(-E_v/k_B T) \tag{1.17}$$

or

$$n \simeq N \exp(-E_v/k_B T) \qquad \text{for } N \gg n \tag{1.18}$$

Equation (1.18) shows that the density of vacancies increases exponentially with temperature. For example, assuming $E_v = 1$ eV and $N = 10^{23}$ cm^{-3}, the density of vacancies at $T =$ 1200 K is equal to 4.5×10^{18} cm^{-3}.

A similar procedure may be employed to derive the expressions for the density of Frenkel and Schottky defects in a crystal. For Schottky defects, this is given by

$$n \simeq N \exp(-E_s/k_B T) \tag{1.19}$$

where E_s is the activation energy for creating a Schottky defect. For Frenkel defects, we obtain

$$n \simeq (NN')^{1/2} \exp(-E_f/2k_B T) \tag{1.20}$$

where E_f is the activation energy of a Frenkel defect. Note that N' is the density of interstitial sites. In general, it is found that $E_s > E_f > E_v$. For example, for aluminum, E_v was found equal to 0·75 eV and $E_s \sim 3$ eV.

It is interesting to note that Frenkel defects may be created by nuclear bombardment, high-energy electron and proton irradiations, or by ion implantation damage. In fact, the radiaton damage created by high-energy particle bombardment in a solid is concerned almost entirely with the creation and annihilation of Frenkel defects. The Frenkel defects may be eliminated or reduced via thermal or laser annealing process. Both thermal and laser annealing procedures are widely used in semiconductor material processing and device fabrication. Recently, rapid thermal annealing (RTA) and laser annealing techniques have also been used extensively by semiconductor industry for removing damages created by radiation, ion implantation, and device processing.

Foreign impurity is another type of point defect which deserves special mention. Both the substitutional and interstitial impurity defects may be introduced by doping the host crystal with foreign impurities using thermal diffusion or ion implantation. It is common practice to use foreign impurities to modify the electrical conductivity and the conductivity types (i.e., n- or p-type) of a semiconductor. Foreign impurities may either occupy a regular lattice site or reside in an interstitial site of the host crystal. As will be discussed in Chapter 5, both the shallow- and deep-level impurities may play a very important role in controlling the physical and electrical properties of a semiconductor. Finally, point defects may also be created by quenching the crystal at high temperatures or by severe deformation of the crystal through hammering or rollling.

1.7.2. Line and Surface Defects

Another type of crystal defects known as line defects may also be created in both single- and poly-crystalline solids. The most common type of line defects created in a crystalline solid is the dislocations. Dislocations are lattice defects created in a crystal which can be best described in terms of partial internal slip. There are two types of dislocations which are commonly observed in a crystalline solid. They are the edge and screw dislocations. The creation of these dislocations and their physical properties are discussed next.

Edge Dislocation. An edge dislocation can be best described by imagining a perfect crystal which is cut open along line AO shown in Fig. 1.12a; the plane of the cut is perpendicular to that of the page. An extra monolayer crystal plane of depth AO is then inserted in the cut and the crystal lattice is repaired as best it can be, leaving a line perpendicular to the plane of the paper and passing through the point O, around which the crystal structure is severely distorted. The distortion of a crystal lattice can be created by the partial insertion of an extra plane of atoms into the crystal. This distortion is characterized by a line defect. The local expansion (known as the dilatation) around the edge dislocation can be described by a simple expression, which reads

$$d = \left(\frac{b}{r}\right) \sin \theta \tag{1.21}$$

where b is the Burgers vector (a measure of the strength of distortion caused by dislocation), r is the radial distance from a point in the crystal to the dislocation line, and θ is the angle between r and the slip plane. The sign of dilatation is positive for expansion and negative for compression. The Burgers vector for an edge dislocation is perpendicular to the dislocation line and lies in the slip plane.

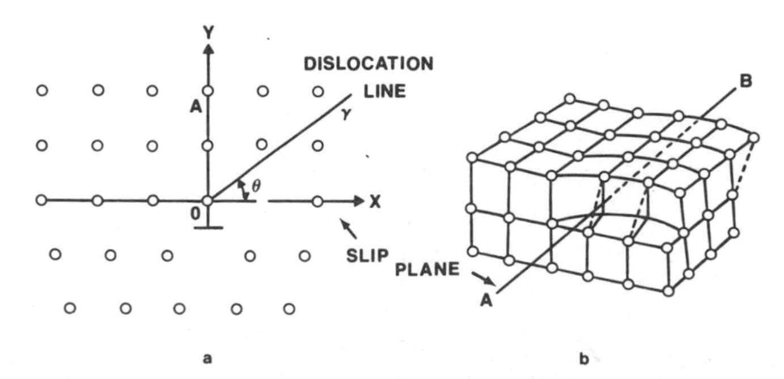

FIGURE 1.12. (a) Edge dislocation and (b) screw dislocation in a crystal lattice.

Screw Dislocation. The second type of dislocation, known as the screw dislocation, is shown in Fig. 1.12b. As shown in the horizontal plane of Fig. 1.12b, the screw dislocation is produced by cutting the crystal partially through and pushing the upper part of the crystal one lattice spacing over. A line of distortion is clearly shown along the edge of the cut. This line is usually called the screw dislocation. In contrast to the type of distortion surrounding an edge dislocation, the atoms near the center of a screw dislocation are not in dilatation, but are on a twisted or sheared lattice. It is noted that in the screw dislocation the relative displacements of the two halves of the crystal are in the direction of the dislocation line rather than normal to it. Again, the Burgers vector can be used to specify the amount of displacement that has occurred.

Dislocations may be created in a number of ways. For example, plastic deformation creates dislocations and consequently creates damage in the lattice. The dislocations themselves introduce defect levels in the forbidden gap of a semiconductor. For semiconductors with a diamond lattice structure the dislocation velocity depends exponentially on the temperature and hence the dislocation generation requires that plastic deformation in a semiconductor occurs at very high temperatures. The density of dislocations, which is defined by the number of dislocation lines intersected by a plane of unit area, can be counted by using either the etch-pit or the X-ray diffraction technique. In the etch-pit technique, the sample is first polished and then chemically etched. Conical pits are formed at places where dislocation lines intersect the crystal surface, and the number of etch-pits are counted. In the X-ray diffraction technique, the observed spread of angle θ in the Bragg diffraction pattern is a measure of the dislocation density.

If the density of dislocation is sufficiently high (e.g., $>10^7$ cm^{-2}), then the electrical and mechanical properties of a crystal may be affected by the dislocations. For example, the electrical conductivity of a semiconductor measured parallel and perpendicular to the dislocation line can be quite different when the density of these dislocation lines is very high. It is noted that a dislocation line may be considered as a line charge, which can trap the minority carriers and affect the minority carrier lifetime in a semiconductor. In a pure silicon or germanium crystal, the dislocation density may range from a few hundred to several tens of thousands per cm^2, depending on the conditions of crystal growth and heat treatment. In general, if the dislocation density is less than 10^6 cm^{-2}, then its effect on the electrical properties of a semiconductor is negligible. Semiconductors can be produced with zero or few dislocations per unit area. In fact, dislocation-free germanium and silicon single crystals have been routinely grown by the present-day crystal pull technology. However, for polycrystalline materials, the dislocation density is usually very high, and thus dislocations play a much more important role in a polycrystalline semiconductor than in a single crystalline semiconductor. Its effects on the minority carrier lifetimes and the majority carrier mobility of a polycrystalline material are also more pronounced than for a single crystalline semiconductor.

A surface defect is another type of defect which can affect the performance of a semiconductor device. Typical example of a surface defect is the grain boundaries in a polycrystalline semiconductor. In general, an array of edge dislocations can be formed near the grain boundaries of any two subregions of the polycrystalline material. Grain boundaries often play an important role in influencing the electrical and transport properties of a polycrystalline semiconductor. For example, depending on the heat treatment used during and after the film growth, the grain size of a thin-film polycrystalline silicon grown by low-pressure chemical vapor deposition (LPCVD) technique may vary from a few hundred angstroms to a few tens of micrometers. On the other hand, for a bulk polycrystalline silicon material, the grain size may vary from a few milimeters to a few centimeters. Polycrystalline silicon thin films prepared by LPCVD technique are widely used for interconnects and thin-film resistors in the silicon integrated circuits. Bulk polycrystalline silicon material and thin-film polycrystalline materials from II–VI semiconductor compounds such as CdS and $CuInSe_2$ have been used in fabricating low-cost solar cells for terrestrial power generation.

PROBLEMS

1.1. Show that the maximum proportion of the available volume which may be filled by hard spheres for the following lattice structures is given by
(a) Simple cubic: $\pi/6$.
(b) Body-centered cubic: $\sqrt{3}\pi/8$.
(c) Face-centered cubic: $\sqrt{2}\pi/6$.
(d) Hexagonal-closed-packed: $\sqrt{2}\pi/6$.
(e) Diamond: $\sqrt{3}\pi/16$.

1.2. Explain why the following listed lattices are not Bravias lattices.
(a) Base-centered tetragonal.
(b) Face-centered tetragonal.
(c) Face-centered rhombohedral.

1.3. Show that a crystal lattice can not have an axis with fivefold rotational symmetry.

1.4. Construct a primitive cell for a body-centered cubic (BCC) and a face-centered cubic (FCC) lattice.

1.5. (a) Show that a diamond lattice structure is made up of two interpenetrating face-centered cubic lattices.
(b) If the cubic edge (or the lattice constant) of a diamond lattice is equal to 3.56 Å. Calculate the distance between the nearest neighbors and the total number of atoms per unit cell.
(c) Repeat for silicon (cubic edge $a = 5.43$ Å) and gemanium (cubic edge $a = 5.62$ Å).

Show that a body-centered tetragonal lattice with $a = \sqrt{2}b$ has the symmetry of a face-centered cubic lattice.

1.7. Find the number of nearest neighbors and the primitive lattice vectors for a diamond lattice structure.

1.8. Show that the reciprocal lattice of a body-centered cubic lattice is a face-centered cubic lattice.

1.9. Draw the crystal planes for the following lattice structures:
(a) (200), (222), (311) planes for a cubic crystal.
(b) $(10\bar{1}0)$ plane of a hexagonal crystal. [*Hint*: the Miller indices for a hexagonal lattice are represented by (a_1, a_2, a_3, c).]

1.10. Show that the first Brillouin zone of a diamond lattice structure is enclosed by eight $\{111\}$ and six $\{200\}$ planes.

1.11. (a) Give the total number of planes for $\{100\}$, $\{110\}$, $\{111\}$, and $\{200\}$ of a cubic lattice.
(b) Find the normal distance from the origin of the unit cell to the planes listed in (a).

1.12. Show that in a hexagonal closed-packed lattice structure, the length of the c-axis is equal to $\sqrt{8/3}a$, where a is the length of one side of the hexagonal base plane.

1.13. Draw the first four Brillouin zones of a two-dimensional square lattice, and show that the area of each zone is identical.

1.14. Show that the density of the Schottky and Frenkel defects in a crystal are given respectively by Eqs. (1.19) and (1.20).

BIBLIOGRAPHY

F. J. Blatt, *Physics Propagation in Periodic Structures*, 2nd ed., Dover, New York (1953).
L. Brillouin, *Wave Propagation in Periodic Structures*, 2nd ed., Dover, New York (1953).
M. J. Buerger, *Elementary Crystallography*, Wiley, New York (1963).
A. J. Dekker, *Solid State Physics*, 6th ed., Prentice-Hall, Englewood Cliffs (1962).
B. Henderson, *Defects in Crystalline Solids*, Crane Russak & Co., New York (1972).

C. Kittel, *Introduction to Solid State Physics*, 5th ed., Wiley, New York (1976).
T. L. Martin, Jr. and W. F. Leonard, *Electrons and Crystals*, Brooks & Cole, California (1970).
J. P. McKelvey, *Solid State and Semiconductor Physics*, Harper & Row, New York (1966).
F. C. Phillips, *An Introduction to Crystallography*, Longmans, Green & Co., London (1946).
J. C. Phillips, *Bonds and Bands in Semiconductors*, Academic Press, New York (1973).
L. Pauling, *The Nature of the Chemical Bond*, Cornell University Press, Ithaca, New York (1960).
J. C. Slater, *Quantum Theory of Molecules and Solids*, Vol. 2, McGraw-Hill, New York (1965).
R. W. G. Wyckoff, *Crystal Structures*, 2nd ed., Interscience Publishers, New York (1963).

2

Lattice Dynamics

2.1. INTRODUCTION

In this chapter we are concerned with the thermal properties and lattice dynamics of a crystalline solid. Under thermal equilibrium conditions, the mass centers or the nuclei of the atoms in a solid are not at rest but instead they vibrate with respect to their mean equilibrium position. In fact, many thermal properties of a solid are determined by the amplitude and the phase factor of these atomic vibrations. For example, the specific heat of an insulator is due entirely to its lattice vibrations. Solid argon, which is perhaps the simplest solid of all, consists of a regular array of neutral atoms with tightly bound closed-shell electrons. These electrons are held together primarily by the van der Waal force, and hence interact only with their nearest-neighbor atoms. The physical properties of such a solid are due entirely to the thermal vibrations of its atoms with respect to their equilibrium positions. Therefore, the specific heat for such a solid is constituted entirely from its lattice vibrations. On the other hand, the specific heat for metals is dominated by the lattice specific heat at high temperatures, and by the electronic specific heat at very low temperatures. The most important effect of the lattice vibrations on metals or semiconductors is that they are the main scattering centers which would limit the carrier mobility in these materials. In fact, the interaction between the electrons and the lattice vibrations is usually responsible for the temperature dependence of the resistivity and carrier mobility in an undoped semiconductor. Furthermore, such interactions also play an important role in the thermoelectric effects of metals and semiconductors.

According to the classical Dulong and Petit law, the lattice specific heat for a solid is constant, and equals 5.96 cal/mol · °C. The Dulong and Petit law prevails for most solids at high temperatures but fails at very low temperatures. The lattice specific heat can be derived from the classical statistics as follows.

Let us consider a solid with N identical atoms which are bounded together by an elastic force. If each atom has three degrees of freedom, then there will be $3N$ degrees of freedom for the N atoms to produce $3N$ independent vibration modes, each with the same vibration frequency. According to classical statistics, the mean energy for each lattice vibration mode is $k_B T$, and hence the total energy U for $3N$ vibration modes in a solid is equal to $3Nk_B T$. Thus, the lattice specific heat under constant volume condition is given by

$$C_v = \frac{dU}{dT} = 3Nk_B = 3R \tag{2.1}$$

where R $(=Nk_B)$ is the ideal gas constant, and k_B is the Boltzmann constant $(=1.38 \times 10^{-23}$ joule/K). As an example, for an ideal gas system, by substituting $N = 6.025 \times 10^{23}$ atoms/g · mol (Avogadro's number) into Eq. (2.1) one finds that C_v is equal to

5.96 cal/mol · °C. This value is in good agreement with the experimental data for most solids at high temperatures. However, Eq. (2.1) fails to predict correctly the lattice specific heat for most solids at very low temperatures. This is due to the fact that at very low temperatures, atoms in a solid are no longer vibrating independently with one another. Instead, the lattice vibration modes can be considered as quasi-continuum with a broad spectrum of vibration frequencies.

In Section 2.2, we derive equations for the dispersion relations of a one-dimensional monatomic linear chain and a diatomic linear chain. The dispersion relation (i.e., frequency versus wave vector) for a three-dimensional lattice is derived in Section 2.3. The concept of phonons is introduced in Section 2.4. In Section 2.5, we derive the phonon density of states function, and present the lattice spectra for some metals and semiconductors. The Debye model for predicting the lattice specific heat of a solid is described in Section 2.6. In Section 2.7 we describe the elastic constants and velocity of sound for a compound semiconductor such as GaAs.

2.2. THE ONE-DIMENSIONAL LINEAR CHAIN

To understand the physical properties associated with lattice vibrations in a solid, it is useful to consider two simple cases, namely, the one-dimensional monatomic linear chain and the diatomic linear chain. This is discussed separately as follows:

Monatomic Linear Chain. For a one-dimensional monatomic linear chain, there is one atom per unit cell. If only the nearest-neighbor interaction is considered, then the linear chain can be represented by a string of identical masses connecting to one another by the same massless spring, as is illustrated in Fig. 2.1. In this case, the equation of motion for the atomic displacement can be easily derived by using Hooke's law. According to this classical law, the force acting on the nth atom with mass m is given by

$$F_n = m\frac{\partial^2 u_n}{\partial t^2} = -\beta(u_n - u_{n+1}) - \beta(u_n - u_{n-1})$$

$$= -\beta(2u_n - u_{n+1} - u_{n-1}) \tag{2.2}$$

where β is the force constant and u_n denotes the displacement of the nth atom. The solution of Eq. (2.2) has the form of a travelling wave which can be expressed by

$$u_n = u_q e^{i(naq - \omega t)} \tag{2.3}$$

where q is the wave vector ($q = 2\pi/\lambda$) of the lattice wave, a is the lattice constant, and n is an integer. Note that u_q denotes the amplitude function of the lattice wave, which is also a

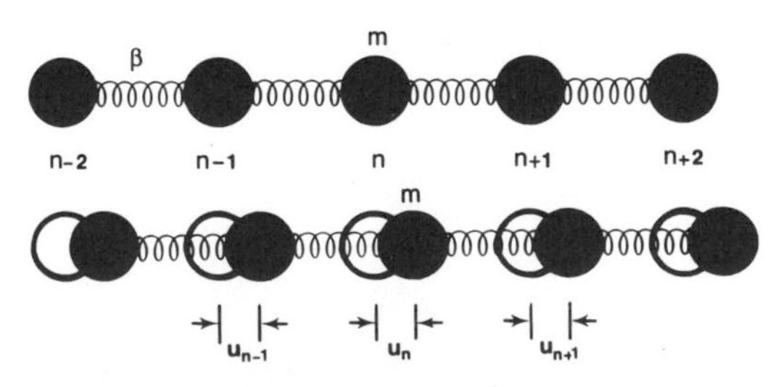

FIGURE 2.1. Lattice vibration of a one-dimensional monatomic linear chain. a is the lattice constant, β is the force constant, and u_n is the displacement of the nth atom from its equilibrium position.

function of wave vector q, By substituting Eq. (2.3) into Eq.(2.2), we obtain

$$m\omega^2 = -2\beta(\cos qa - 1) \tag{2.4}$$

Equation (2.4) is the solution for a simple harmonic oscillator which has a dispersion relation given by

$$\omega = 2\sqrt{\frac{\beta}{m}}\sin(qa/2) = \omega_m \sin(qa/2) \tag{2.5}$$

where $\omega_m = 2(\beta/m)^{1/2}$ is the maximum frequency of the lattice vibration modes. Figure 2.2 shows the dispersion relation for a one-dimensional monatomic linear chain obtained from Eq. (2.5). This dispersion curve has a period of $2\pi/a$.

The dispersion relation given by Eq. (2.5) for a one-dimensional monatomic linear chain exemplifies several fundamental physical properties of lattice dynamics in a solid. First, all the possible lattice vibration modes are limited by the allowed values of wave vector q which fall in the range of $-\pi/a \leq q \leq \pi/a$. This range is known as the first Brillouin zone for this dispersion curve. There are n independent wave vectors within the first Brillouin zone representing the n (i.e., $n = N$, where N is the total number of atoms) independent vibrational modes. Each atomic displacement contributes to one lattice vibration mode. The maximum wave number $q_{\max}$, which occurs at the zone boundary, is given by

$$q_{\max} = \pi/a \simeq 10^8 \text{ cm}^{-1} \tag{2.6}$$

where a is the lattice constant. Since frequency ω is a periodic function of wave vector q in the q-space, for any given wave vector q' outside the first Brillouin zone there is a corresponding wave vector q in the first Brillouin zone which can be obtained by translational operation (i.e., $q' = q \pm K$, where K is the reciprocal lattice vector). This translational symmetry operation has been discussed in detail in Chapter 1. At the zone boundary, the solution of Eq. (2.3) does not represent a traveling wave but a standing wave. Thus, at the zone boundaries $q_{\max} = \pm(n\pi/a)$, and u_n is given by

$$u_n = u_{q_{\max}} e^{i(n\pi - \omega t)} = u_{q_{\max}} e^{-i\omega t}\cos(n\pi) \tag{2.7}$$

Equation (2.7) shows that at the zone boundaries $\cos n\pi = \pm 1$, depending on whether n is an even or an odd integer. This implies that the vibration modes for the alternate atoms are out

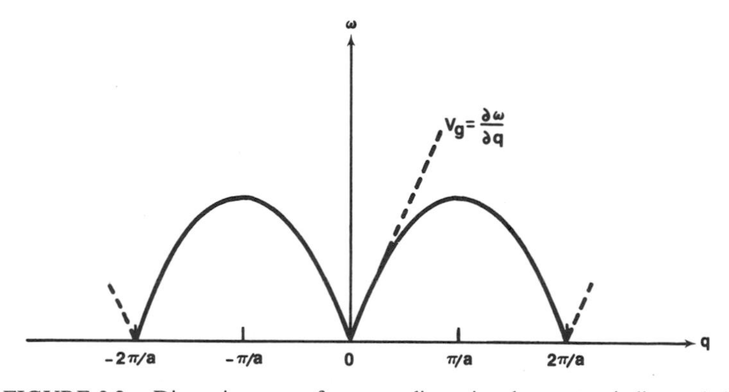

FIGURE 2.2. Dispersion curve for a one-dimensional monatomic linear chain.

of phase at the zone boundary. The group velocity of the lattice wave packet is defined by

$$v_g = \frac{d\omega}{dq} \tag{2.8}$$

The solution of Eqs. (2.5) and (2.8) yields an expression for the group velocity, which reads

$$v_g = (\beta/m)^{1/2} a \cos(qa/2) \tag{2.9}$$

From Eq. (2.9), it is noted that at the zone boundaries where $q_{\max} = \pm \pi/a$, the group velocity v_g is equal to zero. Thus, the lattice wave is represented by a standing wave packet at the zone boundaries, and the incident and the reflected lattice waves are equal in amplitude but travel in opposite directions.

In the long-wavelength limit (i.e., for $qa \to 0$), Eq. (2.5) reduces to

$$\omega = (\beta/m)^{1/2} aq \tag{2.10}$$

which shows that for $qa \to 0$, the vibration frequency of the lattice wave is directly proportional to the wave vector, q. This corresponds to the well known property of ordinary elastic waves in a continuum medium. In this case, the group velocity and the phase velocity are equal, and their values can be determined from the slope of the ω versus q plot at small q, as shown in Fig. 2.2. If $a = 3$ Å and $v_s = 10^5$ cm/sec, then $(\beta/m)^{1/2} \simeq 3 \times 10^{12}\ \text{sec}^{-1}$, and the maximum vibration frequency $\omega_{\max}$ that a lattice can support is equal to $6 \times 10^{12}\ \text{sec}^{-1}$; this value falls in the infrared regime of the optical spectrum.

Diatomic Linear Chain. We shall next consider the case of one-dimensional diatomic linear chain. Figure 2.3 shows a one-dimensional diatomic linear chain which contains two types of atoms of different masses per unit cell. The atoms are equally spaced, but with different masses placed in alternate positions along the linear chain. If we assume that only nearest-neighbor interactions are important, then the force constant between these two types of atoms will be the same throughout the entire linear chain. Therefore, there are two atoms per unit cell with masses of m_1 and m_2. Using Hooke's law, the equations of motion for the $2n$th and $(2n+1)$th atoms of this one-dimensional diatomic linear chain can be written as

$$m_1 \frac{\partial^2 u_{2n}}{\partial t^2} = \beta(u_{2n+1} + u_{2n-1} - 2u_{2n}) \tag{2.11}$$

$$m_2 \frac{\partial^2 u_{2n+1}}{\partial t^2} = \beta(u_{2n+2} + u_{2n} - 2u_{2n+1}) \tag{2.12}$$

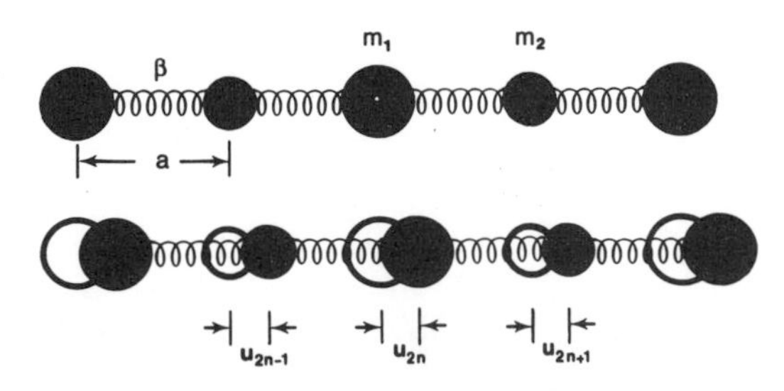

FIGURE 2.3. A diatomic linear chain in equilibrium position and in the displaced position, assuming $m_1 > m_2$.

Solutions of Eqs. (2.11) and (2.12) can be expressed by

$$u_{2n} = u_a\, e^{i(2nqa - \omega t)} \tag{2.13}$$

$$u_{2n+1} = u_0\, e^{i(2n+1)qa - i\omega t} \tag{2.14}$$

where u_{2n} and u_{2n+1} are the displacements for the 2nth and $(2n + 1)$th atoms, respectively. Now, by substituting Eqs. (2.13) and (2.14) into Eqs. (2.11) and (2.12) we obtain

$$2\beta u_a - m_1\omega^2 u_a - 2\beta u_0 \cos qa = 0 \tag{2.15}$$

$$2\beta u_0 - m_2\omega^2 u_0 - 2\beta u_a \cos qa = 0 \tag{2.16}$$

Equations (2.15) and (2.16) will have a nontrivial solution if and only if the determinant for the coefficients of u_a and u_0 in both equations is set equal to zero. Thus, the frequency ω must satisfy the secular equation given by

$$\begin{vmatrix} (2\beta - m_1\omega^2) & -2\beta\cos(qa) \\ -2\beta\cos(qa) & (2\beta - m_2\omega^2) \end{vmatrix} = 0 \tag{2.17}$$

Solving Eq. (2.17) for ω yields

$$\omega^2 = \beta\left\{\left(\frac{1}{m_1} + \frac{1}{m_2}\right) \pm \left[\left(\frac{1}{m_1} + \frac{1}{m_2}\right)^2 - \frac{4\sin^2(qa)}{m_1 m_2}\right]^{1/2}\right\} \tag{2.18}$$

Using the same argument as in the case of the monatomic linear chain, one finds that the allowed values of q for the diatomic linear chain is given by

$$|q| = \frac{n\pi}{Na} \tag{2.19}$$

where N is the total number of unit cells in the linear chain and n is an integer. Since the period of the diatomic linear chain is equal to $2a$, the first Brillouin zone is defined by

$$\frac{-\pi}{2a} \le q \le \frac{\pi}{2a} \tag{2.20}$$

which is a factor of 2 smaller than the first Brillouin zone of the monatomic linear chain. Figure 2.4 shows the dispersion curves (i.e., ω versus q) for a one-dimensional diatomic linear

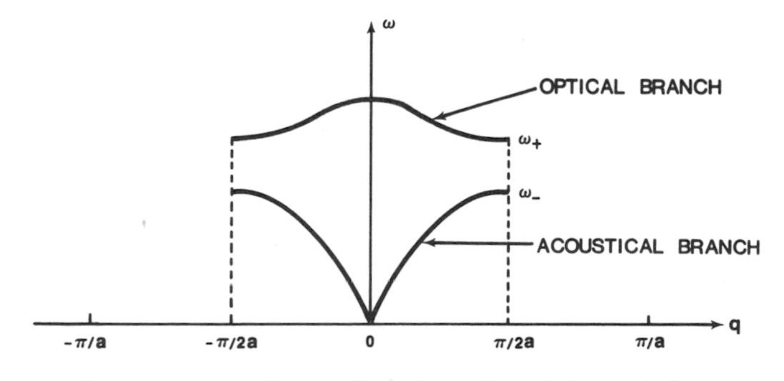

FIGURE 2.4. Dispersion curves of a one-dimensional diatomic linear chain, assuming $m_1 > m_2$.

chain with $m_1 > m_2$. The upper curve shown in Fig. 2.4 corresponds to the plus sign given by Eq. (2.18), and is called the optical branch. The lower curve, which corresponds to the minus sign in Eq. (2.18), is known as the acoustical branch. The lattice vibration modes in the optical branch can usually be excited by the infrared optical radiation, which has frequencies ranging from 10^{12} to 10^{14} Hz. For the acoustical branch, the lattice vibration modes can be excited if the crystal is connected to an acoustical wave transducer that produces pressure waves throughout the crystal. In general, the dispersion curves for a solid with two atoms per unit cell contain both the acoustical and optical branches. For example, the dispersion curves for an alkali-halide crystal such as NaCl consist of both the acoustical and optical branches, corresponding to the positively (i.e., Na^+) and negatively (i.e., Cl^-) charged ions in the crystal.

The physical insights for the dispersion curves of a diatomic linear chain can be best interpreted by considering the two limiting cases, namely, the long-wavelength limit (i.e., $qa \to 0$) and near the zone boundary (i.e., $q \to \pi/2a$). These are now discussed.

Acoustical Branch. For $qa \to 0$, using the minus sign in Eq. (2.18) for the acoustical branch, we obtain

$$\omega = \left[\frac{2\beta}{(m_1 + m_2)}\right]^{1/2} aq \tag{2.21}$$

Equation (2.21) shows that in the long-wavelength limit, ω is directly proportional to the wave vector q. This result is identical to the monatomic linear chain discussed in the previous section. Furthermore, from Eq. (2.16), and for $qa \to 0$, one can show that the ratio of the amplitude of two different mass atoms is given by

$$\frac{u_a}{u_0} = 1 \tag{2.22}$$

which shows that in the long-wavelength limit, atoms at odd and even lattice sites are moving in phase with equal amplitude. From the above analysis, it is obvious that in the long-wavelength limit the dispersion curve in the acoustical branch for a diatomic linear chain is identical to that of a monatomic linear chain if the masses of the two types of atoms are identical (i.e., $m_1 = m_2 = m$).

Optical Branch. The optical branch for the diatomic linear chain is shown by the upper curve of Fig. 2.4. In the long-wavelength limit, where qa approaches zero, we obtain

$$\omega = \left[\frac{2\beta(m_1 + m_2)}{m_1 m_2}\right]^{1/2} \tag{2.23}$$

From Eqs. (2.15) and (2.16) and for $qa \to 0$, the ratio of the amplitude for two different mass atoms is given by

$$\frac{u_a}{u_0} = -\frac{m_2}{m_1} \tag{2.24}$$

Equation (2.23) shows that in the optical branch with $qa \to 0$, the frequency ω is constant and independent of the wave vector q. Thus, the lattice vibration mode at $qa = 0$ represents a standing wave. The ratio of the amplitude factor given by Eq. (2.24) reveals that lattice vibration modes of alternate masses are out of phase, and the amplitude is inversely proportional to the mass ratio of the alternate atoms. This can be best explained by considering the alkali-halide crystal. The two types of ions in an alkali-halide crystal (e.g., NaCl) are oppositely charged, and hence will experience opposing forces when an electric field is applied to the crystal. As a result, the motion of atoms in alternate lattice sites are out of phase with each other and have an amplitude ratio inversely proportional to their mass ratio. If an electromagnetic wave with frequencies corresponding to the frequencies of the optical lattice vibration modes is applied to the crystal, resonant absorption takes place. Since the frequencies in which the lattice vibration modes are excited in this branch usually fall in the infrared spectral range, it is referred to as the optical branch.

Another feature of the dispersion curves shown in Fig. 2.4 is the appearance of a forbidden gap between $\omega_- = (2\beta/m_1)^{1/2}$ and $\omega_+ = (2\beta/m_2)^{1/2}$ at the zone boundary (i.e., $q_{\max} = \pi/2a$). The forbidden region corresponds to frequencies in which lattice waves cannot propagate through the linear chain without attenuation. This can be easily verified by substituting a value of ω which falls in the forbidden region of the dispersion curve shown in Fig. 2.4 into Eq. (2.18). Under this condition, the wave vector q becomes a complex number, and the lattice wave with this frequency is attenuated when it propagates through the linear chain. It is interesting to note that a similar situation also exists in the energy band theory of a semiconductor in which a forbidden band gap exists between the valence band and the conduction band. This will be discussed in detail in Chapter 4.

2.3. DISPERSION RELATION FOR A THREE-DIMENSIONAL LATTICE

The dispersion relation for the one-dimensional linear chain derived in Section 2.2 can be easily extended to the two- and three-dimensional lattices by considering the lattice vibration modes as simple harmonic oscillators. As frequently encountered in quantum mechanics, the displacement of atoms can be expressed in terms of the normal coordinates and normal modes of quantum oscillators. According to quantum mechanics, the lattice vibration modes for the atomic vibrations can be represented by harmonic oscillators, with each vibration mode having its own characteristic frequency ω and wave vector q. According to quantum theory, the equation of motion for a three-dimensional harmonic oscillator is given by

$$\ddot{Q}_{q,s} + \omega^2(q, s)Q_{q,s} = 0 \tag{2.25}$$

where $Q_{q,s}$ is the normal coordinate and $\omega(q, s)$ denotes the normal frequency. Note that both $Q_{q,s}$ and $\omega(q, s)$ are functions of wave vector q and polarization index s (i.e., $s = 1, 2, 3$).

In general, if a crystal contains only one atom per unit cell, then there are three possible polarizations for each wave vector q, one longitudinal and two transverse modes of polarization. In the longitudinal mode of lattice vibration, the motion of atoms is along the direction of wave propagation, while for the transverse modes the motion of atoms is in the plane perpendicular to the direction of wave propagation. If a crystal contains N atoms per unit cell, then the index n varies from 1 to $3N$. For example, if $N = 2$ and $s = 1$, 2, and 3, then there are three polarizations (i.e., $s = 1, 2, 3$; one longitudinal and two transverse modes) in the acoustic branch, and three polarizations in the optical branch (i.e., $s = 4$, 5, and 6).

Figure 2.5a, b, and c show the measured lattice dispersion curves for silicon, GaAs, and aluminum, respectively, obtained from the inelastic slow neutron experiment.[1-3] The dispersion curves for these materials are strongly dependent on the crystal orientations. This is due to the fact that lattice vibration modes depend strongly on crystal symmetry and atomic spacing along a particular crystal orientation. For example, the atomic spacing for silicon crystal along the (100) axis is different from that along the (111) and (110) axes. As a result, the dispersion relations for a silicon lattice are different along (100), (110), and (111) orientations. A similar situation applies to GaAs and aluminum. In general, the dispersion curves for most solids can be determined by using the inelastic slow neutron experiment. In this experiment, the energy losses of a slow neutron due to scattering by the lattice vibrations and the change of wave vector during scattering can be determined experimentally by using conservation of energy and conservation of momentum. A slow neutron atom impinging upon

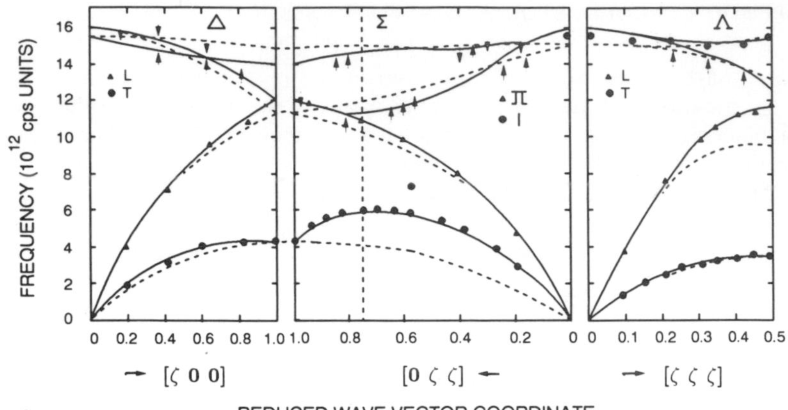

a

REDUCED WAVE VECTOR COORDINATE

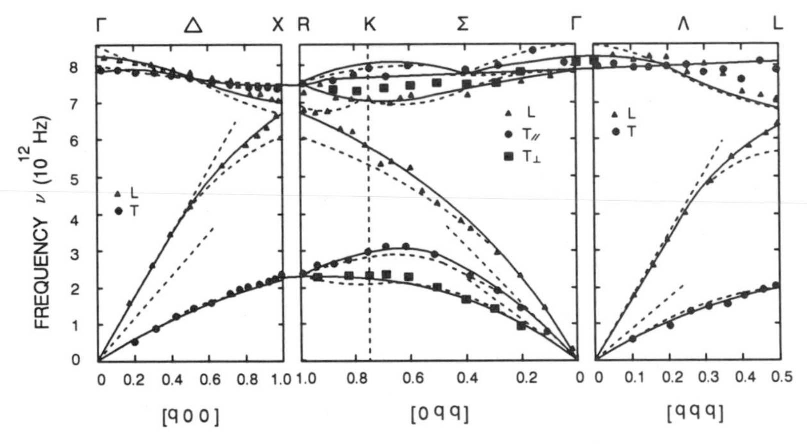

b

REDUCED (DIMENSIONLESS) WAVE-VECTOR, q

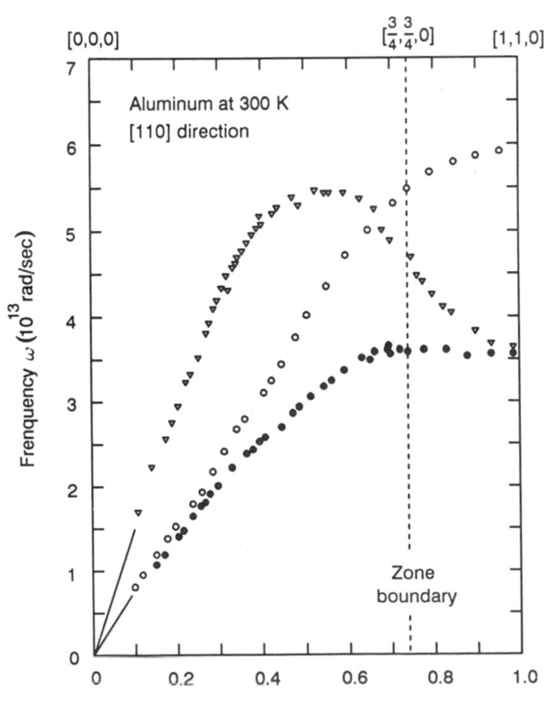

c

Reduced wave number ζ ,dimensionless

a crystal sees the crystal lattice mainly by interacting with the nuclei of the atoms. The momentum conservation for scattering a slow neutron by a lattice vibration mode can be described by

$$k = k' \pm q + K \tag{2.26}$$

where k is the wave vector of the incident neutron, k' is the wave vector of the scattered neutron, q is the phonon wave vector, and K is the reciprocal lattice vector. The plus sign in Eq. (2.26) denotes the creation of a phonon, while the minus sign is for phonon annihilation. Note that we have introduced here the terminology "phonon" for the quantized lattice vibration. The concept of phonon will be discussed further in the next section.

The conservation of energy for scattering of a slow neutron by a lattice atom is given by

$$\frac{\hbar^2 k^2}{2M_n} = \frac{\hbar'^2 k'^2}{2M_n} \pm \hbar\omega_q \tag{2.27}$$

where $\hbar\omega_q$ is the phonon energy. In Eq. (2.27), the plus sign is for phonon emission and the minus sign is for phonon absorption. To determine the dispersion relation from Eqs. (2.26) and (2.27) one needs to find the energy gain and loss of the scattered neutrons as a function of the scattering direction (i.e., $k - k'$) from the slow neutron experiment. This method has been widely used for determining the phonon spectra in the crystalline solids such as metals and semiconductors.

2.4. CONCEPT OF PHONONS

The dispersion relations derived in the previous sections for the one-dimensional monatomic and diatomic linear chains are based on Hooke's law. The results of this classical approach provide a good insight concerning the physical properties of lattice waves which contribute to the specific heat of a crystalline solid. However, it is inconvenient to use the wave concept of lattice vibrations to deal with the problems of interactions between electrons and lattice waves in a crystalline solid, such as scattering of electrons by the lattice waves in a semiconductor or a metal. In fact, it is common practice to use the quantum mechanical approach to solve the problem of scattering of electrons by the lattice vibrations in a solid. In the framework of quantum mechanics, each lattice vibration mode is quantized, and can be treated as a quantum oscillator with a characteristic frequency ω and a wave vector q. This quantized lattice vibration mode is usually referred to as the "phonon", analogous to referring to "photon" as a quantum unit of the electromagnetic radiation. Therefore, it is appropriate to introduce here the concept of "phonons" to represent the quantized lattice vibration modes.

In Section 2.3, we introduced the normal coordinates and normal modes to describe the quantum oscillators for the three-dimensional lattice vibration modes. A quantized lattice vibration mode is represented by a harmonic oscillator, each with a characteristic wave vector q, frequency ω, and polarization index s. According to quantum theory, the energy of a harmonic oscillator is given by

$$E_n = (n + 1/2)\hbar\omega \tag{2.28}$$

FIGURE 2.5. (a) Dispersion curves for silicon along (100) and (111) orientations, (b) for GaAs along (111), (110), and (100) orientations, and (c) for aluminum. After Dolling,[1] Waugh and Dolling,[2] Wallis,[3] by permission.

where $n = 0, 1, 2, 3, \ldots$ and ω is the characteristic frequency of the quantum oscillator. Using Eq. (2.28), the phonon energy can be written as

$$E_n(q, s) = (n_{q,s} + 1/2)\hbar\omega(q, s) \tag{2.29}$$

where $n_{q,s} = 1/[\exp(\hbar\omega/k_B T) - 1]$ is the average phonon occupation number, which can be derived by using Bose–Einstein statistics, as will be discussed later in Chapter 3. The quantity $\hbar\omega/2$ on the right-hand side of Eq. (2.29) represents the zero-point phonon energy (i.e., $n_{q,s} = 0$). It should be noted that the zero-point energy does not affect the distribution function of phonons in any way, nor does it contribute to the average internal energy and the specific heat of a solid at temperatures above absolute zero degree kelvin. A large value of $n_{q,s}$ in Eq. (2.29) corresponds to phonons with large amplitude, and *vice versa*. In the dispersion curves shown previously, the acoustical branch consists of both the longitudinal acoustical (LA) phonons and the transverse acoustical (TA) phonons, while the longitudinal and transverse optical (LO and TO) phonons form the optical branch.

It is noteworthy that phonon scatterings are usually the dominant scattering mechanisms which control the carrier mobilities in an intrinsic or a lightly doped semiconductor. The role of phonon scatterings on the transport properties and carrier mobilities in a semiconductor will be discussed further in Chapter 7 and Chapter 8.

2.5. THE DENSITY OF STATES AND LATTICE SPECTRUM

The density-of-states function for phonons in a crystalline solid can be derived by using the periodic boundary conditions of the crystal lattice. For a three-dimensional cubic lattice, if the length of each side of the cubic cell is equal to L, then the density-of-states function can be derived by using the periodic boundary conditions over the N^3 atoms within the cube cell with a volume of L^3. Values of the phonon wave vector q are determined by using the three-dimensional periodic boundary conditions given by

$$e^{i(q_x x + q_y y + q_z z)} = e^{i[q_x(x + L) + q_y(y + L) + q_z(z + L)]} \tag{2.30}$$

which is reduced to

$$e^{i(q_x + q_y + q_z)L} = 1 \tag{2.31}$$

From Eq. (2.31) we obtain $q_x, q_y, q_z = 0, \pm 2\pi/L, \pm 4\pi/L; \ldots; N\pi/L$. Therefore, there is one allowed value of $\boldsymbol{q}$ per unit volume $(2\pi/L)^3$ in q-space (i.e., momentum space). To find a general expression of the phonon density-of-states function $D(\omega)$ for a three-dimensional crystal lattice, the total number of states per unit volume with frequencies between ω and $\omega + d\omega$ can be expressed by

$$D(\omega)\, d\omega = \left(\frac{L}{2\pi}\right)^3 \int_{\text{shell}} d^3q \tag{2.32}$$

The integrand of Eq. (2.32) represents the total number of states available within a spherical shell in q-space with frequencies varying between ω and $\omega + d\omega$, and $(2\pi/L)^3$ is the volume of the unit cell in q-space. As shown in Fig. 2.6, dS_ω is the area element on the constant frequency surface in q-space, and d^3q is the volume element which can be expressed by

$$d^3q = dS_\omega \, dq_\perp \tag{2.33}$$

Now substituting Eq. (2.33) into Eq. (2.32) and using the relation that $dq_\perp = d\omega/|\nabla_q\omega|$, we obtain a general expression for the phonon density-of-states function, which is given by

$$D(\omega) = \left(\frac{L}{2\pi}\right)^3 \int_{\text{shell}} \frac{dS_\omega}{|\nabla_q\omega|} \tag{2.34}$$

or

$$D(\omega) = \left(\frac{L}{2\pi}\right)^3 \int \frac{dS_\omega}{v_g} \tag{3.35}$$

where $v_g = |\nabla_q\omega|$ is the group velocity. Note that integration of Eq. (2.34) is carried out over the constant frequency surface in q-space.

It is clear that an expression for the phonon density-of-states function can be derived from Eq. (2.34) provided that the dispersion relation between ω and q is known. Figure 2.7a shows a plot of phonon density of states as a function of frequency for copper, and Fig. 2.7b shows the corresponding density-of-states function derived by using a dispersionless relation between ω and q (i.e., assuming $\omega = u_s q$, where u_s is the velocity of sound) in Eq. (2.34).

The phonon density-of-states curve shown in Fig. 2.7a for a copper crystal was obtained from the numerical analysis of the measured dispersion curve. In general, if the constant frequency surface in q-space is spherical, then in the long-wavelength limit, the density-of-states function $D(\omega)$ is proportional to the square of frequency. Note that Eq. (2.34) may be modified to find the density-of-states function for electrons in the conduction band or for holes in the valence band of a semiconductor. The modifications include taking into account the spin degeneracy factor due to the Pauli exclusion principle and replacing the frequency of phonons by the energy of electrons or holes in Eq. (2.34). This will be discussed further in Chapter 4.

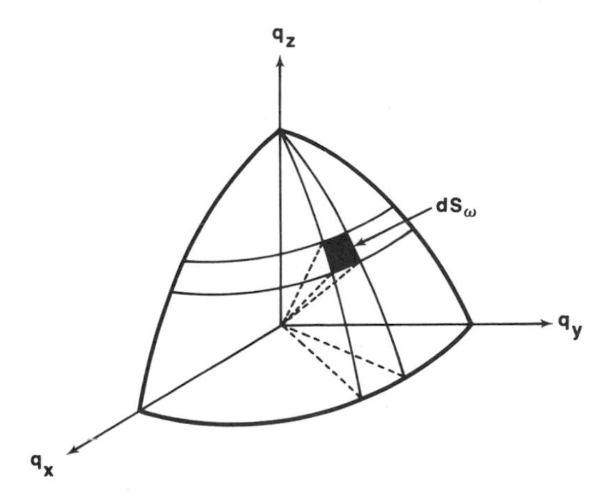

FIGURE 2.6. Constant frequency surface in q-space; $dS_\omega \, dq$ is the volume element between two surfaces of constant frequency ω and $\omega + d\omega$.

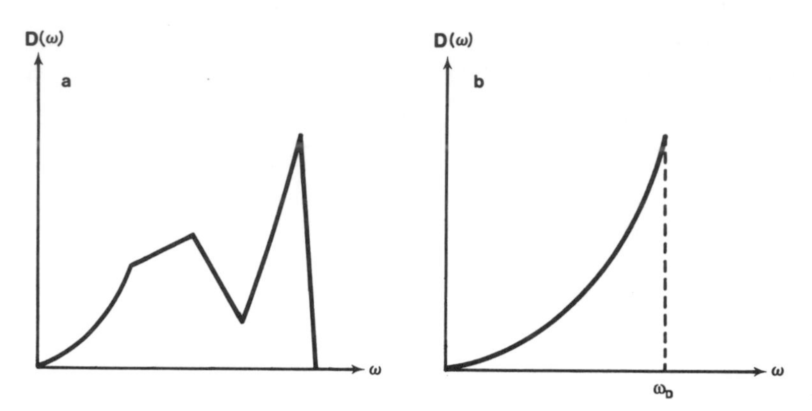

FIGURE 2.7. (a) Density of phonon states versus frequency for copper. (b) Debye lattice spectrum, where ω_D is the Debye cutoff frequency.

2.6. LATTICE SPECIFIC HEAT

The classical Dulong and Petit law described in Section 2.1 fails to predict correctly the temperature dependence of the lattice specific heat of solids at very low temperatures. The reason for this failure arises from the fact that the Dulong and Petit law does not consider all the lattice vibration modes, particularly the long-wavelength phonons which are the most important vibration modes at low temperatures. In deriving the lattice specific heat, Debye uses a continuum model to account for all the possible lattice vibration modes. This assumption is valid as long as the wavelength of phonons is large compared to the interatomic spacing. In this respect, a solid is considered as a continuous medium to the lattice phonons. Furthermore, the number of vibration modes is limited by the total number of constituent atoms, which is equal to N. Therefore, for N atoms each with three degrees of freedom, the total number of vibration modes would be equal to $3N$. In other words, the frequency spectrum corresponding to a perfect continuum is cut off so as to comply with a total of $3N$ vibration modes. The Debye cutoff frequency, ω_D, corresponds to a maximum frequency which a transverse and a longitudinal vibration mode can support. We next discuss the lattice specific heat based on the Debye model.

To derive the lattice specific heat of a crystalline solid, we need to find the total internal energy due to the thermal vibrations of lattice atoms. Using Eq. (2.29) for the average phonon energy and ignoring the zero-point energy, the total energy for the lattice phonons can be expressed by

$$U = \int_0^{\omega_D} \frac{D(\omega)\hbar\omega\, d\omega}{(e^{\hbar\omega/k_BT} - 1)} \tag{2.36}$$

where $D(\omega)$ is the density of states per unit frequency given by Eq. (2.34). To find the solution of Eq. (2.36), the expression for $D(\omega)$ and the dispersion relation between ω and q must first be derived. In the Debye model, it is assumed that the solid under consideration is an isotropic dispersionless continuum, and thus the relation between ω and q can be written as

$$\omega = v_g q = v_p q = v_s q \tag{2.37}$$

where v_g, v_p, and v_s denote the group velocity, phase velocity, and the velocity of sound in a solid, respectively. From Eq. (2.37), it is noted that both the group and phase velocities are equal to the velocity of sound in a dispersionless continuum medium. In an isotropic medium, the phase velocity of phonons is independent of the direction of wave vector q. To derive the phonon density-of-states function, let us consider the spherical shell in q-space shown in Fig. 2.6. Equations (2.34) through (2.37) can be employed to determine the number of vibration modes within this spherical shell for frequencies between ω and $\omega + d\omega$, which is given by

$$D(\omega)\,d\omega = \left(\frac{3V\omega^2}{2\pi^2 v_s^3}\right) d\omega \tag{2.38}$$

where $V = L^3$ is the volume of the cubic cell. A typical Debye spectrum calculated from Eq. (2.38) is shown in Fig. 2.7b. A factor of 3 is included in Eq. (2.38) to account for the three components of polarization (i.e., two transverse and one longitudinal) per wave vector. In general, the propagation velocities for the transverse-mode phonons and the longitudinal-mode phonons are not equal (i.e., $v_t \neq v_l$), and hence Eq. (2.38) must be replaced by

$$D(\omega)\,d\omega = \left(\frac{V}{2\pi^2}\right)\left(\frac{2}{v_t^3} + \frac{1}{v_l^3}\right)\omega^2\,d\omega \tag{2.39}$$

The Debye cutoff frequency ω_D can be obtained by integrating Eq. (2.39) for $D(\omega)$ and by using the fact that there are $3N$ total vibration modes in the crystal. Thus, the total number of vibration modes is given by

$$\int_0^{\omega_D} D(\omega)\,d\omega = 3N \tag{2.40}$$

which yields

$$\omega_D = (6\pi^2 n)^{1/3} v_s \tag{2.41}$$

where $n = N/V$ is the number of atoms per unit volume and

$$v_s = \left[\frac{1}{3}\left(\frac{2}{v_t^3} + \frac{1}{v_l^3}\right)\right]^{-1/3} \tag{2.42}$$

is the average velocity of phonons in the solid. The total energy of phonons can be obtained by substituting Eq. (2.38) into Eq. (2.36) and integrating Eq. (2.36) from $\omega = 0$ to $\omega = \omega_D$, to yield

$$\begin{aligned} U &= \left(\frac{3V}{2\pi^2 v_s^3}\right)\int_0^{\omega_D} \frac{\hbar\omega^2\,d\omega}{e^{\hbar\omega/k_B T} - 1} \\ &= \left(\frac{3Vk_B^4T^4}{2\pi^2\hbar^3 v_s^3}\right)\int_0^{x_m} \frac{x^3\,dx}{e^x - 1} \end{aligned} \tag{2.43}$$

where $x = \hbar\omega/k_B T$, $x_m = \hbar\omega_D/k_B T = T_D/T$, and $T_D = \hbar\omega_D/k_B$ is the Debye temperature. The lattice specific heat per unit volume can be derived from Eq. (2.43) by differentiating the total energy U with respect to the temperature, which is

$$C_v = \frac{dU}{dT} = 9Nk_B\left(\frac{T}{T_D}\right)^3 \int_0^{T_D/T} \frac{e^x x^4\,dx}{(e^x - 1)^2} \tag{2.44}$$

Note that an analytical expression cannot be obtained from Eq. (2.44) that is valid over the entire temperature range. However, an approximate expression can be derived for two limiting cases, namely, for $T \gg T_D$ and $T \ll T_D$. This is discussed as follows:

1. At high temperatures (i.e., for $T \gg T_D$ or $x \ll 1$): In this case, Eq. (2.44) reduces to

$$C_v = 3Nk_B = 3R \tag{2.45}$$

which is identical to the result given by the classical Dulong and Petit law.

2. At low temperatures (i.e., for $T \ll T_D$ or $x \gg 1$): In this case, the upper limit of the integral in Eq. (2.44) may be replaced by infinity, and the integral is given by

$$\int_0^\infty \frac{x^3\,dx}{(e^x - 1)} = \frac{\pi^4}{15} \tag{2.46}$$

Now, by substituting Eq. (2.46) into Eq. (2.43) and differentiating the total energy U with respect to T we obtain the lattice specific heat as

$$C_v = \left(\frac{12\pi^4}{5}\right) R \left(\frac{T}{T_D}\right)^3 \tag{2.47}$$

Equation (2.47) shows that the lattice specific heat is proportional to T^3 at low temperatures. The theoretical prediction given by Eq. (2.47) is in good agreement with the observed temperature behavior of the lattice specific heat for both the semiconductors and insulators at low temperatures. The reason for this good agreement may be attributed to the fact that the Debye model has taken into account the contribution of long-wavelength acoustic phonons to the lattice specific heat, which is important at low temperatures. Figure 2.8 shows a comparison of the lattice specific heat versus temperature predicted by the Debye model and by the Dulong and Petit law.

Although the Debye model generally gives a correct prediction of the lattice specific heat for both insulators and semiconductors over a wide range of temperatures, the Debye temperature used in the theoretical fitting of the experimental data varies from material to material. For example, the Debye temperature is 640 K for silicon and 370 K for germanium.

In spite of the success of the Debye model for predicting the correct temperature behavior of lattice specific heat of semiconductors and insulators over a wide range of temperatures, it fails, however, to predict the correct temperature dependence of the specific heat of metals at very low temperatures. The reason for its failure stems from the fact that the electronic specific

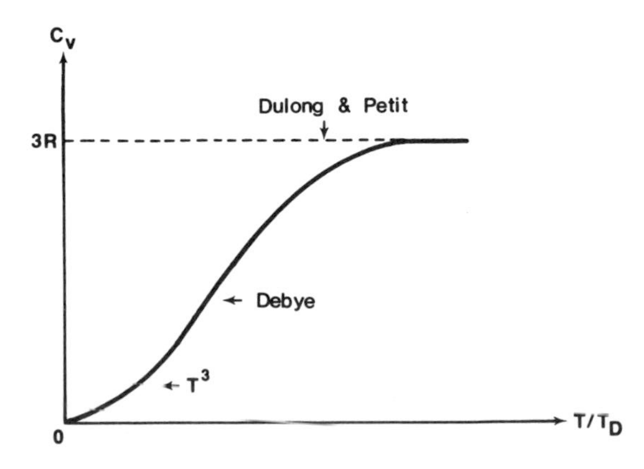

FIGURE 2.8. Lattice specific heat C_v versus normalized temperature T/T_D, as predicted by the Debye model and the Dulong and Petit law.

heat, which is contributed by the total kinetic energy of electrons in a metal, becomes dominant at very low temperatures. In fact, the specific heat of a metal is controlled by the electronic specific heat rather than by the lattice specific heat at very low temperatures. It can be shown by using Fermi–Dirac statistics that the electronic specific heat for a metal varies linearly with temperature, which is in good agreement with the experimental observation at low temperatures. Thus, the total specific heat for a metal consists of the lattice specific heat and the electronic specific heat, which can be expressed by

$$C_v = C_l + C_e = \alpha T^3 + \beta T \tag{2.48}$$

where C_l and C_e denote the lattice and electronic specific heat, respectively. Both α and β are constants which can be determined from the C_v/T versus T^2 plot. From the slope of this plot we can determine the constant α, and the intercept, when extrapolated to $T = 0$ K, yields the constant β. It is noted that at low temperatures, the second term in Eq. (2.48) dominates and the specific heat of metals varies linearly with temperature. Derivation of the electronic specific heat for a metal can be carried out by using Fermi–Dirac statistics. This can be found in Problems 3.6 and 3.7 of Chapter 3.

2.7. ELASTIC CONSTANTS AND VELOCITY OF SOUND

Another important thermal property that needs to be discussed here is associated with the stress–strain relation in a semiconductor. The elastic constant is an important physical parameter when dealing with scattering of electrons by phonons in a semiconductor. In this section we describe the stress–strain relation when a small stress is applied to a semiconductor. The tensor relation of stress to strain in this case involves the second-order elastic moduli. These elastic moduli can be determined accurately by using the ultrasonic speed of sound measurements along different crystal orientations. As an example, Fig. 2.9 shows the pressure

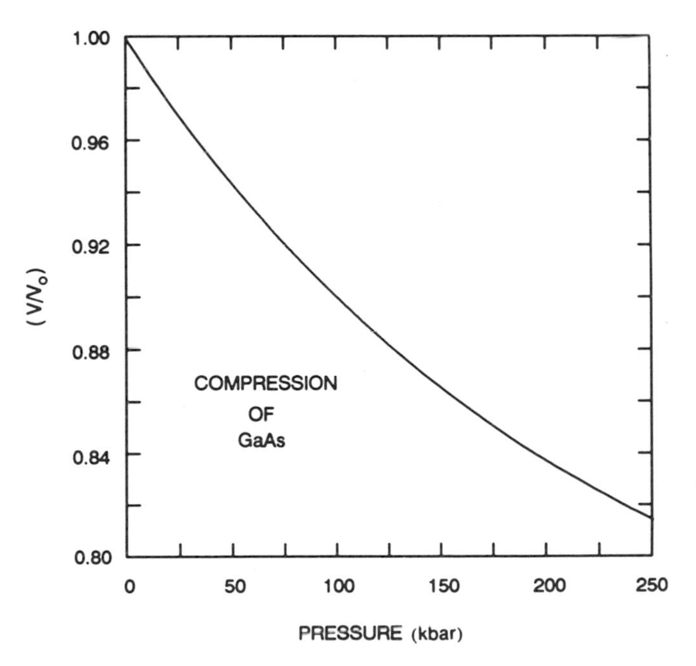

FIGURE 2.9. Pressure–volume relation for a GaAs crystal, as calculated from the bulk modulus for zero pressure. After McSkimin,[4] by permission.

dependence of the specific volume for a single crystal GaAs, as calculated from the bulk modulus at zero pressure.[4] The curve was calculated by assuming that the bulk modulus is linearly dependent on the pressure.

Since GaAs is a cubic crystal with zinc blende lattice structure, the small-stress adiabatic elastic response tensor can be simplified to only three independent second-order moduli, namely, c_{11}, c_{12}, and c_{44}. The speed of long-wavelength longitudinal and transverse acoustic waves can be determined from the density of the crystal, ρ, and the elastic constants c_{11}, c_{12}, c_{44} for any crystal orientations and polarizations. Figure 2.10a, b, and c show the plots of three elastic constants versus lattice constant for several III–V compound semiconductor materials.[5] The solid lines in this figure are obtained by applying the least-squares fit to the equation

$$c_{ij} = A_{ij}a + B_{ij} \tag{2.49}$$

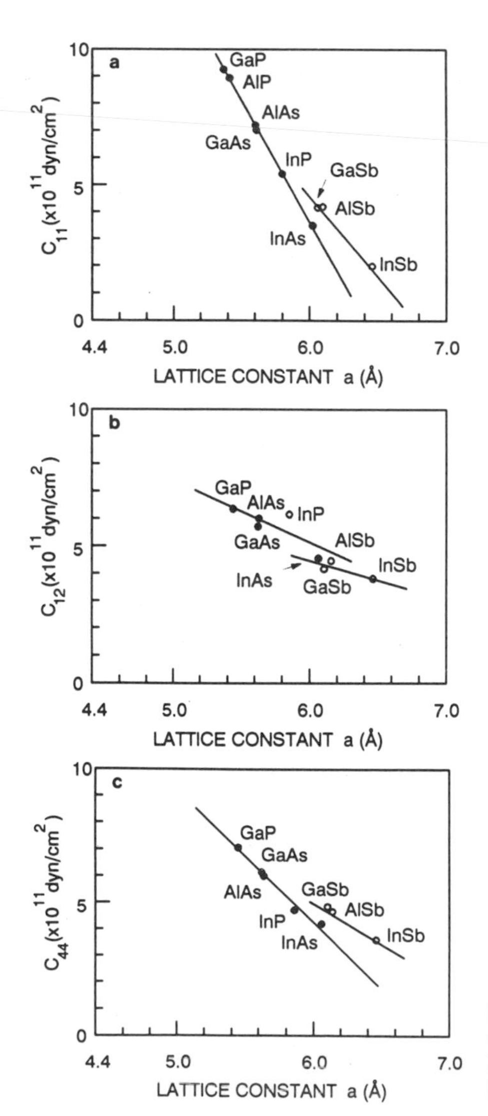

FIGURE 2.10. Elastic stiffness constants c_{11}, c_{12}, and c_{44} versus lattice constant a (Å) for some III–V binary compound semiconductors. After de Launay,[5] by permission.

where A_{ij} and B_{ij} are the fitting constants and a is the lattice constant in Å. It is clearly shown that a good fitting is obtained with the same crystal structure for these III–V compound semiconductors. It should be noted that the elastic constants for III–V binary compound semiconductors have been extensively studied. The elastic constants for III–V ternary compounds such as $Al_xGa_{1-x}As$ and $In_xGa_{1-x}As$ are directly proportional to the alloy mole fraction x. Furthermore, it is also found that the lattice constant a in many III–V compounds is also varied linearly with the alloy mole fraction. The interpolation scheme of Eq. (2.49) should provide reliable elastic parameters for the III–V ternary compounds such as AlGaAs and InGaAs.

The velocity of sound v_s in a crystalline solid can be determined if the density ρ and the stiffness constants c_{ij} of the semiconductors are known. The velocity of sound can be related to these two constants by

$$v_s = \left(\frac{c_{ij}}{\rho}\right)^{1/2} \tag{2.50}$$

Equation (2.50) shows that the velocity of sound is proportional to the square root of the elastic constants. Expressions for the velocity of sound as a function of crystal density ρ and elastic constants c_{ij} for phonons propagating in the [100], [110], and [111] directions of a GaAs crystal are given respectively by

$$\begin{aligned} v_L &= [c_{11}/\rho]^{1/2} \quad &\text{for [100] axis} \\ v_T &= [c_{44}/\rho]^{1/2} \quad &\text{for [100] plane} \end{aligned} \tag{2.51}$$

and

$$\begin{aligned} v_L &= [(c_{11} + 2c_{12} + 4c_{44})/3\rho]^{1/2} \quad &\text{for [111] axis} \\ v_T &= [(c_{11} - c_{12} + c_{44})/\rho]^{1/2} \quad &\text{for [111] plane} \end{aligned} \tag{2.52}$$

where subscripts L and T denote the longitudinal and transverse components, respectively. For propagation along the [110] axis, the velocity of sound can be calculated by using the expressions

$$\begin{aligned} v_L &= [(c_{11} + c_{12} + 2c_{44})/2\rho]^{1/2} \quad &\text{for [110] axis} \\ v_{T1} &= [c_{44}/\rho]^{1/2} \quad &\text{for [001] plane} \\ v_{T2} &= [(c_{11} - c_{12})/2\rho]^{1/2} \quad &\text{for } [1\bar{1}0] \text{ plane} \end{aligned} \tag{2.53}$$

where v_L is the velocity of longitudinal-mode acoustic phonons, v_{T1} is the velocity for the fast transverse-mode acoustical phonons, and v_{T2} is the velocity for the slow transverse-mode acoustical phonons. For a GaAs crystal, v_{T1} is equal to v_{T2} for phonons propagating along the [100] direction. On the other hand, v_{T1} is larger than v_{T2} for phonons propagating along the [110] direction. Using Eqs. (2.51) through (2.53) and the elastic constants shown in Fig. 2.10, we can calculate the velocities of sound for the III–V compound semiconductors. As an example, using values of c_{ij} shown in Fig. 2.10 and value of ρ for GaAs, we found that $v_L = 5.65 \times 10^5$ cm/sec and $v_T = 4 \times 10^5$ cm/sec for phonon propagation along the [100] axis and on the [001] plane, respectively.

PROBLEMS

2.1. (a) Considering only the nearest-neighbor interaction, find the dispersion relation for the diatomic linear chain of a silicon lattice along the (111) crystal axis. Note that the masses of the silicon atoms are identical, and the positions of the nearest-neighbor atoms are located at $(0, 0, 0)$, $(a/4, a/4, a/4)$, and (a, a, a) inside the unit cell. Assume that the force constant is equal to β.

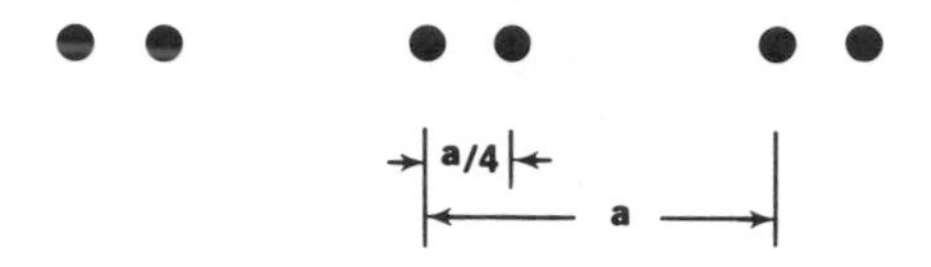

(b) Plot the dispersion curves from the result obtained in (a).
(c) Sketch the atomic displacement for the longitudinal and transverse optical lattice vibration modes at $q = \pi/4\sqrt{3}a$, $\pi/2\sqrt{3}a$, and $\pi/\sqrt{3}a$.

2.2. Using the Einstein model, derive the lattice specific heat for a three-dimensional crystal lattice. (*Hint*: The Einstein model is similar to the Debye model except that it assumes a single vibration frequency ω_E for all lattice phonons. Use this assumption to derive the average phonon energy and the lattice specific heat.)

2.3. (a) For a one-dimensional monatomic linear chain with fixed end boundary condition, show that the density-of-states function is given by

$$D(\omega) = \frac{L}{\pi} \cdot \frac{dq}{d\omega}$$
$$= \left(\frac{2L}{\pi a}\right) \cdot \frac{1}{(\omega_m^2 - \omega^2)^{1/2}}$$

[*Hint*: Use the dispersion relation given by Eq. (2.5) to derive $D(\omega)$.]
(b) Apply the Debye model to this one-dimensional linear chain, and show that the density-of-states function can be expressed by

$$D(\omega) = \frac{L}{\pi v_s}$$

where L is the length of the linear chain.
(c) Plot $D(\omega)$ versus ω for (a) and (b), and explain the difference.

2.4. Apply the Debye model to derive the specific heat for a one-dimensional monatomic linear chain with only nearest-neighbor interaction, and show that at low temperatures the specific heat varies linearly with temperature. What is the cutoff frequency in this case?

2.5 (a) Write down the equations of motion for a one-dimensional linear chain of identical masses which are connected to each other by springs of two difference force constants, β_1 and β_2, in alternating positions. Find the dispersion relation from the solution for this linear chain.
(b) Plot the dispersion curve for the linear chain given in (a).

2.6. (a) Write down the equation of motion for a two-dimensional square lattice with spacing a and atomic mass M. The nearest-neighbor force constant is given by β.
(b) Assume that the solution of (a) is given by

$$u_{lm} = u(0) \exp[i(lk_x a + mk_y a - \omega t)]$$

where u_{lm} denotes the displacement normal to the plane of the square lattice for the atom in the lth column and mth row. Find the dispersion relation in (a).
(c) Plot the dispersion curve for a square lattice based on the result obtained in (b).

2.7. Derive the density-of-states function $D(\omega)$ in a one-dimensional linear chain of length L carrying $N + 1$ particles at a separation of a for the following cases:
(a) The particles $s = 0$ and $s = N$ at the ends of the linear chain are held fixed (i.e., fixed boundary condition).

(b) The linear chain is allowed to form a ring, so that the periodic boundary condition can be applied to the problem [i.e., $u(sa) = u(sa + L)$].
(c) What are the allowed values of the wave vector q in cases (a) and (b)?

2.8. One method of determining the phonon spectra in a solid is by using the slow neutron scattering experiment. Give an example to explain this technique in determining the phonon spectra of a crystal. [See, for example, the paper by A. D. B. Woods, B. N. Brockhouse, R. A. Lowley, and W. Cochran, *Phys. Rev.* **131**, 1025 (1963); see also *Phys. Rev.* **119**, 980 (1960).]

REFERENCES

1. G. Dolling, *Proceedings of Symposium on Inelastic Scattering in Solids and Liquids, II*, Chalk River, P. 37, IAEA, Vienna (1963).
2. J. L. T. Waugh and G. Dolling, *Phys. Rev.* **132**, 2410 (1963).
3. R. F. Wallis (ed.), *Lattice Dynamics*, p. 60, Pergamon Press, New York (1965).
4. H. J. McSkimin, *J. Appl. Phys.* **38**, 2362 (1967).
5. J. de Launay, in: *Solid State Physics*, Vol. 2 (F. Scitz and D. Turnbull, eds.), p. 219, Academic Press, New York (1956).

BIBLIOGRAPHY

L. Brillouin, *Wave Propagation in Periodic Structures*, 2nd ed., McGraw-Hill, New York (1946).
M. Born and K. Huang, *Dynamical Theory of Crystal Lattices*, Oxford University Press, London (1954).
W. Cochran, *Dynamics of Atoms in Crystals*, Crane, Russak, New York (1975).
C. Kittel, *Introduction to Solid State Physics*, 5th ed., Wiley, New York (1975).
A. A. Maradudin, E. W. Montroll, G. H. Weiss, and I. P. Ipatova, *Theory of lattice dynamics in the harmonic approximation*, in: *Solid State Physics*, 2nd ed., Suppl. 3 (1971).
S. S. Mitra, in: *Solid State Physics*, Vol. 13 (F. Seitz and D.Turnbull, eds.), p. 1, Academic Press, New York (1956).
J. M. Ziman, *Electrons and Phonons*, Oxford University Press, London (1960).

3

Semiconductor Statistics

3.1. INTRODUCTION

In this chapter we present the three basic statistics which are commonly used in the derivation of distribution functions for gas molecules, photons and phonons, electrons in a metal, and electrons and holes in a semiconductor Without using these basic statistics, it is impossible to deal with the problems of interactions of a large number of particles in a solid. Since a great deal of physical insight can be obtained and learned from statistical analysis of the particle distribution functions in a solid, it is appropriate for us to devote this chapter to finding the distribution functions associated with different statistical mechanics for particles such as gas molecules, photons, phonons, electrons, and holes. The three basic statistics which govern the distribution of particles in a solid are: (1) Maxwell–Boltzmann (M–B) statistics, (2) Bose–Einstein (B–E) statistics, and (3) Fermi–Dirac (F–D) statistics. The M–B statistics are also known as the classical statistics, since they apply only to particles with weak interactions among themselves. In the M–B statistics, the number of particles in each quantum state is not restricted by the Pauli exclusion principle. Particles such as gas molecules in an ideal gas system and electrons and holes in a dilute semiconductor are examples which obey the M–B statistics. The B–E and F–D statistics are known as the quantum statistics because their distribution functions are derived based on quantum mechanical principles. Particles which obey the B–E and F–D statistics usually have a much higher density and stronger interaction among themselves than the classical particles. Particles which obey the B–E statistics, including photons and phonons, are called Bosons, while particles which obey the F–D statistics, including electrons and holes in a degenerate semiconductor or electrons in a metal, are known as Fermions. The main difference between the F–D and the B–E statistics is that the occupation number in each quantum state for the Fermions is restricted by the Pauli exclusion principle, while Bosons are not subjected to the restriction of the exclusion principle. The Pauli exclusion principle states that no more than two electrons with opposite spin degeneracy can occupy the same quantum state. Therefore, the total number of particles with the same spin should be equal to or less than the total number of quantum states available for occupancy.

The M–B statistics and velocity distrbution function for the ideal gas molecules are depicted in Section 3.2. Section 3.3 presents the F–D statistics, the physics aspect of the F–D distribution function, and its derivative for the free-electron case. Section 3.4 deals with the B–E statistics and the distribution function of phonons and photons. The blackbody radiation formula is also discussed. Finally, the distribution functions for electrons in the shallow donor states and holes in the shallow acceptor states inside the forbidden gap of a semiconductor are derived in Section 3.5.

3.2. MAXWELL–BOLTZMANN (M–B) STATISTICS

In this section, the M–B distribution function for the classical noninteracting particles such as ideal gas molecules and electrons in a dilute semiconductor is derived. Let us first consider an isolated system which contains N distinguishable particles with a total energy of E. Let us then consider the problem of distributing N particles among the q energy levels. If the system consists of $E_1, E_2, \ldots, E_i, \ldots, E_q$ energy levels with $n_1, n_2, \ldots, n_i, \ldots, n_q$ particles in each corresponding energy level, then there are two constraints imposed upon these N particles. These two constraints (i.e., conservation of energy and conservation of particles) can be expressed by

$$C_1(n_1, n_2, \ldots, n_i, \ldots, n_q) = N = \sum_{i=1}^{q} n_i \tag{3.1}$$

and

$$C_2(n_1, n_2, \ldots, n_i, \ldots, n_q) = E = \sum_{i=1}^{q} n_i E_i \tag{3.2}$$

To derive the classical M–B distribution function, the framework of quantum states and energy levels in a noninteracting system is retained and the Pauli exclusion principle is neglected. Therefore, there is no limitation on the number of particles which can be put into a quantum state at a given energy level in the system. In order to derive the distribution function for particles in a noninteracting system, we first analyze the problem of distributing N_1 and N_2 balls in two boxes, and then extend this result to the problem of particle distribution in a solid.

If $W(N_1, N_2)$ represents the total number of independent ways of arranging N_1 and N_2 balls in box 1 and box 2, respectively, then $W(0, N_2)$ is the total number of ways of making box 1 empty and box 2 full. There is only one possible arrangement for this case, and hence $W(0, N_2) = 1$. Next, we consider the case of arranging 1 ball in box 1 and $(N - 1)$ balls in box 2. In this case, there are N different ways of putting 1 ball in box 1, and hence the total number of ways of arranging 1 ball in box 1 and $(N - 1)$ balls in box 2 is given by $W(1, N_2) = N$. Next, let us consider the case of arranging 2 balls in box 1 and $(N - 2)$ balls in box 2. The number of ways of arranging the first ball in box 1 is N, and the number of ways of arranging the second ball in box 1 is $(N - 1)$. Thus, the total number of ways of arranging 2 balls in box 1 and $(N - 2)$ balls in box 2 is $W(2, N - 2) = N(N - 1)/2!$, where 2! is included to account for the permutation between two identical balls. Similarly, we can extend this procedure to the distribution of N_1 balls in box 1 and N_2 balls in box 2. Thus, we can write the total number of ways of arranging N_1 and N_2 balls in boxes 1 and 2 as

$$W(N_1, N_2) = \frac{N(N-1)(N-2)(N-3)\cdots(N-N_1+1)}{N_1!} = \frac{N!}{N_1!N_2!} \tag{3.3}$$

Extending the above procedure, the total number of ways of arranging $N_1, N_2, N_3, \ldots, N_q$ balls in boxes $1, 2, 3, \ldots, q$ can be expressed by

$$W(N_1, N_2, N_3, \ldots, N_q) = \frac{N!}{N_1!N_2!N_3!\cdots N_q!} = \frac{N!}{\prod_{i=1}^{q} N_i!} \tag{3.4}$$

where $N = N_1 + N_2 + N_3 + \ldots + N_q$ is the total number of balls available for distribution in q boxes.

The above results can be applied to find the M–B distribution function for particles in a solid. Next, let us consider the distribution of n particles among the $E_1, E_2, E_3, \ldots, E_q$ energy levels in a solid. If we assume that there are $g_1, g_2, g_3, \ldots, g_q$ degenerate quantum states and $n_1, n_2, n_3, \ldots, n_q$ partices in each corresponding energy level $E_1, E_2, E_3, \ldots . E_q$, then the distribution function for the n particles among the q energy levels each with g_q degenerate states is similar to the distribution of N balls in q boxes discussed above. The only difference is that in this case there are an additional g_i quantum states in each E_i energy level (where $i = 1, 2, 3, \ldots, q$). If the quantum state in each energy level is nondegenerate, then the total number of ways of arranging $n_1, n_2, \ldots, n_q$ particles in the $E_1, E_2, E_3, \ldots, E_q$ energy levels is given by Eq. (3.4). If there are g_i degenerate quantum states in each energy level E_i, then it is necessary to count the total number of ways of arranging n_i particles in each of the g_i quantum states in the ith energy level. For example, the number of ways of arranging n_1 particles in g_1 quantum states in the energy level E_1 is given by $(g_1)^{n_1}$. Similarly, there are $(g_2)^{n_2}$ ways of arranging n_2 particles among the g_2 quantum states in the E_2 energy level. Therefore, the total number of ways of arranging $n_1, n_2, \ldots, n_q$ particles among the g_1, g_2, g_q quantum states in the $E_1, E_2, \ldots, E_q$ energy levels is given by

$$\begin{aligned} W(n_1, n_2, \ldots, n_i, \ldots, n_q) &= \frac{n!(g_1)^{n_1}(g_2)^{n_2} \ldots (g_q)^{n_q}}{(n_1!n_2! \ldots n_q!)} \\ &= n! \prod_{i=1}^{q} \left(\frac{(g_i)^{n_i}}{n_i!} \right) \end{aligned} \tag{3.5}$$

Taking the natural logarithm on both sides of Eq. (3.5), we obtain

$$\ln W(n_1, n_2, \ldots, n_q) = \ln n! + \sum_{i=1}^{q} (n_i \ln g_i - \ln n_i!) \tag{3.6}$$

Since values of n_i, g_i, and n are much much larger than unity, we can employ Stirling's approximation (i.e., $\ln x! \simeq x \ln x - x$, for $x \gg 1$) in Eq. (3.6), and the result yields

$$\ln W \simeq (n \ln n - n) + \sum_{i=1}^{q} (n_i \ln g_i - n_i \ln n_i + n_i) \tag{3.7}$$

From thermodynamics, the entropy of a solid is equal to $k_B \ln W$, where k_B is the Boltzmann constant, and $\ln W$ is given by Eq. (3.7). Furthermore, the most probable distribution function for particles in a solid can be obtained by maximizing the entropy of the system. Therefore, the distribution function for the particles described above can be obtained by maximizing Eq. (3.7) with respect to n_i. This is carried out by using the method of Lagrangian multipliers. From Eqs. (3.1), (3.2), and (3.7), we obtain

$$\begin{aligned} \frac{d \ln W}{dn_i} &= \frac{d}{dn_i} \left[n \ln n + \sum_{i=1}^{g} \left\{ n_i \ln \left(\frac{g_i}{n_i} \right) \right\} \right] \\ &= \alpha \frac{dC_1}{dn_i} + \beta \frac{dC_2}{dn_i} \end{aligned} \tag{3.8}$$

or

$$\ln \left(\frac{g_i}{n_i} \right) - 1 = \alpha + \beta E_i \tag{3.9}$$

where α and β are constants to be determined. From Eq. (3.9), the distribution function of classical particles can be written as

$$f(E_i) = \frac{n_i}{g_i} = \exp[-(1 + \alpha + \beta E_i)] \tag{3.10}$$

If we drop the index i from Eq. (3.10), then the M–B distribution function can be expressed by

$$f(E) = A \exp(-\beta E) \tag{3.11}$$

where A and β are constants to be determined. Both A and β can be determined from the distribution of gas molecules in an idea gas system, which obeys the M–B statistics. For example, if there are N monatomic gas molecules which interact only through collision processes, then the energy of such particles is purely kinetic and may be written as

$$E = \frac{mv^2}{2} = \left(\frac{m}{2}\right)(v_x^2 + v_y^2 + v_z^2) \tag{3.12}$$

By substituting Eq. (3.12) into Eq. (3.11), the M–B distribution function for an ideal gas molecule system can also be written in terms of the velocity distribution function, which is given by

$$N(v) = 4\pi A \ e^{-\beta mv^2/2} v^2 \tag{3.13}$$

The velocity distribution function $N(v)$ given by Eq. (3.13) represents the number of particles in the system whose velocities lie in a range dv about v. Using Eq. (3.13), the number of particles with velocities between v and $v + dv$ in velocity space is given by

$$dN = N(v)\, dv = 4\pi v^2 A \ e^{-\beta mv^2/2}\, dv \tag{3.14}$$

The total number of particles in velocity space is obtained by integrating Eq. (3.14) from zero to infinity, and the result yields

$$N = \int dN = \int_0^\infty 4\pi v^2 A \ e^{-\beta mv^2/2}\, dv \tag{3.15}$$

The total energy of the gas molecules in such a system is given by

$$U = \int_0^\infty \left(\frac{mv^2}{2}\right) 4\pi v^2 A \ e^{-\beta mv^2/2}\, dv \tag{3.16}$$

To determine constants A and β from Eqs. (3.15) and (3.16), the kinetic energy of the gas molecules must be determined first. This energy may also be obtained independently by using the equipartition law, which shows that the kinetic energy for a gas molecule in an ideal gas system is equal to $(3/2)k_BT$. Therefore, the total energy for an ideal gas system containing N gas molecules is given by

$$U = (\tfrac{3}{2}) N k_B T \tag{3.17}$$

Solving Eqs. (3.15) through (3.17) yields

$$A = N\left(\frac{m}{2\pi k_B T}\right)^{3/2} \quad \text{and} \quad \beta = \frac{1}{k_B T} \tag{3.18}$$

The velocity distribution function for a classical particle is obtained by substituting the expressions for A and β given by Eq. (3.18) into Eq. (3.13), and the result yields

$$N(v) = 4\pi N\left(\frac{m}{2\pi k_B T}\right)^{3/2} v^2 e^{-mv^2/2k_B T} \tag{3.19}$$

Figure 3.1 shows the plot of $N(v)$ versus v for three different temperatures. The average velocity for a classical particle which obeys the M–B statistics can be obtained by using the expression

$$\langle v \rangle = \frac{\int_0^\infty vN(v)\,dv}{\int_0^\infty N(v)\,dv} \tag{3.20}$$

where $N(v)$ is given by Eq. (3.19). A general expression for the average of velocity to the nth power, $\langle v^n \rangle$, in velocity space is given by

$$\langle v^n \rangle = \frac{\int_0^\infty v^n N(v)\,dv}{\int_0^\infty N(v)\,dv} \tag{3.21}$$

where $n = 1, 2, 3, \ldots$. Note that both the average velocity $\langle v \rangle$ and the average kinetic energy (i.e., $\sim\langle v^2 \rangle$) of a classical particle in the velocity space can be calculated by using Eq. (3.21). Table 3.1 lists some definite integrals which may be used to calculate the average velocity to the nth power for electrons or holes in a nondegenerate semiconductor or for gas molecules in an ideal gas system.

3.3. FERMI–DIRAC (F–D) STATISTICS

The M–B statistics described in Section 3.2 are applicable for the noninteracting particles which are assumed to be distinguishable. However, in reality it is usually impossible to distinguish electrons in a metal or in a degenerate semiconductor due to the extremely high

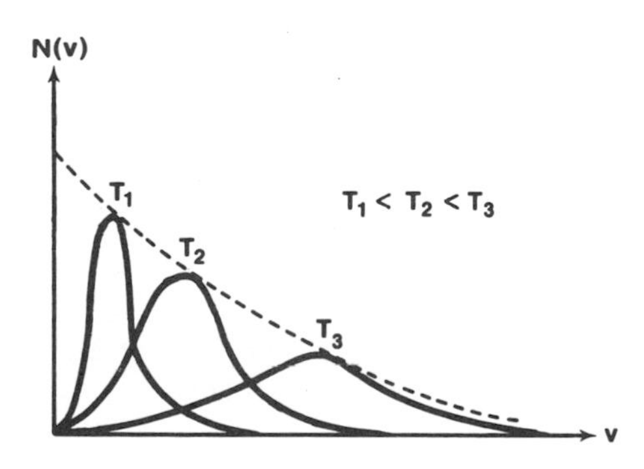

FIGURE 3.1. Maxwell–Boltzmann velocity distribution functon for three different temperatures.

TABLE 3.1. Some Frequently Used Integrals for the M–B Statistics

$$\int_0^\infty e^{-x^2}\,dx = (\pi/4a)^{1/2} = (\pi k_B T/2m)^{1/2}$$

$$\int_0^\infty x\,e^{-ax^2}\,dx = 1/2a = k_B T/m$$

$$\int_0^\infty x^2\,e^{-ax^2}\,dx = (\pi/16a^3)^{1/2} = (\sqrt{\pi}/4)(2k_B T/m)^{3/2}$$

$$\int_0^\infty x^3\,e^{-ax^2}\,dx = 1/2a_2 = (k_B T/2m)^2$$

$$\int_0^\infty x^4\,e^{-ax^2}\,dx = (3/8a^2)(\pi/a)^{1/2} = (3\sqrt{\pi}/8)(2k_B T/m)^{5/2}$$

$$\int_0^\infty x^5\,e^{-ax^2}\,dx = 1/a^3 = (2k_B T/m)^3$$

$$a = m/2k_B T.$$

density of electrons in these materials (i.e., $n_0 \geq 10^{19}\ \text{cm}^{-3}$). To apply statistics method to particles in such a system, an additional constraint imposed by the Pauli exclusion principle from quantum mechanics must be considered. According to the Pauli exclusion principle, no more than one electron with the same spin is allowed per quantum state in a degenerate electron system. The F–D distribution function, which takes into account the Pauli exclusion principle, may be used in finding the electron densities in a metal or in a heavily doped semiconductor. It is interesting to note that several important physical phenomena which cannot be explained properly by using the classical M–B statistics at very low temperatures come as a direct result of the F–D statistics.

To derive the F–D distribution function, three basic constraints must be considered. These are the conservation of particles, conservation of energy, and Pauli exclusion principle, which are given by

$$\sum_{i=1}^{q} n_i = n \tag{3.22}$$

$$\sum_{i=1}^{q} n_i E_i = E \tag{3.23}$$

and

$$n_i \leq g_i \tag{3.24}$$

where n_i and g_i denote the number of particles and degenerate quantum states in the ith energy level, respectively. The total energy of the system is assumed equal to E, and the total number of particles in the system under consideration is n. Equation (3.24) is the additional constraint imposed by the Pauli exclusion principle. To derive the F–D distribution function, it is appropriate to first consider the distribution of particles in the ith energy level. If there are n_i particles and g_i quantum states ($n_i \leq g_i$) in the E_i energy level, then in the g_i quantum states there are g_i ways of arranging the first particle, $(g_i - 1)$ ways of arranging the second particle, $(g_i - 2)$ ways of arranging the third particle and so on. Thus, the total number of ways of arranging n_i particles in the g_i quantum states in the E_ith energy level is

$$g_i(g_i - 1)(g_i - 2)\cdots(g_i - n_i + 1) = \frac{g_i!}{(g_i - n_i)!} \tag{3.25}$$

Since all the n_i particles are indistinguishable, permutation among themselves cannot be counted as independent arrangements. Thus, Eq. (3.25) must be modified to account for the permutation of n_i particles, and the result is given by

$$W(n_i) = \frac{g_i!}{n_i!(g_i - n_i)!} \tag{3.26}$$

Using the above procedure, the total number of independent ways of arranging $n_1, n_2, n_3, \ldots, n_q$ particles among $g_1, g_2, g_3, \ldots, g_q$ quantum states at $E_1, E_2, E_3, \ldots, E_q$ energy levels, with no more than one particle per quantum state with the same spin, is given by

$$\begin{aligned} W(n_1, n_2, \ldots, n_i, \ldots, n_q) &= \frac{g_1!g_2!g_3! \ldots g_q!}{(n_1!n_2! \ldots n_q!)(g_1 - n_1)!(g_2 - n_2)! \ldots (g_q - n_q)!} \\ &= \prod_{i=1}^{q} \left[\frac{g_i!}{n_i!(g_i - n_i)!}\right] \end{aligned} \tag{3.27}$$

Taking the natural logarithm on both sides of Eq. (3.27), we obtain

$$\ln W(n_1, n_2, \ldots, n_q) = \sum_{i=1}^{q} \ln\left[\frac{g_i!}{n_i!(g_i - n_i)!}\right] \tag{3.28}$$

By using Stirling's approximation on the right-hand side of Eq. (3.28), we obtain

$$\ln W \simeq \sum_{i=1}^{q} [g_i \ln g_i - n_i \ln n_i - (g_i - n_i) \ln (g_i - n_i)] \tag{3.29}$$

The most probable distribution function of F–D statistics is obtained by differentiating Eq. (3.29) with respect to n_i and using the method of Lagrangian multipliers on the two constraints given by Eqs. (3.22) and (3.23), as described in the preceding section. The result yields

$$\frac{d \ln W}{dn_i} = \ln\left[\frac{(g_i - n_i)}{n_i}\right] = \eta + \beta E_i \tag{3.30}$$

From Eq. (3.30), we obtain the F–D distribution function which reads

$$f(E_i) = \frac{n_i}{g_i} = \frac{1}{1 + e^{(\eta + \beta E_i)}} \tag{3.31}$$

where $\eta = -E_f/k_BT$ is the reduced Fermi energy, and $\beta = 1/k_BT$. Dropping the index i in Eq. (3.31), the Fermi–Dirac distribution function can be expressed by

$$f_0(E) = \frac{1}{1 + e^{(E-E_f)/k_BT}} \tag{3.32}$$

To explain the physical significance of the F–D distribution function given by Eq. (3.32), one must refer to Fig. 3.2a, which shows the F–D distribution function $f_0(E)$ versus energy E for three different temperatures, and to Fig. 3.2b. which shows a plot of $df_o(E)/dE$ versus E at temperature T. As shown in Fig. 3.2a, at $T = 0$ K, $f_0(E) = 1$ for $E < E_f$, and $f_0(E) = 0$, for $E > E_f$. This means that the probability of finding a particle with $E < E_f$ is equal to unity, which implies that all the quantum states below the Fermi level are completely occupied at $T = 0$ K. On the other hand, the probability of finding a particle with energy greater than the Fermi energy (i.e., $E > E_f$) is zero, which implies that all the quantum states above the Fermi level are empty at $T = 0$ K. These results are in sharp contrast to the results predicted by the classical M–B statistics, which show that the kinetic energy for electrons is zero at $T = 0$ K. The fact that the kinetic energy of electrons is not zero at $T = 0$ K can be explained by using the F–D statistics. It is clear from the F–D distribution function that even at $T = 0$ K, electrons will fill all the quantum states up to the Fermi level, E_f. In fact, we can show that the average kinetic energy of electrons in a metal at $T = 0$ K is equal to $(\frac{3}{5})E_f(0)$, where $E_f(0)$ is the Fermi energy at $T = 0$ K. As discussed previously, this result is in sharp contrast to the zero kinetic energy predicted by the classical M–B statistics.

As shown in Fig. 3.2a, for $T > 0$ K, $f_0(E) = \frac{1}{2}$ at $E = E_f$, which shows that the probability of finding an electron at the Fermi level is 50%. For $E < E_f$, f_0 is greater than $\frac{1}{2}$, and is smaller than $\frac{1}{2}$ for $E > E_f$. The results show that for $T > 0$ K, the quantum states below the Fermi level are partially empty, and the quantum states above the Fermi level are partially filled. It is these partially filled states (electrons) and partially empty quantum states (holes) which are responsible for the electronic conduction in a semiconductor as well as in a metal. This can also be explained from Fig. 3.2b, which shows that df_0/dE is a delta function centered around E_f, and only those quantum states which are a few k_BT above and below the Fermi level will contribute to the electrical conduction process in a semiconductor.

To find the Fermi energy and the electron density in a metal at $T = 0$ K, it is necessary to know the density-of-states function $g(E)$ in the conduction band. This is due to the fact that the density of electrons in an energy band depends on the availability of the quantum states in that energy band as well as the probability of a quantum state being occupied by an electron. The density of quantum states in an energy band can be derived by using the density-of-states function for phonons derived in Section 2.5. The only difference between these two systems is that, in deriving the electron density-of-states function, the spin degeneracy (i.e., Pauli exclusion principle) must be considered. Thus, by taking into account the spin degeneracy of electrons, the density of quantum states per unit energy interval between the constant energy surfaces of E and $E + dE$ in k-space can be found from Eq. (2.34) by simply replacing the

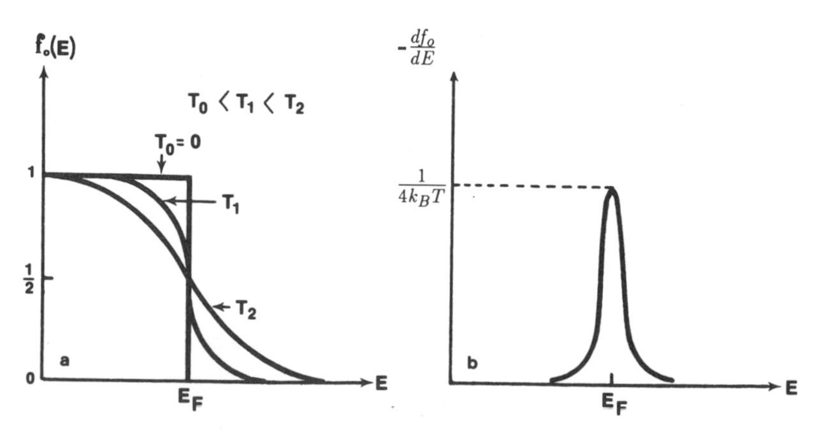

FIGURE 3.2. Fermi–Dirac distribution function $f(E)$ and its derivative, $\partial f(E)/\partial E$, versus energy E.

phonon frequency ω by the electron energy E, and the phonon wave vector q by the electron wave vector k, which yields

$$g(E) = 2\left(\frac{1}{2\pi}\right)^3 \int \frac{dS_k}{|\nabla_k E|} \tag{3.33}$$

where dS_k is the surface element in k-space, and $\nabla_k E$ is the gradient of energy which is related to the group velocity of electrons. The factor 2 on the right-hand side of Eq. (3.33) accounts for spin degeneracy. For the free-electron case, the density-of-states function for a parabolic energy band can be derived as follows: The total energy of free electrons is given by

$$E = \frac{\hbar^2 k^2}{2m_0} \tag{3.34}$$

Thus, using Eq. (3.34), we obtain $\nabla_k E = \hbar^2 k/m_0$ and $\int dS_k = 4\pi k^2$. Now, substituting these results into Eq. (3.33) and carrying out the integration of Eq. (3.33), we obtain the density-of-states function per unit volume, which reads

$$g(E) = \left(\frac{4\pi}{h^3}\right)(2m_0)^{3/2} E^{1/2} \tag{3.35}$$

Equation (3.35) shows that for the free-electron case the density-of-states function $g(E)$ for a parabolic energy band with spherical constant energy surface is proportional to the square root of the energy, as shown in Fig. 3.3. This result can also be applied to describe the density-of-states functions in the conduction band and the valence band of a semiconductor provided that the free-electron mass in Eq. (3.35) is replaced by either the electron or hole effective mass in the conduction or valence bands.

The equilibrium electron density in a metal can be calculated by using Eqs. (3.32) and (3.35). The density of electrons with energy between E and $E + dE$ in the conduction band can be expressed by

$$dn = f_0(E) g(E) dE = \left(\frac{4\pi}{h^3}\right)(2m_0)^{3/2} \frac{E^{1/2} dE}{1 + e^{(E-E_f)k_BT}} \tag{3.36}$$

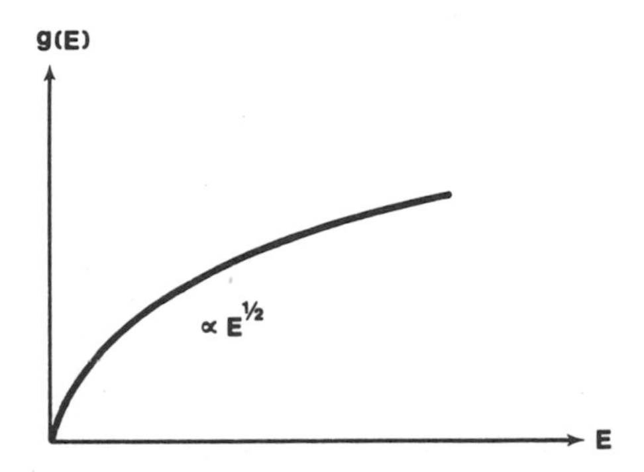

FIGURE 3.3. Density of quantum states $g(E)$ versus energy E for the free-electron case.

Thus, the total density of electrons in the conduction band of a metal is obtained by integrating Eq. (3.36) from $E = 0$ to $E = \infty$, which yields

$$n_0 = \int dn = \left(\frac{4\pi}{h^3}\right)(2m_0)^{3/2} \int_0^{\infty} \frac{E^{1/2}dE}{1 + e^{(E-E_f)/k_BT}} \tag{3.37}$$

At zero degree Kelvin (i.e., $T = 0$ K), Eq. (3.37) reduces to

$$\begin{aligned} n_0 &= \int_0^{E_f(0)} \left(\frac{4\pi}{h^3}\right)(2m_0)^{3/2}E^{1/2}\,dE \\ &= \left(\frac{8\pi}{3h^3}\right)(2m_0)^{3/2}E_f(0)^{3/2} \end{aligned} \tag{3.38}$$

Thus, the Fermi energy at $T = 0$ K can be obtained by solving Eq. (3.38), and the result yields

$$E_f(0) = \left(\frac{h^2}{8m_0}\right)\left(\frac{3n_0}{\pi}\right)^{2/3} \tag{3.39}$$

Note that in Eq. (3.39), n_0 is the free-electron density, m_0 is the free-electron mass, and h is the Planck constant. Now, by substituting the value of $n_0 = 1 \times 10^{22}$ cm^{-3} and $m_0 = 9.1 \times 10^{-31}$ kg into Eq. (3.39), we found that the Fermi energy for a metal at $T = 0$ K was around 10 eV. This implies that the kinetic energy of electrons can still be quite large even at $T = 0$ K (see Problem 3.3). As stated before, this result is in sharp contrast with the prediction given by the classical M–B statistics, which shows that the kinetic energy is equal to zero at $T = 0$ K.

In a metal, both the electron density and the Fermi energy are a very weak function of temperature. For example, using F–D statistics, we can show that the first-order correction in the temperature dependence of Fermi energy is given by

$$E_f(T) = E_f(0)\left[1 - \frac{\pi^2}{12}\left(\frac{T}{T_f}\right)^2\right] \tag{3.40}$$

where $E_f(0)$ is given by Eq. (3.39), and $T_f = E_f(0)/k_B$ is the Fermi temperature. Using Eq. (3.40), we found that for $E_f(0) = 10$ eV, the Fermi temperature T_f is equal to 11,600 K. Thus, it is clear that the Fermi energy in a metal indeed depends very weakly on temperature. In fact, if the values of $E_f(0)$ and T_f given above were used in Eq. (3.40) the second term in the square bracket of Eq. (3.40) would indeed be negligibly small for metals. On the contrary, both the Fermi energy and the carrier density for a nondegenerate semiconductor are in general a strong function of temperature. This will be discussed further in Chapter 5.

3.4. BOSE–EINSTEIN (B–E) STATISTICS

In this section the distributon function for photons and phonons is derived using the Bose–Einstein (B–E) statistics. The general characteristics of photons and phonons include: (i) they are indistinguishable and each with a quantized energy value and wave number, and (ii) the occupation number in each quantum state for these particles is not restricted by the Pauli exclusion principle. To derive the B–E distribution function, consider a linear array of n_i particles and $g_i - 1$ partitions which are necessary to divide these particles into g_i quantum states. It is not difficult to see that the number of ways of arranging the n_i particles among g_i states is equal to the number of independent permutations of particles and partitions. Since there are a total of $n_i + g_i - 1$ particles plus partitions, they can be arranged linearly in $(n_i + g_i - 1)!$ ways. However, permutations of particles among themselves or partitions among themselves do not count as an independent arrangement; we must take into account the number of ways of permuting particles among themselves (i.e., $n_i!$) and the number of ways of permuting partitions among themselves $(g_i - 1)!$. Therefore, the total number of independent ways of arranging n_i particles among g_i quantum states at the ith energy level can be written as

$$W_i = \frac{(n_i + g_i - 1)!}{n_i!(g_i - 1)!} \tag{3.41}$$

Unlike the F–D statistics, n_i can be greater than g_i in the B–E statistics. Therefore, the total number of independent ways of arranging $n_1, n_2, \ldots, n_q$ particles among $E_1, E_2, \ldots, E_q$ energy levels with $g_1, g_2, \ldots, g_q$ quantum states pertaining to each corresponding energy level is given by

$$\ln W(n_1, n_2, \ldots, n_q) = \sum_{i=1}^{q} \ln\left[\frac{(n_i + g_i - 1)!}{n_i!(g_i - 1)!}\right] \tag{3.42}$$

Now differentiating Eq. (3.42) with respect to n_i and using Stirling's approximation and the Lagrangian multipliers, we obtain

$$\frac{d \ln W}{dn_i} \simeq \ln\left[\left(\frac{g_i}{n_i}\right) + 1\right] = \alpha + \beta E_i \tag{3.43}$$

Now solving Eq. (3 43), we obtain the B–E distribution function, which reads

$$f(E_i) = \frac{n_i}{g_i} = \frac{1}{e^{(\alpha + \beta E_i)} - 1} \tag{3.44}$$

where α and β are constants to be determined. Since the numbers of particles in each quantum state is not restricted by the Pauli exclusion principle, the constant α is equal to zero, and β is given by $1/k_BT$. Thus, dropping index i in Eq. (3.44), the B–E distribution function is reduced to

$$f(E) = \frac{1}{e^{h\nu/k_BT} - 1} \tag{3.45}$$

where $E = h\nu$ is the energy of a photon or a phonon, h is Planck's constant, and ν is the phonon frequency. It is useful to introduce here Planck's blackbody radiation formula which describes the photon distributon over a wide range of optical spectrum. According to Planck's blackbody radiation law, the number of photons per unit volume having frequencies between ν and $\nu + d\nu$ is given by

$$Q_{eq} d\nu = \frac{(8\pi\nu^2/c^3)d\nu}{e^{h\nu/k_BT} - 1} \tag{3.46}$$

A comparison of Eqs. (3.45) and (3.46) reveals that the denominators in both equations are identical, implying that the distribution function for photons given by Planck's blackbody radiation law is indeed consistent with the prediction given by the B–E statistics. It is noted that similar to the F–D distribution function, the B–E distribution function will reduce to the classical M–B distribution function at high temperatures or for $h\nu \geq 4k_BT$ in which the lattice phonons are dominated by optical phonons.

3.5. STATISTICS IN THE SHALLOW-IMPURITY STATES

The F–D distribution function derived in Section 3.3 is applicable for both the electrons and holes in a semiconductor. However, the distribution of electrons in a shallow-donor state or holes in a shallow-acceptor state inside the forbidden gap of a semiconductor is somewhat different from that in the conduction or the valence band states. The difference stems mainly from the fact that, in the forbidden gap, a shallow-donor state can be occupied by an electron with either spin-up or spin-down. Once a donor impurity state is occupied by an electron, it becomes neutral. As a result, no additional electron is allowed to occupy that donor impurity state. The distribution function of electrons in this state can be derived as follows:

Let us consider a hydrogenic shallow-donor impurity state located a few k_BT below the conduction band edge in the forbidden band gap of a semiconductor with density N_d and ionization energy E_d. If W_d is the total number of ways of arranging n_d electrons in the N_d donor states with one electron per state, then there are $2N_d$ ways of putting the first electron in the N_d donor states taking into account the spin degeneracy of electrons. Similarly, the number of ways of arranging the second electron in $(N_d - 1)$ donor states is equal to $2(N_d - 1)$, since there are only $(N_d - 1)$ empty donor states available for electron occupancy. This procedure can be repeated until either all the n_d electrons are used up, or until there are only 2 $(N_d - n_d + 1)$ ways of arranging the last electron in the remaining donor impurity state. Since electrons are indistinguishable, the permutations among themselves do not count as independent arrangements. Therefore, the total number of ways of distributing n_d electrons among N_d donor states is given by

$$W_d = 2(N_d)2(N_d - 1)2(N_d - 2) \cdots 2(N_d - n_d + 1)/n_d!$$

$$= \frac{2^{n_d} N_d!}{n_d!(N_d - n_d)!} \tag{3.47}$$

A comparison of Eqs. (3.47) and (3.26) reveals that an extra factor of 2^{n_d} is included in Eq. (3.47) to account for the Coulombic nature of the shallow-donor impurity states. Now applying the same procedure as described in the previous sections to Eq. (3.47), the distribution function of electrons in a shallow-donor state can be written as

$$f(E_d) = \frac{n_d}{N_d} = \frac{1}{1 + \frac{1}{2}\, e^{(E_d - E_f)/k_BT}} \tag{3.48}$$

From Eq. (3.48), it is noted that a degenerate factor of $\frac{1}{2}$ is added to the exponential term of the denominator. This degenerate factor is to account for the Coulombic interaction between the electron and the ionized donor impurity state. A more generalized expression for the electron distribution function in a shallow-donor state can be obtained by replacing the factor $\frac{1}{2}$ in Eq. (3.48) by a degenerate factor, g_D^{-1}, where g_D is the degeneracy factor of the shallow-donor state. Values of g_D may vary between 2 and 12 depending on the nature of the shallow-donor states and the conduction band structure of a semiconductor.

The above result can also be applied to find the distribution function of holes in the shallow-accepor states. By using the similar procedure for electrons, it can be shown that the distribution function of holes in a shallow-acceptor state of the semiconductor is given by

$$f(E_a) = \frac{p_a}{N_a} = \frac{1}{1 + 2\, e^{(E_f - E_a)/k_BT}} \tag{3.49}$$

Note that Eqs. (3.48) and (3.49) are the distributon functions for electrons and holes in the shallow-donor states and shallow-acceptor states, respectively. Both expressions can be used in calculating the electron and hole densities in the shallow-donor and shallow-acceptor states of a semiconductor.

The M–B and F–D distribution functions and the density-of-states function derived in this chapter are very important for calculating the density of electrons and holes in a semiconductor. While the M–B statistics are applicable for both the ideal gas system and the nondegenerate semiconductors, the F–D statistics must be used in metals and degenerate semiconductors. On the other hand, the B–E statistics are used primarily for calculating the average energy and density of phonons and photons.

PROBLEMS

3.1. Using Eq. (3.20) show that the average velocity of a classical particle is given by

$$\langle v \rangle = (8kT/\pi m_0)^{1/2}$$

3.2. Using the M–B statistics, find the average kinetic energy of classical particles. What is the root-mean-square value of the velocity? The average kinetic energy of a classical practice can be obtained by using the following equation:

$$\langle E \rangle = \frac{\int_0^\infty Eg(E)\exp(-E/k_BT)\,dE}{\int_0^\infty g(E)\exp(-E/k_BT)\,dE}$$

where $g(E)$ is the density-of-states function given by Eq. (3.35).

3.3. Using the F–D statistics and Problem 3.2, derive a general expression for the average kinetic energy of electrons, and show that at $T = 0$ K, the average energy of electrons $\langle E \rangle$ is equal to $(\frac{3}{5})E_f(0)$. Compare the average kinetic energies predicted by both the M–B and F–D statistics at $T = 0$ K, and explain the physical meanings of their difference.

3.4. Calculate the Fermi energy (in eV) for electrons in sodium and copper at $T = 0$ K. Assuming one electron per atom, calculate the equivalent Fermi temperature for each case.

3.5. Plot the F–D distribution function, $f(E)$, and its derivative, $\partial f(E)/\partial E$, as a function of electron energy for $T = 0, 300, 600$, and 1000 K.

3.6. Derive Eq. (3.40). *Hint*: the temperature dependence of Fermi energy for a metal can be derived by considering the following integral:

$$I = \int_0^\infty f(E)(dG(E)/dE)\,dE$$

where $f(E)$ is the F–D distribution function, and $G(E)$ is a well-behaved function which vanishes at $E = 0$. Using integration by parts and Taylor's series expansion in $G(E)$ with respect to $E = E_f$, the above integral reduces to

$$I = G(E_f) + (\pi^2/6)(k_BT)^2(d^2G(E)/dE^2)|_{E_f} + \cdots$$

where

$$G(E) = \int_0^E g(E)\,dE$$

and

$$\frac{dG(E)}{dE} = g(E)$$

$$\frac{d^2G(E)}{dE^2} = \frac{dg(E)}{dE}$$

$g(E)$ is the density-of-states function given by Eq. (3.35). It is noted that the temperature dependence of free-electron density, n_0, in a metal can be expressed by

$$n_0 = \int_0^\infty f(E)g(E)\,dE$$

$$= \int_0^{E_f} g(E)\,dE + \left(\frac{\pi^2}{6}\right)(k_B T)^2 \left(\frac{dg(E)}{dE}\right)\bigg|_{E_f}$$

and at $T = 0$ K

$$n_0 = \int_0^{E_f(0)} g(E)\,dE$$

3.7. Using the results given in Problem 3.6, show that the electronic specific heat for a metal is given by

$$C_e = \frac{dU}{dT} = \left(\frac{\pi^2 n_0 k_B}{2}\right)\left(\frac{T}{T_f}\right)$$

where U is the average energy of electrons at any given temperature and is given by

$$U = \int_0^\infty E g(E) f(E)\,dE$$

$$= U_0 + \left(\frac{\pi^2}{6}\right)(k_B T)^2 g(E_f(0))$$

3.8. Derive Eq. (3.48) and Eq. (3.49).

BIBLIOGRAPHY

E. Band, *An Introduction to Quantum Statistics*, D. Van Nostrand, Princeton, New Jersey (1955).
J. S. Blakemore, *Semiconductor Statistics*, Pergamon Press, New York (1960).
R. W. Gurney, *Introduction to Statistical Mechanics*, McGraw-Hill, New York (1949).
J. P. McKelvey, *Solid State and Semiconductor Physics*, Chapter 5, Harper & Row, New York (1966).
R. C. Tolman, *The Principles of Statistical Mechanics*, Oxford University Press, London (1938).
S. Wang, *Solid State Electronics*, Chapter 1, McGraw-Hill, New York (1966).

4

Energy Band Theory

4.1. INTRODUCTION

In this chapter the one-electron energy band theories for the crystalline solids are presented. The importance of energy band theories for a crystalline solid is due to the fact that many important physical and optical properties of a solid can be readily explained by using its energy band structure. In general, the energy band structure of a solid can be constructed by solving the one-electron Schrödinger equation for electrons in a crystalline solid which contains a large number of interacting electrons and atoms. To simplify the difficult task of solving the Schrödinger equation for the many-body problem in a crystal, the effects that arise from the motion of atomic nuclei must be neglected (i.e., it is assumed that the nuclei are at rest in equilibrium positions at each lattice site). Under this condition, the nuclear coordinates enter the problem only as a constant parameter. However, even though the problem has been confined as a purely electronic one, there are still the many-electron problems in the system which cannot be solved explicitly. Therefore, it is necessary to apply additional approximations to solving the Schrödinger equation for electrons in a crystalline solid.

One of the most fruitful methods developed for solving the many-electron problem in a crystal is the one-electron approximation. In this method the total wave function of electrons is chosen as a linear combination of the individual wave functions, where each wave function involves only the coordinates of one electron. It is this approximation which forms the basic framework for calculating the energy band structure of a solid. This method may be described by assuming that each electron sees, in addition to the potential of the fixed charges (i.e., positive ions), only some average potential due to the charge distribution of the rest of the electrons in the crystal. Therefore, the movement of each electron is essentially independent of the other electrons throughout the crystal lattice. By means of the one-electron approximation, the solution of the many-electron problem is reduced to: (1) finding equations which are satisfied by the one-electron wave functions, and (2) obtaining adequate solutions for the electron wave functions and energy in the crystal under consideration.

Section 4.2 discusses some basic constraints imposed on the electron wave functions which may be attributed to the translational symmetry of the crystal. For example, suitable electron wave functions in a crystal must obey the Bloch theorem. According to this theorem, the electron wave functions in a periodic crystal consist of a plane wave modulated by a Bloch function which has the same periodicity as the crystal potential. Section 4.3 presents the Kronig–Penney model for the one-dimensional periodic crystal lattice. Section 4.4 describes the nearly-free electron approximation for a three-dimensional crystal lattice. This method can be used to find the electronic energy states for the outer-shell valence electrons in which the periodic potential of the crystal can be treated as a small perturbation. Section 4.5 depicts the tight-binding approximation [or the linear combination of atomic orbits (LCAO)]. The method is applicable to calculations of the electronic states of the inner-shell core electrons of a crystalline solid. Section 4.6 presents the energy band structures for some important elemental and compound semiconductors. The energy band calculations for a semiconductor are usually carried out by using a more rigorous and sophisticated method, such as the pseudopotential technique, than those depicted above. Section 4.7 describes the effective mass concept for electrons and holes in a semiconductor. Finally, a brief discussion of the miniband structure and the density of states in a superlattice is presented in Section 4.8.

4.2. THE BLOCH–FLOQUET THEOREM

The Bloch–Floquet theorem states that the most generalized solution for a one-electron Schrödinger equation in a periodic crystal lattice is given by

$$\phi_k(r) = u_k(r)\, e^{ik\cdot r} \tag{4.1}$$

where $u_k(r)$ is the Bloch function which has the same spatial periodicity of the crystal potential. The one-electron, time-independent Schrödinger equation for which $\phi_k(r)$ is a solution is given by

$$-\left(\frac{\hbar^2}{2m}\right)\nabla^2\phi_k(r) + V(r)\phi_k(r) = E_k\phi_k(r) \tag{4.2}$$

where $V(r)$ is the periodic crystal potential, which arises from the presence of ions at their regular lattice sites, and has the periodicity of the crystal lattice given by

$$V(r) = V(r + R_j) \tag{4.3}$$

Note that R_j is the translational vector in the direct lattice defined by Eq. (1.3). To prove the Bloch theorem, it is necessary to consider the symmetry operation which translates an eigenfunction in a periodic crystal lattice via the translational vector R_j. This translational operation can be expressed by

$$T_j f(r) = f(r + R_j) \tag{4.4}$$

The periodicity of a crystal lattice can be verified from the fact that $f(r)$ is invariant under the symmetry operations of T_j. Since the translational operator T_j commutes with the Hamiltonian H, it follows that

$$T_j H\phi_k = HT_j\phi_k \tag{4.5}$$

Since ϕ_k is an eigenfunction of T_j, we may write

$$T_j\phi_k(r) = \phi_k(r + R_j) = \sigma_j\phi_k(r) \tag{4.6}$$

where σ_j is a phase factor and an eigenvalue of T_j. The phase factor σ_j can be expressed by

$$\sigma_j = e^{ik\cdot R_j} \tag{4.7}$$

where k is the wave vector of electrons, which could be a complex number in a periodic crystal lattice. If we perform two successive translational operations (i.e., T_jT_i) on the wave function ϕ_k, we obtain from Eq. (4.6) the following relationship

$$T_jT_i\phi_k = T_j\sigma_i\phi_k = e^{ik\cdot(R_i + R_j)}\phi_k(r) \tag{4.8}$$

From Eq. (4.1), the Bloch function $u_k(r)$ can be written as

$$u_k(r) = e^{-ik\cdot r}\phi_k(r) \tag{4.9}$$

From Eqs. (4.4), (4.6), and (4.7), we obtain

$$\begin{aligned} T_j u_k(r) &= u_k(r + R_j) = T_j[e^{-ik\cdot r}\phi_k(r)] \\ &= e^{-ik\cdot(r + R_j)} T_j \phi_k(r) \\ &= e^{-ik\cdot(r + R_j)}\, e^{ik\cdot R_j}(r) \\ &= e^{-ik\cdot r}\phi_k(r) = u_k(r) \end{aligned} \tag{4.10}$$

Thus, using the symmetry operations given in Eq. (4.10) we have shown that

$$u_k(r + R_j) = u_k(r) \tag{4.11}$$

which shows that the Bloch function $u_k(r)$ has indeed the same periodicity in space as the crystal potential $V(r)$. Therefore, the general solution of Eq. (4.2) is given by Eq. (4.1). From Eq. (4.1), it is shown that the electron wave function in a periodic lattice is a plane wave modulated by the Bloch function. The Bloch function $u_k(r)$ is invariant under translational operations. It should be pointed out here that the detailed shape of $u_k(r)$ depends on the electron energy E_k and the crystal potential $V(r)$ of a particular crystalline solid. Thus, the Bloch theorem described in this section can be applied to finding the electron wave functions and energy band structures (i.e., E_k versus k relation) for any crystalline solids with periodic potential.

4.3. THE KRONIG–PENNEY MODEL

In this section, the one-electron Schrödinger equation is used to solve the electron wave functions and energy bands of a one-dimensional (1-D) periodic lattice. The periodic potential $V(x)$ for such a lattice is shown in Fig. 4.1a. The Kronig–Penney model shown in Fig. 4.1b replaces the periodic potential of a 1-D periodic lattice with a delta function at each lattice site. In this model, it is assumed that $V(x)$ is zero everywhere except at the atomic site, where

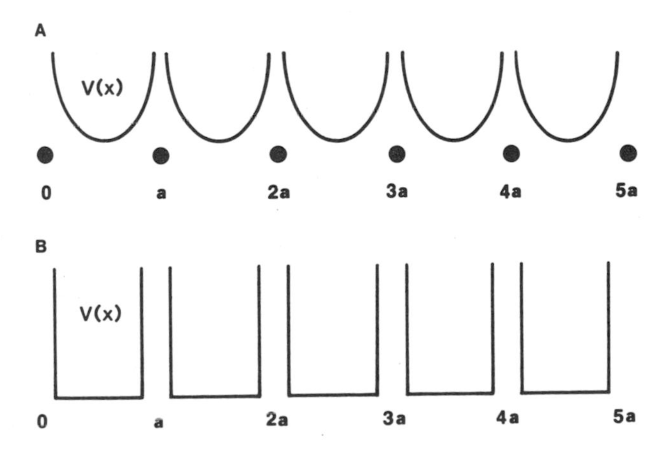

FIGURE 4.1. (A) A one-dimensional periodic potential distribution. (B) The Kronig–Penney model for a one-dimensional periodic lattice.

it approaches infinity in such a way that the integral of $V(x)\,dx$ remains finite and equal to a constant C. Inside the potential barrier, the electron wave functions must satisfy the one-electron Schrödinger equation, which is given by

$$\left(\frac{\hbar^2}{2m}\right)\frac{d^2\phi_k}{dx^2} + [E - V(x)]\phi_k = 0 \tag{4.12}$$

where $V(x)$ is the periodic potential with period a. According to the Bloch–Floquet theorem discussed in the previous section, the general solution for Eq. (4.12) can be written as

$$\phi_k(x) = u_k(x)\,e^{ik\cdot x} \tag{4.13}$$

where $u_k(x)$ is the Bloch function. Between the potential barriers (i.e., $0 < x < a$), $V(x) = 0$, and Eq. (4.12) becomes

$$\frac{d^2\phi_k}{dx^2} + k_0^2\phi_k = 0 \tag{4.14}$$

where

$$k_0^2 = \frac{2mE}{\hbar^2} \tag{4.15}$$

k_0 is the wave vector of free electrons. Since the solution of electron wave functions given by Eq. (4.13) is valid everywhere in the periodic lattice, we can substitute Eq. (4.13) into Eq. (4.14) to obtain an equation for the Bloch function $u_k(x)$, namely,

$$\frac{d^2u_k}{dx^2} + 2ik\frac{du_k}{dx} + (k_0^2 - k^2)u_k = 0 \tag{4.16}$$

This is a linear differential equation with constant coefficients, and the roots of its characteristic equation are equal to $-i(k \pm k_0)$. Thus, the general solution of Eq. (4.16) for u_k can be expressed by

$$u_k(x) = e^{-ik\cdot x}(A\cos k_0x + B\sin k_0x) \tag{4.17}$$

where A and B are constants which can be determined from the periodic boundary conditions. The first boundary condition is obtained by noting that both $u_k(r)$ and $\phi_k(r)$ are invariant under translational operation. Thus, we can write

$$u_k(0) = u_k(a) \tag{4.18}$$

where a is the period of the crystal potential, $V(r)$. To calculate the change in the slope of electron wave functions across the infinitely thin potential barrier at the atomic site, we can integrate Eq. (4.12) from $x = 0_-$ on the left-hand side of the potential barrier to $x = 0_+$ on the right-hand side of the potential barrier at $x = 0$. This yields

$$\int_{0_-}^{0_+}\left\{\frac{d^2\phi_k}{dx^2} + \left(\frac{2m}{\hbar^2}\right)[E - V(x)]\phi_k\right\}dx = 0 \tag{4.19}$$

or

$$\phi_k'(0_+) - \phi_k'(0_-) = \left(\frac{2m}{\hbar^2}\right) C\phi_k(0) \tag{4.20}$$

Equation (4.20) is obtained by noting that as $x \to 0$ inside the potential barrier, integration of $E\,dx$ over the barrier width is reduced to zero, and the change in the slope of electron wave functions (ϕ_k') across the potential barrier is given by Eq. (4.20). Now, upon substituting Eq. (4.13) into Eq. (4.20), we obtain

$$u_k'(0_+) - u_k'(0_-) = \left(\frac{2mC}{\hbar^2}\right) u_k(0) \tag{4.21}$$

By replacing $0_+ = 0$ and $0_- = a$ in Eq. (4.21), we obtain the second boundary condition for $u_k(r)$ in the form

$$u_k'(0) = u_k'(a) + \left(\frac{2mC}{\hbar^2}\right) u_k(0) \tag{4.22}$$

Note that the first derivative of $u_k(r)$ is identical on the left-hand side of each potential barrier shown in Fig. 4.1b. Now substituting the two boundary conditions given by Eqs. (4.18) and (4.22) into Eq. (4.17), we obtain two simultaneous equations for A and B:

$$A(e^{-ika}\cos k_0 a - 1) + B(e^{-ika}\sin k_0 a) = 0 \tag{4.23}$$

and

$$A\left[-ik(1 - e^{-ika}\cos k_0 a) + \left(e^{-ika}k_0 \sin k_0 a - \frac{2mC}{\hbar^2}\right)\right]$$
$$+ B[k_0 + e^{-ika}(ik\sin k_0 a - k_0 \cos k_0 a)] = 0 \tag{4.24}$$

In order to obtain a nontrivial solution for Eqs. (4.23) and (4.24), the determinant of the coefficients for A and B in both equations must be set equal to zero. This yields

$$\begin{vmatrix} [e^{-ika}\cos k_0 a - 1] & e^{-ika}\sin k_0 a \\ -ik(1 - e^{-ika}\cos k_0 a) + \left(e^{-ika}k_0 \sin k_0 a - \dfrac{2mC}{\hbar^2}\right) & [k_0 + e^{-ika}(ik\sin k_0 a - k_0\cos k_0 a)] \end{vmatrix} = 0 \tag{4.25}$$

The latter equation gives

$$\cos ka = \left(\frac{P}{k_0 a}\right)\sin k_0 a + \cos k_0 a \tag{4.26}$$

where $P = mCa/\hbar^2$, and C is defined by

$$C = \lim_{\substack{V(x)\to\infty \\ \Delta x \to 0}} [V(x)\Delta x] \tag{4.27}$$

Equation (4.26) has a real solution for the electron wave vector k if the magnitude of the right-hand side of Eq. (4.26) lies between -1 and $+1$. Figure 4.2 shows a plot of the right-hand side of Eq. (4.26) versus k_0a for a fixed value of P. It is noted that the solution of Eq. (4.26) consists of a series of alternate allowed and forbidden regions with the forbidden regions becoming smaller and smaller as the value of k_0a becomes larger. The physical meaning of this plot is discussed below.

First, we notice that the magnitude of P is closely related to the binding energy of electrons in the crystal. For example, if P is equal to zero, then we have the free-electron case, and the energy of electrons is given by Eq. (4.15). On the other hand, if P approaches infinity, then the energy of electrons becomes independent of k. This corresponds to the case of an isolated atom. In this case, the values of electron energy are determined by the condition that $\sin k_0a$ in Eq. (4.26) must be set equal to zero as P approaches infinity, which implies $k_0a = n\pi$. Thus, the electron energy for this case is given by

$$E_n = \frac{\hbar^2 k_0^2}{2m_0} = \frac{n^2\pi^2\hbar^2}{2ma^2} \tag{4.28}$$

where n is an integer. In this case, electrons are completely bound to the atom and their energy levels become discrete. When P approaches infinity, the energy spectrum is characterized by the alternate allowed and forbidden energy regions as shown in Fig. 4.2. The allowed regions are the regions in which the magnitude of the right-hand side in Eq. (4.26) lies between -1 and $+1$, while the forbidden regions are the regions in which the absolute magnitude of the right-hand side exceeds unity.

Figure 4.3 shows the plot of electron energy as a function of P. As shown in this figure, the origin where $P = 0$ corresponds to the free-electron case, and the energy of electrons is continuous in k-space. In the region where P has a finite value, the energy of electrons may be characterized by a series of allowed (shaded area) and forbidden regions. As P approaches infinity, the energy of electrons becomes discrete (or quantized), which corresponds to the case of an isolated atom (i.e., period $a \to \infty$).

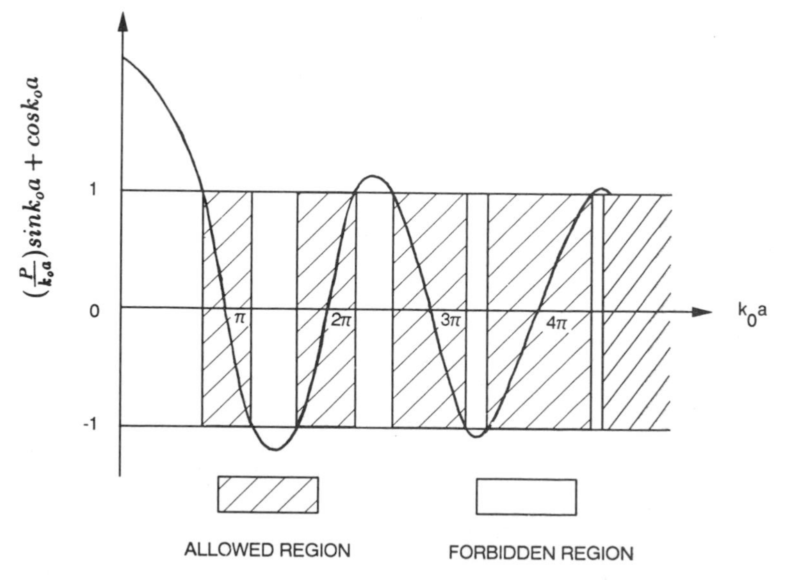

FIGURE 4.2. A plot of the magnitude of the right-hand side of Eq. (4.26) versus k_0a for a one-dimensional periodic lattice.

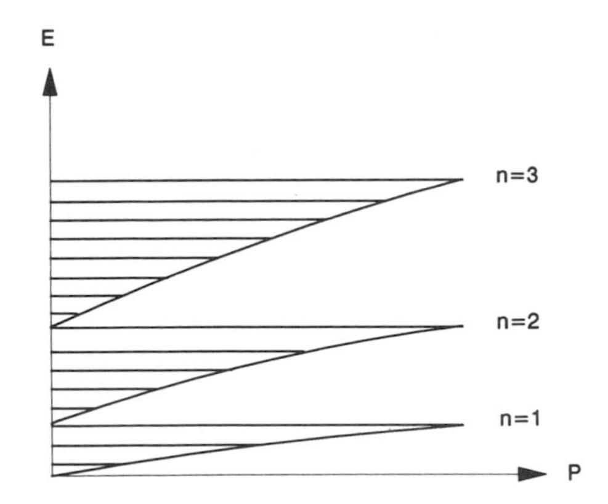

FIGURE 4.3. Energy versus P for a one-dimensional periodic lattice.

Based on the Kronig–Penney model discussed above, the energy band diagram for a 1-D periodic lattice can be represented by Fig. 4.4, which is plotted in the extended zone scheme. The values of wave vector k are given by $-n\pi/a, \ldots, -\pi/a, 0, +\pi/a, \ldots, n\pi/a$. The first Brillouin zone, known as the unit cell of the reciprocal lattice, is defined by the wave vectors with values varying between $-\pi/a$ and $+\pi/a$. Figure 4.4 illustrates two important physical aspects of the energy band diagram: (1) at the zone boundaries where $k = \pm n\pi/a$ and $n = 1, 2, 3, \ldots$, an energy discontinuity occurs, and (2) while the allowed energy band width increases with increasing electron energy, the width of the forbidden gaps decreases as electron energy increases.

If the energy band diagram (E versus k curves) is plotted within the first Brillouin zone, then it is called the reduced zone scheme. The reduced zone scheme (i.e., $-\pi/a \leq k \leq \pi/a$) is more often used than the extended zone scheme, since for any values of wave vector k' in the higher zones there is a corresponding wave vector k in the first Brillouin zone. The relation between k' and k can be obtained through the translational symmetry operation, which is given by

$$k' = k + 2n\pi/a \tag{4.29}$$

where k' represents the wave vector in the higher zones, k is the corresponding wave vector

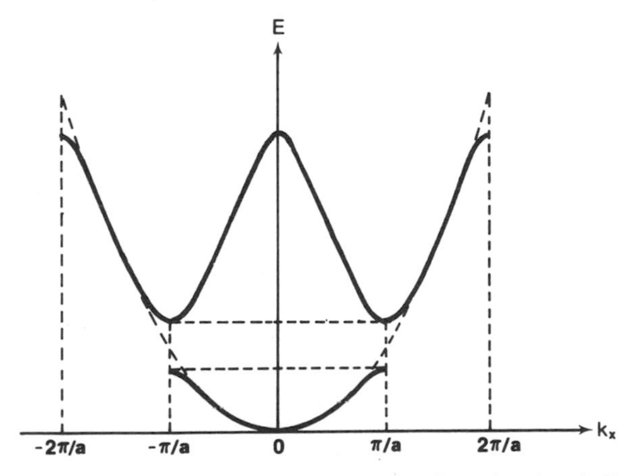

FIGURE 4.4. The energy band diagram for a one-dimensional periodic potential.

in the first Brillouin zone, $n = 1, 2, 3, \ldots$ is an integer, and a denotes the lattice constant of the crystal.

The Kronig–Penney model described above can be employed to construct the energy band diagram of an isolated silicon atom and an artificial one-dimensional periodic silicon lattice. Figure 4.5a and b show the discrete energy level schemes for such an isolated silicon atom and the energy band diagram for a one-dimensional silicon lattice, respectively. As shown in Fig. 4.5a, electrons in the 3*s* and 3*p* shells are known as the valence electrons while electrons in the 1*s*, 2*s*, and 2*p* orbits are called the core electrons. When the valence electrons are excited into the conduction band, the conductivity of a semiconductor increases. Figure 4.5b shows the energy band scheme for a one-dimensional silicon lattice. It is noteworthy that as the spacing of silicon atoms reduces to a few angstroms, the discrete energy levels shown in Fig. 4.5a broadens into energy bands, and each allowed energy band is separated by a forbidden bandgap. In this energy band scheme the highest filled band (i.e., 3*s* and 3*p* states for silicon) is called the valence band, while the lowest empty band is called the conduction band. In a semiconductor, a forbidden bandgap always exists between the conduction and the valence bands, while in metals the energy bands are usually continuous. For most semiconductors, the forbidden-bandgap energies may vary between 0.1 and 3.0 eV.

The main difference in the energy band scheme between the one-dimensional (1-D) and the two- or three-dimensional (3-D) crystal lattice is that, in the 1-D case, an energy discontinuity always exists at the zone boundary, and hence the energy band is characterized by a series of alternate allowed and forbidden bands. However, in the 3-D case, the energy band discontinuity may or may not exist since the values of k_{max} at the zone boundaries along different crystal orientations may be different, as is clearly illustrated in Fig. 4.6. This will lead to an overlap of energy states at the zone boundaries and hence the possible disappearance of the bandgap in the 3-D energy band diagram. It should be emphasized that the electron wave functions in a 3-D periodic crystal lattice are of the Bloch type and can be described by Eq. (4.1). In the next section we shall present the nearly-free electron approximation for calculating the energy bands of valence electrons in a crystalline solid. It should be noted that this method can only provide a qualitative description of the energy band schemes for the valence electrons in a 3-D crystal lattice. To obtain true energy band structures for semiconductors and metals, a more rigorous and sophisticated method, such as the pseudopotential technique, is needed for energy band calculations in these materials.

4.4. THE NEARLY-FREE ELECTRON APPROXIMATION

In the preceding section, it was shown that if the parameter P in Eq. (4.26) is small compared to k_0a, then the behavior of electrons in the 1-D periodic lattice should resemble

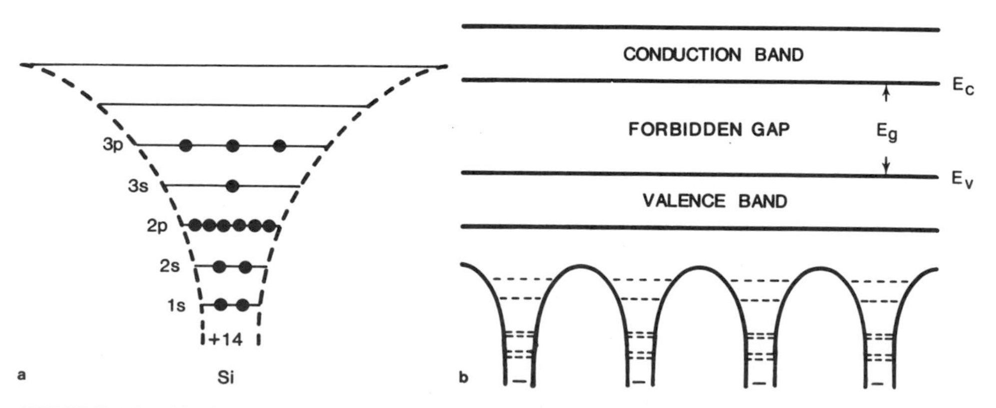

FIGURE 4.5. (A) Energy band diagrams for an isolated silicon atom, and (B) a one-dimensional silicon lattice.

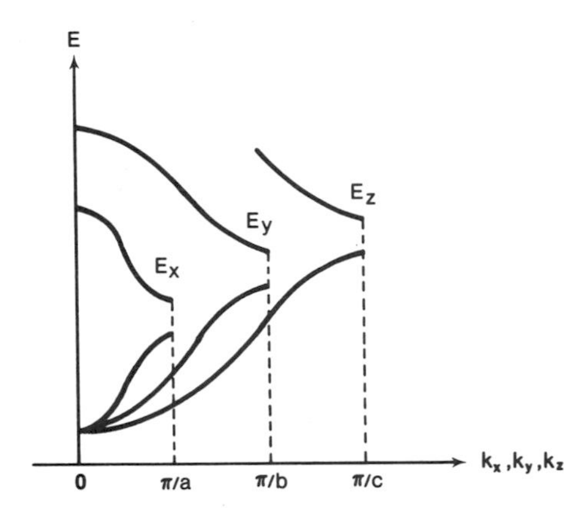

FIGURE 4.6. Energy band diagram in the reduced zone scheme for a three-dimensional (3-D) rectangular lattice, assuming $a = b = c$.

that of the free-electron case in which the energy band is continuous in k-space. In a semiconductor, the outer-shell valence electrons are loosely bound to the atoms, and the effect of the periodic crystal potential on the electron wave functions can be treated as a perturbing potential. In this case, the nearly-free electron approximation can be applied to the valence electrons.

In order to apply the nearly-free electron approximation to a 3-D crystal lattice, the periodic potential must be treated as a small perturbation. In doing so, we assume that the perturbing potential is small compared to the average energy of electrons. The problem can then be solved by using the stationary perturbation theory. From quantum mechanics, the stationary perturbation theory can be derived by using the first- and second-order approximations for the time-independent Schrödinger equation. In this approximation, it is assumed that the total Hamiltonian H consists of two parts, H_0 and H', with H_0 being the unperturbed Hamiltonian and H' the perturbed Hamiltonian. Thus, we can write

$$H = H_0 + aH' \qquad \text{for } a \leq 1 \tag{4.30}$$

The unperturbed one-electron Schrödinger equation is given by

$$H_0 \phi_{n0} = E_{n0} \phi_{n0} \tag{4.31}$$

where ϕ_{n0} and E_{n0} are the unperturbed eigenfunction and eigenvalue, respectively. The perturbed Schrödinger equation is given by

$$H\phi_n = E_n \phi_n \tag{4.32}$$

The solutions of Eq. (4.32) can be expressed in terms of a power series expansion, which reads

$$\phi_n = \phi_{n0} + a\phi_{n1} + a^2\phi_{n2} + \cdots \tag{4.33}$$

$$E_n = E_{n0} + aE_{n1} + a^2 E_{n2} + \cdots \tag{4.34}$$

The new perturbed wave functions ϕ_{nj} ($j = 1, 2, 3, \ldots$) can be expressed in terms of a linear combination of the unperturbed wave functions ϕ_{n0}:

$$\phi_{nj} = \sum_{l=0}^{\infty} b_{lj} \phi_{l0} \tag{4.35}$$

Now, substituting Eqs. (4.33) and (4.34) into Eq. (4.32) and equating the coefficients with equal exponent of a on both sides, we obtain

$$H_0\phi_{n1} + H'\phi_{n0} = E_{n0}\phi_{n1} + E_{n1}\phi_{n0} \tag{4.36}$$

$$H_0\phi_{n2} + H'\phi_{n1} = E_{n0}\phi_{n2} + E_{n1}\phi_{n1} + E_{n2}\phi_{n0} \tag{4.37}$$

Note that the relations given by Eqs. (4.36) and (4.37) are valid for $a \leq 1$. For simplicity we set a equal to 1. Consequently, the first-order correction in energy, E_{n1}, and wave function, ϕ_{n1}, can be obtained by multiplying both sides of Eq. (4.36) by an unperturbed conjugate wave function ϕ^*_{m0} and then integrating the equation over the entire volume. The result yields

$$\int \phi^*_{m0}\left[H_0\left(\sum_{l=0}^{\infty} b_{l1}\phi_{l0}\right) + H'\phi_{n0}\right] dr^3 = \int \phi^*_{m0}\left[E_{n0}\left(\sum_{l=0}^{\infty} b_{l1}\phi_{l0}\right) + E_{n1}\phi_{n0}\right] d^3r \tag{4.38}$$

By integrating Eq. (4.38) with the aid of the orthonormality of the wave functions ϕ_{n0} and the hermitian property of H_0, we obtain

$$b_{m1}E_{m0} + \int \phi^*_{m0}H'\phi_{n0}\, d^3r = E_{n0}b_{m1} \qquad \text{for } m \neq n \tag{4.39}$$

$$E_{n1} = \int \phi^*_{n0}H'\phi_{n0}\, dr^3 = = H'_{nn} \qquad \text{for } m = n \tag{4.40}$$

The solution of the latter two equations is

$$b_{m1} = \frac{H'_{mn}}{(E_{n0} - E_{m0})} \qquad \text{for } m \neq n \tag{4.41}$$

$$E_{n1} = 0 \qquad \text{for } m = n \tag{4.42}$$

In Eq. (4.41), H'_{mn} is called the matrix element defined by the second term on the left-hand side of Eq. (4.39). Thus, the new wave function ϕ_n with the first-order correction due to perturbation becomes

$$\begin{aligned} \phi_n &= \phi_{n0} + \phi_{n1} \\ &= \phi_{n0} + \sum_{\substack{m=0 \\ m \neq n}}^{\infty} \frac{H'_{mn}\phi_{m0}}{(E_{n0} - E_{m0})} \end{aligned} \tag{4.43}$$

Equation (4.43) can be used in deriving the wave functions of valence electrons in a periodic crystal lattice using the nearly-free electron approximation. In order to find the lowest-order correction in the electron energy resulting from the perturbation, it is usually necessary to carry out the expansion to the second-order correction given by Eq. (4.34). The reason for requiring the second-order correction in energy calculations is because the perturbed Hamiltonian H' has a vanishing diagonal matrix element such that the first-order correction in energy is equal to zero (i.e., $E_{n1} = 0$). This can be explained by the fact that the perturbed Hamiltonian H' is usually an odd function of the coordinates, and hence H'_{nn} is equal to zero. From Eq. (4.35), the electron wave functions for the first- and second-order corrections are given respectively by

$$\phi_{n1} = \sum_{l=0}^{\infty} b_{l1}\phi_{l0} \tag{4.44}$$

and

$$\phi_{n2} = \sum_{l=0}^{\infty} b_{l2}\phi_{l0} \tag{4.45}$$

Now, substituting Eqs. (4.44) and (4.45) into Eq. (4.37) and using the same procedures as described above for the first-order correction, we obtain the second-order correction in energy, which reads

$$E_{n2} = \sum_{\substack{m=0\\ m\neq n}}^{\infty} \frac{|H'_{nm}|^2}{(E_{n0} - E_{m0})} \tag{4.46}$$

Thus, using Eq. (4.46), the expression for the electron energy corrected to the second order is given by

$$E_n = E_{n0} + \sum_{\substack{m=0\\ m\neq n}}^{\infty} \frac{|H'_{nm}|^2}{(E_{n0} - E_{m0})} \tag{4.47}$$

Equations (4.43) and (4.47) are the new wave functions and energies of electrons derived from the quantum-mechanical time-independent stationary perturbation theory. The above results may be applied to the nearly-free electron approximation for finding the wave functions and energies of the outer-shell electrons of a crystalline solid. As mentioned earlier, the valence electrons in a semiconductor are loosely bound to the atoms, and hence the periodic crystal potential seen by these valence electrons can be treated as a perturbing potential. In this case, the expressions given by Eqs. (4.43) and (4.47) may be employed to derive the wave functions and energies of the valence electrons. The unperturbed one-electron Schrödinger equation is given by

$$\left(-\frac{\hbar^2}{2m_0}\right)\nabla^2\phi_k^0 = E_k^0\phi_k^0 \tag{4.48}$$

which has the solutions

$$\phi_k^0 = \sqrt{\frac{1}{NV}}\, e^{ik\cdot r} \tag{4.49}$$

and

$$E_k^0 = \frac{\hbar^2 k^2}{2m_0} \tag{4.50}$$

where N is the total number of unit cells in the crystal, V is the volume of the unit cell, ϕ_k^0 is the free-electron wave functions, and E_k^0 is the free-electron energy. The pre-exponential factor given by Eq. (4.49) is the normalization constant. The one-electron Schrödinger equation in the presence of a periodic crystal potential can be written as

$$\left(-\frac{\hbar^2}{2m}\right)\nabla^2\phi_k + V(r)\phi_k = E_k\phi_k \tag{4.51}$$

The crystal potential $V(r)$ can be expressed in terms of the Fourier expansion in the reciprocal space, and is given by

$$V(r) = \sum_{K_j} v(K_j)\, e^{-iK_j\cdot r} \tag{4.52}$$

where K_j is the reciprocal lattice vector and $v(K_j)$ is the Fourier coefficient.

The new electron wave functions and energies can be obtained by finding the matrix element $H_{k'k}$ due to the periodic crystal potential $V(r)$ using the results of perturbation theory described above. From Eqs. (4.49) and (4.52), the matrix element due to periodic potential $V(r)$ is given by

$$H_{k'k} = \langle k'|V(r)|k\rangle$$

$$= \left(\frac{1}{NV}\right)\int e^{-ik'\cdot r}\left(\sum_{K_j} v(K_j)\, e^{-iK_j\cdot r}\right) e^{ik\cdot r}\, d^3r \tag{4.53}$$

Note that the integral on the right-hand side of Eq. (4.53) will vanish unless $k - k' = K_j$. Thus, Eq. (4.53) is reduced to

$$H_{k'k} = v(K_j) \tag{4.54}$$

Substituting Eq. (4.54) into (4.43) yields the new electron wave functions, which read

$$\phi_k = \sqrt{\frac{1}{NV}}\, e^{ik\cdot r}\left[1 + \sum_{K_j} \frac{v(K_j)\, e^{-iK_j\cdot r}}{(E_k^0 - E_{k'}^0)}\right] \tag{4.55}$$

where $K_j = k - k'$ is the reciprocal lattice vector.

It is interesting to note that the term inside the square brackets on the right-hand side of Eq. (4.55) has the periodicity of the crystal potential $V(r)$, and hence may be designated as the Bloch function $u_k(r)$. Thus, the electron wave functions given by Eq. (4.55) is indeed the Bloch wave functions given by Eq. (4.1).

The expression for the electron energy can be derived in a similar manner by substituting Eq. (4.54) into Eq (4.47), and the result yields

$$E_k = E_k^0 + \sum_{K_j} \frac{|v(K_j)|^2}{(E_k^0 - E_{k'}^0)} \tag{4.56}$$

We note that the expressions for the electron wave functions and the energy given by Eqs. (4.55) and (4.56) would become infinity if $E_k^0 = E_{k'}^0$, and hence the perturbation approximation is no longer valid. This condition usually occurs at the zone boundaries, and the energy equation corresponding to this condition is given by

$$E_k^0 = \frac{\hbar^2 k^2}{2m} = \frac{\hbar^2 (k - K_j)^2}{2m} = E_{k'}^0 \tag{4.57}$$

or

$$k \cdot K_j = \frac{|K_j|^2}{2} \tag{4.58}$$

Here we have used the relation $k' = k - K_j$ in Eq. (4.57). Equation (4.58) is exactly the Bragg diffraction condition in a crystalline solid, which occurs at the zone boundaries. Failure of the perturbation theory at the zone boundaries is due to the fact that the periodic crystal potential $V(r)$ at zone boundaries is not small at all, and hence cannot be treated as a small perturbing potential. In fact, the Bragg diffraction condition results in a very severe perturbation of electron wave functions at the zone boundaries. Therefore, to find a proper solution for the electron energy and wave functions at the zone boundaries, it is necessary to reconstruct a new perturbed wave function which is a linear combination of an incident and a diffracted plane wave. Using the linear combination of the incident and diffracted plane waves, we can construct a new electron wave function at the zone boundary, which reads

$$\phi_k^0 = A_0\, e^{ik\cdot r} + A_1\, e^{ik'\cdot r} \tag{4.59}$$

where $k' = k - K_j$. Substituting Eq. (4.59) into Eq. (4.51) yields

$$\left\{\frac{\hbar^2 k^2}{2m} + [V(r) - E_k]\right\} A_0\, e^{ik\cdot r} + \left\{\frac{\hbar^2 k'^2}{2m} + [V(r) - E_k]\right\} A_1\, e^{ik'\cdot r} = 0 \tag{4.60}$$

Now, multiplying Eq. (4.60) by $e^{-ik\cdot r}$ and integrating the equation over the entire space, we obtain

$$A_0(E_k^0 - E_k) - A_1 v^*(K_j) = 0 \tag{4.61}$$

where $E_k^0 = \hbar^2 k^2/2m_0$, and

$$v^*(K_j) = \int_0^\infty e^{-ik\cdot r} V(r)\, e^{ik'\cdot r}\, d^3r \tag{4.62}$$

Similarly, if we multiply Eq. (4.60) by $e^{-ik'\cdot r}$ and integrate over the entire space, we obtain

$$A_0 v(K_j) - A_1(E_k - E_{k'}^0) = 0 \tag{4.63}$$

where $E_{k'}^0 = \hbar^2 k'^2/2m_0$, and

$$v(K_j) = \int_0^\infty e^{-ik'\cdot r} V(r)\, e^{ik\cdot r}\, d^3r \tag{4.64}$$

A nontrivial solution exists for Eqs. (4.61) and (4.63) only if the determinant for the coefficients of A_0 and A_1 is equal to zero. Thus, we can write

$$\begin{vmatrix} (E_k^0 - E_k) & -v^*(K_j) \\ v(K_j) & -(E_k - E_{k'}^0) \end{vmatrix} = 0 \tag{4.65}$$

Now, solving Eq. (4.65) for E_k, we obtain

$$E_k = \tfrac{1}{2}\{(E_k^0 + E_{k'}^0) \pm [(E_k^0 - E_{k'}^0)^2 + 4v^*(K_j)\cdot v(K_j)]^{1/2}\} \tag{4.66}$$

Equation (4.66) shows that a forbidden gap exists at zone boundaries, and the width of the forbidden gap is determined by the magnitude of the Fourier coefficient, $|v(K_j)|$, of the

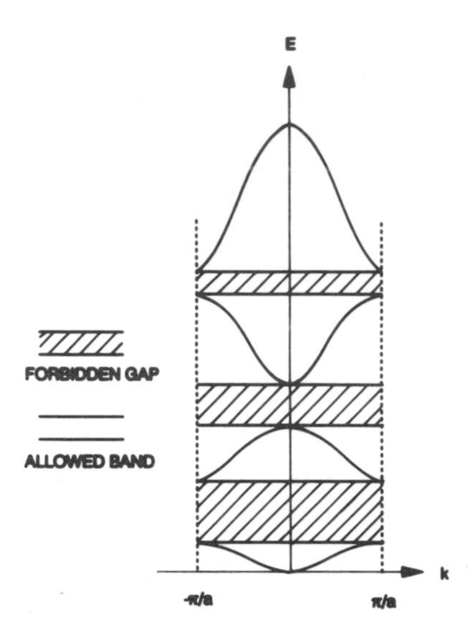

FIGURE 4.7. The energy band diagram in a reduced zone scheme showing the discontinuity of the energy at the zone boundaries.

periodic crystal potential. In general, the energy bandgap will increase with increasing magnitude of the Fourier coefficient $|v(K_j)|$. Figure 4.7 shows a schematic plot of the energy band diagram in the reduced zone scheme derived from the nearly-free electron approximation. It is interesting to note that the energy band scheme derived from this approximation is similar to that obtained from the Kronig–Penney model for the 1-D periodic lattice. Furthermore, the electron wave functions derived from the nearly-free electron approximation are indeed satisfied by the Bloch condition. The results show that, except at the zone boundaries where an energy discontinuity (or a bandgap) occurs, the energy band scheme derived from the nearly-free electron approximation resembles that of the free-electron case [i.e., $v(K_j) = 0$] discussed earlier.

The nearly-free electron approximation presented in this section provides a qualitative description of the electronic states for the outer-shell valence electrons of a three-dimensional crystal lattice. However, in order to obtain true energy band structures for a real crystal, a more rigorous and sophisticated method, such as the pseudopotential or the orthogonalized plane wave method, must be employed in the energy band calculations. These methods have been widely used in the energy band calculations of semiconductors.

4.5. THE TIGHT-BINDING (LCAO) APPROXIMATION

We shall next discuss the energy band calculations in a crystalline solid using the tight-binding approximation or the linear-combination-of-atomic-orbits (i.e., LCAO) method. This method, which was first proposed by Bloch, is often applied to calculations of the electronic states of core electrons in a crystalline solid. It is generally known that core electrons are tightly bound to the individual atoms which interact with one another within the crystal lattice. In this case, the construction of electron wave functions is acieved by using the LCAO method, and the energy bands of electrons are calculated for the corresponding periodic crystal potential. The atomic orbitals are centered on one of the constituent atoms of the crystal. The resulting wave functions are then substituted into the Schrödinger equation, and the energy values are calculated by a procedure similar to that of the nearly-free electron approximation described in the previous section. In order to apply the LCAO method to core electrons in a crystalline solid, the solution for the free atomic orbital wave functions must be obtained first. This is discussed next.

If $\phi_n(r - R_j)$ represents the atomic orbital wave functions centered at the lattice site R_j, then the wave functions of the crystal orbits ϕ_k corresponding to the wave vector k may be represented by a Bloch sum, which reads

$$\phi_k(r) = \sum_j C_j(k)\phi_n(r - R_j) \tag{4.67}$$

The summation in Eq. (4.67) extends over all the constituent atoms of the crystal. The coefficient $C_j(k)$, which satisfies the Bloch condition, can be expressed by

$$C_j(k) = e^{ik \cdot R_j} \tag{4.68}$$

Substituting Eq. (4.68) into Eq. (4.67) yields

$$\phi_k(r) = \sum_j e^{ik \cdot r} e^{-ik \cdot (r - R_j)} \phi_n(r - R_j) = e^{ik \cdot r} U_{k,n}(r) \tag{4.69}$$

To satisfy the Bloch condition, the summation given by Eq. (4.69) must have the periodicity of the crystal lattice.

The LCAO method is clearly an approximation to the true crystal orbitals. This method is adequate when the interatomic spacing is large enough such that overlapping among the atomic orbital wave functions $\phi_n(r - R_j)$ is negligible. Thus, the LCAO method is most suitable for the tightly bound core electrons, and is frequently referred to as the tight-binding approximation. We shall use this method to derive the wave functions and energy band scheme for the core electrons of a crystalline solid.

If $\phi_n(r - R_j)$ denotes a set of atomic orbital wave functions which satisfy the free-atom Schrödinger equation, then we can write

$$-\left(\frac{\hbar^2}{2m^*}\right)\nabla^2\phi_n(r - R_j) + V_{n0}(r - R_j)\phi_n(r - R_j) = E_{n0}\phi_n(r - R_j) \tag{4.70}$$

where $V_{n0}(r - R_j)$ is the free atomic potential. The wave functions for the crystal orbitals may be expressed in terms of a Bloch sum:

$$\begin{aligned}\phi_k(r) &= \left(\frac{1}{NV}\right)^{1/2} e^{ik \cdot r} e^{-ik(r - R_j)} \phi_n(r - R_j) \\ &= \left(\frac{1}{NV}\right)^{1/2} e^{ik \cdot r} u_k(r)\end{aligned} \tag{4.71}$$

where $u_k(r)$ is the Bloch function. In Eq. (4.71), the atomic wave functions are being normalized (i.e., N represents the total number of atoms in the crystal). The factor $(1/NV)^{1/2}$ is the normalization constant for the Bloch sum if overlapping of the atomic orbitals centered at different atomic sites is negligible. Therefore, Eq. (4.71) is a good approximation for the crystal orbitals, provided that the energy levels of the atomic orbits are nondegenerate and overlapping between orbital wave functions in neighboring atoms is negligible. This condition can be expressed by

$$\int \phi_n^*(r - R_j)\phi_n(r - R_i)\, dr^3 = \delta_{ij} \tag{4.72}$$

where $i \neq j$, and $\delta_{ij} = 0$. Now, substituting Eq. (4.71) into Eq. (4.70), multiplying Eq. (4.70) by $\phi_n^*(r - R_i)$, and integrating the result over the entire space, we obtain

$$E_k = \int \phi_k^*(r) H \phi_k(r)\, dr^3$$

$$= \left(\frac{1}{NV}\right)\left\{\int \sum_{ij} e^{ik\cdot(R_j - R_i)} \phi_n^*(r - R_i)\left[-\frac{\hbar^2\nabla^2}{2m} + V_{n0}(r - R_j)\right]\phi_n(r - R_j)\, dr^3\right.$$

$$\left. + \int \sum_{ij} e^{ik\cdot(R_j - R_i)} \phi_n^*(r - R_i) V'(r - R_j) \phi_n(r - R_j)\, dr^3\right\} \tag{4.73}$$

Equation (4.73) can be simplified via Eq. (4.72), and the result is given by

$$E_k = E_{n0} - \alpha_n - \sum_{R_{ij}} \beta_n(R_{ij})\, e^{ik\cdot R_{ij}} \tag{4.74}$$

where $R_{ij} = R_j - R_i$, and

$$E_{n0} = \left(\frac{1}{NV}\right)\int \phi_n^*\left[-\frac{\hbar^2\nabla^2}{2m} + V_{n0}\right]\phi_n\, d^3r \tag{4.75}$$

$$\alpha_n = -\int \phi_n^2(r - R_i) V'(r - R_j)\, dr^3 \tag{4.76}$$

$$\beta_n = -\int \phi_n^*(r - R_i) V'(r - R_j) \phi_n(r - R_j)\, dr^3 \tag{4.77}$$

$$V(r - R_j) = V_{n0}(r - R_j) + V'(r - R_j) \tag{4.78}$$

As shown in Fig. 4.8, $V_{n0}(r - R_j)$ is the unperturbed atomic potential centered at R_j, and $V'(r - R_j)$ is the perturbed crystal potential due to atoms other than that of the *j*th atom.

In general, the atomic orbital wave functions ϕ_n falls off exponentially with distance r, and hence overlapping of each atom's orbital wave function ϕ_n is assumed to be negligibly small. Therefore, it is expected that the contribution to each β_n will come from a rather restricted range of r. Furthermore, it is also expected that β_n will decrease rapidly with increasing distance between atomic centers. Figure 4.8 illustrates that the potential $V'(r - R_j)$, which plays the role of the perturbing potential, is practically zero in the vicinity of R_j. In the

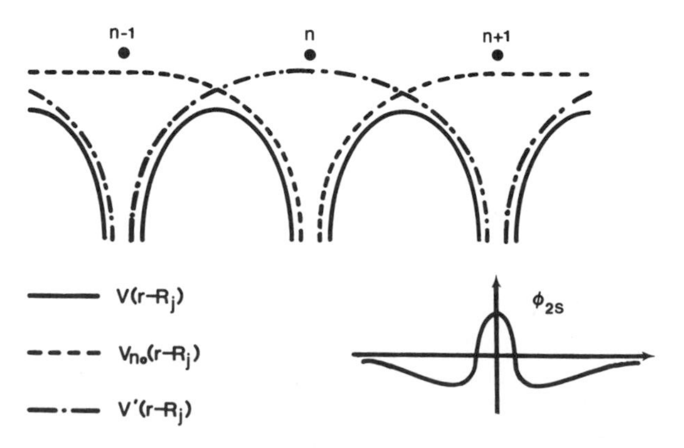

FIGURE 4.8. The crystal potential used in the tight-binding approximation.

following, we shall apply the LCAO method to derive the energy band structures of the s-like states for a simple cubic lattice and a body-centered cubic lattice.

4.5.1. The Simple Cubic Lattice

We shall first apply the LCAO method to calculations of the energy band structure of the s-like states for a simple cubic lattice. In a simple cubic lattice, there are six nearest-neighbor atoms located at an equal distance a from any chosen atomic site. Therefore, the value of $\beta_n(a)$, given by Eq. (4.77), is the same for all six nearest-neighbor atoms. Since the perturbing potential $V'(r)$ is negative, and the atomic wave functions are of the same sign in the region of overlapping, values for both α_n and $\beta_n(a)$ will be positive. Thus, the energy-wave vector (E versus k) relation can be derived by substituting $R_{ij} = (a, 0, 0)$, $(0, a, 0)$, $(0, 0, a)$, $(-a, 0, 0)$, $(0, -a, 0)$, $(0, 0, -a)$ into Eq. (4.74) to yield

$$
\begin{aligned}
E_k &= E_0 - \alpha_n - \beta_n(e^{ik_xa} + e^{ik_ya} + e^{ik_za} + e^{-ik_xa} + e^{-ik_ya} + e^{-ik_za}) \\
&= E_{n0} - \alpha_n - 2\beta_n(a)(\cos k_xa + \cos k_ya + \cos k_za) \qquad (4.79)
\end{aligned}
$$

Equation (4.79) shows the E–k relation for the s-like states of a simple cubic lattice. Figure 4.9a and b show the energy band diagrams plotted in the k_x direction and the k_x–k_y plane, respectively, as calculated from Eq. (4.79). The width of the energy band for this case is equal to 12 $\beta_n(a)$. It is of interest that the shape of the E–k plot is independent of the values of α_n or β_n used, but depends only on the geometry of the crystal lattice. Two limiting cases deserve special mention, namely, near the top of the band and near the bottom of the band. First, let us consider the case in which the value of k is very small (e.g., near the bottom of the band). In this case, the cosine terms in Eq. (4.79) may be expanded for small ka [i.e., $\cos ka \simeq (1 - k^2a^2/2)$]. If only the first-order term is retained, then the energy E is found to vary with k^2. This result is identical to the free-electron case. Under this condition, the E–k relation may be simplified to

$$E_k = E_{n0} - \alpha_n - 6\beta_n(a) + \beta_n(a)k^2a^2 \qquad (4.80)$$

From Eq. (4.80) the electron effective mass m^* in the region where ka is very small may be written as

$$m^* = \frac{\hbar^2}{d^2 E_k/dk^2} = \frac{\hbar^2}{2\beta_n(a)a^2} \qquad (4.81)$$

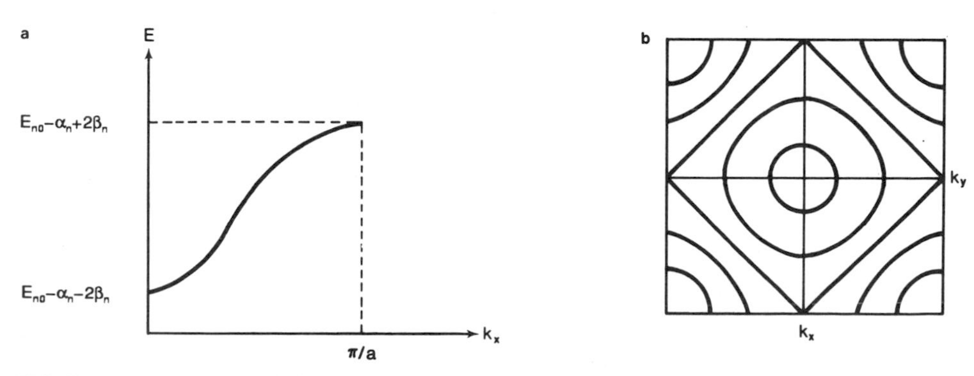

FIGURE 4.9. Energy band diagram for the s-like states of a simple cubic lattice: (a) one-dimensional and (b) two-dimensional energy band diagrams.

which shows that the constant energy surface near the bottom of the band is spherical, and the effective mass of electrons is a scalar quantity. Similarly, the E–k relation near the top of the band (i.e., $k = \pi/a$) can be obtained by expanding E_k in Eq. (4.78) at $k_x = k_y = k_z = \pi/a$, which yields

$$E_k = C + \frac{\hbar^2 k^2}{2m^*} \tag{4.82}$$

where C is a constant, and m^* is given by

$$m^* = -\frac{\hbar^2}{2\beta_n(a)a^2} \tag{4.83}$$

This shows that the electron effective mass m^* is negative near the top of the band. In fact, the effective masses given by Eqs. (4.81) and (4.83) represent the curvatures of the E versus k plot near the bottom and top of the bands, respectively. The effective mass is an important physical parameter in that it measures the curvature of the E–k energy band diagram. It is noteworthy that a positive m^* means that the band is bending upward, and a negative m^* implies that the band is bending downward. Moreover, an energy band with a large curvature corresponds to a small effective mass, and an energy band with a small curvature denotes a large effective mass. The effective mass concept is important since the mobility of electrons in a band is inversely proportional to it. For example, by examining the curvature of the energy band diagram near the bottom of the conduction band we can obtain qualitative information concerning the effective mass and the mobility of electrons in the conduction band. A detailed discussion of the effective masses for electrons (or holes) in the bottom (or top) of an energy band will be given in Section 4.7.

4.5.2. The Body-Centered Cubic Lattice (the s-like states)

For a body-centered cubic (BCC) lattice, there are eight nearest-neighbor atoms for each chosen atomic site [i.e., located at $R_{ij} = (\pm a/2, \pm a/2, \pm a/2)$]. If we substitute these values in Eq. (4.74), we obtain the E–k relation for the s-like states of the BCC crystal in the form

$$E_k = E_{n0} - \alpha_n - 8\beta_n \cos(k_x a/2)\cos(k_y a/2)\cos(k_z a/2) \tag{4.84}$$

In Eq. (4.84), values of k must be confined to the first Brillouin zone in order to have nondegenerate energy states. Using Eq. (4.84), the two-dimensional constant-energy contour plotted in the first quadrant of the k_x–k_y plane for the s-like states of a body-centered cubic lattice is shown in Fig. 4.10. While the energy surfaces are spherical near the zone center and the zone boundaries, the constant-energy contours depart considerably from the spherical shape for the intermediate values of k. For small values of k and for large k near the zone boundaries, the electron energy E is proportional to k^2, and the effective mass of electrons can be derived from Eq. (4.84) to yield

$$m^* = \frac{\hbar^2}{8a^2\beta_n(a)} \tag{4.85}$$

From Eq. (4.84), it can be shown that the total width of the allowed energy band for s-like states in a BCC crystal lattice is equal to $16\beta_n(a)$.

It is clear from the above examples that the tight-binding approximation (or LCAO method) is most suitable for calculating the energy bands of the core electrons, such as the s-like states in the cubic crystals.

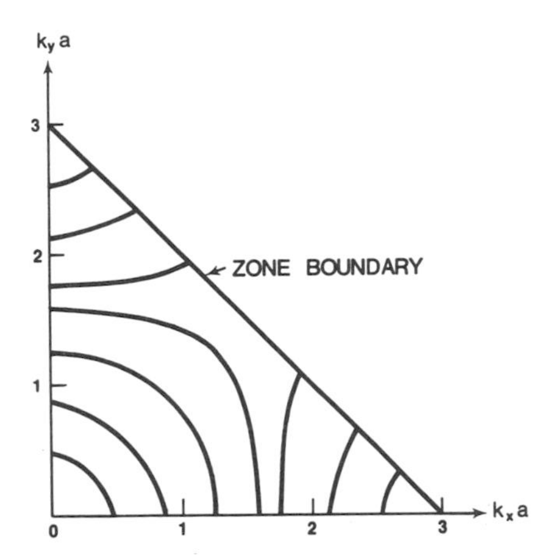

FIGURE 4.10. Constant-energy contours for a two-dimensional body-centered cubic (BCC) lattice. The E–k relation is given by $E_k = E_{n0} - \alpha_n - 8\beta_n \cos(k_x a/2)\cos(k_y a/2)$ and $k_z = 0$.

4.6. ENERGY BAND STRUCTURES FOR SEMICONDUCTORS

Due to the rapid development of semiconductor technology (e.g., silicon and III–V compound semiconductors), calculations of the energy band structures for both the elemental (Si, Ge) and compound semiconductors (GaAs, InP, etc.) have been reported extensively in the literature. As a result a very large amount of information has been obtained about the band structures of these semiconductors from both theoretical and experimental sources. In most cases theoretical calculations of the energy band structure for these materials are guided by the experimental data from the optical absorption and photoemission experiments in which the fundamental absorption process is closely related to the density of states and the transition from the initial to the final states of the energy bands. The energy band structures for some elemental and compound semiconductors calculated from the pseudopotential method are discussed in this section. In general, the exact calculations of the energy band structure for semiconductors are much more complex than those of the nearly-free electron approximation and the tight-binding approximation described in this chapter. In fact, both of these approximations can only provide a qualitative description of the energy bands in a crystalline solid. For semiconductors, the two most commonly used methods for calculating the energy band structures are the pseudopotential and the orthogonalized plane wave methods. They are discussed briefly below.

The main difficulty of band calculations in a real crystal is that the only functions which satisfy the boundary conditions imposed by the Bloch theorem in a simple manner are plane waves, but plane wave expressions do not converge readily in the interior of an atomic cell. The pseudopotential method is based on the concept of introducing a pseudopotential for a crystal which will lead to the same energy levels as the real crystal potential but not the same wave functions. The pseudopotential technique can greatly improve the convergence of the series of the plane waves which represent pseudowave functions of electrons in the crystal. In many cases it is convenient to choose the pseudo-potential to be a constant within the ion core. The parameters of the pseudopotential can be determined from the spectroscopic data for the individual atoms. Results of the empirical pseudopotential energy band calculations for some elemental and compound semiconductors with diamond and zinc blende structures are shown in Fig. 4.11.[1] Figure 4.12 shows the various symmetry points displayed at the zone center (Γ) and along the (100) axis (X) and (111) axis (L) inside the first Brillouin zone of a diamond lattice. The first symmetry point, Γ, is the symmetry point located at the Brillouin zone center. The conduction band minimum and the valence band maximum located at the Γ

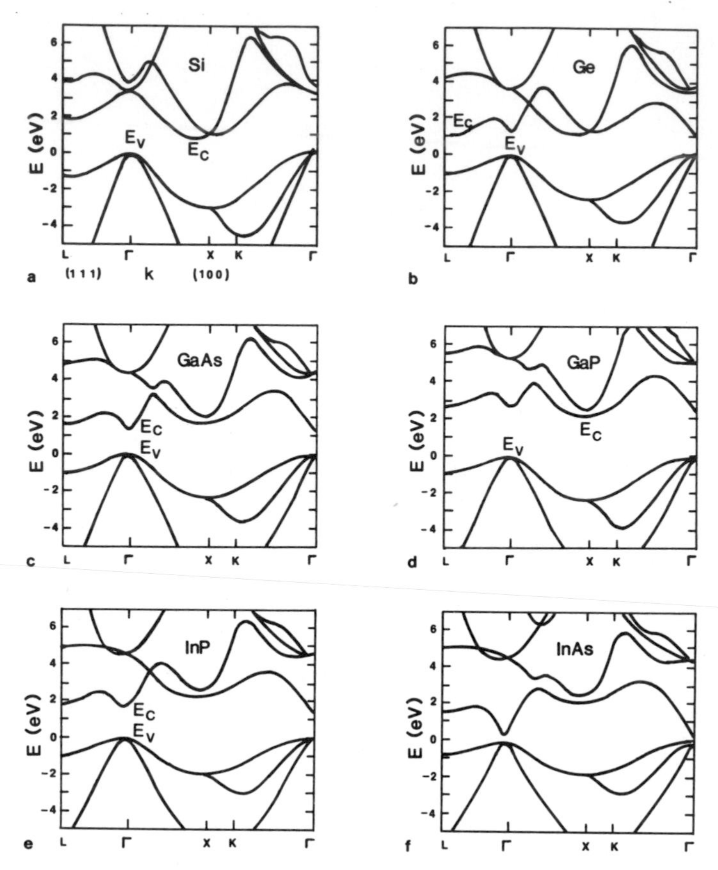

FIGURE 4.11. The energy band structures for some semiconductors with diamond and zinc blende structures. After Cohen and Bergstrasser,[1] by permission.

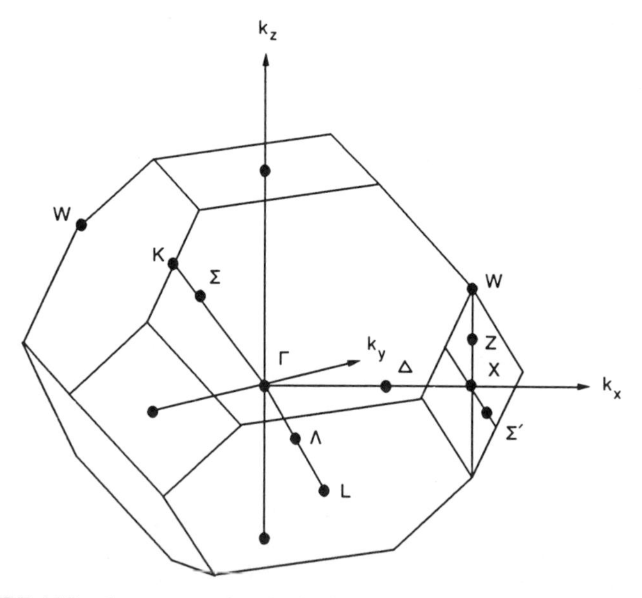

FIGURE 4.12. Symmetry points in the first Brillouin zone of a diamond lattice.

point are denoted by E_c and E_v, respectively. While the conduction band is defined as the lowest empty band, the valence band is defined to be the highest filled band at $T = 0$ K. In most semiconductors, there exists a forbidden gap between these two bands, and the values of the energy bandgap may vary from 0.1 to about 3.0 eV for most semiconductors. If the conduction band minimum and the valence band maximum are located at the same k value in the first Brillouin zone, such as the Γ point at the zone center, then the semiconductor is called a direct bandgap semiconductor. Most of the III–V compound semiconductors, including GaAs, InP, InAs, and InSb, belong to this category. Direct bandgap semiconductors have been widely used for photonic device applications because their band structure allows for direct optical transitions. They are also widely used in high-speed and high-frequency device applications due to the small electron effective mass and high electron mobility in these materials. If the conduction band minimum and the valence band maximum are not located at the same k value in the first Brillouin zone, then the semiconductor is referred to as an indirect bandgap semiconductor. Elemental semiconductors such as silicon and germanium belong to this category. Table 4.1 lists the energy bandgaps and the effective masses of electrons and holes for some elemental and compound semiconductors.

The conduction band of a diamond or a zinc blende crystal usually consists of several subbands or satellite bands. For example, the conduction band minimum of germanium crystal is located at the zone boundaries along the {111} axes, while for silicon it is located near the zone boundaries along the {100} axes; these are shown in Fig. 4.11b and c, respectively. It is noted that the constant energy surfaces for electrons in silicon and germanium are ellipsoidal energy surfaces, while the constant energy surface near the conduction band minimum is

TABLE 4.1. The Energy Bandgaps and the Effective Masses for Elemental and Compound Semiconductors at 300 K[a]

Element	E_g (eV)	Electron mass $\left(\frac{m_e^*}{m_0}\right)$	Hole mass $\left(\frac{m_h^*}{m_0}\right)$
Si	1.12	$m_t^* = 0.19$	$m_{lh}^* = 0.16$
		$m_l^* = 0.97$	$m_{hh}^* = 0.50$
Ge	0.67	$m_t^* = 0.082$	$m_{lh}^* = 0.04$
		$m_l^* = 1.6$	$m_{hh}^* = 0.30$
GaAs	1.43	0.068	$m_{lh}^* = 0.074$
			$m_{hh}^* = 0.62$
AlAs	2.16	$m_l = 2.0$	$m_{lh}^* = 0.15$
			$m_{hh}^* = 0.76$
GaP	2.26	$m_l = 1.12$	$m_{hh}^* = 0.79$
		$m_t^* = 0.22$	$m_{lh}^* = 0.14$
GaSb	0.72	0.045	$m_{hh}^* = 0.62$
			$m_{hl}^* = 0.074$
InAs	0.33	0.023	$m_{hh}^* = 0.60$
			$m_{hl}^* = 0.027$
InP	1.29	0.08	$m_{hh}^* = 0.85$
			$m_{hl}^* = 0.089$
InSb	0.16	0.014	$m_{hh}^* = 0.60$
			$m_{hl}^* = 0.027$
CdS	2.42	0.17	0.60
CdSe	1.70	0.13	0.45
CdTe	1.50	0.096	0.37
ZnSe	2.67	0.14	0.60
ZnTe	2.35	0.18	0.65
ZnS	3.68	0.28	—
PbTe	0.32	0.22	0.29

[a] m_t^* denotes transverse effective mass, m_l^* longitudinal effective mass, m_{lh}^* light-hole mass, m_{hh}^* heavy-hole mass, and m_0 free-electron mass (9.1×10^{-31} kg).

spherical for GaAs and other III–V compounds. Figure 4.13 shows a more detailed band structure of GaAs calculated from the pseudopotential method.[2] The Γ conduction band minimum is located at the zone center, the L-conduction band valleys are located at $(2\pi/a)$ $(1/2, 1/2, 1/2)$ along the (111) axes, and the X-conduction band valleys are located at the zone boundaries along the (100) axes. The separation between the L-valley and the Γ-band minimum is equal to 0.29 eV. The valence band maxima of heavy- and light-hole bands are located at the Γ point in the Brillouin zone center. Therefore, both silicon and germanium are indirect bandgap semiconductors, while GaAs, InP, and InAs are direct bandgap semiconductors. For silicon, the conduction band minima consist of six ellipsoids of constant-energy surfaces along the {100} axes with the center of each ellipsoidal energy surface located about three-fourths the distance from the zone center to the zone boundary. For germanium, the conduction band minima consist of eight ellipsoidal constant-energy surfaces along the {111} axes with the center of each ellipsoid located at the zone boundary. Therefore, for germanium there are eight half-ellipsoids of conduction band valleys inside the first Brillouin zone. For GaAs, the constant-energy surface of the Γ conduction band minimum is spherical, and is located at the zone center. The energy versus wave vector (i.e., E versus k) relation for electrons near the bottom of the conduction band can be expressed by

$$E(k) = E_c + \frac{\hbar^2 k^2}{2m_n^*} \tag{4.86}$$

for a spherical constant-energy surface, and

$$E(k) = E_c + \frac{\hbar^2}{2}\left(\frac{k_l^2}{m_l} + \frac{k_t^2}{m_t}\right) \tag{4.87}$$

for an ellipsoidal energy surface, where m_l and m_t denote the longitudinal and transverse effective masses of electrons, respectively.

The valence bands of silicon, germanium, and gallium–arsenide crystals consist of the heavy-hole and light-hole bands which are degenerate at $k = 0$. In addition, a spin–orbit split-off band is located at a few tens of meV below the top of the valence bands. This can be best

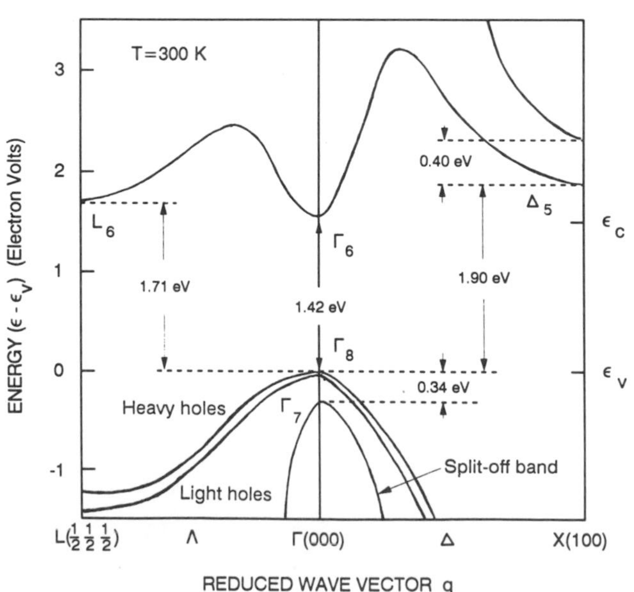

FIGURE 4.13. Detailed energy band diagram for a GaAs crystal calculated from the pseudopotential method, showing both the conduction and valence bands along the (100) and (111) crystal orientations. After Chelikowski and Cohen,[2] by permission.

explained through examination of the band structure shown in Fig. 4.13 for a GaAs crystal. In this figure, it is apparent that the heavy-hole and light-hole bands are degenerate at the top of the valence band and can be represented by a parabolic band with different curvatures. The valence band with a smaller curvature (i.e., a larger hole effective mass) is usually referred to as the heavy-hole band, and the valence band with a larger curvature (i.e., a smaller hole effective mass) is known as the light-hole band. The effective masses for the light-hole and heavy-hole bands of Si, Ge, and GaAs are also listed in Table 4.1. The energy versus wave vector relation (i.e., E versus k) for the heavy- and light-hole bands near the top of the valence bands is nonparabolic and can be expressed by

$$E(k) = E_v - \frac{\hbar^2 k^2 s(k)}{2m_p^*} \tag{4.88}$$

where $s(k)$ is given by

$$s(k) = A \pm [B^2 + C^2(k_x^2k_y^2/k^4 + k_x^2k_z^2/k^4 + k_y^2k_z^2/k^4)]^{1/2} \tag{4.89}$$

with A, B, and C constants (see Problem 4.10); the plus and minus signs correspond to the heavy-hole and light-hole bands, respectively. Note that the constant-energy surfaces near the top of the valence bands are warped.

Another interesting and technologically important feature of the III–V compound semiconductors is their ability to form lattice-matched ternary or quaternary compound semiconductor epitaxial layers on either the GaAs or InP substrate (e.g., $Al_xGa_{1-x}As$ on GaAs and $In_xGa_{1-x}As$ on InP). By forming ternary or quaternary (e.g., $In_xGa_{1-x}As_yP_{1-y}$) alloy compound semiconductors, it is possible to change many important physical and electrical properties of the compound semiconductors, such as the bandgap energy and the electron mobility. Furthermore, many interesting devices can be fabricated by using the binary/ternary superlattice and multiquantum well structures (e.g., $Al_xGa_{1-x}As$/GaAs). This feature is extremely important for many applications in detectors, lasers, and high-speed devices using III–V compound semiconductor epitaxial layers grown by MOCVD and MBE techniques. Figure 4.14 shows the energy bandgap versus lattice constant for a number of II–VI and III–V binary compound semiconductors.[3] The solid lines denote the direct bandgap materials and the dashed lines the indirect bandgap materials. A mixture of AlP/GaP (i.e., to form $Al_xGa_{1-x}P$), AlAs/GaAs (i.e., $Al_xGa_{1-x}As$), InP/GaAs/InAs (i.e., $In_xGa_{1-x}As_yP_{1-y}$), and

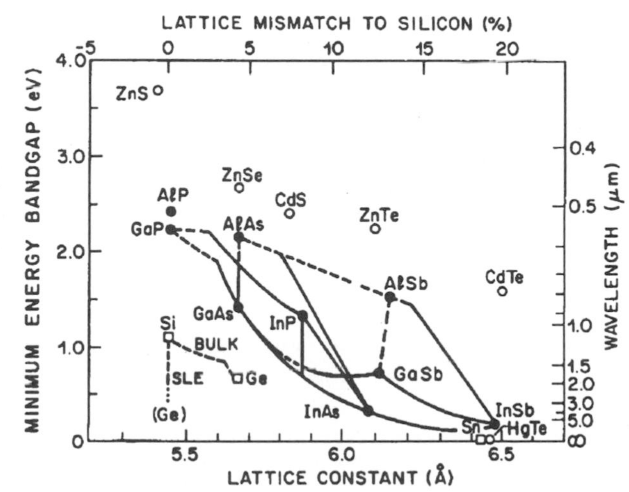

FIGURE 4.14. The energy bandgap versus lattice constant for III–V binary compound semiconductors. Solid lines denote the direct bandgap materials and dashed lines the indirect bandgap materials. Vertical lines are for the lattice-matched ternary compound semiconductors on a selected binary compound semiconductor substrate. After Hansen,[3] reprinted by permission of John Wiley & Sons Inc.

AlSb/GaSb (i.e., $Al_xGa_{1-x}Sb$) along the vertical line of Fig. 4.14 would yield a lattice-matched epilayer grown on GaP, GaAs, InP, and GaSb substrates, respectively. By tailoring the energy bandgap of these III–V alloy systems, it is possible to produce detectors and lasers with wavelengths covering the visible to infrared spectral range.

The energy band structures presented in this section are extremely important for understanding the optical, electrical, and physical properties of the semiconductor materials and devices. In fact, we shall use the energy band diagrams presented in this section to explain the physical and transport properties of a wide variety of semiconductor devices to be discussed in Chapters 10 through 15.

4.7. THE EFFECTIVE MASS CONCEPT

In Section 4.1, we have shown that the wave function of electrons in a periodic crystal is a plane wave modulated by the Bloch function, $u_k(r)$. For the time-dependent electron wave functions, this can be written as

$$\phi_k(r, t) = u_k(r)\, e^{i(k\cdot r - \omega t)} \tag{4.90}$$

Since the wave function for a Bloch-type wave packet extends over the entire crystal lattice, the group velocity for such an electron wave packet is given by

$$v_g = \frac{d\omega}{dk} = \left(\frac{1}{\hbar}\right)\nabla_k E(k) \tag{4.91}$$

Note that the relation $E(k) = \hbar\omega$ was used in Eq. (4.91). According to Eq. (4.91), the group velocity of the electron wave packet is perpendicular to the constant-energy surface at a given wave vector k in k-space. The group velocity is determined by the gradient of energy with respect to wave vector k, as given by Eq. (4.91).

If a Lorentz force $\mathscr{F}$—which may be due to either an electric field or a magnetic field—is applied to the electrons inside a crystal, then the wave vector of electrons will change with the applied Lorentz force according to the following relation:

$$\mathscr{F} = -q(\mathscr{E} + v_g \times \mathscr{B}) = \hbar\left(\frac{dk}{dt}\right) = \hbar\dot{k} \tag{4.92}$$

where $\mathscr{E}$ is the electric field and $\mathscr{B}$ is the magnetic field. The product $\hbar k$ is usually referred to as the crystal momentum. Equation (4.92) shows that an external applied force acting on an electron tends to change the crystal momentum or the electron wave vector in a crystal lattice. The electron effective mass in a crystal lattice can thus be defined by

$$\mathscr{F} = m_e^* a = m_e^*\left(\frac{dv_g}{dt}\right) \tag{4.93}$$

The solution of Eqs. (4.90), (4.91), and (4.92) yields an expression for the acceleration of electrons under a Lorentz force, namely,

$$a = \frac{dv_g}{dt} = \left(\frac{1}{\hbar}\right)\left(\frac{d\nabla_k E}{dk}\right)\left(\frac{dk}{dt}\right) = \left(\frac{1}{\hbar^2}\right)\left(\frac{d^2E}{dk^2}\right)\cdot \mathscr{F} \tag{4.94}$$

From Eqs. (4.93) and (4.94) we can derive an expression for the reciprocal effective mass tensor of the electrons whose component is given by

$$(m_e^*)_{ij}^{-1} = \left(\frac{1}{\hbar^2}\right)\left(\frac{d^2E(k)}{dk_i\, dk_j}\right) \tag{4.95}$$

where $i, j = 1, 2, 3$ are the indices used to specify the crystal orientations. From Eq. (4.95), it is seen that the reciprocal effective mass is directly proportional to the curvature of the energy versus wave vector plot (i.e., E versus k diagram). For example, a larger curvature near the conduction band minimum implies a small electron effective mass, and *vice versa*.

Another important concept to be presented here is concerned with holes in the valence band of a semiconductor. A hole in the valence band marks the absence of a valence electron or the creation of an empty state in the valence band. Furthermore, the motion of a hole can be regarded as the motion of a missing electron in the valence band. Since most of the holes reside near the top of the valence band maximum in which the curvature of the E versus k diagram is always negative (which implies a negative electron effective mass), it is appropriate to replace the missing electrons by the positively charged holes. This arrangement greatly simplifies the treatment of electronic conduction in the valence band of a semiconductor. By using the concept of holes, which has a positive effective mass and a positive charge, the inverse hole effective mass can be derived from the following expression:

$$\frac{dv_g}{dt} = -\left(\frac{1}{m_e^*}\right)\mathscr{F} = \left(\frac{1}{\hbar^2}\right)\nabla_{k'}^2 E_{k'} \cdot \mathscr{F} = \left(\frac{1}{m_h^*}\right)\mathscr{F} \tag{4.96}$$

which yields

$$\frac{1}{m_h^*} = \left(\frac{1}{\hbar^2}\right)\nabla_{k'}^2 E_{k'} \tag{4.97}$$

where $\mathscr{F}$ is the Lorentz force experienced by a hole. Therefore, a hole in the valence band may be considered as a particle with a positive charge q and a positive effective mass m_h^*. Figure 4.15 shows the electrons near the bottom of the conduction band and holes near the top of the valence band.

The effective mass concept presented above is particularly useful for describing the transport properties of a semiconductor. In a semiconductor, most of the electrons reside near the bottom of the conduction band and holes are near the top of the valence band. If the energy

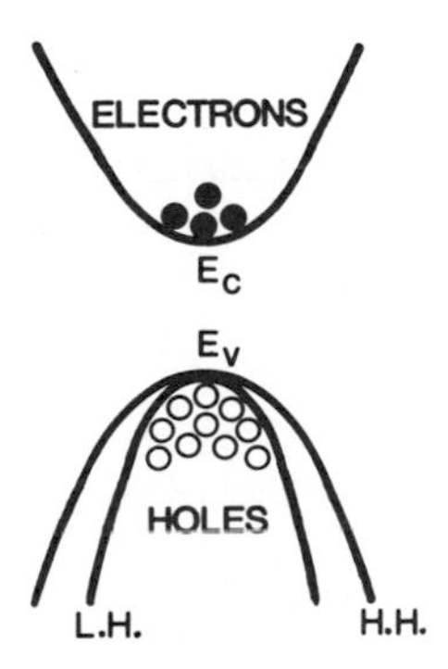

FIGURE 4.15. Electrons near the bottom of the conduction band and holes near the top of the valence bands, respectively.

band structures near the bottom of the conduction band and the top of the valence band have spherical constant-energy surfaces, then the effective masses for both electrons and holes are given by a scalar quantity. If we assume that both the conduction band minimum and the valence band maximum are located at the zone center (i.e., Γ) for $k = 0$, then the E–k relation can be expressed by

$$E_k = E_c + \frac{\hbar^2 k^2}{2m_e^*} \tag{4.98}$$

for electrons in the conduction band, and

$$E_{k'} = E_v - \frac{\hbar^2 k'^2}{2m_h^*} \tag{4.99}$$

for holes in the valence bands. Both the heavy-hole and light-hole bands degenerate into a single band at the top of the valence band edge.

Equations (4.98) and (4.99) can be used to describe the E–k relation for electrons near the bottom of conduction band and holes near the top of the valence bands. These relations are valid for direct bandgap semiconductors such as GaAs, InP, and InAs, in which the constant-energy surfaces near the conduction band minimum and the valence band maximum are spherical. If the constant-energy surface near the band edge is nonspherical, then an effective mass tensor given by Eq. (4.95) should be used instead. For silicon and germanium, the constant-energy surface near the bottom of the conduction band is ellipsoidal, and the electron effective mass may be expressed in terms of its transverse and longitudinal effective masses (i.e., m_t^* and m_l^*). Both these masses can be determined by using the cyclotron resonance experiment performed at very low temperature. Table 4.1 lists the effective masses of electrons and holes for some practical semiconductors. By introducing the effective-mass concept for electrons in the conduction band and holes in the valence bands, we can treat both the electrons and holes as quasi-free particles, which in turn greatly simplify the mathematics of solving the carrier transport problems in a semiconductor.

4.8. ENERGY BAND STRUCTURE AND DENSITY OF STATES IN A SUPERLATTICE

In this section we discuss the band structure and the density of states in a heterostructure superlattice. With the availability of the sophisticated molecular beam epitaxy (MBE) and metal–organic chemical vapor deposition (MOCVD) growth techniques, it is now possible to grow high-quality III–V semiconductor epitaxial layers composed of alternating material systems (e.g., AlGaAs/GaAs, InAlAs/InGaAs) with monoatomic layer thickness. As a result, extensive studies of the fundamental properties of superlattices, such as energy band structures and carrier transport in the growth direction of the superlattice layers, have been undertaken in recent years. Novel devices such as laser diodes, detectors, and modulators using quantum well/superlattice structures have emerged in the past few years. Unlike the three-dimensional (3-D) system in which the size of the sample in the x, y, z directions is much larger than the de Broglie wavelength, λ, the thickness of a two-dimensional (2-D) system along the growth direction is equal to or less than λ. For a GaAs crystal, this corresponds to a layer thickness of 250 Å or less at 300 K. In a 2-D system, carrier confinement occurs along the growth direction in which the layer thickness is comparable to the de Broglie wavelength, but retains quasi-free electron behavior within the plane of the superlattice.

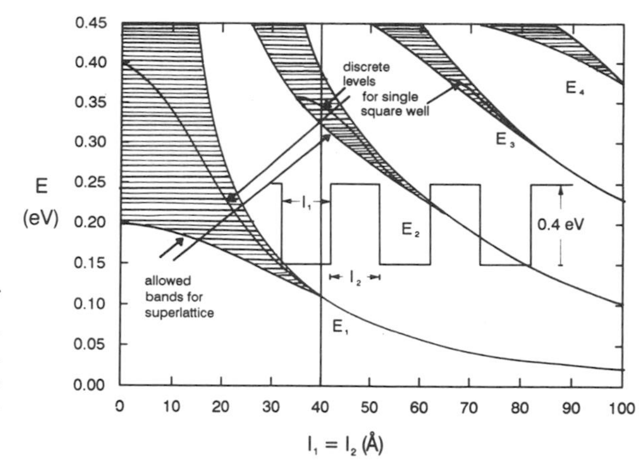

FIGURE 4.16. Calculated widths of minibands and intermittent gaps as a function of the period length for a symmetrical well/barrier heterostructure (e.g., AlGaAs/GaAs). After Esaki,[4] by permission.

A superlattice structure is formed when thin layers (e.g., ≤ 250 Å for GaAs) of a larger-bandgap semiconductor (e.g., AlGaAs) and a smaller-bandgap semiconductor (e.g., GaAs) are grown alternatively on a conducting or a semiinsulating substrate. The periodic structure formed by alternating deposition of thin epitaxial layers of two different-bandgap materials produces a periodic potential similar to the 1-D Kronig–Penney potential discussed in Section 4.3. The potential barrier is formed by the larger-bandgap material while the potential well is formed by the smaller-bandgap material. The energy band diagram for the superlattice is similar to that of free electrons exposed to a periodic crystal potential, except that now the periodic potential is imposed on Bloch electrons with an effective mass m_n^*. Depending on the width of the superlattice, the energy states inside the quantum well could be discrete bound states or minibands. Figure 4.16 shows the calculated widths of minibands and intermittent gaps as a function of the period length (i.e., $l = l_1 + l_2$) for a symmetrical barrier/quantum well structure with a barrier height of 0.4 eV.[4] It is noteworthy that for an equal barrier/well width (i.e., $l_1 = l_2 = 40$ Å) superlattice, the lowest band is rather narrow and lies 100 meV above the bottom of the quantum well. The second miniband extends from 320 to 380 meV, and higher bands overlap above the top of the potential barrier.

Figure 4.17 shows (a) the first and second minibands inside the conduction band of a superlattice along the growth direction (i.e., the z direction), (b) minibands and minigaps in

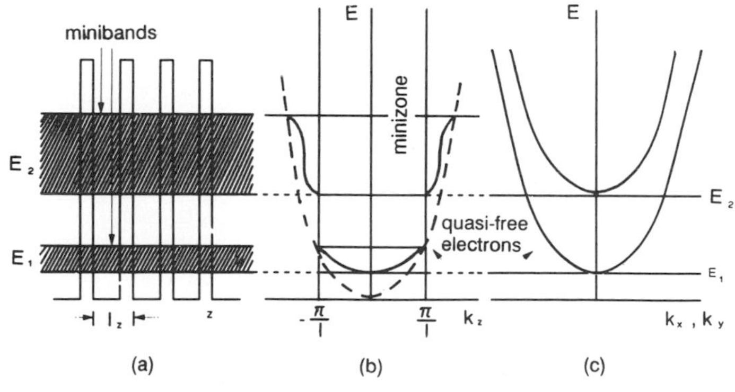

FIGURE 4.17. (a) Minibands in the growth direction (i.e., z) of the superlattice layers, (b) minibands and gaps in the k_z direction (perpendicular to the superlattice layer) inside the Brillouin zone, and (c) energy dispersion curves in the k_x and k_y directions (i.e., in the plane of the superlattice layer), which show the continuous states within the Brillouin zone for the E_1 and E_2 levels.

the k_z direction inside the Brillouin zone, and (c) energy (E_1 and E_2) versus wave vector k in the k_x and k_y directions (i.e., in the plane of the superlattice). It is seen that within the conduction band, we observe a subband structure of minibands across the potential barrier and the quantum well; the higher minibands extend beyond the height of potential barriers. The lower minibands inside the well are separated by the minigaps in the direction of superlattice periodicity (i.e., the z direction). Within the plane of the superlattice layers (i.e., the x–y plane), the electron wave functions experience only the regular periodic lattice potential. Therefore, the energy dispersion relations (i.e., E versus k_x and k_y) are similar to that of the unperturbed crystal lattice except for mixing the states in the z direction, which results in lifting the lowest-energy states at $k = 0$ above E_c of the bulk well material as shown in Fig. 4.17b and c. The second miniband results in a second shifted parabola along the k_x and k_y directions. It is seen that the E versus k relation in the k_x–k_y plane is continuous, while a minigap between the first and second minibands appears in the direction perpendicular to the superlattice (k_z). Formation of the miniband in a superlattice can be realized when the wave functions of carriers in the neighboring quantum wells of a multilayered heterostructure overlap significantly. The energy levels broaden into minibands with extended Bloch states. These minibands are expected to lead to the transport of carriers perpendicular to the superlattice layers which include tunneling, resonant tunneling, ballistic and miniband transport.

Calculations of energy band structures in a superlattice can be carried out by several methods. These include the pseudopotential, tight-binding (LCAO), and envelop-function (i.e., $k \cdot p$) methods. Among these methods, the envelop-function approach is most widely used due to its simplicity. With several refinements, this method can become quite effective in dealing with many problems such as band mixing, the effects of external fields, impurities, and exciton states. A detailed description of the envelop-function approximation for calculating the energy bands in the superlattice heterostructure devices has been given by Altarelli.[5,6]

We shall discuss next the density of states in the minibands of a superlattice. It is shown in Fig. 4.18 that the density-of-states function has a staircase character (dashed steps) for the isolated quantum wells (i.e., the barrier width is much larger than the well width).[4] In this case, each level can be occupied by the number of electrons given by its degeneracy multiplied by the number of atoms in the quantum well. Thus, the two-dimensional (2-D) density of states in each discrete level can be described by

$$g(E) = \frac{nm_n^* k_B T}{\pi \hbar^2} \tag{4.100}$$

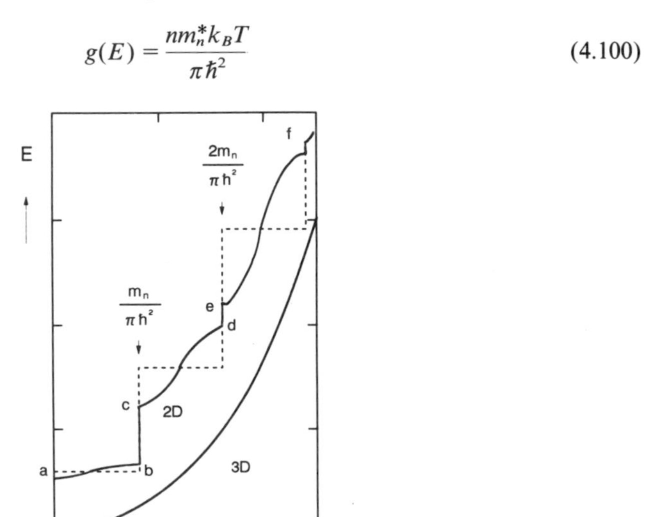

FIGURE 4.18. Staircase density of states for the isolated 2-D quantum wells (dashed line), the superlattice (distorted solid line), and the 3-D system with a parabolic band. After Esaki,[4] by permission.

where $n = 1, 2, 3, \ldots$ and $g(E)$ is measured in cm^{-2}. When significant overlap occurs, tunneling becomes possible and each level splits into minibands, and the staircase behavior (dashed line) changes shape as shown by the solid curly line in Fig. 4.18. For comparison, the density-of-states function for a 3-D system is also included in Fig. 4.18 for a parabolic band.

The concept of minibands and the density of states in a superlattice described in this section are very useful for understanding the device operation using quantum well/superlattice structures, as will be discussed further in Chapters 12 and 15.

PROBLEMS

4.1. Using the nearly-free electron approximation for a one-dimensional crystal lattice and assuming that the only nonvanishing Fourier coefficients of the crystal potential are $v(\pi/a)$ and $v(-\pi/a)$ in Eq. (4.56), show that near the band edge at $k = 0$, the dependence of electron energy on the wave vector k is given by

$$E_k = E_0 + \frac{\hbar^2 k^2}{2m^*}$$

where $m^* = m_0[1 - (32m_0^2 a^4/h^4\pi^4)v(\pi/a)^2]^{-1}$ is the effective mass of the electron at $k = 0$.

4.2. The E–k relation of a simple cubic lattice given by Eq. (4.79) is derived from the tight-binding approximation. Show that near $k = 0$ this relation can be expressed by

$$E_k = E_{n0} + \frac{\hbar^2 k^2}{2m^*}$$

where $m^* = \hbar^2/2\beta_n(a)a^2$.
For $k \simeq \pi/a$, show that the E-k relation can be expressed by

$$E_k = E_{n0} + \frac{\hbar^2 k^2}{2m^*}$$

where $m^* = -\hbar^2/2\beta_n(a)a^2$.

4.3. If the conductivity and the density-of-states effective masses of electrons are defined respectively by

$$m_{cn}^* = 3(1/m_l^* + 2/m_t^*)^{-1} \qquad \text{and} \qquad m_{dn}^* = v^{2/3}(m_l^* m*_t^2)^{1/3}$$

where m_l^* and m_t^* denote the longitudinal and transverse effective masses, respectively, find the conductivity effective mass m_{cn}^* and the density-of-states effective mass m_{dn}^* for silicon and germanium. Given: $m_t^* = 0.19m_0$, $m_l^* = 0.97m_0$, $v=6$ for silicon; and $m_t^* = 0.082m_0$, $m_l^* = 1.64m_0$, $v=4$ for germanium.

4.4. Explain why most of the III–V compound semiconductors, such as GaAs, InP, and InSb, have smaller electron effective masses than that of silicon and germanium.

4.5. Sketch the constant-energy contours for a two-dimensional square lattice using the expression derived from the tight-binding approximation

$$E(k) = E_0 + B\cos(k_x a/2)\cos(k_y a/2)$$

4.6. Derive expressions for the group velocity (v_g), acceleration (dv_g/dt), and the effective mass (m^*) of electrons using the E–k relation for the two-dimensional square lattice described in Problem 4.5. If $\cos(k_y a/2) = 1$, plot E, v_g, dv_g/dt, and m^* versus k for the one-dimensional crystal lattice.

4.7. If the E–k relation for a simple cubic lattice corresponding to an atomic state derived by the tight-binding approximation is given by

$$E(k) = E_0 - E_0' - 2E'(\cos k_1 a + \cos k_2 a + \cos k_3 a)$$

derive the expressions for (a) the group velocity, (b) acceleration, and (c) the effective mass tensor.

4.8. Repeat Problem 4.7 for a body-centered cubic lattice (s-like state). (See Eq. (4.84).)

4.9. Using the tight-binding approximation, derive the E–k relation for the s-like states in a face-centered cubic lattice.

4.10. The E–k relation near the top of the valence band maximum for silicon and germanium is given by

$$E(k) = -\left(\frac{\hbar^2}{2m}\right)\{Ak^2 \pm [B^2k^4 + C^2(k_1^2k_2^2 + k_2^2k_3^2 + k_3^2k_1^2)]^{1/2}\}$$

where E is measured from the top of the valence band edge. Plus refers to the heavy-hole band and minus to the light-hole band.

	A	B	C
Ge	13.1	8.3	12.5
Si	4.0	1.1	4.1

Using the values of A, B, and C for germanium and silicon given by the above table, plot the constant-energy contours for the heavy-hole and light-hole bands in silicon and germanium.

4.11. Plot the energy bandgap for the E_Γ, E_L, and E_X conduction minima versus temperature (T) for GaAs for $0 < T < 1000$ K. Given:

$$E_\Gamma(T) = 1.519 - \frac{5.405 \times 10^{-4}T^2}{(T + 204)}$$

$$E_L(T) = 1.815 - \frac{6.05 \times 10^{-4}T^2}{(T + 204)}$$

$$E_X(T) = 1.981 - \frac{4.60 \times 10^{-4}T^2}{(T + 204)} \qquad \text{(eV)}$$

4.12. Plot the energy bandgap for the E_Γ, E_L, and E_X conduction minima as a function of pressure (P) for GaAs for $0 < P < 50$ bar. At what pressure P will GaAs become an indirect bandgap material? Given:

$$E_\Gamma = E_\Gamma(0) + 0.0126P - 3.77 \times 10^{-5}P^2 \qquad \text{(eV)}$$

$$E_L = E_L(0) + 0.0055P$$

$$E_X = E_X(0) - 0.0015P$$

4.13. Referring to the paper by J. R. Chelikowsky and M. L. Cohen, *Phys. Rev.* *B***14**(2), 556–582 (1976), describe briefly the pseudopotential method for calculating the energy band structures of semiconductors with diamond and zinc blende structures.

4.14. Plot the energy, group velocity, and inverse effective mass of electrons versus the wave vector in the first Brillouin zone of a one-dimensional crystal lattice, using the relation $E = \hbar^2k^2/2m_0$.

REFERENCES

1. M. L. Cohen and Bergstrasser, *Phys. Rev.* **141**, 789–796 (1966).
2. J. R. Chelikowsky and M. L. Cohen, "Nonlocal Pseudopotential Calculations for the Electronic Structure of Eleven Diamond and Zinc-Blende Semiconductors," *Phys. Rev. B* **14**(2), 556 (1976).
3. M. Hansen, *Constitution of Binary Alloys*, McGraw-Hill, New York (1958).
4. L. Esaki, in: *The Technology and Physics of Molecular Beam Epitaxy* (E. M. C. Parker, ed.), p. 143, Plenum Press, New York (1985).
5. M. Altarelli, *Phys. Rev. B* **32**, 5138 (1985).
6. M. Altarelli, in: *Heterojunctions and Semiconductor Superlattices* (G. Allen *et al.*, eds.), Springer-Verlag, Berlin (1986).

BIBLIOGRAPHY

F. J. Blatt, *Physics of Electronic Conduction in Solids*, McGraw-Hill, New York (1968).
R. H. Bube, *Electronic Properties of Crystalline Solids*, Academic Press, New York (1974).
J. Callaway, *Energy Band Theory*, Academic Press, New York (1964).
J. Callaway, *Quantum Theory of the Solid State*, Part A & B, Academic Press, New York (1974).
C. Kittel, *Introduction to Solid State Physics*, 5th ed., Wiley, New York (1976).
R. Kubo and T. Nagamiya, *Solid State Physics*, McGraw-Hill, New York (1969).
K. Seeger, *Semiconductor Physics*, 3rd ed., Springer-Verlag, Berlin/Heidelberg (1985).
J. C. Slater, *Quantum Theory of Molecules and Solids*, Vols. 1, 2, and 3, McGraw-Hill, New York (1963).
S. Wang, *Solid State Electronics*, McGraw-Hill, New York (1966).
J. M. Ziman, *Principles of the Theory of Solids*, Cambridge University Press, Cambridge (1964).

5

Equilibrium Properties of Semiconductors

5.1. INTRODUCTION

In this chapter, we describe the equilibrium properties of semiconductors. The fact that electrical conductivity of a semiconductor can be readily changed by many orders of magnitude through the incorporation of foreign impurities has made the semiconductor one of the most intriguing and unique electronic materials among all the crystalline solids. The invention of germanium and silicon transistors in the early 1950s and the silicon integrated circuits in the 1960s as well as the development of microprocess chips in the late 1970s and 1980s has indeed transformed semiconductors into the most important and indispensable electronic material of modern times.

Unlike metals, the electrical conductivity of a semicondcutor can be changed by several orders of magnitude by simply doping it with foreign impurities or by using external excitations (e.g., by photoexcitation). At low temperatures, a pure semiconductor may behave as a perfect electrical insulator since its valence band is completely filled with electrons and the conduction band is completely empty. However, as the temperature rises, a fraction of the valence electrons will be excited into the conduction band by the thermal energy so creating free holes in the valence band. As a result, the electrical conductivity will increase rapidly with increasing temperature. Thus, even a pure semiconductor may become a good electrical conductor at high temperatures. In general, the semiconductors may be divided into two categories: one is the pure undoped (or as grown) semiconductor, which is usually referred to as the intrinsic semiconductor, and the other is the doped semiconductor, which is also called the extrinsic semiconductor. Another distinct difference between a metal and a semiconductor is that the electrical conduction in a metal is carried out by valence electrons, while the electrical conduction in a semiconductor may be caused by electrons, holes, or both carriers. The electrical conduction of an intrinsic semiconductor is due to both electrons and holes, while the electrical conduction for an extrinsic semiconductor is usually dominated by either electrons or holes, depending on whether the material is doped by the shallow donors or by the shallow acceptor impurities.

To understand the conduction mechanisms in a semiconductor, the equilibrium properties of a semiconductor are first examined. A unique feature of semiconductor materials is that their physical and transport parameters depend strongly on temperature. For example, the intrinsic carrier concentration in a semiconductor depends exponentially on temperature. Other physical parameters, such as carrier mobility, resistivity, and the Fermi level in a nondegenerate semiconductor, are likewise a strong function of temperature. In addition, both the shallow-level and deep-level impurities may also play an important role in controlling the physical and electrical properties of a semiconductor. For example, the equilibrium carrier concentration of a semiconductor is controlled by the shallow-level impurities, and the minority carrier lifetimes are usually closely related to defects and deep-level impurities in a semiconductor.

In Section 5.2, we derive general expressions for the electron density in the conduction band and the hole density in the valence band for the cases of a single spherical energy band and multivalley conduction bands. The equilibrium properties of an intrinsic semiconductor are discussed in Section 5.3. In Section 5.4, we present the equilibrium properties of an extrinsic semiconductor. The conversion of the conduction mechanism from the intrinsic to n-type (electrons) or p-type (holes) conduction by doping the semiconductor with shallow donor or shallow acceptor impurities are discussed in this section. Section 5.5 deals with the physical properties of a shallow dopant impurity. Using a hydrogenic impurity model the ionization energy of a shallow impurity level is derived in this secton. In Section 5.6, the Hall effect and the electrical conductivity of a semiconductor are depicted. Finally, the heavy doping effects, such as carrier degeneracy and bandgap narrowing in a degenerate semiconductor, are discussed in Section 5.7.

5.2. DENSITIES OF ELECTRONS AND HOLES IN A SEMICONDUCTOR

General expressions for the equilibrium densities of electrons and holes in a semiconductor can be derived by using the Fermi–Dirac (F–D) distribution function and the density-of-states function described in Chapter 3. For undoped and lightly doped semiconductors, the Maxwell–Boltzmann (M–B) distribution function is used instead of the F–D distribution function. If we assume that the constant-energy surfaces near the bottom of the conduction band and the top of the valence band are spherical, then the equilibrium distribution functions for electrons in the conduction band and holes in the valence band may be described in terms of the F–D distribution function. The F–D distribution function for electrons in the conduction band is given by

$$f_n(E) = \frac{1}{[1 + e^{(E-E_f)/k_BT}]} \tag{5.1}$$

and the F–D distribution function for holes in the valence band can be expressed by

$$f_p(E) = \frac{1}{[1 + e^{(E_f-E)/k_BT}]} \tag{5.2}$$

The density-of-states function derived in Chapter 2 for phonons may be applied to the electrons in the conduction band and holes in the valence band, provided that the Pauli exclusion principle is taken into account (i.e., including the spin degeneracy factor). Assuming a parabolic band for both the conduction and valence bands, the density-of-states function for the conduction band states can be written as

$$g_n(E - E_c) = \left(\frac{4\pi}{h^3}\right)(2m_n^*)^{3/2}(E - E_c)^{1/2} \tag{5.3}$$

and the density of states for the valence band states is given by

$$g_p(E_v - E) = \left(\frac{4\pi}{h^3}\right)(2m_p^*)^{3/2}(E_v - E)^{1/2} \tag{5.4}$$

Figure 5.1 shows a plot of f_n, f_p, g_n, g_p, and the products $f_n g_n$ and $f_p g_p$ as a function of energy for the conduction and valence bands for $T > 0$ K, respectively. The hatched area denotes the electron density in the conduction band and the hole density in the valence band. The conduction band edge is leveled by E_c, the valence band edge is designated by E_v, and the energy bandgap is represented by E_g. The equilibrium electron density n_0 in the conduction band can be obtained by integrating the product $g_n f_n$ (i.e., the electron density per unit energy interval) with respect to energy over the entire conduction band using Eqs. (5.1) and (5.3), which yields

$$\begin{aligned} n_0 &= \int_{E_c}^{\infty} f_n g_n \, dE \\ &= \left(\frac{4\pi}{h^3}\right)(2m_n^*)^{3/2} \int_{E_c}^{\infty} \frac{(E - E_c)^{1/2} \, dE}{[1 + e^{(E - E_f)/k_B T}]} \\ &= \left(\frac{4\pi}{h^3}\right)(2m_n^* k_B T)^{3/2} \int_0^{\infty} \frac{\varepsilon^{1/2} \, d\varepsilon}{[1 + e^{(\varepsilon - \eta)}]} \\ &= N_c \mathscr{F}_{1/2}(\eta) \end{aligned} \tag{5.5}$$

where

$$N_c = 2(2\pi m_n^* k_B T/h^2)^{3/2} \tag{5.6}$$

is the effective density of the conduction band states, and

$$\mathscr{F}_{1/2}(\eta) = \left(\frac{2}{\sqrt{\pi}}\right) \int_0^{\infty} \frac{\varepsilon^{1/2} \, d\varepsilon}{[1 + e^{(\varepsilon - \eta)}]} \tag{5.7}$$

is the Fermi integral of order one-half; $\varepsilon = (E - E_c)/k_B T$ is the reduced energy, m_n^* is the density-of-states effective mass of electrons, and $\eta = -(E_c - E_f)/k_B T$ is the reduced Fermi energy. Equation (5.5) is a general expression for the equilibrium electron density in the conduction band, and is valid for the entire dopant density range. Since the Fermi integral given by Eq. (5.7) can only be evaluated by numerical integration or through the use of the Fermi integral table, it is a common practice to use simplified expressions for calculating carrier density over a certain range of dopant densities in the heavily doped semiconductors. The following approximations are valid for a specified range of reduced Fermi energies

$$\begin{aligned} \mathscr{F}_{1/2}(\eta) &\simeq e^{n} && \text{for } \eta < -4 \\ &\simeq \frac{1}{(e^{-\eta} + 0{\cdot}27)} && \text{for } -4 < \eta < 1 \\ &\simeq \left(\frac{4}{3\sqrt{\pi}}\right)\left(\eta^2 + \frac{\pi^2}{6}\right)^{3/4} && \text{for } 1 < \eta < 4 \\ &\simeq \left(\frac{4}{3\sqrt{\pi}}\right)\eta^{3/2} && \text{for } \eta > 4 \end{aligned} \tag{5.8}$$

The expression for N_c given by Eq. (5.6) may be rewritten as

$$N_c = 2.5 \times 10^{19}(T/300)^{3/2}(m_n^*/m_0)^{3/2} \tag{5.9}$$

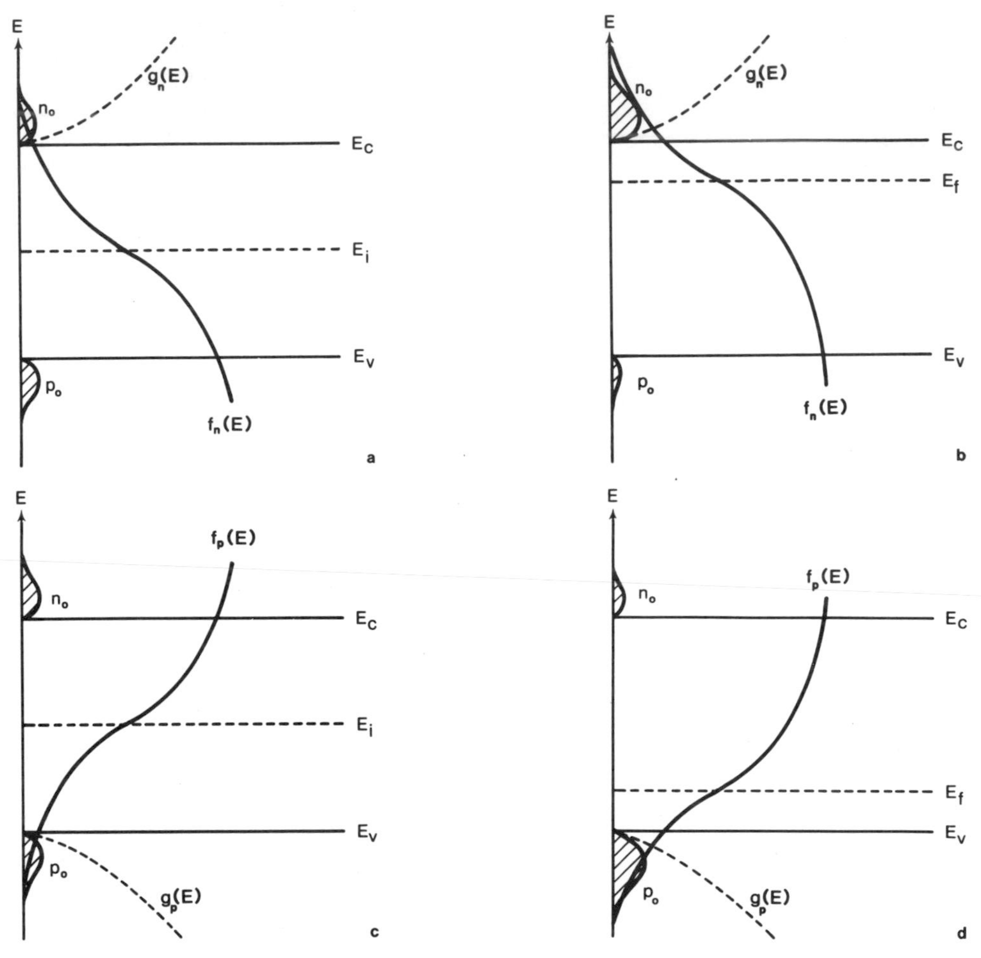

FIGURE 5.1. The Fermi–Dirac distribution function and the density-of-states function for electrons and holes in the conduction and valence bands of a semiconductor respectively, for $T > 0$ K.

where m_0 is the free-electron mass, and m_n^* is the density-of-states effective mass for electrons in the conduction band. For $\eta \leq -4$ (i.e., the Fermi level is $4k_BT$ below the conduction band edge), the Fermi integral of order one-half becomes an exponential function of η, as predicted by the M–B statistics. In this case, the classical M–B statistics prevail, and the semiconductor is referred to as the nondegenerate semiconductor. The density of electrons for the nondegenerate case can be simplified to

$$n_0 = N_c\, e^{-(E_c - E_f)/k_BT} = N_c e^{\eta} \tag{5.10}$$

Equation (5.10) is valid for the intrinsic or lightly doped (i.e., nondegenerate) extrinsic semiconductors. For silicon, Eq. (5.10) is valid for dopant densities up to 10^{19} cm^{-3}. However, for dopant densities higher than this value, Eq. (5.5) must be used instead. A simple rule of thumb for checking the validity of Eq. (5.10) is that n_0 should be three to four times smaller than N_c. Figure 5.2 shows the electron density versus the reduced Fermi energy as calculated by Eqs. (5.5) and (5.10). It is evident from this figure that the two curves are nearly coincided for $\eta \leq -4$, but deviate considerably from one another for $\eta \geq 0$ (i.e., the degenerate case).

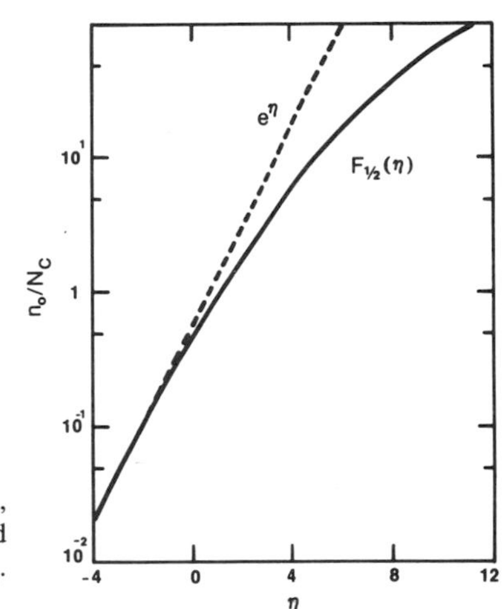

FIGURE 5.2. The normalized electron density, n_0/N_c, versus the reduced Fermi energy, η. The solid line is obtained from Eq. (5.5) and the dashed line is calculated from Eq. (5.10).

The hole density in the valence band can be derived, in a similar way from Eqs. (5.2) and (5.4). The result yields

$$p_0 = \left(\frac{4\pi}{h^3}\right)(2m_p^*)^{3/2} \int_{-\infty}^{E_v} \frac{(E_v - E)^{1/2}\, dE}{[1 + e^{(E_f - E)/k_B T}]}$$

$$= N_v \mathscr{F}_{1/2}(-\eta - \varepsilon_g) \tag{5.11}$$

where $N_v = 2(2\pi m_p^* k_B T/h^2)^3/^2$ is the effective density of the valence band states, m_p^* is the density-of-states effective mass for holes in the valence band, and $\varepsilon_g = (E_c - E_v)/k_B T$ is the reduced bandgap. For the nondegenerate case with $(E_f - E_v) \geq 4k_B T$, Eq. (5.11) becomes

$$p_0 = N_v\, e^{(E_v - E_f)/k_B T} = N_V e^{-\eta - \varepsilon_g} \tag{5.12}$$

which shows that the equilibrium hole density depends exponentially on the temperature and the reduced Fermi energy and energy bandgap

The results derived above are applicable to a single-valley semiconductor with a constant spherical energy surface near the bottom of the conduction band and the top of the valence band edge. III–V compound semiconductors such as GaAs, InP, and InAs, which have a single constant spherical energy surface near the conduction band minima (i.e., Γ-band), fall into this category. However, for elemental semiconductors such as silicon and germanium, which have multivalley conduction band minima, the scalar density-of-states effective mass used in Eq. (5.6) must be modified to account for the multivalley nature of the conduction band minima. This is discussed next.

The constant-energy surfaces near the conduction band minima for Si, Ge, and GaAs are shown in Figs. 5.3a, b, and c, respectively.[1] For silicon, there are six conduction band minima located along the {100} axes, while there are eight conduction band minima located at the zone boundaries of the first Brillouin zone along the {111} axes for germanium. Furthermore, the constant-energy surfaces near the bottom of the conduction bands are ellipsoidal for Si and Ge and spherical for GaAs. If we assume that there are ν conduction band minima, then the total density of electrons in the ν conduction band minima is given by

$$n_0' = \nu n_0 = \nu N_c \mathscr{F}_{1/2}(\eta) = N_c' \mathscr{F}_{1/2}(\eta) \tag{5.13}$$

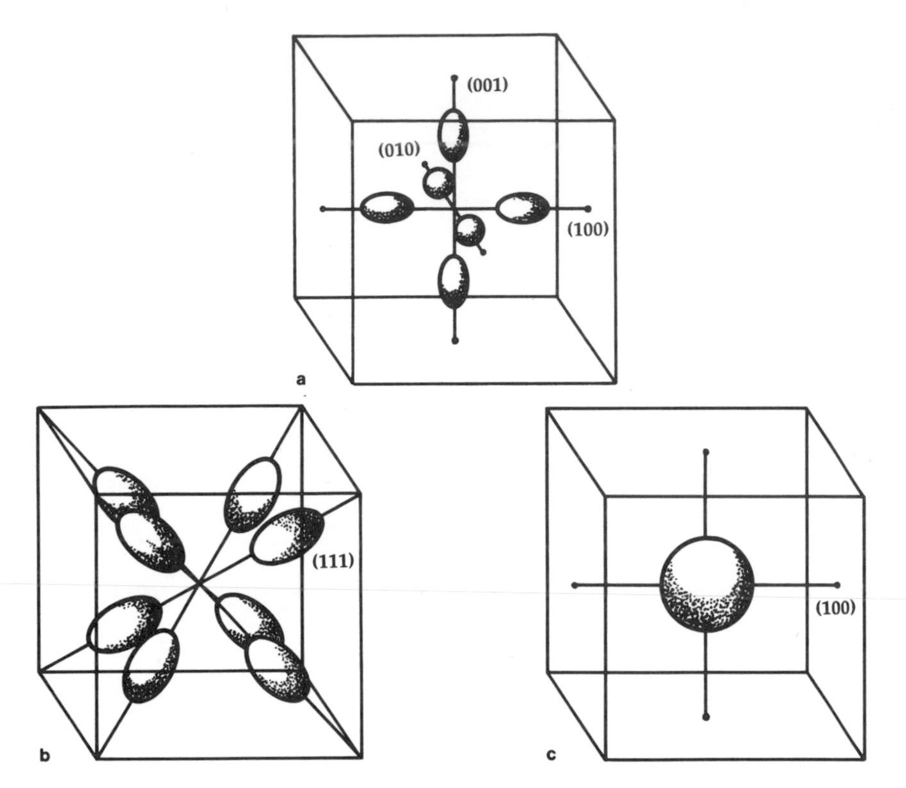

FIGURE 5.3. Constant-energy surfaces near the conduction band edges for (a) Si, (b) Ge, and (c) GaAs.

and

$$N_c' = \nu N_c = 2(2\pi m_n^* \nu^{2/3} k_B T/h^2)^{3/2} = 2(2\pi m_{dn}^* k_B T/h^2)^{3/2} \tag{5.14}$$

is the effective density of the conduction band states for a multivalley semiconductor with an ellipsoidal constant-energy surface. The density-of-states effective mass of electrons m_{dn}^* given in Eq. (5.14) can be expressed in terms of m_t and m_l by

$$m_{dn}^* = v^{2/3}(m_t^2 m_l)^{1/3} \tag{5.15}$$

where m_t and m_l are the transverse and longitudinal effective masses of electrons along the minor and major axes of the constant ellipsoidal energy surface, respectively. Values of m_t and m_l can be determined by using the cyclotron resonant experiment. ν denotes the number of conduction band valleys in the semiconductor (e.g., $\nu = 6$ for Si and 4 for Ge). For silicon crystal with $\nu = 6$, $m_t = 0.19m_0$ and $m_l = 0.98m_0$, m_{dn}^* is found to be $1.08m_0$. Values of m_t, m_l, m_{dn}^*, and N_c' for Si, Ge, and GaAs at 300 K are listed in Table 5.1.

Calculations of the hole densities in the valence bands for Si, Ge, and GaAs are slightly different from that of the electron densities in the conduction band. This is so because the valence band structures for these semiconductors are similar, consisting of a heavy-hole band and a light-hole band as well as the split-off band, as shown in Fig. 5.4. For these semiconductors, the constant-energy surface near the top of the valence bands is nonspherical and warped. For simplicity, we have assumed that the constant-energy surface near the top of the valence

TABLE 5.1. Conduction and Valence Band Parameters for Silicon, Germanium, and GaAs

Parameters	Ge	Si	GaAs
Conduction band			
ν	4	6	1
m_t/m_0	0.082	0.19	—
m_l/m_0	1.64	0.98	—
m_{dn}^*/m_0	0.561	1.084	0.068
N_c' (cm⁻3)	1.03×10^{19}	2.75×10^{19}	3.67×10^{17}
Valence band			
m_L/m_0	0.044	0.16	0.082
m_H/m_0	0.28	0.49	0.45
m_{dp}^{m*}/m_0	0.29	0.55	0.47
N_v' (cm^{-3})	5.42×10^{18}	128×10^{19}	7.0×10^{18}

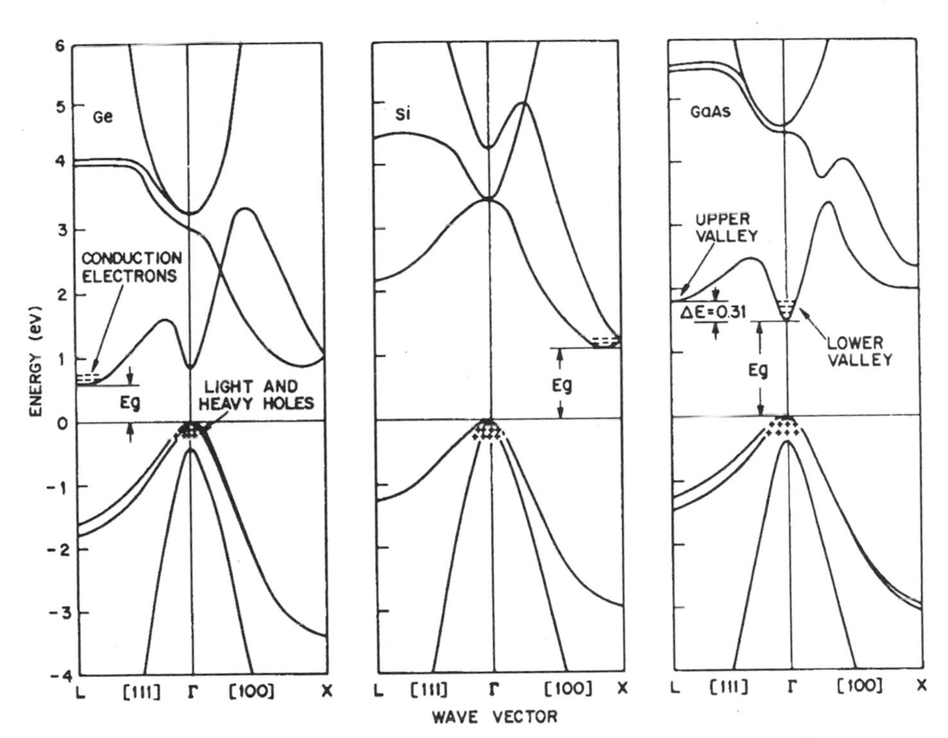

FIGURE 5.4. Energy band structures for Si, Ge, and GaAs along the (111) and (100) axes. For the valence bands, H represents the heavy-hole band and L denotes the light-hole band. Note that both bands are degenerate at $k = 0$. After Sze,[1] reprinted by permission of John Wiley & Sons Inc.

bands is spherical, and by neglecting the split-off band the hole density in the light- and heavy-hole bands can be expressed as

$$p_0 = N_v' \mathscr{F}_{1/2}(-\eta - \varepsilon_g) \tag{5.16}$$

where $N_v' = 2(2\pi m_{dp}^* k_B T/h^2)^{3/2}$ is the effective density of the valence band states, and

$$m_{dp}^* = (m_H^{3/2} + m_L^{3/2})^{-2/3} \tag{5.17}$$

is the hole density-of-states effective mass; m_H is the heavy-hole mass and m_L is the light-hole mass. Values of m_H, m_L, m_{dp}^*, and N_v' for Ge, Si, and GaAs are also listed in Table 5.1.

5.3. INTRINSIC SEMICONDUCTORS

A semiconductor may be considered as an intrinsic semiconductor if its thermally generated carrier density (i.e., n_i) is much larger than the background dopant or residual impurity densities. At $T = 0$ K, an intrinsic semiconductor behaves like an insulator because the conduction band states are empty and the valence band states are completely filled. However, as the temperature increases, some of the electrons from the valence band states are excited into the conduction band states by thermal energy, leaving behind an equal number of holes in the valence band. Therefore, the intrinsic carrier density can be expressed as

$$n_i = n_0 = p_0 \tag{5.18}$$

where n_0 and p_0 denote equilibrium electron and hole density, respectively. Substituting Eqs. (5.5) and (5.11) into (5.18) yields the intrinsic carrier density

$$\begin{aligned} n_i &= N_c \mathscr{F}_{1/2}(\eta) = N_v \mathscr{F}_{1/2}(-\eta - \varepsilon_g) \\ &= (N_c N_v)^{1/2} [\mathscr{F}_{1/2}(\eta) \mathscr{F}_{1/2}(-\eta - \varepsilon_g)]^{1/2} \end{aligned} \tag{5.19}$$

In the nondegenerate case, Eq. (5.19) becomes

$$\begin{aligned} n_i &= (N_c N_v)^{1/2}\, e^{-E_g/2k_B T} \\ &= 2.5 \times 10^{19} (T/300)^{3/2} (m_{dn}^* m_{dp}^* / m_0^2)^{3/4}\, e^{-E_g/2k_B T} \end{aligned} \tag{5.20}$$

A useful relationship between the square of the intrinsic carrier density and the product of electron and hole densities, valid for the nondegenerate case, is known as the law-of-mass-action equation, and is given by

$$n_0 p_0 = n_i^2 = K_i(T) \tag{5.21}$$

From Eqs. (5.20) and (5.21) it is noted that $n_0 p_0$ product depends only on the temperature, energy bandgap, and the effective masses of electrons and holes. Equation (5.21) is very useful since it allows one to calculate the minority carrier density in an extrinsic semiconductor when the majority carrier density is known (i.e., $p_0 = n_i^2/n_0$ for an extrinsic n-type semiconductor).

The intrinsic carrier density depends exponentially on both the temperature and energy bandgap of the semiconductor. For example, at $T = 300$ K, values of the energy bandgap for GaAs, Ge, and Si are given by 1.42, 0.67, and 1.12 eV, and the corresponding intrinsic carrier densities are 2.1×10^6, 2.5×10^{13}, and 1.4×10^{10} cm^{-3}, respectively. Thus, it is clear that a 0.1 eV increase in energy bandgap can lead to a decrease in the intrinsic carrier density by nearly one order of magnitude. This result has a very important practical implication in semiconductor device applications, since the saturation current of a p–n junction diode and a bipolar junction transistor varies with the square of the intrinsic carrier density. Therefore, p–n junction devices fabricated from a large-bandgap semiconductor such as GaAs are expected to have a much lower saturation current than that of a small-bandgap semiconductor such as silicon, and hence are more suitable for high-temperature applications. The intrinsic carrier concentrations for some III–V compound semiconductors such as InP ($E_g = 1.34$ eV), InAs ($E_g = 0.36$ eV), and InSb ($E_g = 0.17$ eV) at 300 K are found to equal 1.2×10^8, 1.3×10^{15}, and 2.0×10^{16} cm^{-3}, respectively.

The Fermi level of an intrinsic semiconductor may be obtained by solving Eqs. (5.10) and (5.12) for the nondegenerate case, which yields

$$\begin{aligned} E_f &= \frac{(E_c + E_v)}{2} + \left(\frac{k_B T}{2}\right) \ln(N_v/N_c) \\ &= E_I + \left(\frac{3}{4}\right) k_B T \ln(m_{dp}^*/m_{dn}^*) \end{aligned} \tag{5.22}$$

where E_I is known as the intrinsic Fermi level, which is located in the middle of the forbidden gap at $T = 0$ K. As the temperature rises from $T = 0$ K, the Fermi level, E_f, will move toward the conduction band edge for $m_{dp}^* > m_{dn}^*$, and toward the valence band edge for $m_{dp}^* < m_{dn}^*$. This is illustrated in Fig. 5.5. The energy bandgap of a semiconductor can be determined from the slope ($= -E_g/2k_B$) of the semilog plot of intrinsic carrier density versus inverse temperature. The intrinsic carrier density may be determined by using either the Hall effect measurement on a bulk semiconductor or the high-frequency capacitance-voltage measurement on a Schottky barrier or a p–n junction diode. The intrinsic carrier densities for Si, Ge, and GaAs as a function of temperature are shown in Fig. 5.6. The energy bandgaps for these materials are determined from the slope of $\ln(n_i T^{3/2})$ versus $1/T$ plot. The energy bandgap determined from n_i is known as the thermal bandgap of the semiconductor. On the other hand, the energy bandgap of a semiconductor can also be determined by using the optical absorption measurement near the cutoff wavelength (i.e., near the absorption edge) of the semiconductor. The energy bandgap thus determined is also referred to as the optical bandgap of the semiconductor. A small difference between these two bandgap values is expected.

Since the energy bandgap for most semiconductors decreases with increasing temperature, a correction of E_g with temperature is necessary when the intrinsic carrier density is calculated from Eq. (5.20). In general, the variation of energy bandgap with temperature can be calculated by using an empirical formula given by

$$E_g(T) = E_g(0) + \alpha T \tag{5.23}$$

where $E_g(0)$ is the energy band gap at $T = 0$ K and $\alpha = dE_g/dT$ is a negative quantity for most semiconductors. For example, $E_g(0) = 1.21$ eV and $\alpha = -2.4 \times 10^{-4}$ eV/K for silicon; $E_g(0) = 1.53$ eV and $\alpha = -4.3 \times 10^{-4}$ eV/K for GaAs.

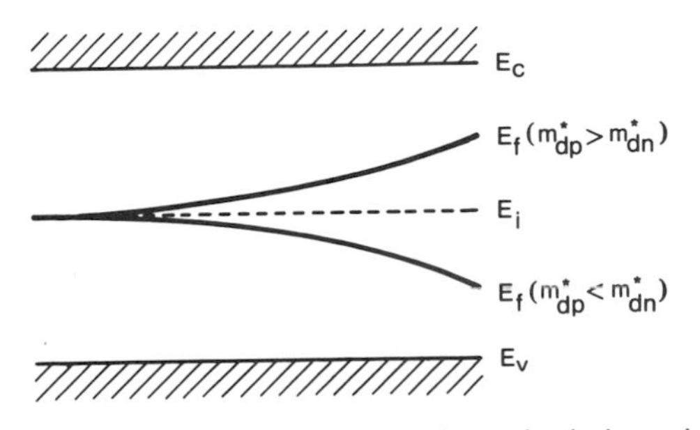

FIGURE 5.5. Fermi level *vs* temperature for an intrinsic semiconductor.

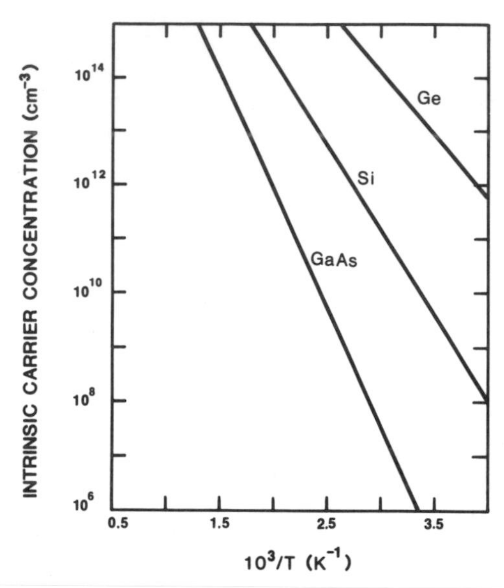

FIGURE 5.6. Intrinsic carrier concentration as a function of temperature for Si, Ge, and GaAs crystals.

5.4. EXTRINSIC SEMICONDUCTORS

As discussed in Section 5.3, the electron–hole pairs in an intrinsic semiconductor are generated by thermal excitation. Therefore, for intrinsic semiconductors with an energy bandgap on the order of 1 eV or higher, the intrinsic carrier density is usually very small at low temperatures (i.e., for $T < 100$ K). As a result, the resistivity for these intrinsic semiconductors is expected to be very high at low temperatures. This is indeed the case for Si, InP, GaAs, and other large-bandgap semiconductors (e.g., $\rho \sim 10^4\ \Omega \cdot$ cm for Si, and $10^7\ \Omega \cdot$ cm for GaAs at 300 K). It should be noted that semi-insulating substrate with resistivity greater than 10^7 ohm · cm can be readily obtained for the undoped and Cr-doped GaAs as well as Fe-doped InP materials. However, high-resistivity semi-insulating substrates are still unattainable for silicon and germanium due to the smaller bandgaps inherited in these materials.

The most important and unique feature of a semiconductor is the fact that its electrical conductivity can be readily changed by many orders of magnitude by simply doping the semiconductor with a shallow donor or a shallow acceptor impurity. By incorporating the dopant impurities into a semiconductor the electron or hole density will increase porportionally to the shallow donor or acceptor impurity concentrations. For example, if a shallow donor impurity density of 10^{16} cm^{-3} is added into a silicon crystal, an equal amount of free-electron density will appear in the conduction band at room temperature. This is illustrated in Fig. 5.7 for a silicon single crystal.

Figure 5.7a shows an intrinsic silicon crystal with covalent bond structure. In this case, each silicon atom shares the four valence electrons reciprocally with its neighboring atoms to form covalent bonds. The covalent structure also applies to other group-IV elements in the periodic table, such as germanium and diamond crystals. Figure 5.7b shows the substitution of a silicon atom by a group-V element such as phosphorus, arsenic, or antimony. In this case, an extra electron from the group-V atom is added to the host silicon lattice. Since this extra electron is loosely bound to the substitutional impurity atom (i.e., with ionization energy of a few tens meV), it can be easily excited into the conduction band via thermal energy, and hence contributes to the electrical conduction at 300 K. If the conduction process is dominated by electrons in the conduction band, then it is called an n-type semiconductor. The dopant

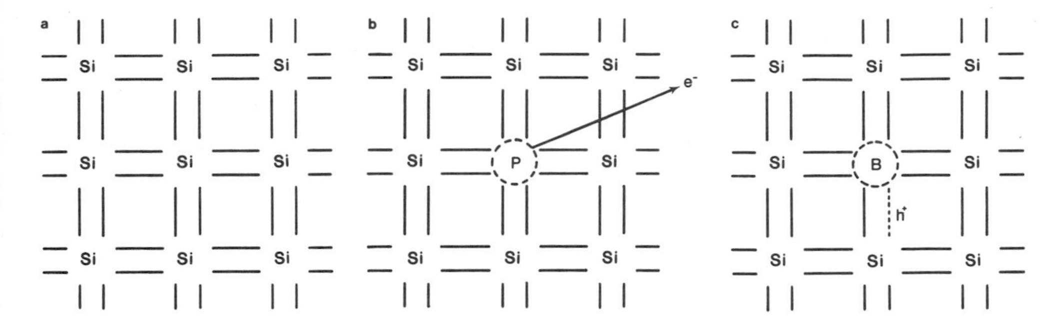

FIGURE 5.7. A covalent bond model for intrinsic and extrinsic silicon.

impurity which contributes an extra electron to the host semiconductor is called the shallow donor impurity. Thus, group-V elements in the periodic table are usually referred to as the shallow donor impurities for the group-IV elemental semiconductors such as Si and Ge. If a group-III element is introduced into a group-IV elemental semiconductor, then there is a deficiency of one electron for each replacement of the host atom by a group-III impurity atom, leaving an empty state (or creation of a hole) in the valence band, as is shown in Fig. 5.7c. In this case, the conduction process is dominated by holes in the valence bands, and the semiconductor is called a p-type semiconductor. The group-III elements including boron, gallium, and aluminum are common dopant impurities for producing p-type doping in the group-IV semiconductors. Therefore, group-III elements are the shallow acceptor impurities for the elemental semiconductors.

For III–V compound semiconductors (such as GaAs, GaP, GaSb, InP, InAs, InSb, and so on) and II–VI compounds (e.g., CdS, CdTe, ZnS, ZnSe, etc.), the control of n-type or p-type doping is more complicated than for the elemental semiconductors For example, in principle, n-type GaAs may be obtained if the arsenic atoms in the arsenic sublattices are replaced by group-VI elements such as tellurium (Te) or selenium (Se), or if the gallium atoms are replaced by group-IV elements such as germanium (Ge), silicon (Si), or tin (Sn). A p-type GaAs can be obtained if arsenic atoms are replaced by group-IV elements such as Ge or Si, or if gallium atoms are replaced by group-II elements such as zinc (Zn) or berillium (Be). However, in practice, Te, Se, Sn, and Si are frequently used as n-type dopants, and Zn or Be are often used as p-type dopants for GaAs or InP material. As for II-VI compound semiconductors, it is even more difficult to produce n- or p-type semiconductors by simply using the doping technique cited above due to the high density of native defects and the nonstoichiometric nature of the II–VI semiconductors. For example, while CdS and ZnSe always show n-type conduction, ZnS and ZnTe always show p-type conduction. The dopant impurities used in controlling the conductivity type of a semiconductor usually have very small ionization energies (i.e., a few tens of meV), and hence these impurities are often referred to as the shallow donors or shallow acceptors. These shallow-level impurities are usually fully ionized at room temperature for most semiconductors.

Figure 5.8a–c shows the energy band diagrams and the impurity levels for Si, Ge, and GaAs, respectively. The energy levels shown in the forbidden gap of silicon and germanium include all the shallow donor and acceptor impurities from group-III elements (e.g., B, Al, Ga) and group-V (e.g., P, As, Sb) elements, and the deep impurity states from the normal metals (Au, Cu, Ag) and transition metals (Fe, Ni, Co). The shallow-level impurities are used mainly for controlling the carrier concentration and the conductivity of the semiconductor, while the deep-level impurities are used to control the recombination and hence the minority carrier lifetime in a semiconductor. As an example, gold is a deep-level acceptor impurity and an effective recombination center in n-type silicon; it has an acceptor level of ionization energy $E_c - 0.55$ eV below the conduction band edge and a donor level of ionization energy

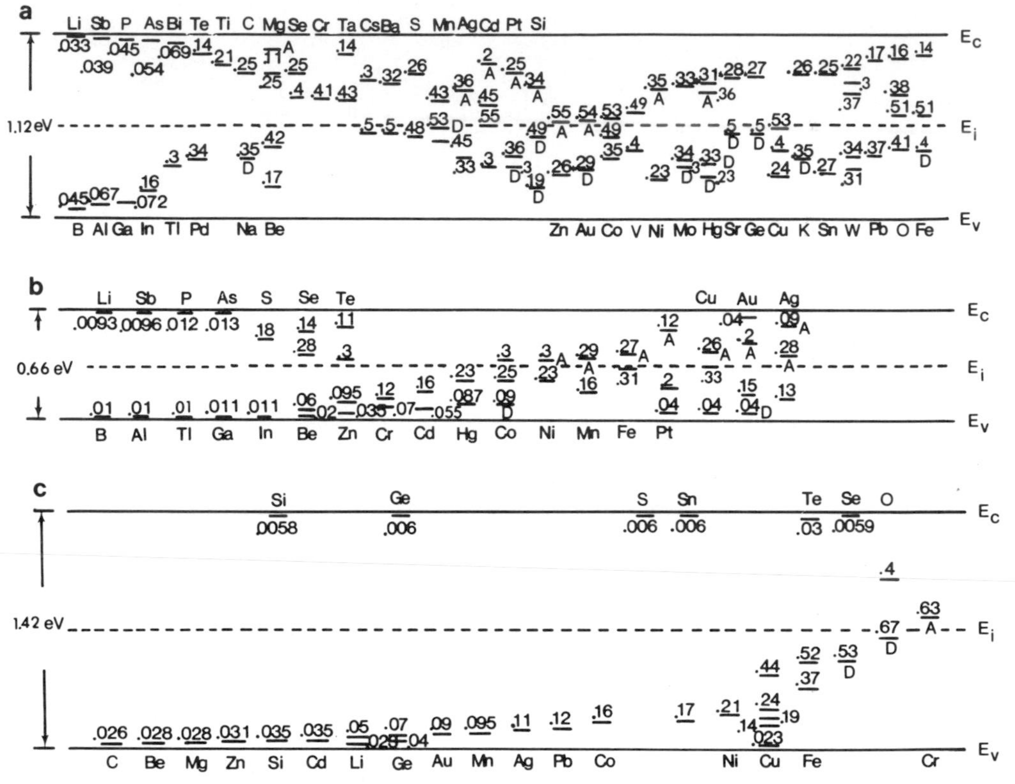

FIGURE 5.8. Ionizaton energies for various impurity levels in (a) Si, (b) Ge, and (c) GaAs. After Sze,[1] reprinted by permission of John Wiley & Sons Inc.

$E_v + 0.35$ eV above the valence band edge in the forbidden gap. Since the gold acceptor level is the most effective midgap recombination center in silicon, gold impurity has been used extensively for controlling the minority carrier lifetimes and hence the switching times in silicon devices.

The temperature behavior of equilibrium carrier density in an extrinsic semiconductor can be determined by solving the charge neutrality equation, using the expressions for the electron and hole densities in the conduction and valence band states as well as in the impurity states derived in the previous section. For an extrinsic semiconductor, if both the donor and acceptor impurities are present in the host semiconductor, then the charge neturality condition in thermal equilibrium is given by

$$\rho = 0 = q(p_0 - n_0 + N_D - n_D - N_A + p_A) \tag{5.24}$$

where p_0 and n_0 are the equlibrium hole and electron densities in the valence and conduction bands, while N_D and N_A are the donor and acceptor impurity densities, respectively. Note that n_D and p_A are the electron and hole densities in the shallow donor and shallow acceptor states, respectively, and are given by

$$n_D = \frac{N_D}{[1 + g_D^{-1}\, e^{(E_D - E_f)/k_B T}]} \tag{5.25}$$

and

$$p_A = \frac{N_A}{[1 + g_A^{-1}\, e^{(E_f - E_A)/k_B T}]} \tag{5.26}$$

where g_D and g_A denote the ground-state degeneracy factors for the shallow donor and shallow acceptor states.

In general, the temperature dependence of the carrier density and the Fermi level for an extrinsic semiconductor can be predicted by using Eqs. (5.24) through (5.26). For a nondegenerate n-type semiconductor, assuming $N_D \gg N_A$ and $N_A \gg p_A$, by substituting Eqs. (5.10), (5.12), and (5.25) into Eq. (5.24), we obtain.

$$N_c\, e^{-(E_c - E_f)/k_B T} = N_v\, e^{(E_f - E_v)/k_B T} - N_A + \frac{N_D}{1 + g_D e(E_f - E_D)/k_B T} \tag{5.27}$$

Equation (5.27) is known as the charge neutrality equation for the n-type extrinsic semiconductor. The Fermi level E_f can be determined by solving Eq. (5.27) using the iteration procedure. However, simple analytical solutions can be obtained from Eq. (5.27) in three different temperature regimes for which simplification can be introduced in Eq. (5.27). The three temperature regimes, which include the intrinsic, exhaustion, and deionization regimes, are discussed separately below.

The Intrinsic Regime. At very high temperatures when the thermal generated carrier densities in both the conduction and valence bands are much greater than the background dopant densities (i.e., $n_i \gg (N_D - N_A)$), the semiconductor becomes an intrinsic semiconductor. In the intrinsic regime (i.e., high-temperature regime), Eq. (5.24) reduces to

$$n_0 = p_0 = n_i \tag{5.28}$$

where

$$n_i = (N_v N_c)^{1/2}\, e^{-E_g/2k_B T} \tag{5.29}$$

is the intrinsic carrier density. In this regime, the intrinsic carrier density is much larger than the net dopant density of the semiconductor. For a silicon specimen with a dopant density of 1×10^{16} cm^{-3}, the temperature corresponding to the onset of the intrinsic regime is $T \geq 800$ K. For a germanium specimen with the same dopant density, this occurs at $T \geq 600$ K. Figure 5.6 shows the plots of intrinsic carrier density versus temperature for Si, Ge, and GaAs. Note that the energy bandgap can be deduced from the plot of $\ln(n_i T^{-3/2})$ versus $1/T$, using Eq. (5.29). It is noted that for the same dopant density, the intrinsic regime for GaAs will occur at a much higher temperature than that of Si or Ge because of the larger energy bandgap for GaAs.

The Exhaustion Regime. In the exhaustion regime, the donor impurities in an n-type semiconductor are completely ionized, and hence the electron density is equal to the net dopant density. Thus, the electron density can be expressed by

$$n_0 = (N_D - N_A) = N_c\, e^{-(E_c - E_f)/k_B T} \tag{5.30}$$

From Eq. (5.30), the Fermi level E_f can be written as

$$E_f = E_c - k_B T \ln[N_c/(N_D - N_A)] \tag{5.31}$$

Equation (5.31) is valid only for the temperature regime in which all the shallow donor impurities are ionized. As the temperature decreases, the Fermi level moves toward the donor level, and a fraction of the donor impurities becomes neutral. This phenomenon is known as the carrier freeze-out, which usually occurs in the shallow donor level at very low temperatures. This temperature regime is referred to as the deionization regime. We shall discuss this regime next.

The Deionization Regime. In the deionization regime, the thermal energy of electrons is usually too small to excite all the electrons in the shallow donor impurity levels into the conduction band, and hence a fraction of the donor levels will be filled by the electrons while some of the donor impurities will still remain ionized. The kinetic equation which governs the transition of electrons between the donor level and the conduction band is given by

$$n_0 + (N_D - n_D) \rightleftharpoons N_D^0 \tag{5.32}$$

or

$$\frac{n_0(N_D - n_D)}{N_D^0} = K_D(T) \tag{5.33}$$

where N_D^0 denotes the neutral donor density (i.e., $n_D = N_D^0$), and $K_D(T)$ is a constant which depends only on temperature. The charge-neutrality condition in this temperature regime (assuming $p_A \ll N_A$ and $(E_f - E_D) \gg k_BT$) is given by

$$n_0 = p_0 + (N_D - n_D) - N_A \tag{5.34}$$

and

$$(N_D - n_D) = N_D g_D^{-1} e^{(E_D - E_f)/k_BT} \tag{5.35}$$

Substituting Eq. (5.35) into Eq. (5.33) and assuming $N_D^0 = \mathrm{N}_D$, we obtain

$$K_D(T) = g_D^{-1} N_c e^{-(E_c - E_D)/k_BT} \tag{5.36}$$

The solution of Eqs. (5.33) and (5.34) yields

$$K_D(T) = \frac{n_0(N_D - n_D)}{N_D^0} \simeq \frac{n_0(n_0 + N_A)}{(N_D - N_A)} \tag{5.37}$$

Equation (5.37), which assumes that $n_0 \gg p_0$ and $N_D - n_D \gg n_0$, relates the carrier density to the temperature in the deionization regime. This equation may be used to determine the ionization energy of a shallow donor impurity level and the dopant compensation ratio in an extrinsic semiconductor. Two limiting cases which may be derived from Eq. (5.37) are discussed below.

The Lightly Compensated Case. For this case with $N_D \gg N_A$ and $N_A \ll n_0$, Eq. (5.37) becomes

$$K_D = \frac{n_0^2}{(N_D - N_A)} \tag{5.38}$$

The solution of Eqs. (5.36) and (5.38) yields

$$n_0 = [(N_D - N_A)N_c g_D^{-1}]^{1/2} e^{-(E_c - E_D)/2k_BT} \tag{5.39}$$

where $g_D = 2$ is the degeneracy factor of the shallow donor level for the present case. Equation (5.39) shows that the electron density increases exponentially with increasing temperature. Thus, from the plot of $\ln(n_0 T^{-3/2})$ versus $1/T$ in the deionization regime, we can determine the ionization energy of a shallow donor level. For the lightly compensated case, the activation energy deduced from the slope of this plot is equal to one-half of the ionization energy of the shallow donor level [i.e., $=(E_c - E_D)/2$].

The Highly Compensated Case. For this case (i.e., $N_D > N_A \gg n_0$), Eq. (5.37) becomes

$$n_0 = K_D(N_D - N_A)/N_A \tag{5.40}$$

The solution of Eqs. (5.36) and (5.40) yields

$$n_0 = [(N_D - N_A)/N_A] N_c g_D^{-1}\, e^{-(E_c - E_D)/k_B T} \tag{5.41}$$

From Eq. (5.41), it is noted that the activation energy determined from the slope of the $\ln(n_0)$ versus $1/T$ plot for the highly compensated case is equal to the ionization energy of the shallow donor impurity level.

Figure 5.9 shows the plot of $\ln n_0$ versus $1/T$ for two n-type silicon samples with different dopant impurity densities and compensation ratios. The result clearly shows that the sample with a higher impurity compensation ratio has a larger slope than that with a smaller impurity compensation ratio. From the slope of this plot we can determine the activation energy of the shallow donor impurity level. Therefore, by measuring the carrier density as a function of temperature over a wide range of temperature, we can determine simultaneously the values of N_D, N_A, E_g, E_D, or E_A for an extrinsic semicoductor. The above analysis is valid for an n-type extrinsic semiconductor. A similar analysis can also be conducted for a p-type extrinsic semiconductor.

The resistivity and Hall effect measurements are often used to determine the carrier concentration, carrier mobility, energy bandgap, and activation energy of shallow impurity levels as well as the compensation ratio of shallow impurities in a semiconductor. In addition to these measurements, the Deep-Level Transient Spectroscopy (DLTS) and Photoluminescence (PL) methods are also widely used to characterize both the deep-level defects and the shallow-level impurities in a semiconductor. Therefore, by performing these experiments, detailed information concerning the equilibrium properties of a semiconductor can be obtained.

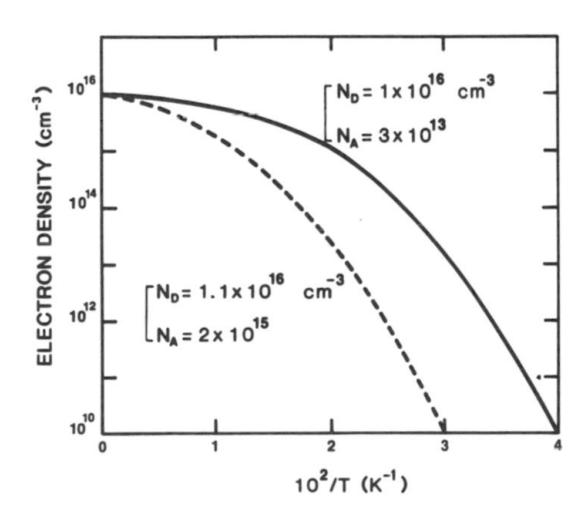

FIGURE 5.9. Electron density versus $10^2/T$ for two n-type silicon samples with different impurity compensations.

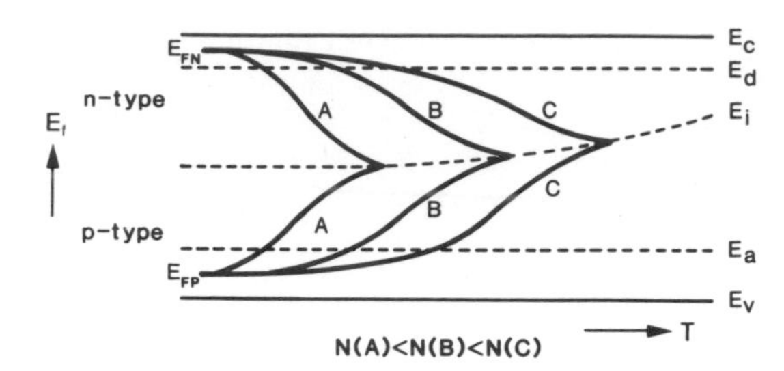

FIGURE 5.10. Fermi level as a function of temperature for n-type and p-type silicon of different compensation ratios, with m_{dp}^* and m_{dn}^*.

Figure 5.10 shows the plot of Fermi level versus temperature for both n- and p-type silicon with different degrees of impurity compensation ratio. As the temperature increases, the Fermi level may move either toward the conduction band edge or toward the valence band edge, depending on the ratio of the electron effective mass to the hole effective mass. As shown in this figure if the hole effective mass is greater than the electron effective mass, then the Fermi level will move toward the conduction band edge at high temperatures. On the other hand, if the electron effective mass is larger than the hole effective mass, then the Fermi level will move toward the valence band edge at high temperatures.

5.5. IONIZATION ENERGY OF A SHALLOW IMPURITY LEVEL

Figure 5.8 shows the ionization energies of the shallow-level and deep-level impurities observed in Ge, Si, and GaAs materials. In general, the ionization energies for the shallow donor impurity levels in these materials are only a few tens meV below the conduction band edge and a few tens meV above the valence band edge for the shallow acceptor impurity levels. The ionization energy of a shallow-level impurity may be determined by using the Hall effect, photoluminescence, or photoconductivity method. For silicon and germanium, the most commonly used shallow donor impurities are phosphorus and arsenic, and the most commonly used shallow acceptor impurity is boron. For GaAs and other III–V compounds, silicon, tellurium, and tin are the common shallow donor impurities used as n-type dopants, while zinc and berrilium are commonly used as p-type dopants.

Shallow impurity levels in a semiconductor may be treated within the framework of the effective mass model, which asserts that the electron is only loosely bound to a donor atom by a spherically symmetric Coulombic potential, and hence can often be treated as a hydrogenic impurity. Although the ionization energy of a shallow impurity may be calculated by solving the Schrödinger equation for the bound electron states associated with a shallow impurity atom, this procedure is a rather complicated one. Instead of solving the Schrödinger equation to obtain the ionization energy of a shallow impurity state, the simple Bohr's model for the hydrogen atom may be applied to estimate the ionization energy of a shallow impurity level in a semiconductor. Although Bohr's model may be oversimplified, it offers some physical insights concerning the nature of the shallow impurity states in a semiconductor. It is interesting to note that the ionization energy of a shallow impurity level calculated from the modified Bohr's model agrees reasonably well with the experimental data for many semiconductors. In the hydrogen-like impurity model, the ionization energy of a shallow impurity state depends only on the effective mass of electrons and the dielectric constant of the semiconductor.

To calculate the ionizaton energy of a shallow impurity level, let us consider the case of a phosphorus donor atom in a silicon host crystal as shown in Fig. 5.7b. Each silicon atom shares four valence electrons reciprocally with its nearest-neighbor atoms to form a covalent bond. The phosphorus atom, which replaces a silicon atom, has five valence electrons. Four of its five valence electrons in the phosphorus atom are shared by its four nearest-neighbor silicon atoms while the fifth valence electron is loosely bound to the phosphorus atom. Although this extra electron of phosphorus ion is not totally free, it has a small ionization energy, which enables it to break loose relatively easily from the phosphorus atom and become free in the host semiconductor. Therefore, we may regard the phosphorus atom as a fixed ion with a positive charge surrounded by an electron with a negative charge. If the ionizaton energy of this bound electron is small, then its orbit will be quite large (i.e., much larger than the interatomic spacing). Under this condition, it is reasonable to treat the bound electron as being imbedded in a uniformly polarized medium whose dielectric constant is given by the macroscopic dielectric constant of the host semiconductor. This assumption resembles that of a hydrogen atom imbedded in a uniform continuous medium with a dielectric constant equal to unity. Therefore, so long as the dielectric constant of the semiconductor is large enough such that the Bohr radius of the shallow-level impurity ground state is much larger than the interatomic spacing of the host semiconductor, the modified Bohr model can be used to treat the shallow impurity states in a semiconductor.

To apply the Bohr model to a phosphorus impurity atom in a silicon crystal, two parameters must be modified. First, the free-electron mass which is used in a hydrogen atom must be replaced by the electron effective mass m^*. Second, the relative permittivity in free space must be replaced by the dielectric constant of silicon, which is equal to 11.7. If we assume that the orbit of a bound electron around a phosphorus atom is circular, then the allowed values of the angular momentum, L_n, are subject to the quantum restriction (i.e., L_n is equal to a multiple integer of Planck's constant h). Under this condition, the angular momentum is given by

$$L_n = m_e^* v_n r_n = nh \tag{5.42}$$

where r_n is the radius of the electron orbit, v_n is the velocity of electrons in the orbit, and n is the quantum number ($n = 1, 2, 3, \ldots$). For a stable orbit, the radius r_n of the orbit must satisfy the condition that the Coulombic attractive force between the bound electron and the phosphorus ion is equal to the centrifugal force required to hold the electron in a circular orbit,

$$F_n = \frac{q^2}{4\pi \varepsilon_0 \varepsilon_s r_n^2} = \frac{m_e^* v_n^2}{r_n} \tag{5.43}$$

Solving Eqs. (5.42) and (5.43) for r_n and v_n yields

$$r_n = \frac{4\pi \varepsilon_0 \varepsilon_s n^2 h^2}{m_e^* q^2} \tag{5.44}$$

and

$$v_n = \frac{q^2}{4\pi \varepsilon_0 \varepsilon_s nh} \tag{5.45}$$

where ε_0 is the permittivity of free space; ε_s is the dielectric constant of the semiconductor and q is the electronic charge. The kinetic energy of the bound electron is given by

$$\text{K.E.} = \tfrac{1}{2} m_e^* v_n^2 = \frac{m_e^* q^4}{32(\pi \varepsilon_0 \varepsilon_s nh)^2} \tag{5.46}$$

and the potential energy is

$$\text{P.E.} = \frac{-q^2}{4\pi\varepsilon_0\varepsilon_s r_n} = \frac{-m_e^* q^4}{(4\pi\varepsilon_0\varepsilon_s nh)^2} \tag{5.47}$$

Therefore, the total energy of the bound electron is equal to the sum of Eqs. (5.46) and (5.47), namely

$$E_i = \text{P.E.} + \text{K.E.} = \frac{-m_e^* q^4}{32(\pi\varepsilon_0\varepsilon_s nh)^2} \tag{5.48}$$

Equation (5.48) represents the ionization energy of a shallow impurity level in a semiconductor, as derived from the modified Bohr's hydrogenic model. The ground-state ionization energy may be obtained by setting $n=1$ in Eq. (5.48), which yields

$$E_i = \frac{-m_e^* q^4}{32(\pi\varepsilon_0\varepsilon_s h)^2} = -13.6(m_e^*/m_0)\,\varepsilon_s^{-2} \quad \text{eV} \tag{5.49}$$

The Bohr radius for the ground state of the shallow impurity level is obtained by setting $n=1$ in Eq. (5.44), and the result is

$$r_1 = \frac{4\pi\varepsilon_0\varepsilon_s g^2}{m_e^* q^2} = 0.53(m_0/m_e^*)\,\varepsilon_s \quad \text{Å} \tag{5.50}$$

Equation (5.49) shows that the ionization energy of a shallow impurity level is inversely proportional to the square of the dielectric constant. On the other hand, Eq. (5.50) shows that the Bohr radius varies linearly with the dielectric constant and inversely with the electron effective mass. For silicon, using the values of the electron effective mass $m_e^* = 0.26m_0$ and the dielectric constant $\varepsilon_s = 11.7$, the ionization energy calculated from Eq. (5.49) is found to be 25.8 meV, and the Bohr radius calculated from Eq. (5.50) is 24 Å. For germanium, with $m_e^* = 0.12m_0$ and $\varepsilon_s = 16$, the calculated value for E_i is found to be 6.4 meV, and the Bohr radius is equal to 71 Å. The above results clearly illustrate that the Bohr radii for the shallow impurity states in both silicon and germanium are indeed much larger than the interatomic spacing of silicon and germanium. The calculated ionization energies for the shallow impurity states in Si, Ge, and GaAs are generally smaller than the measured values shown in Fig. 5.8. However, the agreement should improve for the excited states of the shallow impurity levels.

5.6. HALL EFFECT, HALL MOBILITY, AND ELECTRICAL CONDUCTIVITY

As discussed earlier, carrier density (i.e., n_0 or p_0) and carrier mobility (μ_n or μ_p) are two key parameters which govern the transport and electrical properties of a semiconductor. Both parameters are usually determined by using the Hall effect and resistivity measurements.

The Hall effect was discovered by Edwin H. Hall in 1879 during an investigation of the nature of the force acting on a conductor carrying a current in a magnetic field. Hall found that when a magnetic field is applied at right angles to the direction of current flow, an electric field is set up in a direction perpendicular to both the direction of the current and the magnetic

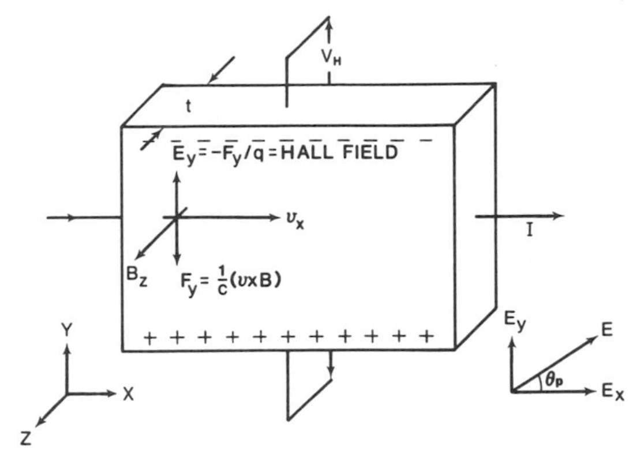

FIGURE 5.11. Hall effect for a p-type semiconductor bar. The Hall voltage V_H, the applied electric field $\mathscr{E}$, and the applied magnetic field B_z are mutually perpendicular to one another. For an n-type sample, electrons are deflected in the y-direction to the bottom of the sample.

field. To illustrate the Hall effect in a semiconductor, Fig. 5.11 shows the Hall effect for a p-type semiconductor bar and the polarity of the induced Hall voltage.

As shown in Fig. 5.11, the Hall effect is referred to the phenomenon in which a Hall voltage (V_H) is developed in the y-directon when an electric current (J_x) is applied in the x-direction and a magnetic field (B_z) is applied in the z-direction of a semiconductor bar. The interaction of a magnetic field in the z-direction with the electron motion in the x-direction produces a Lorentz force along the negative y-direction, which is counterbalanced by the Hall voltage developed in the y-direction. This can be written as

$$q\mathscr{E}_y = -qB_z v_x = -B_z J_x/n_0 \tag{5.51}$$

where B_z is the magnetic flux density in the z-direction and n_0 is the electron density. The electric field component $\mathscr{E}_x$ parallel to the current density J_x in the x-direction is given by

$$J_x = n_0 q \mu_n \mathscr{E}_x \tag{5.52}$$

For the small magnetic field case, the angle between the current density J_x and the induced Hall field $\mathscr{E}_y$ is given by

$$\tan\theta_n \simeq \theta_n = \frac{\mathscr{E}_y}{\mathscr{E}_x} = -B_z\mu_n \tag{5.53}$$

where θ_n is called the Hall angle for electrons. The Hall coefficient R_H is defined by

$$R_H = \left.\frac{\mathscr{E}_y}{B_z J_x}\right|_{J_y=0} = \frac{V_H W}{B_z I_x} \tag{5.54}$$

where V_H is the Hall voltage and W is the width of the semiconductor bar. Solving Eqs. (5.51) and (5.54) yields

$$R_{Hn} = -\frac{1}{n_0 q} \tag{5.55}$$

Equation (5.55) shows that the Hall coefficient is inversely proportional to the electron density. The minus sign in Eq. (5.55) is for electron conduction. Thus, from the measured Hall coefficient, we can calculate the electron density in an n-type semiconductor. Equation (5.55) is valid as long as the relaxation time constant τ is independent of electron energy. If

τ is a function of electron energy, then a generalized expression for the Hall coefficient must be used, namely,

$$R_{Hn} = \frac{-\gamma_n}{qn_0} \tag{5.56}$$

where $\gamma_n = \langle\tau\rangle^2/\langle\tau^2\rangle$ is the Hall factor; its value may vary between 1.18 and 1.93, depending on the types of scattering mechanisms involved. Detailed calculations of the Hall factor for different scattering mechanisms will be discussed further in Chapter 7.

Similarly, for a p-type semiconductor, the Hall coefficient can be expressed by

$$R_{Hp} = \frac{\gamma_p}{qp_0} \tag{5.57}$$

where p_0 is the hole density and γ_p is the Hall factor for holes in a p-type semiconductor. Equation (5.57) shows that the Hall coefficient for a p-type semiconductor is positive since a hole has a positive charge. Values of the Hall factor for a p-type semiconductor may vary between 0.8 and 1.9, depending on the types of scattering mechanisms involved. This will also be discussed further in Chapter 7.

For an intrinsic semiconductor, both electrons and holes are expected to participate in the conduction process, and the mixed conduction prevails. Thus, the Hall coefficient is contributed to by both electrons and holes, and can be expressed by

$$R_H = \frac{\mathscr{E}_y}{B_z J_x} = \frac{R_{Hn}\sigma_n^2 + R_{Hp}\sigma_p^2}{(\sigma_n + \sigma_p)^2} = \frac{(p_0\mu_p^2 - n_0\mu_n^2)}{q(p_0\mu_p + n_0\mu_n)^2} \tag{5.58}$$

where R_{Hn} and R_{Hp} denote the Hall coefficients for n- and p-type conduction given by Eqs. (5.56) and (5.57) respectively. Note that σ_n and σ_p are the electrical conductivities for the n- and p-type semiconductors, respectively. It is interesting to note from Eq. (5.58) that if $p_0\mu_p^2 = n_0\mu_n^2$, then the Hall coefficient $R_H = 0$ results. This situation may in fact occur in an intrinsic semiconductor as one measures the Hall coefficient as a function of temperature over a wide range of temperature in which the conduction process may change from n- to p-type conduction at a certain temperature.

From the above analysis, it is clear that the Hall effect and resistivity measurements are important experimental tools for analyzing the equilibrium properties of a semiconductor. It allows one to determine the majority carrier density, the conductivity mobility, the ionization energy of a shallow impurity level, the type of conduction, the energy bandgap, and the impurity compensation ratio in a semiconductor.

Electrical conductivity is another important physical parameter which will be discussed next. The performance of a semiconductor device is closely related to the electrical conductivity in a semiconductor. The electrical conductivity for an extrinsic semiconductor is equal to the product of electronic charge, carrier density, and carrier mobility, and can be expressed by

$$\sigma_n = qn_0\mu_n \qquad \text{for n-type} \tag{5.59}$$

$$\sigma_p = qp_0\mu_p \qquad \text{for p-type} \tag{5.60}$$

For an intrinsic semiconductor, the electrical conductivity is given by

$$\sigma_i = q(\mu_n + \mu_p)n_i \tag{5.61}$$

where μ_n and μ_p denote the electron and hole mobilities, respectively, and n_i is the intrinsic carrier density.

Since both the electrical conductivity and Hall coefficient are measurable quantities, the product of these two parameters, known as the Hall mobility, can also be obtained experimentally. From Eqs. (5.56), (5.57), (5.59), and (5.60) the Hall mobilities for an n-type and a p-type semiconductor are obtained, which read

$$\mu_{Hn} = R_{Hn}\sigma_n = \gamma_n \mu_n \qquad \text{for n-type} \tag{5.62}$$

and

$$\mu_{Hp} = R_{Hp}\sigma_p = \gamma_p \mu_p \qquad \text{for p-type} \tag{5.63}$$

where γ_n and γ_p denote the Hall factor for the n- and p-type semiconductors, respectively. The ratio of Hall mobility to conductivity mobility is equal to the Hall factor, which depends only on the scattering mechanisms.

5.7. HEAVY DOPING EFFECTS IN A DEGENERATE SEMICONDUCTOR

As discussed earlier, the electrical conductivity of a semiconductor may be changed by many orders of magnitude by simply doping the semiconductor with shallow donor or acceptor impurities. However, when the dopant density is greater than 10^{19} cm^{-3} (i.e., for the cases of silicon and germanium), the semiconductor becomes degenerate, and thus changes in the fundamental physical properties of the semiconductor result. The heavy doping effects in a degenerate semiconductor include the broadening of the shallow impurity level in the forbidden gap from a discrete level into an impurity band, the shrinkage of the energy bandgap, the formation of a band tail at the conduction and valence band edges, and the distortion of the density-of-states function from its square-root dependence on the energy. All these phenomena are referred to as the heavy doping effects in a degenerate semiconductor. In a heavily doped semiconductor, the Fermi-Dirac statistics, rather than the Maxwell–Boltzmann statistics, must be employed in calculating the carrier density and other transport coefficients in such a material.

There are two dominant heavy doping effects in a degenerate semiconductor which will be considered here. The first important consideration is that the Fermi–Dirac statistics must be used to calculate the carrier density in a degenerate semiconductor. This concept was discussed in Chapter 3. The second important effect is known as the bandgap narrowing effect. It is well known that the heavy doping effect is a very difficult and complicated physical problem, and existing theories for dealing with the heavy doping effects are inadequate. Due to the existence of the heavy doping regime in various silicon devices and integrated circuits, most of the theoretical and experimental studies on heavy doping effects conducted in recent years have been focused on degenerate silicon. The results of these studies on the bandgap narrowing effect and carrier degeneracy for heavily doped silicon are discussed next.

Measurements of bandgap narrowing as a function of doping density in heavily doped silicon samples have been widely reported in the literature. A semiempirical formula, based on the stored electrostatic energy of majority–minority carrier pairs, has been derived for the bandgap reduction. The bandgap narrowing, ΔE_g, for an n-type silicon is given by[2]

$$\Delta E_g = \left(\frac{3q^2}{16\pi\varepsilon_0\varepsilon_s}\right)\left(\frac{q^2 N_D}{\varepsilon_s\varepsilon_0 k_B T}\right)^{1/2} \tag{5.64}$$

At room temperature, the bandgap narrowing versus donor concentration for an n-type silicon given by Eq. (5.64) becomes

$$\Delta E_g = 22.5(N_D/10^{18})^{1/2} \quad \text{meV} \tag{5.65}$$

where N_D is the donor density. Using Eq. (5.65), a bandgap reduction of 225 meV is obtained at a dopant density of 10^{20} cm^{-3} for n-type silicon. This value appears to be larger than the measured value reported for silicon. Figure 5.12 shows a plot of bandgap narrowing versus donor density for silicon at 300 K, and a comparison of the calculated values of ΔE_g from Eq. (5.65) with experimental data.

Another important physical parameter to be considered here is the $n_0 p_0$ product, which is equal to the square of the effective intrinsic carrier density, n_{ie}^2. The $n_0 p_0$ product is an important parameter in a heavily doped p$^+$–n junction diode and in the emitter region of a bipolar junction transistor. It relates the bandgap narrowing effect to the saturation current density in the heavily doped bipolar junction devices. To explain this, let us consider the square of the intrinsic carrier density in a heavily doped n-type semiconductor which is given by

$$\begin{aligned} n_{ie}^2 = n_0 p_0 &= N_c N_v \, e^{-E_g'/k_BT} \mathscr{F}_{1/2}(\eta)\, e^{-\eta} \\ &= n_i^2 \, e^{\Delta E_g/k_BT} \mathscr{F}_{1/2}(\eta)\, e^{-\eta} \end{aligned} \tag{5.66}$$

where $E_g' = E_g - \Delta E_g$ is the effective bandgap of a heavily doped n-type semiconductor. Equation (5.66) is obtained by solving Eqs. (5.12) and (5.5). Thus, when the bandgap narrowing effect is considered, the $n_0 p_0$ product (or n_{ie}^2) is found to be much larger for a degenerate than for a nondegenerate semiconductor.

In summary, we have discussed in this chapter the important physical and electrical properties of a bulk semiconductor under equilibrium conditions. Key physical parameters, such as thermal equilibrium carrier densities, shallow- and deep-level impurities and their ionization energies, carrier mobilities, and energy bandgap, have been derived and discussed. Table 5.2 lists some important physical parameters for the elemental and compound semiconductors. The importance of these physical parameters on the transport properties of a semiconductor and device performance will be discussed further in Chapter 7 and subsequent chapters dealing with various electronic and photonic devices.

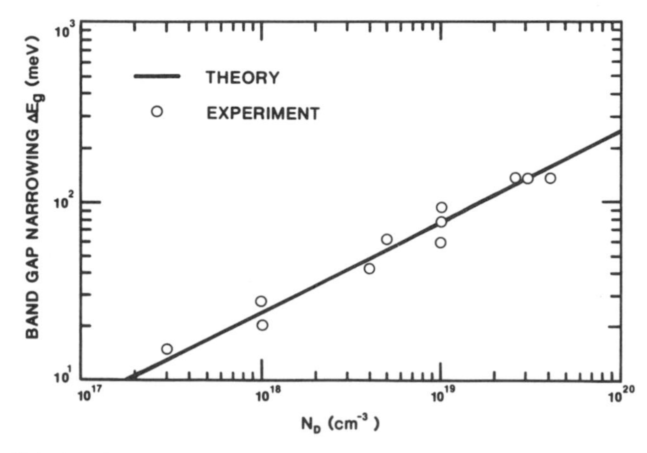

FIGURE 5.12. Calculated and measured values of bandgap narrowing versus donor density in n-type silicon.

TABLE 5.2. Important Physical Parameters for Some Practical Semiconductors at 300 K[a]

Semiconductor	Bandgap (eV)	Effective mass (m^*/m_0)		Carrier mobility ($cm^2/v \cdot sec$)	
		m_n^*	m_p^*	μ_n	μ_p
Ge	0.67	$m_l = 1.64$ $m_t = 0.082$	$m_H = 0.28$ $m_L = 0.04$	3,900	1,900
Si	1.12	$m_l = 0.98$ $m_t = 0.19$	$m_H = 0.49$ $m_L = 0.16$	1,500	450
GaN	3.36	0.60	0.19	380	
GaP	2.25	0.82	0.60	110	75
GaAs	1.43	0.067	0.082	8,500	400
InP	1.35	0.077	0.64	4,600	150
GaSb	0.75	0.042	0.40	5,000	850
InAs	0.36	0.023	0.40	33,000	460
InSb	0.17	0.0145	0.40	80,000	1,250
ZnS	3.68	0.40	—	165	10
ZnSe	2.67	0.17	0.60	102	16
CdS	2.42	0.21	0.80	340	50
CdSe	1.70	0.13	0.45	800	
CdTe	1.56	—	—	1,050	100
PbTe	0.31	0.17	0.20	6,000	7,000

[a] m_l denotes longitudinal mass, m_t transverse mass, m_H heavy-hole mass, m_L light-hole mass, μ_n electron mobility, and μ_p hole mobility.

PROBLEMS

5.1. Consider an n-type silicon doped with phosphorus impurities. The resistivity of this sample is 10 ohm cm at 300 K. Assuming that the electron mobility is equal to 1350 $cm^2/V \cdot s$ and the density-of-states effective mass for electrons is $m_{dn}^* = 1.065 m_0$:
   (a) Calculate the density of the phosphorus impurities, assuming full ionization at 300 K.
   (b) Determine the location of the Fermi level relative to the conduction band edge, assuming that $N_A = 0$ and $T = 300$ K
   (c) Determine the location of the Fermi level if $N_A = 0.5 N_D$ and $T = 300$ K.
   (d) Find the electron density, n_0, at $T = 20$ K, assuming that $N_A = 0$, $g_D = 2$, and $E_c - E_D = 0.044$ eV.
   (e) Repeat (d) for $T = 77$ K.

5.2. If the temperature dependence of the energy bandgap for InAs is given by

$$E_g = 0.426 - 3.16 \times 10^{-4} T^2 (93 + T)^{-1} \quad \text{eV}$$

and the density-of-states effective masses for electrons and holes are given respectively by $m_n^* = 0.002 m_0$ and $m_p^* = 0.4 m_0$, plot the intrinsic carrier density, n_i, as a function of temperature for $200 < T < 700$ K.

5.3. If the dielectric constant for GaAs is 12, and the electron effective mass is $0.086\, m_0$, calculate the ionization energy and the radius of the first Bohr orbit from Bohr's model given in the text. Repeat for an InP crystal.

5.4. Plot the Fermi level as a function of temperature for a silicon specimen with $N_D = 1 \times 10^{16}$ cm^{-3} and compensation ratios of $N_D/N_A = 0.1$, 0.5, 2, 10.

5.5. Show that the expressions given by Eqs. (5.10) and (5.12) for electron and hole densities can be written in terms of the intrinsic carrier density, n_i, as follows:

$$n_0 = n_i\, e^{(E_f - E_i)/k_B T} \quad \text{and} \quad p_0 = n_i\, e^{(E_i - E_f)/k_B T}$$

5.6. Consider a semiconductor specimen. If it contains a small density of shallow donor impurity level such that $k_BT \ll (E_c - E_D) \ll (E_D - E_v)$, show that at $T = 0$ K the Fermi level is located halfway between E_c and E_D, assuming that E_D is completely filled at $T = 0$ K.

5.7. Derive an expression for the Hall coefficient of an intrinsic semiconductor in which conduction is due to both electrons and holes. Specify the condition under which the Hall coefficient vanishes.

5.8. (a) When a current of 1 mA and a magnetic field intensity of 10^3 gauss are applied to an n-type semiconductor bar of width 1 cm and thickness 1 mm, a Hall voltage of 1 mV is developed across the sample. Calculate the Hall coefficient and the electron density in this sample.
(b) If the electrical conductivity of this sample is equal to 2.5 mho · cm^{-1}, what is the Hall mobility? If the Hall factor is equal to 1.18, what is the electron conductivity mobility?

5.9. (a) Using the charge neutrality equation given by Eq. (5.24), derive a general expression for the hole density versus temperature, and discuss both the lightly compensated and heavily compensated cases at low temperatures (i.e., the deionization regime). Assume that the degeneracy factor g_A is equal to 4. Plot p_0 versus T for the case $N_A = 5 \times 10^{16}$, $N_D = 10^{14}\,\text{cm}^{-3}$, and $E_A = 0.044$ eV.
(b) Repeat for the case $N_D = 0.5N_A$; $N_A = 10^{16}\,\text{cm}^{-3}$.
(c) Plot the Fermi level versus temperature.

5.10. Taking into account the Fermi statistics and the bandgap narrowing effects, show that the n_0p_0 product is given by Eq. (5.66) for a heavily doped n-type semiconductor:

$$n_{ie}^2 = n_0p_0 = n_i^2 \exp(\Delta E_g/kT)\mathscr{F}_{1/2}(\eta)\exp(-\eta)$$

where n_i is the intrinsic carrier concentration for the nondegenerate case, ΔE_g is the bandgap narrowing, $\mathscr{F}_{1/2}(\eta)$ is the Fermi integral of order one-half, and $\eta = -(E_c - E_f)/kT$ is the reduced Fermi energy. Using the above equation, calculate the values of n_{ie}^2 and ΔE_g for an n-type degenerate silicon for $\eta = 1$ and 4 (assuming full ionization) and $T = 300$ K. Given:

$$\Delta E_g = 22.5(N_D/10^{18})^{1/2} \quad \text{meV}$$

$$\mathscr{F}^{1/2}(\eta) = (4/3\sqrt{\pi})(\eta^2 + \pi^2/6)^{3/4}$$

$$N_c = 2.75 \times 10^{19}\,\text{cm}^{-3}$$

$$N_v = 1.28 \times 10^{19}\,\text{cm}^{-3}$$

5.11. Using Eq. (5.66), plot n_{ie}^2 versus N_D for n-type silicon with N_D varying from 10^{17} to $10^{20}\,\text{cm}^{-3}$.

5.12. Plot the ratios n_Γ/n_0, n_L/n_0, n_X/n_0 versus temperature (T) for a GaAs crystal for $0 < T < 1000$ K. The electron effective masses for the Γ-, X-, and L-conduction band minima are given respectively by: Γ-band, $m_\Gamma = 0.0632\,m_0$; L-band, $m_l \approx 1.9m_0$; $m_t \approx 0.075m_0$; $m_L = (16m_lm_t^2)^{1/3} = 0.56m_0$; X-band: $m_l \approx 1.9m_0$; $m_t \approx 0.19m_0$; $m_X = (9m_lm_t^2)^{1/3} = 0.85m_0$; $n_0 = n_\Gamma + n_X + n_L$; m_L and m_X are the density-of-states effective masses for the L- and X-bands, and

$$n_\Gamma = N_c^\Gamma \exp[(E_F - E_c)/k_BT] = 2(2\pi m_\Gamma k_BT/h^2)^{3/2}\exp[\eta/k_BT]$$

$$n_X = 2(2\pi m_X k_BT/h^2)^{3/2}\exp[(\eta - \Delta_{\Gamma X})/k_BT]$$

$$n_L = 2(2\pi m_L k_BT/h^2)^{3/2}\exp[(\eta - \Delta_{\Gamma L})/k_BT]$$

where $\eta = E_F - E_c$, $\Delta_{\Gamma X} = E_X - E_\Gamma = 0.50$ eV, and $\Delta_{\Gamma L} = \text{E}_L - \text{E}_\Gamma = 0.33$ eV.

5.13. Using the one-dimensional Schrödinger equation, derive the expressions of quantized energy states for (i) an infinite square well (with well width $a = 100$ Å), (ii) triangular well, and (iii) parabolic well. Assuming that the quantization occurs in the z-direction and the potential energy for the three cases is given by (i) $U(z) \to \infty$, (ii) $U(z) = q\mathscr{E}z$ (where $\mathscr{E}$ is the electric field inside the triangular well), and (iii) $U(z) = (m^*\omega^2/2)z^2$, calculate the energy levels of the ground state and the first excited state of (i) and (ii). Given: $m^* = 0.067m_0$, $a = 100$ Å and $\mathscr{E} = 10^5$ v/cm. (Answer:

$$(i)\ E_r = \frac{\hbar^2\pi^2}{8m^*a^2}(r+1)^2,$$

$r = 0, 1, 2, \ldots, E_0 = 56$ meV, $E_1 = 224$ meV,

$$\text{(ii)}\ E_r = \left(\frac{\hbar^2 q^2 \mathscr{E}^2}{2m^*}\right)^{1/3} \left[\frac{3\pi}{2}(r + 3/4)\right]^{2/3},$$

$E_0 = 87$ meV, $E_1 = 153$ meV, and (iii) $E_r = \hbar\omega(r + 1/2)$.)

REFERENCE

1. S. M. Sze, *Physics of Semiconductor Devices*, 2nd ed., Wiley, New York (1981).
2. H. P. D. Lanyon and R. A. Tuft, *Bandgap Narrowing in Heavily-Doped Silicon*, International Electron Device Meeting, p. 316, IEEE Tech. Dig. (1978).

BIBLIOGRAPHY

J. S. Blakemore, *Semiconductor Statistics*, Pergamon Press, New York (1962).
F. J. Blatt, *Physics of Electronic Conduction in Solids*, McGraw-Hill (1968).
R. H. Bube, *Electronic Properties of Crystalline Solids*, Academic Press, New York (1973).
V. I. Fistul', *Heavily Doped Semiconductors*, Plenum Press, New York (1969).
A. F. Gibson and R. E. Burgess, *Progress in Semiconductors*, Wiley, New York (1964).
N. B. Hannay, *Semiconductors*, Reinhold, New York (1959).
D. C. Look, *Electrical Characterization of GaAs Materials and Devices*, Wiley, New York (1989).
J. P. McKelvey, *Solid-State and Semiconductor Physics*, 2nd ed., Harper & Row, New York (1966).
K. Seeger, *Semiconductor Physics*, 3rd ed. Springer-Verlag, New York (1973).
W. Shockley, *Electrons and Holes in Semiconductors*, D. Van Nostrand, New York (1950).
M. Shur, *Physics of Semiconductor Devices*, Prentice-Hall, New York (1990).
R. A. Smith, *Semiconductors*, 2nd ed., Cambridge University Press, Cambridge (1956).
C. T. Wang, *Introduction to Semiconductor Technology*, Wiley, New York (1990).
H. F. Wolf, *Semiconductors*, Wiley, New York (1971).

6

Excess Carrier Phenomenon in Semiconductors

6.1. INTRODUCTION

The generation of excess carriers in a semiconductor may be accomplished by either electrical or optical means. For example, electron–hole pairs are created in a semiconductor when photons with energies exceeding the bandgap energy of the semiconductor are absorbed. Similarly, minority carrier injection can be achieved by applying a forward bias voltage across a p–n junction diode or a bipolar junction transistor. The inverse process to the generation of excess carriers in a semiconductor is recombination. The annihilation of excess carriers generated by optical or electrical means in a semiconductor may take place via different recombination mechanisms. Depending on the ways in which the energy of an excess carrier is removed during a recombination process, there are three basic recombination mechanisms which are responsible for carrier annihilation in a semiconductor. These include: (1) nonradiative recombination (i.e., the multiphonon process), (2) band-to-band radiative recombination, and (3) Auger band-to-band recombination. The first recombination mechanism, known as the nonradiative or multiphonon recombination process, is usually the predominant recombination process for indirect bandgap semiconductors such as silicon and germanium. In this process, recombination is accomplished via a deep-level recombination center in the forbidden gap, and the energy of the excess carriers is released via phonon emission. The second recombination mechanism, band-to-band radiative recombination, is usually the predominant process occurring in direct bandgap semiconductors such as GaAs and InP. In this case, band-to-band recombination of electron–hole pairs is accompanied by the emission of a photon. The third recombination mechanism, Auger band-to-band recombination, is usually the predominant recombination process occurring in the heavily doped semiconductors or small-bandgap semiconductors such as InSb. The Auger process can also become the predominant recombination process under high injection conditions. Unlike the nonradiative and radiative recombination processes, which are two-particle processes, Auger band-to-band recombination is a three-particle process which involves two electrons and one hole for an n-type semiconductor, or one electron and two holes for a p-type semiconductor. For an n-type semiconductor, Auger recombination is accomplished first via electron–electron collisions in the conduction band, and followed by electron–hole recombination in the valence band. Based on the principle of detailed balance, the rate of recombination is equal to the rate of generation of excess carriers under thermal equilibrium condition, and hence a charge-neutrality condition prevails throughout the semiconductor specimen.

Equations governing the recombination lifetimes for the three basic recombination mechanisms cited above are derived in Sections 6.2, 6.3, and 6.4, respectively. The continuity equations for the excess carrier transport in a semiconductor are described in Section 6.5, and the charge-neutrality equation will be discussed in Section 6.6. The Haynes–Shockley experiment and minority carrier drift mobility are depicted in Secton 6.7. Section 6.8 presents methods of determining minority carrier lifetimes in a semiconductor. The surface states and surface recombination mechanisms in a semiconductor are discussed in Section 6.9. Finally, the deep-level transient spectroscopy (DLTS) technique for characterizing deep-level defects in a semiconductor is described in Section 6.10.

6.2. NONRADIATIVE RECOMBINATION: SHOCKLEY–READ–HALL MODEL

In the nonradiative recombination process, the recombination of electron–hole pairs may take place at the localized trap states in the forbidden gap of the semiconductor. This process involves the capture of electrons or holes by the trap states, followed by their respective recombination with holes in the valence band or electrons in the conduction band. When electron–hole pairs recombine, energy is released via phonon emission. The localized trap states may be created by deep-level impurities (e.g., transition metals or normal metals such as Fe, Ni, Co, W, Au, etc.), or by radiation- and process-induced defects such as vacancies, interstitials, antisite defects and their complexes, dislocation, and grain boundaries. The nonradiative recombination process in a semiconductor can be best described by the Shockley–Read–Hall model.[1,2] This is discussed next.

Figure 6.1 illustrates the energy band diagram for the Shockley–Read–Hall model. In this figure, the four transition processes for the capture and emission of electrons and holes via a localized recombination center are shown. A localized deep-level trap state may be in one of the two charge states differing by one electronic charge. Therefore, the trap could be in either a neutral or a negatively charged state, or in a neutral or a positively charged state. If the trap state is neutral, then it can capture an electron from the conduction band. This capture is illustrated in Fig. 6.1a. In this case, the capture of electrons by an empty neutral trap state is accomplished through the simultaneous emission of phonons during the capture process. Figure 6.1b shows the emission of an electron from a filled trap state. In this illustration, the electron gains its kinetic energy from the thermal energy of the host lattice. Figure 6.1c shows the capture of a hole from the valence band by a filled trap state, and Fig. 6.1d shows the emission of a hole from the empty trap state to the valence band.

The rate equations which describe the Shockley–Read–Hall (SRH) model can be derived from the four emission and capture process shown in Fig. 6.1. In deriving the SRH model, it

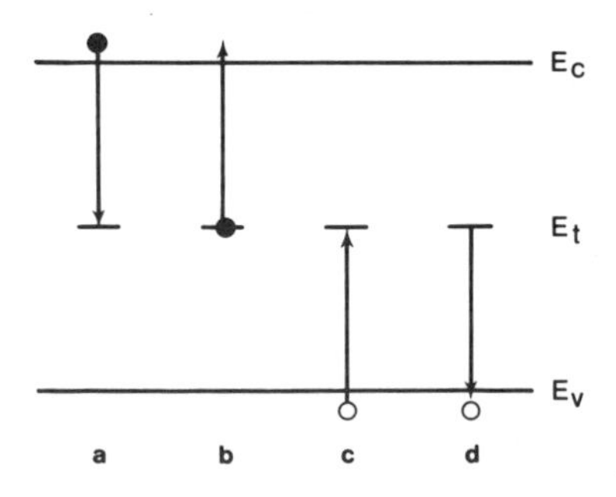

FIGURE 6.1. Capture and emission of an electron and a hole via a deep-level trap. The Shockley–Read–Hall model: (a) electron capture, (b) electron emission, (c) hole capture, and (d) hole emission.

is assumed that the semiconductor is nondegenerate and that the density of trap states is small compared to the majority carrier density. When the specimen is in thermal equilibrium, f_t denotes the probability that a trap state located at E_t in the forbidden gap is occupied by an electron. Using the Fermi–Dirac statisics described in Chapter 3, the distribution function f_t of a carrier at the trap state is given by

$$f_t = \frac{1}{1 + e^{(E_t - E_f)/K_B T}} \tag{6.1}$$

The physical parameters used in the SRH model are defined as follows:

U_{cn} is the electron capture probability per unit time per unit volume ($cm^{-3} \cdot sec^{-1}$).
U_{en} is the electron emission probability per unit time per unit volume.
U_{cp} is the hole capture probability per unit time per unit volume.
U_{ep} is the hole emission probability per unit time per unit volume.
c_n and c_p are the electron and hole capture coefficient (cm^3/sec).
e_n and e_p are the electron and hole emission rates (sec^{-1}).
N_t is the trap density (cm^{-3}).

In general, the rate of electron capture probability is a function of the density of conduction electrons, capture cross section, and density of the empty traps. However, the rate of electron emission probability depends only upon the electron emission rate and the density of traps being filled by the electrons. Thus, the expressions for U_{cn} and U_{en} can be written as

$$U_{cn} = c_n n N_t (1 - f_t) \tag{6.2}$$

$$U_{en} = e_n N_t f_t \tag{6.3}$$

Similarly, U_{cp}, the rate of hole capture probability, and U_{ep}, the rate of hole emission probability are given by

$$U_{cp} = c_p p N_t f_t \tag{6.4}$$

$$U_{ep} = e_p N_t (1 - f_t) \tag{6.5}$$

According to the principle of detailed balance, the rates of emission and capture at a trap level are equal in thermal equilibrium. Thus, we can write

$$U_{cn} = U_{en} \quad \text{for electrons} \tag{6.6}$$

$$U_{cp} = U_{ep} \quad \text{for holes} \tag{6.7}$$

The solution of Eqs. (6.2) through (6.7) yields

$$e_n = c_n n_0 (1 - f_t)/f_t \tag{6.8}$$

$$e_p = c_p p_0 f_t/(1 - f_t) \tag{6.9}$$

The product of Eqs. (6.8) and (6.9) is given by

$$e_n e_p = c_n c_p n_0 p_0 = c_n c_p n_i^2 \tag{6.10}$$

From Eq. (6.1) we can write

$$(1 - f_t)/f_t = e^{(E_t - E_f)/k_B T} \tag{6.11}$$

Now solving Eqs. (6.8), (6.9), and (6.11), we obtain

$$e_n = c_n n_1 \tag{6.12}$$

$$e_p = c_p p_1 \tag{6.13}$$

where n_1 and p_1 denote the electron and hole densities, respectively, when the Fermi level coincides with the trap level. Expressions for n_1 and p_1 are given by

$$n_1 = n_0\, e^{(E_t - E_f)/k_B T} \tag{6.14}$$

$$p_1 = p_0\, e^{(E_f - E_t)/k_B T} \tag{6.15}$$

Equations (6.14) and (6.15) yield

$$n_1 p_1 = n_0 p_0 = n_i^2 \tag{6.16}$$

Under steady-state conditions, the net rate of electron capture per unit volume may be found by solving Eqs. (6.2) through (6.16), which yields

$$U_n = U_{cn} - U_{en} = c_n N_t[n(1 - f_t) - n_1 f_t] \tag{6.17}$$

Similarly, the net rate of hole capture per unit volume may be written as

$$U_p = U_{cp} - U_{ep} = c_p N_t[p f_t - p_1(1 - f_t)] \tag{6.18}$$

The excess carrier lifetime under steady-state conditions is defined by the ratio of the excess carrier density and the net capture rate, and is given by

$$\tau_n = \frac{\Delta n}{U_n} \quad \text{for electrons} \tag{6.19}$$

$$\tau_p = \frac{\Delta p}{U_p} \quad \text{for holes} \tag{6.20}$$

For the small injection case (i.e., $\Delta_n \ll n_0$ and $\Delta p \ll p_0$), the charge-neutrality condition requires that

$$\Delta n = \Delta p \tag{6.21}$$

Under steady-state conditions, if we assume that the net rates of electron and hole capture via a recombination center are equal, then we can write

$$U = U_n = U_p \tag{6.22}$$

From Eqs. (6.19) through (6.22), it is noted that the electron and hole lifetimes are equal (i.e., $\tau_n = \tau_p$) for the small injection case. The electron distribution function, f_t, at the trap level can be expressed in terms of the electron and hole capture coefficients as well as electron and hole densities. By solving Eqs. (6.17), (6.18), and (6.21), we obtain

$$f_t = \frac{(c_n n + c_p p_1)}{c_n(n + n_1) + c_p(p + p_1)} \tag{6.23}$$

A general expression for the net recombination rate can be obtained by substituting Eq. (6.23) into Eq.(6.17) or (6.18), and the result is

$$U = U_n = U_p = \frac{(np - n_i^2)}{\tau_{p0}(n + n_1) + \tau_{n0}(p + p_1)} \tag{6.24}$$

where

$$\tau_{p0} = \frac{1}{c_p N_t} \tag{6.25}$$

and

$$\tau_{n0} = \frac{1}{c_n N_t} \tag{6.26}$$

where $c_p = \sigma_p\langle v_{th}\rangle$ and $c_n = \sigma_n\langle v_{th}\rangle$ denote the hole and electron capture coefficients, respectively, σ_p and σ_n are the hole and electron capture cross sections, respectively, $\langle v_{th}\rangle = (3k_BT/M^*)^{1/2}$ is the mean thermal velocity of electrons or holes, τ_{p0} is the minority hole lifetime for a strong *n*-type semiconductor, and τ_{n0} is the minority electron lifetime for a strong p-type semiconductor. Now by solving Eqs. (6.17) through (6.26), we obtain a general expression for the excess carrier lifetime, which is given by

$$\tau_0 = \frac{\Delta n}{U_n} = \frac{\Delta p}{U_p} = \frac{\tau_{p0}(n_0 + n_1 + \Delta n)}{(n_0 + p_0 + \Delta n)} + \frac{\tau_{n0}(p_0 + p_1 + \Delta p)}{(n_0 + p_0 + \Delta p)} \tag{6.27}$$

where $n = n_0 + \Delta n$ and $p = p_0 + \Delta p$ denote the nonequilibrium electron and hole densities, n_0 and p_0 are the equilibrium electron and hole densities, and Δn and Δp denote the excess electron and hole densities, respectively.

For the small injection case (i.e., $\Delta n \ll n_0$ and $\Delta p \ll p_0$), the excess carrier lifetime given by Eq. (6.27) reduces to

$$\tau_0 = \frac{\tau_{p0}(n_0 + n_1)}{(n_0 + p_0)} + \frac{\tau_{n0}(p_0 + p_1)}{(n_0 + p_0)} \tag{6.28}$$

which shows that under small injection conditions, τ_0 is independent of the excess carrrier density or injection. It is interesting to note that for an n-type semiconductor with $n_0 \gg p_0$, n_1 and p_1, Eq. (6.28) is reduced to

$$\tau_0 = \tau_{p0} \tag{6.29}$$

Similarly, for a p-type semiconductor with $p_0 \gg n_0$, p_1 and n_1, Eq. (6.28) becomes

$$\tau_0 = \tau_{n0} \tag{6.30}$$

Both Eqs. (6.29) and (6.30) show that the excess carrier lifetime in an extrinsic semiconductor is dominated by the minority carrier lifetime. Therefore, the minority carrier lifetime is a key physical parameter for determination of the excess carrier recombination in an extrinsic semiconductor under low injection conditions.

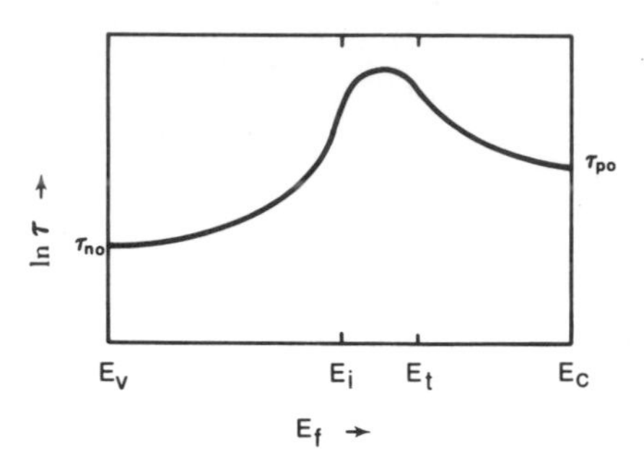

FIGURE 6.2. Dependence of the excess carrier lifetime on the Fermi level.

For the high injection case, with $\Delta n = \Delta p \gg n_0, p_0$, Eq. (6.27) becomes

$$\tau_h = \tau_{p0} + \tau_{n0} \tag{6.31}$$

which shows that, in the high injection limit, the excess carrier lifetime τ_h reaches a plateau (or saturation) and becomes independent of the injection. In general, it is found that in the intermediate injection range the excess carrier lifetime may depend on the injected carrrier density. Furthermore, the excess carrier lifetime is also found to be dependent on n_1 and p_1, which in turn depend on the Fermi level and the dopant density. This is clearly illustrated in Fig. 6.2.

Based on the above discussions, the minority carrier lifetime is an important physical parameter which is directly related to the recombination mechanisms in a semiconductor. A high-quality semiconductor with few defects usually has a long minority carrier lifetime. On the other hand, a poor-quality semiconductor will have a very short minority carrier lifetime and a large defect density. The minority carrier lifetime plays an important role in the performance of many semiconductor devices. For example, the switching speed of a bipolar junction transistor and the conversion efficiency of a p–n junction solar cell may be controlled by the minority carrier lifetime of a semiconductor.

It should be noted that the SRH model described above is applicable for nonradiative recombination process via a single deep-level recombination center in the forbidden gap of a semiconductor. Treatment of the recombination process via multiple deep-level centers can be found in the classical paper by Sah and Shockley.[(3)]

6.3. BAND-TO-BAND RADIATIVE RECOMBINATION

The band-to-band radiative recombination in a semiconductor is the inverse process of optical absorption. Emission of photons due to band-to-band radiative recombination is a common phenomenon observed in a direct bandgap semiconductor such as GaAs, InP, or InAs. In a nondegenerative semiconductor, the rate at which electrons and holes are annihilated via band-to-band radiative recombination is proportional to the product of electron and hole densities in the conduction and valence bands, respectively. In thermal equilibrium, the rate of band-to-band recombination is equal to the rate of thermal generation. This equilibrium rate equation can be expressed by

$$R_0 = G_0 = Bn_0 p_0 = Bn_i^2 \tag{6.32}$$

where B is the rate of radiative capture probability, which can be derived from the optical absorption process by using the principle of detailed balance. Under steady-state conditions, the rate of band-to-band radiative recombination is given by

$$r = Bnp \tag{6.33}$$

where $n = n_0 + \Delta n$ and $p = p_0 + \Delta p$. The net recombination rate is obtained by solving Eqs. (6.32) and (6.33). Thus, the net recombination rate is given by

$$U_r = r - G_0 = B(np - n_i^2) \tag{6.34}$$

The radiative lifetime, τ_r, due to band-to-band recombination is obtained by solving Eqs. (6.34) and (6.21), and the result yields

$$\tau_r = \frac{\Delta n}{U_r} = \frac{1}{B(n_0 + p_0 + \Delta n)} \tag{6.35}$$

From Eq. (6.35), it is noted that τ_r is inversely proportional to the majority carrier density. Under small injection conditions, Eq. (6.35) can be simplified to

$$\tau_{r0} = \frac{1}{B(n_0 + p_0)} \tag{6.36}$$

which shows that, for the small injection case, the band-to-band radiative lifetime is inversely proportional to the majority carrier density. For the intrinsic case (i.e., $n_0 = p_0 = n_i$), the radiative lifetime τ_{ri} due to band-to-band recombination given by Eq. (6.35) is reduced to $1/(2Bn_i)$.

In the high injection limit, $\Delta n = \Delta p \gg n_0, p_0$, and Eq. (6.35) becomes

$$\tau_{rh} = \frac{1}{B\Delta n} \tag{6.37}$$

This equation shows that the band-to-band radiative lifetime under high injection conditions is inversely proportional to the excess carrier density, and is independent of the majority carrier density in the semiconductor.

Since band-to-band radiative recombination is the inverse process of optical absorption, an analytical expression for the radiative recombination capture rate, B, can be derived from the fundamental optical absorption process using the principle of detailed balance.

In a direct bandgap semiconductor, the fundamental absorption process is usually dominated by the vertical transition. It will be shown in Chapter 9 that the energy dependence of the fundamental optical absorption coefficient for a direct bandgap semiconductor can be expressed by

$$\alpha_d = \left(\frac{2^{3/2}q^2}{3nm_0ch^2}\right)(m_r^{3/2} + m_0 m_r^{1/2})(h\nu - E_g)^{1/2} \tag{6.38}$$

where n is the index of refraction, E_g is the energy bandgap, $m_r^{-1} = (m_e + m_h)/m_e m_h$ is the reduced electron and hole effective mass, and m_0 is the free-electron mass. Equation (6.38) shows that, for $h\nu \geq E_g$, the optical absorption coefficient for a direct bandgap semiconductor is proportional to the square root of the photon energy.

In order to correlate the rate of capture probability coefficient, B, to the optical absorption coefficient α_d, we can treat the semiconductor as a blackbody radiation source and use the principle of detailed balance under thermal equilibrium conditions. From Eq. (6.32) B may be determined by setting the rate of radiative recombination equal to the rate of total blackbody radiation absorbed by the semiconductor due to band-to-band recombination, which is given by

$$B_d n_i^2 = \int \frac{n^2 \alpha E^2 \, dE}{(\pi^2 q^2 h^3)(e^{E/k_B T} - 1)} \tag{6.39}$$

The right-hand side of Eq. (6.39) is obtained from the Planck blackbody radiation formula. Solving Eqs. (6.38) and (6.39) we obtain the rate of capture probability B_d for the direct transition, which reads

$$B_d = \left(\frac{E_g}{n_i}\right)^2 (2\pi)^{3/2} \left(\frac{hq^2}{3m_0^2 c^2}\right) \eta(1 + m_0/m_r) \left(\frac{m_0}{m_e + m_h}\right)^{3/2} (k_B T)^{-3/2} (m_0 c^2)^{-1/2} \tag{6.40}$$

It is important to note from Eq. (6.40) that B_d is inversely proportional to the square of the intrinsic carrier density, which leads to an exponential dependence of B_d on temperature (i.e., B_d decreases exponentially with temperature). This implies that the band-to-band radiative recombination lifetime is a strong function of temperature. Table 6.1 lists the values of B_d calculated from Eq. (6.40) for GaSb, InAs, and InSb. The results are found to be in reasonable agreement with the published data for these materials.

A similar calculation of the capture probablity for the indirect transition involving the absorption and emission of phonons in an indirect bandgap semiconductor yields the capture probability coefficient, B_i, which is given by

$$B_i = \left(\frac{4\pi h^3}{m_0^3 c^3}\right) (A\mu^2) \left(\frac{m_0^2}{m_e m_h}\right)^{3/2} E_g^2 \coth(\theta/2T) \tag{6.41}$$

where A and μ are adjustable parameters used to fit the measured absorption data.

A comparison of Eqs. (6.40) and (6.41) reveals that while B_d varies with $\exp(E_g/k_B T) T^{-3/2}$, B_i depends less strongly on temperature [i.e., it is proportional to $\coth(\theta/2T)$]. For a direct bandgap semiconductor in which recombination is via band-to-band radiative transition, values of B_d can be quite high (i.e., 3×10^{-11} cm^3/sec). On the other hand, for indirect transitions, values of B_i are found to be much smaller than B_d (i.e., 10^3 to 10^4 times smaller). Table 6.1 lists the calculated values of B_d and B_i and the radiative lifetimes for some semiconductors at $T = 300$ K.

TABLE 6.1. Band-to-Band Radiative Recombination Parameters for Some Elemental and Compound Semiconductors at 300 K

Semiconductors	E_g (eV)	n_i (cm^{-3})	B (cm^3/sec)	τ_i	τ_0 [a]
Si	1.12	1.5×10^{10}	2.0×10^{-15}	4.6 h	2500 μsec
Ge	0.67	2.4×10^{13}	3.4×10^{-15}	0.61 sec	150 μsec
GaSb	0.71	4.3×10^{12}	1.3×10^{-11}	9 msec	0.37 μsec
InAs	0.31	1.6×10^{15}	2.1×10^{-11}	15 μsec	0.24 μsec
InSb	0.18	2×10^{16}	4×10^{-11}	0.62 μsec	0.12 μsec
PbTe	0.32	4×10^{15}	5.2×10^{-11}	2.4 μsec	0.19 μsec

[a] Assuming n_0 or $p_0 = 10^{17}$ cm^{-3}.

6.4. BAND-TO-BAND AUGER RECOMBINATION

As discussed in the previous section, band-to-band radiative recombination is the inverse process of fundamental optical absorption. In a similar manner, Auger recombination is the inverse process of impact ionization. Band-to-band Auger recombination is a three-particle process which involves either electron–electron collisions in the conduction band followed by recombination with holes in the valence band, or hole–hole collisions in the valence band followed by recombination with electrons in the conduction band. These two recombination processes and their inverse processes are shown schematically in Fig. 6.3.

For small-bandgap semiconductors such as InSb, the minority carrier lifetime is usually controlled by band-to-band Auger recombination, and energy loss is carried out either by electron–electron colllisions or hole–hole collisions and subsequent Auger recombination.

To derive the band-to-band Auger recombination lifetime, the rate of Auger recombination in equilibrium conditions can be written as

$$R_a = G_0 = C_n n_0^2 p_0 + C_p p_0^2 n_0 \tag{6.42}$$

Under nonequilibrium conditions, the Auger recombination rate is given by

$$r = C_n n^2 p + C_p p^2 n \tag{6.43}$$

Therefore, the net Auger recombination rate under steady-state conditions is obtained from Eqs. (6.42) and (6.43), and the result is

$$U_A = r - G_0 = C_n(n^2 p - n_0^2 p_0) + C_p(p^2 n - p_0^2 n_0) \tag{6.44}$$

where C_n and C_p are the capture probability coefficients when the third carrier is either an electron or a hole. Both C_n and C_p can be calculated from their inverse process, namely, the impact ionization. In thermal equilibrium, the rate at which carriers are destroyed via Auger recombination is equal to the generation rate averaged over the Boltzmann distribution function in which the electron–hole pairs are created by impact ionization. Thus, we can write

$$C_n n_0^2 p_0 = \int_0^\infty P(E)(dn/dE)\, dE \tag{6.45}$$

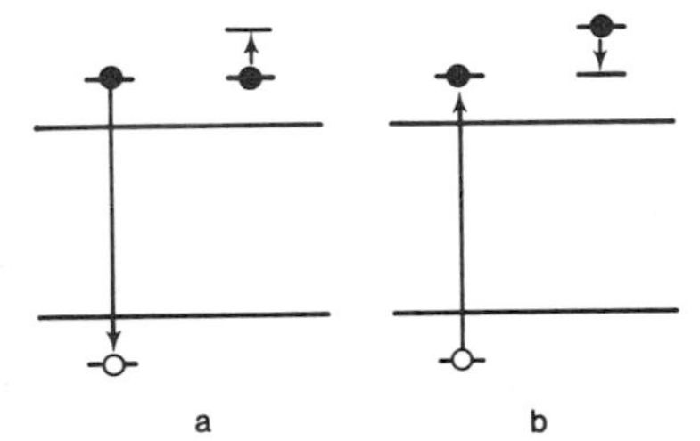

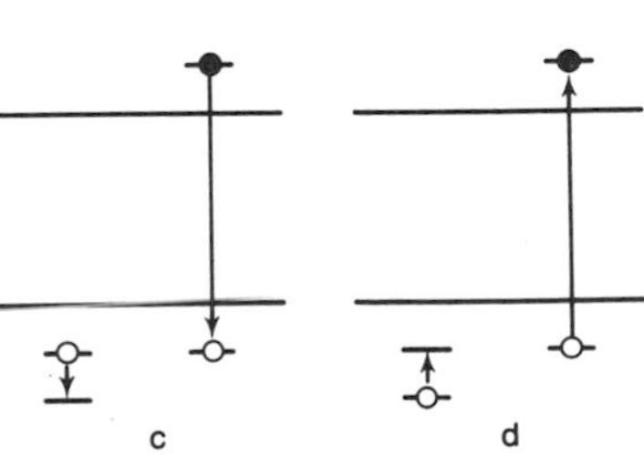

FIGURE 6.3. Auger recombination and its inverse process which shows the annihilation and creation of an electron–hole pair: (a) annihilation of an electron–hole pair by electron–electron collisions, (b) creation of an electron–hole pair by electron impact ionization, (c) destruction of an electron–hole pair by hole–hole collisions, and (d) creation of an electron–hole pair by hole impact ionization.

where $P(E)$ is the probability per unit time that an electron with energy E will make an ionizing collision, and is given by

$$P(E) = (mq^4/2h^3)G(E/E_t - 1)^s \tag{6.46}$$

where $G < 1$ is a parameter which is a complicated function of the band structure of the semiconductor. The exponent s is an integer which is determined by the symmetry of the crystal in momentum space at a threshold energy E_t. The value of E_t for impact ionization is roughly equal to $1.5E_g$, where E_g is the energy bandgap of the semiconductor. By substituting Eq. (6.46) into Eq. (6.45), we obtain

$$n_i^2 C_n = \left(\frac{s}{\sqrt{\pi}}\right)\left(\frac{mq^4}{h^3}\right) G \left(\frac{k_B T}{E_t}\right)^{(s-1/2)} e^{(-E_t/k_B T)} \tag{6.47}$$

Equation (6.47) shows that the Auger capture coefficient C_n for electrons depends exponentially on both the temperature and energy bandgap of the semiconductor. The Auger lifetime may be derived from Eq. (6.44), and the result yields

$$\tau_A = \frac{\Delta n}{U_A} = \frac{1}{n^2 C_n + 2n_i^2(C_n + C_p) + p^2 C_p} \tag{6.48}$$

If C_n is assumed equal to C_p, then Eq. (6.48) shows that τ_A has a maximum value if $n = p = n_i$ (i.e., $\tau_i = 1/6n_i^2 C_n$). For an extrinsic semiconductor, τ_A is inversely proportional to the square of the majority carrier density. For the intrinsic case, however, the Auger lifetime can be obtained from Eqs. (6.47) and (6.48) for $s = 2$,

$$\tau_i = \frac{1}{3n_i^2(C_n + C_p)} = 3.6 \times 10^{-17}(E_t/k_B T)^{3/2} e^{(E_t/k_B T)} \tag{6.49}$$

which shows that the intrinsic Auger lifetime is an exponential function of the temperature and energy bandgap ($E_t \sim 1.5E_g$). Note that the temperature dependence of the Auger lifetime in an extrinsic semiconductor is not as strong as in an intrinsic semiconductor. However, due to the strong temperature dependence of the Auger lifetime, it is possible to identify the Auger recombination process by analyzing the measured lifetime versus temperature in a semiconductor. For a heavily doped semiconductor, Eq. (6.48) predicts that the Auger lifetime is inversely proportional to the square of the majority carrier density. The Auger recombination mechanism has been found to be the dominant recombination process for many degenerate semiconductors as well as in small-bandgap semiconductors. Values of Auger recombination coefficients for silicon and germanium are given by: $C_n = 2.8 \times 10^{-31}$ and $C_p = 10^{-31}$ cm^6/sec for silicon, $C_n = 8 \times 10^{-32}$ and $C_p = 2.8 \times 10^{-31}$ cm^6/sec for germanium. Using these values, the intrinsic Auger lifetime for silicon is 4.48×10^9 sec at 300 K, and is equal to 1.61×10^3 sec for germanium. Thus, the Auger recombination is a very unlikely recombination process for intrinsic semiconductors (with the exception of small-bandgap semiconductors). However, the Auger recombination lifetime for n-type silicon reduces to about 10^{-8} sec at a doping density of 10^{19} cm^{-3}.

Under high injection conditions, Auger recombination may also become the predominant recombination process. In this case, the Auger lifetime is given by

$$\tau_{Ah} = \frac{1}{\Delta n^2(C_n + C_p)} = \left(\frac{3n_i^2}{\Delta n^2}\right)\tau_i \tag{6.50}$$

where τ_i is the intrinsic Auger lifetime given by Eq. (6.49).

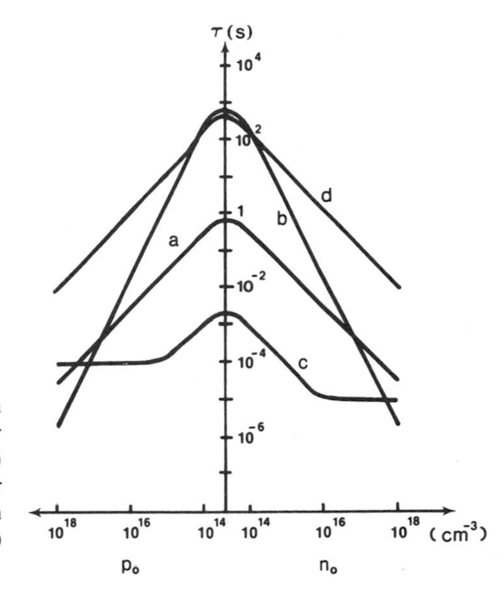

FIGURE 6.4. A comparison of the recombination lifetimes at $T = 300$ K for a germanium crystal for the cases when recombination is dominated by: (a) band-to-band radiative recombination, (b) Auger band-to-band recombination, (c) multiphonon process (Shockley–Read–Hall model), and (d) impurity-to-band Auger recombination.

As an example, let us consider a germanium specimen. If the injected carrier density is assumed equal to $10^{18}\ \mathrm{cm}^{-3}$ and $E_t = 1.0$ eV, then the Auger lifetime, as calculated from Eq. (6.50), is found to be 1 μsec. For small-bandgap semiconductors, we expect the Auger recombination to be the predominant recombination process even at smaller injection level. Additional discussions on the Auger recombination and band-to-band radiative recombination mechanisms in semiconductors can be found in a special issue of *Solid State Electronics* edited by P. T. Landsberg and A. F. W. Willoughby.[4]

In order to obtain an overall picture of the various recombination processes taking place in a semiconductor, Figs. 6.4 and 6.5 show the excess carrier lifetimes due to different recombination mechanisms as a function of the majority carrier density for a germanium and a GaSb crystal. From these two figures, a significant difference in the dominant recombination

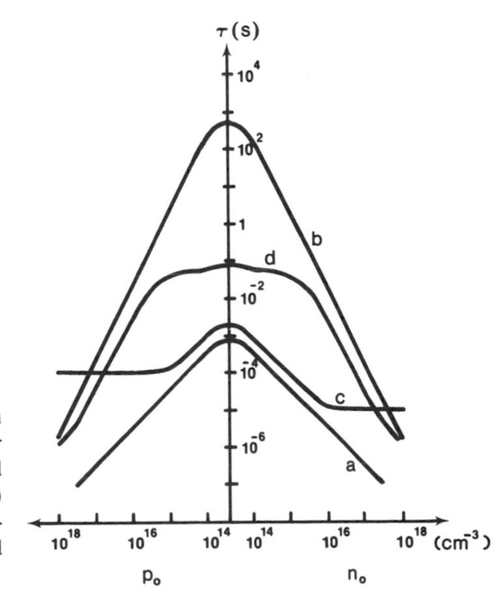

FIGURE 6.5. A comparison of the recombination lifetimes at 300 K for a direct-gap semiconductor such as GaSb when the recombination is dominated by: (a) band-to-band radiative recombination, (b) Auger band-to-band recombination, (c) multiphonon process, and (d) Auger impurity-to-band recombination.

mechanism is observed between these two materials. The difference in the dominant recombination process in germanium and GaSb may be attributed to the fact that germanium is an indirect bandgap semiconductor while GaSb is a direct bandgap semiconductor. For germanium, the Shockey–Read–Hall recombination process is expected to be the predominant process for most doping densities (except in the heavy doping range), while for GaSb the band-to-band radiative recombination is expected to be the predominant process for the medium to lightly doped case.

As discussed above, Auger recombination is usually the predominant recombination process for small-bandgap semiconductors, heavily doped semiconductors, and for semiconductors operating at very high temperatures or under very high injection conditions. As shown in Fig. 6.4, the minority carrier lifetime for an indirect bandgap semiconductor is usually controlled by the nonradiative multiphonon recombination process via a deep-level trap, as predicted by the Shockley–Read–Hall model. As for a direct bandgap semiconductor, the minority carrier lifetime is usually limited by the band-to-band radiative recombination process except in the high doping regime, as is shown in Fig. 6.5.

6.5. BASIC SEMICONDUCTOR EQUATIONS

The spatial and time-varying function of the excess carrier phenomena in a semiconductor under nonequilibrium conditions may be analyzed by using the basic semiconductor equations. These equations contain the diffusion and drift terms (for both electrons and holes) as well as the recombination and generation terms. There are two continuity equations for the excess carriers in a semiconductor: one for electrons and one for holes. As will be discussed later, both the steady-state and transient effects can, in principle, be solved from these two continuity equations.

In a semiconductor, the electron–hole pairs may be created by either thermal or optical means and annihilated by the different recombination processes discussed in the previous sections. In thermal equilibrium, the rate of generation must be equal to the rate of recombination. Otherwise, space charge will be built up within the semiconductor specimen. The nonequilibrium condition is established when an external excitation is applied to the semiconductor specimen. For example, excess electron–hole pairs are created when photons with energies exceeding the bandgap energy (i.e., $h\nu \geq E_g$) are absorbed by the semiconductor. The continuity equations for both electrons and holes under nonequilibrium condition are given respectively by

$$\frac{dn}{dt} = \frac{1}{q}\nabla \cdot J_n - \frac{n}{\tau_n} + g_T \tag{6.51}$$

and

$$\frac{dp}{dt} = -\frac{1}{q}\nabla \cdot J_p - \frac{p}{\tau_p} + g_T \tag{6.52}$$

where $n = \Delta n + n_0$ and $p = \Delta p + p_0$ are the nonequilibrium electron and hole densities, respectively, g_T is the total generation rate, n/τ_n and p/τ_p are the rates of recombination for electrons and holes, τ_n and τ_p are the electron and hole lifetimes, while J_n and J_p are the electron and hole current densities, respectively. In general, the excess electron–hole pairs may be created by thermal generation and external excitation. Thus, we can write the total generation rate as

$$g_T = G_{th} + g_E \tag{6.53}$$

where G_{th} is the thermal generation rate and g_E is the external generation rate. According to the principle of detailed balance, under thermal equilibrium, the rate of generation must be equal to the rate of recombination. Thus, in thermal equilibrium, we can write

$$G_{th} = R_0 = \frac{n_0}{\tau_n} = \frac{p_0}{\tau_p} \tag{6.54}$$

The continuity equations for the excess electrons and holes can be obtained by substituting Eqs. (6.53) and (6.54) into Eqs. (6.51) and (6.52), and the results are given by

$$\frac{\partial \Delta n}{\partial t} = \frac{1}{q} \nabla \cdot J_n - \frac{\Delta n}{\tau_n} + g_E \tag{6.55}$$

$$\frac{\partial \Delta p}{\partial t} = -\frac{1}{q} \nabla \cdot J_p - \frac{\Delta p}{\tau_p} + g_E \tag{6.56}$$

The electron and hole current densities in a semiconductor consist of two components, namely, the drift and diffusion currents. These two components are given respectively by

$$J_n = q\mu_n n\mathscr{E} + qD_n \nabla n \tag{6.57}$$

$$J_p = q\mu_p p\mathscr{E} - qD_p \nabla p \tag{6.58}$$

where $\mathscr{E}$ is the electric field; μ_n and μ_p are the electron and hole mobilities, and D_n and D_p are the electron and hole diffusivities, respectively. The first term on the right-hand side of Eqs. (6.57) and (6.58) is called the drift component while the second term is the diffusion component. The total current density is equal to the sum of electron and hole current densities, which is given by

$$J_T = J_n + J_p \tag{6.59}$$

In thermal equilibrium, both the electron and hole current densities are equal to zero, and from Eqs. (6.57) and (6.58) we obtain

$$D_n = -\left(\frac{n_0}{|\nabla n_0|}\right)\mu_n \mathscr{E} \tag{6.60}$$

$$D_p = \left(\frac{p_0}{|\nabla p_0|}\right)\mu_p \mathscr{E} \tag{6.61}$$

The electric field in a bulk semiconductor can be related to the electrostatic potential by

$$\varepsilon = -\nabla\Phi \tag{6.62}$$

where Φ denotes the electrostatic potential. If a concentration gradient due to the nonuniform impurity profile exists in the semiconductor, then a chemical potential term must be added to the electrostatic potential term given in Eq. (6.62). This is usually referred to as the electrochemical potential or the Fermi potential. The equilibrium carrier density for both electrons and holes can also be expressed in terms of the intrinsic carrier density and the electrostatic potential through the use of Maxwell–Boltzmann statistics; they are given by

$$n_0 = n_i \, e^{q\Phi/k_BT} \tag{6.63}$$

$$p_0 = n_i \, e^{-q\Phi/k_BT} \tag{6.64}$$

where Φ is the electrostatic potential measured relative to the intrinsic Fermi level which is constant throughout the semiconductor, and n_i is the intrinsic carrier density. Solving Eqs. (6.60) through (6.64) yields the relationships btween μ_n and D_n, and μ_p and D_p in thermal equilibrium:

$$D_n = \left(\frac{k_BT}{q}\right)\mu_n \tag{6.65}$$

$$D_p = \left(\frac{k_BT}{q}\right)\mu_p \tag{6.66}$$

Equations (6.65) and (6.66) are the well known Einstein relations for the diffusivity–mobility of electrons and holes under equilibrium condition. They show that the ratio of diffusivity and mobility of electrons and holes in a semiconductor is equal to k_BT/q. This relation is valid for the nondegenerate case. For heavily doped semiconductors, Fermi statistics should be used instead, and Eqs. (6.65) and (6.66) must be modified to account for the degeneracy effect (see Problem 6.5).

In addition to the five basic semiconductor equations discussed above, Poisson's equation should also be included. This equation, which relates the divergence of the electric field to the charge density in a semiconductor, is given by

$$\nabla \cdot \mathscr{E} = \frac{\rho}{\varepsilon_0 \varepsilon_s} = \left(\frac{q}{\varepsilon_0 \varepsilon_s}\right)(N_D^+ - N_A^- + p - n) \tag{6.67}$$

where N_D and N_A denote the ionized donor and acceptor impurity densities, respectively, and ε_s is the dielectric constant of the semiconductor. Equations (6.55) through (6.59) and Eq. (6.67) are known as the six basic semiconductor equations which are being used to solve the spatial and time-dependent problems associated with the steady-state and transient behavior of the excess carriers in a semiconductor. Examples of using these equations to solve the excess carrier phenomena in a semiconductor are given in Sections 6.7 and 6.8.

6.6. CHARGE-NEUTRALITY CONDITIONS

In a homogeneous semiconductor, charge neutrality is maintained under thermal equilibrium conditions and Eq. (6.67) is equal to zero. However, a departure from the charge-neutrality condition may arise from one of the following two sources: (1) a nonuniformly doped semiconductor with fully ionized impurities in thermal equlibrium conditions, and (2) unequal densities of electrons and holes arising from carrier trapping under nonequilibrium conditions. In both situations, an electrochemical potential (i.e., the quasi-Fermi potential) and a built-in electric field may be established within the semiconductor. In this section, a nonuniformly doped semiconductor will be considered.

From Eq. (6.63), the electrostatic potential for an n-type semiconductor can be expressed by

$$\Phi = \left(\frac{k_B T}{q}\right) \ln\left(\frac{N}{n_i}\right) \tag{6.68}$$

where $N = N_D - N_A$ is the net dopant density, which is a function of position in a nonuniformly doped semiconductor.

Now, by substituting Eqs. (6.62), (6.63), and (6.64) into Eq. (6.67) we obtain

$$\nabla^2 \Phi = \left(\frac{2qn_i}{\varepsilon_0 \varepsilon_s}\right)[\sinh(q\Phi/k_B T) - (N/2n_i)] \tag{6.69}$$

Equation (6.69) can be rewritten as

$$\begin{aligned} \nabla^2 \phi &= \left(\frac{2q^2 n_i}{k_B T \varepsilon_0 \varepsilon_s}\right)(\sinh(\phi\ ) - \sinh(\phi_0)) \\ &= \left(\frac{4q^2 n_i}{k_B T \varepsilon_0 \varepsilon_s}\right)\left[\cosh\left(\frac{\phi + \phi_0}{2}\right)\sinh\left(\frac{\phi - \phi_0}{2}\right)\right] \end{aligned} \tag{6.70}$$

In Eq. (6.70), the normalized electrostatic potential ϕ is defined by

$$\phi = \Phi/(k_B T/q) \tag{6.71}$$

and

$$\sinh(\phi_0) = \frac{N}{2n_i} \tag{6.72}$$

The physical significance of Eq. (6.70) can best be illustrated by considering the one-dimensional case in which the impurity density N is only a function of x in the semiconductor. If $(\phi - \phi_0) \ll 1$ (i.e., a small inhomogeneity in the semiconductor) in Eq. (6.70), then we obtain

$$\cosh\left(\frac{\phi + \phi_0}{2}\right) \simeq \cosh(\phi_0) = [1 + \sinh(\phi_0)^2]^{1/2} \simeq \frac{N}{2n_i} \tag{6.73}$$

and

$$\sinh\left(\frac{\phi - \phi_0}{2}\right) \simeq \frac{(\phi - \phi_0)}{2} \tag{6.74}$$

Now substituting Eqs. (6.73) and (6.74) in Eq. (6.70), the one-dimensional Poisson's equation is reduced to

$$\frac{\partial^2 \phi}{\partial x^2} \cong \frac{\partial^2(\phi - \phi_0)}{\partial x^2} \simeq \left(\frac{q^2 N}{k_B T \varepsilon_0 \varepsilon_s}\right)(\phi - \phi_0) \tag{6.75}$$

which has a solution given by

$$\phi - \phi_0 \cong e^{-x/L_D} \tag{6.76}$$

where

$$L_D = \sqrt{\frac{k_B T \varepsilon_0 \varepsilon_s}{q^2 N}} \tag{6.77}$$

is known as the extrinsic Debye length. The physical meaning of L_D is that it is a characteristic length used to determine the distance in which a small variation of the potential can smooth itself out in a homogeneous semiconductor.

Equation (6.76) predicts that in an extrinsic semiconductor under thermal equilibrium, no significant departure from the charge-neutrality condition is expected over a distance greater than a few Debye lengths. It will be left as an exercise for the reader to show that L_D in Eq. (6.77) for an n-type semiconductor can also be expressed by

$$L_{D_n} = \sqrt{D_n \tau_d} \tag{6.78}$$

where $\tau_d = \varepsilon_0 \varepsilon_s / \sigma$ is the dielectric relaxation time and σ is the electrical conductivity.

6.7. THE HAYNES–SHOCKLEY EXPERIMENT

In this section we shall give an example to illustrate how the basic semiconductor equations described in Section 6.6 can be applied to solve the spatial- and time-dependent excess carrier phenomena in a semiconductor. Let us first consider a uniformly doped n-type semiconductor bar in which N electron–hole pairs are generated instantaneously at a point $x = 0$, at $t = 0$. If we assume that the semiconductor bar is infinitely long in the x-direction (see Fig. 6.7), then the continuity equation given by Eq. (6.55) for the excess holes under a constant applied electric field can be reduced to a one-dimensional equation, which is given by

$$\frac{\Delta p}{\partial t} = D_p \frac{\partial^2 \Delta p}{\partial x^2} - \mu_p \mathscr{E} \frac{\partial \Delta p}{\partial x} - \frac{\Delta p}{\tau_p} \tag{6.79}$$

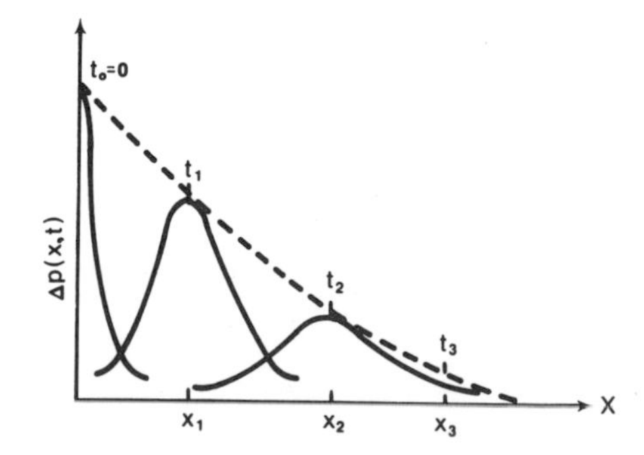

FIGURE 6.6. Spatial and time dependence of the excess hole density in an n-type extrinsic semiconductor bar under a constant applied electric field.

Equation (6.79) is obtained by substituting J_p, given by Eq. (6.58), into Eq. (6.56) and assuming that the external generation rate, g_E, is zero. The solution for Eq. (6.79) is given by

$$\Delta p(x, t) = \left[\frac{N e^{-t/\tau_p}}{(4\pi D_p t)^{1/2}}\right] \exp[-(x - \mu_p \mathscr{E} t)^2/4D_p t] \tag{6.80}$$

From Eq. (6.80), we see that the initial value of $\Delta p(x, 0)$ is zero except at $x = 0$, where $\Delta p(x, 0)$ approaches infinity. Thus, the initial hole concentration distribution corresponds to a Dirac delta function. For $t > 0$, the distribution of $\Delta p(x, t)$ has a Gaussian shape. The half-width of $\Delta p(x, t)$ will increase with time and its maximum amplitude will decrease with distance along the direction of the applied electric field, with a drift velocity $v_d = \mu_p \mathscr{E}$. The total excess carrier density injected at time t into the semiconductor is obtained by integrating Eq. (6.80) with respect to x from $-\infty$ to $+\infty$, thereby giving the following equation:

$$\Delta p(t) = \int_{-\infty}^{+\infty} \Delta p(x, t)\, dx = N e^{-t/\tau_p} \tag{6.81}$$

Equation (6.81) shows that $\Delta p(t)$ decays exponentially with time, with a time constant equal to the hole lifetime, τ_p. Figure 6.6 shows the spatial and time dependence of the excess carrier density in an n-type extrinsic semiconductor under a constant applied electric field. As shown in this figure, in order to maintain the original injection hole density profile, a large hole lifetime τ_p is needed. This implies that the semiconductor specimen should be of high quality with a very low defect density. Figure 6.7 shows the schematic diagram of the Haynes–Shockley experiment for measuring both the diffusivity and drift mobility of minority carriers in a semiconductor. In this experiment, P_1 and P_2 denote the injection and collector contacts for the minority carriers (i.e., holes in the present case), and V_1 and V_2 are the voltages applied to the respective contacts in order to create a uniformed electric field along the specimen and to provide a reverse bias voltage to the collector contact. The injection of minority carriers at contact P_1 can be achieved by using either an electric pulse generator or a pulse laser. An oscilloscope is used to display the pulse shape at contacts P_1 and P_2 and to measure the time delay of minoity carriers traveling between the injecting and collecting contacts.

The Haynes–Shockley experiment is described as follows. At $t = 0$, holes are injected at point P_1 of the sample in the form of a pulse of very short duration (on the order of a few microseconds or less). After this initial hole injection, the excess holes will move along the direction of the applied electric field (i.e., the x-direction) and are collected at contact P_2. This collection results in a current flow and a voltage drop across the load resistor R. The time elapsed between the initial injection pulse at P_1 and the arrival of the collection pulse at P_2 is a measure of the drift velocity of holes in the n-type semiconductor bar. In addition to the drift motion along the direction of the applied electric field, the hole density is also

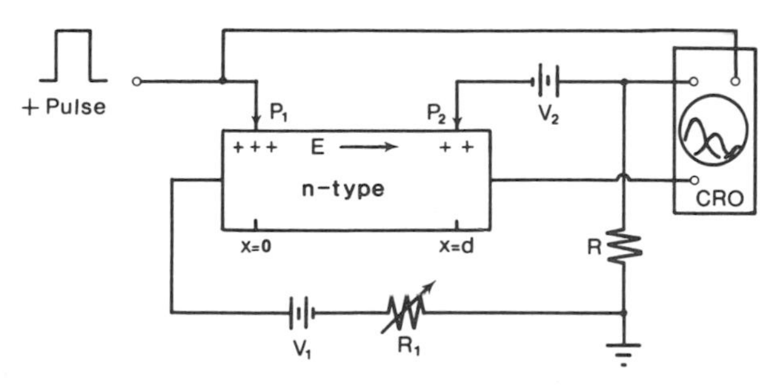

FIGURE 6.7. Schematic diagram for the Haynes–Shockley experiment.

TABLE 6.2. Drift Mobilities in Silicon and Germanium Measured at 300 K by the Haynes–Shockley Experiment

Silicon	Germanium
$\mu_n = 1350 \pm 100$ (cm^2/V · s)	$\mu_p = 3900 \pm 100$
$\mu_p = 480 \pm 15$	$\mu_p = 1900 \pm 50$

dispersed and broadened because of the diffusion effect. This explains why the pulse shown on the right-hand side of Fig. 6.6 is not as sharp as the initial injection pulse shown at $x = 0$.

The steps involved in determining the values of μ_p and D_p from the Haynes–Shockley experiment and Eq. (6.80) are discussed as follows: If t_0 is the time required for the peak of the hole pulse to move from contact P_1 to contact P_2 when an electric field is applied to the specimen, then the distance that the hole pulse traveled is given by

$$d = v_d t_0 = \mu_p t_0 \left(\frac{V_a}{l} \right) \tag{6.82}$$

where d is the distance between the injection and collecting contacts of the specimen; V_a is the applied voltage across the ample of length l. If values of d and t_0 are known, then the hole drift mobility μ_p can be easily calculated from Eq. (6.82) (i.e., $\mu_p = dl/V_a t_0$).

The hole diffusion constant D_p can be determined from the width of the Gaussian distribution function $\Delta p(x, t)$. The output voltage V_R of the hole pulse will drop to 0.367 of its peak value when the second exponential factor on the right-hand side of Eq. (6.80) is equal to unity. Thus, one obtains

$$(d - \mu_p \mathscr{E} \Delta t)^2 = 4 D_p \Delta t \tag{6.83}$$

If t_1 and t_2 denote the two delay time constants which satisfy Eq. (6.83) and $\Delta t = t_2 - t_1$, then D_p can be determined from the following expression:

$$D_p = (\mu_p \mathscr{E})^2 (\Delta t)^2 / 16 t_0 \tag{6.84}$$

The approximation given above is valid as long as the exponential factor, $(-t/\tau_p)$, given by Eq. (6.80) does not change appreciably over the measured time interval, Δt. In practice, the diffusion constants for electrons and holes are determined from the electron and hole mobilities by using the Einstein relations given by Eqs. (6.65) and (6.66). Values of the electron and hole drift mobilities for silicon and germanium determined by the Haynes–Shockley experiment at room temperature are listed in Table 6.2.

6.8. MINORITY CARRIER LIFETIMES AND PHOTOCONDUCTIVITY EXPERIMENT

In this section, measurements of the minority carrier lifetime in a semiconductor by the transient photoconductivity decay method will be depicted. The theoretical and experimental aspects of the transient photoconductivity effect in a semiconductor are discussed. As shown in Fig. 6.8, if the semiconductor bar is illuminated by a light pulse which contains photons with energies greater than the bandgap energy of the semiconductor, then electron–hole pairs will be generated in the specimen. The creation of excess carriers by the absorbed photons will result in a change of the electrical conductivity in the semiconductor bar. This phenomenon is known as the photoconductivity effect in a semiconductor. If the light pulse is abruptly turned off at $t = 0$, then the photoconductivity of the specimen will decay exponentially with time and gradually return to its original value under dark conditions. The time constant of the photoconductivity decay is controlled by the lifetimes of the minority carriers. By measuring the photoconductivity decay time constant, we can determine the minority carrier lifetime in a semiconductor specimen.

The problem of the transient photoconductivity-decay experiment for the excess hole density in an n-type semiconductor can be solved by using Eq. (6.56). As shown in Fig. 6.8, assuming that the light pulse is impinging along the y-direction of the sample, the spatial- and time-dependent excess hole density for $t \geq 0$ can be written as

$$\frac{\partial \Delta p}{\partial t} = D_p \frac{\partial^2 \Delta p}{\partial y^2} - \frac{\Delta p}{\tau_p} \tag{6.85}$$

Equation (6.85) is obtained from Eq. (6.56) by assuming that the light pulse is uniformly illuminated in the x–z plane of the specimen such that its diffusion components $\partial^2 \Delta p / \partial x^2$ and $\partial^2 \Delta p / \partial z^2$ are negligible compared to the diffusion component in the y-direction. The electric field is also assumed to be small such that the drift term in Eq. (6.56) can be neglected. As shown in Fig. 6.8, the boundary conditions at the top and bottom surfaces are given respectively by

$$D_p \frac{\partial \Delta p}{\partial y} = -s_b \Delta p \qquad \text{at } y = d \tag{6.86}$$

and

$$D_p \frac{\partial \Delta p}{\partial y} = s_f \Delta p \qquad \text{at } y = 0 \tag{6.87}$$

where s_b and s_f denote the surface recombination velocities at the top and bottom surfaces of the specimen, respectively. The values of s_b and s_f depend critically on the surface treatment. In addition to the boundary conditions given by Eqs. (6.86) and (6.87), the initial and final conditions are obtained by assuming that

$$\Delta p(y, t = 0) = \Delta p_0 = \text{constant} \tag{6.88}$$

$$\Delta p(y, t \to \infty) = 0 \tag{6.89}$$

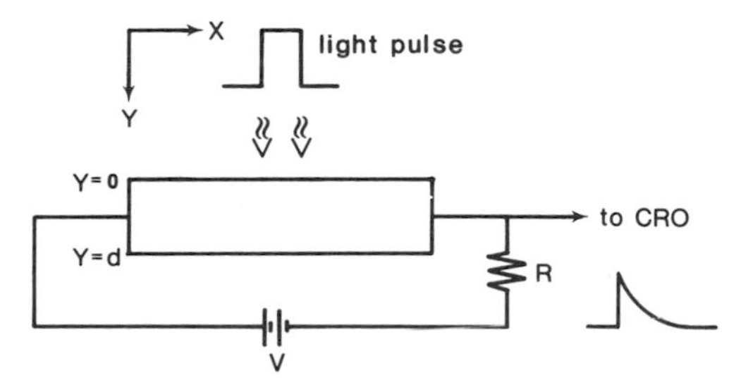

FIGURE 6.8. Photoconductivity-decay experiment for the minority carrier lifetime measurement in a semiconductor.

Since Eq. (6.85) is a homogeneous linear partial differential equation for $\Delta p(y, t)$, its solution can be written as the product of two independent functions of t and y:

$$\Delta p(y, t) = A\, e^{-(b^2 D_p + 1/\tau_p)t} \cos(by) \tag{6.90}$$

It is noteworthy that Eq. (6.90) does not satisfy the boundary conditions imposed by Eqs. (6.88) and (6.89). Therefore, the most general solution for Eq. (6.85) corresponding to an arbitrary initial condition at $t = 0$ can be expressed in terms of a series sum of the solution given by Eq. (6.90) such that

$$\Delta p(y, t) = \sum_{n=0}^{\infty} A_n\, e^{-(b_n^2 D_p + 1/\tau_p)t} \cos((b_n y) \tag{6.91}$$

Substituting Eq. (6.91) into Eq. (6.86) for $y = d$ yields the boundary condition

$$\sin(b_n d) = s/D_p b_n \tag{6.92}$$

The solutions for the surface recombination velocity s can be obtained graphically for different values of b_n (i.e., for $n = 0, 1, 2, \ldots$). The coefficient A_n in Eq. (6.91) can be determined from the initial condition given by Eq. (6.88). Furthermore, we can assume that at $t = 0$

$$\Delta p(y, 0) = \sum_{n=0}^{\infty} A_n \cos(b_n y) = \Delta p_0 = \text{constant} \tag{6.93}$$

Multiplication of Eq. (6.93) by $\cos(b_m y)$ and integrating both sides of the equation from $y = 0$ to $y = d$ yield

$$\int_0^d \Delta p_0 \cos(b_m y)\, dy = \int_0^d \sum_{n=0}^{\infty} A_n \cos(b_n y) \cos(b_m y)\, dy \tag{6.94}$$

If a set of functions of $\cos(b_m y)$ and $\cos(b_n y)$ is orthogonal for $0 < y < d$, then the integration on the right-hand side of Eq. (6.94) will vanish, except for the term with $n = m$. Thus, we obtain

$$A_n = \frac{4\Delta p_0 \sin(b_n d)}{2b_n d + \sin(2b_n d)} \tag{6.95}$$

From Eqs. (6.91) and (6.95) we can derive a general solution for Eq. (6.85) which satisfies the boundary and initial conditions given by Eqs. (6.86) through (6.89). Therefore, the general solution for $\Delta p(y, t)$ is

$$\Delta p(y, t) = 4\Delta p_0\, e^{-t/\tau_p} \sum_{n=0}^{\infty} \left[\frac{\sin(b_n d) \cos(b_n y)}{[2b_n d + \sin(2b_n d)]}\right] e^{-b_n^2 D_p t} \tag{6.96}$$

Using Eq. (6.96), the transient photoconductivity can be expressed by

$$\begin{aligned}\Delta\sigma(t) &= q\mu_p(b + 1) \int_0^d \Delta p(y, t)\, dy \\ &= 4q\mu_p(b + 1)\Delta p_0\, e^{-t/\tau_p} \sum_{n=0}^{\infty} \left[\frac{\sin^2(b_n d)}{b_n[2b_n d + \sin(2b_n d)]}\right] e^{-b_n^2 D_p t} \end{aligned} \tag{6.97}$$

or

$$\Delta\sigma(t) = \sum_{m}^{\infty} C_m e^{-t/\tau_m} \tag{6.98}$$

where

$$C_m = \frac{4q\mu_p(b+1)\Delta p_0 \sin^2(b_m d)}{b_m[2b_m d + \sin(2b_m d)]} \tag{6.99}$$

and

$$\tau_m^{-1} = \tau_p^{-1} + b_m^2 D_p \tag{6.100}$$

Equation (6.98) shows that the transient photoconductivity is represented by a summation of infinite terms, each of which has a characteristic amplitude C_m and decay time constant τ_m, where $m = 0, 1, 2, \ldots$.

Since b_0 (i.e., $m = 0$) is the zeroth-order mode and the smallest member of the set b_m, the time constant $\tau_0^{-1} = (\tau_p^{-1} + b_0^2 D_p)$ must be larger than any other higher-order modes. The fact that the higher-order modes will die out much more quickly than the fundamental mode (i.e., $m = 0$) after the initial transient (i.e., for $t > 0$) implies that the decay time constant will be dominated by the zeroth-order mode. Therefore, the minority carrier lifetime can be determined from the photoconductivity-decay experiment using Eq. (6.98) for $m = 0$. Figure 6.9 shows a plot of $\Delta\sigma(t)$ versus t for a semiconductor specimen. From the slope of this photoconductivity-decay curve, we obtain the zeroth-order decay mode time constant, which is

$$\tau_0^{-1} = \tau_p^{-1} + b_0^2 D_p \tag{6.101}$$

The first term on the right-hand side of Eq. (6.101) denotes the inverse bulk hole lifetime, while the second term represents the inverse surface recombination lifetime (to account for the effect of surface recombination). If the surface recombination velocity is small, then the photoconductivity-decay time constant is equal to the bulk lifetime. However, if the surface recombination term in Eq. (6.101) is much larger than the bulk lifetime term, then one can determine the surface recombination velocity from Eq. (6.101) by measuring the effective lifetimes of two samples with different thicknesses and similar surface treatment.

Since the minority carrier lifetime is an important physical parameter for modeling the silicon devices and integrated circuits, it is important to determine the minority carrier lifetimes

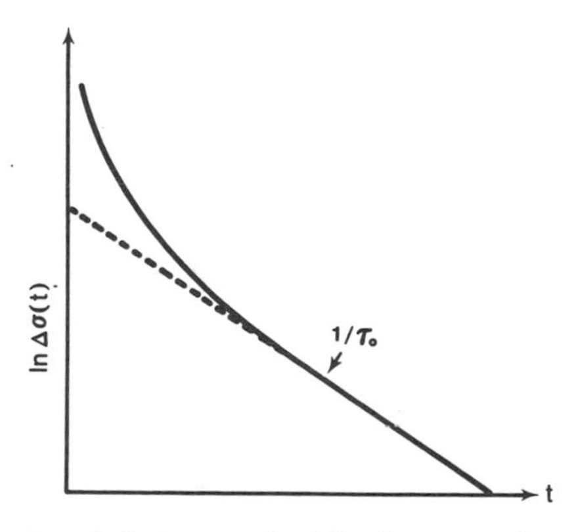

FIGURE 6.9. A typical photoconductivity-decay curve in a semiconductor.

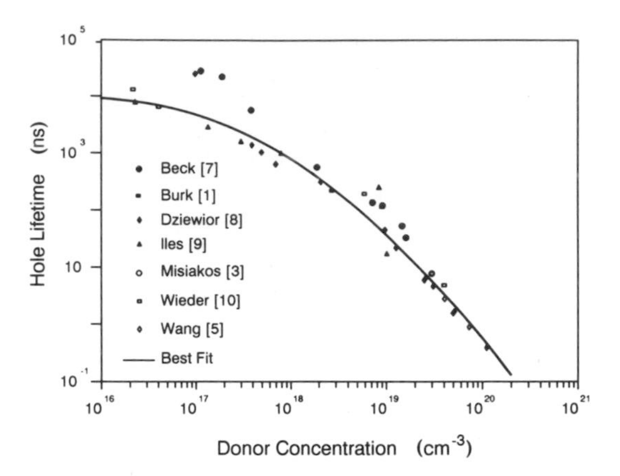

FIGURE 6.10. Measured and best-fit hole lifetimes versus donor concentrations in n-type silicon. After Law *et al.*,[5] by permission, © IEEE–1990.

versus doping concentrations in silicon materials. Figures 6.10 and 6.11 show the measured minority carrier lifetimes as a function of doping concentrations in both n- and p-type silicon, as reported recently by Law *et al.*[5] The effective carrier lifetime is modeled using a concentration-dependent Shockley–Read–Hall (SRH) lifetime τ_{srh} and a band-to-band Auger recombination lifetime τ_A to calculate the total effective lifetime by Mathiessen's rule, which is given by

$$\tau^{-1} = \tau_{srh}^{-1} + \tau_A^{-1} \tag{6.102}$$

where

$$\tau_{srh} = \frac{\tau_0}{1 + N_D/N_{ref}} \tag{6.103}$$

and

$$\tau_A = \frac{1}{C_A N_I^2} \tag{6.104}$$

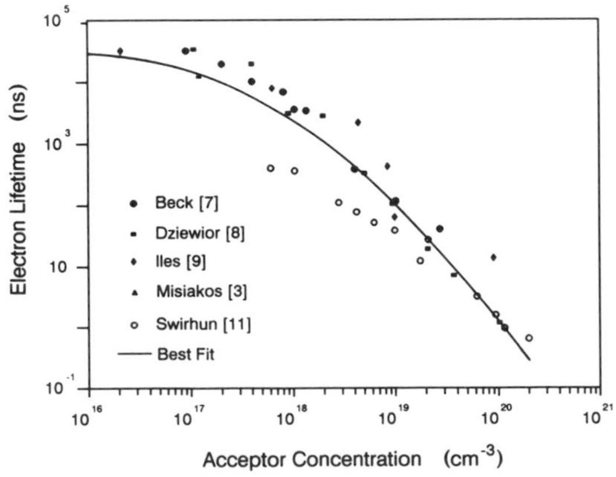

FIGURE 6.11. Measured and best-fit electron lifetimes versus acceptor concentrations in p-type silicon. After Law *et al.*,[5] by permission, © IEEE–1990.

Figure 6.10 shows the measured hole lifetimes as a function of the donor density for n-type silicon. The solid line is the best-fit curve using Eqs. (6.102) through (6.104). The values of parameters used in the fitting of this curve are given by $\tau_0 = 10\ \mu s$, $N_{ref} = 10^{17}\ cm^{-3}$, and $C_A = 1.8 \times 10^{-31}\ cm^6/sec$. Figure 6.11 shows the measured electron lifetimes as a function of the acceptor density for p-type silicon. The solid line is the best-fit curve using values of $\tau_0 = 30\ \mu s$, $N_{ref} = 10^{17}\ cm^{-3}$, and $C_A = 8.3 \times 10^{-32}\ cm^6/s$.

6.9. SURFACE STATES AND SURFACE RECOMBINATION VELOCITY

It is well known that a thin natural oxide layer can be easily formed on a freshly cliffed or chemically polished semiconductor surface when it is exposed to air. As a result, an oxide–semiconductor interface usually exists at an unpassivated semiconductor surface. In general, due to a sudden termination of the periodic structure at the semiconductor surface and the lattice mismatch in the crystallographic structure at the semiconductor–oxide interface, defects are likely to form at the interface which will create discrete or continuous energy states within the forbidden gap of the semiconductor. Figure 6.14 illustrates the energy band diagram for an oxide-semiconductor interface having surface states in the forbidden gap of the semiconductor.

In general, there are two types of surface states which are commonly observed in a semiconductor surface, namely, slow surface states and fast surface states. In a semiconductor surface, the density of slow states is usually much higher than the density of fast states. Furthermore, these surface states can be either positively or negatively charged. To maintain surface charge neutrality, the bulk semiconductor near the surface must supply an equal amount of opposite electric charges. As a result, the carrier density near the surface is different from that of the bulk semiconductor. Due to the slow surface states, the carrier densities at the semiconductor surface not only can change, but may vary so drastically that the surface conductivity type may convert to the opposite type of the bulk. In other words, if the bulk semiconductor is n-type with $n_0 \gg p_0$, then the hole density p_s at the semiconductor–oxide interface may become much larger than the electron density (i.e., $p_s \gg n_s$) such that the surface is inverted to p-type conduction. This is illustrated in Fig. 6.12a in which an inversion layer is formed at semiconductor surface. On the other hand, if the surface electron density is much greater than the surface hole density and bulk electron density (i.e., $n_s > p_s$ and $n_s > n_0$), then an accumulation layer is formed at the semiconductor surface, as shown in Fig. 6.12b. Therefore, the slow surface states at the oxide–semiconductor interface play an important role in controlling the conductivity type of the semiconductor surface.

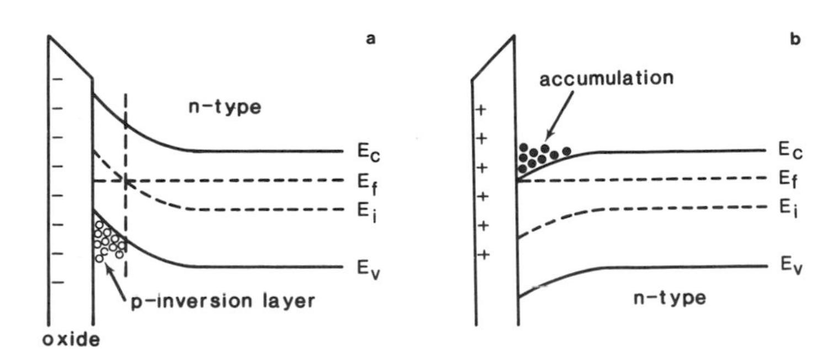

FIGURE 6.12. Potential barrier created at an n-type semiconductor surface: (a) negatively charged slow states and the inversion layer, and (b) positively charged slow states and the accumulation layer.

Figure 6.13 shows both the slow and fast surface states commonly observed in a semiconductor surface. The fast surface states are created either by termination of the periodic lattice structure in the bulk (i.e., creation of dangling bonds at the semiconductor surface) or by lattice mismatch and defects at the oxide–semiconductor interface. These surface states are in intimate electrical contact with the bulk semiconductor, and can reach a state of equilibrium with the bulk within a relatively short period of time (of the order of microseconds or less), and thus are referred to as the fast surface states.

Another type of surface state, usually referred to as the slow state, exists inside the thin oxide layer near the oxide–semiconductor interface. This type of surface state may be formed by either chemisorbed ambient ions or defects in the oxide region (e.g., sodium ions or pin holes in the SiO_2 layer). Carriers transporting from such a state to the bulk semiconductor either have to overcome the potential barrier due to the large energy gap of the oxide, or tunnel through the thin oxide layer. Such a charge transport process involves a large time constant, typically of the order of seconds or more, and hence these states are usually called the slow states.

The concept of surface recombination velocity is discussed next. The Shockley–Read–Hall model derived earlier for dealing with nonradiative recombination in the bulk semiconductor may also be used to explain the recombination in a semiconductor surface. It is noted that a mechanically roughened surface, such as a sand-blasted surface, will have a very high surface recombination velocity while a chemically etched surface will have a much lower surface recombination velocity. Undoubtedly, the fast surface states play an important role in controlling the recombination of excess carriers at the semiconductor surface. For example, GaAs has a very large surface state density and hence a very high surface recombination velocity, while an etched silicon surface has a much lower surface recombination velocity than that of GaAs.

Figure 6.14 shows the energy band diagram for an n-type semiconductor with fast states present at the surface. The energy level introduced by the fast surface states is denoted by E_t, while ϕ_s and ϕ_b denote the surface and bulk electrostatic potentials, respectively. The equilibrium electron and hole densities (n_s and p_s) at the surface can be expressed in terms of the bulk carrier densities, which are given by

$$n_s = n\, e^{-q(\phi_b - \phi_s)/k_B T} \tag{6.105}$$

and

$$p_s = p\, e^{q(\phi_b - \phi_s)/k_B T} \tag{6.106}$$

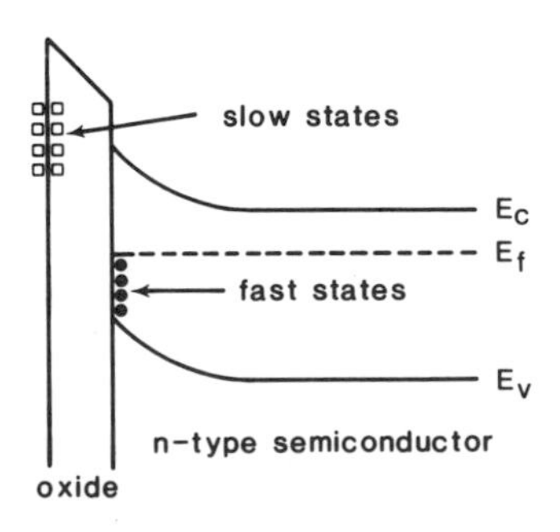

FIGURE 6.13. Energy band diagram of a semiconductor surface in the presence of a thin natural oxide layer and two types of surface states. □ denotes the slow surface states and ● denotes the fast surface states.

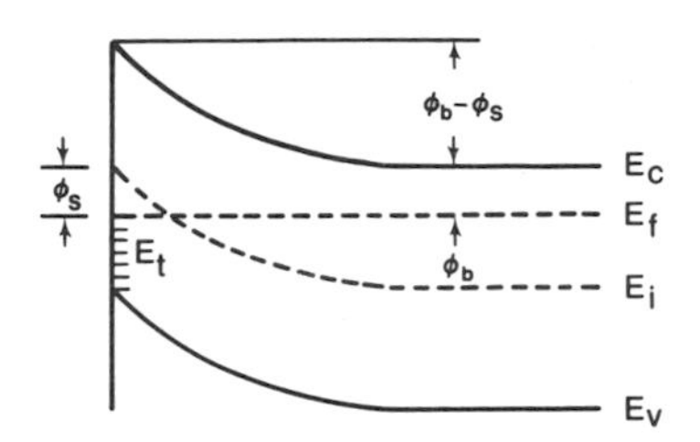

FIGURE 6.14. Energy band diagram of a semiconductor surface showing the fast surface states in the forbidden gap of the vacuum–semiconductor interface; ϕ_s denotes the surface potential and ϕ_b is the bulk potential.

with

$$n_s p_s = np = (n_0 + \Delta n)(p_0 + \Delta p) \tag{6.107}$$

where n and p are the nonequilibrium electron and hole densities, and Δn and Δp are the excess electron and hole densities respectively.

In Eq. (6.24), if n is replaced by surface electron density n_s and p by surface hole density p_s, we obtain an equation for the surface recombination rate,

$$U_s = \frac{N_{ts} c_n c_p (n_s p_s - n_i^2)}{c_p(p_s + p_1) + c_n(n_s + n_1)} \tag{6.108}$$

where

$$n_1 = n_i \, e^{(E_t - E_i)/k_B T} \tag{6.109}$$

and

$$p_1 = n_i \, e^{(E_i - E_t)/k_B T} \tag{6.110}$$

E_t being the energy level of the fast surface states and N_{ts} the surface state density per unit area (cm^2). Thus, the surface recombination rate U_s has the dimensions of cm$^{-2} \cdot$ sec^{-1}. The surface recombination velocity v_s can be defined by

$$v_s = \frac{U_s}{\Delta n} = \frac{U_s}{\Delta p} \tag{6.111}$$

where U_s is given by Eq. (6.108). For the small injection case (i.e., $\Delta n \ll n_0$), the surface carrier densities n_s and p_s can be approximated by their respective equilibrium carrier densities p_{s0} and n_{s0}. Solving Eqs. (6.105) and (6.106) yields

$$n_s \cong n_{s0} = n_i \, e^{q\phi_s/k_B T} \tag{6.112}$$

and

$$p_s \cong p_{s0} = n_i \, e^{-q\phi_s/k_B T} \tag{6.113}$$

where $\phi_s = (E_f - E_{is})/q$ is the surface potential; E_f is the Fermi energy and E_{is} is the intrinsic Fermi level at the semiconductor surface. The solution of Eqs. (6.108) through (6.113) yields an expression for the surface recombination velocity:

$$v_s = U_s/\Delta n = \frac{N_{ts}c(p_0 + n_0)/2n_i}{\cosh[(E_t - E_I - q\phi_0)/k_BT] + \cosh[q(\phi_s - \phi_0)/k_BT]} \tag{6.114}$$

where

$$\phi_0 = (k_BT/2q)\ln(c_p/c_n) \tag{6.115}$$

and

$$c = (c_pc_n)^{1/2} \tag{6.116}$$

is the average capture rate.

Equation (6.114) shows that the surface recombination velocity v_s is directly proportional to the surface state density N_{ts} and the capture rate. It also depends on the surface potential ϕ_s. As the ambient conditions at the surface change, the values of ϕ_s also change accordingly. This fact explains why a stable surface is essential for the operation of a semiconductor device. The surface recombination velocity is closely related to the surface state density. For example, a high surface state density (e.g., $N_{ts} > 10^{13}$ cm^{-2}) in a GaAs crystal also leads to a high surface recombination velocity ($v_s > 10^6$ cm/sec) in this material. For silicon crystal, the surface state density along the (100) surface can be less than 10^{10} cm^{-2} and 10^{11} cm^{-2} along the (111) surface; as a result the surface recombination velocity for a chemically polished silicon surface can be less than 10^3 cm/sec. Therefore, careful preparation of the semiconductor surface is essential for achieving a stable and high-performance semiconductor device.

6.10. DEEP-LEVEL TRANSIENT SPECTROSCOPY (DLTS) TECHNIQUE

As discussed earlier, deep-level defects play an important role in determining the recombination and trapping mechanisms (and hence the minority carrier lifetimes) in a semiconductor. Therefore, it is essential to develop a sensitive experimental tool for characterizing the deep-level defects in a semiconductor. The deep-level transient spectroscopy (DLTS) experiment—a high-frequency transient capacitance technique—is the most sensitive technique for defect characterization in a semiconductor. For example, by performing the DLTS thermal scan from 77 K to around 450 K we can observe the emission spectrum of all the deep-level traps (both majority and minority carrier traps) in the forbidden gap of a semiconductor as positive or negative peaks on a flat baseline as a function of temperature. The DLTS technique offers advantages such as high sensitivity, easy to analyze, and the capability of measuring traps over a wide range of depths in the forbidden gap. By properly changing the experimental conditions, we can measure defect parameters which include: (1) minority and majority carrier traps, (2) activation energy of deep-level traps, (3) trap density and trap density profile, (4) electron and hole capture cross sections, and (5) type of potential well associated with each trap level. In addition, the electron and hole lifetimes can also be calculated from these measured defect parameters. Therefore, by carefully analyzing the DLTS data, all the defect parameters associated with the deep-level defects in a semiconductor can be determined. We shall next discuss the theoretical and experimental aspects of the DLTS technique.

The DLTS experiment can be performed by using a variety of test structures including the Schottky barrier, p–n junction, bipolar junction transistor, MOSFETs, and MOS capacitor structure. The DLTS technique is based on the transient capacitance change associated with the thermal emission of charge carriers from a trap level to thermal equilibrium after an initial nonequilibrium condition in the space-charge region of a Schottky barrier diode or a p–n junction diode. The polarity of the DLTS peak depends on the capacitance change after trapping of the minority or majority carriers. For example, an increase in the trapped minority carriers in the junction space charge region (SCR) of a p–n diode would result in an increase in the junction capacitance of the diode. In general, a minority carrier trap will produce a positive DLTS peak, while a majority carrier trap would display a negative DLTS peak. For a p^+–n junction diode, the SCR extends mainly into the n-region, and the local charges are due to positively charged ionized donors. If a forward bias is applied, the minority carriers (i.e., holes) will be injected into this SRC region. Once the minority holes are trapped in a defect level, the net positive charges in the SCR will increase. This in turn will reduce the width of SCR, and causes a positive capacitance change. Thus, the DLTS signal will have a positive peak. Similarly, if electrons are injected into the SCR and captured by the majority carrier traps, then the local charge density in the SCR is reduced and the depletion layer width is widened, which results in a decrease in the junction capacitance. Thus, the majority carrier trapping will result in a negative DLTS peak. The same argument can be applied to an n^+–p junction diode.

The peak height of a DLTS signal is directly related to the density of a trap level, which in turn is proportional to the change of junction capacitance $\Delta C(0)$ due to carrier emission from the trap level. Therefore, the defect density N_t can be calculated from the capacitance change $\Delta C(0)$ (or the DLTS peak height). If $C(t)$ denotes the transient capacitance across the depletion layer of a Schottky barrier diode or a p–n junction diode, then using abrupt junction approximation we can write

$$\begin{aligned} C(t) &= A\left[\frac{q\varepsilon_0\varepsilon_s(N_d - N_t e^{-t/\tau})}{2(V_{bi} + V_R + k_BT/q)}\right]^{1/2} \\ &= C_0\left[1 - \left(\frac{N_t}{N_d}\right)e^{-t/\tau}\right]^{1/2} \end{aligned} \tag{6.117}$$

where τ is the thermal emission time constant; $C_0 = C(V_R)$ is the junction capacitance measured at a quiescent reverse bias voltage, V_R. If we use the binomial expansion in Eq. (6.117) and assume that $N_t/N_d \ll 1$, then $C(t)$ can be simplified to

$$C(t) \simeq C_0\left[1 - \left(\frac{N_t}{2N_d}\right)e^{-t/\tau}\right] \tag{6.118}$$

For $t = 0$, we obtain

$$N_t \simeq (2\Delta C(0)/C_0)N_d \tag{6.119}$$

where $\Delta C(0) = C_0 - C(0)$ is the net capacitance change due to thermal emission of electrons from the trap level, and $C(0)$ is the capacitance measured at $t = 0$; $\Delta C(0)$ can be determined from the DLTS measurement. It is seen that both the junction capacitance C_0 and the background dopant density N_d are determined from the high-frequency C–V measurements. Therefore, the defect concentration N_t can be determined from Eq. (6.119) by using DLTS and high-frequency (1 MHz) C–V measurements.

The decay time constant of the capacitance transient in the DLTS thermal scan is associated with a specific time constant, which is equal to the reciprocal of the emission rate. For a given electron trap, the emission rate e_n is related to the capture cross section and the activation energy of the electron trap by

$$e_n = (\sigma_n < v_{th} > N_c/g)\, e^{(E_c - E_t)/k_BT} \tag{6.120}$$

where E_t is the activation energy of the electron trap, $\langle v_{th} \rangle$ is the average thermal velocity, N_c is the effective density of conduction band states, and g is the degeneracy factor. The electron capture cross section σ_n, which depends on temperature, can be expressed by

$$\sigma_n = \sigma_0\, e^{-\Delta E_b/k_BT} \tag{6.121}$$

where σ_0 is the capture cross section when temperature approaches infinity, and ΔE_b is the activation energy of the capture cross section. Now substituting σ_n given by Eq. (6.121) into Eq. (6.120) and using the fact that N_c is proportional to $T^{-3/2}$ and $\langle v_{th} \rangle$ is proportional to $T^{-1/2}$, the quantity e_n given in Eq. (6.120) can be reduced to

$$\begin{aligned} e_n &= BT^2\, e^{(E_c - E_t - \Delta E_b)/k_BT} \\ &= BT^2\, e^{(E_c - E_m)/k_BT} \end{aligned} \tag{6.122}$$

where B is a proportionality constant which is independent of temperature. From Eq. (6.122), it is seen that the electron thermal emission rate e_n is an exponential function of the temperature. The change of capacitance transient can be derived from Eq. (6.118), which yields

$$\Delta C(t) = C_0 - C(t) \simeq C_0(N_t/2N_d)\, e^{-t/\tau} = \Delta C(0)\, e^{-t/\tau} \tag{6.123}$$

where $\tau = e_n^{-1}$ is the reciprocal emission time constant.

The experimental procedures for determining the activation energy of a deep-level trap in a semiconductor are described as follows. The first step of the DLTS experiment is to choose the rate windows t_1 and t_2 in a dual-gated integrator of a boxcar averager which is used in the DLTS system, and measures the capacitance change at a preset t_1 and t_2 rate window. This can be written as

$$\Delta C(t_1) = \Delta C(0)\, e^{-t_1/\tau} \tag{6.124}$$

$$\Delta C(t_2) = \Delta C(0)\, e^{-t_2/\tau} \tag{6.125}$$

The DLTS scan along the temperature axis is obtained by taking the difference of Eqs. (6.124) and (6.125), which produces a DLTS spectrum given by

$$S(\tau) = \Delta C(0)(e^{-t_1/\tau} - e^{-t_2/\tau}) \tag{6.126}$$

The maximum emission rate, $\tau_{\max}^{-1}$, can be obtained by differentiating $S(\tau)$ with respect to τ and setting $dS(\tau)/d\tau = 0$, which yields

$$\tau_{\max} = \frac{(t_1 - t_2)}{\ln(t_1/t_2)} \tag{6.127}$$

Note that $S(\tau)$ reaches its maximum value at a characteristic temperature T_m corresponding to the maximum emission time constant $\tau_{\max}$. The emission rate is related to this $\tau_{\max}$ value by $e_n = 1/\tau_{\max}$ for each t_1 and t_2 rate window setting. By changing the values of the rate window t_1 and t_2 in the boxcar gated integrator, a series of DLTS scans with different values of e_n and T_m can be obtained. From these DLTS thermal scans we can obtain an Arrhenius

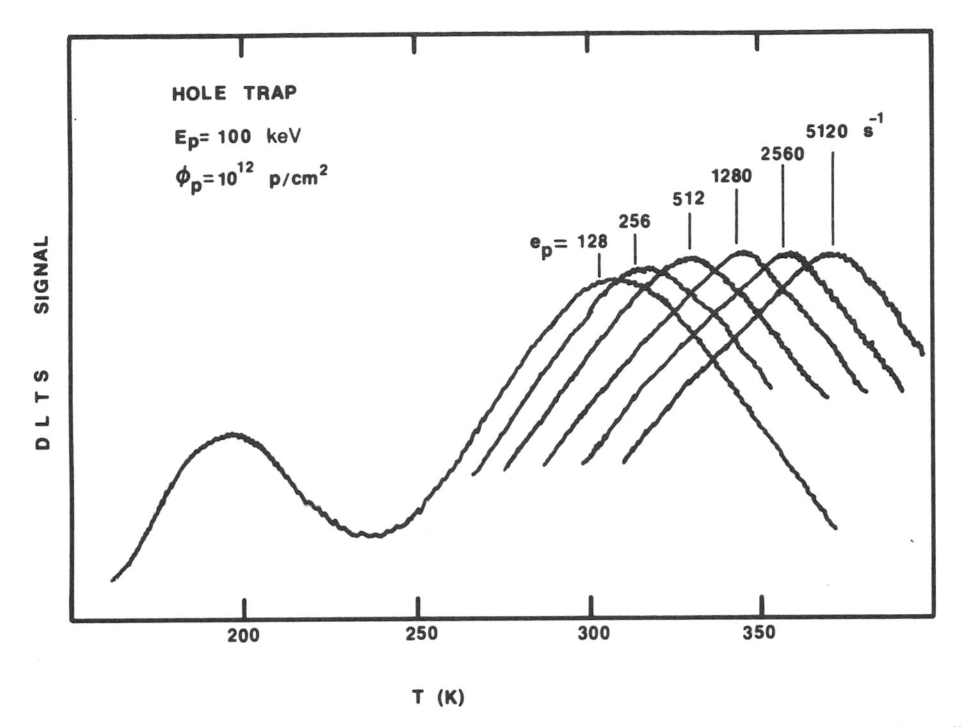

FIGURE 6.15. DLTS scans of the hole traps observed in a 100 keV proton irradiated GaAs solar cell with a proton fluency of 10^{12} cm^{-2}. Six different DLTS scans were performed for the second hole trap observed at a higher temperature. After Li *et al.*,[6] by permission, © IEEE–1980.

plot of $e_n T^2$ versus $1/T$ for a specific trap level, as is shown in Fig. 6.15.[6] The activation energy of the trap level can be calculated from the slope of this Arrhenius plot. Figure 6.16 shows the DLTS scans of electron and hole traps observed in a 290-keV proton irradiated GaAs p–n junction diode.[6] Three electron traps and three hole traps were observed in this sample. Figure 6.17 shows the DLTS scans of a hole trap versus annealing time for a thermally annealed (170°C) Sn-doped InP grown by the liquid encapsulated Czochralski (LEC) technique and the trap density versus annealing time for this sample.[7]

From the above description it is clearly shown that the DLTS technique is indeed a powerful tool for characterizing the deep-level defects in a semiconductor. It allows a quick inventory of all deep-level defects in a semiconductor, and is widely used for defect characterization in a semiconductor.

6.11. SURFACE PHOTOVOLTAGE (SPV) TECHNIQUE

Another characterization method, known as the surface photovoltage (SPV) technique, for measuring the minority carrier diffusion length in a semiconductor wafer has gained popularity in recent years. The SPV method is an attractive technique since it is a steady-state, contactless optical technique. No junction preparation or high-temperature processing are needed for this method. The minority carrier lifetime can also be determined from the SPV measurements by using the relation $\tau = L^2/D$, where L is the minority carrier diffusion length. The SPV technique has been widely used for determining the minority carrier diffusion length in silicon, GaAs, and InP. The basic theory and experimental details of the SPV method are discussed next.

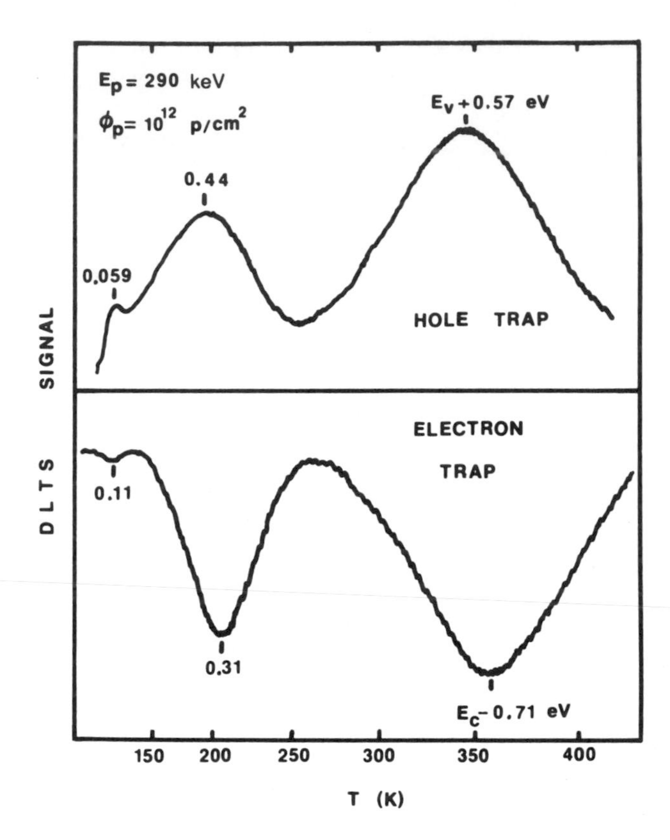

FIGURE 6.16. DLTS scans of electron and hole traps for a 290-keV and 10^{12} p/cm^2 proton irradiated AlGaAs/GaAs p–n junction solar cell. After Li *et al.*,[6] by permission, © IEEE–1980.

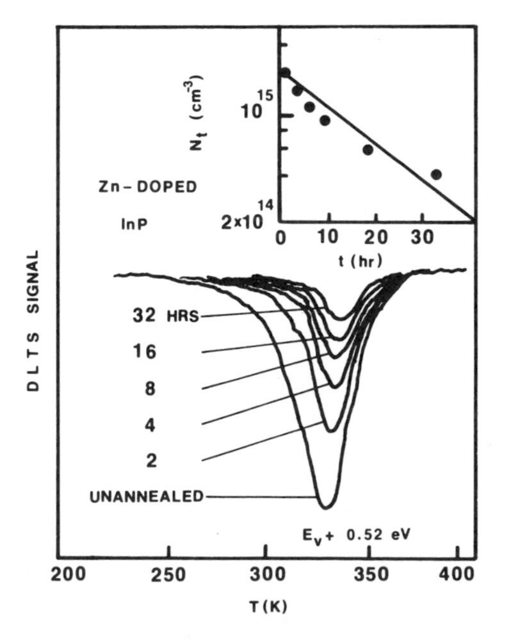

FIGURE 6.17. DLTS scans of a hole trap versus annealing time for a Zn-doped InP specimen annealed at 200°C. After Li *et al.*,[7] by permission.

When a semiconductor specimen is illuminated by chopped monochromatic light with its photon energy greater than the bandgap energy of the semiconductor, a surface photovoltage (SPV) is induced at the semiconductor surface as the photogenerated electron–hole pairs diffuse into the specimen along the direction of incident light. The SPV signal is capacitively coupled into a lock-in amplifier for amplification and measurement. The light intensity is adjusted to produce a constant SPV signal at different wavelengths of the incident monochromatic light. The light intensity required to produce a constant SPV signal is plotted as a function of the reciprocal absorption coefficient for each wavelength near the absorption edge. The resultant linear plot is extrapolated to zero light intensity and intercepts the horizontal axis at $-1/\alpha$, which is equal to the minority carrier diffusion length.

The SPV signal developed at the illuminated surface of a semiconductor specimen is a function of the excess minority carrier density injected into the surface space-charge region. The excess carrier density is in turn dependent on the incident light intensity, the optical absorption coefficient, and the minority carrier diffusion length. Thus, an accurate knowledge of the absorption coefficient versus wavelength is required for the SPV method. In general, the SPV signal for an n-type semiconductor may be written as

$$V_{spv} = f(\Delta p) \tag{6.128}$$

where

$$\Delta p = \frac{\eta I_0(1-R)}{D_p/L_p + s_1} \frac{\alpha L_p}{1+\alpha L_p} \tag{6.129}$$

is the excess hole density, η is the quantum efficiency, I_0 is the light intensity, R is the reflection coefficient, D_p is the hole diffusion coefficient, s_1 is the front surface recombination velocity, α is the optical absorption coefficient, and L_p is the hole diffusion length. Equation (6.128) holds if $\alpha^{-1} \gg L_p$, $n \gg \Delta p$, and $\alpha d > 1$ (where d is the thickness of the specimen).

If η and R are assumed constant over the measured wavelength range, the incident light intensity I_0 required to produce a constant SPV signal is directly proportional to the reciprocal absorption coefficient α^{-1} and can be written as

$$I_0 = C(\alpha^{-1} + L_p) \tag{6.130}$$

where C is a constant, independent of the photon wavelength. The linear plot of I_0 versus α^{-1} is extrapolated to zero light intensity and the negative intercept value is the effective hole diffusion length.

Figure 6.18 shows the relative photon intensity I_0 versus the inverse absorption coefficient α^{-1} for an n-type InP specimen.[8] The negative intercept yields $L_p = 1.4\,\mu$m. SPV measurements have been widely used in determining the minority carrier diffusion lengths in silicon wafers, with minority carrier diffusion lengths in undoped silicon material exceeding 100 μm.

PROBLEMS

6.1. Consider an n-type silicon sample with a dopant density of 2×10^{15} cm^{-3}. If the specimen is illuminated by a mercury lamp with variable intensity, plot the excess carrier lifetimes as a function of the excess carrier density for Δn varying from 2×10^{13} to 5×10^{16} cm^{-3}. It is assumed that the recombination of excess carriers is dominated by the Shockley–Read–Hall process, $\tau_{n0} = \tau_{p0} = 1 \times 10^{-8}$ sec, $n_0 \gg p_0$, and $n_0 \gg n_1, p_1$.

6.2. Consider a gold-doped silicon sample. There are two energy levels for the gold impurity in silicon. The gold acceptor level is located at 0.55 eV below the conduction band edge and the

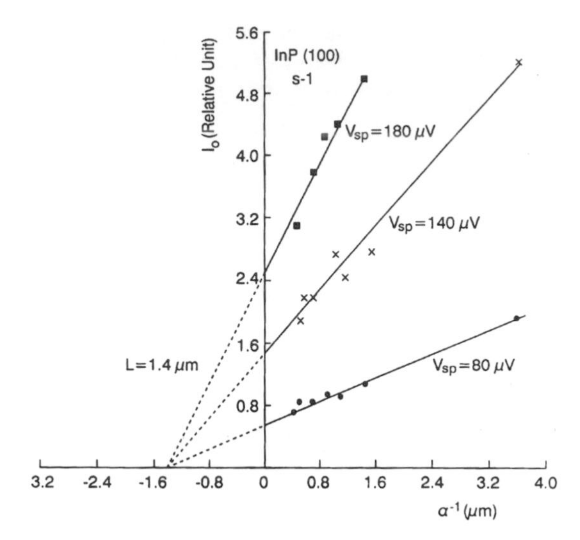

FIGURE 6.18. Relative light intensity I_0 versus the inverse absorption coefficient for an InP specimen. After Li,[8] by permission.

gold donor level is 0.35 eV above the valence band edge. If the electron capture rate C_n for the gold acceptor center is assumed equal to $5 \times 10^{-8}\ \text{cm}^3 \cdot \text{sec}^{-1}$, and the hole capture rate C_p is $2 \times 10^{-8}\ \text{cm}^3 \cdot \text{sec}^{-1}$, and the density of the gold acceptor center, N_{Au}, is equal to $5 \times 10^{15}\ \text{cm}^{-3}$:

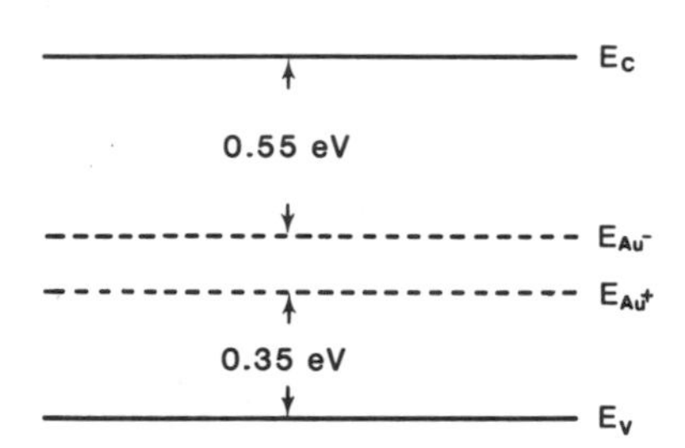

(a) Compute the electron and hole lifetimes in this sample.
(b) If the temperature dependence of the electron emission rate is given by

$$e_n = A_m(T/300)^m \exp[-(E_c - E_{Au}^-)/k_BT]$$

find a solution for e_n when $m = 0$ and 2.
(c) Calculate e_p from (a) and (b).

6.3. The kinetics of recombination, generation, and trapping at a single energy level inside the forbidden bandgap of a semiconductor have been considered in detail by Shockley and Read [*Phys. Rev.* **87**, 835 (1952)]. From the appendix of this paper derive an expression for the excess carrier lifetime for the case when a large trap density is present in the semiconductor [also see Hall's paper, *Proc. IEEE* **106B**, 923 (1959)].

6.4. Plot the radiative lifetime for a GaAs sample as a function of excess carrier density, Δn, at $T = 300$ K, for $\Delta n/n_i = 0$, 1, 3, 10, 30 and $n_0/n_i = 10^{-2}$, 10^{-1}, 1, 10, 10^2. $n_i = 1 \times 10^7\ \text{cm}^{-3}$ is the intrinsic carrier density, and the generation rate G_r is assumed equal to $10^7\ \text{cm}^{-3} \cdot \text{sec}^{-1}$.

6.5. Show that the Einstein relation (i.e., D_n/μ_n) for an n-type degenerate semiconductor is equal to $(k_BT/q)\mathcal{F}_{1/2}(\eta)/\mathcal{F}_{-1/2}(\eta)$. Plot D_n/μ_n versus dopant density for n-type silicon at 300 K.

6.6. Derive an expression for the extrinsic Debye length for nondegenerate and degenerate semiconductors, and calculate the Debye lengths L_{D_n} for an n-type silicon sample with $N_D = 10^{14}$, 10^{15}, 10^{16}, 10^{17}, 10^{18}, and $10^{19}\ \text{cm}^{-3}$.

6.7. Plot the energy band diagram for a p-type semiconductor surface under (a) inversion, (b) accumulation, and (c) depletion conditions.

6.8. Calculate the surface recombination velocity versus surface state density for an n-type silicon with $N_{ts} = 10^9$, 10^{10}, 10^{11}, and $10^{12}\ \text{cm}^{-2}$. Assume that $E_T = E_c - 0.5$ eV, $C_n = C_p = 10^{-8}\ \text{cm}^3 \cdot \text{sec}^{-1}$, $n_0 = 10^{16}\ \text{cm}^{-3}$, $n_0 \gg p_0$, and $T = 300$ K.

6.9. From the paper, "Fast Capacitance Transient Apparatus: Application to Zn- and O-centers in GaP p–n Junctions," by D. V. Lang, *J. Appl. Phys.* **45**, 3014–3022 (1974), describe the electron emission and capture processes in a Zn-O doped GaP p–n diode and their correlation to the DLTS thermal scans. Explain under what conditions the DLTS theory described in this paper fails.

6.10. Using the Arrhenius plot (i.e., e_p/T^2 versus $1/T$) find the activation energy of the second hole trap (located at a higher temperature) shown in Fig. 6.15.

REFERENCES

1. W. Shockley and W. T. Read, *Phys. Rev.* **87**, 835 (1952).
2. R. N. Hall, *Phys. Rev.* **87**, 387 (1952).
3. C. T. Sah and W. Shockley, *Phys. Rev.* **109**, 1103 (1958).
4. P. T. Landsberg and A. F. W. Willoughby (eds.), *Proceeding of the International Conference on Recombination Mechanisms in Semiconductors, in Solid State Electronics* **21**, 1273 (1978).
5. M. E. Law, E. Solley, M. Liang, and D. E. Burk, "Self-Consistent Model of Minority Carrier Lifetime, Diffusion Length, and Mobility," *IEEE Elec. Dev. Lett.* **12**, 40 (1991).
6. S. S. Li, W. L. Wang, P. W. Lai, and R. Y. Loo, "Deep-Level Defects, Recombination Mechanisms, and Performance Characteristics of Low Energy Proton Irradiated AlGaAs/GaAs Solar Cells," *IEEE Trans. Electron Devices*, **ED-27**, 857 (1980).
7. S. S. Li, W. L. Wang, and E. H. Shaban, "Characterization of Deep-Level Defects in Zn-doped InP," *Solid State Commun.* **51**, 595 (1984).
8. S. S. Li, "Determination of Minority Carrier Diffusion Length in InP by Surface Photovoltage Measurement," *Appl. Phys. Lett.* **29**, 126 (1976).

BIBLIOGRAPHY

J. S. Blakemore, *Semiconductor Statistics*, Pergamon Press, New York (1962).

R. H. Bube, *Photoconductivity of Solids*, Wiley, New York (1960).

P. T. Landsberg, *Solid State Physics in Electronics and Telecommunications*, Academic Press, London (1960).

A. Many and R. Bray, "Lifetime of Excess Carriers in Semiconductors," in *Progress in Semiconductors*, Vol. 3, pp. 117–151, Heywood and Co., London (1958).

J. P. McKelvey, *Solid State and Semiconductor Physics*, 2nd ed., Chapter 10, Harper & Row, New York (1982).

W. Shockley, *Electrons and Holes in Semiconductors*, D. Van Nostrand, New York (1950).

R. A. Smith, *Semiconductors*, Cambridge University Press, London (1961).

7

Transport Properties of Semiconductors

7.1. INTRODUCTION

In this chapter we are concerned with the carrier transport phenomena in a semiconductor under the influence of applied external fields. Different galvanomagnetic, thermoelectric, and thermomagnetic effects created by the applied electric and magnetic fields as well as the temperature gradient in a semiconductor will be discussed in this chapter. The transport coefficients associated with the galvanomagnetic, thermoelectric, and thermomagnetic effects in a semiconductor are derived from the Boltzmann transport equation using the relaxation time approximation, which assumes that all the collision processes are elastic and can be treated in terms of a unique relaxation time. This approximation is valid as long as the collision process is elastic. The elastic scattering requires that the energy change of charged carriers (i.e., electrons or holes) during the scattering process must be small compared to the energy of the charged carriers, and the relaxation time must be a scalar quantity. In this approximation, the relaxation time is represented by a unique time constant τ, which may depend on both the energy of the charged carriers and the temperature. Typical examples of elastic scattering processes include the scattering of electrons by longitudinal acoustical phonons, ionized impurities, and the neutral impurities in a semiconductor. In the event that the relaxation time approximation fails, the solutions for the Boltzmann transport equation can be obtained by using the variational principles.

The effect of an applied electric field, magnetic field, or a temperature gradient on the electrons in a semiconductor is to change the distribution function of electrons from its equilibrium value. As shown in Chapter 5, in the absence of an external field, the distribution of electrons in a semiconductor or a metal under equilibrium conditions may be described by the Fermi–Dirac distribution function,

$$f_0(E) = \frac{1}{1 + e^{(E - E_f)/k_B T}} \tag{7.1}$$

Equation (7.1) shows that in thermal equilibrium the electron distribution function $f_0(E)$ depends not only on the electron energy but also on the Fermi energy E_f, a many-body parameter, and temperature T. However, under the influence of external fields, $f_0(E)$ given in Eq. (7.1) is no longer valid for the nonequilibrium electron distribution function in a semiconductor. This can be best explained by considering the case in which an electric field or a magnetic field is applied to the semiconductor specimen. When an electric field or a magnetic field is applied to the semiconductor, the Lorentz force will tend to change the wave vector of electrons [i.e., $\mathscr{F} = -q(\mathscr{E} + \mathbf{v} \times \mathbf{B}) = \hbar\, dk/dt$] along the direction of the applied fields. As a result, the distribution function will be modified by the changing wave vector of electrons under the influence of the Lorentz force. Furthermore, since f_0 depends on both the energy and the temperature as well as the electron concentration, we expect that the nonequilibrium distribution function of electrons will also be a function of the position in space when a temperature gradient or a concentration gradient is presented across the semiconductor specimen.

To illustrate the effect of external forces on the electron distribution function, Fig. 7.1a and b show the two-dimensional distribution functions of electrons in the presence of an applied electric field and a temperature gradient, respectively. As shown in Fig. 7.1a, when an electric field is applied along the x-direction, the change of electron wave vector in the x-direction is given by $\Delta k_x = -q\mathscr{E}_x\tau/\hbar$, where τ is the mean relaxation time of electrons and $\mathscr{E}_x$ is the applied electric field in the x-direction. In this case, the nonequilibrium distribution function as a whole is shifted to the right by an amount equal to Δk_x from its equilibrium position. Note that the shape of the nonequilibrium distribution function remains unchanged from its equilibrium case. The fact that the shape of the electron distribution function in k-space does not change due to the electric field can be explained by the force acting on each quantum state k. Since the Lorentz force F_x $(=q\mathscr{E}_\S)$ due to the electric field is equal to $\hbar\dot{k}_x$, the rate of change of k_x is the same for all electrons. Consequently, if there is no relaxation mechanism to restore the distribution function to equilibrium, an applied electric field can only cause the distribution function to drift, unaltered in shape, along the k_x-direction at a constant velocity $v_x = \hbar\dot{k}_x/m^*$ and the change in the crystal momentum due to this drift is $\Delta k_x = q\mathscr{E}_\S\tau/\hbar$.

On the other hand, the shift in the perturbed distribution function is quite different when a temperature gradient is applied to a semiconductor. In this case, the nonequilibrium distribution function is shifted to the right by an amount equal to Δk_x for those electrons with energies greater than the Fermi energy, and to the left by the same amount Δk_x for those electrons with energies less than the Fermi energy [i.e., $f(k_x, k_y, k_z) = f_0(k_x + \Delta k_x, k_y, k_z)$], as is illustrated in Fig. 7.1b. The physical mechanisms which cause this shift can be explained by introducing a Taylor expansion of $(E - E_f)$ about the Fermi energy $E_f = \hbar^2 k_f^2/2m^*$. If we

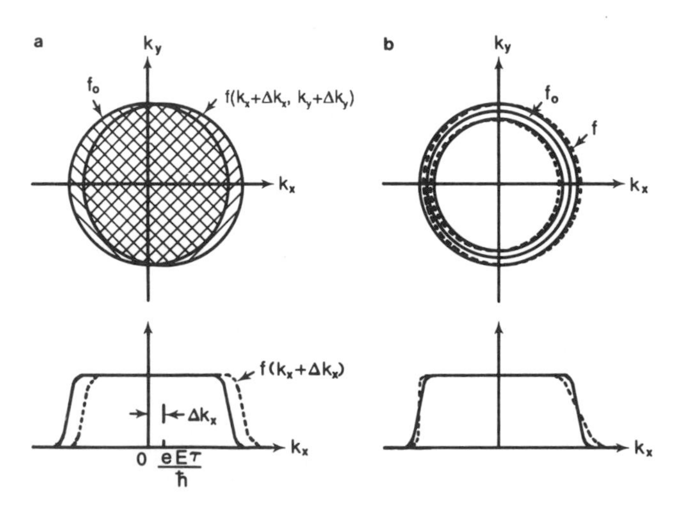

FIGURE 7.1. The effect of an applied electric field and a temperature gradient on the electron distribution function in a semiconductor.

assume $|E - E_f| \ll E_f$, we can replace $(E - E_f)$ by $(\hbar^2 k_f/m^*)(k - k_f)$, and obtain

$$\Delta k_x = \frac{\tau \hbar k_f}{m^* T}(k - k_f)\frac{\partial T}{\partial x}$$

This result shows that the distribution of quantum states on the Fermi surface (i.e., at $k = k_f$) is not affected by the temperature gradient (i.e., $\Delta k_x = 0$). On the other hand, those quantum states corresponding to energies in excess of the Fermi energy have their centers shifted in k-space opposite to the direction of the temperature gradient, while states of lower energies have their centers shifted in the same direction as the temperature gradient.

Section 7.2 describes the various types of galvanomagnetic, thermoelectric, and thermomagnetic effects in a semiconductor. These include the electrical conductivity, Hall effect, Seebeck and Pelter effects, Nernst and Ettinghousen effects, and the magnetoresistance effect. In Section 7.3, we derive the Boltzmann transport equation for the steady-state case. Expressions for the electrical conductivity, electron mobility, Hall coefficient, magnetoresistance, Nernst and Seebeck coefficients are derived in Section 7.4 for n-type semiconductors. Transport coefficients for the mixed conduction case (e.g., intrinsic semiconductors) are depicted in Section 7.5. In Section 7.6 we present some experimental data on transport coefficients for germanium, silicon and III–V compound semiconductors.

7.2. GALVANOMAGNETIC, THERMOELECTRIC, AND THERMOMAGNETIC EFFECTS

In this section, we discuss various types of galvanomagnetic, thermoelectric, and thermomagnetic effects in a semiconductor. These effects are associated with the transport of electrons (for n-type) or holes (for p-type) in a semiconductor when an external electric field, a magnetic field, or a temperature gradient is applied separately or simultaneously to a semiconductor specimen. The transport coefficients to be described here are the electrical conductivity, thermal conductivity, Hall coefficient, Seebeck coefficient, Nernst coefficient, and the magnetoresistance for an n-type semiconductor. Transport coefficients derived for an n-type semiconductor can also be applied to a p-type semiconductor provided that the positive charge and the effective mass of holes are used instead. It is noted that for nondegenerate semiconductors, Boltzmann statistics is used in the derivation of transport coefficients, while Fermi statistics is used in deriving the transport coefficients for degenerate semiconductors and metals.

7.2.1. *Electrical Conductivity* σ_n

In this section we discuss the current conduction in an n-type semiconductor. In this case, the electrical current conduction is due to electrons. When a small electric field is applied to the specimen, the electrical current density can be related to the electric field by using Ohm's law, which reads

$$J_n = \sigma_n \mathscr{E} = q\mu_n n \mathscr{E} \tag{7.2}$$

where $\sigma_n = q\mu_n n$ is the electrical conductivity, μ_n is the electron mobility, and n denotes the electron density.

The electrical current density can also be expressed in terms of the electron density and electron drift velocity v_d along the direction of the applied electric field. Thus, we can write

$$J_n = qnv_d \tag{7.3}$$

where q is the electronic charge. By comparing Eqs. (7.2) and (7.3) one finds that the electron drift velocity is related to the electric field by

$$v_d = \mu_n \mathscr{E} \tag{7.4}$$

where μ_n is the low-field electron drift mobility, which is defined as the electron drift velocity per unit electric field strength. For metals, μ_n can be expressed in terms of the mean collision time τ and the electron effective mass m^*,

$$\mu_n = \frac{q\tau}{m^*} \tag{7.5}$$

From Eq. (7.2) and Eq. (7.5), the electrical conductivity σ_n for a metal can be expressed in terms of the mean collision time and the electron effective mass:

$$\sigma_n = qn\mu_n = \frac{q^2 n\tau}{m^*} \tag{7.6}$$

In collision processes, the transition probability for electron collision is directly related to the density of collision centers, and the collision rate is inversely proportional to the collision time constant. For example, in the case of electron–phonon scattering, the number of scattering centers is equal to the phonon population in thermal equilibrium. At high temperatures the average phonon density is proportional to temperature T. Consequently, at high temperatures, the collision time τ varies as $1/T$, and hence the electrical conductivity σ_n will vary inversely with temperature T. This prediction is consistent with the experimental observation of electrical conductivity versus temperature in metals.

For n-type semiconductors, Eqs. (7.5) and (7.6) are still valid provided that the free-electron mass is replaced by the conductivity effective mass of electrons in the conduction bands m_n^*, and τ is replaced by the average relaxation time $\langle\tau\rangle$. In general, the electron density in a semiconductor is a strong function of temperature, and the relaxation time may depend on both the energy and temperature. A general expression for the current density in an n-type semiconductor can be derived as follows. From Eq. (5.3), the density of quantum states $g_n(E)$ for a single-valley semiconductor with a parabolic conduction band can be rewritten as

$$g_n(E) = \left(\frac{4\pi}{h^3}\right)(2m_n^*)^{3/2}(E - E_c)^{1/2} \tag{7.7}$$

Equations (7.3) and (7.7) yield a general expression for the electron current density in the form

$$J_n = -qnv_x = -q\int_0^\infty v_x f(E) g_n(E)\, dE \tag{7.8}$$

where $f(E)$ is the nonequilibrium distribution function of electrons, which may be obtained by solving the Boltzmann transport equation, to be discussed later in Section 7.4. The integration of Eq. (7.8) is carried out over the entire conduction band. The minus sign in Eq. (7.8) stands for electron conduction in an n-type semiconductor. For hole conduction in a p-type

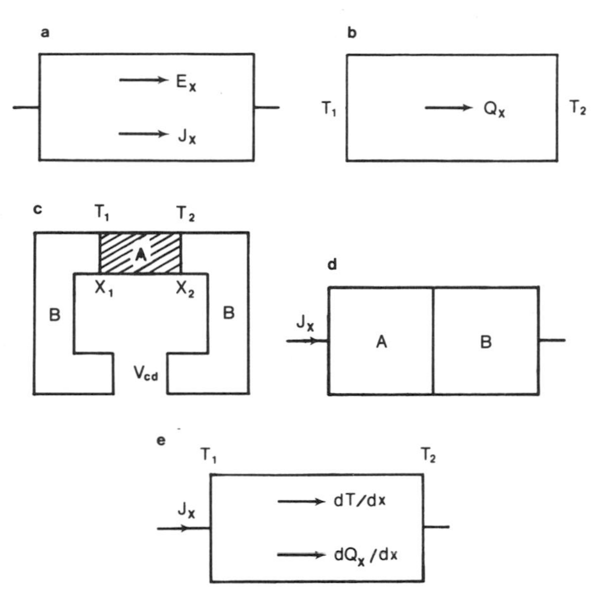

FIGURE 7.2. Longitudinal transport effects in the presence of an electric field or temperature gradient: (a) electrical conductivity, $\sigma = (J_x/\mathscr{E}_x)|_{\Delta T=0}$; (b) electronic thermal conductivity, $K_n = -Q_x/(\partial T/\partial x)|_{J_x=0}$; (c) Seebeck effect, $S_{ab} = V_{dc}/(T_2 - T_1)$; (d) Peltier effect, $\Pi_{ab} = S_{ab}T$; (e) Thomson effect, $\tau_s = (\partial Q_x/\partial x)/I_x(\partial T/\partial x)$.

semiconductor, the plus sign should be used instead. Figure 7.2a shows the applied electric field and the current flow in an n-type semiconductor specimen. The electrical conductivity for an n-type semiconductor is derived in Section 7.4.

7.2.2. *Electronic Thermal Conductivity K_n*

The electronic thermal conductivity is due to the flow of thermal energy carried by electrons when a temperature gradient appears across a semiconductor specimen. As shown in Fig. 7.2b, when a temperature gradient is created in a semiconductor specimen, a heat flux flow will appear across the specimen. The electronic thermal conductivity K_n is defined as the thermal flux density per unit temperature gradient, and can be expressed by

$$K_n = -\left.\frac{Q_x}{(\partial T/\partial x)}\right|_{J_x=0} \tag{7.9}$$

where Q_x is the thermal flux density given by

$$Q_x = nv_x E = \int_0^\infty v_x E f(E) g_n(E)\, dE \tag{7.10}$$

v_x is the thermal velocity of electrons in the x-direction and E is the electron energy. The integration on the right-hand side of Eq. (7.10) is carried out over the entire conduction band. Equations (7.8) and (7.10) are the two basic equations which describe the flow of electric current density and heat flux density for an n-type semiconductor, respectively. All the transport coefficients described in this section can be derived from Eqs. (7.8) and (7.10) provided that the nonequilibrium distribution function $f(k, r)$ is known. The steady-state nonequilibrium distribution $f(k, r)$ can be found by solving the Boltzmann transport equation. It should be noted that in thermal equilibrium both J_n and Q_x, given by Eq. (7.8) and Eq. (7.10), are equal to zero, as $f(E)$ is reduced to $f_0(E)$ (i.e., the Fermi distribution function). Figures 7.2 and 7.3 show plots of electric field, current density, heat flux, temperature gradient, and the resulting galvanomagnetic, thermoelectric, and thermomagnetic effects in a semiconductor. We shall next discuss the various types of galvanomagnetic, thermoelectric, and thermomagnetic effects in a semiconductor. Transport coefficients associated with these effects will be derived in Sections 7.4 and 7.5.

7.2.3. Thermoelectric Coefficients

When a temperature gradient, an electric field, or both is applied across a semiconductor or a metal specimen, three different types of thermoelectric effect can be observed. These include the Seebeck, Peltier, and Thomson effects. The thermoelectric coefficients associated with each of these effects can be defined according to Figs. 7.2c, d, and e, in which two pieces of conductors, A and B, are joined at junctions x_1 and x_2. If a temperature difference ΔT is established between junctions x_1 and x_2, then an open-circuit voltage V_{cd} is developed between terminals "c" and "d". This is known as the Seebeck effect. In this case, the differential Seebeck coefficient, or the thermoelectric power, can be defined by

$$S_{ab} = \frac{V}{\Delta T} = \frac{V_{dc}}{(T_2 - T_1)} \tag{7.11}$$

If the junctions x_1 and x_2 are initially maintained at the same temperature, then by applying a voltage across terminals "c" and "d" one observes an electrical current flow through these two conductors. If the result is a rate of heating at junction x_1, then there will be a cooling at the same rate at junction x_2. This is the well-known Peltier effect, the basic principle for thermoelectric cooling. The differential Peltier coefficient is defined by

$$\Pi_{ab} = \frac{Q_x}{I_x} \tag{7.12}$$

The Thomson effect occurs when an electric current and a temperature gradient are applied simultaneously in the same direction on a semiconductor specimen. In this case, the simultaneous presence of the current flow I_x and the temperature gradient $\partial T/\partial x$ in the direction of current flow will yield a rate of heating or cooling $(\partial Q_x/\partial x)$ per unit length. Thus,

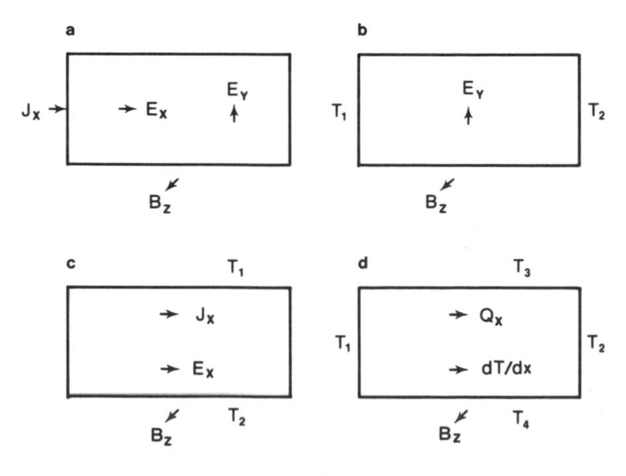

FIGURE 7.3. Galvanomagnetic and thermomagnetic effects in a semiconductor under the influence of an applied electric field, a magnetic field, and a temperature gradient. The polarity shown is for hole conduction: (a) Hall effect, (b) Nernst effect, (c) Ettingshausen effect, (d) Righi–Leduc effect.

the Thomson coefficient can be expressed by

$$\tau_s = \frac{(\partial Q_x/\partial x)}{I_x(\partial T/\partial x)} \tag{7.13}$$

Applying the thermodynamic principles to the thermoelectric effects shown in Figs. 7.2c, d, and e, Thomson derived two important equations, later known as the Kelvin relations, which relate the three thermoelectric coefficients. The two equations are

$$\Pi_{ab} = S_{ab}T \tag{7.14}$$

and

$$\tau_a - \tau_b = T\frac{dS_{ab}}{dT} \tag{7.15}$$

The Kelvin relations, Eqs. (7.14) and (7.15), not only have a sound theorctical basis, but also have been verified experimentally by the measured thermoelectric figure of merit. Equation (7.14) is particularly useful for thermoelectric refrigeration applications. This is due to the fact that the rate of cooling by means of the Peltier effect can be expressed in terms of the Seebeck coefficient, which is a much easier quantity to measure. Furthermore, Eq. (7.15) allows one to account for the influence of the Thomson effect on the cooling power of a thermoelectric refrigerator via the variation in the Seebeck coefficient with temperature.

Since the Thomson coefficient is defined for a single conductor, it is appropriate to introduce here the absolute Seebeck and Peltier coefficients for a single conductor. The differential Seebeck and Peltier coefficients between the two conductors are given by $(S_a - S_b)$ and $(\Pi_a - \Pi_b)$, respectively, where S_a and S_b are the absolute Seebeck coefficients for conductors A and B, and Π_a and Π_b denote the absolute Peltier coefficients for conductors A and B, respectively. Thus, the Kelvin relations given by Eqs. (7.14) and (7.15) can be rewritten for a single conductor as

$$\Pi = ST \tag{7.16}$$

and

$$\tau_s = T\frac{dS}{dT} \tag{7.17}$$

where Π, S, and τ_s denote the absolute Peltier, Seebeck, and Thomson coefficients in a single conductor or semiconductor specimen, respectively. Therefore, using the definitions for the thermoelectric coefficients described in this section, we can derive general expressions for the thermoelectric coefficients by using the Boltzmann transport equation to be discussed in Section 7.4.

7.2.4. Galvanomagnetic and Thermomagnetic Coefficients

When a magnetic field is applied to a semiconductor specimen in addition to an electric field and temperature gradient, the transport phenomena become much more complicated than those without a magnetic field. Fortunately, most of the important galvanomagnetic effects in a semiconductor associated with the applied magnetic fields are those cases when a magnetic field is applied in the direction perpendicular to the electric field or the temperature gradient. These are usually referred to as transverse galvanomagnetic effects, such as the Hall, Nernst, and transverse magnetoresistance effects. In this section, we shall discuss these transverse galvanomagnetic effects in an n-type semiconductor.

The Hall effect is the most well-known galvanomagnetic effect found in a semiconductor. As shown in Fig. 7.3a, when a magnetic field is applied along the z-direction and an electric field is applied in the x-direction, an electric field (known as the Hall field) will be developed in the y-direction of the specimen. The Hall coefficient R_H, under isothermal conditions, can be defined by

$$R_H = \left.\frac{\mathscr{E}_y}{J_x B_z}\right|_{J_y=0} \tag{7.18}$$

where $\mathscr{E}_y$ is the Hall field induced in the y-direction, J_x is the electric current density flow in the x-direction, and B_z is the applied magnetic field in the z-direction.

Figure 7.3a shows a schematic diagram of the Hall effect across a semiconductor specimen under isothermal conditions. It is seen that the electric current density in the direction of the Hall field (i.e., the y-direction) and the temperature gradient across the specimen are assumed equal to zero in this case. The polarity of the Hall voltage depends on the type of charge carriers (i.e., electrons or holes) in the specimen. This is due to the fact that electrons and holes in a semiconductor will experience an opposite Lorentz force if they are moving in the same direction along the specimen. Therefore, the polarity of the Hall voltage will be different for an n-type and a p-type specimen, and the Hall effect measurement is often used for determining the conduction types (i.e., n- or p-type) of a semiconductor in addition to determine the carrier concentration.

If we apply a temperature gradient in the x-direction and a magnetic field in the z-direction, then a transverse electric field will be developed in the y-direction of the specimen; this is illustrated in Fig. 7.3b. This effect is known as the Nernst effect. The Nernst coefficient, which is thermodynamically related to the Ettingshausen coefficient in the same way as the Seebeck coefficient is related to the Peltier coefficient, may be defined by

$$Q_n = \left.\frac{\mathscr{E}_y}{B_z(\partial T/\partial x)}\right|_{J_x=J_y=0} \tag{7.19}$$

where $\mathscr{E}_y$ is the Nernst field developed in the y-direction when a temperature gradient is applied in the x-direction and a magnetic field in the z-direction. Note that the electric current density along the x- and y-directions as well as the temperature gradient along the y-direction are set equal to zero.

It is shown in Fig. 7.3c that if we apply an electric field in the x-direction and a magnetic field in the z-direction, then a temperature gradient will be developed in the y-direction of the specimen. This is known as the Ettingshausen effect. It is this effect which forms the basis of thermomagnetic cooling as a counterpart to thermoelectric cooling by the Peltier effect discussed earlier. The Ettingshausen coefficient P_E can be defined by

$$P_E = \left.\frac{(\partial T/\partial y)}{J_x B_z}\right|_{J_y=\partial T/\partial x=0} \tag{7.20}$$

where $\partial T/\partial y$ is the temperature gradient developed in the y-direction of the specimen. Note that, in defining the Ettingshausen effect, the current density in the y-direction and the temperature gradient in the x-direction are assumed equal to zero.

The Ettingshausen coefficient P_E and the Nernst coefficient Q_n are related by the Bridgeman equation, which is given by

$$P_E K_n = Q_n T \tag{7.21}$$

where K_n is the electronic thermal conductivity defined by Eq. (7.9).

The Righi–Leduc effect refers to the creation of a transverse temperature gradient $\partial T/\partial y$ when a temperature gradient $\partial T/\partial x$ along the x-direction and a magnetic field B_z along the z-direction are applied simultaneously to a semiconductor specimen. The Righi–Leduc coefficient R_L can be expressed by

$$R_L = \frac{(\partial T/\partial y)}{(\partial T/\partial x)B_z} \tag{7.22}$$

Note that the current densities, J_x and J_y, along the x- and y-directions are assumed equal to zero ($J_x = J_y = 0$) in Eq. (7.22).

In order to derive general expressions for the various transport coefficients discussed in this section, we shall next introduce the Boltzmann transport equation for solving the nonequilibrium distribution function $f(E)$ in Eqs. (7.8) and (7.9).

7.3. BOLTZMANN TRANSPORT EQUATION

An analytical expression for the Boltzmann transport equation can be derived for an n-type semiconductor using the relaxation time approximation, which is justified as long as the collision process is elastic. As discussed earlier, elastic scattering refers to the case when the change in electron energy is small compared to the initial energy of electrons during a collision event. Under this assumption the collision term in the Boltzmann equation can be expressed in terms of a unique time constant, known as the relaxation time. The relaxation time can be a function of temperature and energy, depending on the types of scattering mechanism involved, and will be discussed further in Chapter 8.

The transport coefficients for an n-type semiconductor to be derived in this section include electrical conductivity, Hall coefficient, Seebeck coefficient, Nernst coefficient, and magnetoresistance.

According to Liouville's theorem, if $f(k, r, t)$ denotes the nonequilibrium distribution function of electrons at time t, in the volume element $d^3r\, d^3k$, and located at (r, k) in r- and k-space, then $f(k + \dot{k}\, dt, r + \dot{r}\, dt, t + dt)$ is the distribution function at time $(t + dt)$ within the same volume element. The difference between $f(k, r, t)$ and $f(k + \dot{k}\, dt, r + \dot{r}\, dt, t + dt)$ must be balanced out by the collision processes that occurred inside the crystal. Thus, the total rate of change of the distribution function with respect to time in the presence of a Lorentz force or a temperature gradient can be written as

$$\frac{df}{dt} = -\dot{k} \cdot \nabla_k f - \dot{r} \cdot \nabla_r f + \frac{\partial f}{\partial t} + \left.\frac{\partial f}{\partial t}\right|_c \tag{7.23}$$

where $\dot{k} = dk/dt$ and $\dot{r} = dr/dt = v$. The first two terms on the right-hand side of Eq. (7.23) are the external force terms, the third term is the time-dependent term which exists only for the transient case, and the fourth term is the internal collision term, which tends to retard the external force terms. Equation (7.23) is called the Boltzmann transport equation.

In this section, we shall consider only the steady-state case in which the transport coefficients are derived when the time-independent external forces are applied to a semiconductor specimen. In this case, the third term on the right-hand side of Eq. (7.23) is equal to zero. Thus, the steady-state Boltzmann equation given by Eq. (7.23) becomes

$$\dot{k} \cdot \nabla_k f + \dot{r} \cdot \nabla_r f = \left.\frac{\partial f}{\partial t}\right|_c \tag{7.24}$$

In general, the nonequilibrium distribution function $f(k, r)$ can be solved using Eq. (7.24). Thus, the transport coefficients for a semiconductor or a metal can be derived once $f(k, r)$ is found from the Boltzmann equation.

In order to obtain an analytical expression for $f(k, r)$ from Eq. (7.24), it is necessary to assume that the scattering of charge carriers in a semiconductor is elastic so that the relaxation time approximation can be applied to the Boltzmann equation. According to the classical model, electron velocity is accelerated by the applied electric field over a period of time inside the crystal, while its drift velocity drops to zero through the internal collision process. It is, however, more appropriate to consider the way in which the electron system is relaxed toward its equilibrium distribution once the external perturbation is removed. Therefore, if $f(k, r)$ represents the distribution function of electrons under the influence of an applied electric field, and $f_0(E)$ is the thermal equilibrium distribution function, then the collision term given by Eq. (7.24) can be expressed in terms of the relaxation time τ:

$$\left.\frac{\partial f}{\partial t}\right|_c = -\frac{f - f_0}{\tau} \tag{7.25}$$

This equation is the basis of the relaxation time approximation in which the collision term on the right-hand side of Eq. (7.24) is replaced by the ratio of the difference in the nonequilibrium and equilibrium distribution function and the relaxation time constant τ. The relaxation time constant is usually dependent on various scattering mechanisms taking place inside the semiconductor. If the external electric field is removed, then the nonequilibrium distribution function will decay exponentially to its equilibrium value with a time constant τ governed by the internal scattering processes.

For n-type semiconductors, the relaxation time τ depends on the energy of electrons according to the simple power law

$$\tau = aE^s \tag{7.26}$$

where s is a constant whose value depends on the types of scattering mechanism involved. Constant a may or may not be a function of temperature, depending on the types of scattering involved. For example, in a semiconductor in which the ionized impurity scattering is dominated, the value of s is $3/2$ and a is not a function of temperature, while for acoustic phonon scattering the value of s is $-1/2$ and a varies inversely with temperature. For neutral impurity scattering, τ is constant and independent of the electron energy.

7.4. DERIVATION OF TRANSPORT COEFFICIENTS

In this section, the transport coefficients for the single-carrier conduction case are derived for an n-type semiconductor in the presence of an applied electric field, a magnetic field, or a temperature gradient. The transport coefficients such as electrical conductivity, carrier mobility, Hall coefficient, Seebeck and Nernst coefficients, as well as the magnetoresistance, can be derived from Eq. (7.24) using the relaxation time approximation.

To derive the nonequilibrium distribution function from the Boltzmann equation, let us consider an n-type semiconductor. The Lorentz force acting on the electron due to the presence of an electric field and a magnetic field can be expressed by

$$\mathscr{F} = -q(\mathscr{E} + \mathbf{v} \times \mathbf{B}) = \hbar\dot{k} \tag{7.27}$$

or

$$\dot{k} = -\frac{q}{\hbar}(\mathscr{E} + \mathbf{v} \times \mathbf{B}) \tag{7.28}$$

Now on substituting $\dot{k}$ given by Eq. (7.28) into Eq. (7.24), the first term on the left-hand side of Eq. (7.24) becomes

$$-\dot{k} \cdot \nabla_k f = \left(\frac{q}{\hbar}\right)(\mathscr{E} + \mathbf{v} \times \mathbf{B}) \cdot \nabla_k f \tag{7.29}$$

The second term on the left-hand side of Eq. (7.24) is due to the temperature gradient or the concentration gradient present in a semiconductor. Using the relaxation time approximation, the collision term is given by Eq. (7.25). Now substitution of Eqs. (7.29) and (7.25) into Eq. (7.24) yields

$$\left(\frac{q}{\hbar}\right)(\mathscr{E} + \mathbf{v} \times \mathbf{B}) \cdot \nabla_k f - \mathbf{v} \cdot \nabla_r f = \frac{f - f_0}{\tau} \tag{7.30}$$

or

$$\left(\frac{q}{m^*}\right)(\mathscr{E} + \mathbf{v} \times \mathbf{B}) \cdot \nabla_v f - \mathbf{v} \cdot \nabla_r f = \frac{f - f_0}{\tau} \tag{7.31}$$

Equation (7.31) is obtained by substituting $\hbar k = m^* v$ into Eq. (7.30), which is a generalized steady-state Boltzmann equation. To obtain an analytical solution for $f(k, r)$ from Eq. (7.31), certain approximations must be used. Since the equilibrium distribution function f_0 depends only on energy and temperature, the nonequilibrium distribution fucntion f must contain terms which depend only on electron velocity and energy. Therefore, it is appropriate to write a generalized trial solution for Eq. (7.31) in terms of the equilibrium distribution function and a first order correction term which contains both the energy and velocity components. This is given by

$$f = f_0 - \mathbf{v} \cdot \mathbf{P}(E) \frac{\partial f_0}{\partial E} \tag{7.32}$$

where $\mathbf{P}(E)$ is an unknown vector quantity which depends only on electron energy. For the small perturbation case [i.e., $(f - f_0) \ll f_0$], each term in Eq. (7.31) can be expressed by

$$\mathbf{v} \cdot \nabla_r f \simeq \mathbf{v} \cdot \nabla_r f_0 = \mathbf{v} \cdot \nabla_r T \left[\frac{(E_f - E)}{T} \frac{\partial f_0}{\partial E}\right] \tag{7.33}$$

$$\mathscr{E} \cdot \nabla_v f \simeq \mathscr{E} \cdot \nabla_v f_0 = \mathscr{E} \cdot (\nabla_v E) \frac{\partial f_0}{\partial E} = \mathscr{E} \cdot (m^* v) \frac{\partial f_0}{\partial E} \tag{7.34}$$

$$(\mathbf{v} \times \mathbf{B}) \cdot \nabla_v f \simeq - \mathbf{v} \cdot [\mathbf{B} \times \mathbf{P}(E)] \frac{\partial f_0}{\partial E} \tag{7.35}$$

Now, substituting Eqs. (7.33), (7.34), and (7.35) into Eq. (7.31) yields

$$-q\tau \mathbf{v} \cdot \mathscr{E} + \left(\frac{q\tau}{m_n^*}\right)[\mathbf{B} \times \mathbf{P}(E)] \cdot \mathbf{v} + \tau \frac{(E_f - E)}{T} \mathbf{v} \cdot \nabla_r T - \mathbf{v} \cdot \mathbf{P}(E) = 0 \tag{7.36}$$

which is a generalized steady-state Boltzmann equation in the presence of the applied electric fields, magnetic fields, and temperature gradients. Note that Eq. (7.36) can be further simplified by factoring out the velocity component, and the result is

$$\mathbf{P}(E) - \left(\frac{q\tau}{m_n^*}\right)[\mathbf{B} \times \mathbf{P}(E)] = -q\tau\mathscr{E} + \tau\frac{(E_f - E)}{T}\nabla_r T \tag{7.37}$$

In order to obtain a solution for the unknown vector function $\mathbf{P}(E)$ in Eq. (7.37), it is assumed that the applied electric fields and temperature gradients lie in the x–y plane of the semiconductor specimen, and the magnetic field is along the z-axis. Under this assumption, the components for $\mathbf{P}(E)$ in Eq. (7.37) in the x- and y-directions of the specimen are given respectively by

$$P_x(E) + (q\tau/m_n^*)B_z P_y(E) = -q\tau\mathscr{E}_x + \tau\frac{(E_f - E)}{T}\frac{\partial T}{\partial x} \tag{7.38}$$

and

$$P_y(E) - (q\tau/m_n^*)B_z P_x(E) = -q\tau\mathscr{E}_y + \tau\frac{(E_f - E)}{T}\frac{\partial T}{\partial y} \tag{7.39}$$

The solution of Eqs. (7.38) and (7.39) for $P_x(E)$ and $P_y(E)$ yields

$$P_x(E) = \frac{(\beta - \delta\gamma)}{(1 + \delta^2)} \tag{7.40}$$

and

$$P_y(E) = \frac{(\gamma + \delta\beta)}{(1 + \delta^2)} \tag{7.41}$$

where

$$\delta = \frac{q\tau B_z}{m_n^*} = \omega\tau \tag{7.42}$$

$$\beta = \tau\left[-q\mathscr{E}_x + \frac{(E_f - E)}{T}\frac{\partial T}{\partial x}\right] \tag{7.43}$$

$$\gamma = \tau\left[-q\mathscr{E}_y + \frac{(E_f - E)}{T}\frac{\partial T}{\partial y}\right] \tag{7.44}$$

Thus, the transport coefficients for an n-type semiconductor defined in Section 7.2 can be obtained by solving Eqs. (7.32) through (7.44). This is discussed next.

7.4.1. Electrical Conductivity σ_n

Let us consider the cases in which the applied electric field and the current flow are in the x- and y-directions of the specimen. By substituting Eq. (7.32) for f in Eq. (7.8), the electric current density components due to electron conduction along the x- and y-directions defined by Eq. (7.8) are given by

$$J_x = -nqv_x = -\int_0^\infty qv_x f(E)g(E)\,dE$$
$$= q\int_0^\infty v_x^2 P_x(E)g(E)\frac{\partial f_0}{\partial E}\,dE \tag{7.45}$$

and

$$J_y = q\int_0^\infty v_y^2 P_y(E)g(E)\frac{\partial f_0}{\partial E}\,dE \tag{7.46}$$

where $P_x(E)$ and $P_y(E)$ can be obtained by solving Eqs. (7.40) through (7.44) with δ, dT/dx, and dT/dy equal to zero. This yields

$$P_x(E) = -q\tau\mathscr{E}_x \tag{7.47}$$

and

$$P_y(E) = -q\tau\mathscr{E}_y \tag{7.48}$$

In Eqs. (7.45) through (7.48), it is seen that the net current flow vanishes if $P(E)$ is equal to zero. To derive the electrical conductivity, it is assumed that the electron velocity is isotropic within the specimen, and hence the square of the velocity components along x-, y-, and z-directions can be expressed in terms of the kinetic energy of electrons by

$$v_x^2 = v_y^2 = v_z^2 = \frac{2E}{3m_n^*} \tag{7.49}$$

where E is the total kinetic energy of electrons. Equation (7.49) is obtained by noting that the electron kinetic energy is equal to $\frac{1}{2}m_n^* v^2$, where $v^2 = v_x^2 + v_y^2 + v_z^2$ and $v_x = v_y = v_z$. Now by substituting Eqs. (7.47) and (7.49) into Eq. (7.45) we obtain

$$\sigma_n = \frac{J_x}{\mathscr{E}_x} = \left(\frac{-2q^2}{3m_n^*}\right)\int_0^\infty \tau E g(E)\frac{\partial f_0}{\partial E}\,dE \tag{7.50}$$

For a nondegenerate semiconductor, the Fermi–Dirac distribution function given in Eq. (7.1) can be simplified to the classical Maxwell–Boltzmann distribution function, which reads

$$f_0 \simeq \exp[(E_f - E)/k_B T] \tag{7.51}$$

and

$$\frac{\partial f_0}{\partial E} = -\frac{f_0}{k_B T} \tag{7.52}$$

From Eqs. (7.7) and (7.52), the electrical conductivity can be rewritten as

$$\sigma_n = \left(\frac{2q^2}{3m_n^* k_B T}\right)\int_0^\infty \tau E g(E) f_0\, dE$$

$$= \left(\frac{2nq^2}{3m_n^* k_B T}\right)\frac{\int_0^\infty \tau E^{3/2} f_0\, dE}{\int_0^\infty E^{1/2} f_0\, dE}$$

$$= \frac{nq^2\langle\tau\rangle}{m_n^*} \tag{7.53}$$

where

$$\langle\tau\rangle = \frac{\int_0^\infty \tau E^{3/2}\exp(-E/k_B T)\, dE}{\int_0^\infty E^{3/2}\exp(-E/k_B T)\, dE} \tag{7.54}$$

is the mean relaxation time averaged over the electron energy and the M–B distribution function. Equation (7.53) is obtained by using the expression for the electron density, which is given by

$$n = \int_0^\infty f_0 g(E)\, dE = C\int_0^\infty E^{1/2}\exp(-E/k_B T)\, dE \tag{7.55}$$

and the average kinetic energy of electrons is given by

$$\langle E\rangle = \frac{\int_0^\infty E f_0 g(E)\, dE}{\int_0^\infty f_0 g(E)\, dE} = \frac{\int_0^\infty E^{3/2}\exp(-E/k_B T)\, dE}{\int_0^\infty E^{1/2}\exp(-E/k_B T)\, dE} = \frac{3k_B T}{2} \tag{7.56}$$

A generalized expression for the average τ^n is given by

$$\langle\tau^n\rangle = \frac{\int_0^\infty \tau^n E g(E)\, \partial f_0/\partial E\, dE}{\int_0^\infty E g(E)\, \partial f_0/\partial E\, dE} \tag{7.57}$$

where $n = 1, 2, 3, \ldots$, and

$$\langle\tau E^n\rangle = \frac{\int_0^\infty (\tau E^n) E g(E)\, \partial f_0/\partial E\, dE}{\int_0^\infty E g(E)\, \partial f_0/\partial E\, dE} \tag{7.58}$$

where $n = 1, 2, 3, \ldots$. It is noteworthy that the electrical conductivity for an n-type semiconductor given by Eq. (7.53) is similar to that of Eq. (7.6) for a metal. The only difference is that the free-electron mass in Eq. (7.6) is replaced by the electron effective mass, and the constant relaxation time τ is replaced by the mean relaxation time $\langle\tau\rangle$ given by Eq. (7.54).

For a nondegenerate semiconductor, using M–B statistics, Eqs. (7.57) and (7.58) can be expressed by

$$\langle\tau^n\rangle = \frac{\int_0^\infty \tau^n E^{3/2} e^{-E/k_B T}\, dE}{\int_0^\infty E^{3/2} e^{-E/k_B T}\, dE} \tag{7.59}$$

and

$$\langle \tau E^n \rangle = \frac{\int_0^\infty (\tau E^n) E^{3/2} \, e^{-E/k_B T} \, dE}{\int_0^\infty E^{3/2} \, e^{-E/k_B T} \, dE} \tag{7.60}$$

Now, by solving Eqs. (7.51) through (7.60) we obtain an expression for the electrical conductivity of a nondegenerate n-type semiconductor:

$$\sigma_n = \left(\frac{nq^2 \tau_0}{m_n^*} \right) (k_B T)^s \frac{\Gamma_{(5/2+s)}}{\Gamma_{(5/2)}} \tag{7.61}$$

where

$$\Gamma_n = \int_0^\infty x^{n-1} \, e^{-x} \, dx \tag{7.62}$$

is the Gamma function of order n, $\Gamma_n = (n-1)!$, and $\Gamma_{1/2} = \sqrt{\pi}$. Since the electrical conductivity is related to the electron mobility by Eq. (7.6), an expression for the electron mobility can be derived from Eqs. (7.6) and (7.61), and the result yields

$$\mu_n = \left(\frac{q \tau_0}{m_n^*} \right) (k_B T)^s \frac{\Gamma_{(5/2+s)}}{\Gamma_{(5/2)}} \tag{7.63}$$

It will be shown later that for acoustic phonon scattering, the electron mean free path varies as T^{-1} and $s = -1/2$. Consequently, for acoustic phonon scattering, the electron mobility is directly proportional to $T^{-3/2}$. For ionized impurity scattering, with $s = 3/2$ and τ_0 independent of temperature, the electron mobility μ_n varies with $T^{3/2}$. Details of the scattering mechanisms in a semiconductor will be discussed in Chapter 8.

The electrical conductivity given by Eq. (7.53) was derived based on the single-valley model with spherical constant-energy surface for the conduction band. This applies to most of the III–V compound semiconductors such as GaAs and InP. In this case the conductivity effective mass m_n^* is a scalar quantity and isotropic, and n denotes the total carrier concentration in the single spherical conduction band. For multivalley semiconductors such as silicon and germanium, since their crystal structure possesses cubic symmetry, the electrical conductivity remains isotropic. Thus, Eq. (7.53) is still applicable for the multivalley semiconductors provided that the average relaxation time is assumed isotropic and the conductivity effective mass m_n^* is replaced by

$$m_\sigma^* = \left[\frac{1}{3} \left(\frac{1}{m_l} + \frac{2}{m_t} \right) \right]^{-1} = \frac{3m_l}{2K+1} \tag{7.64}$$

where $K = m_l/m_t$ is the ratio of the longitudinal and transverse effective masses of an electron along the two main axes of the ellipsoidal energy surface near the conduction band edge. Values of m_l and m_t can be determined by the cyclotron resonance experiment at 4.2 K. Equation (7.64) is obtained by using the geometrical average of the electron mass along the two main axes of the ellipsoidal energy surface.

7.4.2. Hall Coefficient R_H

The general expression for the Hall coefficient of a nondegenerate n-type semiconductor with a single-valley spherical energy band can be derived from Eqs. (7.43) through (7.46) using the definition given by Eq. (7.18). Let us consider the small magnetic field case in which the δ^2 term in Eqs. (7.40) and (7.41) can be neglected (i.e., $\delta^2 \ll 1$). Thus, by substituting $P_x(E)$ given by Eq. (7.40) into Eq. (7.45) and setting $\partial T/\partial x$ equal to zero, we obtain

$$
\begin{aligned}
J_x &= q\int_0^\infty v_x^2(\beta - \gamma\delta)g(E)\frac{\partial f_0}{\partial E}\,dE \\
&= \left(\frac{2q^2}{3m_n^* k_B T}\right)\int_0^\infty \tau E\left[\mathscr{E}_x - \left(\frac{q\tau B_z}{m_n^*}\right)\mathscr{E}_y\right]g(E)f_0\,dE
\end{aligned}
\tag{7.65}
$$

Similarly, we can write Eq. (7.46) as

$$
\begin{aligned}
J_y &= q\int_0^\infty v_y^2(\gamma + \delta\beta)g(E)\frac{\partial f_0}{\partial E}\,dE \\
&= \left(\frac{2q^2}{3m_n^* k_B T}\right)\int_0^\infty \tau E\left[\mathscr{E}_y + \left(\frac{q\tau B_z}{m_n^*}\right)\mathscr{E}_x\right]g(E)f_0\,dE
\end{aligned}
\tag{7.66}
$$

The solution of Eqs. (7.65), (7.66), and (7.18) yields

$$
\begin{aligned}
R_{Hn} &= \frac{\mathscr{E}_y}{J_x B_z} = -\left(\frac{3k_B T}{2q}\right)\frac{\int_0^\infty \tau^2 E g(E)f_0\,dE}{[\int_0^\infty \tau E g(E) f_0\,dE]^2} \\
&= -\frac{1}{qn}\frac{\langle\tau^2\rangle}{\langle\tau\rangle^2}
\end{aligned}
\tag{7.67}
$$

where $\langle\tau\rangle$ and $\langle\tau^2\rangle$ can be determined from Eq. (7.57). The minus sign in Eq. (7.67) denotes the electron conduction for an n-type semiconductor. For p-type semiconductors, the Hall coefficient is given by

$$
R_{Hp} = \frac{1}{qp}\frac{\langle\tau^2\rangle}{\langle\tau\rangle^2}
\tag{7.68}
$$

which has a positive Hall coefficient due to hole conduction. From Eqs. (7.67) and (7.68), the Hall factor γ can be defined as

$$
\gamma_H = \frac{\langle\tau^2\rangle}{\langle\tau\rangle^2}
\tag{7.69}
$$

If the relaxation time is a function of energy (i.e., $\tau = aE^s$), then the Hall coefficient for a nondegenerate n-type semiconductor is given by

$$
\begin{aligned}
R_{Hn} &= -\frac{1}{nq}\frac{\Gamma_{(2s+5/2)}\Gamma_{(5/2)}}{[\Gamma_{(s+5/2)}]^2} \\
&= -\frac{\gamma_{Hn}}{nq}
\end{aligned}
\tag{7.70}
$$

where Γ_n is the Gamma function defined by Eq. (7.62), and γ_{Hn} is the Hall factor for an n-type semiconductor. In general, the Hall factor can be calculated if the scattering mechanisms in the semiconductor are known. The Hall factor for a p-type semiconductor is defined by the same formula as its n-type counterpart discussed above.

Another important physical parameter, which is usually referred to as Hall mobility, can be obtained from the product of electrical conductivity and Hall coefficient. Equations (7.61) and (7.70) enable this to be expressed by

$$\mu_{Hn} = R_{Hn}\sigma_n = \left(\frac{q\tau_0}{m_n^*}\right)(k_B T)^s \frac{\Gamma_{(2s+5/2)}}{\Gamma_{(s+5/2)}} \tag{7.71}$$

The Hall factor for a nondegenerate semiconductor, which is defined as the ratio of Hall mobility and conductivity mobility, can be obtained from Eq. (7.70)

$$\begin{aligned}\gamma_H &= \frac{\mu_{Hn}}{\mu_n} = \frac{\langle\tau^2\rangle}{\langle\tau\rangle^2} \\ &= \frac{\Gamma_{(2s+5/2)}\Gamma_{(5/2)}}{[\Gamma_{(s+5/2)}]^2}\end{aligned} \tag{7.72}$$

Values of γ_H may vary between 1 and 1.93 depending on the types of scattering mechanism involved. For example, for acoustic phonon scattering with $s = -1/2$ the Hall factor is equal to $3\pi/8$, and for ionized impurity scattering with $s = 3/2$ the Hall factor was found to be $315\pi/512$ ($\sim$1.93). Values of the Hall factor given above are obtained for nondegenerate semiconductors with a single-valley spherical energy band.

For multivalley semiconductors, if the conduction band edge has an ellipsoidal energy surface, then the expression for the Hall factor should be modified to include the mass anisotropic effect. In this case, a "Hall mass factor" a_0 is multiplied by the Hall factor given by Eq. (7.69) for the single-valley model to account for the mass anisotropic effect. Thus, the Hall factor for the case of a multiple conduction valley semiconductor can be expressed by

$$\gamma_H = \frac{\mu_{Hn}}{\mu_n} = \frac{\langle\tau^2\rangle a_0}{\langle\tau\rangle^2} \tag{7.73}$$

where a_0 is known as the "Hall mass factor" which can be described by

$$a_0 = \left(\frac{m_\sigma^*}{m_H^*}\right)^2 = \frac{3K(K+2)}{(2K+1)^2} \tag{7.74}$$

and

$$m_H^* = m_l\sqrt{3/[K(K+2)]} \tag{7.75}$$

m_H^* is the "Hall effective mass" and m_σ^* is the conductivity effective mass defined by Eq. (7.64); K is the ratio of the longitudinal and transverse effective masses of electrons along the two main axes of the constant ellipsoidal energy surface of the conduction band. For germanium $K \simeq 20$ and $a_0 = 0.785$, while for silicon $K = 5.2$ and $a_0 = 0.864$. Thus, the low-field Hall coefficient for n-type silicon and germanium should be multiplied by a Hall mass factor given by Eq. (7.75).

The Hall factor for a semiconductor is equal to unity when neutral impurity scattering is the dominant scattering mechanism. This usually occurs either at extremely low temperatures or for a heavily doped semiconductor. This is due to the fact that, in both cases, the relaxation time is constant and independent of the electron energy. For p-type silicon, the Hall factor may be smaller than unity due to the warped and nonparabolic valence band structures. In general, it is usually difficult to determine an accurate Hall factor from Hall effect measurements. In fact, it is common practice to assume a Hall factor equal to unity so that the carrier density can be readily determined from Hall effect measurements.

7.4.3. *Seebeck Coefficient S_n*

The Seebeck coefficient for a single-valley n-type semiconductor with spherical constant-energy surface can be derived from Eqs. (7.40) and (7.45) by setting both δ and J_x equal to zero, and the result yields

$$S_n = \frac{\mathscr{E}_x}{(\partial T/\partial x)} = \left(-\frac{1}{qT}\right)\left[\frac{\int_0^\infty \tau E^2 g(E)\,\partial f_0/\partial E\,dE}{\int_0^\infty \tau E g(E)\,\partial f_0/\partial E\,dE} - E_f\right]$$

$$= -\left(\frac{1}{qT}\right)\left[\frac{\langle \tau E\rangle}{\langle \tau \rangle} - E_f\right] \tag{7.76}$$

For a nondegenerate semiconductor, the Seebeck coefficient given by Eq. (7.76) can be further simplified by using Eqs. (7.59) and (7.60) and $\tau = aE^s$:

$$S_n = -\left(\frac{1}{qT}\right)[(5/2 + s)k_B T - E_f] \tag{7.77}$$

We note that value of the Seebeck coefficient given by Eq. (7.77) can be determined if the types of scattering mechanism and the position of the Fermi level are known. For example, if acoustic phonon scattering is dominant (i.e., $s = -1/2$), then the Seebeck coefficient is given by

$$S_n = -\left(\frac{1}{qT}\right)(2k_B T - E_f) \tag{7.78}$$

On the other hand, if the ionized impurity scattering (i.e., $s = 3/2$) is dominant, then the Seebeck coefficient becomes

$$S_n = -\left(\frac{1}{qT}\right)(4k_B T - E_f) \tag{7.79}$$

From Eqs. (7.78) and (7.79), it is seen that the Fermi energy for a nondegenerate semiconductor can be determined from the measured Seebeck coefficient provided that the dominant scattering mechanism is known. The negative sign for the Seebeck coefficient shown above is due to the fact that conduction is an n-type semiconductor is carried out by the negative charge carriers (i.e., electrons).

For p-type semiconductors, the sign for the Seebeck coefficient given by Eqs. (7.76) through (7.79) is positive since conduction is carried out by the positive charge carriers (i.e., holes). Expressions for the Seebeck coefficient derived in this section are applicable only for single-valley semiconductors. However, the results can also be applied to multivalley semiconductors such as silicon and germanium provided that the density-of-states effective mass for electrons is modified to account for the multivalley conduction bands.

7.4.4. *Nernst Coefficient Q_n*

We shall next consider the Nernst effect for a nondegenerate n-type semiconductor with a single-valley spherical energy surface in the conduction band. The Nernst coefficient Q_n, defined by Eq. (7.19), can be derived from Eqs. (7.40) through (7.48). Let us consider the low magnetic fields case (i.e., $\mu B \ll 10^8$). Under isothermal conditions, the Nernst coefficient can be derived by setting $J_x = J_y = 0$ and $\partial T/\partial y = 0$. From Eqs. (7.45) and (7.46), we obtain

$$J_x = 0 = q \int_0^\infty v_x^2(\beta - \delta\gamma)g(E)\frac{\partial f_0}{\partial E}\,dE$$

$$= \left(\frac{2q^2}{3m_n^* k_B T}\right)\int_0^\infty \tau E\left[\mathscr{E}_x - \frac{(E_f - E)}{T}\left(\frac{\partial T}{\partial x}\right) - \omega\tau\mathscr{E}_y\right]g(E)f_0\,dE \tag{7.80}$$

and

$$J_y = 0 = \left(\frac{2q^2}{3m_n^* k_B T}\right)\int_0^\infty \tau E\left[\mathscr{E}_y - \frac{(E_f - E)}{T}\,\omega\tau\left(\frac{\partial T}{\partial x}\right) + \omega\tau\mathscr{E}_x\right]g(E)f_0\,dE \tag{7.81}$$

The solution of Eqs. (7.80), (7.81), and (7.19) for Q_n yields

$$Q_n = \frac{\mathscr{E}_y}{B_z(\partial T/\partial x)}$$

$$= \left\{\left(\frac{1}{m^* T}\right)\frac{\int_0^\infty \tau^2 E^2 g(E)f_0\,dE}{\int_0^\infty \tau E g(E)f_0\,dE} - \frac{\int_0^\infty \tau^2 E g(E)f_0\,dE \int_0^\infty \tau E^2 g(E)f_0\,dE}{[\int_0^\infty \tau E g(E)f_0\,dE]^2}\right\}$$

$$= \left(\frac{\mu_n}{qT}\right)\left[\frac{\langle\tau^2 E\rangle}{\langle\tau\rangle^2} - \frac{\langle\tau^2\rangle\langle\tau E\rangle}{\langle\tau\rangle^3}\right] \tag{7.82}$$

where $\mu_n = q\langle\tau\rangle/m_n^*$ is the electron conductivity mobility. By substituting τ, given by Eq. (7.26), into Eq. (7.82) and considering only the nondegenerate case, we obtain the Nernst coefficient as

$$Q_n = \left(\frac{k_B}{q}\right)\mu_n s\,\frac{\Gamma_{(2s+5/2)}\Gamma_{(5/2)}}{[\Gamma_{(s+5/2)}]^2} \tag{7.83}$$

For cases of acoustic phonon scattering ($s = -1/2$) and ionized impurity scattering ($s = 3/2$), Eq. (7.82) reduces to

$$Q_n = -\left(\frac{3\pi}{16}\right)\left(\frac{k_B}{q}\right)\mu_n \qquad \text{for } s = -1/2$$

$$= \left(\frac{945\pi}{1024}\right)\left(\frac{k_B}{q}\right)\mu_n \qquad \text{for } s = 3/2 \tag{7.84}$$

It is of interest that, in contrast to both Hall and Seebeck coefficients, the sign of the Nernst coefficient given by Eq. (7.82) depends only on the types of scattering mechanism rather than on the types of charge carrier. For example, Eq. (7.83) shows that the Nernst coefficient is negative when acoustic phonon scattering (i.e., $s = -1/2$) is dominant, and becomes positive when ionized impurity scattering (i.e., $s = 3/2$) is dominant.

7.4.5. Transverse Magnetoresistance

The transverse magnetoresistance effect corresponds to the change in electrical resistivity when a transverse magnetic field is applied across a semiconductor specimen. For example, if an electric field in the x-direction and a magnetic field in the z-direction are applied simultaneously to a semiconductor specimen, then an increase in resistance with applied magnetic field may be observed along the direction of current flow. The basic equations which govern the magnetoresistance effect in a semiconductor are given by Eqs. (7.40) through (7.46). To derive an expression for the transverse magnetoresistance in a single-valley semiconductor with spherical energy band, let us assume that the sample is subject to isothermal conditions with $\partial T/\partial x = \partial T/\partial y = 0$ and $J_y = 0$. Equations (7.40) and (7.41) then give

$$P_x(E) = \frac{(\beta - \delta\gamma)}{(1 + \delta^2)} = \frac{(\beta - \omega\tau\gamma)}{(1 + \omega^2\tau^2)} \tag{7.85}$$

and

$$P_y(E) = \frac{(\gamma + \delta\beta)}{(1 + \delta^2)} = \frac{(\gamma + \omega\tau\beta)}{(1 + \omega^2\tau^2)} \tag{7.86}$$

where

$$\gamma = -q\tau\mathscr{E}_y, \qquad \beta = -q\tau\mathscr{E}_x, \qquad \delta = \omega\tau = \frac{qB_z\tau}{m_n^*} \tag{7.87}$$

Substitution of Eqs. (7.85) and (7.86) into Eqs. (7.45) and (7.46) yields

$$J_x = \left(\frac{2q^2}{3m_n^* k_B T}\right) \int_0^\infty \left[\frac{(\tau\mathscr{E}_x - \omega\tau^2\mathscr{E}_y)}{(1 + \omega^2\tau^2)}\right] Eg(E) f_0 \, dE \tag{7.88}$$

and

$$J_y = \left(\frac{2q^2}{3m_n^* k_B T}\right) \int_0^\infty \left[\frac{(\tau\mathscr{E}_y + \omega\tau^2\mathscr{E}_x)}{(1 + \omega^2\tau^2)}\right] Eg(E) f_0 \, dE \tag{7.89}$$

Now, by using the single-valley model and solving Eqs. (7.88) and (7.89), we can obtain the transverse magnetoresistance coefficients for a nondegenerate n-type semiconductor for two limiting cases, namely, low and high magnetic fields.

Low Magnetic Field (i.e., $\omega\tau \ll 1$). In this case, the $\omega^2\tau^2$ term shown in the denominator of Eqs. (7.88) and (7.89) is retained. From Eq. (7.89), $\mathscr{E}_y$ may be expressed in terms of $\mathscr{E}_x$ by setting $J_y = 0$, which yields

$$\mathscr{E}_y = -\mathscr{E}_x \left[\frac{\int_0^\infty \omega\tau^2 Eg(E) f_0 \, dE}{\int_0^\infty \tau Eg(E) f_0 \, dE}\right] \tag{7.90}$$

Now, on substituting Eq. (7.90) for $\mathscr{E}_y$ in Eq. (7.88) and using the binomial expansion $(1 + \omega^2\tau^2)^{-1} \simeq (1 - \omega^2\tau^2)$ in Eq. (7.88) for $\omega\tau \ll 1$, we obtain the electrical conductivity for a low magnetic field in the form

$$\begin{aligned} \sigma_n = \frac{J_x}{\mathscr{E}_x} &= \left(\frac{2q^2}{3m_n^* k_B T}\right) \left[\int_0^\infty \tau E(1 - \omega^2\tau^2) g(E) f_0 \, dE \right. \\ &\qquad \left. + \frac{\omega^2 \left(\int_0^\infty \tau^2 Eg(E) f_0 \, dE\right)^2}{\int_0^\infty \tau Eg(E) f_0 \, dE}\right] \\ &= \sigma_0 \left[1 - \omega^2 \left(\frac{\langle\tau^3\rangle}{\langle\tau\rangle} - \frac{\langle\tau^2\rangle^2}{\langle\tau\rangle^2}\right)\right] \end{aligned} \tag{7.91}$$

where $\sigma_0 = nq^2\langle\tau\rangle/m_n^*$ is the electrical conductivity at zero magnetic field.

For a low magnetic field, the electrical conductivity is given by

$$\sigma_n = \frac{1}{\rho_n} = \sigma_0 - \Delta\sigma = \sigma_0\left(1 - \frac{\Delta\sigma}{\sigma_0}\right) \tag{7.92}$$

where ρ_n is the resistivity, which can be expressed by

$$\rho_n = \rho_0\left(1 + \frac{\Delta\rho}{\rho_0}\right) \simeq \sigma_0^{-1}\left(1 + \frac{\Delta\sigma}{\sigma_0}\right) \tag{7.93}$$

The solution of Eqs. (7.91) through (7.93) leads to

$$\begin{aligned}
\frac{\Delta\rho}{\rho_0} = \frac{\Delta\sigma}{\sigma_0} &= \omega^2\left[\frac{\langle\tau^3\rangle}{\langle\tau\rangle} - \frac{\langle\tau^2\rangle^2}{\langle\tau\rangle^2}\right] \\
&= (\sigma_0^2 B_z^2)\left[\left(\frac{1}{nq}\right)\frac{\langle\tau^2\rangle}{\langle\tau\rangle^2}\right]^2\left[\frac{\langle\tau^3\rangle\langle\tau\rangle}{\langle\tau^2\rangle^2} - 1\right] \\
&= R_H^2\sigma_0^2 B_z^2\left[\frac{\langle\tau^3\rangle\langle\tau\rangle}{\langle\tau^2\rangle^2} - 1\right] \\
&= \mu_H^2 B_z^2\left[\frac{\langle\tau^3\rangle\langle\tau\rangle}{\langle\tau^2\rangle^2} - 1\right]
\end{aligned} \tag{7.94}$$

The magnetoresistance coefficient can be deduced from Eq. (7.94), and the result yields

$$\zeta = \left(\frac{\Delta\rho}{\rho_0 B_z^2}\right)\left(\frac{1}{\mu_H^2}\right) = \frac{\langle\tau^3\rangle\langle\tau\rangle}{\langle\tau^2\rangle^2} - 1 \tag{7.95}$$

For the nondegenerate case, using $\tau = \tau_0 E^s$, Eq. (7.95) becomes

$$\zeta = \frac{\Gamma_{(3s+5/2)}\Gamma_{(s+5/2)}}{[\Gamma_{(2s+5/2)}]^2} - 1 \tag{7.96}$$

From the latter equation, one finds that ζ is equal to 0.273 for acoustic phonon scattering (i.e., $s = -1/2$) and 0.57 for ionized impurity scattering (i.e., $s = 3/2$).

For a low magnetic field, the above results show that the transverse magnetoresistance in a nondegenerate semiconductor is directly proportional to the square of the magnetic fields (i.e., B_z^2). The magnetoresistance data obtained for semiconductors with a single spherical energy band were found to be in good agreement with the theoretical prediction at low magnetic fields. It should be noted that for semiconductors with spherical energy bands, the longitudinal magnetoresistance (i.e., $J_n // B$) should vanish under the weak magnetic field condition.

High Magnetic Field (i.e., $\omega\tau \ll 1$). At high magnetic fields, Eqs. (7.88) and (7.89) become

$$J_x = \left(\frac{2q^2}{3m_n^* k_B T}\right)\int_0^\infty E(-\mathscr{E}_y/\omega)g(E)f_0\, dE \tag{7.97}$$

and

$$J_y = \left(\frac{nq^2}{m_n^* \omega^2}\right) [\mathscr{E}_y \langle \tau^{-1} \rangle + \omega \mathscr{E}_x] \tag{7.98}$$

Thus, for $J_y = 0$, we obtain

$$\mathscr{E}_y = -\frac{\omega \mathscr{E}_x}{\langle \tau^{-1} \rangle} \tag{7.99}$$

Now, substituting Eq. (7.99) into Eq. (7.97) yields

$$\sigma_\infty = \frac{J_x}{\mathscr{E}_x} = \left(\frac{nq^2}{m_n^*}\right)\left(\frac{1}{\langle \tau^{-1} \rangle}\right) = \frac{\sigma_0}{\langle \tau \rangle \langle \tau^{-1} \rangle} \tag{7.100}$$

or

$$\frac{\sigma_0}{\sigma_\infty} = \frac{\rho_\infty}{\rho_0} = \langle \tau \rangle \langle \tau^{-1} \rangle \tag{7.101}$$

which shows that the ratio of electrical conductivity at high and zero magnetic fields is equal to a constant whose value depends only on the types of scattering mechanism involved. In general, the transverse magnetoresistance approaches a constant value at very high magnetic fields. For a nondegenerate semiconductor, Eq. (7.101) becomes

$$\frac{\sigma_0}{\sigma_\infty} = \frac{\rho_\infty}{\rho_0} = \frac{\Gamma_{(s+5/2)}\Gamma_{(5/2-s)}}{[\Gamma_{(5/2)}]^2} \tag{7.102}$$

from which we can show that $\rho_\infty/\rho_0 = 1.17$ for $s = -1/2$ and $\rho_\infty/\rho_0 = 3.51$ for $s = 3/2$.

The magnetoresistance coefficients derived above are valid only for the single-valley conduction band with spherical energy surface. For multivalley semiconductors such as silicon and germanium, the situation is much more complicated than that presented above. The effective mass anisotropy (K value) strongly affects the magnetoresistance value. For example, if an electric field and a magnetic field are applied parallel to the x-direction of an n-type germanium in which the conduction valleys are located on the (111) and equivalent axes, the longitudinal magnetoresistance along the (100) direction is given by

$$\frac{\Delta\rho}{\rho B^2} = \frac{q^2}{m_l^2} \frac{2K(K-1)^2}{3(2K+1)} \frac{\langle \tau^3 \rangle}{\langle \tau \rangle} = \mu_{Mn}^2 \frac{2K(K-1)^2}{3(2K+1)} \tag{7.103}$$

where $\mu_{Mn}^2 = (q^2/m_l^2)(\langle \tau^3 \rangle / \langle \tau \rangle)$ is the square of a new mobility associated with the magnetoresistance effect in a semiconductor. For n-type germanium with $K = 20$, the K-dependent factor in Eq. (7.103) has a value equal to 118 while it reduces to zero for $K = 1$. Thus, the magnetoresistance is strongly affected by the effective mass anisotropy for an n-type germanium. For an n-type silicon, if the electric current and magnetic field are applied simultaneously along the (100) direction, the longitudinal magnetoresistance should vanish since σ_{xx} is independent of the magnetic field B.

It is noteworthy that the high magnetic field case discussed above is for the classical limit in that the magnetoresistance coefficient does not exhibit any quantum oscillatory behavior. In the quantum limit, the magnetoresistance exhibits oscillatory behavior at very high magnetic fields. The oscillatory behavior of the magnetoresistance observed in metals is known as the Shubnikov de Haas von Alphan effect. This effect is usually observed in the degenerate electron gas of a metal at very low temperatures (e.g., at 4.2 K) and under very high magnetic fields (e.g., several hundred kilogauss). The de Haas von Alphan effect has been widely used in constructing the energy contour of the Fermi surface in a metal.

In short, the transport coefficients derived in this section are for nondegenerate n-type semiconductors. For p-type semiconductors in which holes are the majority carriers, the expressions for various transport coefficients derived above are still valid provided that the plus sign for the electronic charge q is used and the electron effective mass is replaced by the hole effective mass.

The transport coefficients derived in the preceding section are valid only for nondegenerate n-type semiconductors in which classical Maxwell–Boltzmann statistics were used to obtain the average relaxation time contained in the expressions for the transport coefficients. However, for degenerate semiconductors, Fermi–Dirac statistics should be employed in the derivation of these transport coefficients. As for n-type degenerate semiconductors, the transport coefficients can be derived in a similar way to those discussed in the preceding section provided that Fermi statistics are used in these derivations. This will be left as an exercise for the reader in the problems at the end of this chapter.

7.5. TRANSPORT COEFFICIENTS FOR THE MIXED CONDUCTION CASE

In an intrinsic semiconductor or an extrinsic semiconductor with heavy compensation, both electrons and holes may participate in the conduction process and hence mixed conduction prevails in carrier transport. In this case, the conduction is a two-carrier process contributed by both electrons and holes. Derivation of transport coefficients for the mixed conduction case is much more complicated than that of the single-carrier conduction case discussed in the previous sections. This is discussed next.

7.5.1. *Electrical Conductivity* σ

The electrical conductivity for two-carrier conduction is equal to the sum of single-carrier conductivity due to electrons and holes. Hence

$$\sigma = \sigma_n + \sigma_p = q(n\mu_n + p\mu_p) \tag{7.104}$$

where σ_n and σ_p denote the electron and hole conductivities; μ_n and μ_p are the electron and hole mobilities, while n and p represent the electron and hole densities, respectively. Therefore, the electrical conductivity for the mixed conduction case is obtained by substituting the individual expression for σ_n and σ_p derived in the single-carrier conduction case into Eq. (7.104).

7.5.2. *Hall Coefficient* R_H

The Hall coefficient for the mixed conduction case can be derived as follows. If an electric field is applied in the x-direction and a magnetic field in the z-direction, then a Hall voltage is developed in the y-direction of the specimen. The total current flow due to both electrons and holes in the x-direction is given by

$$J_x = J_{nx} + J_{px} \tag{7.105}$$

where

$$J_x = \sigma\mathscr{E}_x, \qquad J_{nx} = \sigma_n\mathscr{E}_x, \qquad J_{px} = \sigma_p\mathscr{E}_x \tag{7.106}$$

The Hall field developed in the y-direction can be expressed by

$$\mathscr{E}_y = R_H J_x B_z, \qquad \mathscr{E}_{ny} = R_{Hn} J_{nx} B_z, \qquad \mathscr{E}_{py} = R_{Hp} J_{px} B_z \tag{7.107}$$

and

$$\sigma \mathscr{E}_y = \sigma_n \mathscr{E}_{ny} + \sigma_p \mathscr{E}_{py} \tag{7.108}$$

Substitution of Eq. (7.107) into (7.108) now yields

$$\sigma R_H J_x B_z = \sigma_n R_{Hn} J_{nx} B_z + \sigma_p R_{Hp} J_{px} B_z \tag{7.109}$$

By introducing Eq. (7.106) into (7.109), we obtain

$$\sigma^2 R_H \mathscr{E}_x B_z = \sigma_n^2 R_{Hn} \mathscr{E}_{nx} B_z + \sigma_p^2 R_{Hp} \mathscr{E}_{px} B_z \tag{7.110}$$

Thus the Hall coefficient due to both electron and hole conduction can be obtained from Eq. (7.110), which reads

$$R_H = \frac{(R_{Hn}\sigma_n^2 + R_{Hp}\sigma_p^2)}{(\sigma_n + \sigma_p)^2} \tag{7.111}$$

where R_{Hn} and R_{Hp} are the Hall coefficients given by Eqs. (7.68) and (7.70) respectively for n- and p-type conduction; σ_n and σ_p denote the corresponding electron and hole conductivities.

It is interesting to note that the Hall coefficient for the mixed conduction case given above is not equal to the simple summation of the Hall coefficients due to electrons and holes. In fact, Eq. (7.111) shows that the Hall coefficient may become zero if the contribution of the Hall coefficient from electrons (i.e., the first term, negative) is equal to that from holes (i.e., the second term, positive) in the numerator. This phenomenon may be observed in the Hall coefficient versus temperature plot for an intrinsic semiconductor in which changes in conductivity type may occur from n-type to p-type while the Hall coefficient changes from negative to positive at a certain elevated temperature.

7.5.3. Seebeck Coefficient S

The Seebeck coefficient for the mixed conduction case can be derived in a similar way to the Hall coefficient. First we consider the electric current density contributed by both electrons and holes in the presence of an electric field and a temperature gradient in the x-direction. The electric current density due to electrons is given by

$$J_{nx} = \sigma_n\left(\mathscr{E}_x - S_n \frac{\partial T}{\partial x}\right) \tag{7.112}$$

and the electric current density due to holes is

$$J_{px} = \sigma_p\left(\mathscr{E}_x - S_p \frac{\partial T}{\partial x}\right) \tag{7.113}$$

If the temperature gradient is zero, then the total electric current density is

$$J_x = J_{nx} + J_{px} = (\sigma_n + \sigma_p)\mathscr{E}_x \tag{7.114}$$

Note that the electrical conductivity is simply equal to the sum of the conductivities due to electrons and holes. Thus, in the mixed conduction case the thermoelectric effect is obtained from Eqs. (7.112) and (7.113) by setting the total current density equal to zero, which yields

$$(\sigma_n + \sigma_p)\mathscr{E}_x = (S_n\sigma_n + S_p\sigma_p)\frac{\partial T}{\partial x} \tag{7.115}$$

The solution of Eq. (7.115) yields the total Seebeck coefficient for the mixed conduction case which can be expressed in terms of the Seebeck coefficient and the electrical conductivity of electrons and holes obtained for the single-carrier conduction, namely,

$$S = \frac{(S_n\sigma_n + S_p\sigma_p)}{(\sigma_n + \sigma_p)} \tag{7.116}$$

where S_n denotes the Seebeck coefficient for n-type conduction given by Eq. (7.76). The Seebeck coefficient S_p for p-type conduction is similar to Eq. (7.76), except that n is replaced by p, and the plus sign is used instead.

7.5.4. *Nernst Coefficient Q*

When both electrons and holes are present in a semiconductor, the total Nernst coefficient Q is not equal to the simple sum of Q_n and Q_p because in the mixed conduction case an additional temperature gradient is developed in the specimen. This temperature gradient causes a Seebeck voltage to appear across the specimen. The Seebeck voltage in turn creates an electric field, which results in a flow of charge carriers. This current flow can induce a Hall voltage when a transverse magnetic field is applied to the specimen. Therefore, it can be shown that the total Nernst coefficient for the mixed conduction case can be expressed in terms of σ_n, σ_p, S_n, S_p, R_{Hn}, R_{Hp}, Q_n, and Q_p for single-carrier conduction:

$$Q = \frac{(Q_n\sigma_n + Q_p\sigma_p)(\sigma_n + \sigma_p) + (S_n - S_p)\sigma_n\sigma_p(R_{Hn}\sigma_n - R_{Hp}\sigma_p)}{(\sigma_n + \sigma_p)^2} \tag{7.117}$$

It is seen that the total Nernst coefficient for a given material depends on both the location of the Fermi level and the types of scattering mechanism involved. For an intrinsic semiconductor, the Nernst coefficient differs considerably from that of Q_n and Q_p for single-carrier conduction.

The expressions for the transport coefficients given by Eqs. (7.104) through (7.117) are for the mixed conduction case. The transport coefficients derived in the previous section for n- and p-type single-carrier conduction can be incorporated in the transport coefficient formula derived in this section for the mixed conduction case.

7.6. TRANSPORT COEFFICIENTS FOR SOME SEMICONDUCTORS

Measurements of the transport coefficients for elemental and compound semiconductors have been widely published in the literature. Some of these results are discussed in this section. Since silicon, germanium, and GaAs have been studied most extensively in the past, we shall

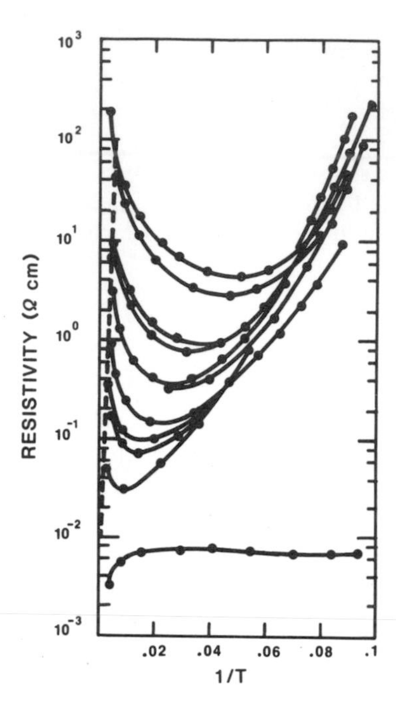

FIGURE 7.4. Resistivity versus inverse absolute temperature for As-doped germanium samples of different donor concentrations (from low 10^{14} for the top curve to 10^{18} cm^{-3} for the bottom curve). After Debye and Conwell,[1] by permission.

present here resistivity, Hall coefficient, Hall mobility, Seebeck coefficient, and magnetoresistance data for these materials.

Figure 7.4 shows resistivity as a function of reciprocal temperature for several n-type As-doped germanium specimens of different doping concentrations.[1] Figure 7.5 shows the corresponding Hall coefficient curves.[1] The results indicate that in the high-temperature regime, values of resistivity for all samples are almost identical and independent of the dopant

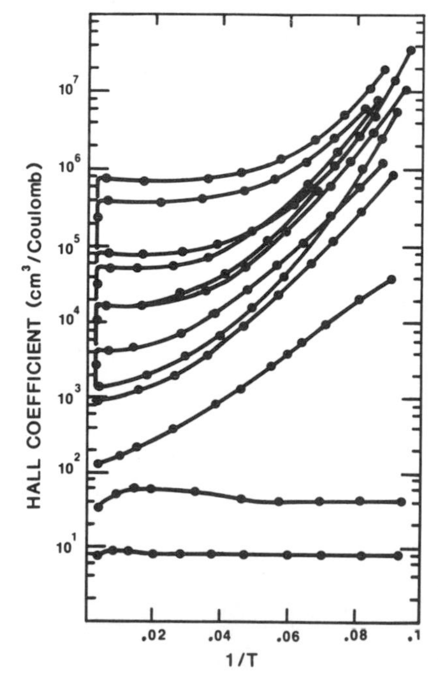

FIGURE 7.5. Hall coefficient versus inverse absolute temperature for As-doped germanium samples of different donor concentrations (from low 10^{14} for the top curve to 10^{18} cm^{-3} for the bottom curve). After Debye and Conwell,[1] by permission.

densities. This is the intrinsic regime, and the carrier concentration is predicted by Eq. (5.20) for the intrinsic semiconductor case. In this regime, the densities of electrons and holes are equal and increase exponentially with temperature with a slope equal to $-E_g/2k_B$. As the temperature decreases, the material becomes an extrinsic semiconductor. In this temperature regime (i.e., the exhaustion regime), all impurity atoms are ionized, and the carrier density is equal to the net dopant density. As the temperature further decreases, carrier freeze-out occurs in the material. This is the so-called deionization regime. In this regime, the Hall coefficient increases again, and from the slope of the Hall coefficient versus temperature curve one can determine the shallow impurity activation energy. At very high doping concentrations, the carrier density becomes nearly constant over the entire temperature range, as is evident by the flatness of the resistivity and Hall coefficient curves shown in Figs. 7.4 and 7.5. The carrier concentration as a function of temperatures can be deduced from Fig. 7.5, and the Hall mobilities are calculated from the product of Hall coefficient and electrical resistivity curves shown in Figs. 7.4 and 7.5.

Measurements of transport coefficient for both n- and p-type silicon have been widely reported in the literature, and some of these results are illustrated in Figs. 7.6 through 7.11.[2-5] Figure 7.6 shows resistivity as a function of temperature for n-type silicon doped with different phosphorus impurity densities (N_D varying from 1.2×10^{14} to 2.5×10^{18} cm^{-3}).[2] Figure 7.7 shows resistivity versus temperature for boron-doped silicon with boron impurity densities varying from 4.5×10^{14} to 3.2×10^{18} cm^{-3}.[3] Excellent agreement between theoretical calculations (solid lines) and experimental data (solid dots) was obtained in both cases. Figure 7.8a and b shows resistivity versus dopant density for both n- and p-type silicon at 300 K.[2,3] The solid line shown in Fig. 7.8a corresponds to theoretical calculations reported by Li, while the dashed line corresponds to experimental data compiled by Irvine.[4] In Fig. 7.8b, the solid line corresponds to theoretical calculations given by Li, the dashed line to Irvinc, while the broken line corresponds to experimental data reported by Wagner.[5]

Resistivity and Hall effect measurements are frequently used to determine the carrier density and mobility in a semiconductor. The mobility determined from the product of electrical conductivity and Hall coefficient is known as the Hall mobility, which is a majority carrier mobility. The drift mobility determined by the Haynes–Shockley experiment is usually referred to as the minority carrier mobility. These two quantities may or may not be equal,

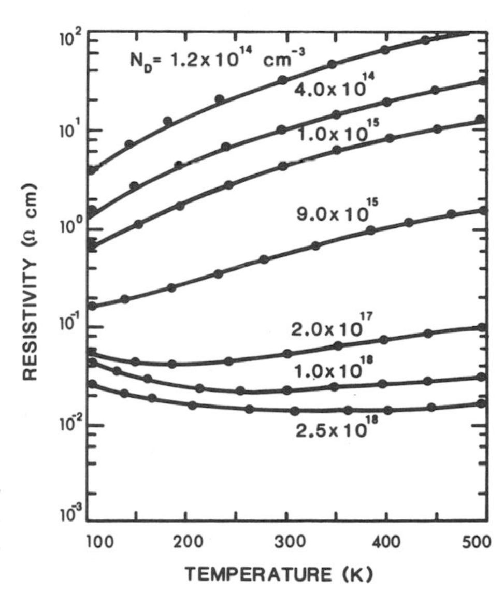

FIGURE 7.6. Resistivity versus temperature for phosphorus-doped silicon. Solid lines denote theoretical calculations and solid dots denote experimental data. After Li,[2] by permission.

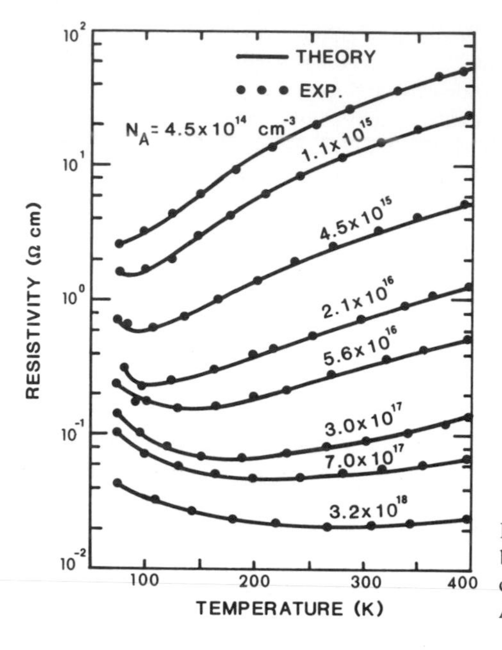

FIGURE 7.7. Resistivity versus temperature for boron-doped silicon. Solid lines denote theoretical calculations and solid dots denote experimental data. After Li,[3] by permission.

depending on the scattering processes and the dopant density of the semiconductor. The ratio of the Hall mobility and conductivity mobility is equal to the Hall factor. Values of the Hall factor may vary between 1 and 1.93, depending on the types of scattering mechanism involved in a semiconductor.

Figures 7.9 and 7.10 show Hall mobility as a function of temperature for both n-type and p-type silicon samples with dopant impurity density as parameter, respectively.[6] The empirical formula for the temperature dependence of Hall mobility for both n- and p-type

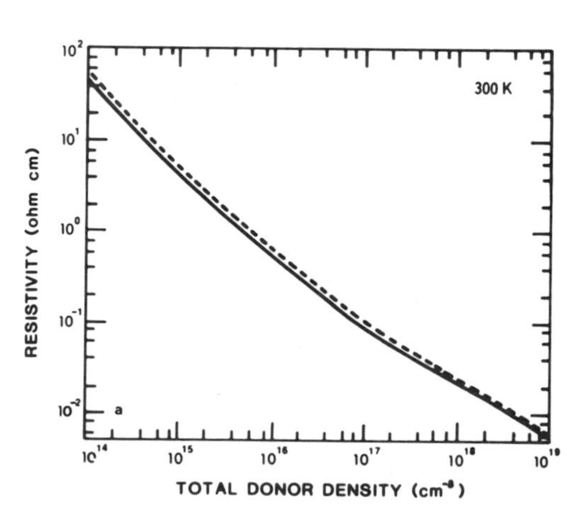

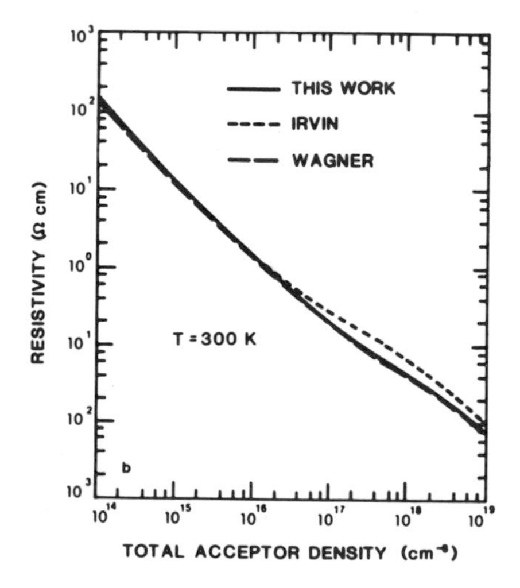

FIGURE 7.8. Resistivity versus dopant density for (a) n-type and (b) p-type silicon at 300 K. Solid lines correspond to calculated values of Li, the dashed line to data published by Irvine,[4] and the broken line to Wager[5]; solid dots correspond to experimental data given by Li.[2,3]

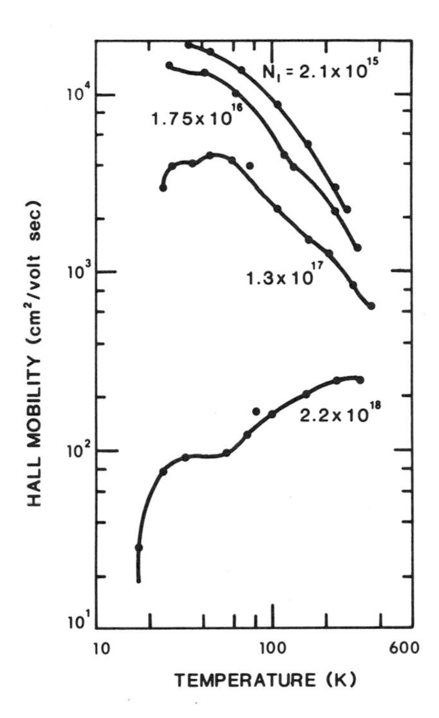

FIGURE 7.9. Hall mobility versus temperature for As-doped silicon samples. After Morin and Maita,[6] by permission.

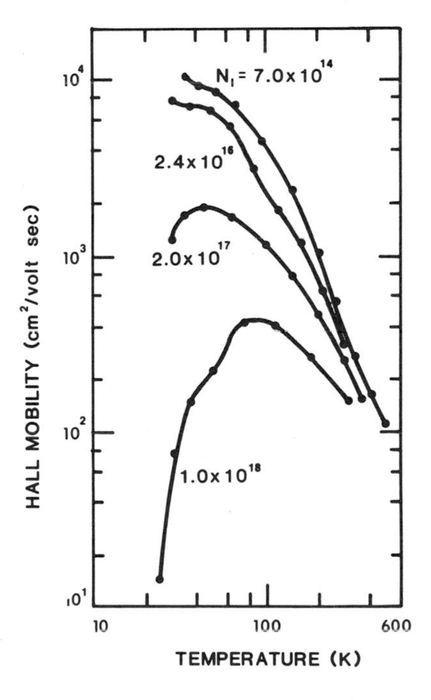

FIGURE 7.10. Hall mobility versus temperature for p-type silicon doped with different boron densities. After Morin and Maita,[6] by permission.

silicon are given respectively by

$$\mu_{Hn} = 5.5 \times 10^6 T^{-3/2} \quad \text{and} \quad \mu_{Hp} = 2.4 \times 10^8 T^{-2.3} \tag{7.118}$$

Equation (7.118) is valid for $T > 100$ K and $N_I < 10^{17}$ cm^{-3}.

Figures 7.11a and b show electron and hole conductivity mobilities as a function of dopant density for n- and p-type silicon, respectively, at $T = 300$ K.[1] The experimental data are deduced from resistivity and junction capacitance–voltage measurements on a specially designed test structure developed at the National Bureau of Standards for accurate determination of the conductivity mobility in silicon. The solid line corresponds to theoretical calculations reported by Li using a more rigorous model than those described here.[2,3] The model takes into account all the scattering mechanisms contributed by acoustic and optical phonons, as well as the ionized and neutral impurities. In addition, intervalley and intravalley phonon scatterings and the effect of the nonparabolic band structure of silicon have also been taken into account in the calculations. The results show excellent agreement between theory and experiment for both n- and p-type silicon over a wide range of dopant densities and temperatures.

We shall next discuss the Seebeck coefficient data for silicon and germanium. The Seebeck coefficient for a nondegenerate n-type semiconductor is given by Eq. (7.77). If the acoustical phonon scattering is dominant, then the Seebeck coefficient for both n- and p-type semiconductors can be expressed by

$$S_{n,p} = \pm\left(\frac{k_B}{q}\right)\left(2 - \frac{E_f}{k_B T}\right) \tag{7.119}$$

where the plus sign is for p-type conduction and the minus sign for n-type conduction. Equation (7.119) shows that the Seebeck coefficient is directly related to the position of the Fermi level in the semiconductor. By measuring the Seebeck coefficient as a function of temperature, the Fermi level can be determined at different temperatures.

Figure 7.12 shows Seebeck coefficient data for n-type germanium[7] and Fig. 7.13 for n- and p-type silicon.[8] For p-type silicon, the Seebeck coefficient changes sign at high temperatures and becomes constant at the onset of the intrinsic regime. It is seen in Fig. 7.12 that the measured and calculated Seebeck coefficients (dashed lines) for germanium are in good agreement for $T > 200$ K. However, the Seebeck coefficient (curves B, C, and D in Fig. 7.12)

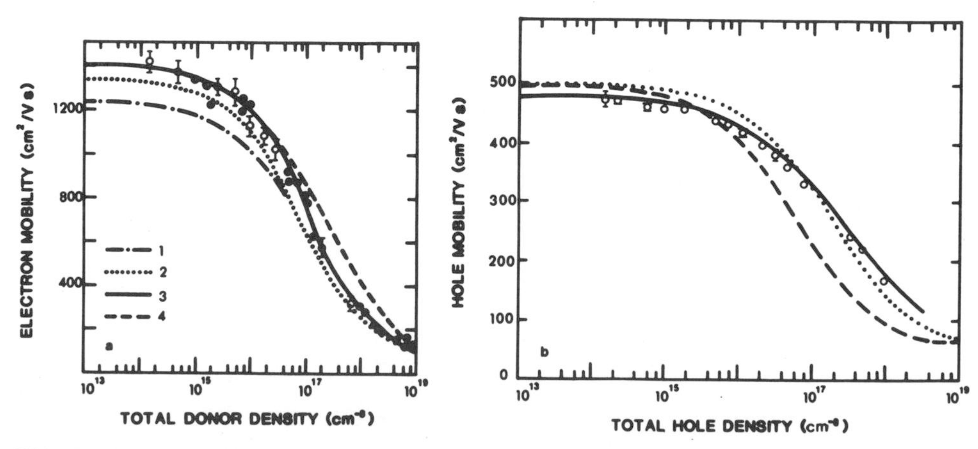

FIGURE 7.11. Electron and hole mobilities versus dopant density for both n- and p-type silicon doped with phosphorus and boron impurities at $T = 300$ K. After Li,[2,3] by permission.

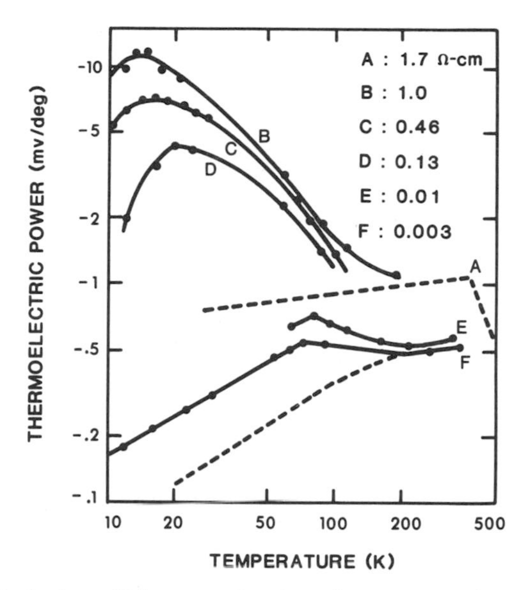

FIGURE 7.12. The Seebeck coefficient as a function of temperature for n-type germanium. After Frederikse,[7] by permission.

increases very rapidly at low temperatures. Such behavior cannot be explained by the theory described above, and the so-called phonon drag effect must be considered in order to explain this anomalous behavior. The phonon drag has a striking effect on the Seebeck coefficient, particularly at low temperatures. This effect can be explained if we assume that the flow of long-wavelength phonons in the presence of a temperature gradient leads to preferential scattering of electrons in the direction of the temperature gradient. It can be shown that the

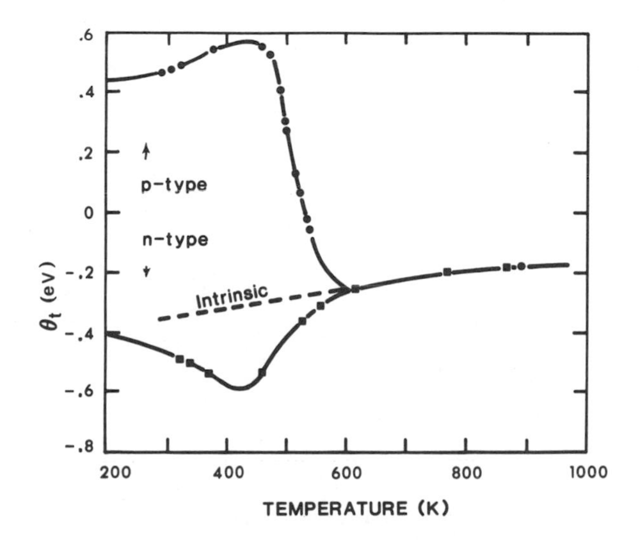

FIGURE 7.13. The Seebeck coefficient as a function of temperature for both n- and p-type silicon. After Geballe and Hull,[8] by permission.

phonon-drag Seebeck coefficient is given by

$$S_{pd} = \pm\left(\frac{xv^2\tau_d}{\mu T}\right) \tag{7.120}$$

where the minus sign denotes n-type and the plus sign p-type; x is the fraction of carrier collisions due to phonons, and τ_d is the relaxation time for loss of momentum from the phonon system.

We note that the phonon-drag Seebeck coefficient has the same sign as the Seebeck coefficient in the absence of the phonon-drag effect. Therefore, the electron and phonon contributions to the Seebeck coefficient reinforce one another at low temperatures. It is seen from Eq. (7.77) that the Seebeck coefficient in a nondegenerate semiconductor can be quite large, of the order of a few mV/K, while for metals the Seebeck coefficient is of the order of

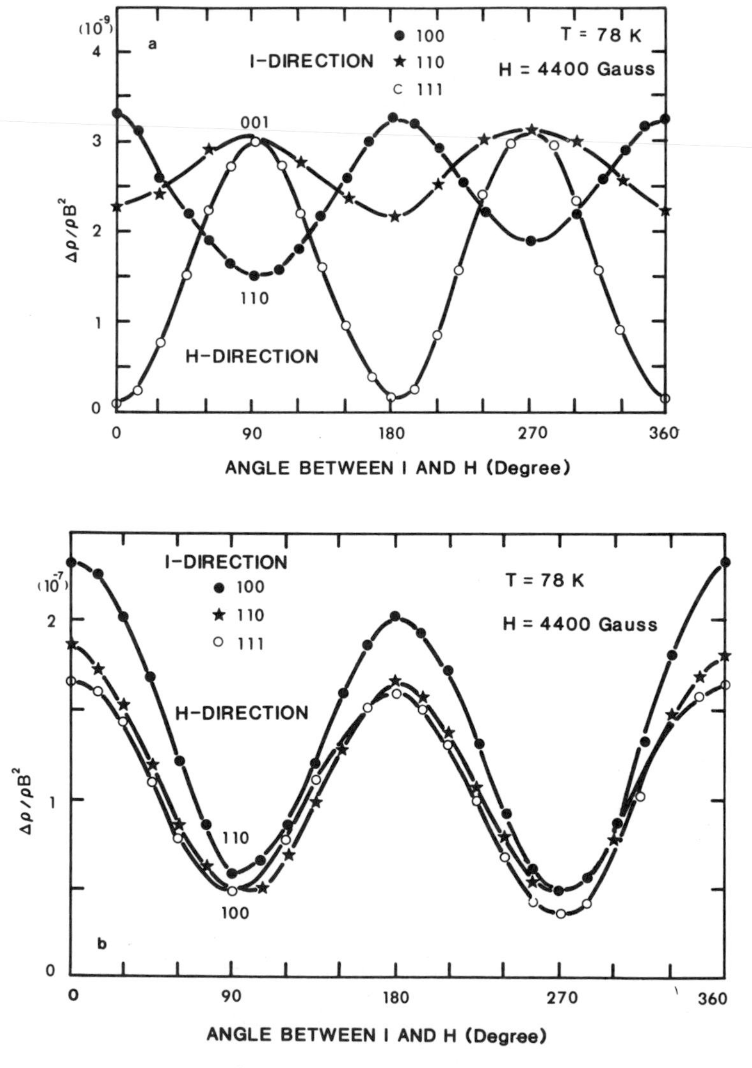

FIGURE 7.14. Variation in $\Delta\rho/\rho B^2$ as the magnetic field H is rotated with respect to the current flow I, for (a) n-type and (b) p-type silicon samples. After Pearson and Herring,[9] by permission.

a few tens μV/K. Thus, at a metal–semiconductor junction, the total Seebeck coefficient is approximately equal to the absolute Seebeck coefficient of the semiconductor.

The transverse magnetoresistance for silicon and germanium is examined. The change in resistivity due to the applied magnetic field is usually referred to as the magnetoresistance effect. The magnetoresistance is defined by

$$\frac{(\rho - \rho_0)}{\rho_0} = \frac{(\sigma_0 - \sigma)}{\sigma_0} \tag{7.121}$$

Early measurements of the magnetoresistance in germanium and silicon were made on polycrystalline materials. Magnetoresistance measurements on single-crystal silicon by Pearson and Herring have produced some interesting results. The effect was found to depend not only on the relative orientations of the current and magnetic field, but also on the crystal orientation. This is illustrated in Fig. 7.14 for an n-type silicon sample.[9] The results show that the theoretical derivation of magnetoresistance based on the assumption that the constant-energy surface is spherical and that σ depends only on the carrier energy are inadequate for the case of silicon. This is due to the fact that the constant-energy surfaces in the conduction and valence band minima of silicon are not exactly spherical. Therefore, refinement of the transport theories presented in this chapter by taking into account various effects cited above is needed in order to obtain an accurate prediction of the experimental results observed in practical semiconductors.

PROBLEMS

7.1. (a) Show that if an electric field is applied in the x-direction, the steady-state nonequilibrium distribution function can be expressed by

$$f(k_x, k_y, k_z) = f_0((k_x - q\mathscr{E}_x\tau/\hbar), k_y, k_z)$$

where f_0 is the equilibrium distribution function.

(b) If a temperature gradient is applied in the x-direction, show that the steady-state nonequilibrium distribution function is given by

$$f(k_x, k_y, k_z) = f_0(k_x + \Delta k_x, k_y, k_z)$$

where $\Delta k_x = (\tau\hbar k_f/m^* T)(k - k_f)(dT/dx)$, k_f being the wave vector of electrons at the Fermi level.

7.2. (a) Plot σ_n and S_n versus η (the reduced Fermi energy) for $-4 \leq \eta \leq 4$, assuming $s = -1/2$ and $\tau = \tau_0 E^s$.

(b) Repeat (a) for $s = +3/2$.

7.3. Using Eq. (7.53) and Fermi–Dirac statistics, show that the electrical conductivity for a degenerate n-type semiconductor is given by

$$\sigma_n = \left(\frac{2nq^2\tau_0}{3m_n^*}\right)(k_B T)^s (s + 3/2)\frac{\mathscr{F}_{(s+1/2)}}{\mathscr{F}_{(1/2)}}$$

where

$$\tau = \tau_0 E^s$$

and

$$\mathscr{F}_r(\eta) = \int_0^\infty \frac{\varepsilon^r d\varepsilon}{[1 + e^{(\varepsilon - \eta)}]}$$

is the Fermi integral of order r; $\varepsilon = E/k_BT$ and $\eta = E_f/k_BT$.

7.4. Show that the seebeck coefficient for a degenerate n-type semiconductor can be expressed by

$$S_n = -\left(\frac{k_B}{q}\right)\left[\frac{(s+5/2)\mathscr{F}_{(s+3/2)}}{(s+3/2)\mathscr{F}_{(s+1/2)}} - \frac{E_f}{k_BT}\right]$$

7.5. If both the electric current and temperature gradient are applied simultaneously to an n-type semiconductor specimen in the x-direction, show that the electric current density and the heat flux density can be expressed by

$$J_x = -nqv_x = -\int_0^\infty qv_x f(E)g(E)\, dE$$

$$= -\left(\frac{2q}{3m_n^*}\right)\int_0^\infty \tau E g(E)\frac{\partial f_0}{\partial E}\left[q\mathscr{E}_x - \frac{(E_f - E)}{T}\frac{\partial T}{\partial x}\right] dE$$

and

$$Q_x = -\left(\frac{2q}{3m_n^*}\right)\int_0^\infty \tau E^2 g(E)\frac{\partial f_0}{\partial E}\left[q\mathscr{E}_x - \frac{(E_f - E)}{T}\frac{\partial T}{\partial x}\right] dE$$

[*Hint*: Solve $P(E)$ from Eq. (7.38) assuming $B = 0$.]

7.6. (a) Using the expressions given in Problem 7.5, show that the electronic thermal conductivity for the degenerate case is given by

$$K_n = -\frac{Q_x}{(dT/dx)} = \left(\frac{n}{m^*T}\right)[\langle\tau E^2\rangle - \langle\tau E\rangle^2/\langle\tau\rangle]$$

(b) Show that, for the nondegenerate case, the expression given by (a) can be reduced to

$$K_n = \left(\frac{n\tau_0}{m^*T}\right)(k_BT)^{s+2}\frac{\Gamma_{(7/2+s)}}{\Gamma_{(5/2)}}$$

7.7. Show that the longitudinal magnetoresistance effect will vanish if the constant-energy surface of the conduction bands is spherical.

7.8. Using Eqs. (7.65) and (7.66) and Fermi–Dirac statistics, derive the Hall coefficient for a degenerate n-type semiconductor and show that the result can be reduced to Eq. (7.70) if nondegenerate statistics is used instead. Derive the Hall factor for a degenerate n-type semiconductor.

7.9. If the total electron mobility of an n-type semiconductor is obtained by using the reciprocal sum of the lattice scattering mobility and the ionized impurity scattering mobility (i.e., $\mu_n^{-1} = \mu_L^{-1} + \mu_I^{-1}$), and $\tau_L = a'T^{-1}E^s$ and $\tau_I = bE^s$, where a and b are constants, derive an expression for the total electron mobility when both the lattice and ionized impurity scatterings dominate in the semiconductor.

7.10. Show that the Seebeck coefficient for the mixed conduction case is given by Eq. (7.116). If lattice scattering is dominant, derive the Hall and Seebeck coefficients from Eqs. (7.104) and (7.116) for a nondegenerate n-type semiconductor.

REFERENCES

1. P. P. Debye and E. M. Conwell, *Phys. Rev.* **93**, 693 (1954).
2. S. S. Li, *The Dopant Density and Temperature Dependence of Electron Mobility and Resistivity in n-Type Silicon*, NBS Special Publication, 400–33 (1977). See also: S. S. Li and R. W. Thurber, *Solid-State Electron.* **20**, 609–616 (1977).
3. S. S. Li, *The Theoretical and Experimental Study of the Temperature and Dopant Density Dependence of Hole Mobility, Effective Mass, and Resistivity in Boron-Doped Silicon*, NBS Special Publication, 400–47 (1979). See also: S. Li, *Solid-State Electron.* **21**, 1109–1117 (1978).
4. J. C. Irvine, *Bell Syst. Tech. J.* **16**, 387 (1962).
5. S. Wagner, *J. Electrochem. Soc.*, **119**, 1570 (1972).
6. F. T. Morin and J. P. Maita, "Hall Mobility of Electrons in N-Type Silicon as a Function of Temperature," *Phys. Rev.* **96**, 28 (1954).
7. H. P. R. Frederikse, *Phys. Rev.* **92**, 248 (1953).
8. T. H. Geballe and G. W. Hull, *Phys. Rev.* **98**, 940 (1955).
9. G. L. Pearson and C. Herring, *Physica* **20**, 975 (1954).

BIBLIOGRAPHY

F. J. Blatt, *Physics of Electronic Conduction in Solids*, McGraw-Hill, New York (1968).

R. H. Bube, *Electronic Properties of Crystalline Solids*, Academic Press, New York (1974).

M. Dresden, "Recent Developments in the Quantum Theory of Transport and Galvanomagnetic Phenomena," *Rev. Mod. Phys.* **33**, 265 (1961).

A. F. Gibson and R. E. Burgess, *The Electrical Conductivity of Germanium*, Wiley, New York (1964).

G. L. Pearson and J. Bardeen, *Phys. Rev.* **75**, 865 (1949).

R. A. Smith, *Semiconductors*, Cambridge University Press, London (1960).

R. K. Willardson and A. C. Beer, *Transport Phenomena, Semiconductors and Semi-metals*, Vol. 10, Academic Press, New York (1975).

A. H. Wilson, *The Theory of Metals*, Cambridge University Press, London (1954).

J. M. Ziman, *Electrons and Phonons*, Oxford University Press, London (1960).

8

Scattering Mechanisms and Carrier Mobilities in Semiconductors

8.1. INTRODUCTION

The relaxation time approximation introduced in Chapter 7 enables us to linearize the Boltzmann transport equation in that the collision term is expressed in terms of the ratio of the perturbed distribution function (i.e., $f-f_0$) and the relaxation time. This approximation allows us to obtain analytical expressions for different transport coefficients in semiconductors. However, detailed physical insights concerning the collision term and the validity of the relaxation time approximation were not discussed in Chapter 7. In fact, the scattering mechanisms associated with the collision term in the Boltzmann equation are so complicated that we shall devote this entire chapter to deriving the relaxation time constants due to different scattering mechanisms in a semiconductor.

The collision term in the Boltzmann transport equation represents the internal relaxation mechanisms which are due to the various scattering processes (e.g., acoustical phonon scattering and ionized impurity scattering) of charge carriers taking place in a semiconductor under the influence of external fields. These scattering mechanisms are responsible for the charge carriers to reach steady-state conditions when an external field is applied to the semiconductor and to return to the equilibrium conditions when the external force is removed from the semiconductor. For elastic scattering, the nonequilibrium distribution function will decay exponentially with time to its equilibrium value after the external force is removed. The time constant associated with this exponential decay is known as the relaxation time or the collision time.

In this chapter, we are concerned with several important scattering mechanisms which play a key role in determining the carrier mobilities in a semiconductor. Using a quantum mechanical treatment, we can derive the relaxation time expressions for a number of important scattering mechanisms which are the dominant scattering sources for charge carriers in the elemental and compound semiconductors. In Section 8.2, the collision term is expressed in terms of the rate of transition probability and the distribution functions for the initial and final states in k-space. The differential scattering cross section, which is defined in terms of the rate of transition probability and the incident flux of the scattering particles, is also introduced in this section. Using the Brooks–Herring (B–H) model, the relaxation time for the ionized impurity scattering is derived in Section 8.3. Section 8.4 describes the neutral impurity scattering, which is an important scattering source at very low temperatures. Using deformation potential theory, the scattering of charge carriers by the longitudinal-mode acoustical phonons is derived in Section 8.5. Section 8.6 is concerned with the scattering of charge carriers by polar and nonpolar optical phonons as well as the intervalley optical phonon scattering in a multivalley semiconductor. Scattering of charge carriers by dislocations is discussed in Section 8.7. Finally, the measured Hall mobility and drift mobility for both electrons and holes for some elemental and compound semiconductors are presented in Section 8.8.

In general, the charge carriers in a semiconductor may be scattered by stationary defects (e.g., impurities and dislocations) and/or by dynamic defects (e.g., electrons, holes, and lattice phonons). Therefore, the transport properties of a semiconductor depend strongly upon the types of scattering mechanism involved. For example, the electrical conductivity of an n-type semiconductor can be expressed in terms of the electron mobility and electron concentration by

$$\sigma_n = n_0 q \mu_n \tag{8.1}$$

where n_0 is the electron concentration, q is the electronic charge, and μ_n is the electron mobility. The electron mobility may be defined in terms of the conductivity effective mass m_σ^* and the relaxation time τ by

$$\mu_n = \frac{q\langle\tau\rangle}{m_c^*} \tag{8.2}$$

where $\langle\tau\rangle$ is the average relaxation time defined by Eq. (7.54). Thus, the electron mobility is directly proportional to the average relaxation time and varies inversely with the conductivity effective mass. Since the average relaxation time given in Eq. (8.2) is directly related to the scattering mechanisms, in order to calculate the carrier mobility, it is necessary to first consider the scattering mechanisms in a semiconductor.

In the relaxation time approximation, the collision term in the Boltzmann equation can be expressed in terms of the perturbed distribution function divided by the relaxation time. From Eq. (7.25), we obtain

$$\left.\frac{\partial f}{\partial t}\right|_c = -\frac{f - f_0}{\tau} \tag{8.3}$$

where f is the nonequilibrium distribution function and f_0 is the equilibrium Fermi–Dirac distribution function. As mentioned earlier, the relaxation time approximation is valid only for the elastic-scattering case. This condition is satisfied so long as the change in energy of the charge carriers before and after each scattering event is small compared to the initial carrier energy. In fact, a generalized expression for the collision term given by Eq. (8.3) can be formulated in terms of the rate of transition probability $P_{kk'}$ and the nonequilibrium distribution function $f(k, r)$. This generalized expression is given by

$$-\left.\frac{\partial f_{k'}}{\partial t}\right|_c = \sum_{k'} [P_{kk'} f_{k'}(1 - f_k) - P_{k'k} f_k(1 - f_{k'})] \tag{8.4}$$

where $P_{kk'}$ is the rate of transition probability from the final state k' to the initial state k, and $P_{k'k}$ is the rate of transition probability from state k to state k'. The electron distribution function in state k' is denoted by $f_{k'}$, and the electron distribution function in state k is represented by f_k.

The right-hand side of Eq. (8.4) represents the net transitions from state k to state k' summed over all the final states k'. The summation in Eq. (8.4) can be replaced by integration over the entire conduction band if all the quantum states in the band are treated as quasi-continuum. Since the density of quantum states in the conduction band is very large and the spacing between each quantum state is so small, such an assumption is usually valid. Therefore, it is common practice to replace the summation in Eq. (8.4) by an integral, in which case

$$-\left.\frac{\partial f_{k'}}{\partial t}\right|_c = \frac{N\Omega}{(2\pi)^3}\int [P_{kk'}f_{k'}(1-f_k) - P_{k'k}f_k(1-f_{k'})]\, d^3 k'$$

$$= \frac{N\Omega}{(2\pi)^3}\int P_{kk'}(f_{k'} - f_k)\, d^3k \tag{8.5}$$

Here $P_{kk'}$ is assumed equal to $P_{k'k}$, N is the total number of unit cells in the crystal, and Ω is the volume of the unit cell.

It is seen that the collision term given by Eq. (8.5) is a differential-integral equation and cannot be solved analytically without introducing further approximations. In order to derive an analytical expression for Eq. (8.5), it is useful to first consider the small-perturbation case (i.e., low-field case). In this case, the nonequilibrium distribution function $f(k, r)$ can be expressed in terms of the equilibrium distribution function f_k^0 and a first-order perturbing distribution function f_k^1, which reads

$$f_k = f_k^0 + f_k^1 + \cdots$$

$$f_{k'} = f_{k'}^0 + f_{k'}^1 + \cdots \tag{8.6}$$

where f_k^0 and $f_{k'}^0$ are the Fermi–Dirac distribution functions in the k and k' states, while f_k^1 and $f_{k'}^1$ denote the first-order correction terms in the k and k' states, respectively.

If we assume that the scattering is elastic, then the energy change during scattering processes is small compared to the average electron energy. Under this condition, the average energy of electrons in the initial and final states can be assumed equal, namely, $E_{k'} = E_k$. Therefore, the equilibrium distribution functions for the initial and final states are identical, and the collision term can be simplified to

$$-\left.\frac{\partial f_{k'}}{\partial t}\right|_c = \frac{f_{k'}^1}{\tau} = \frac{N\Omega}{(2\pi)^3}\int P_{kk'}(f_{k'}^1 - f_k^1)\, d^3k' \tag{8.7}$$

Thus, the inverse relaxation time τ^{-1} can be written as

$$\frac{1}{\tau} = \frac{N\Omega}{(2\pi)^3}\int P_{kk'}\left(1 - \frac{f_k^1}{f_{k'}^1}\right) d^3k' \tag{8.8}$$

Furthermore, if we assume that the scattering process is isotropic, then the ratio of f_k^1 and $f_{k'}^1$ can be expressed in terms of $\cos\theta'$, where θ' is the angle between the k and k' states (see Fig. 8.2b). Under this condition, Eq. (8.8) reduces to

$$\frac{1}{\tau} = \frac{N\Omega}{(2\pi)^3}\int P_{kk'}(1 - \cos\theta')\, d^3k' \tag{8.9}$$

which shows that the scattering rate, τ^{-1} of the charge carriers for isotropic elastic scattering depends only on the angle between the k and k' states and the rate of transition probability $P_{kk'}$.

In order to derive the relaxation time for a specific scattering process, both the rate of transition probability and the differential scattering cross section must be determined first. This will be discussed next.

8.2. DIFFERENTIAL SCATTERING CROSS SECTION

In the present treatment, it is assumed that the scattering of charge carriers is confined within a single energy band (e.g., electrons in the conduction band and holes in the valence band), as illustrated in Fig. 8.1a. Other important scattering processes, such as intervalley (i.e., for multivalley semiconductors such as Si and Ge) and interband (heavy-hole and light-hole bands) scatterings, are also shown in Figs. 8.1b and c, respectively.

The intraband and intravalley scatterings shown in Fig. 8.1a are usually accompanied by the absorption or emission of a longitudinal-mode acoustical phonon, and hence are elastic. However, the interband and intervalley scatterings shown in Figs. 8.1b and c are usually inelastic because the change in electron energy for these scatterings is no longer small compared to the mean electron energy. The intervalley and interband scattering processes are usually accompanied by the absorption or emission of optical phonons.

The rate of transition probability $P_{kk'}$ in a scattering event can be derived from the one-electron Schrödinger equation. The one-electron time-independent Schrödinger equation for the initial unperturbed states is given by

$$H_0\phi_k(r) = E_k\phi_k(r) \tag{8.10}$$

where

$$H_0 = -\frac{\hbar^2\nabla^2}{2m^*} + V(r) \tag{8.11}$$

is the unperturbed Hamiltonian and

$$\phi_k(r) = u_k(r)\exp(ik\cdot r) \tag{8.12}$$

is the initial unperturbed electron wave function.

If a small perturbation (e.g., a small electric field) is applied to the system, then the electron may be scattered from its initial state k into its final state k'. The perturbed Hamiltonian under this condition can be written as

$$H = H_0 + H' \tag{8.13}$$

where H_0 is the unperturbed Hamiltonian given by Eq. (8.11), and H' is the first-order correction due to perturbation. The time-dependent Schrödinger equation under the perturbed

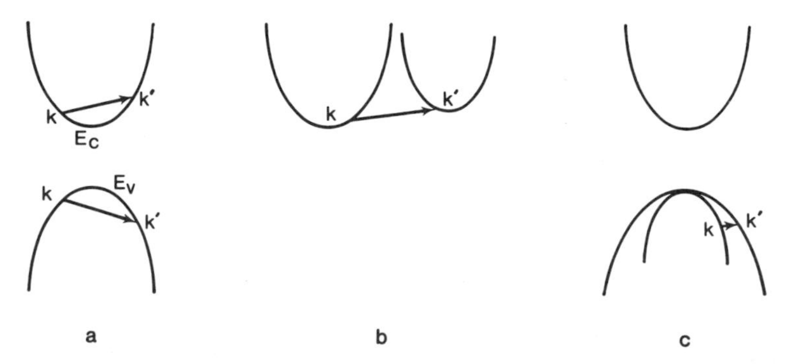

FIGURE 8.1. Scattering of electrons and holes in a semiconductor.

condition is thus given by

$$H\psi_k(r, t) = -i\hbar \frac{\partial \psi_k(r, t)}{\partial t} \tag{8.14}$$

which has a solution given by

$$\psi_k(r, t) = \sum_k a_k(t)\, e^{-iE_k t/\hbar} \phi_k(r) \tag{8.15}$$

where $a_k(t)$ is the time-dependent amplitude function and $\phi_k(r)$ denotes the unperturbed electron wave function defined by Eq. (8.12).

According to time-dependent perturbation theory, the transition probability per unit time from state k to k' can be expressed in terms of the amplitude function $a_k(t)$ by

$$P_{k'k}(t) = \frac{|a_k(t)|^2}{t} \tag{8.16}$$

Similarly, the transition probability per unit time from state k' to k is given by

$$P_{kk'}(t) = \frac{|a_{k'}(t)|^2}{t} \tag{8.17}$$

From the principle of detailed balance, we can assume that $P_{kk'}$ equal to $P_{k'k}$. Using Eqs. (8.14) through (8.17) and the orthogonal properties of electron wave functions, it can be shown from quantum mechanical calculations that the rate of transition probability $P_{kk'}$ in the presence of a step perturbation function (i.e., constant H') is given by

$$P_{kk'} = \frac{|a'_k(t)|^2}{t} = \frac{2\pi}{\hbar} |H_{kk'}|^2 \delta(E_{k'} - E_k) \tag{8.18}$$

where

$$H_{kk'} = \langle k'|H'|k\rangle = \frac{1}{(N\Omega)} \int_{N\Omega} \phi^*_{k'} H' \phi_k \, d^3r \tag{8.19}$$

is the matrix element. In Eq. (8.19), H' denotes the perturbing Hamiltonian, ϕ_k is the electron wave function given by Eq. (8.12), and $\phi^*_{k'}$ is the complex conjugate of $\phi_{k'}$. The function $\delta(E_{k'} - E_k)$ is the Dirac delta function, which is equal to unity for $E_k = E_{k'}$ and vanishes otherwise.

The matrix element $H_{kk'}$, given by Eq. (8.19), has a finite value only if the golden selection (momentum conservation) rule is satisfied (i.e., $k = k'$ for a direct transition and $k' = k \pm q$ for an indirect transition). Calculations of relaxation time can be simplified by introducing a differential scattering cross section $\sigma(\theta', \phi')$ in the relaxation time formula. It is noted that $\sigma(\theta', \phi')$ depends only on θ' if the scattering process is isotropic (i.e., independent of ϕ'). Under this condition, a simple relationship exists between $\sigma(\theta')$ and the rate of transition probability $P_{kk'}$. In general, the differential scattering cross section $\sigma(\theta', \phi')$ is defined as the total number of particles which make transitions from state k to k' per unit solid angle per unit time divided by the incident flux density. This can be written as

$$\sigma(\theta', \phi') = \frac{\dfrac{N\Omega}{(2\pi)^3} P_{kk'} \dfrac{d^3k'}{d\omega}}{\dfrac{v_k}{N\Omega}} = \frac{(N\Omega)^2 P_{kk'}\, d^3k'}{(2\pi)^3 v_k \sin\theta'\, d\theta'\, d\phi'} \tag{8.20}$$

where v_k is the initial particle velocity, $N\Omega$ is the volume of the crystal, and $d\omega = \sin\theta'\, d\theta'\, d\phi'$ is the solid angle between the incident wave vector k and the scattering wave vector k' (see Fig. 8.2).

Let us now consider the case of isotropic elastic scattering. By substituting Eqs. (8.18) and (8.19) into Eq. (8.20), and using the relationships $v_k = v_{k'}$, $k = k'$, and $d^3k' = k'^2 \sin\theta'\, d\theta'\, d\phi'\, dk'$, we can obtain an expression for the differential scattering cross section, which is given by

$$\sigma(\theta') = \frac{(N\Omega)^2 k'^2 |H_{kk'}|^2}{(2\pi\hbar v_{k'})^2} \tag{8.21}$$

The relaxation time τ is related to the total scattering cross section by

$$\tau^{-1} = N_T \sigma_T v_{th} \tag{8.22}$$

where N_T is the density of total scattering centers, σ_T is the total scattering cross section, and v_{th} is the mean thermal velocity $[v_{th} = (3k_BT/m^*)^{1/2}]$. The total scattering cross given by Eq. (8.22) can be derived from the differential scattering cross section given by Eq. (8.21), using the following expression:

$$\sigma_T = 2\pi \int_0^\pi \sigma(\theta')(1 - \cos\theta') \sin\theta'\, d\theta' \tag{8.23}$$

By substituting Eq. (8.21) into Eq. (8.23), the total scattering cross section can be obtained from Eq. (8.23) provided that the perturbing Hamiltonian H', and hence the matrix element $H_{kk'}$, is known. In the following sections, we shall employ Eqs. (8.20) through (8.23) to derive expressions for the relaxation time constants and carrier mobilities in a semiconductor in which scattering of charge carriers is due to either ionized impurities, or neutral impurities, or acoustical phonons.

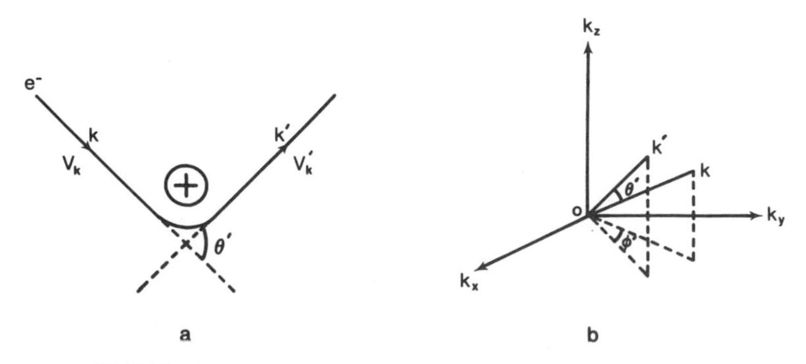

FIGURE 8.2. Scattering of electrons by a positively charged ion.

8.3. IONIZED IMPURITY SCATTERING

Scattering of electrons by an ionized shallow-donor impurity is a typical example of elastic scattering in a semiconductor. This is due to the fact that the mass of a shallow-donor impurity atom is very much larger than that of an electron. As a result, the change of electron energy during such a scattering event is negligible compared to the mean electron energy before the scattering. Therefore, the relaxation time approximation given by Eq. (8.22) is valid in this case. In order to derive the differential scattering cross section and the relaxation time for ionized impurity scattering, the matrix element $H_{kk'}$ and the perturbing Hamiltonian H' due to the shallow impurity potential must be determined first. Let us consider scattering of electrons by a positively charged shallow donor impurity in an n-type semiconductor, as shown in Fig. 8.2a. If the donor impurity is ionized with a single net positive charge, then the potential due to this ionized donor atom, at a large distance from the impurity atom, can be approximated by a bare Coulomb potential

$$V(r) = \frac{q}{4\pi \varepsilon_0 \varepsilon_s r} \tag{8.24}$$

We note that Eq. (8.24) does not consider the screening effect due to the conduction electrons surrounding the rest of the positively charged donor ions in the semiconductor. To take into account the screening effect of these conduction electrons, it is necessary to replace the bare Coulomb potential by a screening Coulomb potential in the derivation of ionized impurity scattering mobility. As shown in Fig. 8.3b, if we include the screening effect of the shallow donor ions by the surrounding conduction electrons, then the screening Coulomb potential (also known as the Yukawa potential) for the ionized impurity atom can be expressed by

$$V'(r) = \frac{q\, e^{-r/\lambda_D}}{4\pi \varepsilon_0 \varepsilon_s r} \tag{8.25}$$

where

$$\lambda_D = \sqrt{\frac{\varepsilon_0 \varepsilon_s k_B T}{q^2 n_0}} \tag{8.26}$$

is the Debye screen length.

In deriving the matrix element for ionized impurity scattering, Conwell and Weisskopf[1] used the bare Coulomb potential given by Eq. (8.24) as the perturbing Hamiltonian, while Brooks and Herring[2] employed the Yukawa potential given by Eq. (8.25) as the perturbing

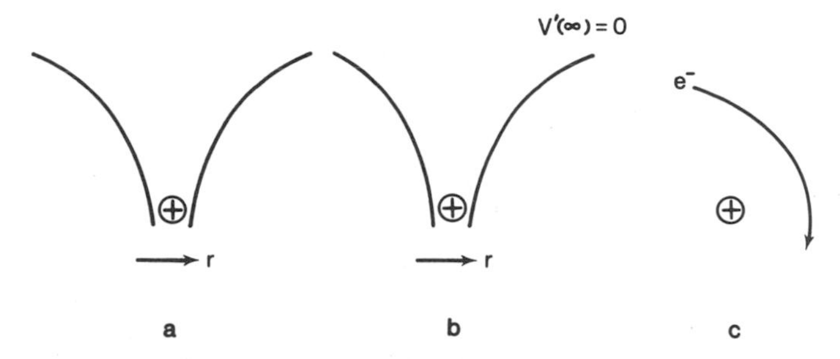

FIGURE 8.3. Potential due to a positively charged donor impurity. (a) Bare Coulombic potential, (b) screening Coulombic potential, (c) trajectory of electron scattering by a positively charged ion.

Hamiltonian. It will be shown later that the relaxation-time formulas derived from both models differ only by a constant and give the same prediction concerning the energy dependence of the relaxation time. Since the Brooks–Herring (B–H) model is based on quantum mechanical principles, and is fundamentally more sound and accurate than the Conwell–Weisskopf (C–W) model, we shall use the B–H model for calculations of ionized impurity scattering mobility. The perturbing Hamiltonian due to the Yukawa potential, given by Eq. (8.25), can be written as

$$H' = qV'(r) = \frac{q^2 e^{-r/\lambda_D}}{4\pi \varepsilon_0 \varepsilon_s r} \tag{8.27}$$

Based on the Bloch theorem, the electron wave functions for the k-state can be expressed by

$$\phi_k(r) = \left(\frac{1}{N\Omega}\right)^{1/2} u_k(r)\, e^{ik\cdot r} \tag{8.28}$$

The matrix element due to the Yukawa potential can be derived from Eqs. (8.19), (8.27), and (8.28), and the result yields

$$\begin{aligned} H_{kk'} &= \frac{1}{N\Omega} \int e^{-ik'\cdot r} \left(\frac{q^2 e^{-r/\lambda_D}}{4\pi \varepsilon_0 \varepsilon_s r}\right) e^{ik\cdot r}\, d^3 r \\ &= \frac{q^2}{2N\Omega \varepsilon_0 \varepsilon_s} \int_0^{\pi} \int_0^{\infty} e^{-iK\cdot r} \left(\frac{e^{-r/\lambda_D}}{r}\right) r^2 \sin\theta_r\, d\theta_r\, dr \\ &= \frac{q^2 \lambda_D^2}{N\Omega \varepsilon_0 \varepsilon_s (1 + K^2 \lambda_D^2)} \end{aligned} \tag{8.29}$$

where $d^3 r = 2\pi r^2 \sin\theta_r\, d\theta_r\, dr$ is the volume element, and

$$K = k' - k = 2|k| \sin\left(\frac{\theta'}{2}\right) \tag{8.30}$$

where K is the phonon wave vector. Figure 8.4 shows the relationship between the incident and scattered wave vectors in real space.

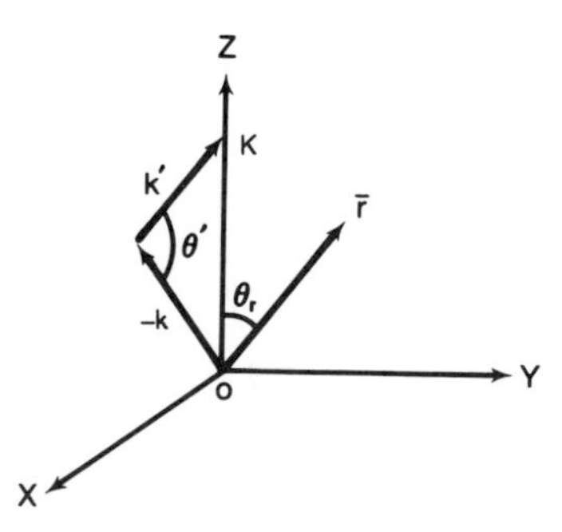

FIGURE 8.4. Coordinates for computing the matrix element of scattering by an ionized impurity, where k is the wave vector of the incident electron, k' is the wave vector of a scattered electron wave, and $K = k' - k = 2|k| \sin(\theta'/2)$.

The differential scattering cross section can be determined by substituting Eq. (8.29) into Eq. (8.21), which yields

$$\sigma(\theta') = \frac{(q^2 m^* \lambda_D^2)^2}{(2\pi\hbar^2 \varepsilon_0 \varepsilon_s)^2 (1 + K^2 \lambda_D^2)^2} = \frac{4\lambda_D^4}{a_B^2 (1 + K^2 \lambda_D^2)^2} \tag{8.31}$$

where

$$a_B = \frac{4\pi \varepsilon_0 \varepsilon_s \hbar^2}{m^* q^2} \tag{8.32}$$

is the Bohr radius for the ground state of the impurity atom.

The relaxation time for ionized impurity scattering can be obtained by substituting Eq. (8.31) into Eqs. (8.22) and (8.23), and the result is given by

$$\begin{aligned}\frac{1}{\tau_I} &= (2\pi N_I v) \int_0^\pi \frac{4\lambda_D^4 (1 - \cos\theta') \sin\theta' \, d\theta'}{a_B^2 [1 + 4\lambda_D^2 k^2 \sin^2(\theta'/2)]^2} \\ &= (2\pi N_I v) \left(\frac{\lambda_D^4}{a_B^2}\right) \left(\frac{1}{k\lambda_D}\right)^4 L(2k\lambda_D)\end{aligned} \tag{8.33}$$

where

$$L(2k\lambda_D) = \ln(1 + 4k^2\lambda_D^2) - \frac{4k^2\lambda_D^2}{(1 + 4k^2\lambda_D^2)} \cong \ln(4k^2\lambda_D^2), \qquad k\lambda_D \gg 1 \tag{8.34}$$

where $L(2k\lambda_D)$ is a slow varying function of temperature and electron density, and $4k^2\lambda_D^2 = 8m^* E \varepsilon_0 \varepsilon_s k_B T / \hbar^2 q^2 n'$, and $n' = n + (N_D - N_A^- - n)(N_A^- + n)/N_D$ are the density of screening electrons surrounding the ionized donor impurities. The integration of Eq. (8.33) is carried out by setting $\sin(\theta'/2) = x$, $1 - \cos\theta' = 2x^2$, $\sin\theta' \, d\theta' = 4x \, dx$, and using tables of integrals.

Equation (8.33) is obtained by using the B–H model, and hence is known as the Brooks–Herring formula for ionized impurity scattering. By employing the relation $E = \hbar^2 k^2 / 2m^*$, and substituting Eqs. (8.26) and (8.32) into Eq. (8.33), we obtain

$$\frac{1}{\tau_I} = \frac{q^4 N_I L(2k\lambda_D)}{16\pi (2m^*)^{1/2} \varepsilon_0^2 \varepsilon_s^2 E^{3/2}} \tag{8.35}$$

This equation shows that for ionized impurity scattering, the relaxation time τ_I is directly proportional to energy $E^{3/2}$. The temperature dependence of τ_I comes only from the variation of $L(2k\lambda_D)$ with T, which is usually very small.

By substituting τ_I, given by Eq. (8.35), into Eq. (8.2), and by averaging τ_I over the energy with the aid of Eq. (7.45), we obtain the ionized impurity scattering mobility μ_I, which reads

$$\mu_I = \frac{q\langle \tau_I \rangle}{m^*} = \frac{64\sqrt{\pi}\,\varepsilon_0^2 \varepsilon_s^2 (2k_B T)^{3/2}}{N_I q^3 m^{*1/2} \ln\left(\dfrac{12 m^* k_B^2 T^2 \varepsilon_0 \varepsilon_s}{q^2 \hbar^2 n'}\right)} \tag{8.36}$$

This equation shows that the ionized impurity scattering mobility μ_I is directly proportional to the temperature to the power of three halves (i.e., $T^{3/2}$). Good agreement has been found between the theoretical prediction given by Eq. (8.36) and mobility data for different semiconductors in which ionized impurity scattering is the dominant scattering mechanism.

Conwell and Weisskopf used the bare Coulomb potential as the perturbing Hamiltonian and derived the relaxation-time formula for ionized impurity scattering in the form

$$\frac{1}{\tau_I'} = \frac{q^4 N_I}{16\pi(2m^*)^{1/2}\varepsilon_0^2\varepsilon_s^2 E^{3/2}} \ln[1 + (2E/E_m)^2] \tag{8.37}$$

where $E_m = q^2/4\pi\varepsilon_0\varepsilon_s r_m$ and $N_I = (2r_m)^{-3}$.

The ionized impurity scattering mobility derived from Eq. (8.37) is given by

$$\mu_I' = \frac{64\sqrt{\pi}\,\varepsilon_0^2\varepsilon_s^2(2k_BT)^{3/2}}{N_I q^3 m^{*1/2} \ln[1 + (12\pi\varepsilon_0\varepsilon_s k_BT/q^2 N_I^{1/3})^2]} \tag{8.38}$$

Equation (8.38) is known as the Conwell–Weisskopf mobility formula for ionized impurity scattering. A comparison of the mobility expressions given by Eqs. (8.36) and (8.38) reveals that both formulas are very similar except that the coefficient inside the logarithmic term is slightly different. It is of interest that both formulas predict the same temperature dependence for the ionized impurity scattering mobility and the same energy dependence for the relaxation time.

8.4. NEUTRAL IMPURITY SCATTERING

Neutral impurity scattering is an important source of resistance in a semiconductor at very low temperatures. As the temperature decreases, carrier freeze-out occurs at the shallow-level impurity centers in an extrinsic semiconductor, and these shallow-level impurities become neutral at very low temperatures. The scattering potential due to a neutral shallow-level impurity center may be described by using a square-well potential which becomes the dominant scattering source for electrons or holes at very low temperatures.

In general, the scattering of charge carriers by neutral shallow-donor or shallow-acceptor impurities can be treated in a similar way to that of scattering of electrons by a hydrogen atom. The neutral impurity in a semiconductor can be represented by a hydrogenic neutral atom immersed in a dielectric medium of the semiconductor whose dielectric constant is determined by the host semiconductor.

Erginsoy[3] derived the neutral impurity scattering mobility for a semiconductor by employing the partial wave technique to obtain the differential scattering cross section. In his derivation, Erginsoy assumed that electron speed was low and elastic scattering prevailed in the semiconductor. Based on his derivation, the total differential scattering cross section for neutral impurity scattering can be written as

$$\sigma_N = \frac{20a_B}{k} \tag{8.39}$$

where a_B is Bohr's radius given by Eq. (8.32). We note that Eq. (8.39) is valid for $E \leq \frac{1}{4}E_i$. From Eqs. (8.22) and (8.39), the relaxation time for neutral impurity scattering can be expressed by

$$\tau_N = (N_N v \sigma_N)^{-1} = \frac{k}{20a_B N_N v} \tag{8.40}$$

When the expression for a_B given by Eq. (8.32) and $k = m^*v/\hbar$ are substituted into Eq. (8.40), we obtain

$$\frac{1}{\tau_N} = \frac{10\varepsilon_0\varepsilon_s N_N h^3}{\pi^2 m^{*2} q^2} \tag{8.41}$$

where N_N denotes the density of neutral impurities. Since the relaxation time for neutral impurity scattering is independent of energy, the mobility due to neutral impurity scattering can be readily obtained by using the relation $\mu_N = q\tau_N/m^*$, which yields

$$\mu_N = \frac{q\tau_N}{m^*} = \frac{\pi^2 m^* q^3}{10\varepsilon_0\varepsilon_s N_N h^3} \tag{8.42}$$

From Eqs. (8.41) and (8.42), it is seen that for neutral impurity scattering the relaxation time is not a function of electron energy, and the mobility is independent of temperature. Experimental results reveal that the mobility data obtained for many semiconductors at very low temperatures do not always agree well with the theoretical predicton given by Eq. (8.42). In fact, experimental results show that carrier mobility is a weak function of temperature for many semiconductors at low temperatures.

8.5. ACOUSTIC PHONON SCATTERING

Scattering of electrons by longitudinal-mode acoustic phonons is discussed in this section. The scattering of electrons by long-wavelength acoustic phonons may be considered as the most important scattering source in intrinsic or undoped semiconductors. This type of scattering process can usually be treated as elastic scattering, since the electron energy is much larger than the phonon energy, and the change in electron energy during the scattering event is small compared to the average energy of electrons. It can be shown that the maximum change of electron energy due to acoustic phonon scattering is given by

$$\Delta E \simeq 4\left(\frac{u_s}{v_{th}}\right)E_e \tag{8.43}$$

where $u_s = 3 \times 10^5$ cm/sec is the velocity of sound in a solid, and v_{th} is the mean thermal velocity of electrons ($\sim 10^7$ cm/sec). Thus, the ratio of phonon energy to mean electron energy as given by Eq. (8.43) is usually much smaller than unity for $T > 100$ K. At very low temperatures, mean electron energy may become comparable to acoustic phonon energy, and the assumption of elastic scattering may no longer be valid. Fortunately, at very low temperatures other types of scattering, such as ionized impurity and neutral impurity scattering, become predominant. It is noteworthy that acoustic phonons may cause scattering in two different ways, either through deformation potential scattering or piezoelectric scattering. An acoustic wave may induce a change in the spacing of neighboring atoms in a semiconductor. This change in atomic spacing may cause the fluctuation of the energy bandgap locally on an atomic scale and is known as the deformation potential. The deformation potential is measured as the change of energy bandgap per unit strain due to the acoustic phonons. This type of scattering is usually the most important scattering source for undoped silicon and germanium crystals at room temperatures.

Piezoelectric scattering is another type of acoustic phonon scattering. This type of scattering is observed in III–V and II–VI compound semiconductors with zinc blende and wurtzite structures. The lack of inversion symmetry in these semiconductors creates a strain-induced microscopic electric-field perturbation, which leads to piezoelectric scattering with emission or absorption of an acoustic phonon. This type of scattering is important for pure III–V and II–VI compound semiconductors at low temperatures. We shall next discuss these two types of acoustic phonon scattering.

8.5.1. Deformation Potential Scattering

In order to derive an expression for the relaxation time of nonpolar acoustic phonon scattering, we shall first discuss the deformation potential technique developed originally by Bardeen and Shockley[4] for calculating the matrix element of longitudinal-mode acoustic phonon scattering. The perturbing Hamiltonian can be obtained from the deformation potential shown in Fig. 8.5. Figure 8.5a shows the change of lattice spacing with respect to its equilibrium position due to lattice vibration. It is seen that thermal expansion and contraction of the lattice with temperature can lead to a change of the conduction and valence band edges or the energy bandgap of the semiconductor, as is shown in Fig. 8.5b. Based on Shockley's deformation potential model, the fluctuation in the conduction band edge due to lattice vibrations may be represented by a deformation potential. Therefore, the perturbing Hamiltonian can be related to the change of crystal volume caused by the lattice phonons and the deformation potential by the relationship

$$H' = \Delta E_c = E_c - E_{c0} = \left(\frac{\Delta E_c}{\Delta V}\right)\Delta V = E_{c1}\left(\frac{\Delta V}{V}\right) \tag{8.44}$$

where E_{c0} is the conduction band edge in thermal equilibrium, and

$$E_{c1} = \frac{\Delta E_c/\Delta T}{\Delta V/V\Delta T} \tag{8.45}$$

is the deformation potential constant. For silicon, $E_{c1} = -16$ eV, and for germanium, $E_{c1} = -9.5$ eV. The ratio $\Delta V/V$ represents the change of the crystal volume to the total crystal volume as a result of temperature change in the semiconductor. Since $\Delta V/V$ can be expanded in terms of a Fourier series in the atomic displacement r_n, we can write

$$\frac{\Delta V}{V} = \nabla_r \cdot r_n \tag{8.46}$$

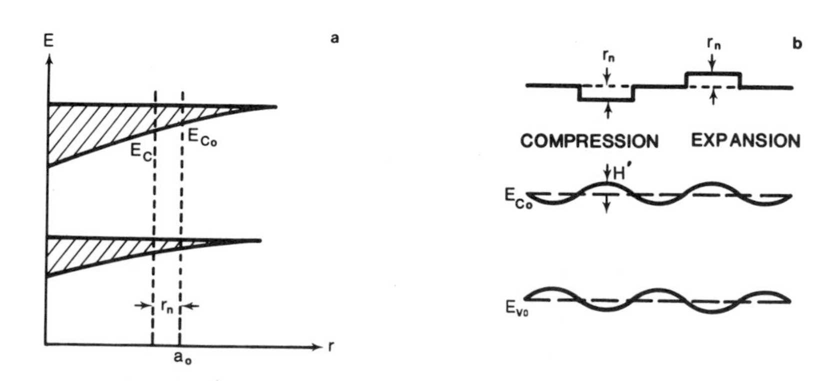

FIGURE 8.5. The change of conduction band edge (deformation potential due to thermal expansion or contraction of the lattice spacing).

where

$$r_n = \sum_{j=1}^{3} (1/N)^{1/2}\zeta_j b_j(q)\, e^{iq\cdot R_{n0}}\, e^{-i\omega t} \tag{8.47}$$

The lattice displacement r_n given by Eq. (8.47) can be expressed in terms of the normal coordinates and normal frequencies in three-dimensional form (i.e., two transverse branches and one longitudinal branch). It is also assumed that only longitudinal-mode acoustic phonon scattering is important in the present case. Therefore, under this condition $\nabla \cdot r_n$ can be expressed by

$$\nabla_r \cdot r_n = \sum_q q_l r_l \tag{8.48}$$

where q_l denotes the wave vector of the longitudinal-mode acoustic phonon and r_l represents the displacement due to longitudinal-mode acoustic phonons. Substituting Eqs. (8.46) and (8.48) for $(\Delta V/V)$ into Eq. (8.44) yields the perturbing Hamiltonian H', which is given by

$$H' = E_{c1}\left(\frac{\Delta V}{V}\right) = E_{c1} \sum_q q_l r_l \tag{8.49}$$

The matrix element $H_{kk'qq'}$ due to this perturbing Hamiltonian is thus given by

$$\begin{aligned} H_{kk'qq'} &= \langle k' n_{q'} | H' | k n_q \rangle \\ &= \int \phi_{k'}^* \phi_{n_q}^* \left(\sum_q E_{c1} q_l r_l \right) \phi_k \phi_{n_q}\, d^3 r\, d^3 r_l \end{aligned} \tag{8.50}$$

where ϕ_{n_q} is the phonon wave function and ϕ_k is the electron wave function (i.e., the Bloch wave). For phonon emission, the solution of Eq. (8.50) for the matrix element can be written as

$$H_{kk'qq'} = \left(\frac{E_{c1} q_l}{N\Omega}\right)\left(\frac{\hbar}{M\omega}\right)^{1/2}\left(\frac{\langle n_q \rangle}{2}\right)^{1/2} \tag{8.51}$$

and for phonon absorption it is given by

$$H_{kk'qq'} = \left(\frac{E_{c1} q_l}{N\Omega}\right)\left(\frac{\hbar}{M\omega}\right)^{1/2}\left[\frac{(\langle n_q \rangle + 1)}{2}\right]^{1/2} \tag{8.52}$$

In Eqs. (8.51) and (8.52), $N\Omega$ is the crystal volume, M is the mass of the atom, and $\langle n_q \rangle$ is the average phonon population density given by

$$\langle n_q \rangle = \frac{1}{e^{\hbar\omega/k_B T} - 1} \simeq \frac{k_B T}{\hbar\omega} \tag{8.53}$$

Equation (8.53) is valid for long-wavelength acoustic phonons (i.e., $k_B T \gg \hbar\omega$). The square of the matrix element due to phonon deformation potential scattering can be obtained from the summation of the square of Eqs. (8.51) and (8.52) and using the dispersionless relation $\omega = u_s q_l$, which yields

$$|H_{kk'qq'}|^2 = \frac{E_{c1}^2 k_B T}{M(u_s N\Omega)^2} \tag{8.54}$$

Now, substitution of $|H_{kk'qq'}|^2$ given by Eq. (8.54) into Eq. (8.21) yields the differential scattering cross section, which reads

$$\sigma_a = \frac{\Omega m^{*2} E_{c1}^2 k_B T}{4\pi^2 \hbar^4 \rho u_s^2} = \frac{\Omega m^{*2} E_{c1}^2 k_B T}{4\pi^2 \hbar^4 c_l} \tag{8.55}$$

where $\rho = M/\Omega$ is the mass density of the atom and Ω is the volume of the unit cell; $c_l = \rho u_s^2$ is the longitudinal elastic constant. For a cubic crystal, $c_l = c_{11}$ for wave propagation along the (100) direction; for a (110) direction $c_l = (c_{11} + c_{12} + c_{44})/2$, and for a (111) propagation direction $c_l = (c_{11} + 2c_{12} + 4c_{44})/3$, where c_{11}, c_{12}, and c_{44} are components of the elasticity tensor (see Chapter 2).

The relaxation time due to longitudinal-mode acoustic phonon scattering can be obtained by substituting Eq. (8.55) into Eqs. (8.23) and (8.22) and then integrating. Hence

$$\frac{1}{\tau_a} = \left(\frac{2\pi v}{\Omega}\right) \int_0^\pi \sigma_a \sin\theta'(1 - \cos\theta')\, d\theta'$$

$$= \frac{m^{*2} v k_B T E_{c1}^2}{\pi \hbar^4 c_l} = \frac{v}{l_a} \tag{8.56}$$

where $l_a = \pi\hbar^4 c_l / m^{*2} E_{c1}^2 k_B T$ is the electron mean free path, which varies inversely with temperature. If we substitute $v = (2E/m^*)^{1/2}$ into Eq. (8.56), then τ_a^{-1} can be rewritten as

$$\frac{1}{\tau_a} = \frac{m_n^{*3/2} k_B T E_{c1}^2 (2E)^{1/2}}{\pi \hbar^4 c_l} \tag{8.57}$$

which shows that τ_a varies with $E^{-1/2}$ and T^{-1}. The electron mobility due to acoustic phonon scattering can be calculated by substituting τ_a from Eq. (8.57) into Eq. (7.64) and averaging over the energy. The result yields

$$\mu_a = \frac{q\langle\tau_a\rangle}{m_c^*} = \left(\frac{2\sqrt{2\pi}\, q\hbar^4 c_l}{3 m_n^{*3/2} m_c^* k_B^{3/2} E_{c1}^2}\right) T^{-3/2} \tag{8.58}$$

where m_c^* is the conductivity effective mass of electrons. For cubic crystals with ellipsoidal constant-energy surfaces, the effective mass product, $m_c m_n^{*3/2}$, satisfies

$$\frac{1}{m_\sigma m_n^{*3/2}} = \frac{1}{3 m_t m_l^{1/2}} \left(\frac{2}{m_t} + \frac{1}{m_l}\right) \tag{8.59}$$

Equation (8.58) predicts that the electron mobility due to longitudinal-mode acoustic phonon scattering is directly proportional to $T^{-3/2}$. Figures 8.7 and 8.8 in Section 8.8 present experimental results for electron mobilities in undoped germanium and silicon crystals that are found to be in good agreement with theoretical predictions for $T < 200$ K. However, at high temperatures, intervalley optical phonon scattering contributes substantially to electron mobility, and hence μ_n varies as T^{-n}, where n lies between 1.5 and 2.7. We note that scattering due to transverse-shear-mode phonons can also be calculated in a similar way to that of longitudinal-mode acoustic phonons.

8.5.2. *Piezoelectric Scattering*

For polar semiconductors (e.g., zinc blende crystals), the bonds are partially ionic, and the unit cell does not possess inversion symmetry. As a result, charged carriers may be scattered by longitudinal-mode acoustic phonons due to piezoelectric scattering. In general, the strain-induced electric field due to the piezoelectric effect can be represented by

$$\mathscr{E}_{pz} = -\left(\frac{e_{pz}}{\varepsilon_0 \varepsilon_s}\right)(\nabla_r r_n) \tag{8.60}$$

where e_{pz} is the piezoelectric constant. Therefore, the perturbing potential due to piezoelectric scattering can be expressed by

$$H' = \frac{e\mathscr{E}_{pz}}{q} = \left(|e| \frac{e_{pz}}{\varepsilon_0 \varepsilon_s q}\right)(\nabla_r r_n) \tag{8.61}$$

where $q = |k' - k| = 2k \sin(\theta'/2) = (2m^* v/\hbar) \sin(\theta'/2)$ is the phonon wave vector. A comparison of Eq. (8.61) with Eq. (8.49) for nonpolar acoustic phonon scattering reveals that, instead of the deformation potential constant E_{c1}, we now have $|e|e_{pz}/\varepsilon_0\varepsilon_s q$, which is not a constant (since q depends on v and θ'). Thus, the matrix element $H_{kk'}$ due to piezoelectric scattering can be written as

$$H_{kk'} = \frac{|e|e_{pz}}{\varepsilon_0 \varepsilon_s q}\left(\frac{k_B T}{2Vc_l}\right)^{1/2} = \left(\frac{e^2 K^2 k_B T}{2V\varepsilon_0 \varepsilon_s q^2}\right)^{1/2} \tag{8.62}$$

which is similar to Eq. (8.54). In Eq. (8.62) we have introduced a dimensionless electromechanical coupling constant, K^2, defined by

$$\frac{K^2}{1 - K^2} = \frac{e_{pz}^2}{\varepsilon_0 \varepsilon_s c_l} \tag{8.63}$$

The left-hand side of Eq. (8.63) will reduce to K^2 if $K^2 \ll 1$. For most polar semiconductors, the value of K^2 is of the order of 10^{-3}.

The transition probability and relaxation time due to piezoelectric scattering can be derived by substituting Eq. (8.62) into Eqs. (8.18) and (8.9), which yields

$$\frac{1}{\tau_{pz}} = \frac{V}{(2\pi)^3} \int 2\left(\frac{2\pi}{\hbar}\right)\left(\frac{e^2 K^2 k_B T}{2V\varepsilon_0 \varepsilon_s q^2}\right)\delta(E_k - E_{k'})k'^2(1 - \cos\theta') \sin\theta' \, d\theta' \, dk' \cdot 2\pi \tag{8.64}$$

where $q^2 = 4k'^2 \sin^2(\theta'/2)$ and $dk' = \hbar^{-1}(m^*/2E)^{1/2} \, dE$. Integration of Eq. (8.64) yields

$$\frac{1}{\tau_{pz}} = \frac{2^{3/2}\pi\hbar^2 \varepsilon_0 \varepsilon_s}{m^{*1/2} e^2 K^2 k_B T} E^{1/2} \tag{8.65}$$

which shows that the relaxation time for piezoelectric scattering is proportional to the square root of the energy. The carrier mobility due to piezoelectric scattering can be derived by using the expression of τ_{pz} given above and Eq. (8.2). For the nondegenerate case, the result is

$$\mu_{pz} = \frac{16\sqrt{2\pi}\hbar^2 \varepsilon_0 \varepsilon_s}{3m^{*3/2} eK^2(k_B T)^{1/2}} \tag{8.66}$$

This equation shows that the piezoelectric scattering mobility depends on $T^{-1/2}$. For a typical III–V compound semiconductor with $\varepsilon_s = 12$, $m^*/m_0 = 0.1$, and $K^2 = 10^{-3}$, we found a mobility value of 1.7×10^5 cm^2/V · sec for piezoelectric scattering at $T = 300$ K. This value is significantly higher than the deformation potential scattering mobility for most polar semiconductors. Therefore, piezoelectric scattering is usually not as important as acoustic phonon scattering due to deformation potential or ionized impurity scattering. Hence, piezoelectric scattering has little practical importance for most III–V compound semiconductors. However, piezoelectric scattering can become important for many II–VI compound semiconductors such as CdS and ZnSe, which have wurtzite crystal structure. For example, ionic and polar crystals, including most of the II–VI compound semiconductors, show a strong piezoelectric effect because wurtzite crystal structure lacks inversion symmetry, and hence the piezoelectric stress tensor is nonvanishing. The microscopic origins of piezoelectricity are due to ionic polarization, strain-dependent ionization, and electronic polarization. It has been suggested that the strain-induced flow of covalent charge between sublattices may be the dominant source of piezoelectricity in II–VI compound semiconductors, since electronic polarization is usually accompanied by acoustic mode phonons in such a crystal. This polarization can lead to a periodic electric perturbation potential which will contribute to electron scattering. The electron mobility due to piezoelectric scattering varies as $T^{-1/2}$, and the effects of piezoelectric scattering may be sufficiently large to be important in determining mobility in a piezoelectric crystal. For example, the temperature dependence of electron mobility for CdS shows that contributions from optical-mode phonon scattering and piezoelectric scattering become dominant at high temperatures. In contrast, for III–V compound semiconductors, piezoelectric scattering becomes important only at low temperatures.

8.6. OPTICAL PHONON SCATTERING

Optical phonon scattering becomes the predominant scattering source at high temperatures or at high electric fields. Both polar and nonpolar optical phonons are responsible for this type of scattering. The scattering of electrons by nonpolar optical phonons may be treated as one type of deformation potential scattering process. Nonpolar optical phonon scattering becomes important for silicon and germanium crystals above room temperatures in which intervalley scattering becomes the dominant process. However, intervalley scattering is generally not important for electrons in the conduction band minima located at the Γ band or along the ⟨100⟩ axes, but is important for conduction band minima located along the ⟨111⟩ (e.g., the Γ band in germanium and the L band in GaAs) crystal axis. Polar optical phonon scattering is the predominant scattering mechanism for ionic or polar crystals such as II–VI and III–V compound semiconductors. For these crystals the motion of negatively and positively charged atoms in a unit cell will produce an oscillating dipole, and the vibrational mode is called the polar optical-mode phonon. Polar optical phonon scattering is associated with atomic polarization arising from displacement caused by optical phonons. This is often the most important scattering mechanism at room temperature for III–V compound semiconductors. Optical phonon scattering is usually an inelastic process that cannot be treated by the relaxation time approximation, because the optical phonon energy is comparable to that of the mean electron energy (i.e., $\hbar\omega \sim k_B T$) at room temperature.

For a multivalley semiconductor such as silicon or germanium, intravalley scattering (i.e., scattering within a single conduction band minimum) near room temperature is usually accompanied by absorption or emission of a longitudinal-mode acoustic phonon. In this case, Eq. (8.58) may be employed to calculate the mobilities in these materials. However, at higher temperatures, intervalley scattering (i.e., scattering from one conduction band minimum to another) may become the dominant scattering mechanism. Intervalley scattering is usually accompanied by absorption or emission of a longitudinal-mode optical phonon. Since the energy of an optical phonon is comparable to that of the average electron energy, scattering of electrons by intervalley optical phonons is generally regarded as inelastic. In this case, the change in electron energy during scattering is no longer small, and so the relaxation time approximation can only be used if certain assumptions are made for this type of scattering.

For silicon and germanium, it is found that, over the temperature range in which intervalley optical phonon scattering is comparable to acoustic phonon scattering, the temperature dependence of electron mobility can be described by an empirical formula given by

$$\mu_n \propto T^{-n} \qquad \text{with } 2.5 > n > 1.5 \tag{8.67}$$

Figure 8.8 in Section 8.8 shows the temperature dependence of electron mobility in silicon at high temperatures. Theoretical calculations of hole mobilities for p-type silicon show that hole mobility varies as $T^{-2.3}$ when both optical and acoustic phonon scatterings become dominant. This result compares favorably with measured data.

In multivalley semiconductors such as silicon and germanium, intervalley scattering becomes important at high temperatures. In this case the scattering of electrons is controlled by nonpolar optical phonons, and the relaxation time is given by[5]

$$\frac{1}{\tau_{0i}} = \left(\frac{m_d^{3/2}}{\tau_0}\right) W\theta_D T^{1/2}\left[\langle n_0 + 1\rangle\left(\mathscr{E}_0 - \frac{\theta_D}{T}\right)^{1/2} + \langle n_0\rangle\left(\mathscr{E}_0 + \frac{\theta_D}{T}\right)^{1/2}\right] \tag{8.68}$$

where θ_D is the Debye temperature, $n_0 = [\exp(\theta_D/T) - 1]^{-1}$ is the phonon distribution function, and W is a constant which determines the relative coupling strength between the electrons and the optical phonons; $W = (D_0 hu_s)^2/2(k_0 a\theta_D)^2$, where D_0^2 is the optical deformation potential constant, $\mathscr{E} = \hbar\omega/k_BT$ is the reduced optical phonon energy, and a is the optical coupling constant. Note that the first term in Eq. (8.68) corresponds to the emission of an optical phonon, and the second term corresponds to the absorption of an optical phonon. Emission of optical phonons is important only when it is energetically possible (i.e., $\mathscr{E}_0 > \theta_D/T$). The mobility due to intervalley optical phonon scattering can be calculated by using Eq. (8.68) to find the average relaxation time $\langle\tau_{0i}\rangle$ and then substituting it into the mobility formula $\mu_{0i} = q\langle\tau_{0i}\rangle/m_c^*$. Based on Eq. (8.68) and the mobility formula, we can expect that the electron mobility due to intervalley optical phonon scattering will increase exponentially with temperature [i.e., $\mu_{0i} \sim \exp(-\theta_D/T)$].

In polar crystals such as II–VI and III–V compound semiconductors, polar optical phonon scattering becomes the dominant scattering mechanism at room temperature. Coupling between the conduction electrons and the optical-mode phonons in a polar crystal such as GaAs is a very effective scattering process. Both perturbation theory and polaron theory have been employed to derive the polar optical phonon scattering mobility. The theoretical expression of electron mobility derived by Petritz and Scanlon for polar optical-mode phonon scattering is given by[6]

$$\mu_{p0} = \frac{8qa_0}{3(2\pi mk_B\Theta)^{1/2}}\left(\frac{1}{\varepsilon_\infty} - \frac{1}{\varepsilon_s}\right)^{-1}\left(\frac{m_0}{m^*}\right)^{1/2}\frac{\chi(Z_0)[\exp(Z_0) - 1]}{Z_0^{1/2}} \tag{8.69}$$

where ε_∞ is the high-frequency dielectric constant, ε_s is the low-frequency dielectric constant $[\varepsilon_s = \varepsilon_\infty(\omega_l/\omega_s)^2]$, $\Theta = \hbar\omega_l/k_B$, $a_0 = \hbar^2/mq^2$, and $Z_0 = \Theta/T$; ω_l is the angular frequency of the longitudinal optical (LO) modes and $\chi(Z_0)$ is a quantity defined by Howarth and Sondheimer.[5] For a pure GaAs crystal, with a longitudinal optical phonon temperature Θ equal to 416 K (i.e., LO phonon energy $\hbar\omega_l \sim 36$ meV), the mobility μ_{p0} is roughly equal to 10,000 $\text{cm}^2/\text{V}\cdot\text{sec}$ at 300 K.

Due to the exponential dependence of μ_{p0} on temperature, the scattering of electrons by polar optical phonons becomes very unlikely at low temperatures. For example, at room temperature, the electron mobility in a lightly to moderately doped GaAs is due mainly to longitudinal acoustic phonon and polar optical phonon scattering, while ionized impurity scattering becomes dominant at low temperatures.

8.7. SCATTERING BY DISLOCATIONS

Dislocations in a semiconductor can act as scattering centers for both electrons and holes. The scattering of electrons by a dislocation center may be attributed to two effects. First, a dislocation center may be viewed as a line charge, and hence has an effect similar to a charged impurity center. Second, the strain field created by the dislocations in a crystal can produce a scattering potential similar to that of a deformation potential. However, it is generally known that scattering by dislocations can become important only if the density of dislocations is greater than 10^8 cm^{-2}.

To deal with scattering of electrons by dislocations, we may regard the dislocation line as a space charge cylinder of radius R and length L, as shown in Fig. 8.6. The probability that an electron is scattered into an angle $d\theta'$ by a dislocation line can be expressed by

$$P_d = \frac{d(b/R)}{d\theta'} = \frac{1}{2}\sin\left(\frac{\theta'}{2}\right) \tag{8.70}$$

where b is the scattering impact parameter. The differential scattering cross section per unit length of dislocation line charge is thus given by

$$\sigma_d(\theta') = R\sin\left(\frac{\theta'}{2}\right) \tag{8.71}$$

The total scattering cross section can be obtained by substituting Eq. (8.71) into Eq. (8.23) and integrating over θ' from 0 to π, which yields

$$\sigma_T = \frac{8R}{3} \tag{8.72}$$

Therefore, the relaxation time due to scattering of electrons by dislocations is given by

$$\tau_d = \frac{1}{N_d\sigma_T v} = \frac{3}{(8N_dRv)} \tag{8.73}$$

The electron mobility due to scattering by dislocations can be obtained directly from Eq. (8.73), which reads

$$\mu_d = \frac{q\tau_d}{m^*} = \left(\frac{3q}{8N_dR}\right)\frac{1}{(3m^*k_BT)^{1/2}} \tag{8.74}$$

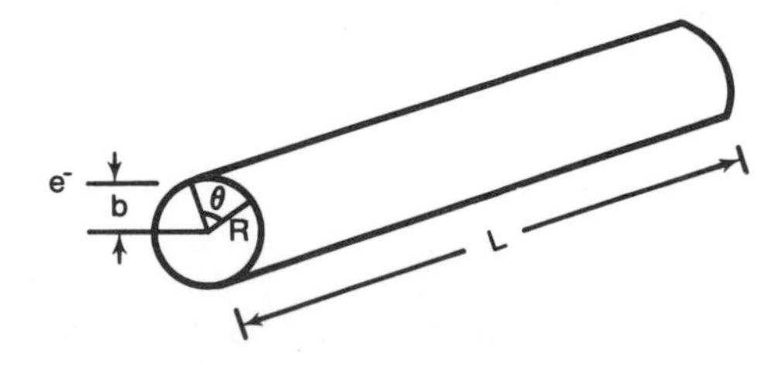

FIGURE 8.6. Scattering of electrons by a dislocation line.

where N_d is the density of dislocation lines. Equation (8.74) shows that the electron mobility due to scattering by dislocations is directly proportional to $T^{-1/2}$. For single-crystal silicon and germanium the dislocation density is usually very low, and hence scattering of electrons by dislocations is usually negligible. It should be pointed out that scattering of carriers by dislocations can also take place by virtue of their surrounding strain fields. The effect of strain fields can be calculated by finding a deformation potential from the known strain field. The scattering due to these strain fields is usually not important for n-type semiconductors, but can become important for p-type semiconductors.

8.8. ELECTRON AND HOLE MOBILITIES IN SEMICONDUCTORS

By using the relaxation time expressions and the mobility formulas derived in the preceding sections for different scattering mechanisms, the electron and hole mobilities in a semiconductor can in principle be calculated over a wide range of temperatures and doping concentrations. However, one must realize that these mobility formulas are derived for the isotropic elastic scattering case. Some modifications may be needed so that these mobility formulas can be applied to practical semiconductors. In general, it is not a simple task to fit theoretical calculations with experimental data for electron and hole mobilities in a semiconductor over a wide range of doping concentrations and temperatures, because in most semiconductors the total carrier mobility is usually controlled by several scattering mechanisms, such as acoustic phonons, optical phonons, and ionized impurities. An exception may exist for ultrapure semiconductors in which longitudinal acoustic phonon scattering may prevail over a wide range of temperatures, and thus allows for direct comparison between theoretical calculations and measured values. In general, electron mobility in a semiconductor due to multiple scattering processes can be calculated by using the expression

$$\mu_n = \frac{q\langle\tau\rangle}{m_c^*} \tag{8.75}$$

where

$$\frac{1}{\tau} = \sum_i \frac{1}{\tau_i} \tag{8.76}$$

and τ_i denotes the relaxation time due to a particular scattering process. For example, if the scattering mechanisms are due to acoustic phonons, optical phonons, ionized impurities, and neutral impurities, then the total scattering time constant can be obtained by employing the reciprocal sum of the relaxation times due to these scattering processes, namely,

$$\tau^{-1} = \tau_a^{-1} + \tau_0^{-1} + \tau_I^{-1} + \tau_N^{-1} \tag{8.77}$$

The electron mobility for the mixed scattering case can be calculated as follows: (1) find the total relaxation time τ due to different scattering mechanisms by using either Eq. (8.76) or Eq. (8.77), (2) calculate the average relaxation time $\langle\tau\rangle$ from Eq. (7.57), and (3) calculate the total electron mobility by using Eq. (8.75). It should be pointed out here that computing the carrier mobility by above procedure could be quite cumbersome and difficult if the relaxation time due to different scattering mchanisms is energy-dependent. In this case it may not be possible to obtain an analytical expression for the average relaxation time, and instead a numerical solution may be needed for the mean relaxation time and the total mobility. However, if the relaxation time due to different scattering mechanisms is independent of energy, then one can use the simplified reciprocal sum formula to obtain the total electron mobility (i.e., $\mu_n^{-1} = \sum_{i=1}^{n} \mu_i^{-1}$).

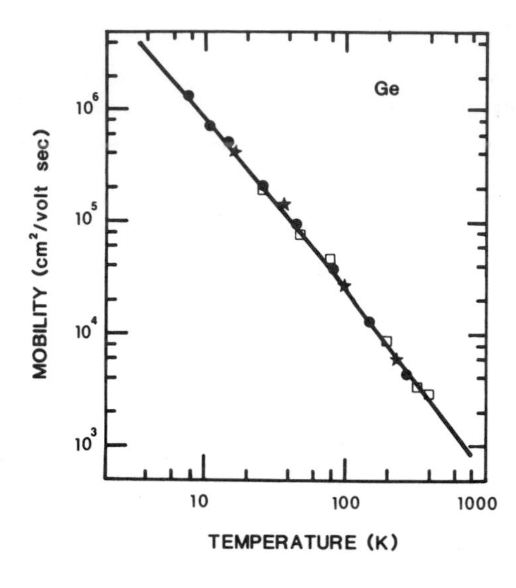

FIGURE 8.7. Comparison of calculated drift mobility of electrons (solid curve) and measured Hall mobility in a pure germanium specimen. The results show that acoustic phonon scattering is the dominant scattering mechanism in this sample. After Rode,[7] p. 83, by permission.

Figures 8.7 through 8.15 show calculated and measured data of electron and hole mobilities versus temperature for pure Ge, Si, GaAs, GaP, InSb, InP, InAs, CdS, and CdTe crystals, respectively. The solid lines are theoretical calculations, while the solid dots are measured data.[7] Table 8.1 lists electron drift mobilities for Ge, Si, GaP, and GaAs measured at 77 and 300 K. A comparison of the mobility data between these semiconductors shows that InSb has the highest and CdS the lowest electron mobility. In general, the electron mobilities for III–V compound semiconductors such as GaAs, InP, and InAs are higher than that of elemental semiconductors such as Si and Ge. Therefore, various electronic and photonic devices fabricated from III–V compound semiconductors are expected to operate at a much higher frequency and speed than that of silicon devices. To facilitate mobility calculations in GaAs due to various scattering mechanisms, Table 8.2 lists some bulk- and valley-dependent material parameters for the GaAs crystal.

In n-type silicon, the important scattering mechanisms for electrons are mainly due to acoustical phonon scattering, and ionized impurity scatterings. At room temperature, longitudinal-mode acoustic phonons are the dominant scattering source for undoped silicon, while ionized impurity scattering becomes important for $N_D \geq 10^{17}$ cm^{-3}. Optical deformation

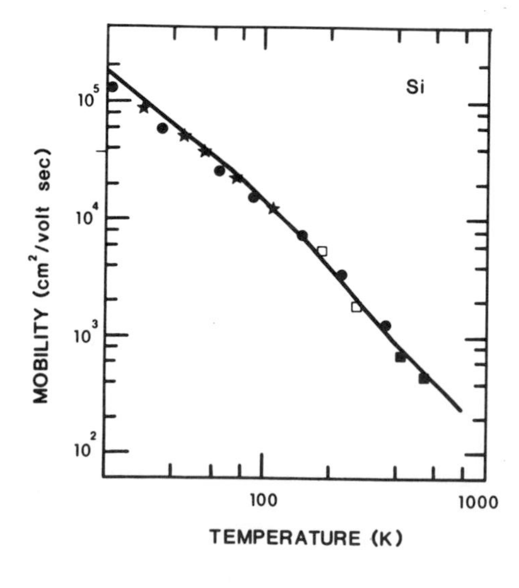

FIGURE 8.8. Comparison of calculated drift mobility of electrons (solid curve) and measured Hall mobility in a pure silicon specimen. The results show that acoustic phonon scattering is dominant for $T < 80$ K, and intervalley scattering becomes comparable to acoustic phonon scattering for $T \geq 300$ K. After Rode,[7] p. 81, by permission.

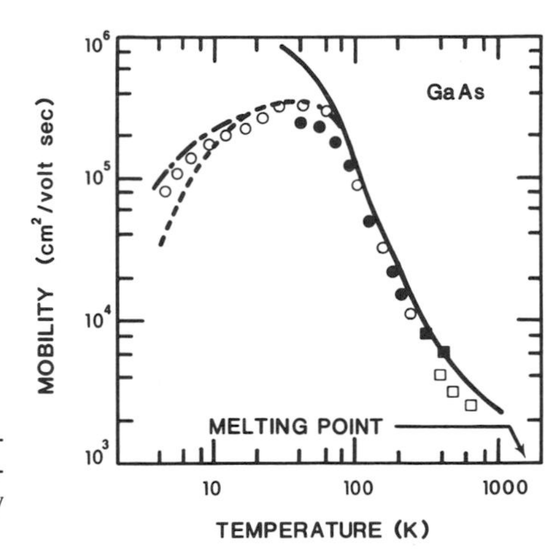

FIGURE 8.9. Comparison of calculated drift mobility of electrons (solid curve) and measured Hall mobility for a pure GaAs crystal. After Rode,[7] by permission.

potential scattering is negligible for electron scattering within a particular conduction band minimum, since the matrix element vanishes due to symmetry. However, scattering between different conduction band minima is significant (i.e., intervalley optical phonon scatterings). The scattering mechanisms in the conduction band of GaAs crystal are different from that of silicon. Due to the spherical symmetry of the electron wave function at the Γ-band, optical deformation potential scattering is zero in this conduction band minimum. Furthermore, due to the small electron effective mass (i.e., $m^* = 0.067m_0$) at the Γ-band minimum, the contribution of acoustic phonon scattering to electron mobility is also negligible in GaAs. As a result, electron mobility in GaAs is much higher than in silicon. The most important scattering mechanisms for GaAs are polar optical phonon scattering, ionized impurity scattering (for $N_D \geq 10^{17}\ \text{cm}^{-3}$), and intervalley optical phonon scattering (at high fields).

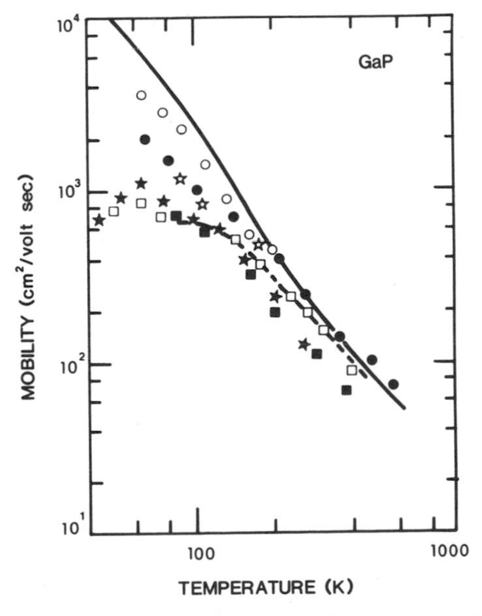

FIGURE 8.10. Comparison of calculated drift mobility of electrons (solid curve) and measured Hall mobility for a pure GaP crystal. After Rode,[7] by permission.

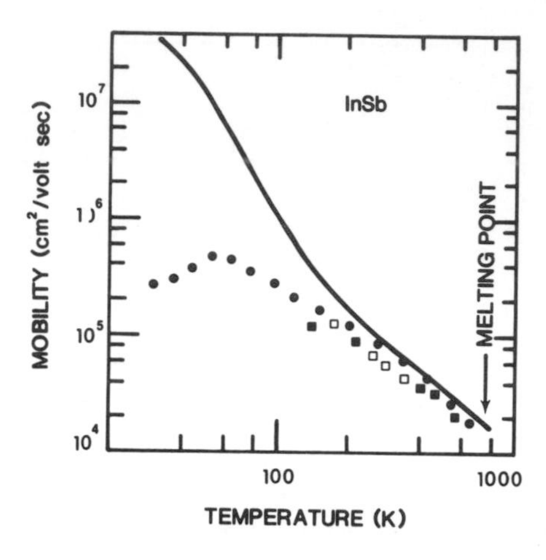

FIGURE 8.11. Comparison of calculated drift mobility of electrons (solid curve) and measured Hall mobility for a pure InSb crystal. After Rode,[7] by permission.

For p-type silicon and GaAs, the valence band maxima for both silicon and GaAs are located at the Γ point (i.e., the zone center), and wave functions of holes do not possess spherical symmetry. Therefore, optical deformation potential scattering is important for holes. In addition, acoustic phonon scattering and ionized impurity scattering are also important in the valence bands for both materials.

8.9. HOT ELECTRON EFFECTS IN A SEMICONDUCTOR

As discussed in Chapter 7, Ohm's law prevails under a low electric field condition and the current density varies linearly with the applied electric field. This can be expressed as

$$J_n = \sigma_n \mathscr{E} = n_0 q \mu_n \mathscr{E} \tag{8.78}$$

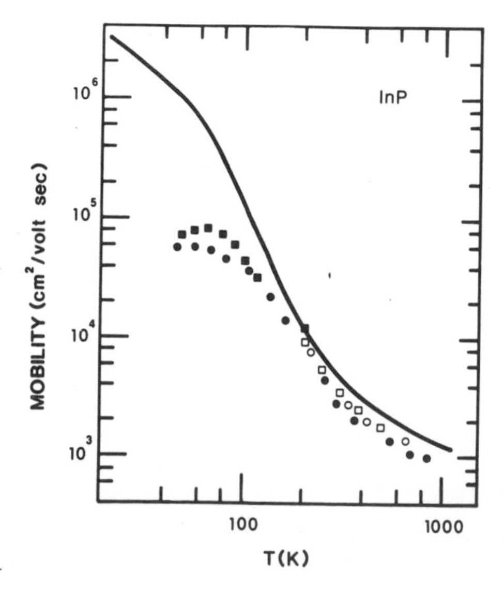

FIGURE 8.12. Comparison of calculated drift mobility of electrons (solid curve) and measured Hall mobility for a pure InP crystal. After Rode,[7] by permission.

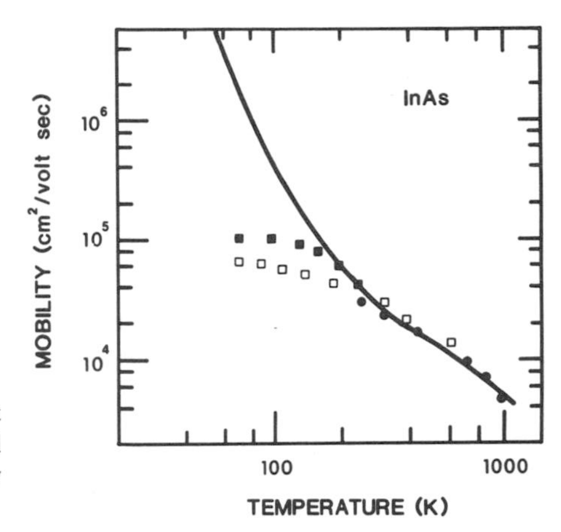

FIGURE 8.13. Comparison of calculated drift mobility of electrons (solid curve) and measured Hall mobility for a pure InAs crystal. After Rode,(7) by permission.

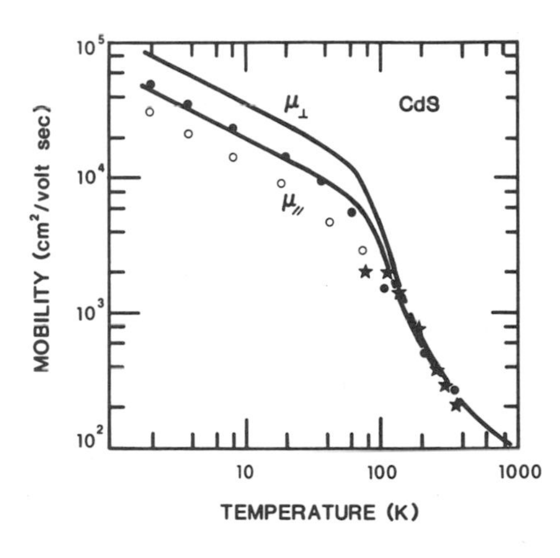

FIGURE 8.14. Comparison of calculated drift mobility of electrons (solid curve) and measured Hall mobility for a pure CdS specimen. After Rode,(7) by permission.

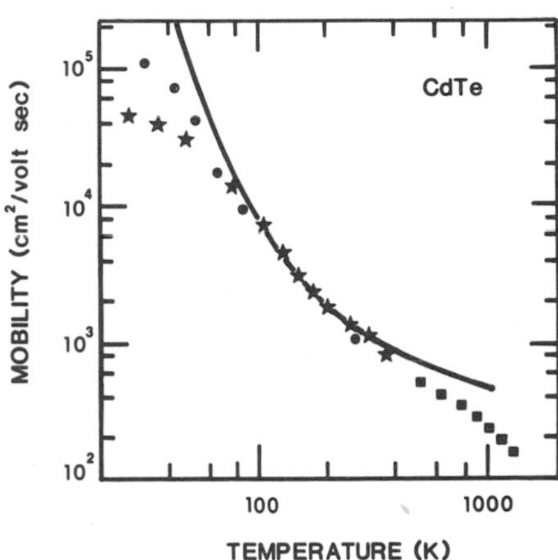

FIGURE 8.15. Comparison of calculated drift mobility of electrons (solid curve) and measured Hall mobility for a pure CdTe specimen. After Rode,(7) by permission.

where σ_n is the electrical conductivity and $\mathscr{E}$ is the applied electric field. As the electric field continues to increase, a point is reached at which the electric current density will no longer vary linearly with the electric field. This means that either the electron density or the electron mobility becomes a function of the electric field. An increase in the electron density is possible

TABLE 8.1. Electron Drift Mobility in Ge, Si, GaP, and GaAs

Material	Ge	Si	GaP	GaAs
$T = 300$ K:	μ_n (cm^2/V · sec)			
Calculated	4080	1580	183	8920
Measured	3800–4200	1350–1450	120–200	3500–9000
$T = 77$ K	μ_n (cm^2/V · sec)			
Calculated	37,400	22,800	4370	2.9×10^5
Measured	35,000–47,000	18,000–24,000		2.2×10^5

TABLE 8.2. Bulk and Valley Material Parameters for Mobility Calculations in GaAs

(a) Bulk parameters			
Density (g/cm)	5.36		
Piezoelectric constant (C/m^2)	0.16		
LO phonon energy (eV)	0.36		
Longitudinal sound velocity (cm/sec)			
Optical dielectric constant	10.92		
Static dielectric constant	12.90		
(b) Valley material parameters	Γ[100]	L[111]	X[100]
Electron effective mass (m^*/m_0)	0.067	0.222	0.58
Energy band gap E_g (eV)	1.43	1.77	1.96
Acoustic deformation potential (eV)	7.0	9.2	9.7
Optical deformation potential (eV/cm)	0	3×10^8	0
Number of equivalent valleys	1	4	3
Intervalley deformation potential constant D (eV/cm)			
Γ	0	1×10^9	1×10^9
L	1×10^9	1×10^9	1×10^8
X	1×10^9	5×10^8	7×10^8

if the electric field is high enough to cause (1) impact ionization (i.e., ionization of other imperfections or crystal atoms upon impact by hot electrons), (2) field ionization (i.e., ionization of imperfections by quantum mechanical tunneling to the nearest band), or (3) electrical injection (i.e., injection of electrons from contacts into the semiconductor). These processes will usually lead to a change of electric current with applied electric field that is faster than that predicted by Eq. (8.70). It will be shown later that these effects are usually observed in a p–n junction diode operating under a large reverse bias condition. Another high-field effect which has been found in many III–V compound semiconductor devices is that, under high fields, the current density will increase with the electric field at a slower rate than that predicted by Eq. (8.79). This effect arises from the decrease in electron mobility with increasing electric field resulting from scattering of electrons by optical phonons under high electric field conditions.

In this section, only the effect of applied electric fields on electron mobility will be considered. The mobility versus electric field relation is derived by assuming that scattering of electrons is dominated only by the longitudinal-mode acoustical phonons.

It is generally known that as the electric field increases, the electrons will gain energy from the applied electric field. Furthermore, scattering of electrons is associated with absorption or emission of phonons. Thus, in order to calculate energy loss by electrons due to phonon scattering, it is necessary to determine the average energy resulting from either absorption or emission of phonons. Thus, in order to calculate energy loss by electrons due to phonon scattering, it is necessary to determine the average energy resulting from either absorption or emission of phonons under high electric field conditions. The electron energy will increase, if there is a net gain in energy due to phonon absorption.

Under high electric field conditions, electron energy can be described in terms of an effective electron temperature, T_e. For a nondegenerate semiconductor, an increase in energy of the order of k_BT represents a large change in the mean electron energy, and an effective electron temperature T_e for such an energetic electron may be defined by the Maxwellian mean velocity, which is given by

$$\langle v \rangle = \left(\frac{k_B T_e}{8\pi m^*} \right)^{1/2} \tag{8.79}$$

If the effective electron temperature defined by Eq. (8.79) is equal to the lattice temperature, then the electron mobility will not be a function of the electric field. On the other hand, if there is a net gain of energy due to the effects of the applied electric field and acoustical phonon scattering, then the electrons will heat up. Under this condition, T_e becomes larger than the lattice temperature of the crystal, and the electric current will no longer vary linearly with the electric field. The derivation of the current–electric field relation under hig-field conditions is quite complicated, and only the relation between electron mobility and the electric field will be discussed in this section.

Under steady-state conditions, the electron mobility as a function of the electric field can be expressed in terms of the effective electron temperature T_e. The low-field electron mobility for longitudinal acoustical phonon scattering is given by

$$\mu_0 = \frac{4ql}{3(2\pi m^* k_B T)^{1/2}} \tag{8.80}$$

where l is the mean free path of electrons and is inversely proportional to the temperature. If the electron mobility under high-field conditions is expressed in terms of the low-field electron mobility μ_0 and the effective electron temperature T_e, then we can write a field-dependent electron mobility as

$$\mu = \mu_0 \left(\frac{T}{T_e} \right)^{1/2} \tag{8.81}$$

The condition for T_e to exceed the lattice temperature T is that $\mu_0 \mathscr{E} > u_s$. This means that the effective electron temperature starts to rise when the drift velocity becomes comparable to the velocity of sound (u_s) in the semiconductor. Since μ_0 is proportional to $T^{-3/2}$ for acoustic phonon scattering, it follows that μ is proportional to $T^{-1}T_e^{-1/2}$. For scattering by acoustic-mode phonons, the effective electron temperature versus electric field can be written as[8]

$$T_e = \left(\frac{T}{2} \right) \left\{ 1 + \left[1 + \left(\frac{3\pi}{8} \right) \left(\frac{\mu_0 \mathscr{E}}{u_s} \right)^2 \right]^{1/2} \right\} \tag{8.82}$$

In the relatively low-field regime, with $\mu_0 \mathscr{E} \ll u_s$, T_e reduces to

$$T_e \simeq T \left[1 + \left(\frac{3\pi}{32} \right) \left(\frac{\mu_0 \mathscr{E}}{u_s} \right)^2 \right] \tag{8.83}$$

Now by substituting Eq. (8.83) into Eq. (8.81) and using a binomial expansion, we obtain the corresponding field-dependent electron mobility, which reads

$$\mu = \mu_0\left[1 - \left(\frac{3\pi}{64}\right)\left(\frac{\mu_0\mathscr{E}}{u_s}\right)^2\right] \tag{8.84}$$

It is seen from Eq. (8.84) that in the intermediate-field regime, the differential mobility $(\mu_0 - \mu)$ varies as the square of the applied electric field.

In the high field regime, with $\mu_0\mathscr{E} \gg u_s$, Eq. (8.82) becomes

$$T_e = T\left[\left(\frac{3\pi}{32}\right)^{1/2}\left(\frac{\mu_0\mathscr{E}}{u_s}\right)\right] \tag{8.85}$$

and the corresponding electron mobility is given by

$$\mu = \left(\frac{32}{3\pi}\right)^{1/4}\left(\frac{\mu_0 u_s}{\mathscr{E}}\right)^{1/2} \tag{8.86}$$

Equation (8.86) predicts that electron mobility at high fields is inversely proportional to the square root of the electric field. Since the drift velocity v_d is equal to the product of electron mobility and electric field, it will increase with the square root of the electric field at high fields.

It is noteworthy that the results obtained above are for the case when scattering of electrons is due to longitudinal acoustical phonons. For such scattering, increasing electron energy with the applied electric field will result in an increase of phonon scattering, which in turn will lead to the reduction of electron mobility with increasing electric field. On the other hand, if scattering is dominated by ionized impurity scattering, then an increase in electron energy with increasing electric field will result in an increase of electron mobility. This is due to the fact that, for ionized impurity scattering, the probability of scattering decreases with increasing electron energy (i.e., $\tau_I^{-1} \sim E^{-3/2}$).

Figure 8.16 shows a plot of electron mobility and drift velocity versus electric field calculated for a typical semiconductor at 300 K.[2] In this figure, it is assumed that scattering of electrons is dominated by longitudinal-mode acoustical phonons at low and intermediate electric fields and by optical-mode phonons at high electric fields. The results clearly show that, for scattering by acoustical phonons, the electron mobility will decrease with the square of the applied electric fields, and the high-field electron mobility will vary inversely with the square root of the electric fields. At very high fields hot electrons will start interacting with optical phonons, which in turn will limit the drift velocity to a saturation value.

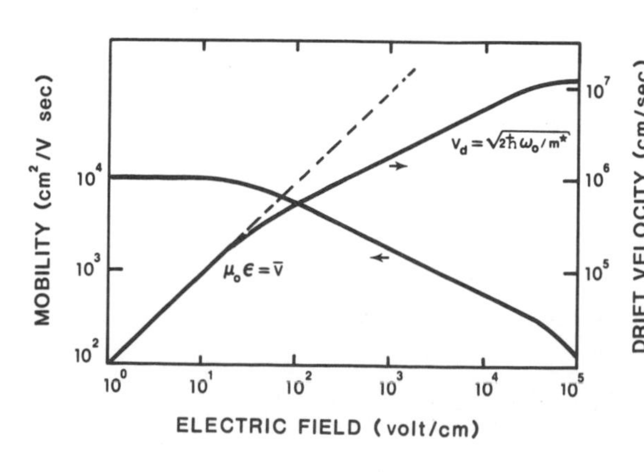

FIGURE 8.16. Electron mobility and drift velocity versus electric field calculated for a typical semiconductor, assuming longitudinal acoustic phonon scattering dominates at low and intermediate fields, and optical phonon scattering at high fields. Values of parameters used in the calculations are: $\mu_0 = 10^4$ cm²/V · sec, $u_s = 2 \times 10^5$ cm/sec, $\hbar\omega_0 = 0.04$ eV, and $m^* = m_0$. After Bube,[8] by permission.

The most widely used method to study hot electron effects in a semiconductor is the Monte Carlo approach. It consists of a simulation of the motion of one or more electrons inside a semiconductor subject to the action of an external applied electric field and of given scattering mechanisms. The basic principle of the Monte Carlo method relies on the generation of a sequence of random numbers with given distribution probabilities. When charge transport is analyzed on submicrometer scales under very high electric field conditions, the conventional semiclassical approach of transport processes in terms of the Boltzmann equation can be substituted by a full quantum mechanical description, namely, the Monte Carlo approach.

A brief description of the general procedure governing the Monte Carlo method is as follows. Consider the case of a cubic semiconductor under a very high electric field $\mathscr{E}$. The simulation starts with a set of given initial conditions with initial wave vector k_0. The duration of the first free flight is determined stochastically from a probability distribution determined by the scattering probabilities. The simulation of all quantities of interest, such as velocity and energy, are recorded. A dominant scattering mechanism is then selected as being responsible for the end of the free flight, according to the relative probabilities of all possible scattering mechanisms. From the transition rate of this scattering mechanism, a new k value after scattering is determined stochastically as the initial state of the new free flight, and the entire process is repeated iteratively. The results of the calculation become more and more accurate as the simulation ends when the quantities of interest are known with the desired precision. A detailed description of this method can be found in a monograph edited by Reggiani.[9] The Monte Carlo method allows one to extract derived phyical information from simulated experiments, and is a powerful tool for analyzing stationary and transient transport effects in semiconductors under high-field conditions. It is particularly useful for analyzing high-field transport properties in submicron devices.

PROBLEMS

8.1. Using Eqs. (8.18), (8.19), and (8.20) derive Eq. (8.21), assuming that $v_k = v_{k'}$, $k = k'$, and $d^3k' = k'^2 \sin\theta'\, d\theta'\, d\phi'\, dk'$.

8.2. Using the Conwell–Weisskopf model, derive Eqs. (8.37) and (8.38) [i.e., $V(r) = q/4\pi\varepsilon_0\varepsilon_s r$ for ionized impurity scattering).

8.3. Show that the maximum change of electron energy due to acoustic phonon scattering is given by Eq. (8.43). Does this satisfy the condition of elastic scattering?

8.4. Calculate the Debye screen lengths for Si, Ge, and GaAs for $N_D = 10^{15}$, 10^{17}, and $10^{19}\ \text{cm}^{-3}$. Given: $\varepsilon_s = 11.7$ for Si, 12 for GaAs, and 16 for Ge.

8.5. If the electron mobility in silicon is due to scattering of acoustic phonons and ionized impurities, show that the mixed scattering mobility can be approximated by

$$\mu_{LI} = \mu_L\left\{1 + \chi^2\left[\mathrm{Ci}(\chi)\cos\chi + \sin\chi\left(\mathrm{Si}(\chi) - \frac{\pi}{2}\right)\right]\right\}$$

where $\chi^2 = 6\mu_L/\mu_I$, and μ_L and μ_I are the lattice and ionized impurity scattering mobilities; $\mathrm{Ci}(\chi)$ and $\mathrm{Si}(\chi)$ are the cosine and sine integrals of χ, respectively. [See the paper by P. P. Debye and E. M. Conwell, *Phys. Rev.* **93**, 693 (1954).]

8.6. Using Eq. (8.58), calculate the electron mobility due to acoustic phonon scattering for pure silicon when $100 \geq T \leq 300\ K$. *Given*: $(m_0/m^*)^{5/2} = 20.4$, $E_{c1} = 12.8$ eV, and $Mu_s^2 = 1.97 \times 10^{12}\ \text{dynes/cm}^2$.

8.7. Using the expression for τ_I given by Eq. (8.35) and $\mu_I = q\langle\tau_I\rangle/m^*$, show that the ionized impurity scattering mobility is given by Eq. (8.36).

8.8. The inverse scattering relaxation time for piezoelectric scattering in a nondegenerate semiconductor with a parabolic band is given by

$$\tau_{pe}^{-1} = \frac{3q^2 \kappa T P^2 m^* d}{6\pi \hbar^3 \varepsilon_0 k'}$$

where k' is the electron wave vector and P is the piezoelectric coefficient. Derive an expression for the piezoelectric scattering mobility, and show that the mobility is proportional to $T^{-1/2}$.

8.9. For a GaAs crystal, the polar optical phonon scattering mobility μ_{p0} given by Eq. (8.69) can be simplified to

$$\mu_{p0} = 5.3 \times 10^3 \left(\frac{\chi(Z_0)[\exp(Z_0) - 1]}{Z_0^{1/2}} \right)$$

where $\Theta = \hbar\omega_l / k_B$, $a_0 = \hbar^2/mq^2$, and $Z_0 = \Theta/T$; ω_l is the angular frequency of the longitudinal optical modes, and $\chi(Z_0)$ is a quantity defined by Howarth and Sondheimer.[5] For pure GaAs, the longitudinal optical phonon temperature Θ is equal to 416 K (i.e., the LO phonon energy $\hbar\omega_l \sim 36$ meV), and $\mu_{p0} \approx 10{,}000\ \text{cm}^2/\text{V} \cdot \text{sec}$ at 300 K.
The ionized impurity scattering mobility, μ_i, is given by

$$\mu_i = \frac{1.5 \times 10^{18}}{N_I[\ln(1 + b) - b/(1 + b)]} T^{3/2}$$

where

$$b = \frac{9.1 \times 10^{13}}{n_0} T^2$$

The piezoelectric scattering mobility is given by

$$\mu_{pz} = 4.89 \times 10^5 \left(\frac{100}{T} \right)^{1/2}$$

Assuming that Matthiessen's rule prevails, the total electron mobility for this GaAs crystal can be approximated by

$$\mu_n^{-1} = \mu_{p0}^{-1} + \mu_i^{-1} + \mu_{pz}^{-1}$$

Using the above expression, plot the electron mobility versus temperature for this GaAs crystal for $100 < T < 600$ K at $N_I = 10^{16}$, 10^{17}, and $10^{18}\ \text{cm}^{-3}$.

REFERENCES

1. E. M. Conwell and V. F. Weisskopf, "Ionized Impurity Scattering in Semiconductors," *Phys. Rev.* **77**, 388–390 (1950).
2. H. Brooks, "Theory of the Electrical Properties of Germanium and Silicon," in: *Advances in Electronics and Electron Physics* (L. Marton, ed.), Vol. 7, pp. 85–182, Academic Press, New York (1955).
3. C. Erginsoy, "Neutral Impurity Scattering in Semiconductors," *Phys. Rev.* 1013–1017 (1956).
4. J. Bardeen and W. Shockley, "Deformation Potential and Lattice Scattering in Semiconductors," *Phys. Rev.* **80**, 72–84 (1950).
5. D. Howarth and E. Sondheimer, *Proc. R. Soc. London, Ser. A* **219**, 53 (1953).
6. R. L. Petritz and W. W Scanlon, "Optical Mode Phonon Scattering in Semiconductors," *Phys. Rev.* **97**, 1620 (1955).

7. D. L. Rode, "Low-Field Electron Transport," in: *Semiconductors and Semimetals* (R. K. Willardson and A. C. Beer, eds.), Vol. 10, Academic Press, New York (1975).
8. R. H. Bube, *Electronic Properties of Crystalline Solids*, Chapter 8, p. 289, Academic Press, New York (1974).
9. R. Reggiani, *Hot Electron Transport in Semiconductors*, Springer-Verlag, New York (1985).

BIBLIOGRAPHY

F. J. Blatt, *Physics of Electronic Conduction in Solids*, McGraw-Hill, New York (1968).

C. Herrings and E. Vogt, "Transport and Deformation Potential Theory for Many-Valley Semiconductors with Anisotropic Scattering," *Phys. Rev.* **101**, 944–961 (1956).

S. S. Li, *The Dopant Density and Temperature Dependence of Electron Mobility and Resistivity in n-type Silicon*, NBS Special Publication, 400–33 (1977).

S. S. Li, *The Dopant Density and Temperature Dependence of Hole Mobility and Resistivity in p-type Silicon*, NBS Special Publication, 400–47 (1979).

K. Seeger, *Semiconductor Physics*, Springer-Verlag, New York (1973).

9

Optical Properties and Photoelectric Effects

In this chapter, we are concerned with fundamental optical properties and internal photoelectric effects in a semiconductor. The optical properties associated with the fundamental and free-carrier absorption processes and internal photoelectric effects such as photoconductive, photovoltaic, and photomagnetoelectric effects in a semiconductor will be depicted. Many fundamental physical properties, such as the energy band structure and recombination mechanisms, can be understood by studying the optical absorption process and photoelectric effects in a semiconductor. Practical applications using internal photoelectric effects such as photovoltaic and photoconductivity effects in a semiconductor have been extensively reported in the literature. Future trends are moving toward the development of various optoelectronic devices and optoelectronic integrated circuits (OEICs) for use in computing, communications, signal processing, and data transmission. In fact, many photonic devices such as LEDs, laser diodes, and photodetectors fabricated from silicon and III–V compound semiconductors have been widely used in optical communications, data transmission, and signal processing systems.

Depending on the energy of incident photons, there are two kinds of optical absorption processes which may occur in a semiconductor. The first kind of absorption process involves the absorption of photons which have energies equal to or greater than the bandgap energy of a semiconductor. This type of optical absorption is called the fundamental or interband absorption process. The fundamental absorption process is usually accompanied by an electronic transition across the forbidden gap, and as a result excess electron–hole pairs are generated in the semiconductor. The absorption coefficient due to interband transition is usually very large. For example, in the ultraviolet (UV) to visible spectral range, typical values of the absorption coefficient for most semiconductors range from $10^6\ \mathrm{cm}^{-1}$ near the UV wavelength to $1\ \mathrm{cm}^{-1}$ near the cutoff wavelength of the semiconductor (e.g., 0.4 to 1.1 μm for silicon). However, the absorption coefficient becomes very small (e.g., less than $1\ \mathrm{cm}^{-1}$) when the photon energies fall below the bandgap energy of the semiconductor. In this case, another type of optical absorption process takes over in the semiconductor. This type only results in electronic transitions within the allowed energy band, and is called the free-carrier absorption process. The fundamental absorption process, which leads to an interband transition, must be treated quantum mechanically, while the free-carrier absorption process can be described adequately by classical electromagnetic wave theory. Finally, absorption of photons with energies below the bandgap energy of the semiconductor may also lead to electronic transitions from localized impurity states to conduction or valence band states. For example, the extrinsic photoconductivity observed at low temperatures is due to the photoexcitation of free carriers from shallow impurity states to conduction or valence band states. Since the energy bandgap varies between 0.1 and 3.0 eV for most semiconductors, the fundamental optical absorption occurs in the visible to infrared spectral regimes. Therefore, most semiconductors are opaque from the ultra-violet (UV) to the infrared spectral range, and become transparent in the far-infrared spectral regime (i.e., for $\lambda > 10\ \mu$m).

In order to better understand the optical absorption processes in a semiconductor, it is necessary to first consider the two optical constants, namely, the index of refraction and the extinction coefficient. These two optical constants may be derived by solving Maxwell's equations for electromagnetic waves in a solid, as will be depicted in Section 9.2. The free-carrier absorption process is presented in Secton 9.3. Section 9.4 deals with the fundamental absorption process in a semiconductor. Internal photoelectric effects, including photoconductive (PC), photovoltaic (PV), and photomagnetoelectric (PME) effects, in a semiconductor, are depicted in Sections 9.5, 9.6, and 9.7, respectively. Section 9.5 will discuss both intrinsic and extrinsic photoconductivity effects in a semiconductor and their practical applications. The internal photovoltaic effect, also known as the Dember effect, is discussed in Section 9.6. Finally, the photomagnetoelectric (PME) effect in a semiconductor is examined in Section 9.7.

9.1. OPTICAL CONSTANTS OF A SOLID

The optical constants, such as the index of refraction and the extinction coefficient, may be derived by solving Maxwell's equations for electromagnetic waves propagating in a solid. It is well known that some solids are transparent while others are opaque, that some solid surfaces are strongly reflective while others tend to absorb optical radiation that falls on them. The optical absorption depends on the wavelength of the incident optical radiation. For instance, most semiconductors show strong absorption from ultraviolet (UV) to near-infrared (IR) wavelengths, and become transparent in the far-infrared spectral regime. Therefore, in order to obtain a better understanding of the optical absorption process in a semiconductor, it is important to derive expressions for these two basic optical constants in the UV to IR spectral range.

The propagation of electromagnetic waves in a solid may be described by the well-known Maxwell's equations, which are given by

$$\nabla \cdot \mathscr{E} = 0 \tag{9.1}$$

$$\nabla \times \mathscr{E} = -\frac{\partial B}{\partial t} \tag{9.2}$$

$$\nabla \cdot B = 0 \tag{9.3}$$

$$\nabla \times H = \sigma \mathscr{E} + \varepsilon_0 \varepsilon_s \frac{\partial \mathscr{E}}{\partial t} \tag{9.4}$$

In free space, the electromagnetic wave equations may be obtained from Eqs. (9.1) through (9.4) using the relations $B = \mu_0 H$, $\sigma = 0$, and $\varepsilon_s = 1$. This approach yields

$$\nabla^2 \mathscr{E} = \mu_0 \varepsilon_0 \frac{\partial^2 \mathscr{E}}{\partial t^2} = \left(\frac{1}{c^2}\right) \frac{\partial^2 \mathscr{E}}{\partial t^2} \tag{9.5}$$

where $c = 1/\sqrt{\varepsilon_0 \mu_0}$ is the velocity of light in free space. Inside the solid, the wave equation can also be obtained by solving Eqs. (9.1) through (9.4), and the result is

$$\nabla^2 \mathscr{E} = \mu_0 \varepsilon_0 \varepsilon_s \frac{\partial^2 \mathscr{E}}{\partial t^2} + \mu_0 \sigma \frac{\partial \mathscr{E}}{\partial t} \tag{9.6}$$

A comparison of Eqs. (9.5) and (9.6) shows that the difference between waves propagating in free space and in a solid is due to the difference in the dielectric constant and the electrical conductivity in both media. It is evident that Eq. (9.6) will reduce to Eq. (9.5) if the dielectric constant ε_s is equal to 1 and the electrical conductivity σ is zero. The first term on the right-hand side of Eq. (9.6) is due to the displacement current, while the second term represents the conduction current. An electromagnetic wave with frequency ω propagating in the z-direction and polarizing in the x-direction can be expressed by

$$\mathscr{E}_x = \mathscr{E}_0 \exp\left[i\omega\left(\frac{z}{v} - t\right)\right] = \mathscr{E}_0 \exp[i(k^* \cdot z - \omega t)] \tag{9.7}$$

where k^* is the complex wave number, and v is the velocity of the electromagnetic waves inside the solid and could also be a complex number. We note that k^* and v are related by

$$k^* = \frac{\omega}{v} \tag{9.8}$$

Now by substituting Eq. (9.7) into Eq. (9.6) we obtain

$$k^{*2} = \frac{\omega^2}{v^2} = \mu_0 \varepsilon_0 \varepsilon_s \omega^2 + i\mu_0 \sigma \omega \tag{9.9}$$

or

$$k^* = \frac{\omega}{v} = \left(\frac{\omega}{c}\right)\left(\varepsilon_s + \frac{i\sigma}{\omega \varepsilon_0}\right)^{1/2} = \left(\frac{\omega}{c}\right) n^* \tag{9.10}$$

where

$$n^* = \left(\varepsilon_s + \frac{i\sigma}{\omega \varepsilon_0}\right)^{1/2} = \varepsilon_s^{*1/2} \tag{9.11}$$

is the complex refractive index of the solid, and ε_s^* is the complex dielectric constant. Quantity n^* can be written as

$$n^* = n + i\kappa_e \tag{9.12}$$

where n is the index of refraction of the media, and κ_e is the extinction coefficient, which is a constant relating the attenuation of the incident electromagnetic wave inside the solid to its penetration depth. For example, if an incident electromagnetic wave propagates into a solid at a distance equal to one wavelength in free space (i.e., $\lambda_0 = 2\pi c/\omega$), then its amplitude is decreased by a factor of $e^{-2\pi\kappa_e}$. It is seen that the fundamental optical absorption coefficient can be defined in terms of κ_e, to be discussed later. From Eqs. (9.11) and (9.12), we obtain the real and imaginary parts of the complex refractive index, given respectively by

$$n^2 - \kappa_e^2 = \varepsilon_s \tag{9.13}$$

and

$$2n\kappa_e = \frac{\sigma}{\omega \varepsilon_0} \tag{9.14}$$

Therefore, the optical properties of a solid, as observed macroscopically, may be described in terms of the complex refractive index n^*. For example, by substituting Eqs. (9.10) and (9.12) into Eq. (9.7), the solution for the electromagnetic waves inside the solid becomes

$$\mathscr{E}_x = \mathscr{E}_0 \exp\left(\frac{-k_e \omega z}{c}\right) \exp\left[i\omega\left(\frac{nz}{c} - t\right)\right] \tag{9.15}$$

which shows that the velocity of an incident electric wave in a solid reduces to c/n, and its amplitude decreases exponentially with distance. The attenuation of incident electric waves is associated with the absorption of electromagnetic energy by the dissipating medium. However, the optical constant commonly measured in a solid is not the extinction coefficient, k_e, but the absorption coefficient, α. The optical absorption coefficient is related to the Poynting vector of the electromagnetic wave energy flow by

$$S(z) = S_0\, e^{-\alpha z} \tag{9.16}$$

where $S(z)$ is the Poynting vector which is proportional to the square of the amplitude of the electric waves (i.e., $\sim|\mathscr{E}^2|$) given by Eq. (9.15). By equating the square of Eq. (9.15) and Eq. (9.16), we obtain the absorption coefficient in the form

$$\alpha = \frac{2k_e\omega}{c} = \frac{4\pi k_e}{\lambda_0} \tag{9.17}$$

where λ_0 is the wavelength of the electromagnetic waves in free space. Thus, the extinction coefficient k_e, can be determined from the absorption coefficient of the semiconductor. It is noteworthy that both the real $(n^2 - k_e^2)$ and imaginary $(2nk_e)$ parts of the complex refractive index are the quantities measured in a solid. In practice, $n^2 - k_e^2$ and $2nk_e$ are obtained by measuring the reflection and transmission coefficients of a solid.

In deriving the reflection coefficient we consider the case of normal incidence, as shown in Fig. 9.1. If $\mathscr{E}_x$ $(\mathscr{H}_y)$ and $\mathscr{E}_x''$ $(\mathscr{H}_y'')$ denote the incident and reflected electric (magnetic) waves, respectively, and $\mathscr{E}_x'$ $(\mathscr{H}_y')$ represents the transmitted electric (magnetic) waves into the solid in the z-direction, then for $z > 0$ we can write

$$\mathscr{E}_x' = \mathscr{E}_0 \exp\left[i\omega\left(\frac{n^*z}{c} - t\right)\right] \tag{9.18}$$

For $z < 0$ (i.e., in free space), the electric wave is equal to the sum of the incident and reflected waves, which can be expressed by

$$\mathscr{E}_x = \mathscr{E}_1 \exp\left[i\omega\left(\frac{z}{c} - t\right)\right] + \mathscr{E}_2 \exp\left[-i\omega\left(\frac{z}{c} + t\right)\right] \tag{9.19}$$

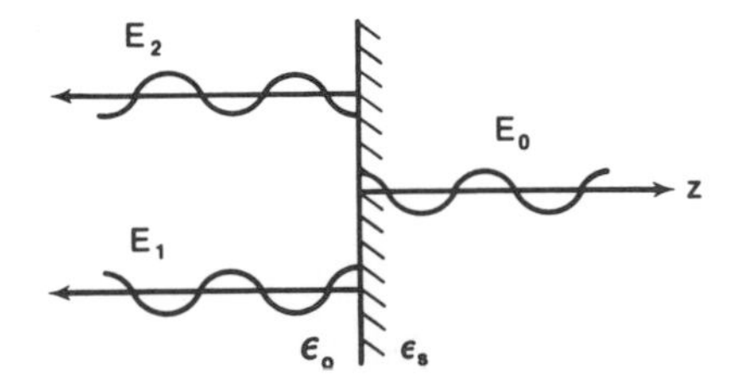

FIGURE 9.1. An electromagnetic wave propagating into a solid under normal incidence.

The magnetic wave components polarized in the y-direction (i.e., the $\mathscr{H}_y$ quantities) may be related to the electric wave components in the x-direction by the characteristic impedance of the medium. For the incident, transmitted, and reflected electromagnetic waves, they are given by

$$\frac{\mathscr{E}_x}{\mathscr{H}_y} = \sqrt{\frac{\mu_0}{\varepsilon_0}} = Z_0, \qquad \frac{\mathscr{E}'_x}{\mathscr{H}'_y} = \sqrt{\frac{\mu'}{\varepsilon}} = Z', \qquad \frac{\mathscr{E}''_x}{\mathscr{H}''_y} = -\sqrt{\frac{\mu_0}{\varepsilon_0}} = -Z_0 \tag{9.20}$$

The boundary conditions at the $z = 0$ plane requires that the tangential components of both $\mathscr{E}_x$ and $\mathscr{H}_y$ be continuous. Thus, we can write

$$\mathscr{E}'_x = \mathscr{E}_x + \mathscr{E}''_x \qquad \text{and} \qquad \mathscr{H}'_y = \mathscr{H}_y + \mathscr{H}''_y \tag{9.21}$$

The solution of Eqs. (9.20) and (9.21) yields

$$\frac{\mathscr{E}''_x}{\mathscr{E}_x} = \frac{Z' - Z_0}{Z' + Z_0} = \frac{\sqrt{\varepsilon_0/\mu_0} - \sqrt{\varepsilon/\mu'}}{\sqrt{\varepsilon_0/\mu_0} + \sqrt{\varepsilon/\mu'}} \tag{9.22}$$

Since $\mathscr{H}''_y$ is reversed in sign from that of $\mathscr{H}_y$, we obtain

$$\frac{\mathscr{H}''_y}{\mathscr{H}_y} = -\frac{\mathscr{E}''_x}{\mathscr{E}_x} = -\frac{Z' - Z_0}{Z' + Z_0} \tag{9.23}$$

The Poynting vector is equal to the product of the electric and magnetic field strengths (i.e., $|S| = \mathscr{E}_x\mathscr{H}_y$, $|S''| = \mathscr{E}''_x\mathscr{H}''_y$), so the reflection coefficient R can be obtained by solving Eqs. (9.22) and (9.23):

$$\begin{aligned} R = \left|\frac{S''}{S}\right| &= \left(\frac{\mathscr{E}''_x}{\mathscr{E}_x}\right)\left(\frac{\mathscr{H}''_y}{\mathscr{H}_y}\right) = \left(\frac{Z' - Z_0}{Z' + Z_0}\right)^2 \\ &= \left[\frac{\sqrt{\varepsilon_0/\mu_0} - \sqrt{\varepsilon/\mu_0}}{\sqrt{\varepsilon_0/\mu_0} + \sqrt{\varepsilon/\mu_0}}\right]^2 = \left(\frac{n' - n_0}{n' + n_0}\right)^2 \end{aligned} \tag{9.24}$$

For nonmagnetic materials, $\mu' = \mu_0$, $\varepsilon = \varepsilon_0\varepsilon_s$, and $n' = n + ik_e$; for free space, $n_0 = 1$. Thus, the absolute value of the reflection coefficient for normal incidence can be written as

$$R = \left|\frac{(n-1)^2 + k_e^2}{(n+1)^2 + k_e^2}\right| \tag{9.25}$$

where n is the index of refraction in free space and k_e is the extinction coefficient of the solid.

The transmission coefficient T, defined as the ratio of the transmission power and the incident power, can be derived in a similar way to that of the reflection coefficient given above. Hence

$$T = \left|\frac{\mathscr{E}'_x\mathscr{H}'_y}{\mathscr{E}_x\mathscr{H}_y}\right| = \frac{4Z_0Z'}{(Z' + Z_0)^2} = \frac{4n_0n'}{(n_0 + n')^2} \tag{9.26}$$

For $n_0 = 1$, the absolute value of the transmission coefficient can be obtained from Eq. (9.26), and the result yields

$$T = \frac{4n}{(n+1)^2 + k_e^2} \tag{9.27}$$

It is seen from Eqs. (9.24) and (9.26) that the sum of the transmission and reflection coefficients is equal to one (i.e., $R + T = 1$).

For normal incidence, it is evident from Eqs. (9.25) and (9.27) that by measuring T and R we can determine both n and k_e. However, for incident angles other than normal, the reflection coefficient will, in general, depend on the polarization, and from observation of different angles of incidence, both n and k_e can be determined if k_e is not too small. If both n and k_e are large, then R will approach unity.

An inspection of Eq. (9.13) reveals that the dielectric constant ε_s can also be determined directly from the refractive index n, provided that k_e is much smaller than 1. Values of n can be found directly from measurements of reflection coefficients if k_e is very small.

There is considerable practical interest in measurements of transmission and reflection coefficients in free space under normal incidence by using a plane-parallel sheet of crystal with refractive index n and thickness d. If I_0, I_t, and I_r denote the incident, transmitted, and reflected wave intensities through the sample, then the normalized transmitted and reflected wave intensities can be expressed, respectively, by

$$\frac{I_t}{I_0} = \frac{(1-R)^2 e^{-\alpha d}(1 + k_e^2/n^2)}{1 - R^2 e^{-2\alpha d}} \tag{9.28}$$

and

$$\frac{I_r}{I_0} = \frac{R(1 - e^{-2\alpha d})}{1 - R^2 e^{-2\alpha d}} \tag{9.29}$$

Equations (9.28) and (9.29) show that both n and k_e can be found by measuring I_t and I_r. For most transmission experiments, it is valid to assume that $k_e^2 \ll n^2$. If the sample thickness d is further chosen to ensure that $R^2 e^{-2\alpha d} \ll 1$, then Eq. (9.29) can be simplified to

$$\frac{I_t}{I_0} = (1-R)^2 \exp(-\alpha d) \tag{9.30}$$

From Eq. (9.30), it is of interest that the optical absorption coefficient α of a semiconductor near the band edge can be determined by measuring the transmission intensity versus wavelength on two thin samples of different thicknesses without knowledge of the reflectance. This is true as long as the reflection coefficients in both samples are assumed identical.

For elemental semiconductors such as Si and Ge, the main contribution to the dielectric constant arises from electronic polarization. However, in compound semiconductors (such as II–VI compounds), both electronic and ionic polarizations contribute to the dielectric behavior. The increase in the degree of ionicity in these compounds relative to the group-IV elements will lead to a significant difference between the static and optical (high-frequency) dielectric constants. The high-frequency dielectric constant ε_s^∞ is equal to n^2. The static or low-frequency dielectric constant ε_s^s can be calculated from the relation

$$\varepsilon_s = \varepsilon_s^\infty \left(\frac{\omega_l}{\omega_t}\right)^2 \tag{9.31}$$

TABLE 9.1. Refractive Indices and Dielectric Constants of Silicon, Germanium, and Some III–V and II–VI Compound Semiconductors

Material	n	ε_s^s	ε_s^∞
Si	3.44	11.8	11.6
Ge	4.00	16	15.8
InSb	3.96	17	15.9
InAs	3.42	14.5	11.7
GaAs	3.30	12.5	10.9
GaP	2.91	10	8.4
CdS	2.30	8.6	5.2
CdSe	2.55	9.2	6.4
CdTe	2.67	9.7	7.1
ZnS	2.26	8.1	5.1
ZnSe	2.43	8.7	5.9

where ω_l and ω_t denote the longitudinal- and transverse-mode optical phonon frequencies, respectively. Table 9.1 lists values of dielectric constants and refractive indices for Si, Ge, and some III–V and II–VI compound semiconductors.

9.2. FREE-CARRIER ABSORPTION PROCESS

When the energy of incident electromagnetic radiation is smaller than the bandgap energy of the semiconductor, excitation of electrons from the valence band into the conduction band will not occur. Instead, the absorption of incident electromagnetic radiation will result in the excitation of lattice phonons and the acceleration of free electrons inside the conduction band. In the conduction band, free-carrier absorption is proportional to the density of conduction electrons. Since free-carrier absorption involves only electronic transitions within the conduction band, we can apply the classical equation of motion to handle the interaction between the electromagnetic wave and the conduction electrons. The equation of motion for an electron due to a time-varying electric wave (i.e., $\mathscr{E}_0 e^{i\omega t}$) of frequency ω and propagating in the z-direction is given by

$$m^* \frac{\partial^2 z}{\partial t^2} + \left(\frac{m^*}{\tau}\right)\frac{\partial z}{\partial t} = q\mathscr{E}_0 e^{i\omega t} \tag{9.32}$$

where τ is the relaxation time and m^* is the effective mass of electrons in the conduction band. The solution of Eq. (9.32) is given by

$$z = \frac{(q\mathscr{E}_0/m^*)\, e^{i\omega t}}{(i\omega/\tau - \omega^2)} \tag{9.33}$$

If the electron density in the conduction band is equal to N_0, then the total polarization P, which is proportional to the product of displacement z and electron density N_0, can be expressed by

$$P = N_0 q z \tag{9.34}$$

The polarizability p^*, which is defined as the polarization per unit electric field, can be written as

$$p^* = \frac{N_0 q z}{\mathscr{E}_0} \tag{9.35}$$

The complex dielectric constant ε_s^* given by Eq. (9.11) is related to the polarizability p^* by

$$\varepsilon_s^* = \varepsilon_s' - i\varepsilon_s'' = n^{*2} = \varepsilon_s + \frac{p^*}{\varepsilon_0} = \varepsilon_s + \frac{(N_0 q^2/m^* \varepsilon_0)}{(i\omega/\tau - \omega^2)} \tag{9.36}$$

Note that the second term on the right-hand side of Eq. (9.36) is due to the contribution of free-carrier absorption. Therefore, from Eq. (9.36), the real and imaginary parts of the complex dielectric constant can be written as

$$\varepsilon_s' = n^2 - k_e^2 = \varepsilon_s - \frac{\tau\sigma_0}{\varepsilon_0(1 + \omega^2\tau^2)} \tag{9.37}$$

and

$$\varepsilon_s'' = 2nk_e = \frac{\sigma_0}{\omega\varepsilon_0(1 + \omega^2\tau^2)} \tag{9.38}$$

where $\sigma_0 = N_0 q^2 \tau/m^*$ is the dc electrical conductivity. In Eqs. (9.37) and (9.38), it is assumed that τ is a constant and independent of energy.

We note that the absorption coefficient α is related to the extinction coefficient k_e via Eq. (9.17). Thus, from Eq. (9.38), we obtain

$$\alpha = \frac{4\pi k_e}{\lambda_0} = \frac{\sigma_0}{nc\varepsilon_0(1 + \omega^2\tau^2)} \tag{9.39}$$

which shows the frequency dependence of the optical absorption coefficient. Two limiting cases, $\omega\tau \gg 1$ and $\omega\tau \ll 1$, will be considered next.

Long Wavelength Limit ($\omega\tau \ll 1$). In this case, the absorption coefficient given by Eq. (9.39) reduces to

$$\alpha = \frac{\sigma_0}{nc\varepsilon_0} \tag{9.40}$$

and the real part of the dielectric constant in Eq. (9.37) is reduced to

$$\varepsilon_s' = \varepsilon_s - \frac{\tau\sigma_0}{\varepsilon_0} \tag{9.41}$$

Equation (9.40) shows that the absorption coefficient is independent of frequency, but depends on temperature through σ_0. For example, for an n-type germanium sample with $\omega\tau = 10^{-12}$ sec this corresponds to a wavelength of about 2 mm. For a lightly doped semiconductor with large dielectric constant, the contribution of σ_0/ε_0 in Eq. (9.41) to ε_s' is quite small. Thus, from Eq. (9.41) it can be shown that ε_s' is equal to ε_s.

For heavily doped semiconductors with large σ_0, $(\tau\sigma_0/\varepsilon_0)$ becomes much larger than ε_s and hence ε_s' becomes negative. This corresponds to the metallic case. If we assume $\varepsilon_s \ll \tau\sigma_0/\varepsilon_0$, then from Eqs. (9.41) and (9.37) we obtain the real part of the dielectric constant:

$$\varepsilon_s' = n^2 - k_e^2 = -\frac{\tau\sigma_0}{\varepsilon_0} \tag{9.42}$$

Similarly, from Eq. (9.38) we obtain the imaginary part of the dielectric constant:

$$\varepsilon_s'' = 2nk_e = \frac{\sigma_0}{\omega\varepsilon_0} \tag{9.43}$$

Now, by combining Eqs. (9.42) and (9.43) we obtain

$$\omega\tau = -\frac{(n^2 - k_e^2)}{2nk_e} \tag{9.44}$$

It is seen that from Eq. (9.44) for $\omega\tau \ll 1$, k_e is equal to n. Thus, by using $n = k_e$ in Eq. (9.43), the refractive index is given by

$$n = \left(\frac{\sigma_0}{2\omega\varepsilon_0}\right)^{1/2} \tag{9.45}$$

The absorption coefficient can be deduced from Eqs. (9.40) and (9.45), and the result yields

$$\alpha = \left(\frac{2\sigma_0\omega}{\varepsilon_0 c^2}\right)^{1/2} \tag{9.46}$$

Equation (9.46) shows that α is proportional to the square root of the frequency. Therefore, in this case, the material exhibits metallic behavior. This corresponds to the well-known skin effect in which the penetration depth of the incident electromagnetic wave, δ, is inversely proportional to the square root of the frequency and the electrical conductivity.

Short-Wavelength Limit ($\omega\tau \gg 1$). This usually occurs in the wavelength regime extending upward from far-infrared toward the fundamental absorption edge of the semiconductor. The short-wavelength free-carrier absorption becomes negligible when the photon energy exceeds the bandgap energy of the semiconductor. To understand the free-carrier absorption process in the short-wavelength limit, we solve Eqs. (9.37) through (9.39) to obtain

$$\varepsilon_s' = \varepsilon_s - \frac{\sigma_0}{\varepsilon_0\omega^2\tau} = n^2 - k_e^2 \tag{9.47}$$

$$\varepsilon_s'' = 2nk_e = \frac{\sigma_0}{\varepsilon_0\omega^3\tau^2} \tag{9.48}$$

$$\alpha = \frac{\sigma_0}{nc\varepsilon_0\omega^2\tau^2} = \frac{N_0 q^3 \lambda_0^2}{4\pi^2 c^3 m^{*2} \mu n \varepsilon_0} \tag{9.49}$$

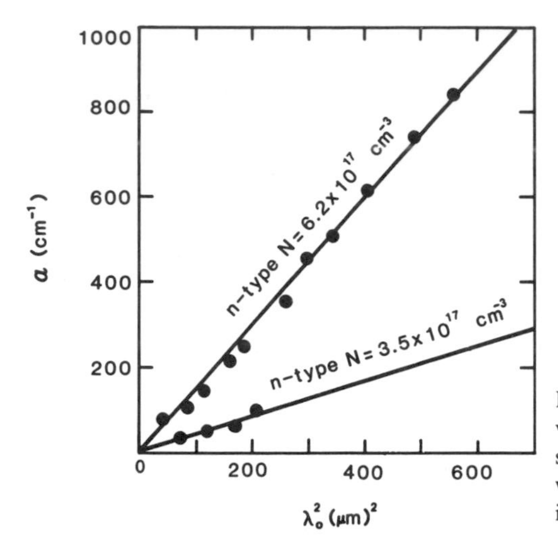

FIGURE 9.2. Optical absorption coefficient versus the square of the wavelength for two InSb specimen with different doping densities over the wavelength range in which free-carrier absorption is dominant. After Moss,[1] by permission.

Equation (9.49) shows that, for $\omega\tau \gg 1$, the absorption coefficient is directly proportional to the square of the wavelength. This prediction has been observed in a number of semiconductors. Figure 9.2 shows the absorption coefficient versus the square of the wavelength for two n-type InSb specimens with different dopant concentrations.[1] The results follow the prediction given by Eq. (9.49).

It is seen from Eq. (9.47) that ε_s' changes sign as ω decreases. The condition for which $\varepsilon_s' = 0$ corresponds to total internal reflection, and the frequency for this to occur is called the plasma resonance frequency, ω_p, which is given by

$$\omega_p = \left(\frac{\sigma_0}{\varepsilon_0 \varepsilon_s \tau}\right)^{1/2} = \left(\frac{N_0 q^2}{m^* \varepsilon_0 \varepsilon_s}\right)^{1/2} \tag{9.50}$$

where ω_p is the frequency at which a classical undamped plasma of free electrons exhibits its normal mode of oscillation. For a germanium sample with $N_0 = 10^{16}\ \text{cm}^{-3}$, $m^* = 0.12\, m_0$, and $\varepsilon_s = 16$, we find that ω_p is equal to 2×10^2 GHz, which falls into the rather difficult millimeter-wavelength regime. In order to observe plasma resonance in the microwave-frequency regime, an ultra-pure crystal should be used, otherwise the experiment must be performed at an extremely low temperature. For a germanium crystal, a carrier concentration of $10^{13}\ \text{cm}^{-3}$ or less is required for plasma resonance to be observed in the microwave-frequency range. For metals, since the electron concentration is very high (i.e., $10^{22}\ \text{cm}^{-3}$), the plasma resonance frequency usually falls in the far-ultraviolet regime, which corresponds to a photon energy of 10 to 20 eV. Free-carrier absorption has been used extensively in determining the relaxation time constant and the conductivity effective mass of electrons and holes in a semiconductor.

9.3. FUNDAMENTAL ABSORPTION PROCESS

A fundamental absorption process takes place when photons with energies greater than the bandgap energy of the semiconductor are absorbed in a semiconductor. This process usually results in the generation of electron–hole pairs in a semiconductor. For most semiconductors, the fundamental absorption process may occur in the ultraviolet (UV), visible, and infrared (IR) wavelength regimes. It is the most important optical absorption process because important photoelectric effects are based on such absorption process to generate electron–hole pairs in a semiconductor.

There are two types of optical transition associated with the fundamental absorption process, namely, direct and indirect band-to-band transitions, as shown in Fig. 9.3a and b. In direct transition only one photon is involved, while in indirect transition additional energy is supplied or released in the form of phonons. The absorption coefficient associated with these two transition processes depends on the probability per unit time that an electron makes a transition from the valence band into the conduction band when an incident photon is absorbed. The transition probability P_i can be calculated by using first-order time-dependent perturbation theory, which is given by

$$P_i = \left(\frac{2\pi}{h}\right)|M_{if}|^2 g_n(E) \tag{9.51}$$

where P_i is the transition probability per unit time from the initial state k_i in the valence band to the final state k_f in the conduction band; M_{if} denotes the matrix element due to perturbation which connects states k_i and k_f of the system, and $g_n(E)$ is the density of final states in the conduction band.

In the present case, the perturbation is due to incident electromagnetic radiation, and the important matrix element corresponding to electric dipole transition is given by

$$\begin{aligned} M_{if} &= \int \psi_i^* \nabla_r \psi_f \, d^3r \\ &= \int u_v^*(k_i, r)\, e^{-ik_i\cdot r} \nabla_r u_c(k_f, r)\, e^{ik_f\cdot r}\, d^3r \\ &= \int u_v^*(k_i, r) \nabla_r u_c(k_f, r)\, e^{i(k_f - k_i)\cdot r}\, d^3r + ik_f \int u_v^*(k_f, r) u_c(k_f, r)\, e^{i(k_f - k_i)\cdot r}\, d^3r \end{aligned} \tag{9.52}$$

where ψ_i and ψ_f denote the electron wave functions in the valence and conduction bands, respectively; $u_v(k_i, r)$ and $u_c(k_f, r)$ are the Bloch functions for the valence and conduction bands, respectively. Both terms on the right-hand side of Eq. (9.52) contain the factor $e^{i(k_f - k_i)\cdot r}$, which oscillates rapidly. Thus, the integrand of Eq. (9.52) will vanish unless

$$k_f = k_i \tag{9.53}$$

which is the condition of momentum conservation for such a transition. In fact, the contribution of photon momentum in Eq. (9.53) has been ignored since it is very small compared to

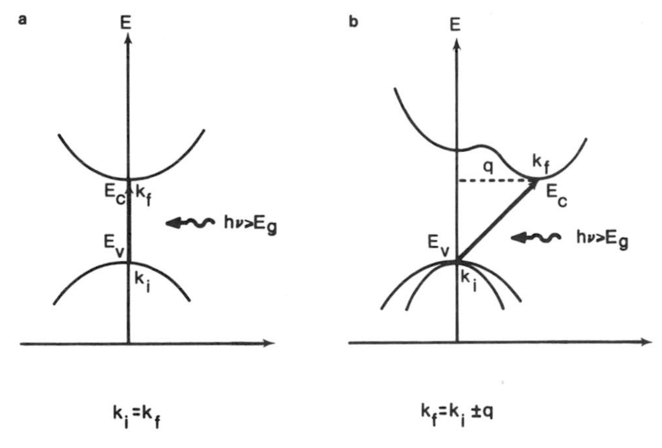

FIGURE 9.3. Direct and indirect transitions associated with the fundamental absorption processing in a semiconductor.

that of the crystal momentum. We note that the first term on the right-hand side of Eq. (9.52) is known as the allowed transition; its value is real and independent of the wave vector k_f of the final state. The second term on the right-hand side of Eq. (9.52) is imaginary, and it depends on the wave vector of the final state, k_f. Transition associated with the second term is called forbidden transition. Since the absorption coefficient is directly related to the rate of transition probability, P_{if}, Eq. (9.52) can be used to derive the absorption coefficient for direct and indirect interband transitions taking place between the valence band and conduction band of a semiconductor.

9.3.1. Direct Transition Process

The direct (or vertical) transition shown in Fig. 9.3a is the dominant absorption process taking place in a direct bandgap semiconductor when the conduction band minimum and the valence band maximum are located at the same k-value in the reciprocal space (i.e., at the Γ point). In order to derive an expression for the absorption coefficient near the conduction band minimum, it is necessary to find the density of final states $g(E_n)$ in Eq. (9.51). We note that electron energy in the conduction band is given by

$$E_n = E_c + \frac{\hbar^2 k^2}{2m_n^*} \tag{9.54}$$

and in the valence band by

$$E_p = E_v - \frac{\hbar^2 k^2}{2m_p^*} \tag{9.55}$$

The photon energy corresponding to such a transition can be written as

$$h\nu = E_n - E_p = E_g + \frac{\hbar^2 k^2}{2m_r^*} \tag{9.56}$$

where E_g is the bandgap energy of the semiconductor and $m_r^* = m_n^* m_p^*/(m_n^* + m_p^*)$ is the reduced effective mass.

Equations (9.54) through (9.56) enable the density of final states for the conduction band with a parabolic band structure to be written in the form

$$g(E_n) = \left(\frac{4\pi}{h^3}\right)(2m_r^*)^{3/2}(h\nu - E_g)^{1/2} \tag{9.57}$$

From Eqs. (9.52) and (9.57), the absorption coefficient for a direct allowed transition can be expressed by

$$\alpha_d^a = K_d^a(h\nu - E_g)^{1/2} \tag{9.58}$$

In the direct allowed transitions, the square of the matrix element (i.e., $|M_{if}|^2$) is independent of the wave vector, and hence K_d^a is a constant, independent of energy.

Equation (9.58) shows that, for an allowed direct optical transition, the absorption coefficient α_d^a varies as $(h\nu - E_g)^{1/2}$. A plot of α_d^2 versus $h\nu$ near the fundamental absorption edge allows the energy bandgap of a semiconductor to be determined. This is illustrated in

Fig. 9.4 for a p-type GaAs[2]; the intercept of the straight line with the horizontal axis yields the bandgap energy E_g of the GaAs.

In the forbidden direct transitions, as given by the second term of Eq. (9.52), the matrix element M_{if} is proportional to k, and hence the optical absorption coefficient in this case can be written as

$$\alpha_d^f = K_d^f(h\nu - E_g)^{3/2} \tag{9.59}$$

The energy $[E = (h\nu - E_g)]$ to the power 3/2 relationship with α_d^f, given by Eq. (9.59), is due to the fact that the transition probability for direct forbidden transitions is proportional to the product of $k^2(\sim E)$ and the density-of-states function $(\sim E^{1/2})$.

9.3.2. Indirect Transition Process

For an indirect bangap semiconductor, the conduction band minimum and valence band maximum are not located at the same k-value in the reciprocal space. Therefore, the indirect optical transition induced by photon absorption is usually accompanied by the simultaneous absorption or emission of a phonon. As illustrated in Fig. 9.3b, conservation of momentum in this case is given by

$$k_f = k_i \pm q \tag{9.60}$$

where k_f and k_i are the wave vectors of electrons for the final and initial states, respectively, and q is the phonon wave vector. The plus sign in Eq. (9.60) corresponds to phonon emission and the minus sign to phonon absorption. Conservation of energy for indirect optical transitions requires that

$$\begin{aligned} h\nu &= E_n - E_p \pm \hbar\omega_q \\ &= E_g + \frac{\hbar^2(k_n - k_c)^2}{2m_n^*} + \frac{\hbar^2 k_p^2}{2m_p^*} \pm \hbar\omega_q \end{aligned} \tag{9.61}$$

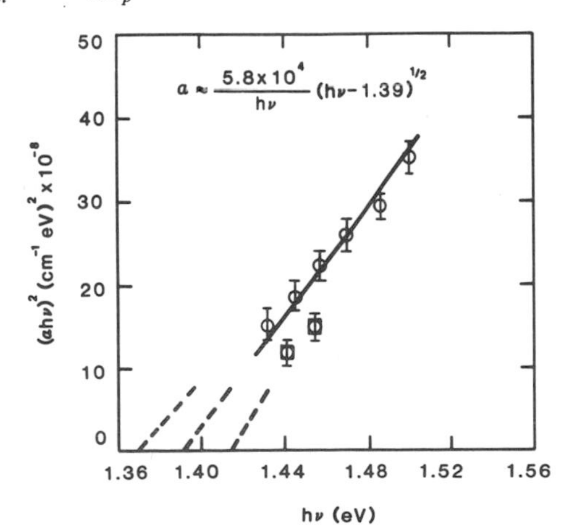

FIGURE 9.4. Direct transition in a p-type GaAs specimen with $N_A = 10^{17}$ cm^{-3} and an absorption coefficient greater than 9×10^3 cm^{-1}. The threshold energy is (1.39 ± 0.02) eV. After Kudman and Seidel,[2] by permission.

where

$$E_n = E_c + \frac{\hbar^2(k_n - k_c)^2}{2m_n^*} \tag{9.62}$$

$$E_p = E_v - \frac{\hbar^2 k_p^2}{2m_p^*} \tag{9.63}$$

and $(k_n - k_c) \ll k_c$.

It is seen from Eq. (9.61) that conservation of energy is achieved by a phonon emission or phonon absorption process. The plus sign in Eq. (9.61) refers to phonon emission and the minus sign to phonon absorption, where k_p denotes the initial state in the valence band, k_c the state at the conduction band minimum, and k_n the final state in the conduction band. Now consider the case in which transition from the k_p state in the valence band is induced by a photon with energy $h\nu$. The density of states in the valence band can be described by

$$g_v(E_p) = A_v E_p^{1/2} \tag{9.64}$$

where $E_p = E_v - \Delta E$, with ΔE a small energy interval in which transitions can take place; $A_v = (4\pi/h^3)(2m_p^*)^{3/2}$.

The density of final conduction band states involving phonon absorption is given by

$$\begin{aligned} g_c(E_n) &= A_c(E_n - E_c)^{1/2} = A_c(h\nu - E_g - E_p + \hbar\omega_q)^{1/2} \\ &= A_c(\Delta E - E_p)^{1/2} \end{aligned} \tag{9.65}$$

where $A_c = (4\pi/h^3)(2m_n^*)^{3/2}$. Equation (9.65) is obtained by solving Eqs. (9.61) through (9.64). In Eq. (9.65), we have used the relation $h\nu = (E_g \pm \hbar\omega_q + \Delta E)$. Therefore, the total effective density of states for transitions involving absorption and emission of a phonon can be expressed by

$$\begin{aligned} g(h\nu) &= \int_0^{\Delta E} g_c(E_n) g_v(E_p)\, dE_p = A_c A_v \int_0^{\Delta E} (\Delta E - E_p)^{1/2} E_p^{1/2}\, dE_p \\ &= K_i^a \Delta E^2 = K_i^a (E_v - E_p)^2 = K_i^a (h\nu - E_g \pm \hbar\omega_q)^2 \end{aligned} \tag{9.66}$$

where the plus sign denotes phonon absorption and the minus sign phonon emission.

The probability of phonon absorption and emission is directly related to the average phonon density, which is given by

$$\langle n_q \rangle = (e^{\hbar\omega_q/k_B T} - 1)^{-1} \tag{9.67}$$

Equations (9.66) and (9.67) yield the optical absorption coefficient due to indirect transitions with phonon absorption in the form

$$\alpha_{ia} = \langle n_q \rangle g(h\nu) = K_{ia} \frac{(h\nu - E_g + \hbar\omega_q)^2}{(e^{\hbar\omega_q/k_B T} - 1)} \tag{9.68}$$

For transitions involving phonon emission, the optical absorption coefficient can be expressed by

$$\alpha_{ie} = \langle n_q + 1 \rangle g(h\nu) = K_{ie} \frac{(h\nu - E_g - \hbar\omega_q)^2}{(1 - e^{-\hbar\omega_q/k_B T})} \tag{9.69}$$

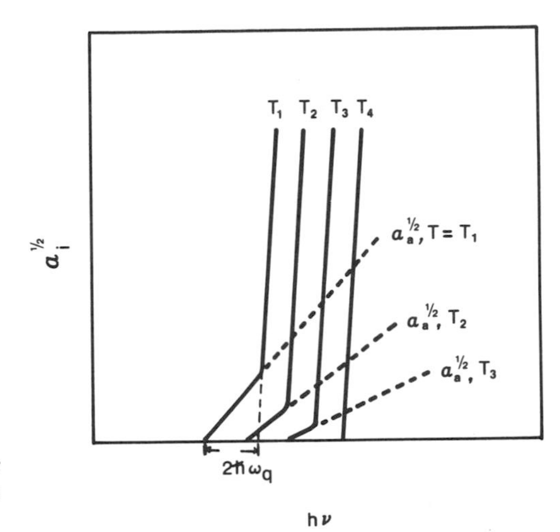

FIGURE 9.5. $\alpha_i^{1/2}$ versus photon energy $h\nu$ for indirect optical transitions with temperature as a parameter. Note that $T_1 > T_2 > T_3 > T_4$.

Therefore, by combining Eqs. (9.68) and (9.69), the absorption coefficient for indirect allowed transitions involving both emission and absorption of a phonon may be written as

$$\alpha_i = \alpha_{ie} + \alpha_{ia}$$
$$= K_i\left\{\frac{(h\nu - E_g - \hbar\omega_q)^2}{(1 - e^{-\hbar\omega_q/k_BT})} + \frac{(h\nu - E_g + \hbar\omega_q)^2}{(e^{\hbar\omega_q/k_BT} - 1)}\right\} \tag{9.70}$$

The first term in Eq. (9.70) is due to phonon emission, while the second term is attributed to phonon absorption. It is clear from Fig. 9.5 that the optical absorption coefficient curve for indirect allowed transitions involving phonon absorption will extend to longer wavelengths than those associated with phonon emission. The optical absorption coefficient for indirect allowed transitions varies with the square of photon energy. A plot of $\alpha_i^{1/2}$ versus $h\nu$ at different temperatures should yield a straight line, and its intercept with the horizontal axis allows one to determine the phonon energy and the energy bandgap of the semiconductor.

A plot of $\alpha_i^{1/2}$ versus $h\nu$ involving the emission and absorption of a phonon for four different temperatures is shown in Fig. 9.5. According to Eq. (9.70), two straight-line segments can be observed in the $\alpha_i^{1/2}$ versus $h\nu$ plot. For small photon energy, only α_{ia} (i.e., associated with phonon absorption) contributes, and the $\alpha_{ia}^{1/2}$ versus $h\nu$ plot intersects the axis at $h\nu = E_g - \hbar\omega_q$. For $h\nu > E_g + \hbar\omega_q$, α_{ie} becomes dominant at lower temperatures. Since the intersection of $\alpha_{ie}^{1/2}$ versus hv occurs at $h\nu = E_g + \hbar\omega_q$, we can determine both the energy bandgap and phonon energy from this plot. Figure 9.6 shows the square root of the absorption coefficient versus photon energy near the fundamental absorption edge of germanium with temperature as a parameter.[3] For this sample the phonon emission process becomes dominant at $T < 20$ K.

Several effects could influence the accuracy of determining the bandgap energy from optical absorption measurements in a semiconductor. The first effect is due to the Burstein shift in a degenerate semiconductor. In a heavily doped n-type semiconductor, the Fermi level lies inside the conduction band. Therefore, in order for photon-generated electrons to make transitions from the valence band into the conduction band, the photon energy must be greater than the bandgap energy of the semiconductor so that electrons can be excited into the empty states above the Fermi level in the conduction band. This shifts the optical absorption edge to a higher energy with increased doping concentration. This problem is particularly severe for small-bandgap semiconductors such as InSb and InAs, since the electron effective masses and densities of states in the conduction band are small for these materials. In calculating the

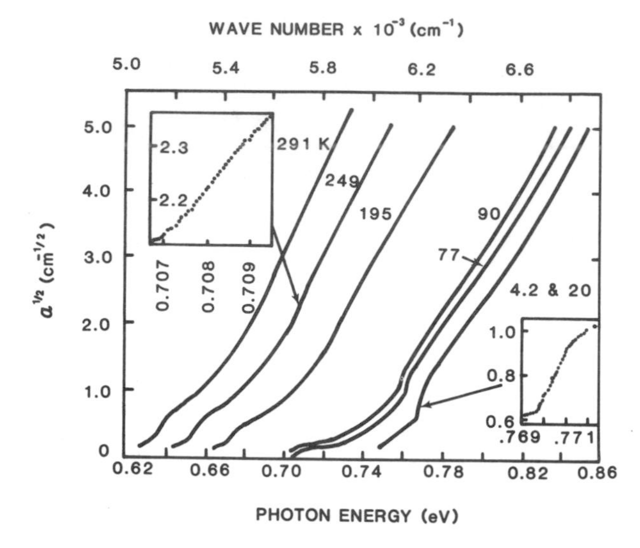

FIGURE 9.6. Square root of the absorption coefficient versus photon energy for a germanium specimen with temperature as a parameter. The inserts show the spectral resolution. After Macfarlane *et al.*,[3] by permission.

Burstein shift, the effect of energy-band nonparabolicity should also be considered. The second effect is related to the formation of impurity band-tail states (or impurity bands) arising from high concentration of shallow impurities or defects, which can merge into the conduction band or the valence band. This effect will result in an exponential absorption edge in the semiconductor. The third effect is associated with exciton formation in the semiconductor. An exciton is an electron–hole pair bound together by Coulombic interaction. Excitons may be free, bound, or constrained to a surface, or associated with a defect complex. The binding energies for excitons are slightly below the conduction band edge, and hence exciton features are sharp peaks just below the absorption edge. Excitons are usually observed at low temperatures and become dissociated into free carriers at room temperature.

It is seen that the absorption coefficient increases rapidly above the fundamental absorption edge (i.e., $h\nu \geq E_g$). In the visible spectral range, values of the absorption coefficient for most semiconductors may vary from 10^3 to 10^6 cm^{-1}. In general, the magnitude of the absorption coefficient represents the degree of interaction between the semiconductor and the incident photons. The internal photoelectric effects in a semiconductor are closely related to the optical absorption coefficient. Experimental results of absorption coefficient versus photon energy for some elemental and compound semiconductors (i.e., Si, Ge, GaAs, GaP, and InSb) are shown in Figs. 9.7 through 9.10.[4–7] Information concerning the optical absorption coefficient versus photon energy is very important for analyzing internal photoelectric effects in a semiconductor.

9.4. THE PHOTOCONDUCTIVE EFFECT

In this section, the photoconductivity effect in a semiconductor is described. In the absence of illumination, the dark conductivity of a semiconductor is given by

$$\sigma_0 = q(n_0\mu_n + p_0\mu_p) \tag{9.71}$$

where n_0 and p_0 denote the densities of electrons and holes in thermal equilibrium, while μ_n and μ_p are the electron and hole mobilities, respectively.

When photons with energies greater than the bandgap energy of a semiconductor are absorbed in a semiconductor, intrinsic photoconductivity results. The absorbed photons create excess electron-hole pairs (i.e., Δn and Δp), and as a result the densities of electrons and holes

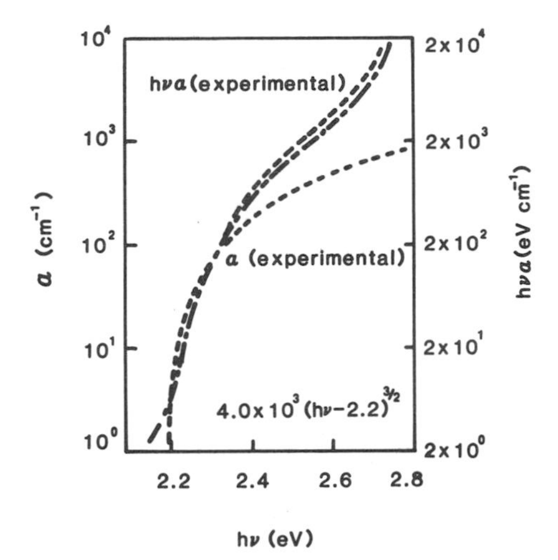

FIGURE 9.7. Absorption coefficient versus photon energy for a GaP sample at room temperature. After Spitzer *et al.*,[4] by permission.

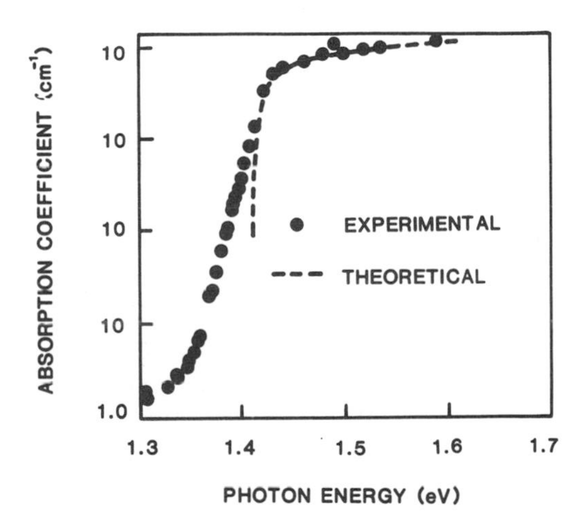

FIGURE 9.8. Absorption coefficient versus photon energy for a GaAs sample at room temperature. After Moss and Hawkins,[5] by permission.

(i.e., n and p) become larger than their equilibrium values of n_0 and p_0 (i.e., $n = n_0 + \Delta n$, $p = p_0 + \Delta p$).

The photoconductivity is defined as the net change in electrical conductivity under illumination and can be expressed by

$$\Delta\sigma = q(\Delta n\mu_n + \Delta p\mu_p) \tag{9.72}$$

where Δn and Δp are the photogenerated excess electron and hole densities, respectively.

In a degenerate semiconductor, Δp and Δn are generally much smaller than p_0 and n_0, and the effect of incident photons can be considered as a small perturbation. However, in an insulator or a nondegenerate semiconductor, values of Δn and Δp can become larger than their equilibrium carrier densities. If the effect of electron or hole trapping is small and the

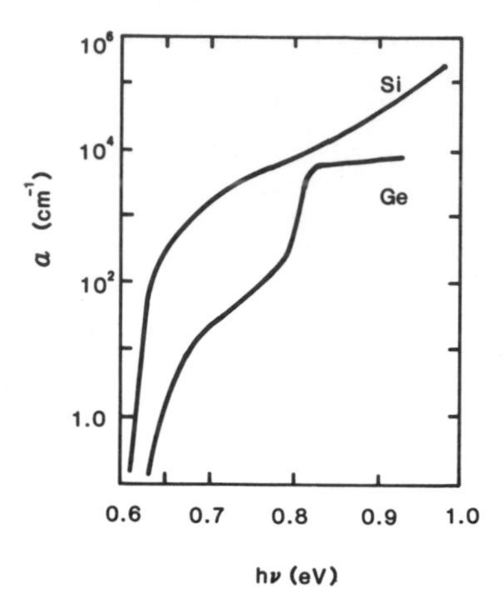

FIGURE 9.9. Absorption coefficient versus photon energy for silicon and germanium crystals measured at 300 K. After Dash and Newman,[6] by permission.

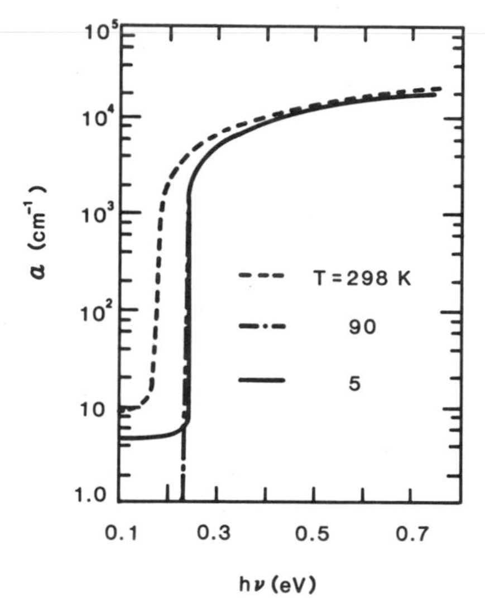

FIGURE 9.10. Absorption coefficient versus photon energy for a pure InSb sample measured at three different temperatures. After Johnson,[7] by permission.

semiconductor remains neutral under illumination (i.e., without building up the space charge in the material), then $\Delta n = \Delta p$ holds throughout the specimen.

Depending on the incident photon energies, there are two types of photoconduction processes which are commonly observed in a semiconductor. One type of photoconduction process is known as intrinsic photoconductivity, in which the excess electron–hole pairs are generated by the absorption of photons with energies exceeding the bandgap energy of the semiconductor (i.e., $h\nu \geq E_g$). This type of photoconduction process is illustrated in Fig. 9.11a. The other type of photoconduction process is known as extrinsic photoconductivity, in which electrons (or holes) are excited from the localized donor (or acceptor) states into the conduction (or valence) band states by the absorption of photons with energy equal to or greater than the activation energy of the donor (or acceptor) levels, but is less than the bandgap energy of the semiconductor (i.e., $E_D \leq h\nu \leq E_g$). This is shown in Fig. 9.11b.

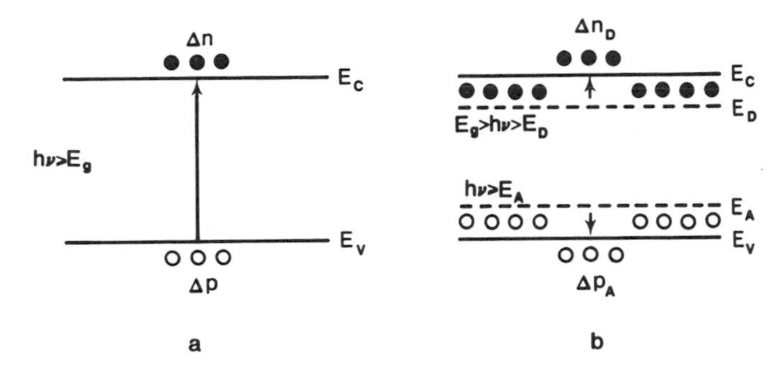

FIGURE 9.11. (a) Intrinsic and (b) extrinsic photoconductivity in a semiconductor.

In intrinsic photoconduction, both the photogenerated electrons and holes are participating in the photoconduction process, and the photoconductivity is described by Eq. (9.72). However, for extrinsic photoconductivity, the photoconduction process usually involves only one type of carrier (i.e., either electrons or holes), and the expression for the extrinsic photoconductivity is given by

$$\Delta\sigma_n = qn_D\mu_n \qquad \text{for n-type} \tag{9.73}$$

and

$$\Delta\sigma_p = qp_A\mu_p \qquad \text{for p-type} \tag{9.74}$$

where n_D and p_A are the photogenerated excess electron and hole densities from the donor and acceptor centers, respectively.

An extrinsic photoconductor usually operates at cryogenic temperatures because at very low temperatures freeze-out occurs for electrons in the conduction band states or for holes in the valence band states. The return of electrons in the shallow-donor states or of holes in the shallow-acceptor states due to the freeze-out effect is the basis for extrinsic photoconductivity. At very low temperatures, the electrical conductivity under the dark condition and background noise are generally very low. If photons with energies of $E_D \leq h\nu \leq E_g$ impinge upon a specimen, then the conductivity of the sample will increase dramatically with the incident photons since electrons (or holes) in the shallow impurity states will be excited into the conduction (or valence) band states by these photons. The sensitivity of extrinsic photoconductivity depends greatly on the density of sensitizing shallow impurity centers and the thickness of the specimen. Extrinsic photoconductivity is most frequently used for long-wavelength infrared detection. For example, a Cu-doped germanium extrinsic photoconductor operating at 4.2 K can be used to detect photons of wavelengths ranging from 2.5 μm to 30 μm, while a Hg-doped germanium photodetector operating at 28 K can be used for 10.6 μm wavelength detection.

We shall next consider intrinsic photoconductivity. Figure 9.12 shows the schematic diagram of a photoconductor under illumination and bias conditions.

In the intrinsic photoconduction process, electron–hole pairs are generated in a semiconductor when photons with energies exceeding the bandgap energy of the semiconductor are absorbed. The rate of generation of electron–hole pairs per unit volume per unit time can be written as

$$g_E = \alpha\phi_0(1 - R) \qquad \text{for } \alpha d \ll 1 \tag{9.75}$$

and

$$g_E = \alpha\phi_0(1 - R)\, e^{-\alpha y} \qquad \text{for } \alpha d \gg 1 \tag{9.76}$$

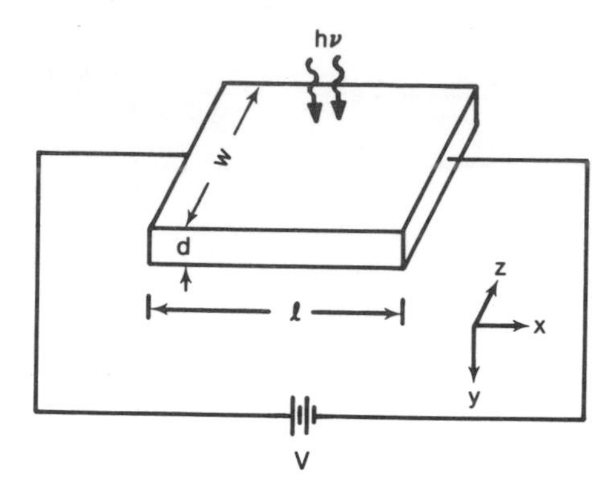

FIGURE 9.12. Photoconductivity process in a semiconductor specimen.

where R is the reflection coefficient of the semiconductor defined by Eq. (9.25), α is the absorption coefficient, ϕ_0 is the photon flux density (i.e., $\phi_0 = I_0/h\nu$), and I_0 is the incident light intensity per unit area (watt/cm^2).

Equation (9.75) is valid for a very thin photoconductor (i.e., $\alpha d \ll 1$) in which photons are uniformly absorbed throughout the sample, while Eq. (9.76) is applicable for a thick specimen (i.e., $\alpha d \gg 1$) in which the photogeneration rate decays exponentially with penetration distance. We shall discuss both cases.

Let us first consider the case of a thin specimen with $\alpha d \ll 1$. Here, the excess electron and hole densities are related to the generation rate g_E by

$$\Delta n = g_E \tau_n \tag{9.77}$$

and

$$\Delta p = g_E \tau_p \tag{9.78}$$

where τ_n and τ_p are the electron and hole lifetimes, respectively. The change of electrical conductance due to the incident photons can be described by

$$\begin{aligned}\Delta G = \Delta\sigma\left(\frac{A}{l}\right) &= q(\Delta n\mu_n + \Delta p\mu_p)\left(\frac{Wd}{l}\right)\\ &= qg_E(\tau_n\mu_n + \tau_p\mu_p)\left(\frac{Wd}{l}\right)\end{aligned} \tag{9.79}$$

or

$$\Delta G = qG_E\frac{(\tau_n\mu_n + \tau_p\mu_p)}{l^2} \tag{9.80}$$

where $G_E = g_E(Wdl)$ is the total volume generation rate (i.e., total number of carriers generated per second), and $A = Wd$ is the cross-section area. If V is the applied voltage and I_{ph} is the photocurrent, then the latter can be written as

$$I_{ph} = V\Delta G = qVG_E\frac{(\tau_n\mu_n + \tau_p\mu_p)}{l^2} = qVG_ES \tag{9.81}$$

where

$$S = \frac{(\tau_n\mu_n + \tau_p\mu_p)}{l^2} \tag{9.82}$$

is the photosensitivity factor. It is seen that value of S is directly proportional to the product $\mu\tau$. This means that in order to obtain a high photosensitivity factor, the carrier lifetimes and mobilities must be as large as possible and the sample length should be as small as possible. As an example, let us consider a silicon photoconductor. If the wavelength of the incident photon is 0.5 μm, the absorption coefficient α is $10^4\ \text{cm}^{-1}$, τ_n is 100 μsec. R is 0.3, and the photon flux density ϕ_0 is $10^{14}\ \text{cm}^{-2}\cdot\text{sec}^{-1}$, then the excess electron density can be calculated by using the following formula:

$$\Delta n = \alpha\phi_0(1 - R)\tau_n = 7 \times 10^{13}\ \text{cm}^{-3} \tag{9.83}$$

which shows that a relatively large density of excess electrons can be generated even with a relatively small incident light intensity. Another parameter which has often been used to evaluate a photoconductor is the photoconductivity gain, G_p. This figure of merit (G_p) is defined as the ratio of the excess carrier lifetime, τ, to the carrier transit time, t_r, across the specimen, and can be expressed by

$$G_p = \frac{\tau}{t_r} = SV \tag{9.84}$$

where $t_r = l/v_d = l^2/\mu V$ is the transit time for the excess carriers to drift across the photoconductor specimen, v_d is the drift velocity, and μ is the carrier mobility. A photoconductivity gain of 10^4 can be readily obtained for a CdS photoconductor.

In the above treatment, loss due to surface recombination has been neglected. For a thin-film photoconductor, the effect of surface recombination can be incorporated into an effective excess carrier lifetime as

$$\frac{1}{\tau'} = \frac{1}{\tau_B} + \frac{1}{\tau_s} \tag{9.85}$$

where τ' is the effective excess carrier lifetime, τ_B is the bulk carrier lifetime, and τ_s is the surface recombination lifetime given by

$$\tau_s = \frac{d}{2s} \tag{9.86}$$

where s is the surface recombination velocity and d the sample thickness. For example, in a chemically polished silicon specimen, with s equal to 500 cm/sec and thickness d equal to 2 μm, the surface recombination lifetime, τ_s, is found equal to 2×10^{-7} sec. Therefore, for a thin-film photoconductor, if τ_s is less than τ_B, then the surface recombination lifetime rather than the bulk lifetime may control the effective excess carrier lifetime.

In general, the photocurrent for a thin-film photoconductor can be derived from Eqs. (9.75) and (9.79), and the result yields

$$I_{ph} = \Delta GV = q\eta(1 - R)\alpha\phi_0\tau'(1 + b)\mu_p\left(\frac{V}{l}\right)(Wd) \tag{9.87}$$

where τ' is given by Eq. (9.85), $b = \mu_n/\mu_p$ is the electron to hole mobility ratio, and η is the quantum efficiency. From Eq. (9.87) it is seen that, for a constant τ', the photocurrent I_{ph} is directly proportional to the light intensity. This is generally true under both low and high injection conditions (i.e., for $\Delta n \ll n_0$ or $\Delta n \gg n_0$). However, for the intermediate injection range (i.e., $\Delta n \simeq n_0$), τ may become a function of the injected excess carrier density, and the photocurrent is no longer a linear function of the light intensity. Depending on the relationship between the excess carrier lifetime and the injected carrier density, a superlinear or sublinear region may exist in the intermediate injection regime.

We shall next consider the case of a thick photoconductor with $\alpha d \gg 1$. Here, the generation rate is given by Eq. (9.76). Due to nonuniform absorption the diffusion of excess carriers along the direction of incident photons plays an important role in this case. As shown in Fig. 9.12, excess carrier densities as a function of distance along the y-direction can be obtained by solving the following continuity equation:

$$D_n \frac{\partial^2 \Delta n}{\partial y^2} - \frac{\Delta n}{\tau_n} = -g_E = -\alpha \phi_0 (1 - R)\, e^{-\alpha y} \tag{9.88}$$

If the electron diffusion length $L_n = (D_n \tau_n)^{1/2}$ is much larger than the sample thickness d, then Eq. (9.88) has the solution

$$\Delta n = \frac{\alpha I_0 (1 - R) \tau_n}{h\nu(\alpha^2 L_n^2 - 1)} \left[\left(\frac{\alpha L_n^2 + s\tau_n}{L_n + s\tau_n} \right) e^{-y/L_n} - e^{-\alpha y} \right] \tag{9.89}$$

We note that Eq. (9.89) is obtained by using the boundary condition

$$D_n \left. \frac{\partial \Delta n}{\partial y} \right|_{y=0} = s \Delta n|_{y=0} \tag{9.90}$$

The photocurrent is obtained by integrating Eq. (9.89) with respect to y from 0 to ∞, and the result yields

$$\begin{aligned} I_{ph} &= \left(\frac{W}{l} \right) V q \mu_p (1 + b) \int_0^\infty \Delta n \, dy \\ &= \frac{q I_0 W L_n \mu_p (1 + b) \tau_n (1 - R) V}{l (L_n + s\tau_n) h\nu} \left[1 + \frac{s\tau_n}{L_n (1 + \alpha L_n)} \right] \end{aligned} \tag{9.91}$$

where l is the sample length. In Eq. (9.91) we have used $y = d$ by $y = \infty$ in the integration. This is valid as long as the sample thickness is much greater than the diffusion length of electrons.

Figure 9.13 shows photocurrent versus wavelength of incident photons for different surface recombination velocities. For $s\tau_n \gg L_n$, I_{ph} reaches a maximum when $\alpha \sim 1/L_n$. However, if the surface recombination velocity is small and $s\tau_n \ll 1$, then the photocurrent will increase

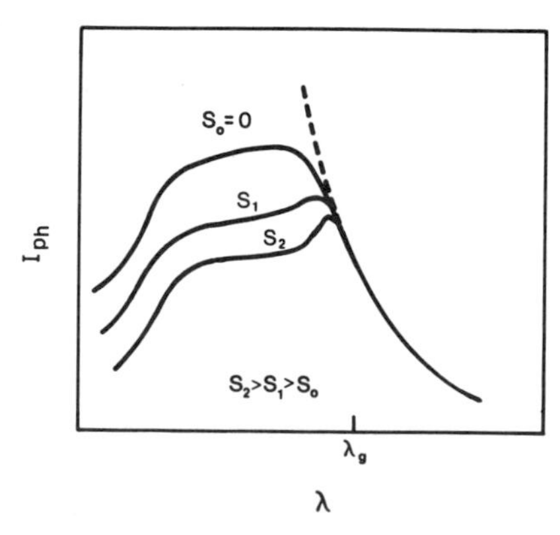

FIGURE 9.13. Relative photoresponse versus wavelength for different surface recombination velocities, s_i.

monotonically with decreasing wavelength. This is shown in Fig. 9.13 for the case $s = 0$. The sharp decrease in photocurrent is usually observed near the absorption edge (i.e., $h\nu \sim E_g$) in which the absorption coefficient decreases exponentially with increasing wavelength. However, in the very-short-wavelength region (i.e., near the UV regime), the absorption coefficient is usually very large (i.e., $\alpha \geq 10^5\ \text{cm}^{-1}$) and $\alpha L_n \gg 1$. Excess carriers are generated near the surface of the photoconductor where excess carrier lifetime is controlled by surface recombination. Therefore, the photocurrent is expected to decrease rapidly in the short-wavelength regime. In order to improve the short-wavelength photoresponse, careful preparation of the semiconductor surface is necessary so that the surface recombination velocity of the photoconductor can be kept low.

9.4.1. *Kinetics of Photoconduction*

Since the photocurrent is directly related to the excess carrier densities generated by the incident photons, a study of photocurrent as a function of light intensity usually yields useful information concerning the recombination mechanisms of the excess carriers in a semiconductor. As an example, let us consider an n-type direct bandgap semiconductor. If the band-to-band radiative recombination dominates the excess carrier lifetimes, then the kinetic equation for the photoconduction process can be expressed by

$$\frac{dn}{dt} = g_E - U \tag{9.92}$$

where g_E is the generation rate of the excess carriers defined by Eq. (9.75), U is the net recombination rate which, for band-to-band radiative recombination, is given by

$$U = B(np - n_i^2) \tag{9.93}$$

In the steady-state case, the carrier generation rate is obtained by solving Eqs. (9.92) and (9.93). The result yields

$$g_E = U = B(np - n_i^2) \tag{9.94}$$

where $n = n_0 + \Delta n$ and $p = p_0 + \Delta p$.

In order to understand the kinetics of photoconduction, we consider two limiting cases, namely, the low- and high-injection cases.

Low-injection case $(\Delta n \ll n_0,\ \Delta p \ll p_0)$. Under the low-injection condition, Eq. (9.94) becomes

$$B\Delta n(p_0 + n_0) = g_E = \frac{\alpha I_0(1-R)}{h\nu} \tag{9.95}$$

or

$$\Delta n = \frac{g_E}{B(n_0 + p_0)} = \frac{\alpha I_0(1-R)}{Bh\nu(n_0 + p_0)} \tag{9.96}$$

In Eq. (9.66) we have assumed that $\Delta n = \Delta p$ (i.e., no trapping), and the charge-neutrality condition prevails. Equation (9.96) shows that Δn is directly proportional to light intensity I_0. In the low-injection case, since I_{ph} varies linearly with Δn, I_{ph} is also a linear function of the light intensity I_0.

High-injection case ($\Delta n \gg n_0$, and $\Delta p \gg p_0$). Under the high-injection condition, Eq. (9.94) becomes

$$g_E = B\Delta n^2 \tag{9.97}$$

or

$$\Delta n = \left(\frac{g_E}{B}\right)^{1/2} = \left[\frac{\alpha I_0(1-R)}{Bh\nu}\right]^{1/2} \tag{9.98}$$

which shows that Δn is directly proportional to the square root of the light intensity. Therefore, under the high-injection condition, when band-to-band radiative recombination is dominant, the photocurrent varies with the square root of the light intensity.

Figure 9.14 shows a p-type extrinsic photoconductor with a deep-acceptor center. The kinetic equation for the photoconduction process in this case is given by

$$\frac{dp}{dt} = e_p(N_A - p) - c_p p^2 + g_E \tag{9.99}$$

where e_p and c_p are the emission and capture rates of holes, respectively, as shown in Fig. 9.14. The first term on the right-hand side of Eq. (9.99) is the rate of spontaneous generation from the neutral acceptor centers. The second term gives the rate of recombination of free holes and ionized acceptor centers. For the steady-state low-injection case, solving Eq. (9.99) yields

$$\Delta p = \frac{g_E}{(e_p + 2c_p p_0)} = g_E \tau_p \tag{9.100}$$

where $\tau_p = 1/(e_p + 2c_p p_0)$. Equation (9.100) is obtained by employing the relationship $e_p(N_A - p_0) = c_p p_0^2$. The results predict a linear relationship between Δp (or I_{ph}) and g_E (or I_0), providing that the hole lifetime is constant and independent of injection.

For the high-injection limit, Eq. (9.99) becomes

$$g_E = c_p \Delta p^2 \tag{9.101}$$

or

$$\Delta p = \left(\frac{g_E}{c_p}\right)^{1/2} \tag{9.102}$$

which shows that Δp is directly proportional to the square root of the generation rate $g_E^{1/2}$.

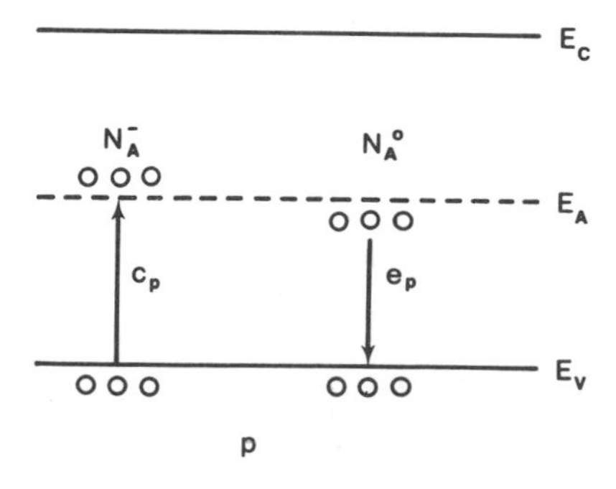

FIGURE 9.14. Kinetics of photoconduction in a p-type semiconductor with a deep-acceptor center.

A typical plot of the photocurrent versus light intensity is shown in Fig. 9.15, which usually consists of a linear, sublinear, and a superlinear region. This type of behavior has been observed in a number of photoconductors such as CdS and other II–VI compound semiconductors.

9.4.2. *Practical Applications of Photoconductivity*

Photoconductors made from high-resistivity semiconductors are often used to detect visible to infrared radiations. Intrinsic photoconductors such as lead sulfide (PbS), lead selenide (PbSe), lead telluride (PbTe), and indium antimonide (InSb), operating at 77 K, are commonly used for 3–5 μm infrared detection, while wide-bandgap photoconductors such as CdS and CdTe are mainly used in the near-UV to visible spectral region. For longer wavelengths (i.e., $\lambda > 10\ \mu$m), extrinsic photoconductors such as Au-, Cu-, Hg-, and Cd-doped germanium photoconductors and CdHgTe detectors are commonly used. Figure 9.16 shows detectivity versus wavelength for various photoconductive and photovoltaic infrared detectors.[8] The detection mechanism for these devices is based on the optical excitation of holes from acceptor impurity centers into the valence band. In order to suppress the competing thermal excitation, extrinsic photoconductors are generally operated in the temperature range between 77 and 4.2 K. Since optical absorption coefficients for extrinsic photoconduction are usually very small (typically of the order of 1 to 10 cm^{-1}), to obtain high quantum efficiency the thickness of the photoconductor along the direction of incident photons should be several millimeters or centimeters long.

In an intrinsic photoconductor, the conduction process is due to band-to-band excitation. Consequently, the absorption coefficients are usually very large (e.g., 10^3 to 10^5 cm^{-1}), and thus the thickness of the photoconductor in the direction of incident light may vary from a few micrometers to a few tens of micrometers. Practical long-wavelength infrared photodetectors operating in 8 to 14 μm have been fabricated from II–IV–VI compound semiconductors such as $Hg_{1-x}Cd_xTe$ and $Pb_{1-x}Sn_xTe$ crystals. A $Hg_{1-x}Cd_xTe$ IR detector with extremely high detectivity [i.e., $\geq 10^{11}$ cm(Hz)$^{1/2}$/W] and operating at 77 K has been achieved for 10.6 μm detection. The energy bandgap for $Hg_{1-x}Cd_xTe$ varies with the mole fraction ratio x (i.e., E_g varies from 1.4 eV to less than 0.1 eV as x varies from 1 to 0). As a result, the $Hg_{1-x}Cd_xTe$ IR detector can be tailored to the desired wavelength by varying the mole fraction x in the material. The main drawbacks of $Hg_{1-x}Cd_xTe$ for long-wavelength IR detection applications are that the quality of this material system is not very good; in particular, long-term stability and doping uniformity across the wafer are some of the key problems that need to be solved.

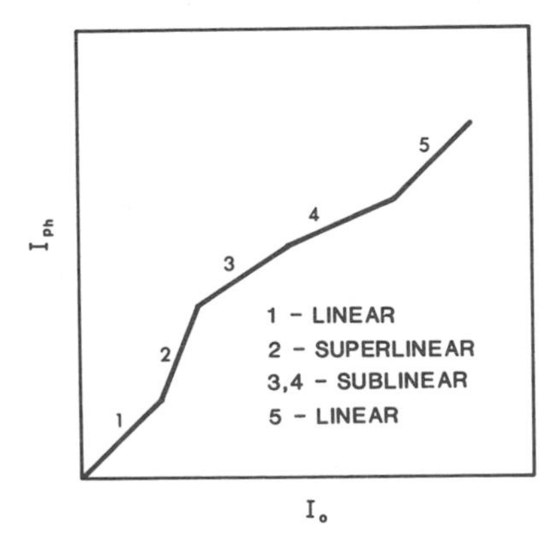

FIGURE 9.15. Photocurrent versus light intensity for a photoconductor, showing the linear, superlinear, and sublinear regions.

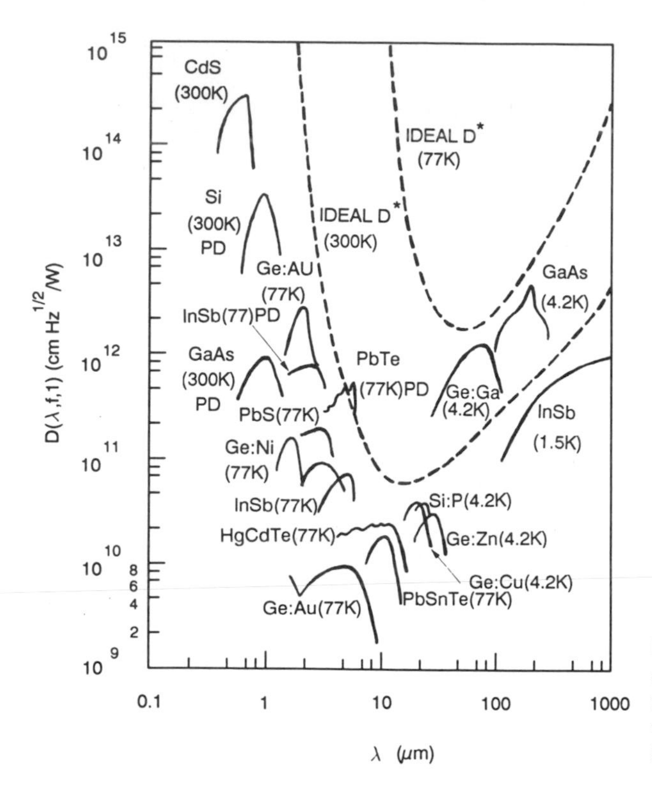

FIGURE 9.16. Spectral dependence of detectivity for some important photoconductive and photovoltaic infrared detectors. After Sze,[8] by permission.

In addition to practical application of photoconductors described above, there are fundamental physical parameters related to excess carrier recombination which can be determined by studying photoconductivity effects in a semiconductor. For example, from a study of photoconductance versus light intensity, we can deduce information concerning recombination and trapping mechanisms in a semiconductor. The minority carrier lifetime or diffusion length for a semiconductor can be determined by using either the steady-state or transient photoconductivity method. This will be discussed further in Section 9.7.

9.5. THE PHOTOVOLTAIC (DEMBER) EFFECT

The internal photovoltaic effect (i.e., the Dember effect) in a semiconductor is depicted in this section. In Fig. 9.12, we show that photons with energies greater than the bandgap energy are impinging on a p-type semiconductor specimen. If the sample thickness is much larger than the inverse absorption coefficient, then a concentration gradient of the photoinjected carriers is established in the direction of incident photons. This will cause electron and hole currents to flow by diffusion along the direction of the incident light. The total diffusion current is equal to zero if the mobilities of electrons and holes are equal in the semiconductor. In general, electron and hole mobilities are not equal in most semiconductors; an unbalanced electron and hole diffusion current will create an internal electric field along the direction of incident photons. This internal electric field will assist hole diffusion and retard electron diffusion. As a result, an internal electric field is established under this condition. This internal electric field is usually referred to as the Dember field. To derive an expression for the Dember field, the components of electron and hole current densities (i.e., J_{ny} and J_{py})

along the direction of incident photons can be written as

$$J_{ny} = qn\mu_n\mathscr{E}_y + qD_n\frac{\partial n}{\partial y} \tag{9.103}$$

and

$$J_{py} = qp\mu_p\mathscr{E}_y - qD_p\frac{\partial p}{\partial y} \tag{9.104}$$

where $n = n_0 + \Delta n$, $p = p_0 + \Delta p$, and $\Delta n = \Delta p$; μ_n and D_n, and μ_p and D_p are related through Einstein's relation, which is given by

$$D_n = \left(\frac{k_BT}{q}\right)\mu_n \quad \text{and} \quad D_p = \left(\frac{k_BT}{q}\right)\mu_p \tag{9.105}$$

The total current density in the y-direction is equal to the sum of J_{ny} and J_{py}, which can be written as

$$J_y = J_{ny} + J_{py} = q(bn + p)\mu_p\mathscr{E}_y + (b-1)qD_p\frac{\partial \Delta n}{\partial y} \tag{9.106}$$

Under open-circuit conditions, $J_y = 0$, and the Dember field is given by

$$\mathscr{E}_y = -\left(\frac{k_BT}{q}\right)\frac{(b-1)}{(bn+p)}\frac{\partial \Delta n}{\partial y} \tag{9.107}$$

where $b = \mu_n/\mu_p$ is the mobility ratio; $\mathscr{E}_y$ given by Eq. (9.107) is the Dember electric field. We note that the field vanishes if $b = 1$. The Dember field developed in the sample will enhance the diffusion of holes and retard the diffusion of electrons. The resulting photovoltage (or Dember voltage) between the front (i.e., illuminated side) and the back side of the sample is obtained by integrating Eq. (9.107) from $y = 0$ to $y = d$. For small injection (i.e., $\Delta n \ll n_o$), the Dember voltage is given by

$$\begin{aligned} V_d &= \int_0^d -\mathscr{E}_y\,dy \\ &= \left(\frac{k_BT}{q}\right)\int_{\Delta n_0}^{\Delta n_d}\frac{-(b-1)}{(bn_0+p_0)}\,dn \\ &= \left(\frac{k_BT}{q}\right)\frac{(b-1)}{(bn_0+p_0)}(\Delta n_0 - \Delta n_d) \end{aligned} \tag{9.108}$$

where Δn_0 is the excess electron density at $y = 0$ and Δn_d is the excess electron density at $y = d$. For a thick sample, $\alpha d \gg 1$ and $\Delta n_d \ll \Delta n_0$. Equation (9.108) becomes

$$V_d = \left(\frac{k_BT}{q}\right)\frac{(b-1)}{(bn_0+p_0)}\Delta n_0 \tag{9.109}$$

The excess electron density Δn_0 at $y = 0$ can be related to the incident light intensity by using the result obtained in the previous section. This is given by

$$\Delta n_0 = \frac{\eta\alpha L_n I_0(1-R)\tau_n}{h\nu(\alpha L_n + 1)(L_n + s\tau_n)} \tag{9.110}$$

Now substituting Eqs. (9.110) into (9.109) yields

$$V_d = \left(\frac{k_B T}{q}\right) \frac{\eta \alpha L_n I_0 (1 - R) \tau_n (b - 1)}{h\nu(\alpha L_n + 1)(L_n + s\tau_n)(bn_0 + p_0)} \tag{9.111}$$

Equation (9.111) shows that the Dember voltage is directly proportional to the light intensity, carrier lifetime, mobility ratio, and the absorption coefficient. In contrast to the photoconductive effect, the photovoltaic effect requires no external applied voltage, and hence can be used to generate electrical power from the photovoltaic effect in a semiconductor. Devices using this internal photovoltaic effect are known as solar cells or photovoltaic cells. Typical photovoltaic devices are formed by using the p–n junction or Schottky barrier structure. Details of photovoltaic devices will be discussed in Chapter 12.

9.6. THE PHOTOMAGNETOELECTRIC EFFECT

The photomagnetoelectric (PME) effect refers to the voltage (or current) developed in a semiconductor specimen as a result of interaction of an applied magnetic field with the diffusion current produced by photogenerated excess carriers. The PME open-circuit voltage (or short-circuit current), the magnetic field, and the incident light are mutually perpendicular to one another; their relative orientations are illustrated in Fig. 9.17.

The development of a PME field in a semiconductor under the influence of a magnetic field and incident light may be explained as follows. Electron–hole pairs generated near the surface of a semiconductor specimen by the incident photons are diffused into the specimen in the direction of the incident light (i.e., the y-direction). The time-invariant magnetic flux density, B, along the z-direction deflecting holes in the positive x-direction and electrons in the negative x-direction results in a net PME short-circuit current flowing in the positive x-direction. Under open-circuit conditions, a PME voltage is developed in the x-direction.

In order to derive an expression for the PME open-circuit voltage or the PME short-circuit current, we consider a semi-infinite semiconductor slab. The following equations hold for the rectangular slab shown in Fig. 9.17. For a small magnetic field (i.e., $\mu B \ll 1$) and

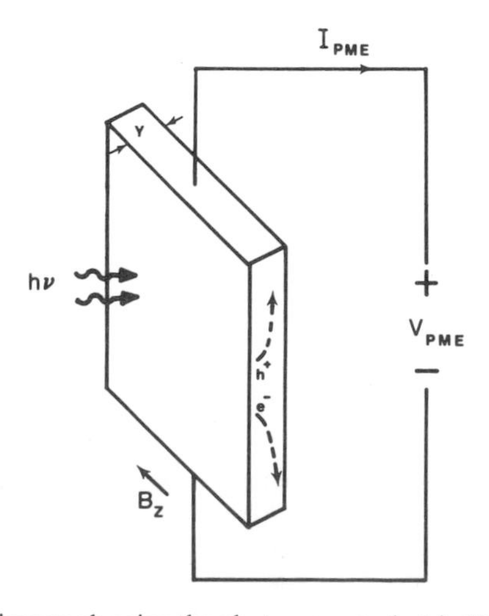

FIGURE 9.17. Schematic diagram showing the photomagnetoelectric (PME) effect in a semiconductor.

small-injection case, the Hall angles are given by

$$\tan\theta_p \simeq \theta_p = \mu_p B \tag{9.112}$$

$$\tan\theta_n \simeq \theta_n = -\mu_n B \tag{9.113}$$

$$\theta = \theta_p - \theta_n = \mu_p(1+b)B \tag{9.114}$$

where $\Delta n = \Delta p \ll n_0$ or p_0, and $b = \mu_n/\mu_p$ is the mobility ratio.

The hole and electron current density can be best described by vector equations with $\hat{i}, \hat{j}, \hat{k}$ denoting unit vectors along the x-, y-, and z-axes, respectively

$$\mathbf{J}_p \simeq J_{py}\hat{\mathbf{j}} + \theta_p J_{py}\hat{\mathbf{j}} \times \hat{\mathbf{k}} \tag{9.115}$$

$$\mathbf{J}_n \simeq J_{ny}\hat{\mathbf{j}} + \theta_n J_{ny}\hat{\mathbf{j}} \times \hat{\mathbf{k}} \tag{9.116}$$

Here J_{py} and J_{ny} are given respectively by

$$J_{py} = q\left(\mu_p p \mathscr{E}_y - D_p \frac{\partial \Delta p}{\partial y}\right) \tag{9.117}$$

and

$$J_{ny} = q\left(\mu_n n \mathscr{E}_y + D_n \frac{\partial \Delta n}{\partial y}\right) \tag{9.118}$$

while the total current density is thus given by

$$\mathbf{J} = \mathbf{J}_p + \mathbf{J}_n = (J_{px} + J_{nx})\hat{\mathbf{i}} + (J_{py} + J_{ny})\hat{\mathbf{j}} \tag{9.119}$$

Since the total current density in the y-direction is zero, Eq. (9.119) reduces to

$$\begin{aligned} J = J_x &= (J_{px} + J_{nx}) \\ &= [q(p\mu_p + n\mu_n)\mathscr{E}_x + \theta_p J_{py} + \theta_n J_{ny}] \end{aligned} \tag{9.120}$$

Equations (9.117) through (9.119) yield

$$J_{py} = -J_{ny} = -qD\left(\frac{\partial \Delta n}{\partial y}\right) \tag{9.121}$$

where $D = D_n(n+p)/(bn+p)$ is the effective diffusion coefficient. Now, on substituting Eq. (9.121) into Eq. (9.120) we obtain

$$\begin{aligned} J_x &= J_{nx} + J_{px} \\ &= q\mu_p(p + nb)\mathscr{E}_x - q(b+1)\mu_p BD\left(\frac{\partial \Delta n}{\partial y}\right) \end{aligned} \tag{9.122}$$

To derive the PME electric field or open-circuit voltage along the x-direction, it is usually not sufficient to assume that $J_x = J_{nx} + J_{px} = 0$. Such a solution would lead to J_x being a function of z. This result is not correct, because for a constant magnetic field the electric field must be irrotational. With $\partial\mathscr{E}_z/\partial x = 0$, it follows that $\partial\mathscr{E}_x/\partial z = 0$. Therefore, the correct boundary condition is given by

$$\int_0^d (J_{nx} + J_{px})\, dy = 0 \tag{9.123}$$

For the small-injection case, with $\Delta n = \Delta p \ll n_0$ or p_0, the PME electric field in the x-direction becomes

$$\begin{aligned}\mathscr{E}_x &= \frac{(b+1)\,BD}{d(n_0 b + p_0)} \int_0^d \frac{\partial \Delta n}{\partial y}\, dy \\ &= -\frac{BD(b+1)}{d(n_0 b + p_0)} (\Delta n_0 - \Delta n_d)\end{aligned} \tag{9.124}$$

The PME short-circuit current can be obtained by setting $\mathscr{E}_x = 0$ in Eq. (9.122) and then integrating the equation from $y = 0$ to $y = d$. The result yields

$$\begin{aligned}I_{PME} &= -W \int_0^d q(b+1)\mu_p BD\left(\frac{\partial \Delta n}{\partial y}\right) dy \\ &= qW(b+1)\mu_p BD(\Delta n_0 - \Delta n_d)\end{aligned} \tag{9.125}$$

For $\alpha d \gg 1$ and $d \gg L_n$, we may assume that $\Delta n_d = 0$ and substitute Eq. (9.110) into Eq. (9.125). The result yields

$$I_{PME} = \frac{qW(b+1)\mu_p I_0(1-R)BL_n^2}{h\nu(L_n + s\tau_n)} \cdot \frac{\alpha L_n}{(1+\alpha L_n)} \tag{9.126}$$

In Eq. (9.126), the quantum yield is assumed equal to unity. It shows that, at low magnetic fields, the PME short-circuit current is directly proportional to the magnetic flux density B. For $\alpha L_n \gg 1$, I_{PME} becomes independent of wavelength.

The ratio of the PME short-circuit current given by Eq. (9.126) to the photocurrent given by Eq. (9.90) is

$$\frac{\left(\dfrac{I_{PME}}{B}\right)}{\left(\dfrac{I_{ph}}{\mathscr{E}_x}\right)} = \frac{\alpha L_n^2}{\left(1 + \alpha L_n + \dfrac{s\tau_n}{L_n}\right)\tau_n} \tag{9.127}$$

If $\alpha L_n \gg 1$, $s\tau_n/L_n$, then Eq. (9.127) reduces to

$$\frac{\left(\dfrac{I_{PME}}{B}\right)}{\left(\dfrac{I_{ph}}{\mathscr{E}_x}\right)} = \frac{L_n}{\tau_n} = \left(\frac{D_n}{\tau_n}\right)^{1/2} \tag{9.128}$$

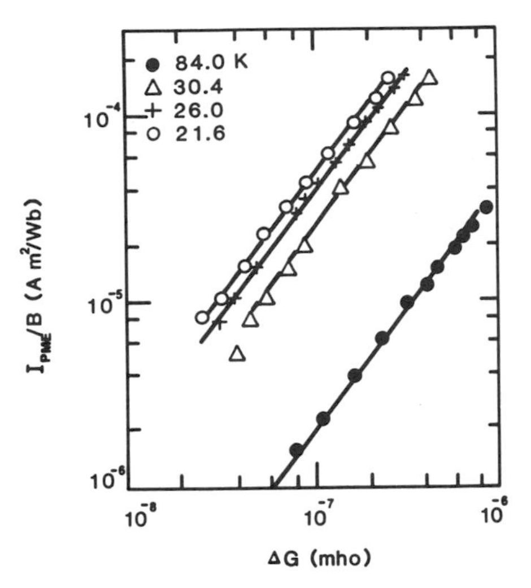

FIGURE 9.18. PME short-circuit current versus photoconductance for an Au-doped silicon specimen with $N_{Au} = 5 \times 10^{16}\ \text{cm}^{-3}$. After Agraz and Li,[9] by permission.

where $L_n = (D_n \tau_n)^{1/2}$ is the electron diffusion length. Equation (9.128) provides a direct means for determining the minority carrier lifetime from PME and PC measurements. The diffusion constant can be determined from mobility data using Einstein relations given by Eqs. (6.60) and (6.61).

The PME effect has been widely used in determining minority carrier lifetimes in a semiconductor. This method is particularly attractive for semiconductors with extremely short carrier lifetimes in which the transient photoconductivity decay method fails. Studies of the PME effect have been carried out in many elemental and compound semiconductors such as Ge, Si, InSb, and GaAs in the past. Figure 9.18 shows PME short-circuit current versus photoconductance for a gold-doped silicon sample measured at different temperatures. The nonlinear relationship between I_{PME} and ΔG is due to the effect of minority carrier trapping in the gold-doped silicon sample.

Figure 9.19a shows PME short-circuit current versus photoconductance over a wide range of light intensity for two Cr-doped n-type GaAs samples. The dependence of lifetime on

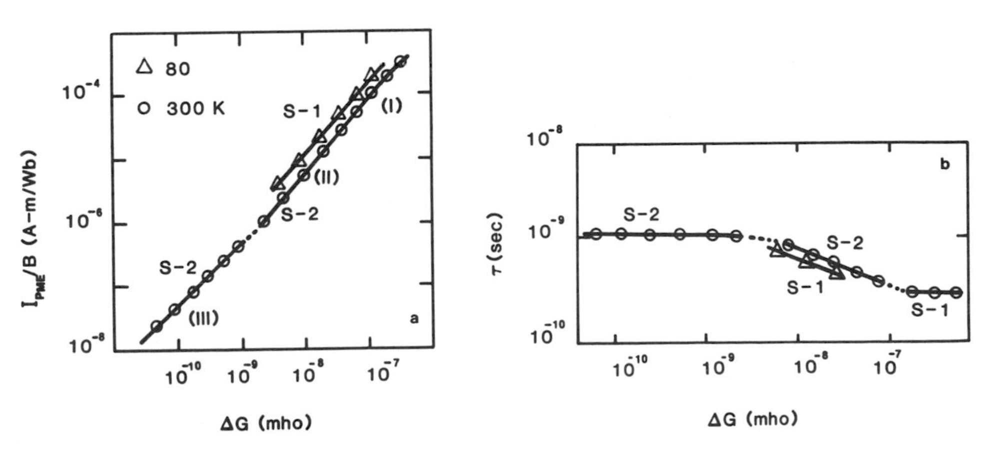

FIGURE 9.19. (a) PME short-circuit current, and (b) lifetime versus photoconductance for a Cr-doped n-type GaAs sample. After Huang and Li,[10] by permission.

photoconductance is also illustrated in Fig. 9.19b. In both low- and high-injection regimes I_{PME} varies linearly with ΔG, while in the intermediate injection range a nonlinear relationship exists between I_{PME} and ΔG.

PROBLEMS

9.1. Consider an n-type silicon specimen: the sample is 0.2 cm thick, 1 cm wide, and 2 cm long. Monochromatic light with wavelength equal to 0.9 μm and intensity of 5×10^{-4} W/cm^2 impinges upon the sample. Assuming that the equilibrium electron density is 10^{16} cm^{-3} and the absorption coefficient $\alpha = 320$ cm^{-1} at $\lambda = 0.9$ μm, find:
 (a) The number of incident photons per second on the sample.
 (b) The depth at which the light intensity is 10% of its value at the surface.
 (c) The number of electron–hole pairs generated per second in this sample assuming a quantum yield of 0.8.
 (d) The photoconductance ΔG, if the electron diffusion length $L_n = 50$ μm and surface recombination at the illuminated surface is zero.

9.2. Repeat Problem 9.1, (a) through (d), assuming that the wavelength of the incident photons is changed to 0.63 μm and the absorption coefficient of silicon is equal to 3×10^3 cm^{-1} at $\lambda =$ 0.63 μm.

9.3. Show that under the high-injection condition (i.e., $\Delta p = \Delta n \gg n_0$ or p_0) and with $\tau_n = \tau_p = \tau_h$ (where τ_h is the carrier lifetime at high injection as defined by the Shockley–Read–Hall model), the following relations prevail:
 (a) $\Delta G = qW\mu_p(1 + b)(D\tau_n)^{1/2}\Delta n_0 l$.
 (b) $I_{PME} = qW\mu_p(1 + b)DB\Delta n_0$.
 (c) $V_{PME} = I_{PME}/\Delta G = (D/\tau_n)^{1/2}Bl$.
 $D = 2D_n/(1 + b)$ is the ambipolar diffusion constant, and Δn_0 is the excess carrier concentration generated at the illuminated surface.

9.4. An n-type CdS photoconductor, which has a dark resistivity of 10^8 ohm · cm, is illuminated by a He–Ne laser ($\lambda = 0.6328$ μm). If the power output of the laser beam is 0.5 mW/cm^2, find:
 (a) The incident photon flux density per second.
 (b) The volume generation rate. Given: $\alpha = 4 \times 10^3$ cm^{-1} at $\lambda = 0{\cdot}6328$ μm and $R = 0$.
 (c) The photogenerated electron density. Given: $\tau_n = 10^{-3}$ sec.
 (d) The photoconductivity. Given: $\mu_n = 400$ cm^2/V · sec.

9.5. Using Eq. (9.91), calculate the photocurrent for an InSb photoconductor for $s = 0$, 100, and 10000 cm/sec. Given: $\tau_n = 10$ μsec. $L_n = 10$ μm, $R = 0.35$, $w/l = 1$, $I_0 = 1$ mW/cm^2, $V = 100$ mV, $\mu_p = 500$ cm^2/V · sec, $b = 100$, $\lambda = 4$ μm, $h\nu = hc/\lambda$.

9.6. Using Eq. (9.111), compute the Dember voltage for the InSb photoconductor given in Problem 9.5 for $\alpha = 10^4$ cm^{-1} and $n_0 = 5 \times 10^{16}$ cm^{-3}. Assuming that p_0 is negligible, the quantum yield $\eta = 0.8$, $s = 0$, 100, and 10^4 cm/sec, and $T = 300$ K.

9.7. If the intensity of incident photons on an n-type semiconductor is such that $\Delta p \gg n_0$, and band-to-band radiative recombination is dominant, show that the photocurrent is directly proportional to the square root of the light intensity.

9.8. If the energy bandgap for the $Al_xGa_{1-x}As$ ternary compound semiconductor is given by:

$$E_g = 1.424 + 1.247x \qquad \text{for } 0 \le x \le 0.45$$

$$E_g = 1.900 + 0.125x + 0.143x^2 \qquad \text{for } 0.45 \le x \le 1.0$$

 (a) Plot E_g versus x for $0 \le x \le 1.0$. Note that $Al_xGa_{1-x}As$ is a direct bandgap material if $x \le 0.45$, and becomes an indirect bandgap for $x > 0.45$.
 (b) Plot the optical absorption coefficient versus λ for $Al_xGa_{1-x}As$ crystal when $x = 0.2$, 0.4, 0.6, and 0.8.
 (c) If a 0.5-mW He–Ne laser beam of wavelength 0.6328 μm is illuminated on an $Al_xGa_{1-x}As$ specimen 10 μm thick, 100 μm wide, and 500 μm long, what is the total volume generation rate in this sample? Assume that the optical absorption coefficient at this wavelength is 5×10^4 cm^{-1}.

REFERENCES

1. T. S. Moss, *Optical Properties of Semiconductors*, Academic Press, New York (1959).
2. I. Kudman and T. Seidel, *J. Appl. Phys.* **33**, 771 (1962).
3. G. G. Macfarlane, T. P. Mclean, J. E. Quarrington, and V. Roberts, *J. Phys. Chem. Solids* **8**, 390 (1959).
4. W. G. Spitzer, M. Gershenzon, C. J. Frosch, and D. F. Gibbs, *J. Phys. Chem. Solids* **11**, 339 (1959).
5. T. S. Moss and T. D. Hawkins, *Infrared Phys.* **1**, 111 (1962).
6. W. C. Dash and R. Newman, *Phys. Rev.* **99**, 1151 (1955).
7. D. L. Greenaway and G. Harbeke, *Optical Properties and Band Structure of Semiconductors*, Pergamon Press, New York (1968).
8. S. M. Sze, *Physics of Semiconductor Devices*, 2nd ed., Wiley–Interscience, New York (1981).
9. A. G. Agraz and S. S. Li, *Phys. Rev.* **2**, 1947 (1970).
10. C. I. Huang and S. S. Li, *J. Appl. Phys.* **44**, 4214 (1973).

BIBLIOGRAPHY

R. H. Bube, *Photoconductivity of Solids*, Wiley, New York (1960).

R. H. Bube, *Electronic Properties of Crystalline Solids*, Academic Press, New York (1974).

O. Madelung, *Physics of III–V Compounds*, Wiley, New York (1964).

T. Moss, *Optical Properties of Semiconductors*, Butterworths, London (1959).

A. Rose, *Concepts of Photoconductivity and Allied Problems*, Wiley, New York (1963).

S. M. Ryvkin, *Photoelectric Effects in Semiconductors*, Consultants Bureau, New York (1968).

R. K. Willardson and A. C. Beer, *Semiconductors and Semimetals*, Vol. 3, Academic Press, New York (1967).

10

Metal–Semiconductor Contacts

10.1. INTRODUCTION

In this chapter, we are concerned with the basic physical principles, electrical properties, and applications of the metal–semiconductor contacts and devices. It is well known that the quality of metal–semiconductor contacts plays an important role in the performance of various semiconductor devices and integrated circuits. For example, good ohmic contacts are essential for achieving excellent electrical characteristics of a semiconductor device, while Schottky (i.e., rectifying) contacts can be used for a wide variety of device applications. In addition to a wide variety of device and circuit applications, Schottky barrier contacts can also be used as test vehicles for investigating the physical and electrical properties of a semiconductor material and its surfaces. For example, a Schottky diode can be used to study bulk defects and interface properties of a metal–semiconductor system. Therefore, it is essential to obtain a better understanding of the fundamental physical and electrical properties of the metal–semiconductor interface so that technologies for preparing good ohmic and Schottky contacts can be developed for device applications.

Two types of metal–semiconductor contacts are commonly encountered in the fabrication of semiconductor devices and integrated circuits. They are the Schottky and ohmic contacts. A Schottky barrier contact exhibits an asymmetrical current–voltage (I–V) characteristic when the polarity of a bias voltage applied to the metal–semiconductor contacts is changed. The ohmic contact, on the other hand, shows a linear I–V characteristic regardless of the polarity of the externally applied voltage. For a good ohmic contact the voltage drop across the metal–semiconductor contact is usually negligible compared to that of the bulk semiconductor material.

Section 10.2 describes the metal work function and the Schottky effect at a metal–vacuum interface. Thermionic emission theory, used to describe carrier transport at a metal–semiconductor contact, is presented in Section 10.3. In Section 10.4, the energy band diagram, the spatial distributions of the space charge, potential, and electric field across the depletion layer of a Schottky barrier diode are derived. Section 10.5 presents diffusion (Schottky) and thermionic emission models for carrier transport in a Schottky barrier diode. Section 10.6 depicts I–V characteristics and fabrication schemes for a metal–Si and metal–GaAs Schottky barrier diode. Section 10.7 describes three commonly used methods for determining the barrier height of a Schottky diode. Methods for effective barrier height enhancement of a metal–semiconductor Schottky contact are discussed in Section 10.8. In Section 10.9, applications of Schottky barrier diodes for photodetectors, microwave mixers, clamped transistors, MESFETs, and solar cells are depicted. Finally, conventional and novel approaches for forming ohmic contacts on a semiconductor are discussed in Section 10.10.

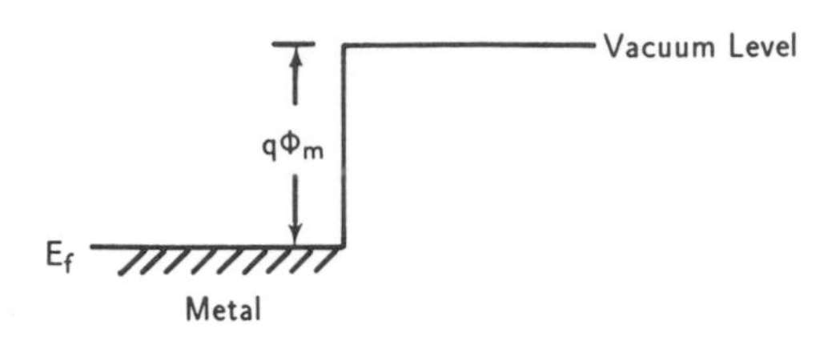

FIGURE 10.1. Energy band diagram at a metal–vacuum interface: ϕ_m is the metal work function and E_f is the Fermi level.

10.2. METAL WORK FUNCTION AND SCHOTTKY EFFECT

The equilibrium energy band diagram for a metal in free space is shown in Fig. 10.1. The energy difference between the vacuum level and the Fermi level is known as the work function of a metal. The work function, ϕ_m, is defined as the minimum energy required for an electron to escape from the metal surface (i.e., Fermi level) into free space at $T = 0$ K. The probability for an electron to escape from the metal surface into the vacuum depends on the velocity of electrons perpendicular to the metal surface. The minimum kinetic energy for an electron to escape from the metal surface into vacuum is given by

$$\tfrac{1}{2}(m_0 v_1^2) \geq q\phi_m \tag{10.1}$$

where v_1 is the electron velocity normal to the metal surface and m_0 is the free-electron mass. If an external electric field is applied to the metal surface, then the so-called Schottky effect or image lowering effect appears. To understand the Schottky effect, let us consider the energy band diagram shown in Fig. 10.2a. When an electric field is applied to the metal surface, electrons that escape from the metal surface will experience two external forces: the image force which arises from the Coulomb attractive force due to the positive image charges induced inside the metal by the escaping electrons, and the Lorentz force due to the applied electric field. The positive image charges create a Coulombic attractive force which tends to pull the escaping electrons back into the metal. The image force can be expressed by

$$F_i = -\frac{q^2}{16\pi\varepsilon_0 x^2} \tag{10.2}$$

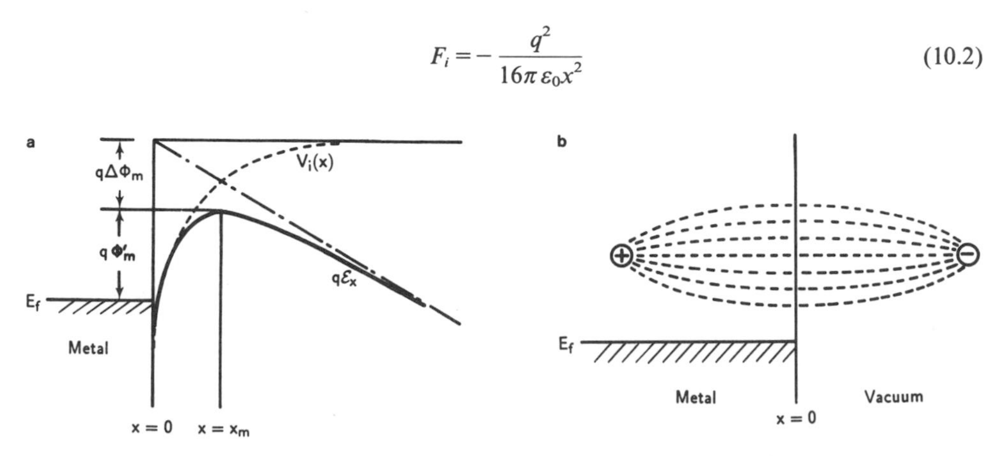

FIGURE 10.2. Schottky (or image lowering) effect at the metal–vacuum interface in the presence of an applied electric field: (a) energy band diagram, showing the applied field ($\mathscr{E}_x$), the image potential $V_i(x)$, and the image lowering potential $\delta\phi_m$; (b) the induced image charge (positive) inside the metal.

The potential energy associated with this image force is given by

$$V_i(x) = -\int_{\infty}^{x} F_i\,dx \tag{10.3}$$

The potential energy due to the applied electric field can be written as

$$V_a(x) = -q\mathscr{E}x \tag{10.4}$$

The total potential energy of the electron is equal to the sum of Eqs. (10.3) and (10.4), namely,

$$V(x) = V_i(x) + V_a(x) = -\frac{q^2}{16\pi\varepsilon_0 x} - q\mathscr{E}x \tag{10.5}$$

The distance at which the maximum potential energy occurs is obtained by differentiating Eq. (10.5) with respect to x and then setting the result equal to zero. This yields

$$x_m = \sqrt{\frac{q}{16\pi\varepsilon_0\mathscr{E}}} \tag{10.6}$$

Now, on substituting Eq. (10.6) into Eq. (10.5), we obtain the maximum potential energy for the electrons,

$$V_m(x_m) = -q\mathscr{E}\sqrt{\frac{q}{4\pi\varepsilon_0\mathscr{E}}} = -2q\mathscr{E}x_m = -q\Delta\phi_m \tag{10.7}$$

It is shown in Fig. 10.2 that the effect of the image force and the applied electric field is to lower the work function of a metal. Therefore, the effective metal work function under the applied electric field can be obtained from Fig. 10.2, and the result is

$$\begin{aligned} q\phi'_m &= q\phi_m + V_m = q\phi_m - q\Delta\phi_m \\ &= q\phi_m - q\sqrt{\frac{q\mathscr{E}}{4\pi\varepsilon_0}} \end{aligned} \tag{10.8}$$

where $q\Delta\phi_m = -V_m$ is the image lowering potential energy. To see the effect of the image lowering effect, we consider two field strengths. If the applied electric field $\mathscr{E}$ is equal to 10^5 V/cm, then x_m is equal to 60 Å and $q\Delta\phi_m = 0.12$ eV. On the other hand, if $\mathscr{E} = 10^7$ V/cm, then $x_m = 6$ Å and $q\Delta\phi_m = 1.2$ eV. Thus, it is obvious that the effective metal work function is greatly reduced at high electric fields as a result of the image lowering effect. Figure 10.3 shows the image lowering potential versus the square root of the applied electric field with the dielectric constant ε_s as parameter. Table 10.1 lists work function data for some metals.

10.3. THERMIONIC EMISSION THEORY

Thermionic emission usually refers to the emission of electrons from a hot metal surface. If the metal is used as a cathode, and all the emitted electrons from the metal surface are collected at the anode of a vacuum diode, then the cathode is in a saturation emission condition. The emitted current density is then called the saturation current density, J_s, and the

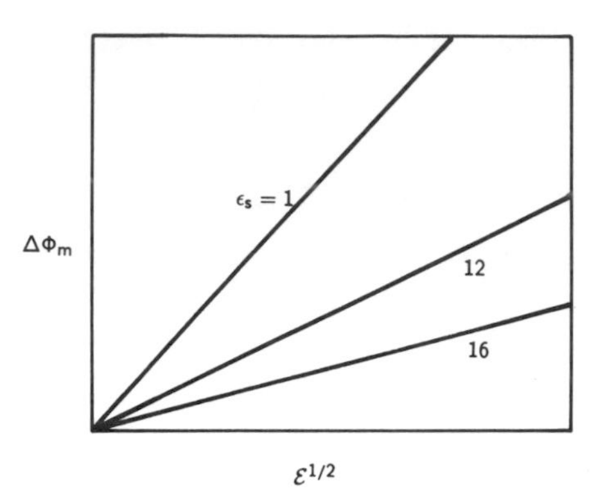

FIGURE 10.3. Image lowering potential versus square root of the applied electric field with dielectric constant ε_s as parameter.

TABLE 10.1. Metal Work Functions for a Clean Metal Surface in Vacuum

Metal(s)	Work function (eV)	Metal	Work function (eV)
Ti, Al, Ag	4.3	Sn	4.4
Au	5.1	W	4.63
Pt	5.7	Mo	4.28
Pd	5.15	Ni	4.5
Mg	3.65	Rh	5.05
Cu	4.45	Hf	4.0

equation which relates this latter quantity to the cathode temperature and the work function of a metal is known as the Richardson equation.

The Richardson equation is derived by considering the geometry of the metal surface shown in Fig. 10.4. The surface is assumed infinite in the x–y plane and the electron mission is normal to the metal surface along the z-axis. The free-electron density in the metal with velocity between (v_x, v_y, v_z) and $(v_x + dv_x, v_y + dv_y, v_z + dv_z)$ is given by

$$dn = 2\left(\frac{1}{2\pi}\right)^3 f(k)\, d^3k = \left(\frac{2m_0^3}{h^3}\right) f(v)\, dv_x\, dv_y\, dv_z \tag{10.9}$$

where m_0 is the free-electron mass and $\hbar k = m_0 v$. Using Maxwell–Boltzmann statistics, the electron distribution function $f(v)$ is given by

$$f(v) = \exp\left[-\frac{m_0(v_x^2 + v_y^2 + v_z^2)}{2k_BT}\right] \tag{10.10}$$

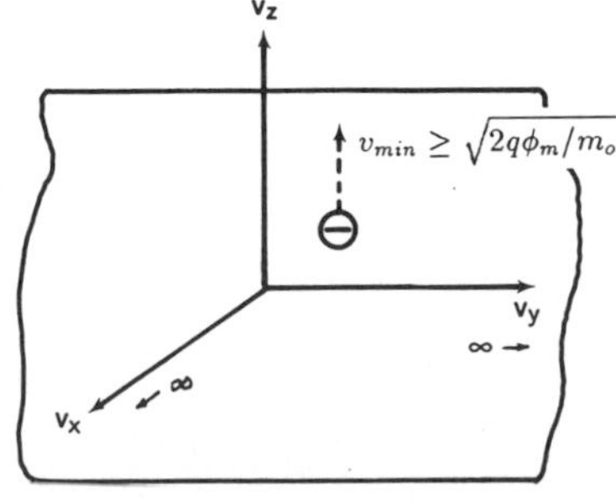

FIGURE 10.4. Thermionic emission of electrons from a metal surface.

Substitution of Eq. (10.10) into Eq. (10.9) yields the thermionic emission current density in the z-direction, namely,

$$J_s = \int q v_z \, dn$$

$$= \left(\frac{2qm_0^3}{h^3}\right) \int_{-\infty}^{\infty} \exp\left(-\frac{m_0 v_x^2}{2k_B T}\right) dv_x \int_{-\infty}^{\infty} \exp\left(-\frac{m_0 v_y^2}{2k_B T}\right) dv_y$$

$$\times \int_{v_{zm}}^{\infty} v_z \exp\left(-\frac{m_0 v_z^2}{2k_B T}\right) dv_z$$

$$= A_0 T^2 \exp\left(-\frac{q\Phi_m}{k_B T}\right) \tag{10.11}$$

where $A_0 = 4\pi q m_0 k_B^2/h^3$ is the Richardson constant, equal to 120 $\mathrm{A/cm^2 \cdot K^2}$ for the free-electron case. In deriving Eq. (10.11), we assume that only electrons with kinetic energies greater than the metal work function (i.e., $v \geq v_{zm}$) can escape from the metal surface.

It is seen from Eq. (10.11) that both the Richardson constant A_0 and the metal work function ϕ_m can be determined from the $\ln(J_s/T^2)$ versus $1/T$ plot; this is illustrated in Fig. 10.5. The intercept of this plot with the ordinate yields A_0, while the slope gives the value of the metal work function.

The image lowering effect must be considered when deriving the thermionic emission current density equation if an electric field is applied to the metal surface. In this case, ϕ_m in the exponent of Eq. (10.11) is replaced by the effective metal work function ϕ_m' given by Eq. (10.8). By replacing ϕ_m' for ϕ_m in Eq. (10.11), we obtain an expression for the effective thermionic current density:

$$J_s' = A_0 T^2 \exp\left(-\frac{q\phi_m'}{k_B T}\right)$$

$$= A_0 T^2 \exp\left(-\frac{q\phi_m}{k_B T}\right) \exp\left[\left(\frac{q}{2k_B T}\right)\left(\frac{q\mathscr{E}}{\pi \varepsilon_0}\right)^{1/2}\right]$$

$$= J_s \exp\left(\frac{4.39\mathscr{E}^{1/2}}{T}\right) \tag{10.12}$$

where the electric field $\mathscr{E}$ is in V/cm. It is seen that Eq. (10.12) takes into account the image lowering effect.

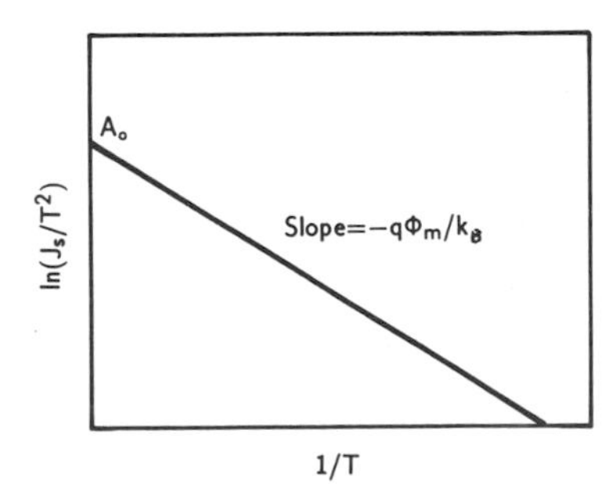

FIGURE 10.5. Plot of $\ln(J_s/T^2)$ versus $1/T$ using Eq. (10.11). The Richardson constant A_0 and the metal work function ϕ_m can be determined from this plot.

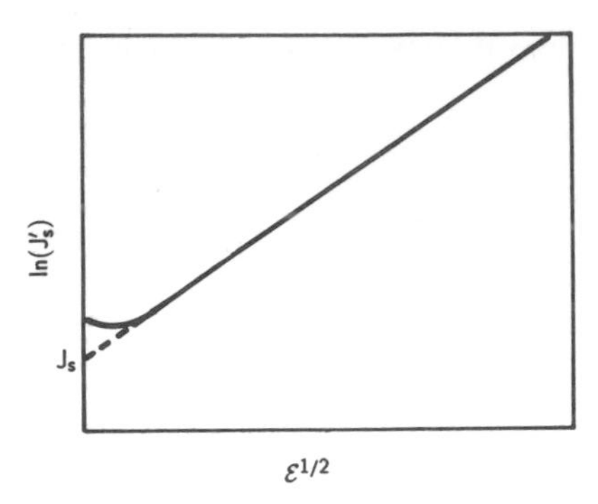

FIGURE 10.6. Plot of $\ln J_s'$ versus $\mathscr{E}^{1/2}$ for a thoriated tungsten metal, assuming $J_s = 1\ \text{A/cm}^2$ at $T = 1873\ \text{K}$.

In general, a plot of $\ln(J_s')$ versus $\mathscr{E}^{1/2}$ should yield a straight line over a wide range of electric fields. However, deviation from linearity is expected at very small electric fields. Figure 10.6 shows a plot of $\ln(J_s')$ versus $\mathscr{E}^{1/2}$ for a thoriated tungsten metal. Using Eq. (10.11) and assuming that $J_s = 1\ \text{A/cm}^2$, $T = 1873\ \text{K}$, and $A_0 = 120\ \text{A/cm}^2 \cdot \text{K}^2$, we obtain the value of ϕ_m, which is equal to 3.2 eV for the thoriated tungsten metal.

10.4. IDEAL SCHOTTKY BARRIER CONTACT

According to the Schottky model, the barrier height for an ideal metal–semiconductor Schottky contact is equal to the difference between the work function of a metal ϕ_m and the electron affinity χ_s of a semiconductor. For a metal/n-type semiconductor Schottky contact, this is given by

$$\phi_{Bn} = \phi_m - \chi_s \tag{10.13}$$

Figure 10.7 shows energy band diagrams for a metal/n-type semiconductor system before and after contact and under various conditions. Figure 10.7a, b, and c denote the cases for $\phi_m > \phi_s$, and Fig. 10.7d, e, and f denote the cases with $\phi_m < \phi_s$. Figure 10.7a and d are before the contact and Fig. 10.7b and e after the contact, assuming the existence of a thin insulating interfacial layer (e.g., $20 \sim 30$ Å) between the metal and semiconductor. Figure 10.7c and f pertain to intimate contact without the interfacial layer. From Fig. 10.7c it is seen that for $\phi_m > \phi_s$, there exists a potential barrier for electrons to cross from the metal to the semiconductor, and the metal–semiconductor contact exhibits a rectifying behavior. However, an ohmic contact is obtained if $\phi_m < \phi_s$, as shown in Fig. 10.7f. For a metal/p-type semiconductor contact, the opposite behavior results. We note that the measured barrier heights for most of the metal–n-type semiconductor contacts do not always follow the simple prediction of Eq. (10.13), because this equation does not consider the interface state density and the image lowering effect. In fact, for most III–V compound semiconductors, due to the high surface state density and Fermi-level pinning at the interface states, the barrier height for Schottky diodes formed on these materials is found to be independent of the metals used. A detailed explanation of this result will be given in Section 10.7.

Similarly, the barrier height for an ideal metal/p-type semiconductor Schottky contact can be expressed by

$$\phi_{Bp} = \frac{E_g}{q} - (\phi_m - \chi_s) = \frac{E_g}{q} - \phi_{Bn} \tag{10.14}$$

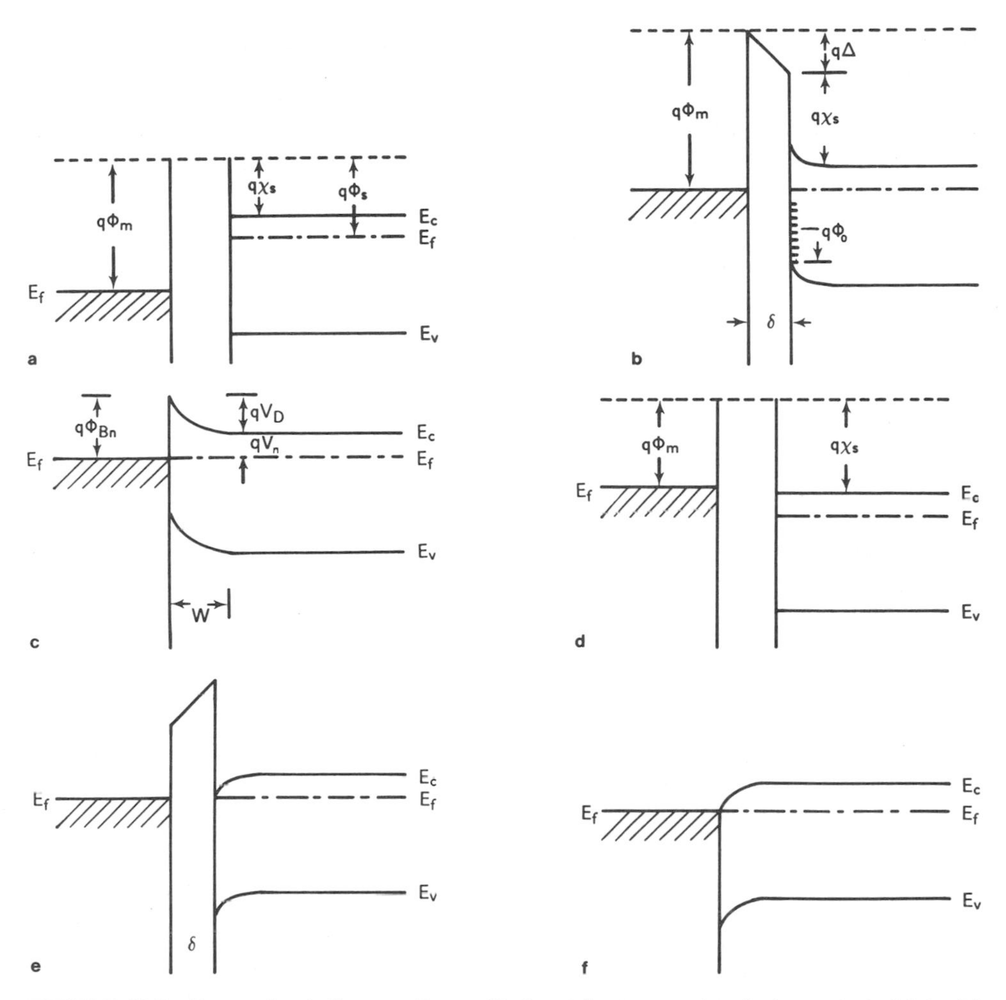

FIGURE 10.7. Energy band diagrams for an ideal metal-n-type semiconductor contact. (a) to (c) $\phi_m > \phi_s$: (a) before contact, (b) in contact with a small air gap and interface states, (c) in intimate contact (rectifying contact), with no interface states; (d) to (f) $\phi_m < \phi_s$: (d) before contact, (c) in contact with a small air gap, and (f) in intimate contact (ohmic contact).

where E_g is the energy bandgap and q is the electronic charge. Equation (10.14) shows that, for a given metal/semiconductor system, the sum of barrier heights for both n- and p-type semiconductors is equal to the bandgap energy of the semiconductor. As shown in Fig. 10.7c, the potential difference $q(\phi_m - \chi_s - V_n)$ is known as the contact potential or diffusion potential, V_D, and can be expressed by

$$V_D = \phi_m - \phi_s = \phi_{Bn} - V_n \tag{10.15}$$

where ϕ_{Bn} is the barrier height and $V_n = (E_c - E_F)/q = (k_B T/q)\ln(N_c/N_D)$ is the Fermi potential.

Equation (10.13) implies that the barrier height for an ideal Schottky diode is equal to the difference between the metal work function and the electron affinity of the semiconductor. This simple relationship holds only when both the interface states and the image lowering effect are negligible. Furthermore, it is noted from Eqs. (10.13) and (10.14) that the sum of the barrier heights for a metal on n-type and p-type semiconductors is equal to the energy bandgap of the semiconductor (i.e., $q\phi_{Bn} + q\phi_{Bp} = E_g$).

The potential and electric field distributions in the space-charge region, the depletion layer width, and the junction capacitance of a Schottky diode can be derived by solving Poisson's equation using proper boundary conditions. The one-dimensional Poisson's equation in the space-charge region of a Schottky diode is given by

$$\frac{d^2V(x)}{dx^2} = -\frac{\rho}{\varepsilon_0\varepsilon_s} \tag{10.16}$$

where ε_s is the dielectric constant of the semiconductor and ε_0 is the permittivity of the free space. The charge density for $0 < x < W$ is given by

$$\rho = q[N_D - n(x)] \tag{10.17}$$

where $n(x)$ is the electron density at $x = W$ and is equal to $n_0 \exp(-qV_D/k_BT)$. We note that $n(x)$ decreases exponentially with distance from the depletion layer edge into the depletion region.

By using a one-sided abrupt junction approximation and assuming that $n(x) = 0$ for $0 < x < W$, we can obtain the spatial distribution of the electric field by integrating Eq. (10.16). This approach leads to

$$\mathscr{E}(x) = -\frac{dV(x)}{dx} = \left(\frac{qN_D}{\varepsilon_0\varepsilon_s}\right)x + C_1 \tag{10.18}$$

where C_1 is a constant determined by the boundary conditions.

The potential distribution can be obtained by integrating Eq. (10.18) to yield

$$V(x) = -\left(\frac{qN_D}{2\varepsilon_0\varepsilon_s}\right)x^2 - C_1x + C_2 \tag{10.19}$$

where C_2 is another constant of integration. The constants C_1 and C_2 are determined by the following boundary conditions:

$$V(0) = -\phi_{Bn} \qquad \text{at } x = 0$$

$$\mathscr{E}(x) = -\frac{dV(x)}{dx} = 0 \qquad \text{at } x = W \tag{10.20}$$

Equations (10.18), (10.19), and (10.20) give

$$C_1 = -\frac{qN_DW}{\varepsilon_0\varepsilon_s} \qquad \text{and} \qquad C_2 = -\phi_{Bn} \tag{10.21}$$

When these latter values of C_1 and C_2 are substituted into Eqs. (10.18) and (10.19), we obtain the spatial distributions of the electric field and potential inside the depletion region. These are, respectively,

$$\mathscr{E}(x) = \left(\frac{qN_D}{\varepsilon_0\varepsilon_s}\right)(x - W) \tag{10.22}$$

and

$$V(x) = -\left(\frac{qN_D}{\varepsilon_0 \varepsilon_s}\right)\left(\frac{x^2}{2} - Wx\right) - \Phi_{Bn} \tag{10.23}$$

The depletion layer width W can be expressed in terms of N_D, V_D, and V_a across the barrier. Hence Fig. 10.8a and Eq. (10.23) yield the potential at $x = W$ in the form

$$V(W) = (V_D - V_a) - \phi_{Bn} = \left(\frac{qN_D W^2}{2\varepsilon_0 \varepsilon_s}\right) - \phi_{Bn} \tag{10.24}$$

or

$$W = \sqrt{\frac{2\varepsilon_0 \varepsilon_s (V_D - V_a)}{qN_D}} \tag{10.25}$$

It is seen from Eq. (10.25) that the depletion layer width is directly proportional to the square root of the applied voltage, and is inversely proportional to the square root of the dopant density of the semiconductor. Furthermore, Eq. (10.25) shows that the depletion layer width decreases with the square root of the forward bias voltage (i.e., for $V_a \geq 0$), and increases with the square root of the reverse bias voltage (i.e., for $V_a < 0$).

In order to derive the depletion layer capacitance, we note that the space charge per unit area, Q_s, in the depletion region (Fig. 10.8d) is given by

$$Q_s = qN_D W = \sqrt{2qN_D \varepsilon_0 \varepsilon_s (V_D - V_a)} \tag{10.26}$$

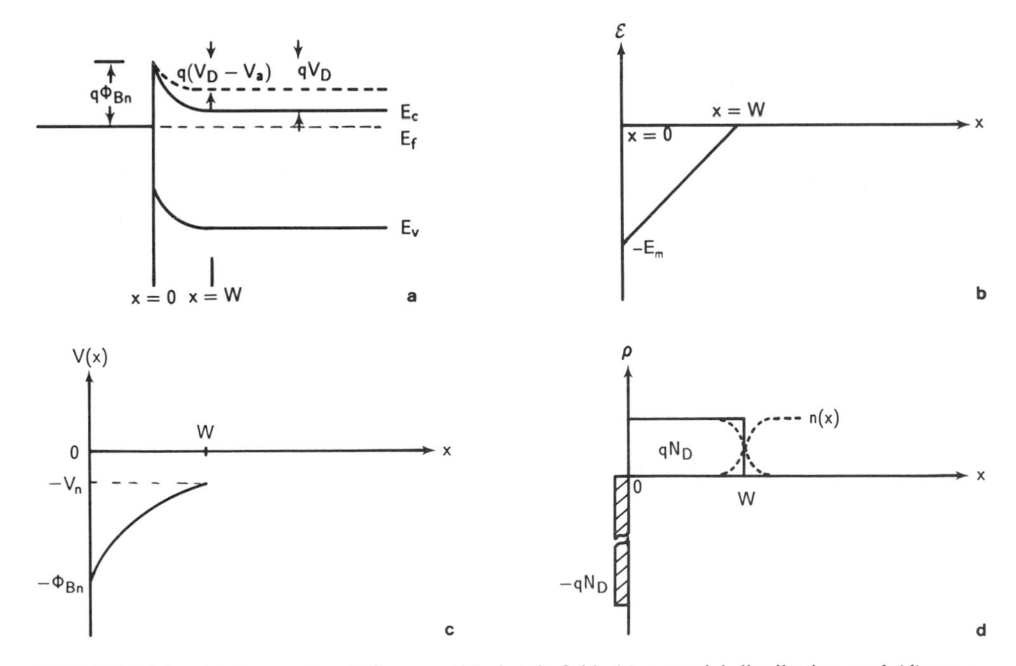

FIGURE 10.8. (a) Energy band diagram, (b) electric field, (c) potential distribution, and (d) space-charge distribution for a metal/n-type semiconductor Schottky barrier diode.

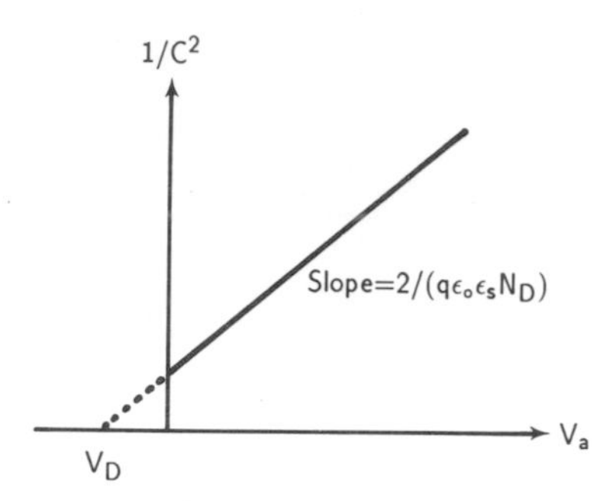

FIGURE 10.9. Square of the inverse capacitance versus the applied voltage for a metal/n-type semiconductor Schottky barrier contact.

The depletion layer capacitance per unit area is obtained by differentiating Eq. (10.26) with respect to the applied voltage V_a to yield

$$C_d = \frac{dQ_s}{dV_a} = \sqrt{\frac{qN_D\varepsilon_0\varepsilon_s}{2(V_D - V_a)}} \tag{10.27}$$

This equation shows that the depletion layer capacitance is inversely proportional to the square root of the applied voltage. Figure 10.8a presents the energy band diagram for a metal/n-type semiconductor Schottky barrier diode in thermal equilibrium (solid line), and under forward bias conditions (dashed line). Figure 10.8b depicts the spatial dependence of the electric field in the depletion region. The maximum electric field, which occurs at $x = 0$, is given by

$$\mathscr{E}_m = -\frac{qN_DW}{\varepsilon_0\varepsilon_s} \tag{10.28}$$

The potential and space-charge distributions in the depletion region are shown in Fig. 10.8c and d, respectively. In Fig. 10.8d the dashed line denotes the actual charge distribution, which shows that at $x = W$ the free-electron density n_0 decreases exponentially with distance as it spreads into the depletion region. The solid line is the abrupt junction approximation, which has been used for the present derivation. The above analysis is valid only for an ideal Schottky diode in which both the surface states and the image lowering effect are ignored. Figure 10.9 shows a plot of $1/C^2$ versus the applied bias voltage, V_a. A linear relation is obtained if N_D is constant throughout the depletion region. Both V_D and N_D can be determined from this plot.

10.5. CURRENT FLOW IN A SCHOTTKY BARRIER DIODE

The metal–semiconductor Schottky barrier diode is a majority-carrier device, because the current flow in such a device is due to the majority carriers (i.e., electrons in an n-type material). This is in contrast to a p–n junction diode in which the minority carriers are responsible for the current conduction. To illustrate the current flow in a Schottky diode, the energy band diagrams and current components for an ideal metal/n-type semiconductor Schottky barrier diode under zero-bias, forward-bias, and reverse-bias conditions are shown

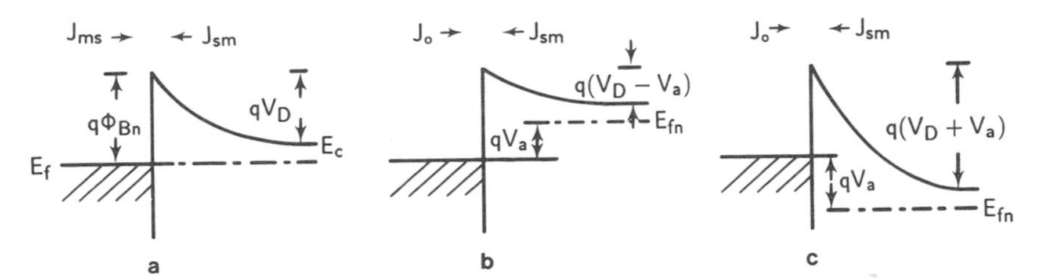

FIGURE 10.10. Energy band diagrams and current components for a Schottky barrier diode under (a) zero bias, (b) forward bias, and (c) reverse bias. J_{sm} denotes the current flow from semiconductor to metal, J_{ms} is the current density from metal to semiconductor, and $J_0 = J_{ms}$ is the saturation current density.

in Figs. 10.10a, b, and c, respectively. Figure 10.10a shows the energy band diagram under zero-bias conditions. The potential barrier for electrons moving from the semiconductor side to the metal side is designated as V_D, while the potential barrier for electrons moving from the metal side to the semiconductor side is ϕ_{Bn}. The net current flow at zero bias is zero.

If a forward-bias voltage V_a is applied to the Schottky diode, then the potential barrier on the semiconductor side of the diode is reduced to $(V_D - V_a)$, as is shown in Fig. 10.10b. However, the barrier height remains relatively unaffected by the applied bias voltage or the doping density of the semiconductor. Therefore, the current flow from the semiconductor to the metal increases drastically under a forward-bias condition, while the current flow from the metal to the semiconductor remains essentially the same. Under a forward-bias condition, the net current flow is controlled by the electron current flow from the semiconductor to the metal, as is shown in Fig. 10.10b. Under a reverse-bias condition, the potential barrier on the semiconductor side increases to $(V_D + V_a)$, and the current flow from the semiconductor to the metal becomes negligibly small compared to the current flow from the metal to the semiconductor. Thus, the net current flow under a reverse-bias condition is controlled by the thermionic emission current flow from the metal to the semiconductor; this is illustrated in Fig. 10.10c.

Carrier transport in a metal–semiconductor Schottky barrier diode can be treated by using the thermionic emission, the diffusion, or the combined thermionic–diffusion model. The current–voltage (I–V) equation derived from these models may be used to predict the current versus temperature or voltage behavior in a Schottky barrier diode. The simple thermionic emission model developed by Bethe and the diffusion model developed by Schottky are the most widely used physical models for predicting current–voltage characteristics of a Schottky barrier diode. In this section, we derive current density equations from both the thermionic emission and the diffusion models. In addition, the combined thermionic–diffusion model developed by Sze and Crowell will be depicted. Finally, the tunneling phenomenon in a highly-doped Schottky diode will also be discussed.

10.5.1. Thermionic Emission Model

The thermionic emission model described in Section 10.3 for electron emission from a hot metal surface into free space can be easily modified for a metal–semiconductor system. The current flow from the semiconductor into the metal is determined mainly by the barrier potential $(V_D - V_a)$ under a forward-bias condition. To overcome this potential barrier, the kinetic energy of electrons in the semiconductor side along the x-direction is given by

$$\tfrac{1}{2}(m_n^* v_{xm}^2) \geq q(V_D - V_a) \tag{10.29}$$

Thus, the electron current density component flowing from the semiconductor to the metal, J_{sm}, may be obtained by integrating the following equation, which is given by

$$
\begin{aligned}
J_{sm} &= \left(\frac{2qm^{*3}}{h^3}\right) \int_{-\infty}^{\infty} \exp\left[\frac{-m^* v_z^2}{2k_B T}\right] dv_z \int_{-\infty}^{\infty} \exp\left[\frac{-m^* v_y^2}{2k_B T}\right] dv_y \\
&\quad \times \int_{v_{xm}}^{\infty} v_x \exp\left[\frac{-m^* v_x^2}{2k_B T}\right] dv_x \\
&= A^* T^2 \exp\left(-\frac{q\phi_{Bn}}{k_B T}\right) \exp\left(\frac{qV_a}{k_B T}\right) \\
&= J_0 \exp\left(\frac{qV_a}{k_B T}\right)
\end{aligned}
\tag{10.30}
$$

where

$$
J_0 = A^* T^2 \exp\left(-\frac{q\phi_{Bn}}{k_B T}\right) \tag{10.31}
$$

is the saturation current density. In Eq. (10.31), $A^* = 4\pi m_n^* q k_B^2 / h^3$ is the effective Richardson constant, m_n^* is the electron effective mass, ϕ_{Bn} is the barrier height, and J_0 is the reverse saturation current density. The current flow from the metal to the semiconductor can be derived from Eq. (10.30) by using the fact that in thermal equilibrium $V_a = 0$, and

$$
J_{ms} = -J_{sm} = -J_0 \tag{10.32}
$$

Thus, the total current flow under a forward-bias condition is equal to the sum of Eqs. (10.30) and (10.32), and can be written as

$$
J = J_{sm} + J_{ms} = J_0 \left[\exp\left(\frac{qV_a}{k_B T}\right) - 1\right] \tag{10.33}
$$

Equation (10.33) is the well-known Schottky diode equation, which predicts an exponential dependence of current density on both temperature and the applied bias voltage. Since the saturation current density J_0 depends exponentially on barrier height, a large barrier height is needed in order to reduce the value of J_0 in a Schottky diode. Several methods of enhancing the effective barrier height of a Schottky diode will be discussed in Section 10.8.

10.5.2. Image Lowering Effect

Similar to the case of a metal–vacuum interface, the image lowering effect also exists at the metal–semiconductor interface, as is shown in Fig. 10.11. By incorporating the image

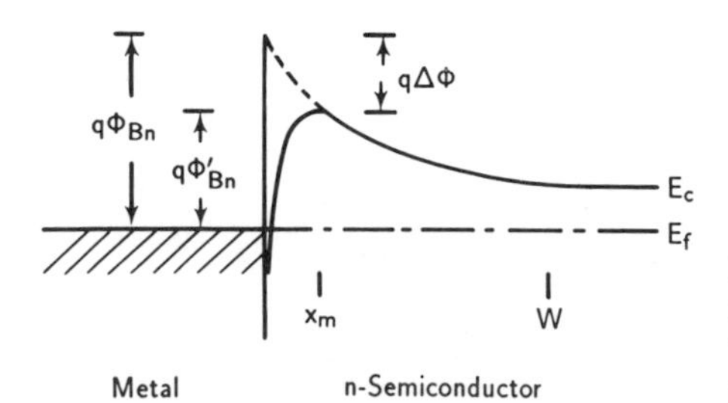

FIGURE 10.11. Energy band diagram for a metal/n-type semiconductor Schottky barrier diode showing the image lowering effect; $q\Delta\phi$ is the image lowering potential.

lowering effect into Eq. (10.31), the saturation current density can be expressed by

$$J_0 = A^* T^2 \exp\left[-\frac{q(\phi_{Bn} - \Delta\phi_m)}{k_B T}\right]$$

$$= A^* T^2 \exp\left(-\frac{q\phi_{Bn}}{k_B T}\right) \exp\left(\frac{q^3 \mathscr{E}_m}{4\pi \varepsilon_0 \varepsilon_s k_B^2 T^2}\right)^{1/2} \tag{10.34}$$

In the depletion region, the maximum field strength $\mathscr{E}_m$ at the metal–semiconductor interface can be derived from Eqs. (10.25) and (10.28), and the result yields

$$\mathscr{E}_m = \sqrt{\frac{2qN_D(V_D - V_a)}{\varepsilon_0 \varepsilon_s}} \tag{10.35}$$

As shown in Fig. 10.8b, the maximum electric field occurs at $x = 0$, and decreases linearly with distance from the metal–semiconductor interface (i.e., $x = 0$) to the edge of the depletion layer (i.e., $x = W$) in the bulk semiconductor. From Eqs. (10.34) and (10.35), it is seen that $\ln J_0$ is directly proportional to $(\mathscr{E}_m)^{1/2}$ or $(V_D - V_a)^{1/4}$ when the image lowering effect is included. This current–voltage behavior has indeed been observed in many metal–semiconductor Schottky barrier diodes.

10.5.3. The Diffusion Model

The Schottky diffusion model is based on the assumption that the barrier height is greater than a few $k_B T$ and that the semiconductor is lightly doped so that the space-charge layer width is greater than the carrier diffusion length. Based on this model, both the drift and diffusion current components should be considered in the depletion region, which can be written as

$$J_n = qn(x)\mu_n \mathscr{E}_x + qD_n \frac{dn(x)}{dx}$$

$$= qD_n\left[\left(\frac{qn(x)}{k_B T}\right)\left(-\frac{dV(x)}{dx}\right) + \frac{dn(x)}{dx}\right] \tag{10.36}$$

We note that the Einstein relation $\mu_n = (q/k_B T)D_n$ and $\mathscr{E}_x = -\,dV(x)/dx$ have been used in Eq. (10.36). Since the total current density J_n in the depletion region is constant and independent of x, we can multiply both sides of Eq. (10.36) by $\exp[-qV(x)/k_B T]$ and then integrate the equation over the entire depletion region (i.e., from $x = 0$ to $x = W$). This yields

$$J_n \int_0^W e^{-qV(x)/k_B T}\, dx = qD_n \int_0^W \left[-\left(\frac{qn(x)}{k_B T}\right)\frac{dV(x)}{dx} e^{-qV(x)/k_B T} + \frac{dn(x)}{dx} e^{-qV(x)/k_B T}\right] dx \tag{10.37}$$

or

$$J_n \int_0^W e^{-qV(x)/k_B T}\, dx = qD_n n(x)\, e^{-qV(x)/k_B T}\Big|_0^W \tag{10.38}$$

The boundary conditions for Eq. (10.38) at $x = 0$ and $x = W$ are given by

$$qV(0) = -q\phi_{Bn} \quad \text{and} \quad qV(W) = -q(V_n + V_a) \tag{10.39}$$

where $qV_n = (E_c - E_f)$ and V_a is the applied voltage. The electron densities at $x = 0$ and $x = W$ are respectively given by

$$n(0) = N_c \exp\left\{-\frac{[E_c(0) - E_f]}{k_B T}\right\} = N_c \exp\left(-\frac{q\phi_{Bn}}{k_B T}\right) \quad \text{and} \quad n(W) = N_c \exp\left(-\frac{qV_n}{k_B T}\right) \tag{10.40}$$

Now on substituting Eqs. (10.39) and (10.40) into Eq. (10.38), we obtain

$$J_n = \frac{(qD_nN_c)[\exp(qV_a/k_BT) - 1]}{\int_0^W \exp[-qV(x)/k_BT]\,dx} \tag{10.41}$$

The integral in the denominator of Eq. (10.41) is evaluated by substituting Eq. (10.23) for $V(x)$ (neglecting the x^2 term) and Eq. (10.25) for W into Eq. (10.41). The result is

$$\begin{aligned} J_n &= \left(\frac{q^2 D_n N_c}{k_B T}\right)\sqrt{\frac{2q(V_D - V_a)N_D}{\varepsilon_0 \varepsilon_s}}\exp\left(-\frac{q\Phi_{Bn}}{k_B T}\right)\left[\exp\left(\frac{qV_a}{k_B T}\right) - 1\right] \\ &= J_0'\left[\exp\left(\frac{qV_a}{k_B T}\right) - 1\right] \end{aligned} \tag{10.42}$$

where

$$J_0' = \left(\frac{q^2 D_n N_c}{k_B T}\right)\sqrt{\frac{2q(V_D - V_a)N_D}{\varepsilon_0 \varepsilon_s}}\exp\left(-\frac{q\Phi_{Bn}}{k_B T}\right) \tag{10.43}$$

is the saturation current density derived from the diffusion model.

A comparison of Eq. (10.43) and Eq. (10.31) reveals that the saturation current density derived from the thermionic emission model is more sensitive to temperature than that of the diffusion model. However, the latter shows a stronger dependence on the applied-bias voltage than the former. We note that the image lowering effect is neglected in Eq. (10.43).

Finally, a synthesis of the diffusion and thermionic emission models has been developed by Crowell and Sze.[1] The so-called thermionic–emission model uses the boundary conditions of the thermionic recombination velocity near the metal–semiconductor interface and considers the effects of electron-optical phonon scattering and quantum-mechanical reflection at the metal–semiconductor interface. The current density equation derived from the thermionic–diffusion model is given by

$$J = \frac{qN_c v_R}{(1 + v_R/v_D)}\exp\left(-\frac{q\Phi_{Bn}}{k_B T}\right)\left[\exp\left(\frac{qV}{k_B T}\right) - 1\right] \tag{10.44}$$

where v_R is the recombination velocity, and v_D is the diffusion velocity associated with electron transport from the depletion layer edge at W to the potential energy maximum at x_m. If v_R is much greater than v_D, then the diffusion process is dominant. On the other hand, if v_D is much greater than v_R, then the pre-exponential factor in Eq. (10.44) is dominated by v_R and the thermionic emission current becomes the predominant current component.

Finally, it is noteworthy that if a metal–semiconductor Schottky contact is formed on a degenerate semiconductor, then the barrier width becomes so thin that the current flow could be dominated by the tunneling process. In this case, the current flow in the diode is determined by the quantum-mechanical tunneling transmission coefficient, and the tunneling current density is proportional to the exponential function of the barrier height, which is given by

$$J_t \sim \exp(-q\phi_{Bn}/E_{00}) \tag{10.45}$$

where $E_{00} = (q\hbar/2)\sqrt{N_D/m^*\varepsilon_0\varepsilon_s}$. From Eq. (10.45), it is seen that the tunneling current will increase exponentially with the square root of the dopant density and decreases with increasing barrier height. Equation (10.45) may be applied to analyze the specific contact resistance for the ohmic contact on a heavily doped semiconductor. This will be discussed further in Section 10.10.

10.6. I–V CHARACTERISTICS OF A SILICON AND A GaAs SCHOTTKY DIODE

In this section, we compare the current–voltage (I–V) characteristics of a Au/n-type Si Schottky diode and a Au/n-type GaAs Schottky diode. The experimental results for both diodes under forward-bias conditions are displayed in Figs. 10.12 and 10.14, respectively.[1] Since the slope of the I–V curve under forward-bias conditions is usually greater than unity, a diode ideality factor 'n' is incorporated into Eq. (10.33) for a practical Schottky barrier diode. This leads to a semiempirical formula for predicting the I–V characteristics of a Schottky diode and is given by

$$J = J_0\left[\exp\left(\frac{qV_a}{nk_BT}\right) - 1\right] \tag{10.46}$$

where J_0 is the saturation current density given by Eq. (10.31). Under a forward-bias condition and for $qV_a \geq 3k_BT$, Eq. (10.46) becomes

$$J_F \cong J_0 \exp\left(\frac{qV_a}{nk_BT}\right) \tag{10.47}$$

For an ideal metal–semiconductor Schottky diode, the diode ideality factor n is equal to unity. Deviation of n from unity may be attributed to a number of factors such as large surface leakage current, high density of bulk recombination centers in the depletion region, and high interface state density as well as high series resistance.

Metal n-type Si Schottky barrier diodes with diode ideality factor n varying from 1.01 to 1.12 have been reported. To achieve near-ideal I–V characteristics for the Si Schottky barrier diodes, various fabrication techniques have been developed. The most widely used techniques including field-plate and guard-ring structures are discussed here. Figure 10.12 shows an Al/n-type Si Schottky barrier diode with a field-plate structure. As shown in Fig. 10.12, a field oxide is grown under the edge of an aluminum Schottky contact. The aluminum is overlaid on top of this field oxide to serve as a field plate. When the Schottky diode is reverse-biased, this field plate keeps the underlying contact surface fully depleted so that the soft breakdown arising from the surface accumulation layer formed around the edge of the metal plate can be eliminated. It is shown in Fig. 10.12 that the I–V characteristics for this diode closely follows the theoretical prediction given by the thermionic emission model for about six decades of current with values of n very close to unity. The intercept of the forward current at zero bias gives a barrier height of 0.70 V. The barrier height deduced from the activation energy plot of ln J_F versus $1/T$ at a fixed forward bias is found equal to 0.69 V. This value is in excellent agreement with the value determined for photoemission excitation of electrons from the metal

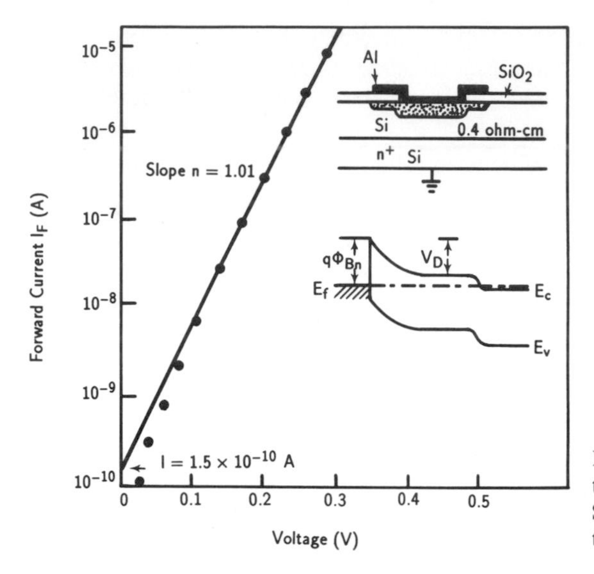

FIGURE 10.12. Forward *I–V* curve and the energy band diagram for an Al/n-type Si Schottky diode with a field-plate structure. After Yu and Mead,[2] by permission.

into the semiconductor. The reverse *I–V* characteristics for the Al/n-type Si Schottky barrier diodes are also displayed in Fig. 10.13 for two different substrate resistivities (i.e., $\rho = 0.4$ and 1.0 ohm · cm). The breakdown voltages for both diodes are presumably limited by the metal edge curvature in the depletion region. Figure 10.14 shows the near-ideal forward *I–V* characteristics for three Au/n-type GaAs Schottky barrier diodes formed on substrates with different orientations.[2]

Another Schottky barrier diode structure with near-ideal *I–V* characteristics, which can be obtained by using a p-type diffused guard-ring structure on an n-type silicon substrate, is shown in Fig. 10.15.[3] A p-type diffused guard-ring structure is extended in the normal planar fashion under the oxide. The Pt–Si Schottky barrier contact formed on n-type silicon inside the p^+ guard ring is in electrical contact with the p-type Si substrate. The doping profile of the p^+ guard ring is tailored in a such way that the breakdown voltage of the p–n junction in the guard-ring region is higher than that of the Schottky barrier contact. In this structure, the region of maximum electric field depends on the depth and profile of the diffused junction.

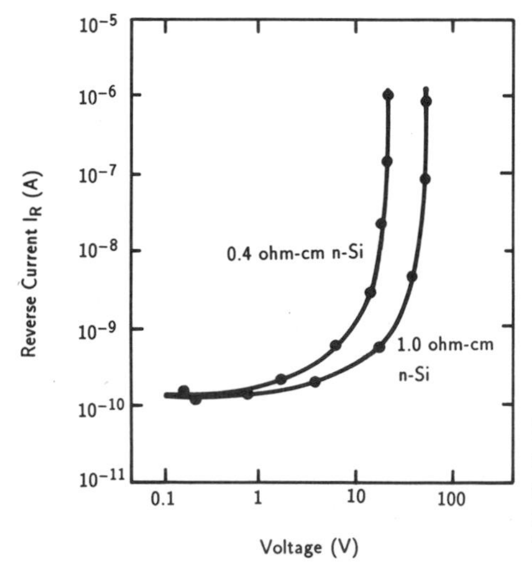

FIGURE 10.13. Reverse *I–V* curves for the Al/n-type silicon Schottky barrier diode with two different substrate resistivities. Al overlaid on SiO_2 is used to control the soft breakdown due to the edge effect. After Yu and Mead,[2] by permission.

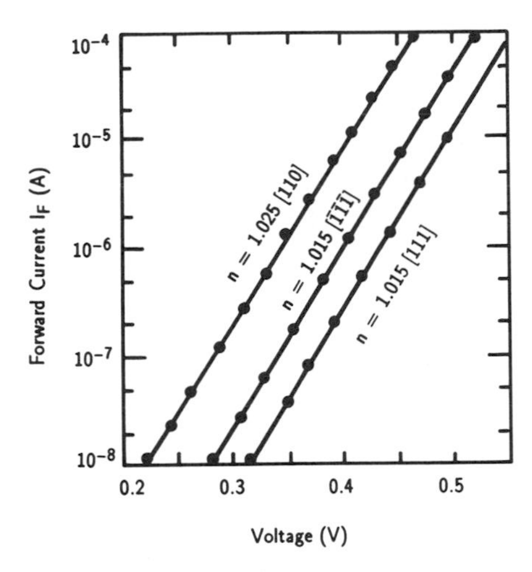

FIGURE 10.14. Forward *I–V* curves for Au/n-type GaAs diodes fabricated on different substrate orientations. After Kahng,[3] by permission.

For an ideal linearly-graded junction, the breakdown voltage is higher than that of a planar junction. Figure 10.16 shows the reverse *I–V* characteristics of a PtSi/n-type Si Schottky diode with a diffused guard-ring structure.[3] The solid line refers to experimental data while the dashed line is calculated from Eq. (10.34) by including the image lowering effect.

In silicon integrated circuits, aluminum is most often used for ohmic contacts and interconnects for silicon devices. In addition, aluminum is also widely used as a gate metal for silicon MOS devices and as Schottky contacts for bipolar transistor circuits. Unfortunately, the aluminum/silicon system has low eutectic temperature (577°C) and interdiffusion occurs at a relatively low temperature (i.e., approximately 400°C). As a result, large leakage current is often observed in the silicon shallow junction bipolar transistors and the n–p junction diodes when aluminum is used for interconnects and ohmic contacts. To overcome this problem, metal silicides with low resistivity and high-temperature stability are required for contacts in silicon ICs. Silicide is a metal–silicon compound which can be formed with a specific ratio of metal/silicon composition. Important silicides for silicon are those of the refractory metals such as Mo, Ti, Ta, and W and the near-noble metals such as Co, Ni, Pt, and Pd. Silicides

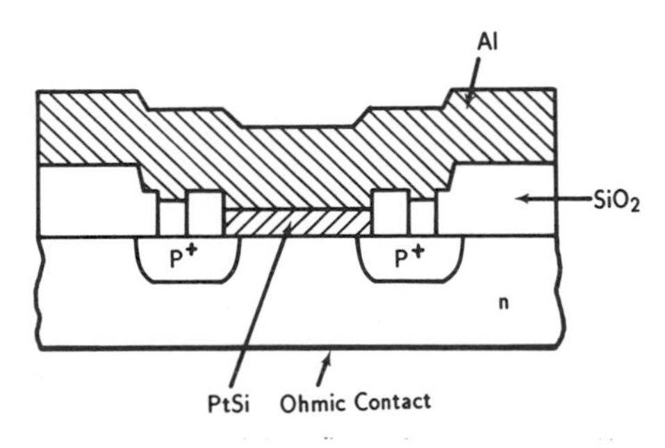

FIGURE 10.15. A PtSi-n-Si Schottky barrier diode with a diffused guard-ring structure. After Lepselter and Sze,[4] by permission.

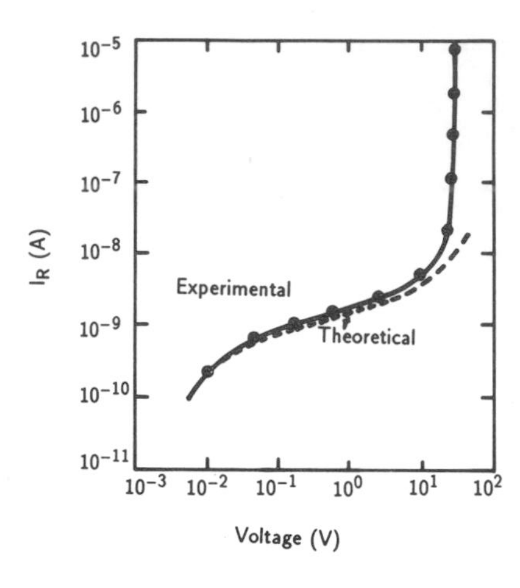

FIGURE 10.16. Comparison of theoretical and measured reverse I–V characteristics for a PtSi/n-type Si Schottky barrier diode shown in Fig. 10.15. After Lepselter and Sze,[4] by permission.

formed from these metals have low resistivity, high eutectic temperature, good adhesive characteristics, and stability. Furthermore, they are easy to form and etch, and can form on oxide. The most stable silicides are the silicon-rich metal disilicides (e.g., $CoSi_2$, $MoSi_2$, $TiSi_2$, and WSi_2), which have eutectic temperatures ranging from 1195 to 1440°C and a resistivity of around 20 to 40 $\mu\Omega \cdot$ cm. The reaction temperatures for these silicides vary from 350 to 650°C. Schottky barrier diodes formed on these silicides have barrier heights varying between 0.58 and 0.67 eV on n-type silicon. The most widely used metal silicide Schottky barrier contact in bipolar circuit applications is the PtSi/n-Si system. The barrier height for a PtSi/n-Si Schottky barrier diode is around 0.90 eV, which is probably the highest achievable barrier height (without barrier height enhancement) for a silicon Schottky barrier diode. In addition, a high-quality PtSi/p-type silicon Schottky barrier diode with barrier height of 0.2 eV or less has also been developed for infrared (3 to 5 μm) photodetector array applications.

In recent years, metal silicides have been widely used at the source or drain contact region of silicon MOS transistors. In process technology, the silicide is formed by depositing the metal onto the exposed silicon area and annealing it to form the silicide. The annealing occurs at temperatures well below the melting point of the silicon, but solid-state interdiffusion takes place and a silicide film forms. For metals deposited on Si, different silicide compounds form under different annealing conditions. For example, Pt deposited on Si, Pt_2Si compound forms at around 300°C, while PtSi forms with further annealing at 450°C. Although PtSi has been widely used in silicon ICs, the PtSi films are not very stable at high-temperature operation and hence require further processing steps. The group of refractory metal silicides of titanium (Ti), tantalum (Ta), molybdenum (Mo), and tungsten (W) has proved stable at high-temperature operation. For example, Ti film deposited on Si forms stable $TiSi_2$ compound after 650°C annealing. $TiSi_2$ has been widely used in VLSI device contacts. In addition, silicide films can also be formed by epitaxial growth. For example, epitaxial silicides, such as $CoSi_2$ and $NiSi_2$, which have cubic crystal structure, have been used as low-resistivity contacts and in novel high-speed device structures, such as metal-based transistors. More recently, epitaxial silicide film of $TiSi_2$ has also been reported for use in VLSI circuits and devices.

10.7. DETERMINATION OF BARRIER HEIGHT

Expressions for the barrier height of ideal metal/n-type and metal/p-type semiconductor Schottky barrier diodes are given by Eqs. (10.13) and (10.14), respectively. However, these equations are valid only when the image lowering effect is negligible and the surface state density is small. However, in most III–V compound semiconductors, the surface state density is usually very high. As a result, it is necessary to include the interface state effect in the barrier height expression. As shown in Fig. 10.7b, the effect of the surface states is expressed in terms of the energy level, $q\phi_0$. This energy level coincides with the Fermi level at the semiconductor surface before the metal–semiconductor contact is formed. In fact, $q\phi_0$ can be considered as a demarcation level in which the energy states below it must be filled in order to satisfy the charge-neutrality condition at the surface. If the surface states become very large, then the Fermi level at the surface will be pinned at $q\phi_0$, and the barrier height for metal–semiconductor contact becomes independent of the metal work function. This has indeed been observed in many Schottky barrier contacts formed on III–V compounds. Cowley and Sze have derived a general expression for the barrier height taking into account effects of image lowering and surface state density. Based on their derivation, the barrier height for a metal–semiconductor Schottky contact is given by[5]

$$\phi_{Bn} = c_2(\phi_m - \chi_s) + (1 - c_2)(E_g/q - \phi_0) - \Delta\phi = c_2\phi_m + c_3 \tag{10.48}$$

where

$$c_2 = \frac{\varepsilon_i \varepsilon_0}{\varepsilon_i \varepsilon_0 + q^2 \delta D_s} \tag{10.49}$$

Here, ε_i is the dielectric constant of the interfacial layer and δ is the thickness of this interfacial layer (see Fig. 10.7b). Equation (10.48) is obtained by assuming that δ is only a few Å thick, and hence ε_i is roughly equal to unity. Since c_2 and c_3 are determined experimentally, we can express $q\phi_0$ and D_s in terms of these two quantities:

$$q\phi_0 = E_g - \frac{c_2\chi_s + c_3 + \Delta\phi}{1 - c_2} \tag{10.50}$$

and the interface state density D_s is given by

$$D_s = \frac{(1 - c_2)\varepsilon_i \varepsilon_0}{c_2 \delta q^2} \tag{10.51}$$

It is of interest that the value of $q\phi_0$ for a wide variety of metals on III–V compound semiconductor Schottky contacts was found to be about one-third of the bandgap energy. Therefore, the barrier height for Schottky diodes formed on n-type semiconductors with very high surface state density is roughly equal to two-thirds of the bandgap energy (i.e., $\phi_{Bn} \sim \frac{2}{3} E_g$). Measurements of barrier heights for many metal/III–V semiconductor Schottky diodes are found to be in good agreement with this prediction. Theoretical calculations and experimental data reveal that the surface state densities for many III–V semiconductors such as GaAs is indeed very high (e.g., $Q_{ss} \geq 10^{13}$ states/cm^2) and the barrier height is found to be independent of the metal work function for Schottky diodes formed on semiconductor materials with high surface state density.

If the interfacial layer is assumed to be only a few Å and $\varepsilon_i = 1$, then the interface state density given by Eq. (10.51) is reduced to

$$D_s \simeq 1.1 \times 10^{13}(1 - c_2)/c_2 \text{ states/cm}^2 \cdot \text{eV} \tag{10.52}$$

For infinite interface state density, $D_s \to \infty$ and $c_2 \to 0$, Eq. (10.48) becomes

$$\phi_{Bn} = E_g/q - \phi_0 - \Delta\phi \tag{10.53}$$

If the interface state density is negligible and only the image lowering effect is considered, then the barrier height is given by

$$\phi_{Bn} = q(\phi_m - \chi_s) - \Delta\phi \tag{10.54}$$

which reduces to Eq. (10.13) when the image lowering effect is neglected.

Experimental results reveal that values of c_2 for Si, GaAs, and GaP are equal to 0.27, 0.09, and 0.27 eV, respectively. The calculated values of $q\phi_0$ for Si, GaAs, and GaP are given by 0.30, 0.54, and 0.67 eV, respectively. Figure 10.17 shows experimental results of the barrier height versus metal work function for Si, GaAs, GaP, and CdS metal/n-type semiconductor contacts.[4] The results clearly show that for GaAs Schottky contacts, the barrier height is nearly independent of the metal work function. This is due to the extremely high interface state density in GaAs crystal, and the barrier height is determined by Eq. (10.48).

We shall next discuss the three most commonly used methods for determining the barrier height of a Schottky diode. They are: (i) current–voltage (I–V), (ii) capacitance–voltage (C–V), and (iii) photoemission methods. These are discussed as follows:

The Current–Voltage Method. The semiempirical formula for the current density of a Schottky barrier diode given by Eq. (10.46) can be rewritten as

$$J = C\exp\left(-\frac{q\Phi_{Bn}}{k_B T}\right)\left[\exp\left(\frac{qV_a}{nk_B T}\right) - 1\right] \tag{10.55}$$

where C is a pre-exponential factor; its value depends on the model employed (i.e., thermionic emission or diffusion model). Typical plots of $\ln J$ versus applied voltage V_a and inverse temperature $1/T$ for a Schottky barrier diode are shown in Figs. 10.18a and b, respectively. The barrier height can be determined either from the saturation current density J_0, shown in Fig. 10.18a or from the $\ln J$ versus $1/T$ plot at a fixed-bias voltage, as shown in Fig. 10.18b. To increase the accuracy of the barrier height determined from the $\ln(J_F/T^2)$ versus $1/T$ plot at a fixed forward-bias voltage, it is important to choose a bias voltage for which the diode

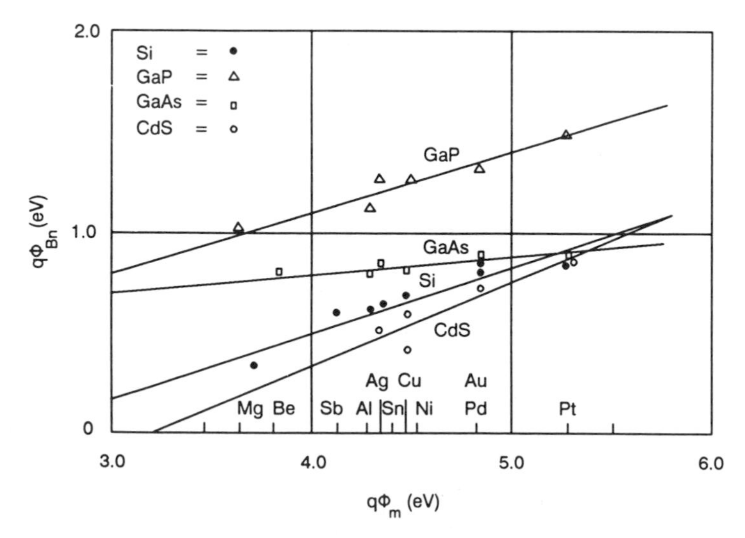

FIGURE 10.17. Schottky barrier heights versus metal work function for metal–Si, GaAs, GaP, and CdS Schottky contacts. After Cowley and Sze,[5] by permission.

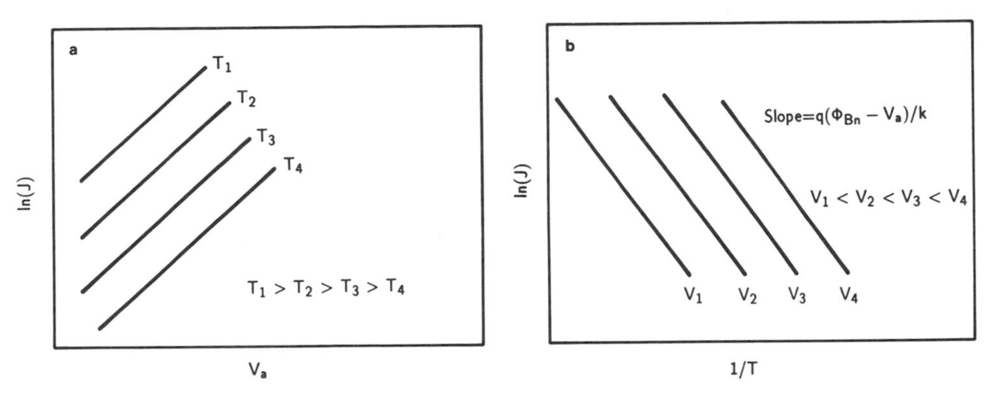

FIGURE 10.18. (a) ln J versus V_a for a Schottky barrier diode at four different temperatures, (b) ln J versus $1/T$ for four different forward-bias voltages.

ideality factor is nearly equal at different temperatures (i.e., the slope for the ln J_F versus V_F curves at T_1, T_2, T_3, and T_4 is the same), as shown in Fig. 10.18a.

The Photoemission Method. The barrier height of a Schottky diode can be determined by measuring the photocurrent versus wavelength of the incident photons near the fundamental absorption edge, as is illustrated in Fig. 10.19a. When photons with energies falling between the barrier height and the bandgap energy of the semiconductor (i.e., $q\phi_{Bn} < h\nu < E_g$) impinge on the Schottky contact, electrons are excited from the metal and injected into the semiconductor. This is illustrated in process (i) of Fig. 10.19b. If the energy of the incident photons exceeds the bandgap energy of the semiconductor (i.e., $h\nu > E_g$), then direct band-to-band excitation occurs and electron–hole pairs are generated in the semiconductor. This is illustrated in process (ii) of Fig. 10.19b. When process (ii) becomes dominant, a sharp increase in photoresponse near the absorption edge is observed, and intrinsic photoconduction becomes the dominant process.

In the photoemission method, the energies of the incident photons fall between the barrier height and the bandgap energy of the semiconductor. According to Fowler's theory, the photocurrent of a Schottky barrier diode produced by photogenerated electrons in the metal [i.e., process (i)] is given by[6]

$$I_{ph} = C(h\nu - q\phi_{Bn})^2 \tag{10.56}$$

which is valid for $(h\nu - q\phi_{Bn}) \geq 3k_BT$ and $q\phi_{Bn} < h\nu < E_g$. From Eq. (10.56), it is seen that the photocurrent is directly proportional to the square of the photon energy. Therefore, a plot of the square root of the photocurrent versus photon energy should yield a straight line. Extrapolation of this straight line to the intercept of the horizontal ($h\nu$) axis yields the barrier

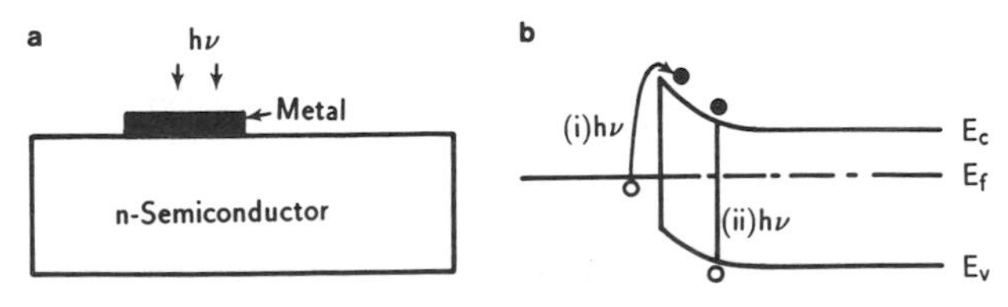

FIGURE 10.19. (a) Illumination from the top surface of a Schottky photodiode. (b) Photoexcitation of electrons: (i) from the metal into the semiconductor with $q\phi_{Bn} < h\nu < E_g$; (ii) inside the semiconductor with $h\nu > E_g$.

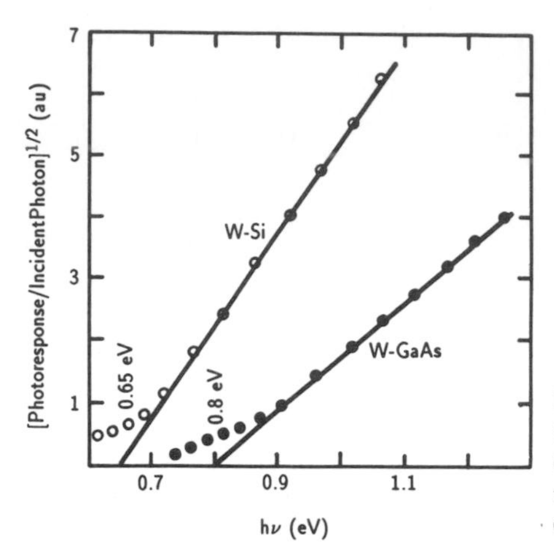

FIGURE 10.20. Square root of photocurrent (in arbitrary units) versus photon energy for a W–Si and a W–GaAs Schottky diode. The intercept of the curves with the horizontal axis yields the barrier height. After Crowell *et al.*,[7] by permission.

height, ϕ_{Bn}. Figure 10.20 shows barrier heights determined by the photoemission method for a W–Si and a W–GaAs Schottky diode.

The Capacitance–Voltage Method. Another method of determining the barrier height of a Schottky diode is by using the capacitance versus voltage (C–V) measurement. From Eq. (10.27), it is seen that for a uniformly doped semiconductor, a plot of C^{-2} versus V should yield a straight line and its intercept with the voltage axis is equal to the diffusion potential V_D. This is illustrated in Fig. 10.21 for a W–Si and a W–GaAs Schottky barrier diode.[5] The diffusion potential determined by the C–V measurement is directly related to the barrier height by

$$\phi_{Bn} = V_D + V_n - q\Delta\phi + \frac{k_B T}{q} \tag{10.57}$$

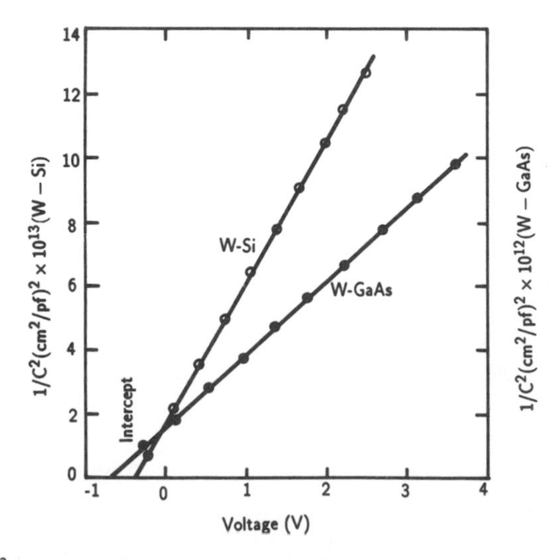

FIGURE 10.21. $1/C^2$ versus applied voltage for a W–Si and a W–GaAs Schottky diode. After Crowell *et al.*,[7] by permission.

TABLE 10.2. Barrier Heights for Some Metal/n-Type Semiconductor Schottky Barrier Diodes

Φ_{Bn} (eV) at 300 K for some n-type semiconductors

Metal	Si	Ge	GaAs	GaP	InP	InAs	InSb
Al	0.55–0.77	0.48	0.80	1.05	—	ohmic	—
Ag	0.56–0.79	—	0.88	1.20	0.54	ohmic	0.18
Au	0.81	0.45	0.90	1.30	0.49	ohmic	0.17
Cu	0.69–0.79	0.48	0.82	1.20	—	—	—
Mo	0.56–0.68	—	—	—	—	—	—
Ni	0.67–0.70	—	—	—	—	—	—
Pd	0.71	—	—	—	—	—	—
Pt	0.90	—	0.86	1.45	—	—	—
PtSi	0.85	—	—	—	—	—	—
W	0.66	0.48	0.71–0.80	—	—	—	—
Ti	0.60	—	0.82	—	—	—	—
In	—	—	0.82	—	ohmic	ohmic	—

where $\Delta\phi$ is the image lowering potential given by

$$\Delta\phi = \sqrt{\frac{q\mathscr{E}_m}{4\pi\varepsilon_0\varepsilon_s}} \tag{10.58}$$

and

$$V_n = \left(\frac{k_B T}{q}\right)\ln\left(\frac{N_c}{N_D}\right) \tag{10.59}$$

is the depth of the Fermi level below the conduction band edge. Thus, knowing V_D, V_n, and $\Delta\phi$, the barrier height ϕ_{Bn} can be determined from the C–V measurement. Table 10.2 lists values of barrier heights for some metal semiconductor Schottky contacts determined by using the three methods described above. Values of barrier heights determined by these methods are generally in good agreement.

10.8. ENHANCEMENT OF EFFECTIVE BARRIER HEIGHT

As discussed in the previous section, the barrier height of an ideal metal–semiconductor Schottky diode is equal to the difference between the metal work function and the electron affinity of the semiconductor. In reality, however, the surface state density of a semiconductor plays an important role in determining the effective barrier height of a Schottky diode. Since only a limited number of metals are suitable for forming Schottky contacts on any semiconductor, it is important to explore alternative methods for enhancing the effective barrier height of a Schottky diode.

We have seen that the effective barrier height of a Schottky diode can be strongly affected by the electric field distribution near the metal–semiconductor interface. Therefore, the barrier height of a Schottky contact can, in principle, be modified by changing the built-in electric field distribution (e.g., by creating a concentration gradient near the semiconductor surface) in a thin region below the metal–semiconductor interface. Evidence of such a dependence has indeed been observed in various Schottky barrier contacts. In fact, the barrier height will decrease if a heavily doped layer of the same dopant type as the substrate is grown on the substrate to form a metal-n^+/n or metal-p^+/p structure. This technique is widely used for making good ohmic contacts in various semiconductor devices. On the other hand, if a thin surface layer of opposite dopant to the substrate is deposited onto the substrate to form a metal–p$^+$/n or metal–n$^+$/p structure, the effective barrier height can be greatly enhanced by using such a structure.

In this section, two different barrier-height-enhancement techniques are described. It was seen earlier that the effective barrier height of a Schottky diode can be enhanced by depositing a very thin epilayer of opposite dopant on the substrate. In such a structure, the barrier height of a metal–p$^+$/n or metal–n$^+$/p Schottky barrier contact is controlled by the thickness and dopant density of the thin semiconductor layer. This thin surface layer can be deposited by using a low-energy ion implantation, molecular beam epitaxy (MBE), or metal–organic chemical vapor deposition (MOCVD) technique. Theoretical and experimental results for metal–p$^+$–n and metal–n$^+$–p silicon Schottky barrier diodes are discussed next.

Figure 10.22a shows the cross-sectional view and Fig. 10.22b the energy band diagram of a metal/p$^+$–n GaAs Schottky barrier diode. We note that the p$^+$–n junction shown in Fig. 10.22a is an abrupt junction structure, and the thickness (W_p) of the p-region is treated as an adjustable parameter. As long as this p-layer remains very thin, the entire p-layer will be fully depleted even at zero-bias conditions. The potential distribution in such a structure can be evaluated by using the depletion approximation. However, if the p-layer becomes too thick, then the p-layer will be partially depleted. As a result, a quasi-neutral p-region will exist, and the structure becomes a conventional metal/p-type Schottky barrier diode in series with a p–n junction diode. We shall exclude such a structure from our present analysis. Therefore, it is important to keep in mind that this structure will work as a Schottky diode only if the deposited thin surface layer remains fully depleted.

In order to analyze the barrier height enhancement in a metal/p–n or metal/n–p Schottky barrier structure, let us introduce χ_s as the electron affinity of the semiconductor, and ϕ_m, ϕ_p, and ϕ_n as the work functions for a metal, p-semiconductor, and n-semiconductor, respectively. If a voltage V is applied to the metal contact, then a potential maximum V_m will appear in front of the metal contact. In this case, $\phi'_{Bn} = \phi_m - \chi_s + V_m(V)$ is the barrier height seen by electrons in the metal, and $\phi'_{Bn} = \phi_m - \phi_n - V + V_m(V)$ is the barrier height seen by electrons in the n-type semiconductor side. If W_n is the width of the space-charge region that extends into the n-region, and x is equal to 0 at the metal contact, then by using the depletion

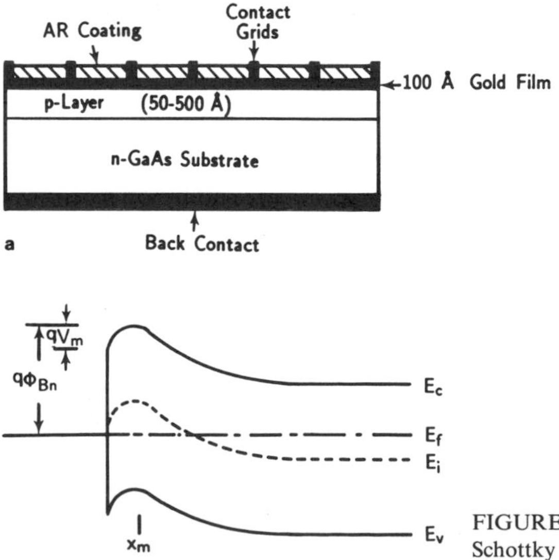

FIGURE 10.22. Schematic diagram of a metal/p–n Schottky barrier diode: (a) cross-sectional view, and (b) energy band diagram showing barrier height enhancement of qV_m.

approximation, Poisson's equation can be written as

$$\frac{d^2\phi}{dx^2} = \frac{qN_a}{\varepsilon_0\varepsilon_s} \qquad \text{for } 0 < x < W_p \tag{10.60}$$

$$\frac{d^2\phi}{dx^2} = -\frac{qN_d}{\varepsilon_0\varepsilon_s} \qquad \text{for } W_p < x < W_p + W_n \tag{10.61}$$

where the boundary conditions are given by

$$\begin{aligned} \phi(x) &= 0 && \text{at } x = 0 \\ \phi(x) &= \phi_m - \phi_n + V_n && \text{at } x = W_p + W_n \end{aligned} \tag{10.62}$$

We note that $\phi(x)$ and $d\phi(x)/dx$ are continuous at $x = W_p$, and

$$\begin{aligned} \left.\frac{d\phi}{dx}\right|_{x=0} &= \frac{q(N_d W_n - N_a W_p)}{\varepsilon_0\varepsilon_s} \\ \left.\frac{d\phi}{dx}\right|_{x=W_p+W_n} &= 0 \end{aligned} \tag{10.63}$$

It can be shown that the solution for $\phi(x)$ when $0 \le x \le W_p$ is given by

$$\phi(x) = \phi_1(x) = \left(\frac{qN_a}{\varepsilon_0\varepsilon_s}\right)\left(\frac{x^2}{2} - xW_p\right) + \left(\frac{qN_d}{\varepsilon_0\varepsilon_s}\right) W_n x \tag{10.64}$$

while if $W_p < x < W_p + W_n$, the solution for $\phi(x)$ is

$$\phi(x) = \phi_2(x) = -\left(\frac{qN_d}{\varepsilon_0\varepsilon_s}\right)\left[\frac{x^2}{2} - x(W_p + W_n)\right] - \frac{q(N_d + N_a)W_p^2}{\varepsilon_0\varepsilon_s} \tag{10.65}$$

The width of the n-region can be obtained by using the second boundary condition, namely,

$$\phi_m - \phi_n + V_n = \frac{1}{2}\left[\frac{qN_d(W_n + W_p)^2}{\varepsilon_0\varepsilon_s}\right] - \frac{1}{2}\left[\frac{q(N_d + N_a)W_p^2}{\varepsilon_0\varepsilon_s}\right] \tag{10.66}$$

If $N_d W_n \ll N_a W_p$, then a potential maximum exists inside the space-charge region of the semiconductor and in front of the metal contact. The position of this potential maximum, x_m, can be determined by setting $d\phi/dx = 0$ at $x = x_m$, which yields

$$\frac{qN_a(x_m - W_p)}{\varepsilon_0\varepsilon_s} + \frac{qN_d W_n}{\varepsilon_0\varepsilon_s} = 0 \tag{10.67}$$

or

$$x_m = W_p - \left(\frac{N_d}{N_a}\right) W_n \tag{10.68}$$

Note that V_m can be obtained by substituting x_m given by Eq. (10.68) into Eq. (10.64), and the result is

$$V_m = -\Delta\phi = \left(\frac{q}{2\varepsilon_0\varepsilon_s N_a}\right)(N_a W_p - N_d W_n)^2 \tag{10.69}$$

Therefore, the effective barrier height for the metal/p$^+$–n Schottky barrier diode shown in Fig. 10.22a is given by

$$\phi'_{Bn} = \phi_m - \chi_s + V_m \tag{10.70}$$

where V_m is given by Eq. (10.69) and W_p is the thickness of the p-layer. The depletion layer width in the n-region, W_n, can be calculated by using the expression

$$W_n = -W_p + (W_p^2 + C)^{1/2} \tag{10.71}$$

where

$$C = \left(\frac{N_a}{N_d}\right) W_p^2 + \frac{2\varepsilon_0\varepsilon_s(\phi_m - \phi_n)}{qN_d} \tag{10.72}$$

and

$$\phi_n = \chi_s + V_n = \chi_s + \left(\frac{k_B T}{q}\right) \ln\left(\frac{N_c}{N_d}\right) \tag{10.73}$$

From the results discussed above, the barrier height enhancement on a metal/p$^+$–n Schottky barrier diode can be calculated by using Eqs. (10.69) through (10.73). Figures 10.23 and 10.24 show theoretical calculations of barrier height enhancement versus dopant density of the p-layer and n-substrate for a Au/p–n GaAs Schottky barrier diode with p-layer thickness as parameter.[8] Figure 10.25a shows a plot of the effective barrier height for a Ti/n–p silicon Schottky barrier diode, and Fig. 10.25b shows the forward I–V characteristics for several Ti/n–p silicon Schottky diodes with different phosphorus implant doses in the n$^+$ surface layer.[9] The barrier height was found to increase from 0.60 eV for a conventional Ti/p-silicon Schottky diode to 0.93 eV for a Ti/n–p silicon Schottky diode fabricated by using a phosphorus implant dose of 1.2×10^{12} cm^{-2} on the p-type silicon substrate. The results show that a significant increase in barrier height was obtained by using this approach. In principle, an effective barrier

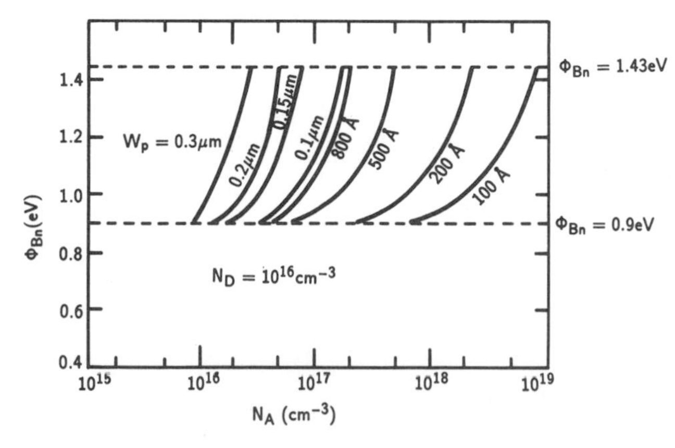

FIGURE 10.23. Calculated barrier heights of a Au/p–n GaAs Schottky barrier diode versus dopant density, N_A, of the p-layer for different p-layer thicknesses. Note: N_d of the n-substrate is fixed at 10^{16} cm^{-3}. After Li,[8] by permission.

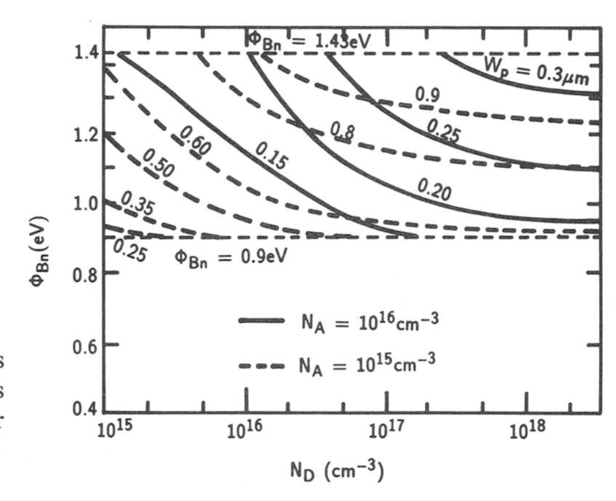

FIGURE 10.24. Calculated barrier heights of a Au/p–n Schottky barrier diode versus dopant density, N_d, of the n-substrate for different p-layer thicknesses and for $N_A = 5 \times 10^{16}$ cm^{-3}. After Li,[8] by permission.

height equal to the bandgap energy of the semiconductor can be obtained by using the structure described in this section, provided that the thickness and dopant density of the thin surface layer are properly chosen for such a Schottky barrier structure. For silicon Schottky barrier diodes, incorporation of such a thin surface layer is usually achieved by using the ion implantation technique, while for Schottky barrier diodes formed on III–V compound semiconductors the thin epilayer can be deposited by using either the molecular beam epitaxy (MBE) or the metal–organic chemical vapor deposition (MOCVD) growth technique. Layer thicknesses from a few tens of angstroms to a few hundreds or thousands of angstroms can be readily deposited onto the GaAs or InP substrates by using either the MBE or the MOCVD growth technique. Therefore, the metal/p–n or metal/n–p Schottky barrier structure described in this section can be regarded as a viable approach for enhancing the effective barrier height of a conventional metal–semiconductor Schottky diode.

Another barrier height enhancement technique using the bandgap engineering approach on a small-bandgap semiconductor such as $In_{0.53}Ga_{0.47}As$ has been reported recently.[10] The technique involves the growth of a thin graded superlattice of n periods on top of the small-bandgap epilayer to create a larger-bandgap surface layer, so that effective barrier height can be enhanced in such a Schottky barrier diode. For example, a high-quality $In_{0.53}Ga_{0.47}As$

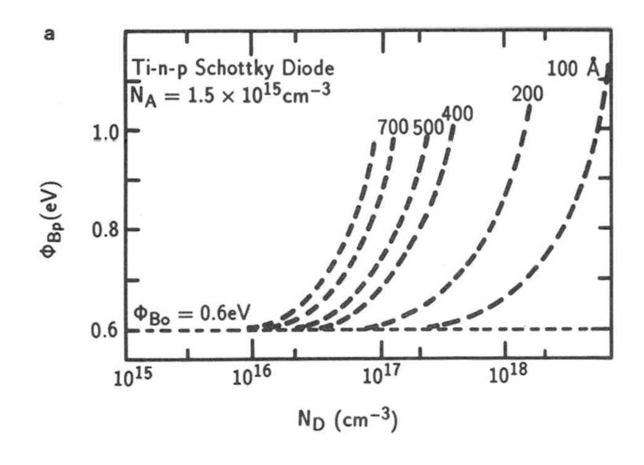

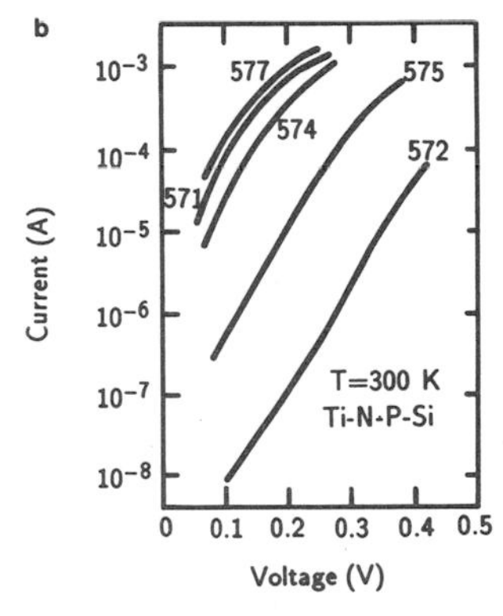

FIGURE 10.25. (a) Calculated barrier heights versus density of the phosphorous implanted layer for different layer thicknesses of a Ti/n–p silicon Schottky barrier diode. Solid dots denote experimental data. (b) Forward-biased I–V curves for a controlled Ti/p-silicon Schottky diode (577) and four other Ti/n–p silicon Schottky barrier diodes with different implant doses. After Li *et al.*,[9] by permission, © IEEE–1980.

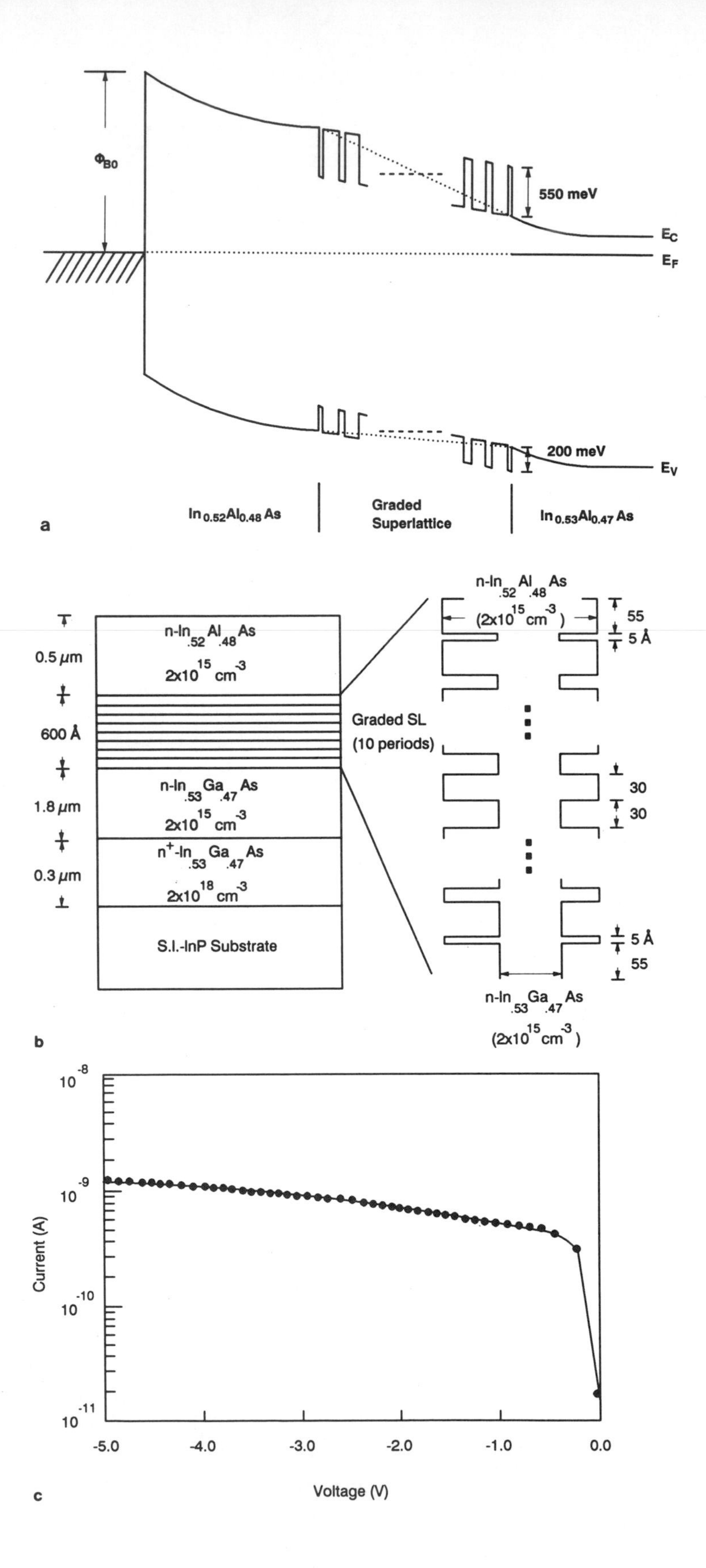

Φ_{B0}
550 meV
E_C
E_F
200 meV
E_V
a
$In_{0.52}Al_{0.48}As$
Graded Superlattice
$In_{0.53}Al_{0.47}As$
0.5 μm
600 Å
1.8 μm
0.3 μm
$n\text{-}In_{.52}Al_{.48}As$
$2x10^{15}\ cm^{-3}$
$n\text{-}In_{.53}Ga_{.47}As$
$2x10^{15}\ cm^{-3}$
$n^+\text{-}In_{.53}Ga_{.47}As$
$2x10^{18}\ cm^{-3}$
S.I.-InP Substrate
Graded SL
(10 periods)
$n\text{-}In_{.52}Al_{.48}As$
$(2x10^{15}\ cm^{-3})$
55
5 Å
30
30
5 Å
55
$n\text{-}In_{.53}Ga_{.47}As$
$(2x10^{15}\ cm^{-3})$
b
Current (A)
10^{-8}
10^{-9}
10^{-10}
10^{-11}
-5.0
-4.0
-3.0
-2.0
-1.0
0.0
Voltage (V)
c

Schottky barrier diode has been fabricated by using a novel graded superlattice structure with 10 periods of $In_{0.52}Al_{0.48}As/In_{0.53}Ga_{0.47}As$ graded superlattice deposited on top of the n-type $In_{0.53}Ga_{0.47}As$ epilayer grown by the MBE technique on an InP substrate. The result shows a barrier height enhancement of 0.41 eV (i.e., from $\phi_{Bn} = 0.3$ eV to 0.71 eV), and near-ideal I–V and C–V characteristics are obtained for this novel Schottky diode. Figure 10.26a shows the energy band diagram of an InAlAs/InGaAs graded superlattice structure formed on an InGaAs Schottky diode, and Fig. 10.26b shows the dimensions of the graded InAlAs/InGaAs superlattice layer structure and doping densities in each region. Figure 10.26c shows the reverse leakage current for a Schottky barrier diode formed on the graded superlattice structure shown in Fig. 10.26b.

10.9. APPLICATIONS OF SCHOTTKY DIODES

Schottky diodes have been widely used for a variety of applications such as solar cells, photodetectors, Schottky-clamped transistors, metal-gate field-effect transistors (MESFETs), modulation-doped field-effect transistors (MODFETS or HEMTs), microwave mixers, and Schottky transistor logic (STL) gate arrays. For example, the exact logarithmic relationship displayed by the I–V curve of a Schottky diode under forward-bias conditions over several decades of current change enables it to be used in the logarithmic converter circuits. A metal–semiconductor Schottky diode can also be used as a variable capacitor in parametric circuits for frequency multiplication. The Schottky barrier solar cell has the potential for use as a low-cost photovoltaic power conversion device for large-scale terrestrial power generation. High-speed Schottky barrier photodetectors covering a broad wavelength range from ultraviolet to visible and into far-infrared spectral regimes have been reported using different metal/semiconductor materials. In this section, some practical applications of Schottky barrier diodes are depicted.

10.9.1. Photodetectors and Solar Cells

A Schottky barrier diode can be used as a high-speed photodetector for low-level light detection or as a solar cell for conversion of sun energy into electricity. To reduce absorption loss in the metal contact of a Schottky barrier photodiode, it is common practice to use either a thin metal film (100 Å or less) or a grating-type structure for the Schottky contact. The reflection loss on a semiconductor surface is minimized by incorporating an antireflection (AR) coating in the front side of the Schottky barrier photodiode, as illustrated in Fig. 10.27a. For the grating-type Schottky barrier photodiode shown Fig. 10.27b, the pattern of metal grids used in the Schottky contact can be generated by using a photolithography technique. Selection of the metal-grid spacing is determined by the operating bias voltage and substrate doping concentration of the diode to ensure that the spacing between the metal grids is fully depleted under operating conditions. For example, in the case of a Au/n-Si Schottky diode with a doping density N_D of 10^{14} cm^{-3}, a spacing of around 10 μm between the metal grids is adequate for creating a fully depleted region between the metal grids of such a Schottky contact. Photodetectors using a semitransparent Schottky contact or a grating-type Schottky contact structure have shown excellent quantum efficiency and high responsivity.

FIGURE 10.26. (a) Energy diagram of an Au/n-InAlAs/n-InGaAs Schottky barrier diode with a 600 Å graded superlattice of InAlAs/InGaAs for barrier height enhancement, (b) dimensions and dopant densities of the structure shown, and (c) reverse I–V characteristics for such a Schottky barrier diode. After Lee *et al.*,[10] by permission.

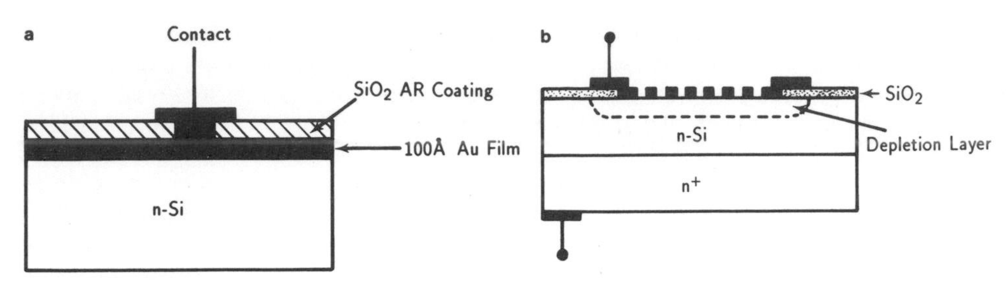

FIGURE 10.27. Schematic diagrams of (a) a conventional Schottky barrier photodiode and (b) a grating-type Schottky barrier photodiode.

In general, there are three modes of detection which can be used in a Schottky barrier photodiode; these are illustrated in Fig. 10.28a, b, and c. The operation of each of these detection modes depends greatly on the incident photon energies, the applied bias voltage, and the breakdown voltage of the photodiode. These are discussed as follows:

For $q\phi_{Bn} < h\nu < E_g$ *and* $V_a \ll V_B$. In this detection mode, electrons are excited from the metal and injected into the semiconductor, as illustrated in Fig. 10.28. In this case, the Schottky barrier photodiode may be used for a variety of applications, which include: (1) infrared (IR) detector, (2) test vehicle to determine barrier height by the photoemission method, and (3) test device for studying bulk defects and interface states in a semiconductor, and hot-electron transport in a metal film. The reason that a Schottky diode can be used for long-wavelength infrared (LWIR) detection is because the barrier height for most Schottky diodes is smaller than the bandgap energy of the semiconductor. As a result, photons with

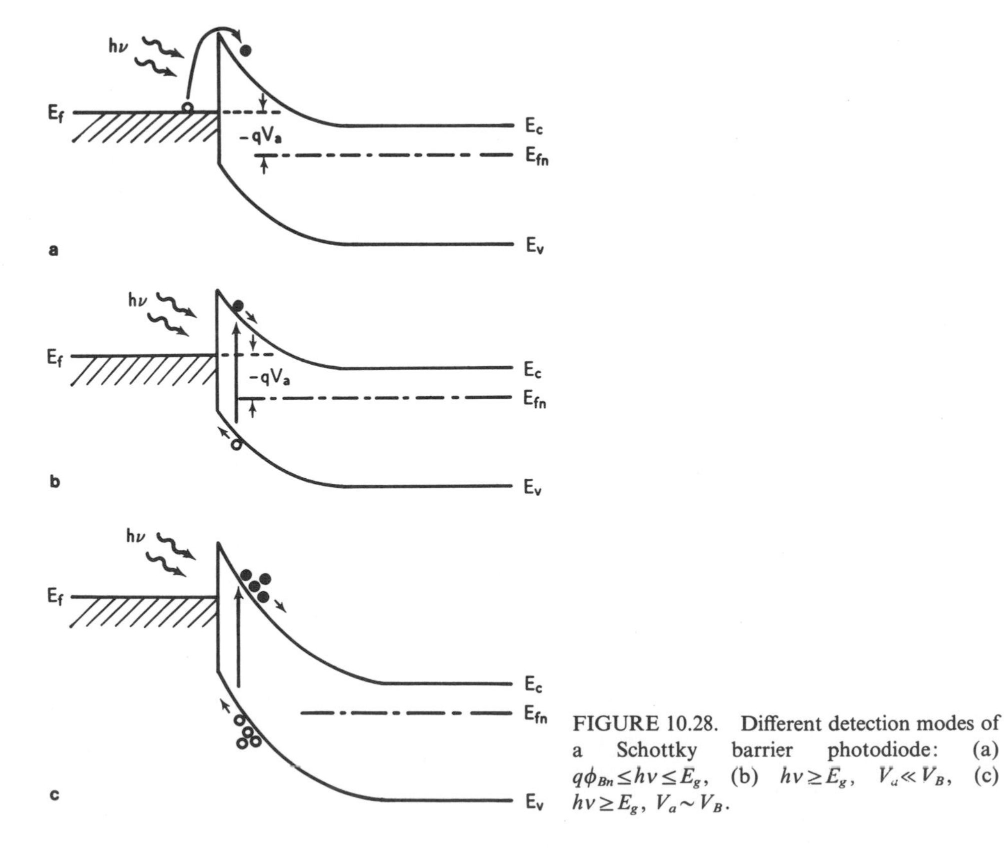

FIGURE 10.28. Different detection modes of a Schottky barrier photodiode: (a) $q\phi_{Bn} \leq h\nu \leq E_g$, (b) $h\nu \geq E_g$, $V_a \ll V_B$, (c) $h\nu \geq E_g$, $V_a \sim V_B$.

energy equal to the barrier height absorbed inside the metal film of a Schottky diode usually fall in the infrared regime. Since the barrier height for an IR Schottky barrier photodiode is low, the reverse leakage current in such a device is expected to be very large at room temperature. Therefore, in order to reduce the reverse leakage current, a Schottky barrier photodetector is usually operated at cryogenic temperatures (e.g., below 77 K). For example, PtSi/p-type Si Schottky barrier photodiode (with barrier height $\phi_{Bp} \simeq 0.2$ eV) arrays integrated with CCD (charge-couple devices) arrays have been developed for 3- to 5-μm IR image-sensor applications. Extending the detection wavelength to 10 μm is possible if the operating temperature for the low-barrier ($\sim$0.1 eV) Schottky barrier photodiode is lowered to 4.2 K.

For $h\nu \geq E_g$ *and* $V_a \ll V_B$. It is shown in Fig. 10.28b that, in this detection mode, electron–hole pairs are generated inside the depletion region of the semiconductor, and the Schottky diode may be operating as a high-speed photodetector. Since the Schottky diode is a majority carrier device, its response speed is limited mainly by the RC time constant and the carrier transit time across the depletion region of the detector. The grating-type Au/n-Si Schottky barrier photodiode shown in Fig. 10.27b has a responsitivity of 0.63 A/W at 0.9 μm and a bandwidth of 1 GHz. The Au/n-GaAs Schottky barrier photodiode shown in Fig. 10.27a has a response speed of less than 100 ps at 0.85 μm. In fact, high-speed GaAs Schottky barrier photodiodes with a 3-dB bandwidth exceeding 100 GHz have been reported recently in the literature. Further discussion of high-speed photodetectors using a Schottky barrier structure will be given in Chapter 12.

For $h\nu \geq E_g$ *and* $V_a = V_B$. In this mode of operation, the Schottky barrier photodiode is in the avalanche mode of detection; this is shown in Fig. 10.28c. When a Schottky diode is operating in the avalanche regime, an internal current gain is obtained. Thus, an avalanche photodiode can provide both high-speed and high-sensitivity detection. A diffused guard-ring structure is usually employed in an avalanche photodiode to eliminate the edge breakdown effect. Internal current gain of several thousands has been accomplished for the germanium and silicon p–i–n avalanche photodiodes; this will be discussed further in Chapter 12.

A Schottky barrier photodiode can also be used as an efficient ultraviolet (UV) light detector. For example, in the UV region the absorption coefficient for most semiconductors is greater than 10^5 cm^{-1}, which corresponds to an effective absorption depth of 0.1 μm or less. Thus, by using a thin metal film and an antireflection (AR) coating film simultaneously on the Schottky barrier structure, efficient collection of photons in the UV spectrum can be achieved. For example, both Ag–ZnS and Au–ZnSe Schottky barrier photodiodes have been developed for UV light detection.

Finally, Schottky barrier solar cells have also been studied extensively for photovoltaic conversion of sunlight into electricity. The Schottky barrier solar cell is easy to fabricate, and has the potential for low-cost and large-scale production. However, due to the low barrier height, Schottky barrier solar cells usually have lower open-circuit voltage and lower conversion efficiency than most p–n junction solar cells. This will be discussed further in Chapter 12.

10.9.2. Schottky-Clamped Transistors

In the saturated switching process of a conventional n–p–n or p–n–p junction transistor, the turnoff speed is limited by the minority carrier storage time in the collector region of the transistor. For a silicon switching transistor, the conventional method of decreasing the storage time is to reduce the minority carrier lifetime by doping the transistor with gold impurities. However, since current gain is also proportional to minority carrier lifetime, the gold-doped silicon transistor will also have a lower current gain. Therefore, it is generally not desirable to dope silicon transistors with gold impurities. To overcome this problem a Schottky barrier diode is usually added to the base–collector junction of the transistor. As shown in Fig. 10.29,

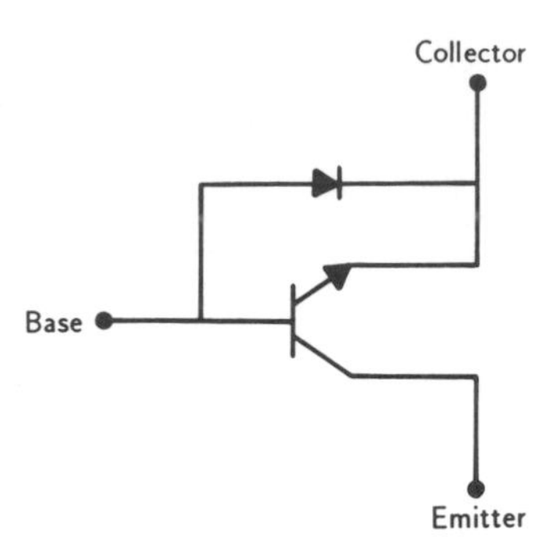

FIGURE 10.29. A Schottky-barrier clamped bipolar transistor. The Schottky diode is connected to the base and collector of a bipolar junction transistor.

the minority carrier storage problem can be virtually eliminated if a Schottky barrier diode is connected to the base–collector junction of the transistor to form a Schottky-clamped transistor. The switching time constant is drastically shortened in this case, since the minority carrier storage in the collector is greatly reduced by the Schottky diode connecting in parallel with the base–collector junction. In the saturation region, the collector junction of the transistor is slightly forward-biased instead of reverse-biased. If the forward voltage drop in the Schottky diode is much lower than the base–collector voltage of the transistor, then most of the excess base current will flow through the Schottky diode. Therefore, the minority carriers are not stored in the collector. Furthermore, the saturation time is greatly reduced when compared to a transistor without a Schottky diode connecting to its base–collector junction. A switching time of less than 1 ns for a silicon Schottky-clamped bipolar transistor has been reported. Recently, low-power and high-speed Schottky-clamped transistor logic (STL) gate arrays have been developed for computer and other custom IC applications. In a STL gate array, two Schottky barrier diodes with different barrier heights are used. For example, in a silicon STL gate array one Schottky diode (e.g., PtSi/n-Si) with large barrier height is connected to the base–collector junction, while another Schottky diode with low barrier height (e.g., TiW/p-Si) is connected to the collector of the transistor. Si STL gate arrays with propagation delay times of less than 1 ns and voltage swings of a few hundred millivolts have been achieved. STL gate arrays fabricated from III–V compound semiconductors such as GaAs, InP, and InGaAs with propagation delay times of a few tens of picoseconds have also been reported.

10.9.3. Microwave Mixers

High-frequency application of Schottky barrier diodes is concerned with low-level signal detection and mixing at microwave frequencies. It has been shown that burnout resistance and noise performance of a Schottky diode is usually superior to that of a point-contact mixer diode.

The frequency response of a Schottky diode is generally superior to that of a p–n junction diode, since it is limited by the RC time constant of the Schottky diode rather than by the minority carrier lifetime as in the case of a p–n junction diode. Using an n–n$^+$ epitaxial wafer for Schottky contact fabrication, both junction capacitance (e.g., $C < 0.1$ pF at $V_a = 0$ and $N_d = 10^{17}$ cm^{-3}) and series resistance of the diode can be reduced to a very low value. This is a direct result of using a very thin n-type epitaxial layer on an n$^+$-silicon substrate, and the resistance drop across the n$^+$-region is negligibly small. A point-contact Schottky diode with barrier contact area of 5 to 10 μm in diameter has been reported. Figure 10.30 shows the

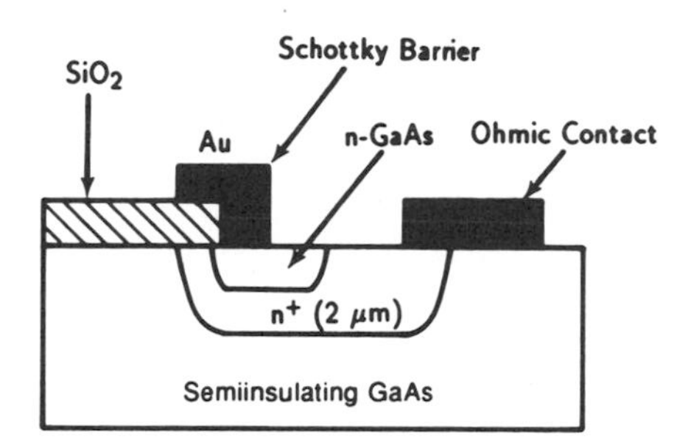

FIGURE 10.30. Cross-section view of a Au/n-type GaAs microwave mixer using a Schottky barrier structure.

geometry of a microwave Schottky barrier diode mixer using a microstrip line configuration. This Schottky diode is capable of excellent mixer performance, either as a discrete component in a waveguide or as a balanced mixer in a microstrip line at carrier frequencies as high as several tens of GHz.

The RC time constant of a Schottky barrier diode can be computed by considering an n–n^+ structure with a barrier contact of radius r, which is large compared to the epilayer thickness. The series resistance of a Schottky diode can be calculated using the expression

$$R_s = \frac{\rho d}{\pi r^2} = \left(\frac{1}{qN_D\mu_n}\right)\left(\frac{d}{\pi r^2}\right) \tag{10.74}$$

and the depletion layer capacitance is calculated from the expression

$$C_d = (\pi r^2)\left(\frac{qN_D\varepsilon_0\varepsilon_s}{2V_D}\right)^{1/2} \tag{10.75}$$

where N_D is the doping density of the n-epitaxial layer and V_D is the diffusion potential. Thus, the RC time constant for such a Schottky diode is given by

$$R_s C_d = \left(\frac{d}{\mu_n}\right)\left(\frac{\varepsilon_0\varepsilon_s}{2qN_D V_D}\right)^{1/2} \tag{10.76}$$

To achieve high-frequency performance in a Schottky diode, the RC time constant of the diode should be kept as small as possible. This requires the use of a very thin epitaxial layer with high carrier mobility and doping density. Other important applications of the Schottky barrier structure include metal gate (i.e., Schottky gate) field-effect transistors (MESFETS) and modulation-doped FETs formed on III–V compound semiconductors, such as GaAs and AlGaAs, and metal–semiconductor IMPATT diodes for microwave power generation.

10.10. OHMIC CONTACTS

Formation of good ohmic contact between a metal and a semiconductor is an extremely important process step for fabricating high-performance semiconductor devices and integrated

TABLE 10.3. Metals for Forming Ohmic Contact in Various Semiconductors

Semiconductor	Metals
Ge (N)	Ag–Al–Sb, Al, Au, Bi, Ai–Au–P, Sb, Sn, Pb–Sn
Ge (P)	Ag, Al, Au, Cu, Ga, Ga–In, In, Al–Pd, Ni, Pt, Sn
Si (N)	Ag, Al, Al–Au, Au, Ni, Pt, Sn, In, Ge–Sn, Sb, Au–Sb
Si (P)	Ag, Al, Al–Au, Au, Ni, Pt, Sn, In, Pb, Ga, Ge
GaAs (N)	Au–Ge (88%, 12%)–Ni, Ag–In (95%, 5%)–Ge, Ag–Sn
GaAs (P)	Au–Zn (84%, 16%), Ag–In–Zn, Ag–Zn
GaP (N)	Ag–Te–Ni, Al, Au–Si, Au–Sn, In–Sn
GaP (P)	Au–In, Au–Zn, Ga, In–Zn, Zn, Ag–Zn
GaAsP (N)	Au–Sn
GaAsP (P)	Au–Zn
GaAlAs (N)	Au–Ge–Ni
GaAlAs (P)	Au–Zn
InAs (N)	Au–Ge, Au–Sn–Ni, Sn
InAs (P)	Al
InGaAs (N)	Au–Ge, Ni
InGaAs (P)	Au–Zn, Ni
InP (N)	Au–Ge, In, Ni, Sn
InSb (N)	Au–Sn, Au–In, Ni, Sn
InSb (P)	Au–Ge
CdS (N)	Ag, Al, Au, Au–In, Ga, In, Ga–In
CdTe (N)	In
CdTe (P)	Au, In–Ni, Indalloy 13, Pt, Rh
ZnSe (N)	In, In–Ga, Pt, InHg
SiC (N)	W
SiC (P)	Al–Si, Si, Ni

circuits. Table 10.3 gives a list of metals which are used in forming ohmic contacts in various semiconductors. Good ohmic contact is a must in order to effectively extract electric current and power from a semiconductor device. In general, an ohmic contact is referred to a noninjecting contact in which the current–voltage relationship under both reverse- and forward-bias conditions is linear and symmetrical. However, in reality, a contact is considered ohmic if the voltage drop across the metal–semiconductor interface is negligible compared to the voltage drop across the bulk semiconductor.

Ohmic contacts can be characterized in terms of the specific contact resistance, R_c, which is defined as the reciprocal of the derivative of the current density with respect to the applied voltage. An expression for the specific contact resistance evaluated at zero bias is given by

$$R_c = \left(\frac{dJ}{dV}\right)^{-1}\Bigg|_{V=0} \tag{10.77}$$

The specific contact resistance defined by Eq. (10.77) is an important figure of merit for evaluating ohmic contacts. In general, current flow through the ohmic contact region of a moderately doped semiconductor is dominated by the thermionic emission process. Thus, the expression for R_c can be derived directly from Eqs. (10.33) and (10.77), which yield

$$R_c = \left(\frac{k_B}{qA^*T}\right)\exp\left(\frac{q\phi_{Bn}}{k_BT}\right) \tag{10.78}$$

This equation shows that in order to achieve a small specific contact resistance, the barrier height of the metal–semiconductor contact should be as small as possible.

For ohmic contact on a heavily doped semiconductor, the field-emission process (i.e., tunneling) dominates the current transport, and hence the specific contact resistance R_c can be expressed by[11]

$$R_c \sim \exp\left[\left(\frac{2\phi_{Bn}}{\hbar}\right)\sqrt{\frac{\varepsilon_0\varepsilon_s m^*}{N_D}}\right] \quad (10.79)$$

which shows that, in the tunneling range, R_c depends strongly on doping density and varies exponentially with $(\phi_{Bn}/\sqrt{N_D})$. The specific contact resistance for the Au–Ge–Ni on an n-type GaAs contact may vary between 10^{-4} and 10^{-7} ohm · cm^2. Figure 10.31a and b show energy band diagrams for a low-barrier-height contact and an ohmic contact on a heavily doped n^{++}/n-semiconductor, respectively. Figure 10.31c shows I–V characteristics of a metal–n^{++}–n-type semiconductor ohmic contact. We note that the specific contact resistance calculated from Eq. (10.79) agrees well with the measured value for the molecular beam epitaxy (MBE) doped ohmic contacts. However, alloyed contacts usually exhibit a linear dependence of ln R_c on N_D^{-1}, and the simple formula given by Eq. (10.79) cannot adequately explain the I–V behavior of the alloyed ohmic contacts.

Formation of ohmic contacts can be achieved in a number of ways. These include: (1) choosing a metal with a lower work function than that of an n-type semiconductor such that the potential barrier between the metal and the semiconductor is small enough for thermionic

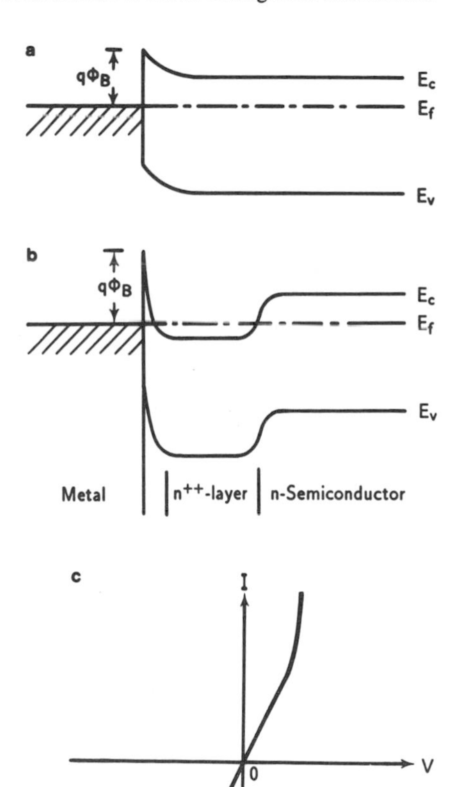

FIGURE 10.31. (a) Energy band diagram for a low-barrier Schottky contact, (b) a metal–n^{++}–n-semiconductor ohmic contact, and (c) the I–V curve for the ohmic contact structure shown in (b).

emission electrons to tunnel through in both directions of the metal–semiconductor contact; (2) depositing a thin and heavily doped epilayer of the same doping type as the substrate to form an n^+/n or p^+/p high–low junction structure on the semiconductor surface; this will reduce the barrier width of the metal–semiconductor contact such that current flow can be achieved by quantum-mechanical tunneling through the thin barrier with low contact resistance; (3) using a graded heterojunction approach to form an ohmic contact (e.g., growing an n^+-InAs/n-GaAs or n^+Ge/n-GaAs structure by the MBE technique); and (4) increasing the density of recombination centers at the semiconductor surface (e.g., by surface roughening) so that the surface will serve as an infinite sink for the majority carriers at the contact.

Techniques for forming ohmic contacts on the semiconductor devices include alloying, electroplating, thermal or E-beam evaporation, sputtering, and ion implantation. In principle, an ohmic contact is formed if the potential barrier between the metal and semiconductor contact is very small. It was seen in Figs. 10.7c and f that, in order to obtain good ohmic contact on an n-type semiconductor, the metal work function should be less than that of the n-type semiconductor. Conversely, to form ohmic contact on a p-type semiconductor, the metal work function must be greater than that of the p-type semiconductor. Unfortunately, for covalent semiconductors such as Ge, Si, and GaAs, formation of ohmic contacts does not always follow the simple rule cited above. For example, the barrier height of an n-GaAs Schottky barrier contact tends to remain nearly constant regardless of the metals used. This is because the surface state density for GaAs is usually very high (e.g., $\geq 10^{13}\ \mathrm{cm}^{-2} \cdot \mathrm{eV}^{-1}$), and as a result the Fermi level at the interface is pinned at the level where peak interface state density occurs. In this case, the barrier height of the Schottky contact is determined by Fermi-level pinning at the surface and is independent of the metal work function.

The most widely used technique for forming ohmic contact on a semiconductor is by first growing a heavily doped thin surface layer to form an n^+/n or p^+/p high–low junction structure on an n- or p-type semiconductor substrate before making ohmic contact. By using such a structure, the barrier width between the metal and the heavily doped n^+ or p^+-layer can be greatly reduced, and hence quantum-mechanical tunneling of charge carriers through such a thin barrier becomes very likely. In this case, the barrier becomes essentially transparent to the charge carriers, and the specific contact resistance is usually very small. The heavily doped layer can be deposited by alloying, thermal diffusion, ion implantation, or epitaxial growth using a suitable dopant impurity on the growth layer. This approach of forming ohmic contact has been quite successful for both silicon and GaAs IC technologies. However, it is not as successful for many large-bandgap semiconductors such as AlGaAs and CdS because these crystals have a tendency to compensate the foreign dopant impurities. This is due to the large density of native defects created by the nonstoichiometric crystal structures in these crystals. Finally, it should be pointed out that ohmic contacts on p-type III–V compound semiconductors are even more difficult to obtain than their n-type counterpart. This is due to the fact that a p-type III–V compound semiconductor surface (e.g., p-type AlGaAs) is much easier to oxidize than an n-type surface during metallization or simply exposure to air.

Another approach which has also been employed for forming ohmic contacts on a semiconductor is introducing a large density of recombination centers at the semiconductor surface before making ohmic contact. The recombination centers at the semiconductor surface may be created by damaging or straining the surface through mechanical lapping, or by introducing impurities at the semiconductor surface which in turn produce effective recombination centers. Plating of metals to such a damaged surface usually results in adequate ohmic contact.

We shall next discuss procedures for forming ohmic contacts on a semiconductor. Various contact-forming techniques, such as alloying, thermal or E-beam evaporation, sputtering, ion implantation, plating, liquid regrowth, and bonding, have been employed. The alloying process is achieved by first placing metal pellets or thin metal foil on the semiconductor surface and then heating the specimen to the eutectic temperature of the metal such that a small portion of the semiconductor is dissolved with the metal. The contact is then cooled, and semiconductor regrowth takes place. This regrowth results in the incorporation of some metal and residual impurities into the metal–semiconductor interface. In this method, impurities in the metal form a heavily doped interfacial layer of the same dopant type as the bulk semiconductor and hence good ohmic contact on the semiconductor can be obtained. However, many large-bandgap semiconductors have contact problems because they cannot be doped heavily enough to form good ohmic contact. Another technique for making ohmic contact is to wet the semiconductor with a metal. In order to promote wetting and good ohmic contact, both the metal and semiconductor surface must be ultraclean. A flux is often used during alloying in order to remove any residual impurities or oxide film such that surface wetting can be enhanced. Care must be taken to identify the difference in thermal expansion coefficients between the semiconductor and metal so that residual stress does not develop in the semiconductor upon cooling from the alloying temperature to room temperature. Liquid regrowth can also be used to form a high–low junction (i.e., n^+/n or p^+/p) on a semiconductor substrate before forming ohmic contact. This technique has been used to form ohmic contact on GaAs Gunn-effect devices.

Ohmic contact on semiconductors can also be created by using electrodeless plating. This method involves the use of a metallic salt (such as $AuCl_3$) which is reduced to a metal at the semiconductor surface by a chemical reducing agent present in the plating solution. Solutions of nickel, gold, and platinum are the most widely used metals for electrodeless plating in semiconductors. It has been found that a nickel film adheres better on a mechanically lapped semiconductor surface than on a chemically polished semiconductor surface.

In silicon integrated circuit (IC) technologies, the most widely used technique for depositing metal films on silicon devices and integrated circuits is by using thermal or E-beam evaporation. The E-beam evaporation is particularly attractive for integrated circuits in which complex contact patterns are formed by using a photolithography technique. The metal contact is deposited by using a heated filament, an electron-beam gun, or by using sputtering. The heated boat or filament technique is the simplest method, but suffers from the disadvantage of being contaminated by the heated container (e.g., tungsten basket or graphite boat). A shutter mechanism is often used to block the initial vapor which may be contaminated with the more volatile impurities. If multiple boats or filaments are used, two or more metals may be evaporated simultaneously with independent control of each evaporating source. In the E-beam system, the electron-beam gun melts the evaporant alone, which serves as its own crucible. This eliminates contamination from the crucible and gives purer deposited metal films. In the sputtering system, positive gas ions bombard the source (cathode), emitting metal atoms. These ejected atoms traverse the vacuum chamber and are deposited onto the semiconductor substrate. The sputtering system has the advantage that the polarity of the system may be reversed so that sputtering may occur from the substrate to a remote anode, thereby cleaning the substrate surface. The back-sputtering approach is particularly useful for removing any residual thin oxide films or other impurities on the substrate surface while it is in the vacuum, and can be followed immediately by depositing metal contact on the substrate.

Standard procedures for metallic contacts in IC fabrication are usually accomplished by a two-step process. For example, ion implantation is first employed to create a thin, heavily doped layer of the same dopant type as that of the bulk substrate to form a p^+/p or n^+/n high–low junction, and is then followed by deposition of metal contacts using either thermal or E-beam evaporation. Finally, thermal annealing is performed on the implanted region to obtain good ohmic contact. In this procedure, careful cleaning of the semiconductor surface is essential to ensure good ohmic contact prior to metal film deposition.

Wire bonding on semiconductor devices and integrated circuits is usually carried out by using either thermal compression or the ultrasonic wire bonder. In thermal compression bonding, both heat and pressure are applied simultaneously to the ball bonder (which contains 1 mil gold or aluminum wire) and the contact pad. In ultrasonic bonding, a combination of pressure and 60 kHz ultrasonic vibration is employed. The ultrasonic vibration gives rise to a scrubbing action which breaks up any thin surface insulating film, and hence intimate contact between the metal and semiconductor can be made. Wire bondings are made by using fine gold or aluminum wires. The advantage of using ultrasonic bonding is that heating is not required, and any previous bondings will not be affected.

Finally, it is worth mentioning that in silicon IC technologies, the chemical vapor deposition (CVD) technique has been widely used for depositing metal contacts as well as heavily-doped polysilicon films for ohmic contacts and interconnects in integrated circuits.

PROBLEMS

10.1. The saturation current density for a thoriated tungsten metal is $1\ A/cm^2$ at 1873 K, and the work function computed from Eq. (10.11) for this metal is 3.2 eV. Assume that $A_0 = 120\ A/cm^2 \cdot K^2$.
 (a) Plot $\ln J_s'$ versus $\mathscr{E}^{1/2}$ for $T = 1873$ K and for $10^4 < \mathscr{E}^{1/2} < 10^7$ V/cm.
 (b) Repeat (a) for $T = 873$ K and 1500 K.

10.2. (a) Draw the energy band diagram for an ideal metal/p-type semiconductor Schottky barrier diode and show that the barrier height for a metal/p-type semiconductor is given by Eq. (10.14).
 (b) Plot the energy band diagrams for an ideal metal/p-type semiconductor Schottky barrier diode for the cases:
  (i) $\phi_m > \phi_s$,
  (ii) $\phi_m < \phi_s$.
  (iii) Explain which of the above cases would form an ohmic or a Schottky contact.

10.3. Using Eq. (10.25), plot the depletion layer width versus reverse bias voltage ($V = 0$ to 20 V) for a Au/n-type silicon Schottky barrier diode for $N_D = 10^{14}$, 10^{16}, and $10^{18}\ cm^{-3}$. Given: $\varepsilon_s = 11.7$, $V_D = \phi_{Bn} - (k_B T/q)\ln(N_D/N_c)$, and $q\phi_{Bn} = 0.81$ eV at $T = 300$ K.

10.4. Taking into account the image lowering effect [i.e., using Eq. (10.34)], plot the saturation current density versus reverse bias voltage for a Au/n-Si and Pt/n-Si Schottky diode. Assume that $q\phi_{Bn} = 0.81$ eV for the Au/n-Si Schottky diode, $q\phi_{Bn} = 0.90$ eV for the Pt/n-Si Schottky diode, and $A^* = 110\ A/cm^2 \cdot K^2$.

10.5. Derive Eqs. (10.30) and (10.42), and compare the results with the current density equation derived from the thermionic-diffusion model reported by C. R. Crowell and S. M. Sze, *Solid State Electronics*, **8**, 979 (1966).

10.6. Design a Au/n-type GaAs Schottky barrier photodiode for detecting a 20-GHz modulated optical signal with center wavelength at 0.84 μm. Show the Schottky barrier structure, and calculate the thickness of the antireflecting coating layer (e.g., SiO_2), the diode area, and the RC time constant of this photodiode. If the incident photosignal has a power intensity of $2\ mW/cm^2$, what is the responsivity of this photodiode? (Hint: Select your own design parameters.)

10.7. If the barrier height of a TiW/p-type Si Schottky barrier diode is equal to 0.55 eV, use Eqs. (10.65) and (10.66) to design a TiW/n^+–p Schottky barrier diode structure to enhance the effective barrier height to 0.90 eV. Assuming that the p-substrate has a dopant density of $1 \times 10^{16}\ cm^{-3}$, calculate the required dopant density and thickness of the n^+-surface layer.

10.8. Assume that the ideality factor n for a Schottky barrier diode is defined by

$$n = \left(\frac{q}{k_B T}\right)\frac{\partial V}{\partial(\ln J)}$$

 (a) Show that

$$n = \left\{1 + \left(\frac{\partial \Delta\phi}{\partial V}\right) + \left(\frac{k_B T}{q}\right)\left[\frac{\partial(\ln A^*)}{\partial V}\right]\right\}^{-1}$$

 (b) What are the possible physical mechanisms which may cause the n value to differ from unity?

10.9. Using the general expressions of the barrier height for a Schottky barrier diode given by Eq. (10.48) and Eqs. (10.49) through (10.51), with $D_s \cong 1.1 \times 10^{13}(1 - c_2)/c_2$ states/cm$^2 \cdot$ eV, where D_s is the interface state density:
   (a) What is the barrier height if D_s approaches infinity? Explain the Fermi-level pinning effect under this condition.
   (b) If $D_s \to 0$, what is the value of c_2 and the expression for ϕ_{Bn}?
   (c) If the values of c_2, c_3, and χ_s for Si, GaAs, and GaP Schottky contacts are given by:

   $c_2 = 0.27, 0.07$, and 0.27 for Si, GaAs, and GaP, respectively,

   $c_3 = -0.66, 0.61$, and -0.07 for Si, GaAs, and GaP, respectively,

   $\chi_s = 4.05, 4.07$, and 4.0 for Si, GaAs, and GaP, respectively,

   calculate D_s, $q\phi_0$, and $q\phi_{Bn}$ for the above Schottky diodes.

10.10. Using Eqs. (10.61) through (10.65), calculate the barrier height enhancement for a Au/p–n GaAs Schottky barrier diode for the cases:
   (a) $N_d = 10^{16}$ cm^{-3}, plot ϕ_{Bn} versus N_a (10^{15} to 5×10^{18} cm^{-3}) for $W_p = 0.3, 0.2, 0.1, 0.05$, and 0.02 μm; $q\phi_{Bn} = 0.9$ eV for an Au/n-GaAs Schottky contact.
   (b) $N_a = 2 \times 10^{17}$ cm^{-3}, calculate and plot ϕ_{Bn} versus N_d and W_n for the values given in (a).

REFERENCES

1. C. R. Crowell and S. M. Sze, "Current Transport in Metal–Semiconductor Barriers," *Solid-State Electron.* **8**, 979 (1966).
2. A. Y. C. Yu and C. A. Mead, *Solid State Electron.* **13**, 97 (1970).
3. D. Kahng, *Bell Syst. Tech. J.* **43**, 215 (1964).
4. M. P. Lepselter and S. M. Sze, *Bell Syst. Tech. J.* **47**, 195 (1968).
5. A. M. Cowley and S. M. Sze, "Surface States and Barrier Height of Metal–Semiconductor Systems," *J. Appl. Phys.* **36**, 3212 (1965).
6. R. H. Fowler, "The Analysis of Photoelectric Sensitivity Curves for Clean Metals at Various Temperatures," *Phys. Rev.* **38**, 45 (1931).
7. C. R. Crowell, J. C. Sarace, and S. M. Sze, "Tungsten–Semiconductor Schottky Barrier Diodes," *Trans. Metall. Soc. AIME* **233**, 478 (1965).
8. S. S. Li, "Theoretical Analysis of a Novel MPN GaAs Schottky Barrier Solar Cell," *Solid-State Electron.* **21**, 435–438 (1977).
9. S. S. Li, C. S. Kim, and K. L. Wang, "Enhancement of Effective Barrier Height in Ti–Silicon Schottky Barrier Diode Using Low Energy Ion-Implantation," *IEEE Trans. Electron Devices* **ED-27**, 1310–1312 (1980).
10. D. H. Lee, S. S. Li, N. J. Sauer, and T. Y. Chang, "A High Quality $In_{0.53}Ga_{0.47}As$ Schottky Barrier Diode Formed by Graded Superlattice of $In_{0.53}Ga_{0.47}As/In_{0.52}Al_{0.48}As$," *Appl. Phys. Lett.* **54**(19), 1863 (1989).
11. F. A. Padovani and R. Stratton, "Field and Thermionic Field Emission in Schottky Barriers," *Solid-State Electron.* **9**, 695 (1966).

BIBLIOGRAPHY

N. Braslau, *Thin Solid Films* **104**, 391 (1983).

P. Chattopadhyay and A. N. Daw, "On the Barrier Height of a Metal–Semiconductor Contact with a Thin Interface Layer," *Solid-State Electron.* **28**, 831 (1985).

M. Heiblum, M. I. Nathan, and C. A. Chang, *Solid-State Electron.* **25**, 185 (1982).

H. K. Henisch, *Semiconductor Contacts: An Approach to Ideas and Models*, Clarendon Press, Oxford, New York (1984).

V. G. Keramidas, *Inst. Phys. Conf. Ser.* No. 45, 396 (1979).

R. S. Muller and T. I. Kamins, *Device Electronics for Integrated Circuits*, Wiley, New York (1977).

A. Piotrowska *et al.*, "Ohmic Contacts to III–V Compound Semiconductors: A Review of Fabrication Techniques," *Solid-State Electron.* **26**, 179 (1983).

E. H. Rhoderick, *Metal–Semiconductor Contacts*, Clarendon Press, Oxford (1978).

V. L. Rideout, "A Review of the Theory and Technology for Ohmic Contacts to Group III–V Compound Semiconductors," *Solid-State Electron.* **18**, 541 (1975).

B. L. Sharma, "Ohmic Contacts to III–V Compound Semiconductors," in: *Semiconductors and Semimetals*, Vol. 15, p. 1, Academic Press, New York (1981).

B. L. Sharma, *Metal–Semiconductor Schottky Barrier Junctions and Their Applications*, Plenum Press, New York (1984).

W. E. Spicer, P. W. Chyre, C. M. Garner, I. Lindau, and P. Pianetta, "The Surface Electronic Structure of III–V Compounds and the Mechanism of Fermi Level Pinning by Oxygen Passivation and Metal Schottky Barriers," *Surf. Sci.* **86**, 763 (1979).

W. E. Spicer, I. Lindau, P. Skeath, C. Y. Su, and P. W. Chyre, "Unified Mechanism for Schottky Barrier Formation and III–V Oxide Interface States," *Phys. Rev. Lett.* **44**, 420 (1980).

S. M. Sze, *Physics of Semiconductor Devices*, 2nd edn., Wiley, New York (1982).

A. Van der Ziel, *Solid State Physical Electronics*, 3rd edn., Prentice-Hall, New York (1976).

C. W. Wilmsen, *Physics and Chemistry of III–V Compound Semiconductor Interfaces*, Plenum Press, New York (1985).

C. Y. Wu, "Interfacial Layer Theory of the Schottky Barrier Diodes," *J. Appl. Phys.* **51**(7), 3786 (1980).

11

p–n Junction Diodes

11.1. INTRODUCTION

In this chapter, we present basic device physics, general characteristics, and operation principles of a p–n junction diode. A p–n junction diode is referred to as a minority carrier device since its current conduction is controlled by the minority carrier diffusion process (i.e., electrons in the p-region and holes in the n-region). This is in contrast to the metal–semiconductor Schottky diode discussed in Chapter 10, which is known as a majority carrier device since current conduction in such a device is due to majority carrier transport.

If a p–n junction diode is formed by using the same type of semiconductor material, but with opposite dopant impurities in the p- and n-regions, then it is referred to as a p–n homojunction diode. On the other hand, if a p–n junction is formed by using two different types of semiconductor materials with different bandgap energies and opposite dopant impurities, then it is referred to as a p–n heterojunction diode. Both the p–n homo- and heterojunction diodes will be considered in this chapter. The p–n junction devices play an important role as active elements in many intergrated circuits. p–n junction structures have been used in a wide variety of practical applications, such as switching diodes, diode rectifiers, solar cells. LEDs, laser diodes, photodetectors, bipolar junction transistors (BJTs), and heterojunction bipolar transistors (HBTs). As for p–n heterojunction diodes most are fabricated from II–VI and III–V compound semiconductors such as n-ZnSe/p-GaAs, p-AlGaAs/n-GaAs, p-Ge/n-GaAs, n-InGaAs/n-InP, and p-InAlAs/n-InGaAs heterojunctions.

Derivations of the charge carrier distribution, built-in potential, electric field, and the potential distribution in the junction space charge region of a p–n junction diode under equilibrium and applied bias conditions are given in Sections 11.2 and 11.3, respectively. The minority carrier distribution and current flow in a p–n junction diode are derived by using the continuity equations presented in Chapter 6. The current–voltage (I–V) and capacitance–voltage (C–V) characteristics under forward and reverse bias conditions are discussed in Section 11.4. The minority carrier storage and transient behavior in a p–n junction diode are discussed in Section 11.5. Section 11.6 describes the avalanche and Zener breakdown phenomena under large reverse bias conditions. Finally, basic device physics and general characteristics of a p–n heterojunction diode are discussed in Section 11.7.

11.2. EQUILIBRIUM PROPERTIES OF A p–n JUNCTION DIODE

A p–n junction diode is formed when an opposite dopant impurity is introduced into a region of the semiconductor substrate by using the alloying, thermal diffusion, or ion-implantation technique. For example, a silicon p–n junction diode is formed when a p-type dopant impurity such as boron (B), aluminum (Al), or gallium (Ga) is introduced into an n-type silicon substrate via the thermal diffusion or ion-implantation process. On the other hand, if an n-type dopant impurity such as phosphorus (P) or arsenic (As) is introduced into a p-type silicon substrate, then a silicon n–p junction diode is formed. For compound semiconductors such as GaAs, InP, InGaAs, and AlGaAs, the p–n junctions can be formed by using different growth techniques such as liquid-phase epitaxy (LPE), vapor-phase epitaxy (VPE), metal–organic chemical vapor deposition (MOCVD), and molecular beam epitaxy (MBE) techniques.

A p–n heterojunction diode can be formed by using two semiconductor materials with different bandgap energies (e.g., p-AlGaAs/n-GaAs, p-InAlAs/n-InGaAs, and p-Ge/n-GaAs). Figure 11.1a and b show energy band diagrams for a p–n junction diode under thermal equilibrium conditions, before and after the intimate contact. We note that the Fermi level is constant across the entire p–n junction under equilibrium conditions. Figure 11.2a shows the charge distribution in the p- and n-quasi-neutral regions as well as in the depletion region of the junction. In general, depending on the dopant impurity profile across the junction, a diffused p–n junction may be approximated by either a step junction or a linearly-graded junction. As shown in Fig. 11.3a, the impurity profile for a step junction changes abruptly across the metallurgical junction of the diode, while the impurity profile for a linearly-graded junction varies linearly with distance across the junction, as is clearly shown in Fig. 11.3b.

The static properties of an abrupt p–n junction diode and a linearly graded p–n junction diode are discussed next. The carrier distribution, built-in potential, electric field, and potential profile in the junction space-charge region of a p–n junction diode can be derived for both abrupt and linearly graded junctions using the Poisson and continuity equations.

In thermal equlibrium, the Fermi level is constant throughout the entire p–n junction, as shown in Fig. 11.1b. Poisson's equation, which relates charge density to electrostatic potential, is given by

$$\frac{d^2V(x)}{dx^2} = -\frac{\rho}{\varepsilon_0\varepsilon_s} = \left(\frac{q}{\varepsilon_0\varepsilon_s}\right)(n - p - N_d + N_a) \tag{11.1}$$

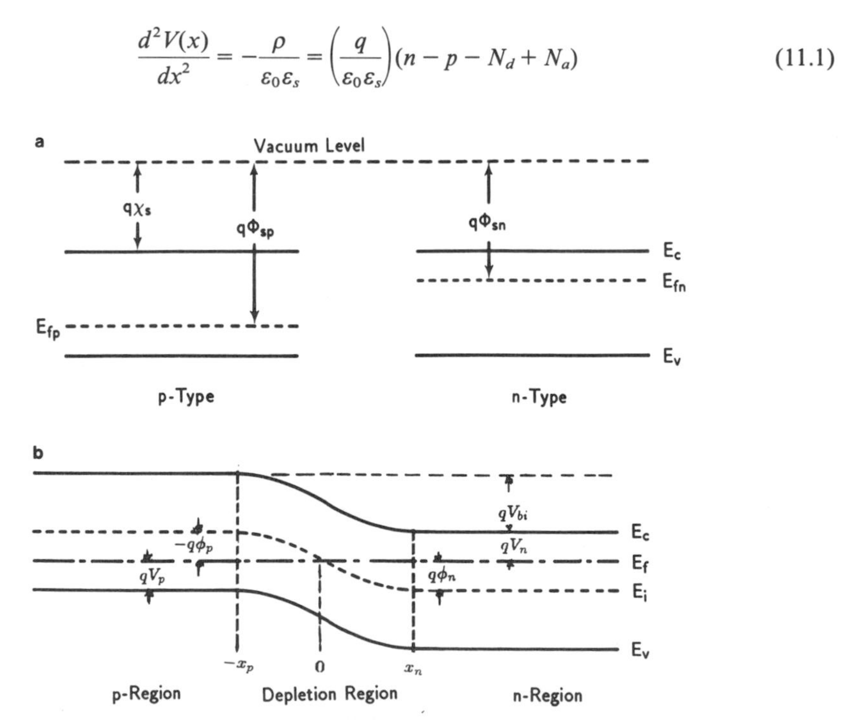

FIGURE 11.1. Energy band diagrams for an isolated n- and p-type semiconductor (a) before contact, and (b) in intimate contact.

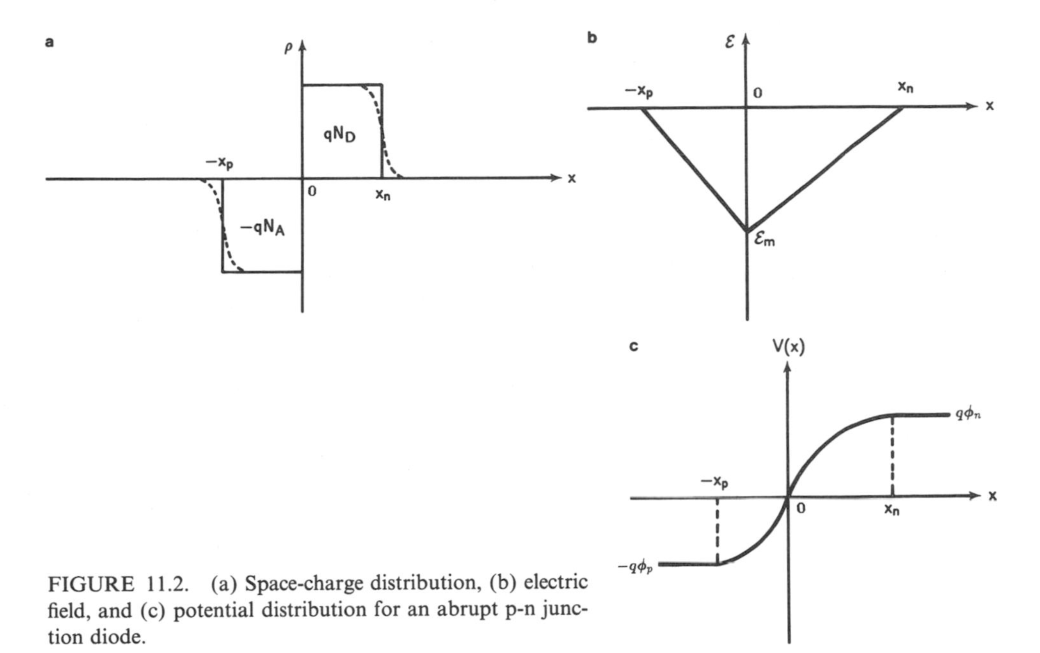

FIGURE 11.2. (a) Space-charge distribution, (b) electric field, and (c) potential distribution for an abrupt p-n junction diode.

The electron and hole densities in the n- and p-regions of the junction can be expressed in terms of the intrinsic carrier density and electrostatic potential, and are given by

$$n = n_i \exp\left(\frac{\phi_n}{V_T}\right) \tag{11.2}$$

and

$$p = n_i \exp\left(\frac{-\phi_p}{V_T}\right) \tag{11.3}$$

where n_i is the intrinsic carrier density and $V_T = k_B T/q$ is the thermal voltage. We use the Fermi level as a reference level (see Fig. 11.1b). Using proper boundary conditions, expressions for the potential, electric field, and charge distribution in the different regions of a p–n junction

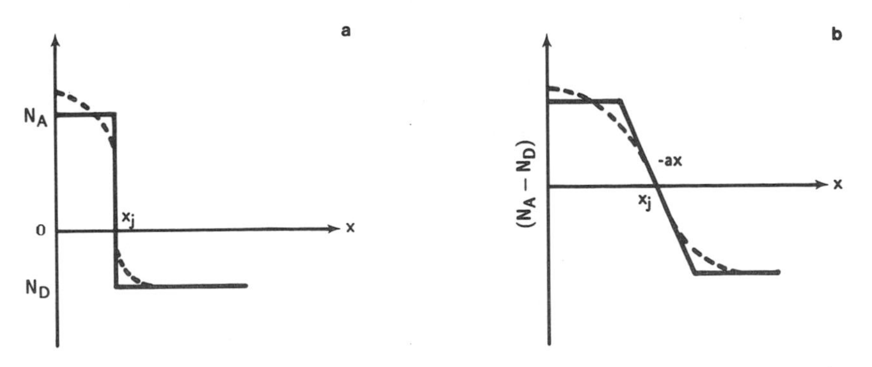

FIGURE 11.3. Impurity profile for (a) a shallow-diffused junction (i.e., an abrupt or step junction) and (b) a deep-diffused junction (i.e., a linearly graded junction).

diode can be derived from Eqs. (11.1) through (11.3). Figure 11.1b illustrates that there are three distinct regions which can be defined in a p–n junction diode, namely, the n- and p-quasi-neutral regions away from the metallurgical junction, and the depletion (or space-charge) region which is occupied by the ionized shallow acceptors in the p-depletion region and the ionized shallow donors in the n-depletion region. In addition to these three distinct regions, a boundary layer of several Debye lengths may also exist in the transition region between the quasi-neutral region and the depletion edge. This transition layer is usually much smaller than the depletion layer width, and hence may be neglected in the diode analysis.

In the quasi-neutral regions, the total charge density is equal to zero so Eq. (11.1) becomes

$$\frac{d^2V(x)}{dx^2} = 0 \tag{11.4}$$

and

$$n - p - N_d + N_a = 0 \tag{11.5}$$

In the n-quasi-neutral region, N_a is assumed equal to zero, and $p \ll n$. The electrostatic potential ϕ_n at the depletion layer edge of the n-quasi-neutral region can be derived by assuming that $N_a = p = 0$ in Eq. (11.5) and then substituting the result into Eq. (11.2). This yields

$$\phi_n = V_T \ln\left(\frac{N_d}{n_i}\right) \tag{11.6}$$

Similarly, the potential distribution at the depletion edge of the p-quasi-neutral region can be written as

$$\phi_p = -V_T \ln\left(\frac{N_a}{n_i}\right) \tag{11.7}$$

Therefore, the built-in or diffusion potential of a p–n junction diode between the n- and p-quasi-neutral regions can be obtained from Eqs. (11.6) and (11.7) in the form

$$V_{bi} = \phi_n - \phi_p = V_T \ln\left(\frac{N_d N_a}{n_i^2}\right) \tag{11.8}$$

where V_{bi} is known as the built-in or diffusion potential of the p–n junction diode in thermal equilibrium. In the depletion region, the free-carrier density is assumed equal to zero (i.e., $n = p = 0$), and hence Eq. (11.1) becomes

$$\frac{d^2V}{dx^2} = \left(\frac{q}{\varepsilon_s \varepsilon_0}\right)(N_a - N_d) \tag{11.9}$$

Equation (11.9) may be used to solve the potential and electric field distribution in the junction space-charge region for both the abrupt and linearly graded p–n junction diodes shown in Figs. 11.2a and 11.3a. The abrupt junction approximation can be applied to the shallow-diffused step junction or an implanted junction, while the linearly graded junction approximation is more suitable for a deep-diffused p–n junction diode.

Figure 11.2a shows the impurity distribution in the space-charge region of an abrupt junction diode. We note that the boundary layer effect (i.e., the spreading of space charges a few Debye lengths into the quasi-neutral regions) shown by the dotted line is neglected in the present analysis. In the depletion region, free carriers are negligible, and Poisson's equation in both the n- and p-space-charge regions are given respectively by

$$\frac{d^2V(x)}{dx^2} = -\frac{qN_d}{\varepsilon_0\varepsilon_s} \quad \text{for } 0 < x < x_n \tag{11.10}$$

and

$$\frac{d^2V(x)}{dx^2} = \frac{qN_a}{\varepsilon_0\varepsilon_s} \quad \text{for } -x_p < x < 0 \tag{11.11}$$

where x_n and x_p denote the depletion layer widths in the n- and p-regions, respectively. The charge-neutrality condition in the depletion region of the junction requires that

$$N_a x_p = N_d x_n \tag{11.12}$$

which shows that the depletion layer width on each side of the junction is inversely proportional to the dopant density. The total depletion layer width W_d of the junction is given by

$$W_d = x_n + x_p \tag{11.13}$$

From Eq. (11.12) it is seen that if N_a is much greater than N_d, then x_n will be much greater than x_p, and the depletion region will spread mostly into the n-region. Thus, if $x_n \gg x_p$, then $W_d \cong x_n$, and we have a one-sided abrupt p–n junction diode. In this case, the depletion layer width on the heavily doped p-region is negligible compared to the depletion layer width on the lightly doped n-region. As a result, we can solve Poisson's equation for the lightly doped side (i.e., the n-region) to obtain basic information on the junction characteristics. Integration of Eq. (11.10) once from x_n to x yields the electric field

$$\mathscr{E} = -\frac{dV(x)}{dx} = \left(\frac{qN_d}{\varepsilon_0\varepsilon_s}\right)(x - x_n) \tag{11.14}$$

which was obtained subject to the boundary condition that $dV(x)/dx = 0$ at $x = x_n$. Since the maximum electric field occurs at $x = 0$, Eq. (11.14) can be expressed in the form

$$\mathscr{E} = \mathscr{E}_m\left(1 - \frac{x}{x_n}\right) \quad \text{for } 0 < x < x_n \tag{11.15}$$

where $\mathscr{E}_m = qN_dx_n/\varepsilon_0\varepsilon_s$ is the maximum field strength at $x = 0$. We note that the electric field is negative throughout the entire depletion region, and varies linearly with distance from $x = 0$ to either side of the junction. As illustrated in Fig. 11.2b, the electric field on the right-hand side of the junction (i.e., the n-region) is negative since the force exerted by the electric field is offset by the electrons diffusing toward the left from the quasi-neutral n-region. Similarly, the electric field in the p-region is also negative in order to retard the diffusion of holes to the right-hand side of the junction. Thus, the electric field for $x < 0$ can be written as

$$\mathscr{E}(x) = -\left(\frac{qN_a}{\varepsilon_s\varepsilon_0}\right)(x + x_p) \quad \text{for } -x_p < x < 0 \tag{11.16}$$

The potential in the n-region can be obtained by integrating Eq. (11.14) once more, and the result yields

$$V(x) = V_n - \left(\frac{qN_d x_n^2}{2\varepsilon_s\varepsilon_0}\right)\left(1 - \frac{x}{x_n}\right)^2 \qquad \text{for } 0 < x < x_n \tag{11.17}$$

where $V_n = (kT/q)\ln(N_c/N_d)$ is the potential difference between the conduction band edge and the Fermi level at the depletion edge of the n-quasi-neutral region. Similarly, the potential in the p-region may be written as

$$V(x) = V_p + \left(\frac{qN_a x_p^2}{2\varepsilon_0\varepsilon_s}\right)\left(1 + \frac{x}{x_p}\right)^2 \qquad \text{for } -x_p < x < 0 \tag{11.18}$$

where $V_p = (kT/q)\ln(N_c/N_a)$ is the potential at the edge of the p-depletion region.

The built-in potential, V_{bi}, which is defined as the total potential change from the quasi-neutral p-region to the quasi-neutral n-region, is equal to $(\phi_n - \phi_p)$, as given by Eq. (11.8). It is noteworthy that most of the potential drop and the depletion region are on the lightly doped side of the junction. The depletion layer width can be obtained by solving Eqs. (11.12), (11.17), and (11.18) at $x = 0$ to yield

$$W_d = x_n + x_p = \left[\left(\frac{2\varepsilon_s\varepsilon_0 V_{bi}}{q}\right)\left(\frac{N_d + N_a}{N_d N_a}\right)\right]^{1/2} \tag{11.19}$$

Equation (11.19) shows that the depletion layer width is controlled by the dopant density of the lightly doped base region, which varies inversely with the square root of the dopant density.

For a linearly graded p–n junction diode, the space-charge distribution in the depletion region is given by

$$N_a - N_d = -ax \tag{11.20}$$

where a is the slope of the dopant impurity density profile (cm^{-4}). Thus, Poisson's equation for a linearly graded p–n junction diode can be expressed by

$$\frac{d^2V(x)}{dx^2} = -\left(\frac{q}{\varepsilon_0\varepsilon_s}\right)ax \tag{11.21}$$

By employing the same procedures as for the step-junction diode described above, we can derive the depletion layer width W_d and built-in potential V_{bi} for a linearly graded p–n junction diode:

$$W_d = \left[\frac{12\varepsilon_0\varepsilon_s V_{bi}}{qa}\right]^{1/3} \tag{11.22}$$

and

$$V_{bi} = 2V_T \ln\left(\frac{aW_d}{2n_i}\right) \tag{11.23}$$

A comparison of Eqs. (11.19) and (11.22) shows that for a linearly graded junction, W_d depends on $(V_{bi}/N_d)^{1/3}$, while for an abrupt junction, it depends on $(V_{bi}/N_d)^{1/2}$.

11.3. p–n JUNCTION UNDER BIAS CONDITIONS

When an external bias voltage is applied to a p–n junction diode, the thermal equilibrium condition is disrupted and a current flow across the junction results. Since the resistance across the depletion region is many orders of magnitude greater than the resistance in the quasi-neutral regions, the voltage drops across both the n- and p-quasi-neutral regions are negligible compared to the voltage drop across the depletion region. Thus, it is reasonable to assume that the voltage applied to a p–n junction diode is roughly equal to the voltage drop across the depletion layer region. The current–voltage (I–V) characteristics of a p–n junction diode under reverse- and forward-bias conditions are discussed next.

The current flow in a p–n junction depends on the polarity of the applied bias voltage. Under forward-bias conditions, the current increases exponentially with applied voltage. Under reverse-bias conditions, the current flow is limited mainly by the thermal generation current and hence depends very little on the applied voltage. Figure 11.4 shows energy band diagrams for a p–n junction diode under: (a) zero-bias, (b) forward-bias, and (c) reverse-bias conditions. As shown in Fig. 11.4b, when a forward-bias voltage V (i.e., positive polarity applied to the p-side and negative to the n-side) is applied to the p–n junction, the potential barrier across the junction will decrease to $(V_{bi} - V)$. In this case, the potential barrier for the majority carriers at the junction is reduced, and the depletion layer width is decreased. Thus, under forward-bias condition a small increase in applied voltage will result in a large increase in current flow across the junction. On the other hand, if a reverse-bias voltage is applied to the junction, then the potential barrier across the junction will increase to $(V_{bi} + V)$, as shown in Fig. 11.4c. Therefore, under a reverse-bias condition the potential barrier for the

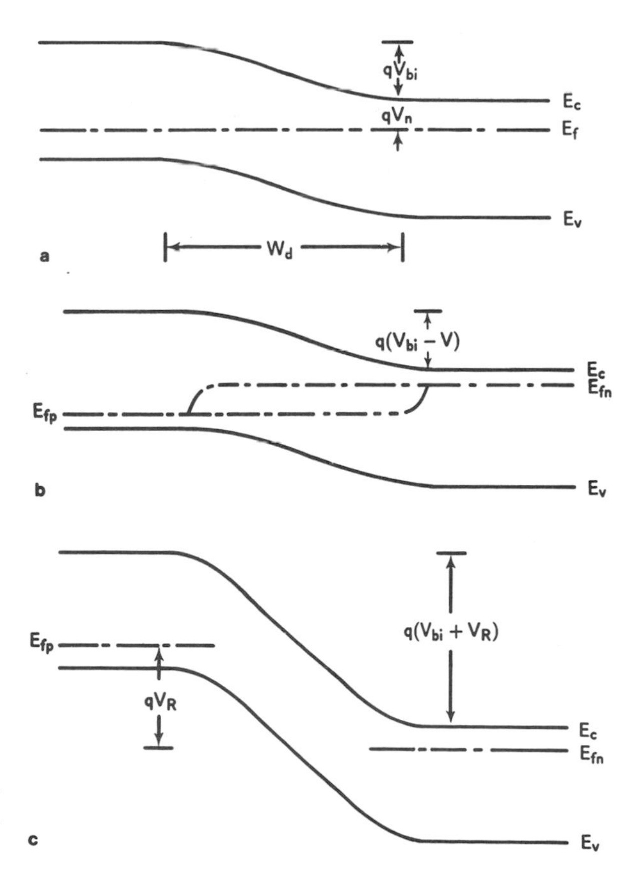

FIGURE 11.4. Energy band diagram of a p–n junction diode under (a) zero-bias, (b) forward-bias, and (c) reverse-bias conditions.

majority carriers and the depletion layer width will increase with increasing reverse-bias voltage. As a result, current flow through the junction becomes very small, and the junction impedance is extremely high.

The abrupt junction approximation is used to analyze the I–V characteristics of a step-junction diode under bias conditions. In the analysis it is assumed that (1) the entire applied voltage drop is only across the junction space-charge region, but is negligible in the quasi-neutral regions; (2) the solution of Poisson's equation obtained under thermal equilibrium conditions can be extended to the applied bias case; and (3) the total potential across the junction space-charge region changes from V_{bi} for the equilibrium case to $(V_{bi} \pm V)$ when the bias voltage is applied to the p–n junction. Thus, the depletion layer width for a step-junction diode under bias conditions is given by

$$W_d = x_n + x_p = \left[\left(\frac{2\varepsilon_0 \varepsilon_s (N_a + N_d)}{q N_a N_d} \right) (V_{bi} \pm V) \right]^{1/2} \tag{11.24}$$

where the plus sign refers to the reverse-bias case and the minus sign to the forward-bias case. Equation (11.24) shows that for a step-junction diode under reverse-bias conditions with $V \gg V_{bi}$, the depletion layer width is proportional to the square root of the applied voltage.

Similarly, for a linearly graded junction, the depletion layer width under bias conditions can be expressed by

$$W_d = \left[\frac{12 \varepsilon_0 \varepsilon_s (V_{bi} \pm V)}{qa} \right]^{1/3} \tag{11.25}$$

The relationship between the maximum electric field and the applied bias voltage in the junction space-charge region of a p–n junction diode can be derived as follows. For a step-junction diode, assuming $W_d \sim x_n$ (i.e., $N_a \gg N_d$ and $x_n \gg x_p$), the maximum electric field at the junction can be derived from Eqs. (11.15) and (11.19), and the result yields

$$\mathscr{E}_m = q N_d x_n / \varepsilon_0 \varepsilon_s \simeq \frac{2(V_{bi} - V)}{W_d} \tag{11.26}$$

Similarly, the maximum electric field versus applied bias voltage for a linearly graded junction is given by

$$\mathscr{E}_m = \frac{3(V_{bi} - V)}{2 W_d} \tag{11.27}$$

We shall next derive the junction capacitance in the space-charge region of the p–n diode. A p–n junction diode can be viewed as a parallel-plate capacitor filled with positive and negative fixed charges due to ionized donor and acceptor impurities in the depletion region, which determine the junction capacitance of the diode. For a step-junction diode with dopant densities of N_a and N_d in the p- and n-regions, respectively, the transition capacitance per unit area may be derived from the total space charge Q_s per unit area on either side of the depletion region. Thus, we can write

$$C_j = \frac{dQ_s}{dV} = \frac{d(q N_a x_p)}{dV} = \frac{d(q N_d x_n)}{dV} \tag{11.28}$$

If we assume that N_a and N_d are constant and independent of position, and use the relations $x_p = (N_d/N_a) x_n$ and $W_d = x_n + x_p$, then the small signal junction capacitance per unit area can be derived from Eq. (11.28) in the form

$$C_j = \sqrt{\frac{q\varepsilon_0\varepsilon_s}{2(1/N_d + 1/N_a)(V_{bi} - V)}} \tag{11.29}$$

For a one-sided step-junction diode (i.e., $N_a \gg N_d$), Eq. (11.29) predicts that the transition capacitance due to fixed charges in the depletion region is directly proportional to the square root of the dopant density, and varies inversely with the square root of the applied reverse-bias voltage for $-V \gg V_{bi}$. Since the transition capacitance per unit area for a one-sided step-junction diode is equal to $\varepsilon_0\varepsilon_s/W_d$, then from Eq. (11.29) we obtain

$$C_j = \frac{dQ_s}{dV} = \frac{\varepsilon_s\varepsilon_0}{W_d} = \left[\frac{q\varepsilon_0\varepsilon_s N_d}{2(V_{bi} \pm V)}\right]^{1/2} \tag{11.30}$$

This equation shows that $1/C_j^2$ varies linearly with the applied voltage V. Thus, a plot of $1/C_j^2$ versus V yields a straight line. The slope of this straight line yields the dopant density of the lightly doped semiconductor, and the intercept at $1/C_j^2 = 0$ on the voltage axis gives the built-in potential V_{bi}. It is noteworthy that if the dopant density is not uniform across the lightly doped semiconductor, the dopant density profile can be determined by using the differential C–V technique similar to that described above, except that the dopant density is determined step by step at a small incremental voltage across the substrate. Figure 11.5 shows a typical plot of $1/C_j^2$ versus V for a step-junction diode. The dopant density of the substrate and the built-in potential can be determined from this plot.

The transition capacitance for a linearly graded junction diode can be derived in a similar way as that of the step-junction diode discussed above. Thus Eq. (11.25) enables the transition capacitance for a linearly graded junction to be expressed by

$$C_j = \frac{dQ_s}{dV} = \frac{\varepsilon_s\varepsilon_0}{W_d}$$

$$= \left[\frac{qa(\varepsilon_s\varepsilon_0)^2}{12(V_{bi} \pm V)}\right]^{1/3} \tag{11.31}$$

Hence the transition capacitance for a linearly graded junction diode is inversely proportional to the cubic root of the applied reverse-bias voltage.

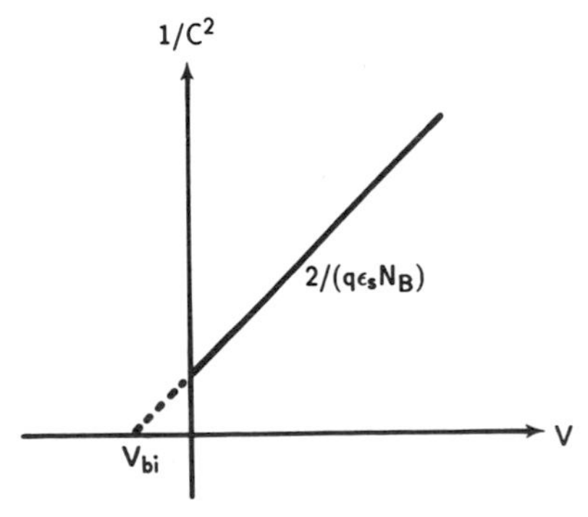

FIGURE 11.5. Inverse of capacitance squared versus applied reverse-bias voltage for a one-sided step-junction diode.

11.4. MINORITY CARRIER DISTRIBUTION AND CURRENT FLOW

In order to derive the current density equations for a p–n junction diode, it is necessary to find the relationship between the minority carrier density and the applied bias voltage at the edges of the depletion layer near both the p- and n-quasi-neutral regions. Figure 11.6 shows a schematic diagram of a p–n junction diode for deriving the current density equations in the n- and p-quasi neutral regions as well as in the depletion region. The cross-section area of the diode perpendicular to the current flow is assumed equal to A. As illustrated in Fig. 11.6, the minority carrier densities at the edge of the quasi-neutral p- (at $x = -x_p$) and n- (at $x = x_n$) regions can be related to the majority carrier densities at the edge of the depletion region under bias conditions. These are given by

$$p_n(x_n) = p_{p0}(-x_p) \exp[-q(V_{bi} - V)/k_BT] \tag{11.32}$$

which is the hole density at the depletion edge of the n-quasi-neutral region, and

$$n_p(-x_p) = n_{n0}(x_n) \exp[-q(V_{bi} - V)/k_BT] \tag{11.33}$$

which is the electron density at the depletion edge of the p-region. We note that $p_{p0}(-x_p) = N_a(-x_p)$ and $n_{n0}(x_n) = N_d(x_n)$ denote the majority carrier densities at the edges of the p-quasi-neutral and n-quasi-neutral regions, respectively. If the applied voltage V is set equal to zero, then $n_p = n_{p0} = n_{n0} \exp(-qV_{bi}/k_BT)$ and $p_n = p_{n0} = p_{p0} \exp(-qV_{bi}/k_BT)$. The above equations are valid only for the low-level injection case. The excess carrier densities at the depletion layer edges under bias conditions can be obtained by solving Eqs. (11.32) and (11.33) in terms of their equilibrium values, and the result yields

$$p_n'(x_n) = p_n(x_n) - p_{n0}(x_n) = p_{n0}(x_n)(e^{qV/k_BT} - 1) \tag{11.34}$$

and

$$n_p'(x_p) = n_p(x_p) - n_{p0}(x_p) = n_{p0}(x_p)(e^{qV/k_BT} - 1) \tag{11.35}$$

If Eqs. (11.34) and (11.35) are used as the boundary conditions, expressions for the minority carrier densities can be derived by solving the continuity equations in the quasi-neutral regions of a p–n junction diode. It is seen from Eqs. (11.32) through (11.35) that the majority carrier density is insensitive to the applied bias voltage, while the minority carrier density is an exponential function of the applied bias voltage. We shall show next that the current flow in a p–n junction diode is in fact governed by the diffusion of the minority carriers across the p–n junction.

The current density equations for a p–n junction diode may be derived by using the continuity equations described in Chapter 6. The one-dimensional (i.e., x-direction only)

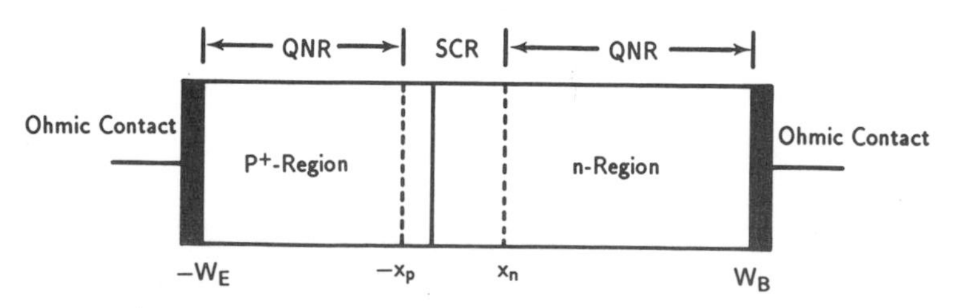

FIGURE 11.6. Schematic diagram of a p–n junction diode showing the dimensions and boundaries of the n- and p-quasi neutral regions and the space-charge region.

continuity equation for the excess holes injected into the n-quasi-neutral region under steady-state conditions is given by

$$D_p \frac{d^2 p_n'}{dx^2} - \frac{p_n'}{\tau_p} = 0 \tag{11.36}$$

which has a solution given by

$$p_n'(x) = A\, e^{-(x-x_n)/L_p} + B\, e^{(x-x_n)/L_p} \tag{11.37}$$

where A and B are constants to be determined by the boundary conditions in Eqs. (11.34) and (11.35), $L_p = (D_p\tau_p)^{1/2}$ is the hole diffusion length, while D_p and τ_p denote the hole diffusion constant and hole lifetime, respectively.

The solutions for the continuity equation given by Eq. (11.36) are determined by examining two special cases, namely, the long-base diode with base width greater than the hole diffusion length (i.e., $W_B \gg L_p$) and the short-base diode with base width smaller than the hole diffusion length (i.e., $L_p \gg W_B$). We shall next derive the current densities in both the n- and p-quasi-neutral regions and the depletion region, as shown in Fig. 11.6 for both long-base and short-base diodes.

For a long-base diode, the base width in the n-quasi-neutral region is much larger than the hole diffusion length. As a result, the excess hole density p_n' will decrease exponentially with increasing distance x, and the constant B in Eq. (11.37) can be set equal to zero. Constant A can be determined by using Eq. (11.34) to determine $p_n'(x)$ at $x = x_n$, and the result yields

$$p_n'(x) = p_{n0}(e^{qV/k_BT} - 1)\, e^{-(x-x_n)/L_p} \tag{11.38}$$

Equation (11.38) is the excess hole density in the n-quasi-neutral region of the p–n junction. Figure 11.7a and b show the distribution of minority carriers in the p- and n-quasi-neutral regions under forward- and reverse-bias conditions, respectively, while Fig. 11.7c and d show the corresponding current densities under forward- and reverse-bias conditions. Excess carriers inside the depletion region are assumed equal to zero. Therefore, the hole current density in the n-quasi-neutral region contributes only via the diffusion of excess holes in this region, i.e.,

$$J_p(x) = -qD_p \frac{dp_n'}{dx} = \left(\frac{qD_p n_i^2}{N_d L_p}\right)(e^{qV/k_BT} - 1)\, e^{-(x-x_n)/L_p} \tag{11.39}$$

In this equation, p_{n0} is replaced by n_i^2/N_d in the pre-exponential factor. As shown in Fig. 11.7c, the hole current density (J_{p1}) has a maximum value at the depletion layer edge, $x = x_n$, and decreases exponentially with x in the n-region. This is a result of recombination of holes with electrons in the n-region before they reach the ohmic contact. Since the total current density (J_{total}) across the diode is invariant under steady-state conditions, the majority electron current (J_{n2}) which supplies electrons for recombination with holes must increase with x away from the junction and reaches a maximum at the ohmic contact in the n-quasi-neutral region. Similarly, the minority electrons injected into the p-quasi-neutral region and the resulting electron current flow (i.e., J_{n1}) in this region can be derived in a similar way as that of the hole current density in the n-quasi-neutral region described above. Thus, the electron current density in the p-quasi-neutral region can be expressed by

$$J_n(x) = \left(\frac{qD_n n_i^2}{N_a L_n}\right)(e^{qV/k_BT} - 1)\, e^{(x+x_p)/L_n} \tag{11.40}$$

which is obtained by assuming that the emitter width is much greater than the electron diffusion length (i.e., $W_E \gg L_n$) in the p-emitter region. We note that x is negative in the p-region and positive in the n-region, and is equal to zero at the metallurgical junction.

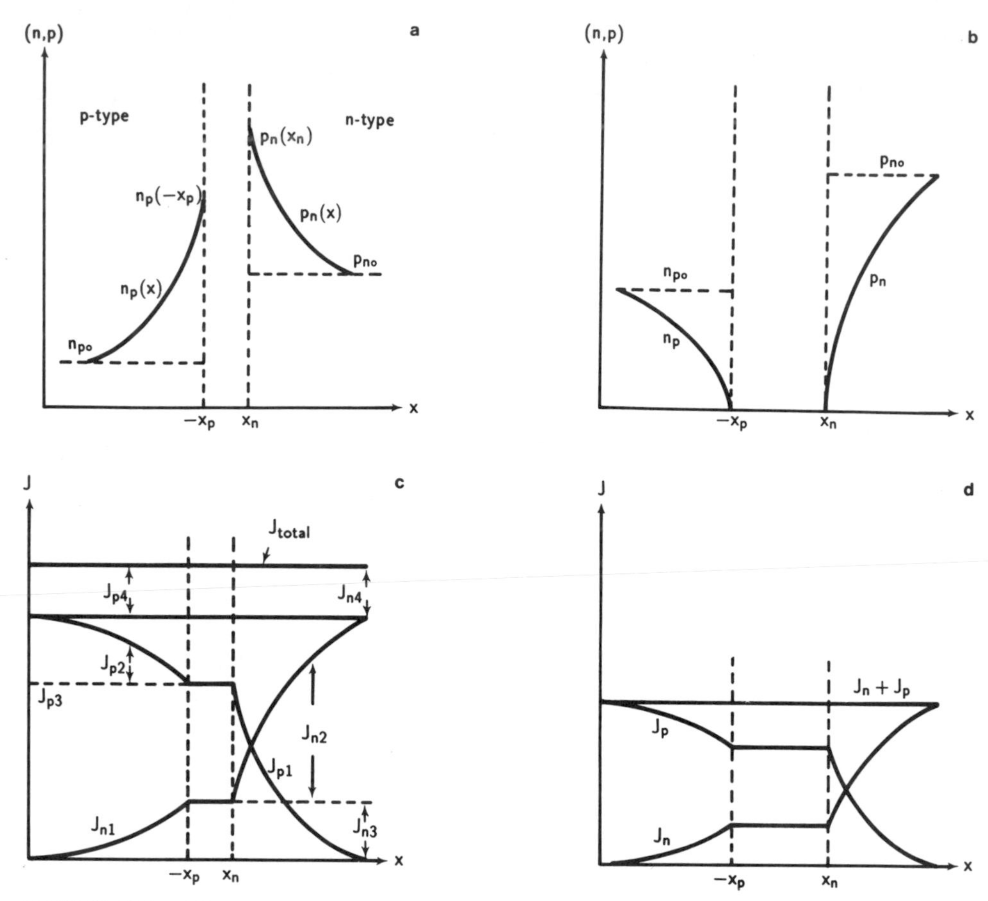

FIGURE 11.7. Minority carrier distribution under (a) forward-bias and (b) reverse-bias conditions; current components under (c) forward-bias and (d) reverse-bias conditions: J_{p1} and J_{n1} are injected minority hole and electron currents; J_{n2} and J_{p2} are majority electron and hole currents recombining with J_{p1} and J_{n1}, respectively; J_{n3} and J_{p3} are electron and hole recombination currents in the space-charge region.

It is seen in Fig. 11.7c that if the recombination current (i.e., J_{n4} or J_{p4}) in the depletion region is neglected, then the total current density in the p–n junction diode can be obtained by adding the injected minority hole current density evaluated at $x = x_n$ and the injected minority electron current density evaluated at $x = -x_p$. This can be written as

$$J = J_{p3}(-x_p) + J_{n1}(-x_p) = J_{p1}(x_n) + J_{n1}(-x_p) = J_0(e^{qV/k_BT} - 1) \tag{11.41}$$

where

$$J_0 = qn_i^2\left(\frac{D_p}{N_d L_p} + \frac{D_n}{N_a L_n}\right) \tag{11.42}$$

is the saturation current density. Since J_0 is proportional to n_i^2, its value depends exponentially on temperature and the energy bandgap of the semiconductor [i.e., $J_0 \sim \exp(-E_g/k_BT)$]. For a Si p–n junction diode, J_0 will double roughly for every 10°C increase in temperature.

We shall next consider a short-base diode, which has a base width W_B and an emitter width W_E much smaller than the minority carrier diffusion lengths (i.e., $W_B \ll L_p$) in the n-base region and (i.e., $W_E \ll L_n$) in the p-emitter region. In this case, the recombination loss in the p- and n-quasi-neutral regions is negligible, and thus all the injected minority carriers are expected to recombine at the ohmic contact regions of the diode. It can be shown that the excess hole density in the n-base region of a short-base diode can be expressed by

$$p'_n(x) = p_{n0}(e^{qV/k_BT} - 1)\left[1 - \frac{(x - x_n)}{W'_B}\right] \tag{11.43}$$

where $W'_B = W_B - x_n$ is the width of the quasi-neutral n-base region. Equation (11.43) is obtained by linearizing the exponential term given in Eq. (11.38), using the boundary condition that $p'_n = 0$ at $x = W_B$. The boundary condition at $x = x_n$ is identical for both the short- and long-base diodes discussed above. Equation (11.43) predicts that the excess hole density in the n-base region decreases linearly with distance x. Thus, the hole current density can be determined from Eq. (11.43), and the result yields

$$J_p = -qD_p \left.\frac{dp'_n}{dx}\right|_{x=x_n} = \left(\frac{qD_p n_i^2}{N_d W'_B}\right)(e^{qV/k_BT} - 1) \tag{11.44}$$

which shows that the hole current density in the n-base region is constant (i.e., the recombination loss in the base region is negligible). If the width of the p-emitter layer is smaller than the electron diffusion length (i.e., $W_E \ll L_n$), then the electron current density in the p^+-emitter region is given by

$$J_n = qD_n \left.\frac{dn'_p}{dx}\right|_{x=-x_p} = \left(\frac{qD_n n_i^2}{N_a W'_E}\right)(e^{qV/k_BT} - 1) \tag{11.45}$$

Since the total current density of a short-base diode is equal to the sum of J_n and J_p, given by Eqs. (11.44) and (11.45), we obtain

$$J = J_n + J_p = qn_i^2\left(\frac{D_n}{N_a W'_E} + \frac{D_p}{N_d W'_B}\right)(e^{qV/k_BT} - 1) \tag{11.46}$$

which shows that the current flow in a short-base diode does not depend on the minority carrier diffusion lengths in the emitter and base regions of the diode, but varies inversely with the width of both regions.

A comparison of the current density equations for a long-base and short-base diode reveals that the pre-exponential factor for the former depends inversely on the minority carrier diffusion length, while the pre-exponential factor for the latter depends inversely on the width of the n- and p-region of the diode. This is easy to understand, since for a long-base diode the widths of the n- and p-regions are much larger than the minority carrier diffusion lengths in both regions. Thus, one expects that the current density in both regions will be controlled by the minority carrier diffusion length. This is not the case for the short-base diode in which little or no recombination loss in the base region is expected. It is seen, however, that both Eqs. (11.41) and (11.46) predict the same exponential dependence of the current on the applied bias voltage under forward-bias conditions and a very small saturation current under reverse-bias conditions. It should be pointed out that under the reverse-bias condition, the saturation current is contributed by the thermal generation currents produced in both the n- and p-quasi-neutral regions of the junction. It is also seen that if one side of the junction is heavily doped, then the reverse saturation current will be determined by the thermal generation current produced in the lightly doped side of the junction. However, if bandgap-narrowing and Auger-recombination effects are taken into account in the heavily doped emitter region, then the saturation current may be dominated by the heavily doped side of the junction.

The ideal diode analysis presented above is based on the assumption that the total current flow in a p–n junction diode is due solely to the diffusion current components produced in the n- and p-quasi-neutral regions. This approximation is valid as long as the recombination current in the junction space-charge region is negligible compared to the diffusion currents produced in the quasi-neutral regions. However, for a silicon p–n junction diode and p–n junction diodes fabricated from III–V compound semiconductors such as GaAs and InP, recombination in the junction space-charge region may be important. In this case, the ideal diode equation described above may become inadequate under small forward-bias conditions, and hence we need to add the recombination current component (i.e., J_{n4} or J_{p4}) generated in the junction space-charge region to the total curent density given by Eq. (11.41) for a long-base diode.

The expression for the generation–recombination current density in the junction space-charge region can be derived by using the Shockley–Read–Hall (SRH) model described in Chapter 6. For simplicity, it is assumed that the electron and hole capture cross sections at the midgap recombination center are equal. Under this condition, the net recombination–generation rate for electrons and holes in the junction space-charge region is given by[1]

$$U = \frac{n_i^2(e^{qV/k_BT} - 1)}{[p + n + 2n_i \cosh(E_t - E_i)/k_BT]\tau_0} \tag{11.47}$$

where E_t is the activation energy of the recombination center and E_i is the intrinsic Fermi level. Note that $np = n_i^2 \exp(qV/k_BT)$ and $\tau_0 = 1/(N_t v_{th} \sigma)$ are used in Eq. (11.47). It is seen that the recombination rate given by Eq. (11.47) is positive under forward-bias conditions, and negative under reverse-bias conditions when the generation process becomes dominant.

The total recombination–generation current in the junction space-charge region can be obtained by integrating the recombination rate over the entire depletion region, in which case

$$J_{gr} = q \int_0^W U\, dx \tag{11.48}$$

Although the latter integration cannot be readily carried out, it is possible to obtain an analytical expression for the recombination current in the junction space-charge region if certain assumptions are made. For example, if we assume that the recombination process is via a midgap trap center [i.e., $E_t = E_i$ and $n = p = n_i \exp(qV/2k_BT)$ for a maximum recombination rate, U_{max}], then the recombination current under forward-bias conditions is given by

$$\begin{aligned} J_r &= \frac{qW'n_i^2(e^{qV/k_BT} - 1)}{2n_i\tau_0(e^{qV/2k_BT} + 1)} \\ &\cong \left(\frac{qW'n_i}{2\tau_0}\right) e^{qV/2k_BT} \end{aligned} \tag{11.49}$$

where $\tau_0 = (\tau_{n0}\tau_{p0})^{1/2}$ is the effective carrier lifetime associated with the recombination of excess carriers in the junction space-charge region of width W'. It is of interest that if we calculate the ratio of the diffusion current and the recombination current components from Eqs. (11.46) and (11.49), we find that the recombination current component is important only in the small forward-bias voltage regime, while the diffusion current becomes the dominant current component at the intermediate forward-bias regime.

Under reverse-bias conditions, the numerator in Eq. (11.47) reduces to $(-n_i^2)$, and thus U becomes negative, which implies a net generation rate inside the junction space-charge

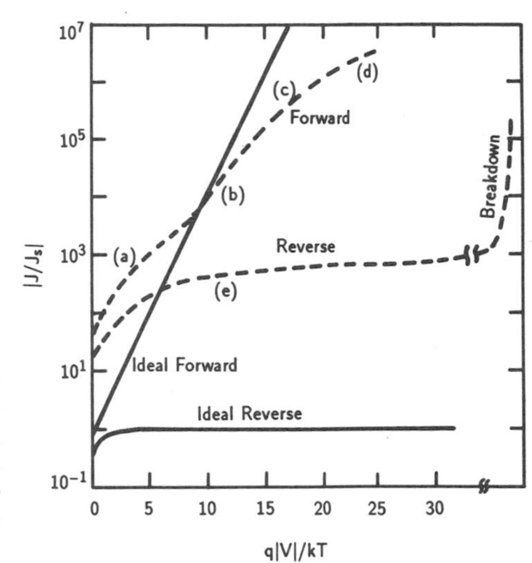

FIGURE 11.8. Current–voltage (I–V) characteristics of a practical silicon p–n diode; (a) generation–recombination (g–r) regime, (b) diffusion regime, (c) high injection regime, (d) series resistance effect, (e) reverse leakage current due to g–r current and surface effects. After Moll,[1] by permission.

region. The generation current can be determined from the product of the maximum generation rate and the depletion layer width W_i, namely,

$$J_g = \frac{qn_iW_i}{2\tau_0} \cong \left(\frac{n_i}{\tau_0}\right)\left[\left(\frac{q\varepsilon_0\varepsilon_s}{2N_d}\right)(V_{bi} + V)\right]^{1/2} \tag{11.50}$$

Equation (11.50) is obtained by assuming that the generation center coincides with the intrinsic Fermi level (i.e., $E_t = E_i$) and the depletion region is dominated by the lightly doped n-region. The result shows that the generation current varies linearly with n_i and the square root of the reverse-bias voltage.

It is noteworthy that the reverse saturation current for a p–n junction diode is in general much smaller than that of a Schottky barrier diode discussed in the previous chapter. This is because the saturation current of a p–n junction diode depends exponentially on the energy bandgap of the semiconductor, while the saturation current of a Schottky diode depends exponentially on the barrier height. Since the barrier height is usually smaller than the energy bandgap, the saturation current of a Schottky diode can be several orders of magnitude higher than that of a p–n junction diode. Furthermore, we also expect that the saturation current of a p–n junction diode will depend much more on temperature than that of a Schottky barrier diode due to the exponential dependence of the current density on both temperature and bandgap energy (or barrier height).

Figure 11.8 shows I–V characteristics of a practical silicon p–n junction diode under forward and reverse-bias conditions.[1] The solid line corresponds to the ideal I–V curve predicted by using the Shockley diode equation, While the dashed line corresponds to the I–V curve for a practical silicon p–n junction diode in which recombination-current and series-resistance effects are also included under forward-bias conditions.

11.5. DIFFUSION CAPACITANCE AND CONDUCTANCE

The transition capacitance derived in Section 11.3 is the dominant junction capacitance under the reverse-bias condition. However, under the forward-bias condition, when a small ac signal superimposed on a dc bias voltage is applied to a p–n junction diode, another capacitance component, known as the diffusion capacitance, becomes the dominant component. This diffusion capacitance is associated with the minority carrier rearrangement in the quasi-neutral regions of the p–n junction under the forward-bias condition. The diffusion capacitance can be derived by considering the following time-varying voltage and current density equations:

$$V(t) = V_0 + v_1 e^{i\omega t} \tag{11.51}$$

and

$$J(t) = J_0 + j_1 e^{i\omega t} \tag{11.52}$$

where V_0 and J_0 denote the dc bias voltage and current density v_1 and j_1 are the amplitude of the small-signal voltage and current density applied to the p–n junction, respectively. The small-signal condition is satisfied if $v_1 \ll k_BT/q$. Thus, when a small ac signal is applied to the junction, the minority hole density in the n-quasi-neutral region can be expressed by

$$p_n = p_{n0} \exp\left[\frac{q(V_0 + v_1 e^{i\omega t})}{k_BT}\right] \tag{11.53}$$

Since $v_1 \ll V_0$, an approximate solution can be obtained by expanding the exponential term of Eq. (11.53), which yields

$$p_n \simeq p_{n0} \exp\left(\frac{qV_0}{k_BT}\right)\left(1 + \frac{qv_1}{k_BT} e^{i\omega t}\right) \tag{11.54}$$

The first term in Eq. (11.54) is the dc component while the second term corresponds to the small-signal component evaluated at the depletion layer edge of the n-quasi-neutral region. A similar expression for the electron density in the p^+-quasi-neutral region can also be derived. Now, substitution of the ac component of p_n given by Eq. (11.54) into the continuity equation yields

$$D_p \frac{\partial^2 \tilde{p}_n}{\partial x^2} - \frac{\tilde{p}_n}{\tau_p} = i\omega \tilde{p}_n \tag{11.55}$$

where

$$\tilde{p}_n = p_{n1} e^{i\omega t} = \left(\frac{p_{n0} q v_1}{k_BT}\right) \exp\left(\frac{qV_0}{k_BT}\right) e^{i\omega t} \tag{11.56}$$

By substituting Eq. (11.56) into (11.55) we obtain

$$\frac{\partial^2 \tilde{p}_n}{\partial x^2} - \frac{\tilde{p}_n}{D_p \tau_p^*} = 0 \tag{11.57}$$

where

$$\tau_p' = \frac{\tau_p}{1 + i\omega\tau_p} \tag{11.58}$$

is the effective hole lifetime which is frequency-dependent. The solution of Eq. (11.57) is given by

$$\tilde{p}_n = p_{n1} \exp[-(x - x_n)/L_p^*] \tag{11.59}$$

where $L_p^* = \sqrt{D_p \tau_p^*}$ is the effective hole diffusion length in the n-quasi-neutral region. Similar to the solution given by Eq. (11.41) for the dc current density, the solution for the ac hole current density is obtained by substituting Eq. (11.59) into Eq. (11.39) and evaluating the result at $x = x_n$. This procedure yields

$$j_p(x_n) = -qD_p \frac{dp_n}{dx}\bigg|_{x = x_n} = \left(\frac{qv_1}{k_B T}\right)\left(\frac{qD_p n_i^2}{N_d L_p'}\right) \exp\left(\frac{qV_0}{k_B T}\right) \tag{11.60}$$

Similarly, the ac electron current density in the p^+-quasi-neutral region can be expressed by

$$j_n(-x_p) = qD_n \frac{dn_p}{dx}\bigg|_{x = -x_p} = \left(\frac{qv_1}{k_B T}\right)\left(\frac{qD_n n_i^2}{N_a L_n'}\right) \exp\left(\frac{qV_0}{k_B T}\right) \tag{11.61}$$

The total ac current density is equal to the sum of $j_p(x_n)$ and $J_n(-x_p)$ given by Eqs. (11.60) and (11.61), respectively, and can be written as

$$j_1 = j_p(x_n) + j_n(-x_p) = \left(\frac{qv_1}{k_B T}\right)\left(\frac{qD_p n_i^2}{N_d L_p'} + \frac{qD_n n_i^2}{N_a L_n'}\right) \exp\left(\frac{qV_0}{k_B T}\right) \tag{11.62}$$

The small-signal admittance of the diode can be obtained from Eq. (11.62) and the result is given by

$$Y = \frac{j_1}{v_1} = G_d + i\omega C_d = \left(\frac{q}{k_B T}\right)\left(\frac{qD_p n_i^2}{N_d L_p^*} + \frac{qD_n n_i^2}{N_a L_n^*}\right) \exp\left(\frac{qV_0}{K_B T}\right) \tag{11.63}$$

where $L_p^* = L_p/\sqrt{1 + i\omega\tau_p}$ and $L_n^* = L_n/\sqrt{1 + i\omega\tau_n}$ denote the effective hole and electron diffusion lengths, respectively. We note that both L_p^* and L_n^* depend on the frequency of the applied ac signal. At very low frequencies, $\omega\tau_{p,n} \ll 1$, the diffusion capacitance and conductance of a p–n diode can be derived from Eq. (11.63), and yield

$$C_{d0} \simeq \left(\frac{q^2 n_i^2}{2k_B T}\right)\left(\frac{L_p}{N_d} + \frac{L_n}{N_a}\right) \exp\left(\frac{qV_0}{k_B T}\right) \tag{11.64}$$

and

$$G_{d0} \simeq \left(\frac{q^2 n^2}{k_B T}\right)\left(\frac{D_p}{N_d L_p} + \frac{D_n}{N_a L_n}\right) \exp\left(\frac{qV_0}{k_B T}\right) \tag{11.65}$$

Equation (11.63) shows that the diffusion capacitance varies inversely with the square root of the frequency and the minority carrier lifetimes, while the conductance increases with the square root of the frequency and the minority carrier lifetime. The small-signal analysis presented above for a p–n junction diode reveals that, under the forward-bias condition, the diffusion capacitance will contribute significantly to the total junction capacitance. It increases exponentially with dc forward-bias voltage. Thus, the equivalent circuit of a p–n junction diode under small-signal operation should include both transition and diffusion capacitances in parallel with diffusion conductance and the series resistance which accounts for the ohmic drop across the contacts and quasi-neutral regions of the diode.

11.6. MINORITY CARRIER STORAGE AND TRANSIENT BEHAVIOR

As shown in the previous section, under forward-bias conditions, electrons are injected from the n-quasi-neutral region into the p-quasi-neutral region, while holes are injected from the p-quasi-neutral region into the n-quasi-neutral region. This will lead to a current flow and minority carrier storage in both the n- and p-quasi-neutral regions of the diode. In this section, the minority carrier storage and transient behavior of a p–n junction diode are depicted.

Although in principle one can predict the transient behavior of minority carriers by solving the continuity equations, it is usually difficult to obtain an analytical solution by this approach. Fortunately, we can solve the problem more readily by using another approach, which will be discussed next.

The total injected minority carrier charge per unit area stored in the n-quasi-neutral region can be found by integrating the excess hole density distribution across the n-quasi-neutral region. For a long-base diode, this is given by

$$\begin{aligned} Q'_p &= q \int_{x_n}^{W_B} p'_n(x)\, dx \\ &= q \int_{x_n}^{W_B} p_{n0}(e^{qV/k_BT} - 1)\, e^{-(x-x_n)/L_p}\, dx \\ &= qL_p p_{n0}(e^{qV/k_BT} - 1) \end{aligned} \tag{11.66}$$

which shows that the minority carrier charge storage is proportional to both the minority carrier diffusion length and the minority carrier density at the depletion layer edge. The stored minority carrier charge given by Eq. (11.66) can be related to the hole injection current density by using Eq. (11.39) for the hole current density in the n-quasi-neutral region evaluated at $x = x_n$, namely

$$Q'_p = \left(\frac{L_p^2}{D_p}\right) J_p(x_n) = \tau_p J_p(x_n) \tag{11.67}$$

This equation shows that the minority carrier (i.e., holes) charge storage in the n-quasi-neutral base region is equal to the product of the hole lifetime and the hole current density. Thus, a long hole lifetime will result in more hole storage in the n-base region. This is expected since the injected holes can stay longer and diffuse deeper into the n-base region when the hole lifetime is long.

Similarly, the minority carrier storage in a short-base diode can be derived by substituting Eq. (11.43) into (11.66), and using Eq. (11.44) for the hole current density. Hence

$$\begin{aligned} Q'_p &= \frac{q(W_B - x_n) p_{n0}}{2} (e^{qV/k_BT} - 1) \\ &= \left[\frac{(W_B - x_n)^2}{2D_p}\right] J_p = \tau_{tr} J_p \end{aligned} \tag{11.68}$$

which shows that for a short-base diode the minority carrier storage is not dependent on the minority carrier lifetime, but instead varies linearly with the average transit time τ_{tr} across the n-base region. The term inside the square brackets of Eq. (11.68) denotes the average transit time for a hole to move through the n-quasi-neutral region.

Another important diode parameter under forward-bias conditions that is associated with the minority carrier storage in the quasi-neutral regions is diffusion capacitance. The diffusion capacitance per unit area for hole storage in the n-quasi-neutral region can be derived by employing the definition $C_d = dQ'_p/dV$, where Q'_p is given by Eqs. (11.67) and (11.68) for long- and short-base diodes, respectively. Thus, we obtain the diffusion capacitance due to hole charge storage in the n-base region as

$$C_d = \left(\frac{q^2 L_p p_{n0}}{k_B T}\right) \exp\left(\frac{qV}{k_B T}\right) \tag{11.69}$$

for the long-base diode, and as

$$C_d = \left(\frac{q^2 (W_B - x_n) p_{n0}}{2k_B T}\right) \exp\left(\frac{qV}{k_B T}\right) \tag{11.70}$$

for the short-base diode. It is seen that the diffusion capacitance is important only under forward-bias conditions, and is negligible under reverse-bias conditions in which transition capacitance becomes the dominant component.

The transient behavior of the minority carrier storage in a p–n junction diode is very important when the diode is used for switching applications, because the switching time of a p–n diode depends on the amount of stored charge which must be injected and removed from the quasi-neutral regions of the diode. For example, one may shorten the switching time by reducing the stored charge in the quasi-neutral regions of the diode. This can be achieved by either reducing the minority carrier lifetime or by limiting the forward current flow in the diode. For switching applications the forward- to reverse-bias transition must be nearly abrupt, and the transit time must be short. In a switching diode the turnoff time is limited by the speed with which the stored holes can be removed from the n-quasi-neutral base region. When a reverse-bias voltage is suddenly applied across a forward-biased junction, the current can be switched in the reverse direction quickly. This is due to the fact that the gradient near the edge of the depletion region can only make a small change in the number of stored holes in the n-quasi-neutral region. Figure 11.9a shows a qualitative sketch of the transient decay of the excess stored holes in a long base p–n diode. Figure 11.9b shows the basic switching circuit and Fig. 11.9c displays the transient response of the current from forward- to reverse-bias conditions. It is seen that the turnoff time constant t_{off} shown in Fig. 11.9c is the time required for the current to reach 10% of the initial reverse current, I_r. This turnoff time can be estimated by considering a forward-bias p^+–n junction diode. In this case, the charge of the stored excess holes in the n-quasi-neutral region is given by

$$Q'_p = qA \int_{x_n}^{W_B} p'_n(x)\, dx \tag{11.71}$$

where W_B is the n-base width, $p'_n(x)$ is the excess hole density in the n-base region, and A is the diode cross-section area. By integrating once the continuity equation for the excess hole density given by Eq. (6.50) from $x = x_n$ to $x = W_B$ and using Eq. (11.71), we obtain

$$I_p(x_n) - I_p(W_B) = \frac{dQ_s}{dt} + \frac{Q_s}{\tau_p} \tag{11.72}$$

Equation (11.72) is known as the charge control equation for a long-base diode. We note that $I_p(W_B)$ in this case can be set equal to zero. Thus, the steady-state forward bias current can be obtained by setting $dQ_s/dt = 0$ in Eq. (11.72), which yields

$$I_f = I_p(x_n) = \frac{Q_{sf}}{\tau_p} \tag{11.73}$$

or

$$Q_{sf} = I_f \tau_p \tag{11.74}$$

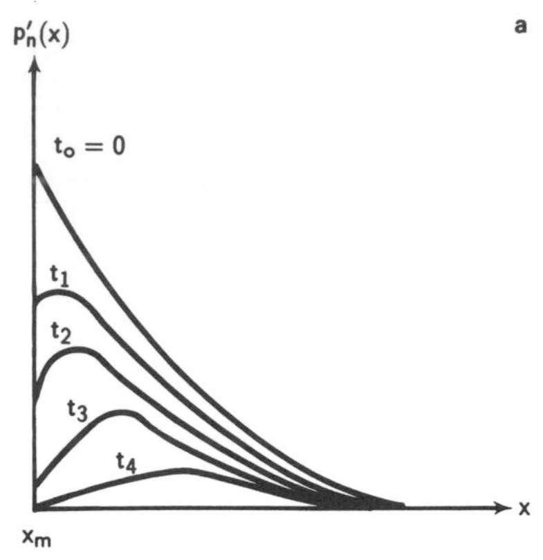

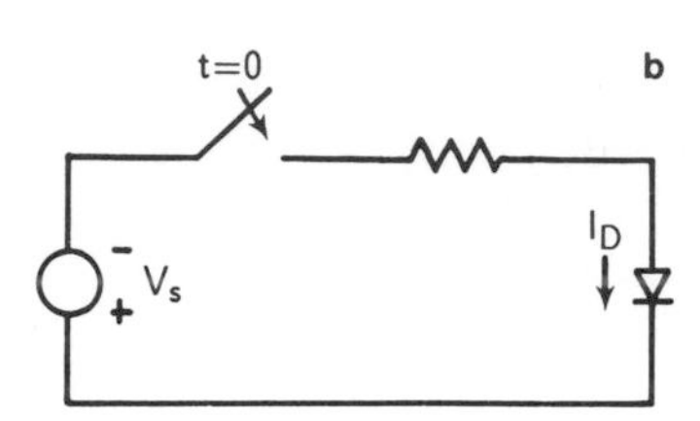

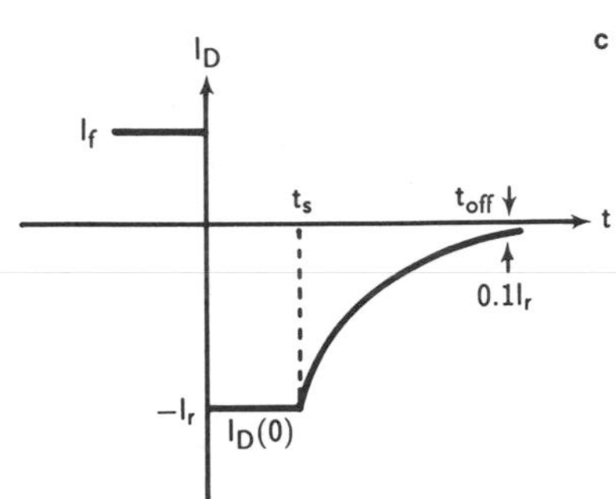

FIGURE 11.9. Transient behavior of a p–n junction diode: (a) transient decay of the minority hole density, (b) basic circuit diagram, and (c) transient response of the current from forward- to reverse-bias conditions.

If the reverse-bias current in the diode during turnoff period is given by I_r, then Eq. (11.72) becomes

$$-I_r = \frac{dQ_s}{dt} + \frac{Q_s}{\tau_p} \tag{11.75}$$

With the aid of Eq. (11.73) as the initial condition, the solution of Eq. (11.75) is the time-dependent storage charge equation, which reads

$$Q_s(t) = \tau_p[-I_r + (I_f + I_r)\, e^{-t/\tau_p}] \tag{11.76}$$

The turnoff time t_{off}, which is the length of time required to remove the minority holes from the n-quasi-neutral region in order to reduce Q_s to zero, can be obtained by solving Eq. (11.76). The result yields

$$t_{off} = \tau_p \ln\left(1 + \frac{I_f}{I_r}\right) \tag{11.77}$$

which shows that the turnoff time or switching time is directly proportional to the minority carrier lifetime and the ratio of the forward current to the reverse current in the diode. Thus, it is evident that the switching speed of a p–n junction diode can be enhanced by reducing the lifetime of the minority carriers. This is why gold impurity is often used as an effective mid-gap recombination center in silicon switching diodes and transistors for reducing the minority carrier lifetime and increasing the switching speed in these devices. Another approach, such as adding a Schottky barrier diode to the collector-base junction of a bipolar junction transistor to form a Schottky clamped bipolar junction transistor, has also been used to reduce the minority carrier storage time in a switching transistor.

11.7. ZENER AND AVALANCHE BREAKDOWNS

In this section, the junction breakdown phenomena in a p–n junction diode will be discussed. As shown in the previous section, the depletion layer width and the maximum electric field in the space-charge region of a p–n junction will increase with increasing reverse-bias voltage. The increase of maximum field strength in the depletion region will eventually lead to junction breakdown phenomena commonly observed in a p–n junction diode. There are two types of junction breakdown commonly observed in a p–n diode: Zener breakdown and avalanche breakdown. Zener breakdown occurs when valence electrons gain sufficient energy from the electric field and then tunnel through the forbidden gap into the conduction band. In this case an electron–hole pair is created by the high reverse-bias voltage, which results in current flow. Avalanche breakdown is different from Zener breakdown in that the electric field is usually much higher. In avalanche breakdown, electrons (or holes) gain sufficient energy from the electric field and then engage in collisions. Between collisions of these high-energy electrons (or holes) they break the covalent bonds in the lattice and thus create more elecron–hole pairs during the collisions. In this process, every electron (or hole) interacting with the lattice will create additional electrons (or holes), and all these electrons can participate in further avalanche collisions. This avalanche process will eventually lead to a sudden multiplication of carriers in the junction space-charge region where the maximum electric field becomes large enough to cause avalanche multiplication. We note that avalanche multiplication (or impact ionization) is probably the most important mechanism in junction breakdown, since the avalanche breakdown voltage imposes an upper limit on the reverse I–V characteristics of a p–n diode as well as other bipolar junction devices. Both the avalanche breakdown and Zener breakdown are nondestructive processes. Values of the breakdown voltage for each of these two processes depend on the junction structure and the doping concentraton of the diode. Figure 11.10 shows critical electric fields for avalanche and Zener breakdowns as a function of doping concentration in a silicon crystal.[2–3] Both of these breakdown phenomena are very important in practical device applications. The physical mechanisms and mathematical derivation of avalanche and Zener breakdowns are discussed next.

Avalanche multiplication (or impact ionization) is the most important mechanism in junction breakdown phenomena, because the avalanche breakdown voltage determines the maximum reverse-bias voltage which can be applied to a p–n junction without destroying the device. For example, the avalanche multiplication mechanism has been used extensively to obtain the internal current gain in an avalanche photodiode or to generate microwave power in an IMPATT diode.

The basic ionization integral which determines the breakdown condition can be derived as follows. Let us consider the case in which impact ionization is initiated by electrons (see Fig. 11.11). The electron current $I_n(0)$ enters on the left-hand side (p-region) of the depletion layer region of width equal to W, at $x = 0$. If the electric field in the depletion region is large enough (i.e., $\mathscr{E} \geq \mathscr{E}_c$), then electron–hole pairs will be created by impact ionization, and the electron current I_n, will increase with distance through the depletion region, reaching a maximum value of $I_n(W) = M_n I_n(0)$ at $x = W$. Similarly, the hole current $I_p(x)$ will increase from

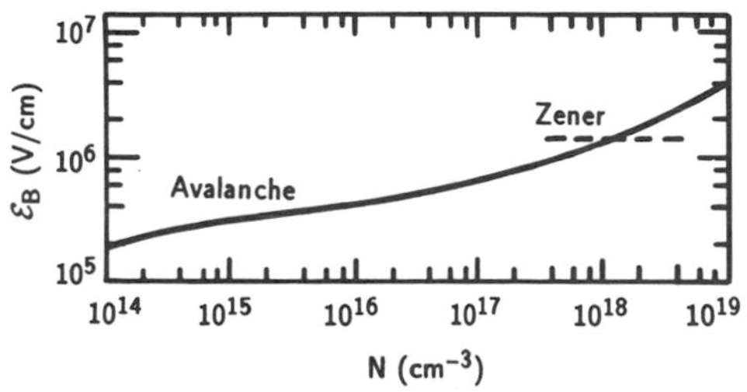

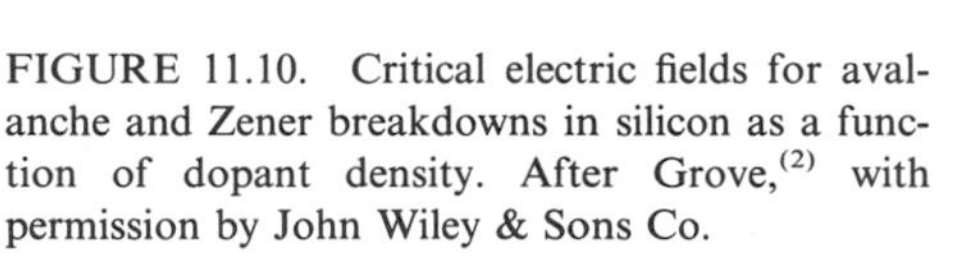
FIGURE 11.10. Critical electric fields for avalanche and Zener breakdowns in silicon as a function of dopant density. After Grove,[2] with permission by John Wiley & Sons Co.

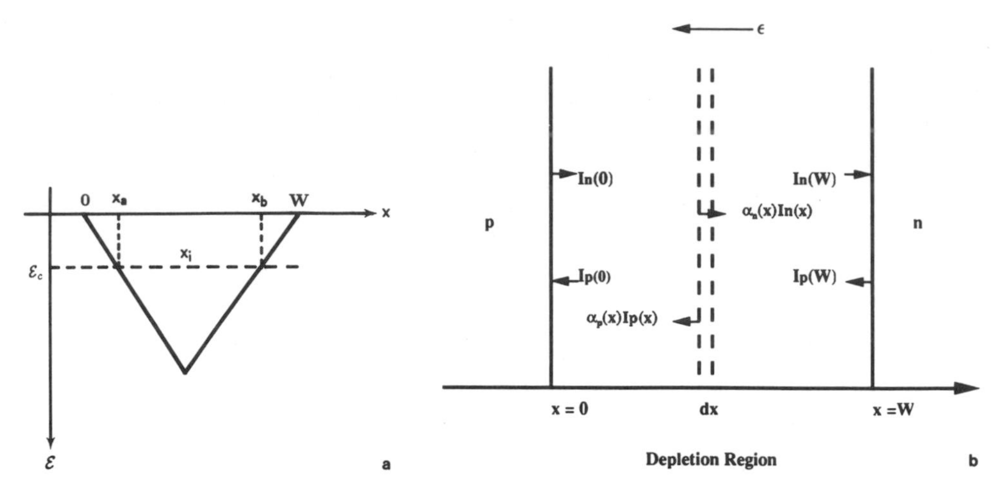

FIGURE 11.11. Schematic representation of (a) the electric field distribution and (b) the avalanche process in the space-charge region showing that ionization occurs in the high-field portion of the space-charge region (i.e., x_i).

$x = W$ to $x = 0$ as it moves through the depletion region from right to left in the junction space-charge region. Figure 11.11b shows current flows due to the avalanche multiplication process for electrons and holes in the depletion region of the diode under a large reverse-bias condition. The total current, $I = I_p(x) + I_n(x)$ is constant under steady-state conditions. The incremental electron current at x is equal to the number of electron–hole pairs generated per second in an interval dx, which is given by

$$d\left(\frac{I_n}{q}\right) = \left(\frac{I_n}{q}\right)\alpha_n\,dx + \left(\frac{I_p}{q}\right)\alpha_p\,dx \tag{11.78}$$

or

$$\frac{dI_n}{dx} - (\alpha_n - \alpha_p)I_n = \alpha_p(I_n + I_p) = \alpha_p I \tag{11.79}$$

where α_n and α_p denote the electron and hole ionization coefficients (cm^{-1}), respectively. If we introduce boundary conditions, that $I_n(0) = I_{n0}$ at $x = 0$ and $I = I_n(W) = M_n I_{n0}$ at $x = W$, then the solution of Eq. (11.79) can be expressed by

$$I_n(x) = \frac{I\{1/M_n + \int_0^x \alpha_p \exp[-\int_0^x (\alpha_n - \alpha_p)\,du]\,dx\}}{\exp[-\int_0^x (\alpha_n - \alpha_p)\,du]} \tag{11.80}$$

where M_n is the multiplication factor of electrons, defined by

$$M_n = \frac{I_n(W)}{I_n(0)} \tag{11.81}$$

Now on solving Eqs. (11.80) and (11.81) we obtain the electron multiplication factor, which reads

$$M_n = \frac{1}{\exp(-\int_0^W (\alpha_n - \alpha_p)\,dx] - \int_0^W \alpha_p \exp[-\int_0^x (\alpha_n - \alpha_p)\,du]\,dx} \tag{11.82}$$

or

$$M_n = \frac{1}{1 - \int_0^W \alpha_n \exp[-\int_0^x (\alpha_n - \alpha_p)\,du]\,dx} \tag{11.83}$$

where Eq. (11.83) is obtained by using the relation

$$\exp\left[-\int_0^W (\alpha_n - \alpha_p)\,dx\right] = 1 - \int_0^W (\alpha_n - \alpha_p) \exp\left[-\int_0^x (\alpha_n - \alpha_p)\,du\right] dx \tag{11.84}$$

The avalanche breakdown voltage is defined by the threshold voltage at which the multiplication factor M_n approaches infinity. Thus, if breakdown is initiated by electrons, then the breakdown condition is obtained from Eq. (11.83) when $M_n \to \infty$, which yields

$$\int_0^W \alpha_n \exp\left[-\int_0^x (\alpha_n - \alpha_p)\,du\right] dx = 1 \tag{11.85}$$

Similarly, if avalanche multiplication is initiated by holes instead of electrons, then the ionization integral given by Eq. (11.83) can be written as

$$\int_0^W \alpha_p \exp\left[-\int_0^x (\alpha_p - \alpha_n)\,du\right] dx = 1 \tag{11.86}$$

Equations (11.85) and (11.86) should produce the same breakdown condition within the depletion region of the diode regardless of whether the avalanche process is initiated by electrons or holes. For semiconductors (e.g., GaP) with equal ionization coefficients (i.e., $\alpha_n = \alpha_p = \alpha$), Eqs. (11.85) and (11.86) can be simplified to

$$\int_0^W \alpha\,dx = 1 \tag{11.87}$$

If the ionization coefficients for both electrons and holes are independent of position in the depletion region, then Eq. (11.82) becomes

$$M_n = \frac{(1 - \alpha_n/\alpha_p)\exp(\alpha_n - \alpha_p)W}{(1 - \exp(\alpha_n - \alpha_p)W)} \tag{11.88}$$

In general, the ionization coefficient is a strong function of the electric field since the energy necessary for an ionizing collision is imparted to the carriers by the electric field. The field-dependent ionization coefficient can be expressed by the empirical formula

$$\alpha = A \exp\left(-\frac{B}{\mathscr{E}}\right) \tag{11.89}$$

where A and B are material constants; $\mathscr{E}$ is the electric field, which can be calculated for each material from the solution of Poisson's equation. For Si, $A = 9 \times 10^5\ \text{cm}^{-1}$ and $B = 18 \times 10^6\ \text{V/cm}$. It is seen that not only does the ionization coefficient vary with the electric field (and hence the position in the depletion region), but the width of the depletion region will also change with the applied bias voltage. It is therefore usually difficult to evaluate the avalanche multiplication factor M from Eqs. (11.85) or (11.86). Instead, an empirical formula for M given by

$$M = \frac{1}{[1 - (V_R/V_B)^n]} \qquad (2 < n < 6) \tag{11.90}$$

is often used. Here, V_R denotes the applied reverse-bias voltage, and V_B is the breakdown voltage defined by

$$V_B = \frac{\mathscr{E}_m W}{2} = \frac{\varepsilon_0 \varepsilon_s \mathscr{E}_m^2}{2qN_B} \tag{11.91}$$

for a one-sided abrupt junction diode, and

$$V_B = \frac{2\mathscr{E}_m W}{3} = \left(\frac{4\mathscr{E}_m^{3/2}}{3}\right)\left(\frac{2\varepsilon_0\varepsilon_s}{qa}\right)^{1/2} \tag{11.92}$$

for a linearly graded junction diode. We note that N_B is the background dopant density in the lightly doped region of the junction, a is the impurity gradient, and $\mathscr{E}_m$ is the maximum electric field in the junction space-charge region. An approximate universal expression for the breakdown voltage as a function of energy bandgap and dopant density for an abrupt p–n junction diode is given by

$$V_B \cong 60\left(\frac{E_g}{1.1}\right)^{1.5}\left(\frac{N_B}{10^{16}}\right)^{-0.75} \tag{11.93}$$

where E_g is the bandgap energy in eV. For a linearly graded junction diode, the breakdown voltage is given by

$$V_B \cong 60\left(\frac{E_g}{1.1}\right)^{1.2}\left(\frac{a}{3 \times 10^{20}}\right)^{-0.4} \tag{11.94}$$

Using Eq. (11.93), the breakdown voltage V_B at room temperature for a silicon p^+/n step-junction diode with $N_d = 10^{16}\ \text{cm}^{-3}$ is found from Eq. (11.93) to be 60 V, and for a GaAs p^+/n diode with similar dopant density the breakdown voltage V_B is around 75 V. Figure 11.12 shows avalanche breakdown voltage versus impurity density for a one-sided abrupt junction and a linearly graded junction diode formed on Ge, Si, GaAs, and GaP, respectively.[3] The dashed line indicates the maximum doping density beyond which the tunneling mechanism will dominate the voltage breakdown characteristics. The Zener breakdown phenomenon in a p–n junction dode is discussed next.

As shown in Fig. 11.10, when the dopant density increases, the width of the space-charge region will decrease and the critical field at which avalanche breakdown occurs will also increase. At very high dopant density, the electric field required for avalanche breakdown to occur exceeds the field strength necessary for Zener breakdown to take place, and hence the latter becomes more likely to occur. To explain the Zener breakdown mechanism, Fig. 11.13a and b show the energy band diagram under reverse-bias conditions and the triangle potential barrier for a heavily doped p^+–n^+ junction diode, respectively. The probability of tunneling

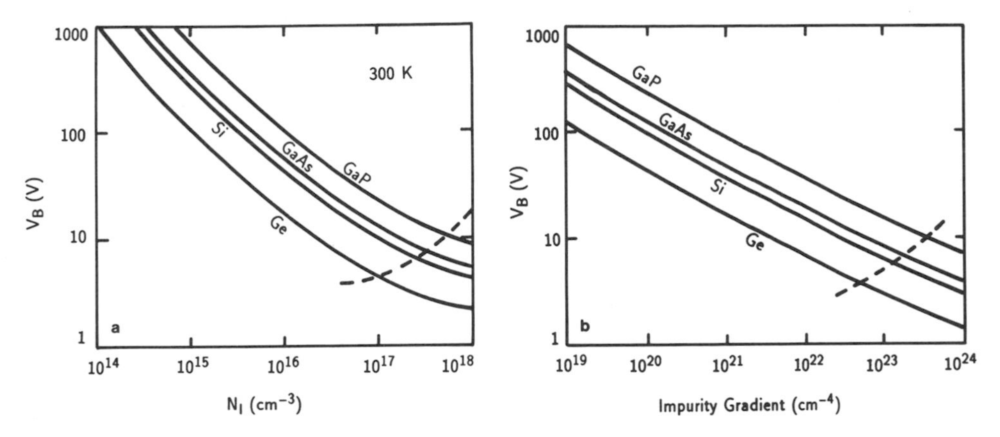

FIGURE 11.12. Avalanche breakdown voltage versus impurity density for (a) a one-sided abrupt junction and (b) a linearly graded diode in Ge, Si, GaAs, and GaP. The dashed line indicates the maximum doping density beyond which the tunneling mechanism will dominate the voltage breakdown characteristics. After Sze and Gibbons,[3] by permission.

from the valence band to the conduction band under high-field conditions may be approximated by considering tunneling through a triangular potential barrier. The height of the energy barrier, $E_B(x)$, decreases linearly from E_g at $x = 0$ to 0 at $x = L$, and the average field is $\mathscr{E} = E_g/qL$. The probability of tunneling, T_x, can be derived from the WKB approximation by using the barrier height in the equation, which reads

$$T_z \cong \exp\left[-2\int_0^L (2m^* E_b/\hbar^2)^{1/2}\, dx\right]$$

$$= \exp(-B/\mathscr{E}) = \exp(-qBL/E_g) \tag{11.95}$$

where

$$B = \frac{4(2m^*)^{1/2} E_g^{3/2}}{3q\hbar} \tag{11.96}$$

Therefore, the Zener tunneling probability decreases exponentially with decreasing electric field or increasing tunneling distance. If n is the number of valence electrons tunneling through

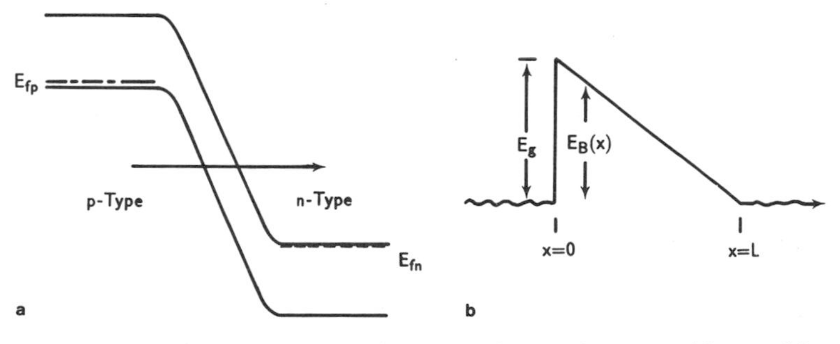

FIGURE 11.13. (a) Energy band diagram of a Zener diode under reverse-bias conditions. (b) The probability of tunneling across the junction is represented by tunneling through a triangle potential barrier.

the barrier, and v_{th} is the thermal velocity of electrons, then the tunneling current can be written as

$$I_t = Aqnv_{th}T_z \tag{11.97}$$

where A is the cross-section area of the diode and T_z is given by Eq. (11.95). Equations (11.95) through (11.97) enable one to estimate the tunneling probability, the tunneling distance, and the electric field for a given tunneling current. It is seen that diodes which exhibit Zener breakdown generally have a lower breakdown voltage than that of avalanche diodes. For example, in a silicon p–n junction diode with doping densities on both sides of the junction greater than 10^{18} cm^{-3}, Zener breakdown will occur at a voltage less than -6 V while avalanche breakdown becomes dominant at a much higher reverse-bias voltage.

11.8. TUNNEL DIODE

An interesting p–n junction structure, known as the tunnel diode or Esaki diode, will be discussed in this section. The tunnel diode was discovered by L. Esaki in 1958, when he observed a negative differential resistance and microwave oscillation in a heavily doped germanium p^+–n^+ junction diode under forward-bias conditions. The current flow in a forward-bias tunnel diode may be attributed to quantum-mechanical tunneling of charged carriers through the thin potential barrier across the junction.

A tunnel diode is formed when the densities of shallow donor and acceptor impurities in both the p^+- and n^+-regions of the junction are in the mid-10^{19} cm^{-3} range. Figure 11.14a shows the energy band diagram of a tunnel diode in the equilibrium condition ($V = 0$), Fig.

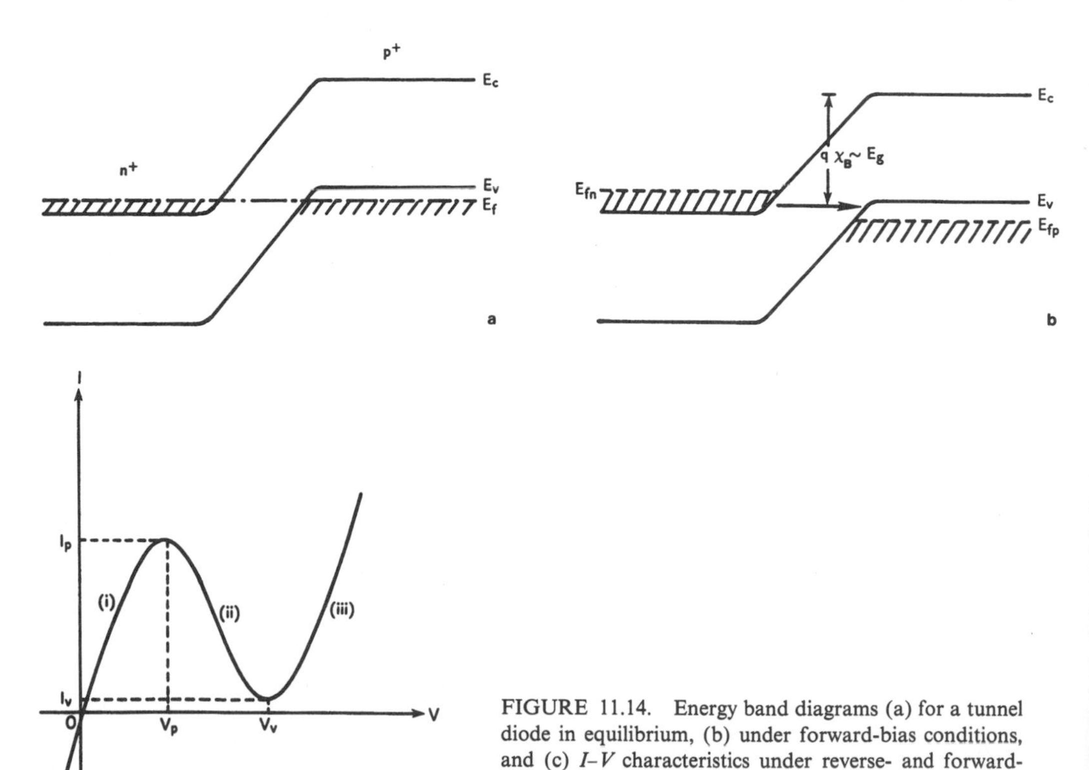

FIGURE 11.14. Energy band diagrams (a) for a tunnel diode in equilibrium, (b) under forward-bias conditions, and (c) I–V characteristics under reverse- and forward-bias conditions.

11.14b presents the energy band diagram under a small forward-bias voltage with a triangle potential barrier of height $q\chi_B \sim E_g$, and Fig. 11.14c displays the forward current–voltage (I–V) characteristics of a tunnel diode. Due to the high doping densities on both sides of the junction, the Fermi levels on either side of the junction are located a few k_BT inside the conduction and valence bands, as shown in Fig. 11.14a. For a tunnel diode, the depletion layer width at zero bias condition is of the order of 50 to 100 Å, which is considerably less than that of a regular p–n junction diode.

The electron tunneling process from the valence band to the conduction band, which is dominated in a tunnel diode under a forward-bias condition, can be explained by using the quantum-mechanical tunneling mechanism. As shown in Fig. 11.14a, at $V = 0$ and $T = 0$ K, the states above the Fermi-level in the conduction band of the n^+-region are empty, and the states below the Fermi level in the valence band of the p^+-region are completely filled. Therefore, under this condition no tunneling of electrons from the conduction band to the valence band will take place, and the tunneling current is equal to zero. This situation will usually prevail even at room temperature. Now, if a forward-bias voltage is applied to the tunnel diode as in Fig. 11.14b, then the quasi-Fermi level in the n-region will move above the quasi-Fermi level in the p^+-region. As a result, it is possible for some of the electrons in the conduction band of the n^+-region to tunnel through the thin potential barrier across the junction into the empty states in the valence band of the p^+-region. The tunneling probability in this case will depend on the thickness of the potential barrier across the junction, which will increase with decreasing barrier thickness.

A typical current–voltage characteristic curve for a tunnel diode under a forward-bias condition is illustrated in Fig. 11.14c, where I_p and V_p denote the peak current and peak voltage, while I_v and V_v are the valley current and valley voltage, respectively. The current–voltage (I–V) characteristics under a forward-bias condition may be divided into three regions: (i) the low-bias (i.e., $V < V_p$) regime, where the current increases monotonically with voltage to a peak value I_p at voltage V_p, (ii) the intermediate-bias (i.e., $V_p < V < V_v$) regime, where the current decreases with increasing voltage to a minimum current I_v at voltage V_v, and (iii) the high bias regime (i.e., $V > V_v$), where the current increases exponentially with applied voltage. In general, the current components contributing to the forward I–V characteristics of a tunnel diode shown in Fig. 11.14c are dominated by the band-to-band tunneling current, the excess current, and the diffusion current. For $V < V_v$ the diffusion current component becomes the dominant current component. In the negative resistance regime [i.e., regime (ii)] the current component is dominated by band-to-band tunneling through the thin triangle potential barrier of the junction.

Tunneling mechanisms and physical insight into a tunnel diode can be understood with the aid of a simple model using the triangle potential barrier shown in Fig. 11.14b under a forward-bias condition. If the barrier height of the triangle potential barrier is assumed equal to the bandgap energy (i.e., $\sim E_g$), and n is the density of electrons in the conduction band available for tunneling, then using the WKB approximation the tunneling probability of electrons across the potential barrier may be written as

$$T_t = \exp\left(-\frac{4\sqrt{2qm_e^*}\, E_g^{3/2}}{3\hbar\mathscr{E}} \right) \tag{11.98}$$

where $\mathscr{E} = E_g/W$ is the electric field across the depletion region of the tunnel diode; W is the depletion layer width. It is seen in Fig. 11.14b that if the states in the valence band of the p^+-region are mostly empty, then the tunneling current due to band-to-band tunneling from the n^+- to the p^+-region is given by

$$I_t = Aqv_{th}nT_t \tag{11.99}$$

where A is the cross-section area of the tunnel diode, v_{th} is the thermal velocity of the tunneling electrons, and T_t is the tunneling probability given by Eq. (11.98).

The tunneling current given by Eq. (11.99) is relatively insensitive to temperature. For example, the peak current of a typical germanium tunnel diode varies by only ±10% over a temperature range from −50 to 100°C. Since the tunneling time across a tunnel diode is very short, the switching speed of a tunnel diode is usually very fast. A wide variety of device and circuit applications using tunnel diodes, including microwave oscillators, multivibrators, low-noise microwave amplifiers, and high-speed logic circuits, have been reported in the literature. In addition to microwave and digital applications, a tunnel diode can also be used as a test vehicle in tunneling spectroscopy for studying fundamental physical parameters such as electron energy states in a solid and excitation modes in a p–n junction device. Finally, it should be noted that since a tunnel diode is a two-terminal device, it is difficult to incorporate such a device structure in many integrated circuit applications.

11.9. p–n HETEROJUNCTION DIODES

A p–n heterojunction structure is formed when two types of semiconductor with opposite dopant impurities and different bandgaps are brought into intimate contact. Typical examples of a p–n heterojunction diode are the n-Ge on p-GaAs and the p-AlGaAs on n-GaAs heterostructures. The heterojunction structures have many important applications ranging from room-temperature injection lasers, light-emitting diodes (LEDs), photodetectors, solar cells, modulation doped field effect transistors (MODFETs or HEMTs), and heterojunction bipolar transistors (HBTs). With recent advances in MOCVD and MBE epitaxial layer growth techniques for III–V compounds and Ge/Si systems, it is now possible to grow high-quality heterojunction structures as well as periodic heterojunction layer structures with layer thicknesses of 100 Å or less for superlattice structures, and greater than 100 Å for multiquantum well (MQW) structures.

Figure 11.15a shows the energy band diagram for an isolated n-Ge and p-GaAs semiconductor in thermal equilibrium, and Fig. 11.15b shows the energy band diagram for an ideal

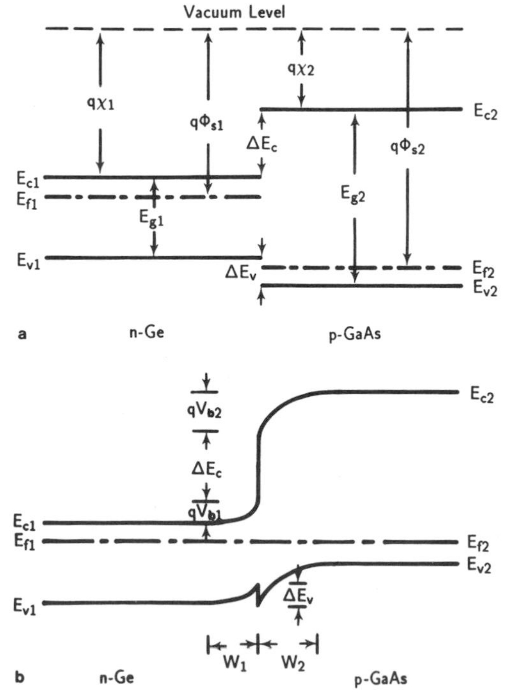

FIGURE 11.15. Energy band diagrams for (a) an isolated n-Ge and p-GaAs semiconductor in equilibrium, and (b) n-Ge and p-GaAs brought into intimate contact to form an n–p heterojunction diode.

n-Ge/p-GaAs heterojunction diode. The energy band diagrams for an ideal n-Ge/n-GaAs, p-Ge/n-GaAs, and p-Ge/p-GaAs heterostructure without considering interface states are illustrated in Figs. 11.16a, b, and c, respectively. Although the energy bandgaps (i.e., $E_g =$ 0.67 eV for Ge and 1.43 eV for GaAs) and dielectric constants (i.e., 16 for Ge and 12 for GaAs) for Ge and GaAs are quite different, the lattice constants for both materials are very similar (5.658 Å for Ge and 5.654 Å for GaAs). As a result, high-quality Ge/GaAs heterojunction structures can be formed in this material system. As shown in Fig. 11.15b and Fig. 11.16 the energy band diagram for a heterojunction diode is considerably more complicated than that of a p–n homojunction due to the presence of energy band discontinuities in the conduction band (ΔE_c) and valence band (ΔE_v) at the metallurgical junction of the two materials. In Fig. 11.15b, subscripts 1 and 2 refer to Ge and GaAs, respectively; the energy discontinuity step arises from the bandgap and work function difference in the two semiconductors. The difference in energy at the conduction band edge is ΔE_c and that in the valence band edge ΔE_v. Based on the Anderson model, the conduction and valence band discontinuities

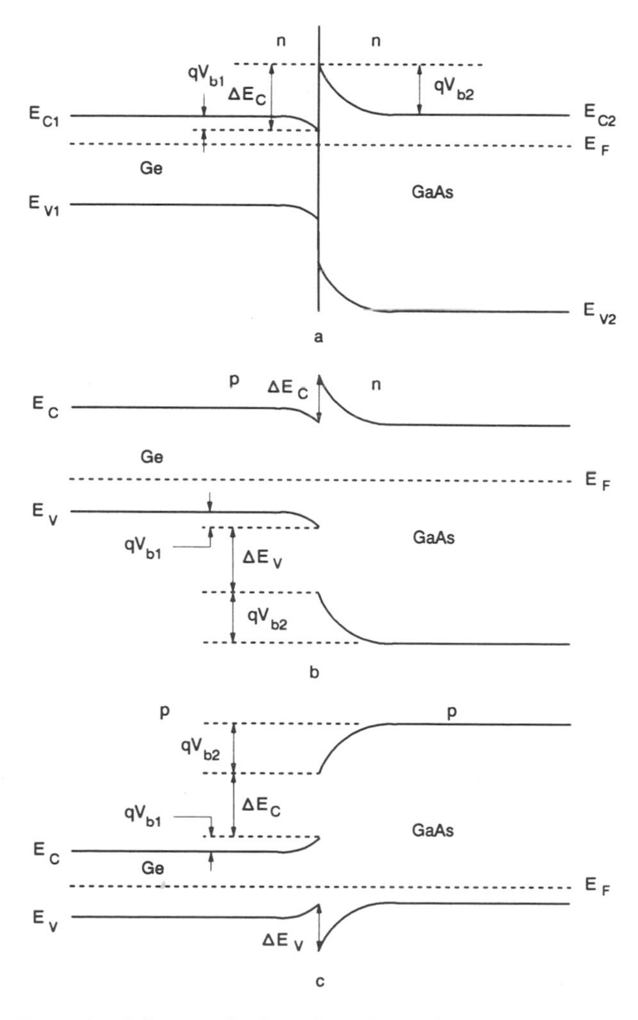

FIGURE 11.16. Energy band diagrams for (a) n-Ge/n-GaAs, (b) p-Ge/n-GaAs, and (c) p-Ge/p-GaAs heterojunction diodes in thermal equilibrium.

(ΔE_c and ΔE_v) can be obtained from the energy band diagram in Fig. 11.15a and is given by

$$\Delta E_c = q(\chi_1 - \chi_2) \tag{11.100}$$

and

$$\Delta E_v = (E_{g1} - E_{g2}) - \Delta E_c \tag{11.101}$$

which shows that the conduction band discontinuity is equal to the difference in the electron affinity of these two materials, and the valence band discontinuity is equal to the bandgap difference minus the conduction band discontinuity. When these two semiconductors are brought into intimate contact, the Fermi level (or chemical potential) must line up in equilibrium. As a result, electrons on the n-Ge side are transferred to the p-GaAs side, and holes on the p-GaAs side are transferred to the n-Ge side until the equilibrium condition is reached (i.e., the Fermi energy is lined up across the heterojunction). As in the case of a p–n homojunction, the redistribution of charges creates a depletion region across each side of the junction, and the energy band diagram for an ideal n-Ge/p-GaAs heterojunction diode in equilibrium is shown in Fig. 11.15b. This figure shows the discontinuities in the conduction band and valence band at the Ge/GaAs interface. The band bending across the depletion region indicates that a built-in potential exists on both sides of the junction. The total built-in potential, V_{bi}, is equal to the sum of the built-in potential on each side of the junction, i.e.,

$$V_{bi} = V_{b1} + V_{b2} \tag{11.102}$$

where V_{b1} and V_{b2} are the band bending potentials in p-Ge and n-GaAs, respectively. We note that the discontinuity in the electrostatic field at the interface is due to the difference in dielectric constants between these two materials. The depletion approximation enables V_{b1} and V_{b2} in the n-Ge and p-GaAs regions to be expressed respectively by

$$V_{b2} = \frac{\varepsilon_2 N_{D1}}{\varepsilon_1 N_{D1} + \varepsilon_2 N_{A2}} V_{bi} \tag{11.103}$$

and

$$V_{b1} = \frac{\varepsilon_2 N_{A2}}{\varepsilon_1 N_{D1} + \varepsilon_2 N_{A2}} V_{bi} \tag{11.104}$$

The solution of Poisson's equation (assuming a Schottky barrier exists across the heterointerface) yields the depletion layer width on either side of the step heterojunction diode under applied bias conditons in the form

$$W_1 = \left[\frac{2N_{A2}\varepsilon_1\varepsilon_2(V_{bi} - V)}{qN_{D1}(\varepsilon_1 N_{D1} + \varepsilon_2 N_{A2})}\right]^{1/2} \tag{11.105}$$

and

$$W_2 = \left[\frac{2_{N_{D1}}\varepsilon_1\varepsilon_2(V_{bi} - V)}{qN_{A2}(\varepsilon_1 N_{D1} + \varepsilon_2 N_{A2})}\right]^{1/2} \tag{11.106}$$

Thus, the total depletion layer width of the heterojunction is given by

$$\begin{aligned} W_d &= W_1 + W_2 \\ &= \left[\frac{2\varepsilon_1\varepsilon_2(V_{bi} - V)(N_{A2}^2 + N_{D1}^2)}{q(\varepsilon_1 N_{D1} + \varepsilon_2 N_{A2})N_{D1}N_{A2}}\right]^{1/2} \end{aligned} \tag{11.107}$$

Equations (11.105) and (11.106) are derived from Poisson's equation in the space-charge region of the heterojunction diode. The two boundary conditions are

$$W_1 N_{D1} = W_2 N_{A2} \tag{11.108}$$

and

$$\varepsilon_1 \mathscr{E}_1 = \varepsilon_2 \mathscr{E}_2 \tag{11.109}$$

where ε_i and $\mathscr{E}_i$ $(i = 1, 2)$ denote the dielectric constants and electric fields in regions 1 and 2, respectively. The relative voltage drop across regions 1 and 2 of the semiconductors is given by

$$\frac{(V_{b1} - V_1)}{(V_{b2} - V_2)} = \frac{N_{A2}\varepsilon_2}{N_{D1}\varepsilon_1} \tag{11.110}$$

The transition capacitance per unit area for the heterojunction diode can be derived from Eq. (11.107), and the result yields

$$C_j = \left[\frac{qN_{D1}N_{A2}\varepsilon_1\varepsilon_2}{2(\varepsilon_1 N_{D1} + \varepsilon_2 N_{A2})(V_{bi} - V)}\right]^{1/2} \tag{11.111}$$

The current–voltage characteristics for an n-Ge/p-GaAs heterojunction diode can be derived by using the energy band diagram in Fig. 11.15b and the thermionic emission theory for a Schottky barrier diode depicted in Chapter 10. Since the relative magnitudes of the current components in a heterojunction are determined by the potential barriers involved, for the n-Ge/p-GaAs heterojunction diode shown in Fig. 11.15b the hole curent from p-GaAs to n-Ge is expected to dominate the current flow because of the low potential barrier ($=V_{b2}$) for hole injection and the high potential barrier ($=V_{b1} + \Delta E_c + V_{b2}$) for electron injection. Therefore, to derive the current–voltage relationship for the n–p heterojunction diode in Fig. 11.15b only the hole current need be considered. At zero bias voltage, the barrier-to-hole flow from p-GaAs to n-Ge is qV_{b2} and in the opposite direction it is $(\Delta E_v - qV_{b1})$. Under thermal equilibrium condition, the two oppositely directed fluxes of holes must be equal since the net current flow is zero. Therefore we can write

$$A_1 \exp[-(\Delta E_v - qV_{b1}/k_B T] = A_2 \exp(-qV_{b2}/k_B T) \tag{11.112}$$

where constants A_1 and A_2 depend on the doping levels and carrier effective masses in the diode.

If we apply a forward-bias voltage V_a across the junction, then the portions of the voltage drops on the two sides of the junction are determined by the relative doping densities and dielectric constants of the materials, and are given by

$$V_2 = K_2 V_a \qquad \text{where } K_2 = \frac{N_{D1}\varepsilon_1}{N_{D1}\varepsilon_1 + N_{A2}\varepsilon_2} \tag{11.113}$$

and

$$V_1 = K_1 V_a \qquad \text{where } K_1 = 1 - K_2 \tag{11.114}$$

The energy barriers are now $(qV_{b2} - V_2)$ on the p-GaAs side and $(\Delta E_v - qV_{b1} - V_1)$ on the n-Ge side of the junction. By using Eq. (11.112), the net hole flux from right to left under forward-bias conditions can be expressed by

$$\phi_p = A_1 \exp(-qV_{b2}/k_B T)[\exp(qV_2/k_B T) - \exp(-qV_1/k_B T)] \tag{11.115}$$

Therefore, if the conduction mechanism is governed by thermionic emission, the current density due to hole injection from p-GaAs to n-Ge for an n–p heterojunction diode shown in Fig. 11.15b is of a similar form to that given by Eq. (11.115), and can be written as

$$J_p = A \exp(-qV_{b2}/k_BT)[\exp(qV_2/k_BT) - \exp(-qV_1/k_BT)]$$

$$\simeq J_0\left(1 - \frac{V_a}{V_{bi}}\right)[\exp(qV_a/k_BT) - 1] \qquad (11.116)$$

where

$$J_0 = \left(\frac{qA^*TV_{bi}}{k_B}\right) e^{-qV_{bi}/k_BT} \qquad (11.117)$$

It is seen that Eq. (11.116) is obtained by using Eq. (11.102) and employing the approximation $e^{q(V_{b1}-V_1)/k_BT} \simeq (q/k_BT)(V_{bi} - V_a)$, while the total applied voltage V_a is equal to the sum of V_1 and V_2. From Eq. (11.116), the current–voltage relationship for an n–p heterojunction diode is somewhat different from that of a metal-semiconductor Schottky diode. The main difference is that the reverse saturation current density is not a constant, but increases linearly with the applied reverse-bias voltage for $V_a \gg V_{bi}$. Under a forward-bias condition, as in the case of a Schottky barrier diode, the current–voltage relationship can be approximated by an exponential dependence of the form e^{-qV_a/nk_BT}, where n is the diode ideality factor. Practical applications of heterojunction structures for a wide variety of devices such as solar cells, LEDs, laser diodes, photodetectors, HEMTs, and HBTs will be discussed further in Chapters 12 and 15. Finally, it should be mentioned that multilayer heterojunction structures such as superlattices and quantum wells as well as modulation-doped heterostructures grown by MBE and MOCVD techniques have been extensively studied for a wide variety of applications. Figure 11.17 shows (a) the energy band diagram for a modulation-doped $Al_{0.3}Ga_{0.7}As$/GaAs heterostructure and a single-period $Al_{0.3}Ga_{0.7}As$/GaAs heterostructure, (b) a comparison of the electron mobilities as a function of temperature in the undoped GaAs quantum well and in several bulk GaAs samples of different doping concentrations, and (c) a comparison of electron mobilities versus temperature in the triangle potential well of the undoped GaAs of a single-period modulation-doped AlGaAs/GaAs heterostructure and in bulk GaAs doped to 10^{17} cm^{-3}. Using a modulation-doping technique, the electron mobility in the undoped GaAs quantum well can be greatly enhanced since the impurity scattering due to ionized donor impurities in the AlGaAs layer can be eliminated in the undoped GaAs quantum well. Typical layer thickness for modulation-doped heterostructure devices is around 100 Å. The modulation-doping technique has been widely used in the heterojunction field-effect transistors (HEMTs or MODFETs) for high-speed device applications. Other applications using superlattice and quantum well heterostructures include resonant tunneling devices and quantum well IR detectors as well as laser diodes and modulators. These devices will be discussed further in Chapters 12 and 15.

11.10. JUNCTION FIELD-EFFECT TRANSISTORS

The junction field-effect transistor (JFET) is a three-terminal device which consists of a source, a gate, and a drain electrode. In a JFET the lateral current flow between the source and drain electrodes is controlled by the applied vertical electric field via a controlled gate, which is formed by a p^+–n junction. Figure 11.18 shows the basic structure of an n-channel junction field-effect transistor formed on a lightly doped n-type epilayer grown on a p-type substrate. The heavily-doped n^+ source and drain regions as well as the p^+-gate region may

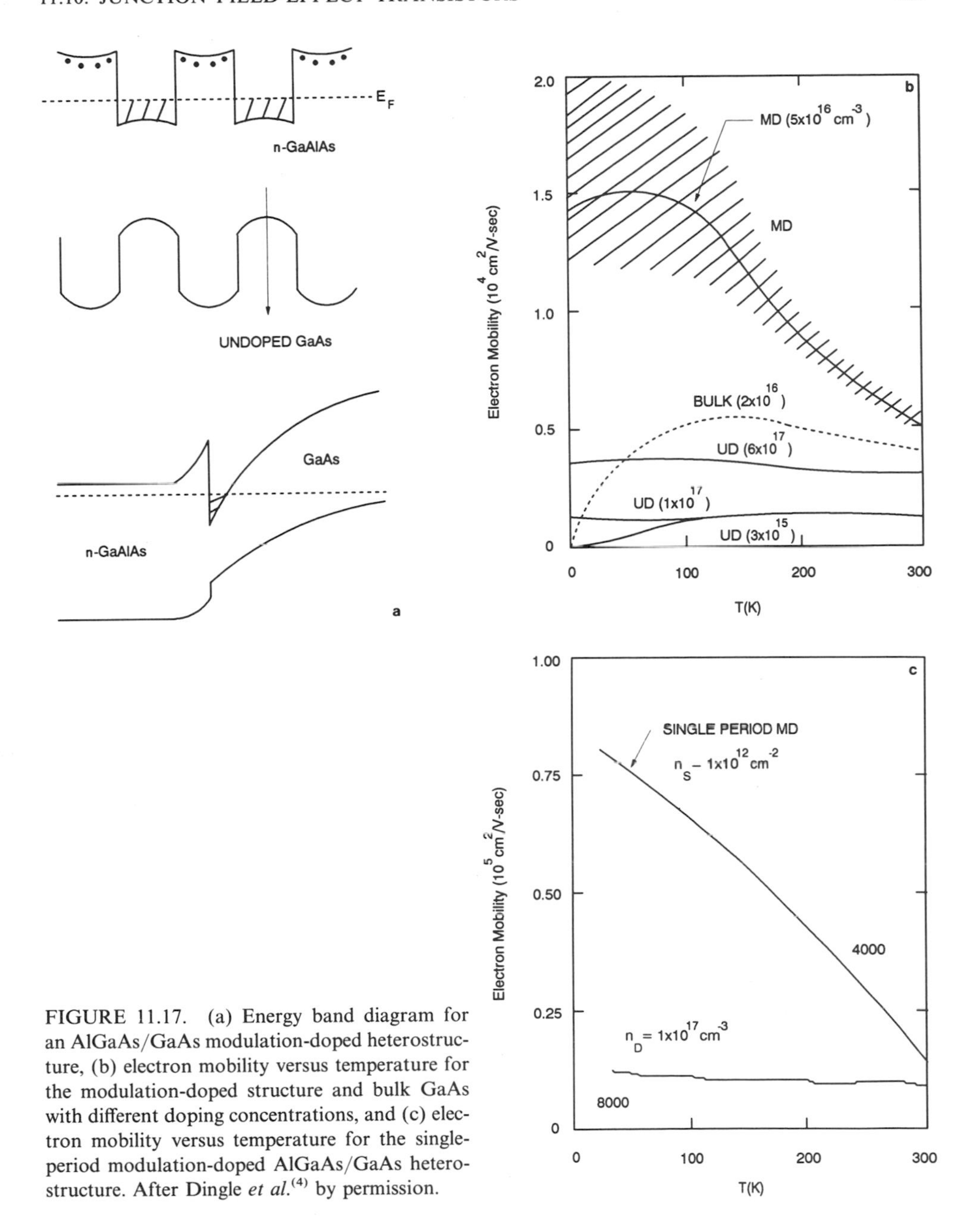

FIGURE 11.17. (a) Energy band diagram for an AlGaAs/GaAs modulation-doped heterostructure, (b) electron mobility versus temperature for the modulation-doped structure and bulk GaAs with different doping concentrations, and (c) electron mobility versus temperature for the single-period modulation-doped AlGaAs/GaAs heterostructure. After Dingle *et al.*[4] by permission.

be formed by using either the thermal diffusion or the ion implantation technique. In IC fabrication, the ion-implantation technique is more widely used since better control of the geometries, dopant densities, and profiles of both the source and drain regions can be obtained using this technique.

Since carrier transport in a JFET is due to the majority carriers in the channel formed between the p^+-gate and the p-substrate, the JFET is also known as a unipolar transistor. The unique feature of a JFET is that the conductivity in the n-channel can be easily controlled by applying a reverse-bias gate voltage to change the depletion layer width in the p^+-gate and

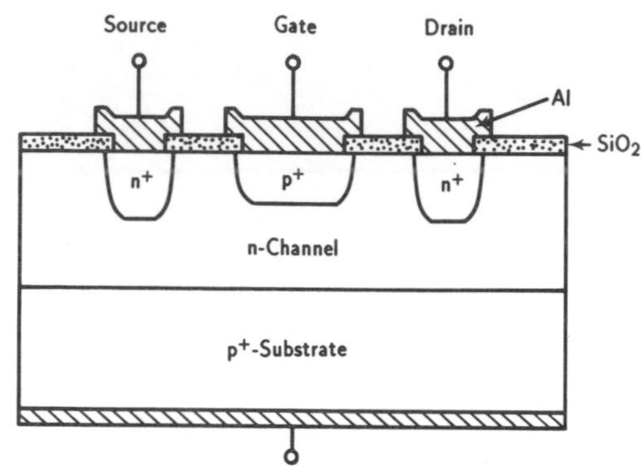

FIGURE 11.18. The device structure of an n-channel junction field-effect transistor (JFET).

the n-channel space-charge regions. If the current flow in the channel is due to electrons, then one has an n-channel JFET. On the other hand, if the current flow in the channel is due to holes, then one has a p-channel JFET (in this case the source and drain electrodes are p^+ dopant and the substrate is n-type).

The dc characteristics of a JFET can be analyzed by considering a one-dimensional JFET structure as shown in Fig. 11.19a and b under different bias conditions. In this figure, L is the channel length between the source and drain, Z is the depth of the channel, $2a$ is the channel width, and the drain current is along the x-direction of the channel length. If the channel length (L) is much greater than the channel width ($2a$), then the change in channel width along the channel is small compared with the channel width. Thus, the electric field in the depletion region of the gate junction is assumed perpendicular to the channel (i.e., along the y-direction), while the electric field inside the neutral n-channel may be assumed to be in the x-direction only. Separation of the electric field in the gate depletion region and the channel

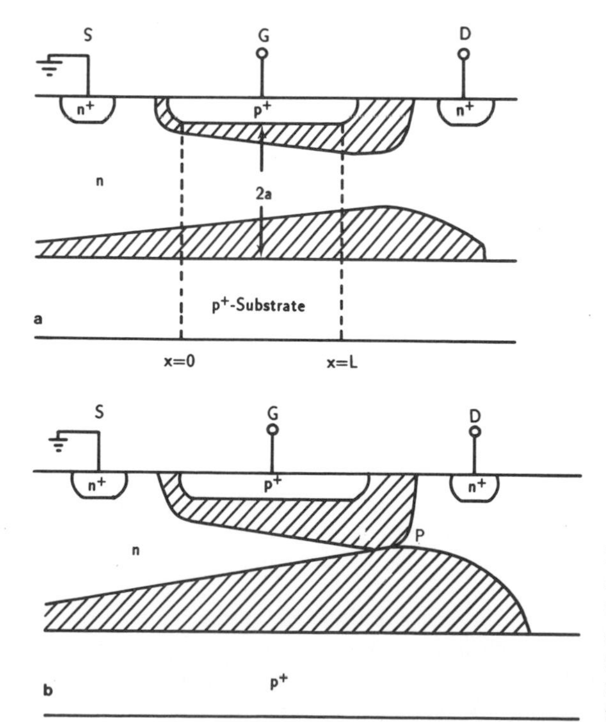

FIGURE 11.19. Schematic diagram of an n-channel JFET showing (a) the source (S), gate (G), and drain (D) regions, the dimensions of the channel and the depletion region (shaded area) in the channel under small gate bias voltage and (b) at pinch-off condition.

gradual-channel approximation, first introduced by Shockley. By assuming a one-sided step junction at the gate region with its dopant density N_A much greater than N_D in the channel, the depletion layer will extend mainly into the channel region. Under normal operating conditions, a reverse-bias voltage is applied across the gate electrode so that free carriers are depleted from the channel and the space-charge region extends into the channel. Consequently, the cross-section area of the channel is reduced, and channel conduction is also reduced. Therefore, the channel current flowing between the source and drain is modulated by the gate voltage. The depletion layer width in this case can be expressed by

$$W_d(x) = \sqrt{\frac{2\varepsilon_0\varepsilon_s[V(x) + V_{bi} - V_g]}{qN_D}} \tag{11.118}$$

The resistance in the channel region can be written as

$$R = \frac{\rho L}{A} = \frac{L}{q\mu_n N_D A} = \frac{L}{2q\mu_n N_D Z(a - W_d)} \tag{11.119}$$

The drain current I_D in the n-channel is due to the drift component only and is given by

$$I_D = q\mu_n n A \varepsilon_x = 2q\mu_n N_D[a - W_d(x)]Z\frac{dV}{dx} \tag{11.120}$$

where $A = 2(a - W_d)Z$ is the cross-section area of the channel. Now, substituting Eq. (11.118) for $W_d(x)$ into Eq. (11.120) and integrating from $x = 0$ to $x = L$ with corresponding voltages from 0 to V_D enable Eq. (11.120) to attain the form

$$I_D \int_0^L \frac{dx}{2q\mu_n N_D Z} = \int_0^{V_D} \left\{\left[a - \left(\frac{2\varepsilon_0\varepsilon_s}{qN_D}\right)(V + V_{bi} - V_g)\right]^{1/2}\right\} dV \tag{11.121}$$

By integrating and rearranging the terms in Eq. (11.121), we obtain

$$I_D = G_0\left\{V_F - \frac{2}{3}\left(\frac{2\varepsilon_0\varepsilon_s}{qa^2N_d}\right)^{1/2}[(V_d + V_{bi} - V_g)^{3/2} - (V_{bi} - V_g)^{3/2}]\right\} \tag{11.122}$$

This equation is a general expression for the current–voltage relationship for a JFET; it can also be expressed in terms of the pinch-off voltage and pinch-off current, namely,

$$I_D = I_p\left\{\left(\frac{V_D}{V_p}\right) - \frac{2}{3}\left[\frac{(V_D + V_g + V_{bi})}{V_p}\right]^{3/2} + \frac{2}{3}\left[\frac{(V_g + V_{bi})}{V_p}\right]^{3/2}\right\} \tag{11.123}$$

where

$$I_p = \frac{q^2\mu_n N_D^2 Z a^3}{\varepsilon_0\varepsilon_s L} \tag{11.124}$$

and

$$V_p = \frac{qN_D a^2}{2\varepsilon_0\varepsilon_s} \tag{11.125}$$

where V_p is the pinch-off voltage (i.e., $V_p = V_D + V_g + V_{bi}$, and $W_d = a$ at $x = L$) and I_p is the pinch-off current.

The I–V characteristics for a JFET can be analyzed in two regions with pinch-off as the boundary condition. At low drain voltages (i.e., $V_D \ll V_g + V_{bi}$), Eq. (11.123) becomes

$$I_D = G_0\left\{1 - \left[\frac{\varepsilon_0\varepsilon_s}{2qN_d(a - W_d)(V_{bi} - V_g)}\right]^{1/2}\right\}V_D \tag{11.126}$$

where $G_0 = 2q\mu_n N_d Z(a - W_d)/L$ is the channel conductance. Equation (11.126) shows that at a given gate voltage, a linear relationship between I_D and V_D exists in this region. The square-root dependence of drain current on gate voltage is a direct result of assuming an abrupt junction for the gate-channel junction. It is seen in Eq. (11.126) that the drain current reaches a maximum value when the gate voltage is zero and decreases with increasing gate voltage. In addition, this equation also predicts zero drain current when the gate voltage is large enough to deplete the entire channel region. If the drain voltage is further increased, the depletion layer width will also increase. Eventually, the two depletion regions touch each other at the drain electrode as shown in Fig. 11.19b. This pinch-off condition occurs when the depletion width W_d is equal to a at the drain elecrode. For a p^+–n junction, solving Eq. (11.118) yields the corresponding value of the drain voltage, which is given by

$$V_{Ds} = \frac{qN_D a^2}{2\varepsilon_0\varepsilon_s} - V_{bi} \qquad \text{for } V_g = 0 \tag{11.127}$$

where V_{Ds} is the saturation drain voltage. The pinch-off condition is reached at this drain voltage, and both the source and drain regions are completely separated by a reverse-bias depletion region. The location of point P in Fig. 11.19b is called the pinch-off point and the corresponding drain current is called the saturation drain current I_{Ds}, that can flow through the depletion region. Beyond the pinch-off point, as V_D is increased further, the depletion region near the drain will expand and point P will move toward the source region. However, the voltage at point P remains the same as V_{Ds}. As a result, since the potential drop in the channel from the source to point P remains unaltered, the current flowing in the channel also stays constant. Therefore, for drain voltages larger than V_{Ds}, the current flowing in the channel is independent of V_D and equals I_{Ds}. Under this condition, the JFET is operating in the saturation regime, and the expression for the saturation current can be deduced from Eq. (11.122) in the form

$$I_{Ds} = G_0\left\{\frac{qN_D a^2}{6\varepsilon_0\varepsilon_s} - (V_{bi} - V_g)\left[1 - \frac{2}{3}\left(\frac{2\varepsilon_0\varepsilon_s(V_{bi} - V_g)}{qN_D a^2}\right)^{1/2}\right]\right\} \tag{11.128}$$

or

$$I_{Ds} = I_p\left\{\frac{1}{3} - \frac{(V_g + V_{bi})}{V_p} + \frac{2}{3}\left[\frac{(V_g + V_{bi})}{V_p}\right]^{3/2}\right\} \tag{11.129}$$

The corresponding drain saturation voltage is given by

$$V_{Ds} = V_p - V_g - V_{bi} \tag{11.130}$$

Therefore, based on the above analysis, the I_D versus V_D curve can be divided into three different regimes, namely, (1) the linear regime at low drain voltages, (2) a regime with less than a linear increase of drain current with drain voltage, and (3) a saturation regime where the drain current remains constant as the drain voltage is further increased. This is illustrated in Fig. 11.20.

The JFETs often operate in the saturation regime within which output drain current does not depend on output drain voltage but depends only on input gate voltage. Under this condition, the JFET may be used as an ideal current source controlled by an input gate voltage. The transconductance of a JFET may be obtained by differentiating Eq. (11.122) with respect to gate voltage, which yields

$$\begin{aligned} g_m &\equiv \left.\frac{\partial I_D}{\partial V_g}\right|_{V_D = \text{constant}} \\ &= G_0\left(\frac{2\varepsilon_0\varepsilon_s}{qN_Da^2}\right)^{1/2}[(V_{bi} - V_g + V_D)^{1/2} - (V_{bi} - V_g)^{1/2}] \\ &= \left(\frac{I_p}{V_p}\right)\left\{1 - \left[\frac{(V_g + V_{bi})}{V_p}\right]^{1/2}\right\} \end{aligned} \tag{11.131}$$

It is found from this equation that in the saturation regime, the transconductance has a maximum value given by

$$g_{ms} = G_0\left\{1 - \left[\frac{2\varepsilon_0\varepsilon_s}{qN_Da^2}(V_{bi} - V_g)\right]^{1/2}\right\} \tag{11.132}$$

The theoretical analysis presented in this section for a JFET involves several simplified assumptions. For example, it is assumed that the depletion layer width is controlled solely by the gate–channel junction and not by the channel–substrate junction. In reality, there will be a variation in potential across the channel–substrate junction along the channel, with maximum potential and depletion width occurring near the drain region. As a result, this simplified assumption may lead to disagreement between theoretical predictions and experimental data on I–V characteristics of a practical JFET. In general, the simple model presented here is valid only for a long-channel JFET. For a short-channel JFET with $L/a < 2$, the saturation mechanism becomes more complex and the above theories require refinement in order to obtain good agreement between theory and experiment.

Based on the above theoretical analysis, it is clear that the dc characteristics of a JFET are usually quite sensitive to dopant density and thickness of the channel region. Therefore,

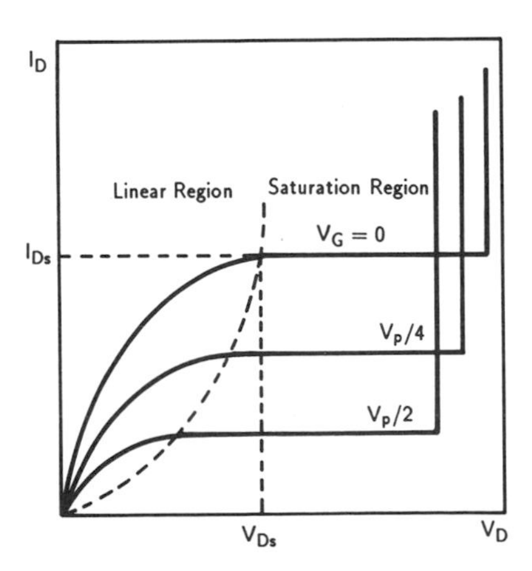

FIGURE 11.20. Output I–V characteristics of a JFET for different gate bias voltages, showing the linear and saturation regions of the device operation.

precise control of thickness and dopant density in the channel region of a JFET is very important. An n-channel silicon JFET can usually be made with excellent control by employing the epitaxial growth technique. As for the source, drain, and gate regions, since the density and location of the dopant impurity in these regions can be controlled better using ion implantation rather than thermal diffusion, the ion-implantation technique is preferable to the thermal-diffusion technique. In fact, both the n-channel and p-gate of a silicon JFET are usually formed by using the ion-implantation technique.

PROBLEMS

11.1. Derive the electric field, potential distribution, and depletion layer width for a linearly graded p–n junction diode, and show that the depletion layer width and the built-in potential under zero-bias conditions are given by Eqs. (11.22) and (11.23), respectively.

11.2. If the impurity gradient a for a Si and GaAs linearly graded p–n junction diode is equal to 10^{22} cm^{-4} at 300 K, calculate the depletion layer width, built-in potential, and breakdown voltage for these two diodes. Repeat for $a = 10^{20}$ cm^{-4}. Calculate and plot junction capacitance versus applied voltage for both diodes.

11.3. The small-signal ac characteristics of a p–n junction diode are important for circuit applications. The diode admittance can be derived by solving the small-signal carrier distribution from the steady-state continuity equation. The small-signal condition is satisfied if the applied ac signal, v_1, is small compared to the thermal voltage, V_T. Draw a small-signal equivalent circuit for the diode by including the circuit elements r_s, G_d, C_d, and C_j, where r_s is the series rsistance caused by the ohmic drop across the neutral semiconductor regions and the contacts, $G_d = I/V_T$ ($V_T = k_BT/q$) is the small-signal conductance, $C_d \cong \tau_p I/2V_T$ is the diffusion capacitance, and C_j is the transition capacitance that arises from the junction space-charge layer. Note that $I = I_p(0)$ is the dc current density for a p$^+$–n junction diode.

11.4. From the results obtained in Problem 11.3, calculate the small-signal conductance and capacitance for a long-base silicon p$^+$–n diode if $N_A = 5 \times 10^{18}$ cm^{-3}, $N_D = 2 \times 10^{16}$ cm^{-3}, $\tau_n = 2 \times 10^{-8}$ sec and $\tau_p = 5 \times 10^{-8}$ sec, $A = 2 \times 10^{-4}$ cm^2, and $T = 300$ K.
(a) For the forward-bias voltages, $V_f = 0.1$, 0.3, 0.5, and 0.7 V.
(b) For the reverse-bias voltages, $V_R = -0.5$, -5, -10, and -20 V.
(c) What is the series resistance of the n-neutral region if the thickness is 2 mm?

11.5. Consider the minority carrier charge storage effect in a long-base p$^+$–n diode.
(a) Show that the turnoff time t_{off} of holes in the n-region is given by

$$t_{off} = \tau_p \ln\left(1 + \frac{I_f}{I_r}\right)$$

The latter equation is obtained from the charge-control equation for a long-base p$^+$–n diode. It can be shown that an exact analysis by solving the time-dependent diffusion equation would yield

$$\operatorname{erf}\left(\frac{t_{off}}{\tau_p}\right)^{1/2} = \frac{I_f}{I_f + I_r}$$

where erf (x) is the error function.
(b) Plot t_{off}/τ_p versus I_r/I_f using the above two expressions.

11.6. Under forward-bias conditions, the space-charge recombination current can be calculated from the Shockley–Read–Hall (SRH) model via a mid-gap recombination center. The recombination rate derived from the SRH model is given by

$$U_r = \frac{n_i^2(e^{qV_a/k_BT} - 1)}{[p + n + 2n_i \cosh(E_t - E_i)/k_BT]\tau_0} \tag{1}$$

where $\tau_0 = 1/N_t\sigma v_{th}$ is the effective carrier lifetime.

(a) Find the conditions for the maximum recombination rate from the above expression.
(b) If the recombination current in the junction space-charge region can be derived from

$$J_r = q \int_{-x_p}^{x_n} U_r \, dx \tag{2}$$

where U_r is given in Eq. (1), and x_n and x_p are the depletion layer widths in the n- and p-regions, respectively, show that the recombination current density can be expressed by

$$J_r = \left(\frac{qW'n_i}{2\tau_0}\right) e^{qV_a/2k_BT}$$

where W' is the portion of the depletion region in which the recombination current is dominant, and $qV_a > 3k_BT$.

11.7. *Consider a long-base silicon* p^+–n *diode. If the diode parameters are given by* $N_A = 10^{19}\ \text{cm}^{-3}$, $N_D = 5 \times 10^{16}\ \text{cm}^{-3}$, $D_p = 4\ \text{cm}^2/\text{sec}$, $D_n = 20\ \text{cm}^2/\text{sec}$, $\tau_p = 10^{-8}$ sec, $\tau_n = 10^{-6}$ sec, $n_i = 1.4 \times 10^{10}\ \text{cm}^{-3}$, and $A = 10^{-4}\ \text{cm}^2$:
(a) Calculate the hole injection current into the n-region for forward-bias voltages of 0.1, 0.3, 0.5, and 0.7 V at 300 K.
(b) Repeat (a) for the electron injection current into the p^+–n region.
(c) What is the total injection current for $V = 0$, 0.3, and 0.5 V if the silicon diode is a short-base diode with p-emitter width $W_E = 5 \times 10^{-5}$ cm and n-base width $W_B = 10^{-3}$ cm?

11.8. If the silicon p–n diode given in Problem 11.7 has a mid-gap recombination center in the junction space–charge region of significant density, calculate the recombination current for the cases $N_t = 10^{13}$, 10^{14}, and $10^{15}\ \text{cm}^{-3}$ and $V = 0.2$, 0.4, and 0.6 V assuming that $\sigma = 10^{-15}\ \text{cm}^2$, $v_{th} = 10^7$ cm/sec, and recombination occurred throughout the entire depletion region. (*Hint*: use the expression for the recombinaton current given by Problem 11.6.)

11.9. The diode parameters for a Ge, Si, and GaAs p^+–n short-base diode are the same as those given by Problem 11.7, except that the intrinsic carrier densities are 2.5×10^{13}, 1.4×10^{10}, and $10^6\ \text{cm}^{-3}$ for Ge, Si, and GaAs at 300 K, respectively. Calculate the total injection current for these diodes. Explain why GaAs is more suitable for high-temperature applications than Si and Ge.

11.10. (a) Construct the energy band diagram for an $Al_xGa_{1-x}As$/GaAs p^+–n heterojunction diode for $x = 0.4$ ($E_g = 1.8$ eV).
(b) If $N_A = 5 \times 10^{18}\ \text{cm}^{-3}$, $N_D = 2 \times 10^{17}\ \text{cm}^{-3}$, $D_n = D_p = 5\ \text{cm}^2/\text{sec}$, and $\tau_n = \tau_p = 10^{-9}$ sec, calculate the injection currents in both regions of the diode assuming $A = 2 \times 10^{-5}\ \text{cm}^2$.
(c) Plot the injection current versus temperature (100 to 400 K) for this diode.

11.11. The onset of Zener breakdown in an abrupt silicon p–n diode takes place when the maximum electric field approaches 10^6 V/cm. If the dopant density in the p-region is $5 \times 10^{19}\ \text{cm}^{-3}$, what would be the dopant density in the n-region in order to have a Zener breakdown voltage of 2 V? Repeat the calculation for $N_A = 10^{20}\ \text{cm}^{-3}$ and $V_{ZB} = 3$ V.

11.12. Plot the energy band diagrams for a p^+-$Al_{0.3}Ga_{0.7}As$/n-GaAs and a n-Ge/p-GaAs heterojunction diode. The bandgap energy for $Al_xGa_{1-x}As$ can be calculated by using the equation $E_g = 1.424 + 1.247x$ eV (for $0 < x < 0.45$). Calculate the values of ΔE_c and ΔE_v for both cases.

REFERENCES

1. J. L. Moll, "The Evolution of the Theory of the Current-Voltage Characteristics of p–n Junctions," *Proc. IRE*, **46**, 1076 (1958).
2. A. S. Grove, *Physics and Technology of Semiconductor Devices*, Chap. 6, p. 192, Wiley, New York (1967).
3. S. M. Sze and G. Gibbons, "Avalanche Breakdown Voltages of Abrupt and Linearly Graded p–n Junctions in Ge, Si, GaAs, and GaP," *Appl. Phys. Lett.* **8**, 111 (1966).
4. R. Dingle, H. L. Stormer, A. C. Gossard, and W. Wiegmann, "Electron Mobilities in Modulation-Doped Semiconductor Heterojunction Superlattices," *Appl. Phys. Lett.* **33**, 665 (1978).

BIBLIOGRAPHY

R. L. Anderson, "Experiments on Ge–GaAs Heterojunctions," *Solid-State Electron.* **5**, 341 (1962).

W. R. Frensky and H. Kroemer, "Theory of the Energy-Band Lineup at an Abrupt Semiconductor Heterojunction," *Phys. Rev. B* **16**, 2642 (1977).

R. N. Hall, *Phys. Rev.* **87**, 387 (1952).

A. G. Miles and D. L. Feucht, *Heterojunctions and Metal–Semiconductor Junctions*, Academic Press, New York (1972).

J. L. Moll, *Physics of Semiconductors*, McGraw-Hill, New York (1964).

R. S. Muller and T. I. Kamins, *Device Electronics for Integrated Circuits*, Wiley, New York (1977).

C. T. Sah, R. N. Noyce, and W. Shockley, "Carrier Generation and Recombination in p–n Junction and p–n Junction Characteristics," *Proc. IRE* **45**, 1228 (1957).

W. Shockey, *Electrons and Holes in Semiconductors*, Van Nostrand, Princeton (1950).

W. Shockley and W. T. Read, *Phys. Rev.* **87**, 835 (1952).

S. M. Sze, *Physics of Semiconductor Devices*, 2nd ed., Wiley, New York (1981).

E. S. Yang, *Microelectronic Devices*, McGraw-Hill, New York (1988).

12

Photonic Devices

12.1. INTRODUCTION

This chapter is concerned with basic device physics, device structures and characteristics, and the operation principles of various photonic devices fabricated from elemental and compound semiconductors. The photonic devices are becoming increasingly important due to a wide variety of applications in the areas of optical communications, optical computing and interconnects, data transmission and signal processing. In addition, recent advances in III–V compound semiconductor technologies have enabled these applications to become a reality. As a result, various photonic devices such as lasers, LEDs, modulators, and photodetectors using III–V semiconductors have been developed for use in optical communications, displays, data transmission, and signal processing. Depending on the device structure and operational mode, photonic devices can in general be divided into three categories: (i) photovoltaic devices (i.e., solar cells) which convert sunlight directly into electricity by generating electron-hole pairs in a solar cell via internal photovoltaic effects, (ii) photodetectors which detect photons or optical signals and convert them into electrical signals via internal photoelectric effects, and (iii) light-emitting diodes (LEDs) or laser diodes which convert electrical energy into coherent or incoherent optical radiation by electrical injection into the junction region of the p–n diode.

The solar cell, which utilizes the internal photovoltaic effect in a semiconductor, will be discussed first. Solar cells may be fabricated by using the p–n junction, Schottky barrier, and MIS (metal–insulator–semiconductor) structures on different semiconductors. The basic principles, cell structure and characteristics, design criteria, and performance limitations for different types of solar cell structures are discussed in Section 12.2. It is of interest that, prior to 1973, the majority of solar cell research had been focussed mainly on developing high-efficiency silicon p–n junction solar cells for space applications. However, in recent years much effort has been shifted toward the development of various low-cost and high-efficiency solar cells for both terrestrial and space power generation as well as for applications in consumer electronics. It is well known that cost and conversion efficiency are the two key factors that determine the compatibility of a photovoltaic system with other types of power generation systems using windmill, geothermal, fossil-fuel, and nuclear power generation for terrestrial applications. In this respect, amorphous and polycrystalline silicon thin-film solar cells appear to have the best chance of meeting both low-cost and high-efficiency criteria for large scale terrestrial power generation. On the other hand, multijunction (or cascade) solar cells using different bandgap materials offer the possibility of achieving a much higher conversion efficiency (e.g., $\geq 30\%$ under air-mass-one (AM1) conditions) than a single-junction solar cell. For example, a three-junction solar cell structure using a large-bandgap semiconductor such as an $Al_xGa_{1-x}As$ or $GaAs_yP_{1-y}$ as the top cell, a GaAs as the middle cell, and an $In_xGa_{1-x}As$ or germanium as the bottom cell could yield a theoretical air-mass-zero (AM0) conversion efficiency exceeding 30%. Furthermore, a high concentrated solar cell using a Fresnel lens to focus sunlight to 10, 500, or 1000 times of one-sun light intensity has been reported. In fact, a silicon concentrated solar cell operating at 100 sun intensity has achieved a remarkable AM1 conversion efficiency of 27%. Large-bandgap III–V compound semiconductor materials such as GaAs and AlGaAs are particularly attractive for concentrator solar cell applications, since they can be operated at a much higher temperature than that of a silicon solar cell. An AM1 conversion efficiency greater than 20% has been achieved for a single-junction solar cell fabricated from Si, GaAs, and InP material systems.

The light-emitting diode (LED) is discussed in Section 12.3. An LED made out of III–V compound semiconductors can emit incoherent light under a forward-bias condition. The basic mechanism of an LED involves the spontaneous emission of photons from the radiative recombination of electron-hole pairs, so converting electrical energy into optical radiation. The random emission of incoherent optical radiation from an LED leads to a broad emission spectral linewidth of 100 Å or greater. LEDs have found a wide variety of applications including optical displays, solid state lamps, optical isolators, data transmission and signal processing, optical computing, and optical communications.

The photodetectors, which employ internal photoelectric effects to detect photons in a semiconductor device, are described in Section 12.4. A p–n junction or a Schottky barrier photodetector can be very fast and sensitive when operated under a reverse-bias condition. If sensitivity is the main concern, then the avalanche photodiode (APD) should be used to obtain the necessary internal current gain. On the other hand, a p–i–n photodiode can offer both sensitivity and speed. In fact, different types of photodetector covering a wide range of wavelengths from ultraviolet (UV) to visible and far infrared (IR) have been developed for a variety of applications. For example, extremely high sensitivity photomultipliers are commercially available for photon counting in the visible to near-IR (0.3 to 0.9 μm) spectral range, while extrinsic photoconductors described in Chapter 9 can extend detection wavelengths into the far-IR spectral regime (e.g., $\lambda>30$ μm). Photodetectors fabricated from narrow bandgap semiconductor materials such as $Cd_xHg_{1-x}Te$ and $Pb_xSn_{1-x}Te$ are used primarily for 8 to 12 μm IR detection, while silicon, GaAs, $In_{0.53}Ga_{0.47}As$ p–i–n, and Schottky barrier photodiodes cover wavelengths from 0.4 to 1.6 μm. Recent development of intersubband transition quantum-well infrared photodetectors (QWIPs) shows that the detectivity of an AlGaAs/GaAs QWIP operating at 77 K is comparable to that of a $Cd_xHg_{1-x}Te$ IR detector in the 8 to 12 μm wavelength range.

Section 12.5 describes basic device physics and structures, general characteristics and performance parameters of a semiconductor laser diode. The conditions for population inversion and oscillation, as well as the threshold current density for a semiconductor laser diode are discussed. Various single- and double-heterostructure laser diodes, formed on III–V semiconductor materials, are also discussed in this section.

12.2. PHOTOVOLTAIC DEVICES

A photovoltaic device, also known as a solar cell, can convert sunlight directly into electricity. The conversion efficiency for a typical silicon p–n junction solar cell is 14 to 16% under AM1 conditions. In order to compute the conversion efficiency of a solar cell, one needs to know the exact incident solar irradiance power under different insolation conditions. Figure 12.1 shows the solar irradiance spectra for two air-mass conditions.[1] The top curve is the solar irradiance spectrum measured outside the earth's atmosphere, and is defined as the AM0 insolation. The irradiant power of the sun under AM0 conditions is 135.3 mW/cm^2 (or 1.353 kW/m^2). The bottom curve is the solar irradiance spectrum measured under AM1 conditions. The AM1 solar spectrum represents the sunlight on the earth's surface when the sun is at its zenith. The total incident power of sunlight under AM1 conditions is 92.5 mW/cm^2.

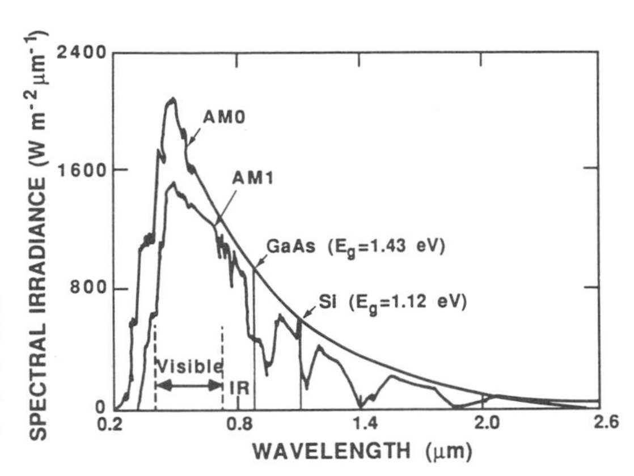

FIGURE 12.1. Solar irradiance versus wavelength under air-mass-zero (AM0) and air-mass-one (AM1) conditions. Also shown are the energy bandgaps and the corresponding cutoff wavelengths for both GaAs and Si. After Thekaekara,[1] by permission.

The AM1.5 condition (i.e., the sun is at 45° above sea level) has an incident power of 84.4 mW/cm^2. This is the most suitable incident solar irradiance power for calculating the conversion efficiency of a solar cell in the terrestrial environment. Since the conversion efficiency will vary under different air-mass conditions, it is important to specify the exact AM_x (i.e., $x = 0$, 1, 1.5, or 2) condition when computing the conversion efficiency of a solar cell.

In this section, different types of solar cells including the p–n junction, Schottky barrier, MIS, p–n heterojunction, and multijunction solar structures are described. We first discuss the generation and collection of photogenerated electron-hole pairs in a p–n junction solar cell, and then derive expressions for the spectral response, short-circuit current, open-circuit voltage, and conversion efficiency for such a solar cell. Formation of front ohmic contact grids and antireflection (AR) coatings for a solar cell is also discussed in this section.

12.2.1. p–n Junction Solar Cells

Since most solar cells use p–n junction structure, we shall first consider the basic cell structure, the dark and photo I–V characteristics, the derivation of photocurrent and the design parameters of such a solar cell. Figure 12.2 shows (a) the cell structure and (b) the cross section view of a typical silicon n^+–p junction solar cell. The basic characteristics of a p–n junction solar cell are analyzed by studying the current–voltage (I–V) behavior under dark and illumination conditions. The dark I–V characteristic is as important as the photo-I–V characteristic in determining the open-circuit voltage, output power, and conversion

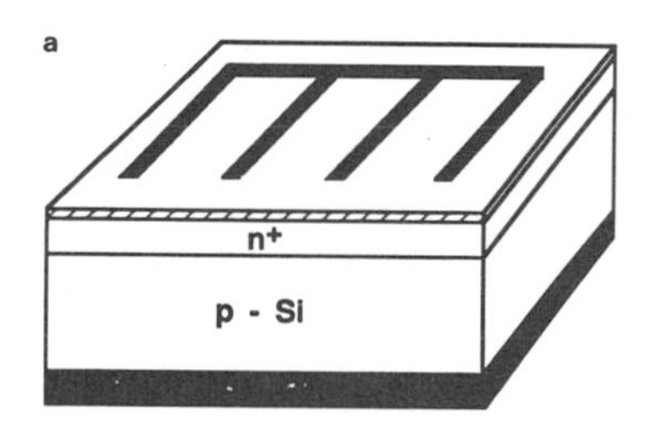

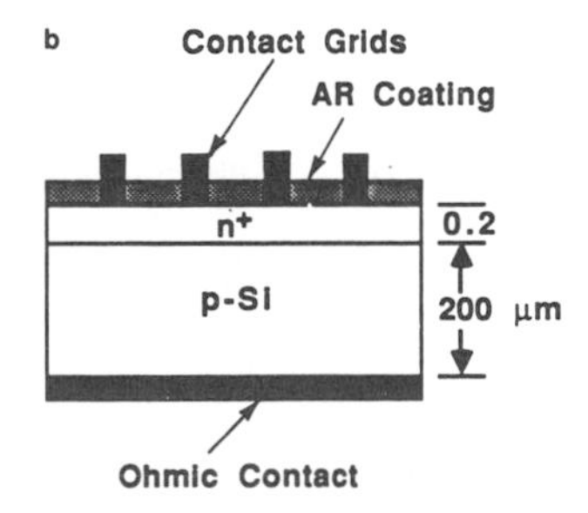

FIGURE 12.2. (a) A typical cell structure and (b) the cross-section view of a silicon n^+–p junction solar cell. Also shown are the front ohmic contact grids and the antireflection (AR) coating layer (e.g., Ta_2O_5).

efficiency of a solar cell. In order to explain the basic operation principles of a p–n junction solar cell, Fig. 12.3 shows the energy band diagrams of a n^+–p junction solar cell under illumination: Fig. 12.3a shows the electron–hole pairs generated by absorbed photons in different regions of the solar cell, and Fig. 12.3b shows the dark current components generated in different regions of the junction. The *I–V* curves under dark and illumination conditions are shown in Fig. 12.3c. We note that the shaded area in the fourth quadrant of the photo-*I–V* curve represents the output power generated from the solar cell. The equivalent circuit for a p–n junction solar cell is shown in Fig. 12.3d, where R_s denotes the series resistance and R_p the shunt resistance. The dark and photo-*I–V* characteristics of a p–n injunction solar cell are discussed next.

Dark I–V Characteristics. As shown in Fig. 12.3b, the dark current of a p–n junction solar cell under forward-bias conditions consists of three components: (i) the injection current (1) due to injection of majority carriers across the p–n junction, (ii) the recombination current (2) due to recombination of electrons and holes via deep-level traps in the junction space-charge region, and (iii) the tunneling current (3) due to multistep tunnelings via deep-level defect states in the junction space-charge region. In a silicon p–n junction solar cell, the injection current is usually the predominant component. However, the recombination current may become the predominant component for solar cells fabricated from low-quality materials such as polycrystalline semiconductor materials. The tunneling current component may become important in a Cu_2S–CdS p–n junction cell or a MIS solar cell.

The injection current, which is due to the injection of holes from the p- into the n-region and electrons from the n- into the p-region of the junction, can be described by the ideal Shockley diode equation, which is given by

$$I_d = I_{01}[\exp(qV/k_BT) - 1] \tag{12.1}$$

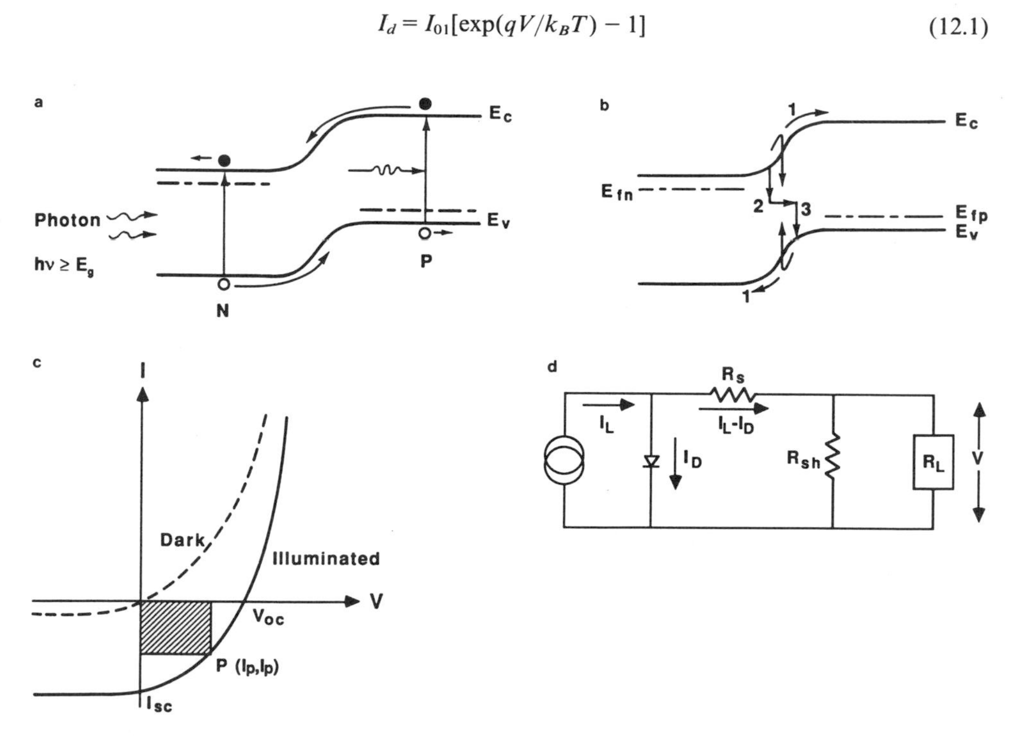

FIGURE 12.3. (a) The energy band diagram of a n^+–p junction solar cell under illumination, (b) the energy band diagram showing (1) the injection current, (2) the recombination current via a deep-level trap, and (3) the trap-assisted tunneling current in the junction space-charge region, (c) *I–V* characteristics under dark and illumination conditions, and (d) the equivalent circuit diagram.

where

$$I_{01} = qn_i^2 A_j \left(\frac{D_p}{L_p N_D} + \frac{D_n}{L_n N_A} \right) \tag{12.2}$$

is the reverse saturation current due to majority electron and hole injection across the p–n junction, A_j is the junction area, n_i is the intrinsic carrier density, D_n and D_p denote the electron and hole diffusion coefficients, while L_n and L_p are the electron and hole diffusion lengths, respectively. Equation (12.1) is obtained by assuming uniform doping in both the n- and p-regions of the solar cell so that a quasi-neutral condition prevails in both regions.

The recombination current in a p–n junction solar cell is due to recombination of electrons and holes via deep-level defect centers inside the junction space-charge region. Based on the Shockley–Read–Hall (SRH) model, this recombination current component can be expressed by

$$I_r = I_{02}[\exp(qV/mk_BT) - 1] \tag{12.3}$$

where

$$I_{02} = \frac{qn_i W A_j}{2\sqrt{\tau_{n0}\tau_{p0}}} \tag{12.4}$$

In Eq. (12.4), W is the depletion layer width, and τ_{n0} and τ_{p0} denote the minority electron and hole lifetimes in the p- and n-regions, respectively. We note that factor m in the exponent may vary between 1 and 2, depending on the location of the recombination centers in the forbidden gap. For example, m equals 2 if the recombination of electron–hole pairs is via a mid-gap recombination center (i.e., $E_t = E_i$), and is less than 2 if the recombination center is not located at the mid-gap or if the multilevel recombination centers exist in the junction space-charge region.

In general, the total dark current of a p–n junction solar cell can be represented by the sum of injection and recombination current components given by Eqs. (12.1) and (12.3), namely,

$$I_D = I_d + I_r \tag{12.5}$$

The main difference between the two current components given by Eq. (12.5) is that the injection current varies with $n_i^2 \exp(qV/k_BT)$, while the recombination current is proportional to $n_i \exp(qV/mk_BT)$. For a typical silicon p–n junction solar cell, values of the injection current density may vary from 10^{-8} to 10^{-12} A/cm^2, while values of the recombination current will depend on the density of recombination centers in the junction space-charge region. In general, the recombination current is important only at low forward-bias regimes and becomes negligible at higher bias voltages. For a high-quality p–n junction solar cell, the recombination current component can usually be neglected in Eq. (12.5). The tunneling current component, which may be important for a MIS solar cell or a CdS solar cell, is not included in Eq. (12.5).

Photo-I–V Characteristics. The photocurrent generated in a p–n junction solar cell under the one-sun condition is discussed next. When photons with energies $h\nu \geq E_g$ impinge upon a p–n junction solar cell, the rate of generation of electron–hole pairs as a function of distance x from the surface of the solar cell is given by

$$g_E(x) = \alpha\phi_0(1 - R)\, e^{-\alpha x} \tag{12.6}$$

where α is the optical absorption coefficient, ϕ_0 is the incident photon flux density per unit bandwidth per second, and R is the reflection coefficient. The photocurrent generated in a solar cell by the incident sunlight can be derived by using the continuity equations described in Chapter 6. For a n^+–p junction solar cell, the spatial distribution of the excess hole density generated in the n-region can be obtained by solving the steady-state continuity equation given by

$$D_p \frac{d^2\Delta p}{dx^2} - \frac{\Delta p}{\tau_p} = -\alpha\phi_0(1 - R)\,e^{-\alpha x} \tag{12.7}$$

The solution of this equation can be obtained by using the boundary conditions at the front surface of the cell where recombination occurs and at the depletion edge in the n^+-quasi-neutral region. The first boundary condition is given by

$$D_p \frac{d\Delta p}{dx} = s_p \Delta p \qquad \text{at } x = 0 \tag{12.8}$$

where $\Delta p = p_n - p_{n0}$ is the excess hole density in the n-region, and s_p is the surface recombination velocity at $x = 0$. The second boundary condition is obtained at the edge of the space-charge region where the excess hole density is assumed equal to zero (i.e., holes swept out by the high electric field in the depletion region). Thus, we can write

$$\Delta p = 0 \qquad \text{at } x = x_j \tag{12.9}$$

where x_j is the junction depth. Now, the solution of Eqs. (12.7) through (12.9) yields

$$\Delta p(x) = \frac{\alpha\phi_0(1 - R)\tau_p}{(\alpha^2 L_p^2 - 1)} \times \left\{ -e^{-\alpha x} + \frac{(s_p + \alpha D_p)\sinh\left(\frac{x_j - x}{L_p}\right) + e^{-\alpha x_j}\left[s_p \sinh\left(\frac{x}{L_p}\right) + \left(\frac{D_p}{L_p}\right)\cosh\left(\frac{x}{L_p}\right)\right]}{s_p \sinh\left(\frac{x_j}{L_p}\right) + \left(\frac{D_p}{L_p}\right)\cosh\left(\frac{x_j}{L_p}\right)} \right\} \tag{12.10}$$

The hole current density at the depletion edge of the n-region is given by

$$\begin{aligned} J_p(\lambda) &= -qD_p \left.\frac{d\Delta p}{dx}\right|_{x = x_j} \\ &= \frac{q\phi_0(1 - R)\alpha L_p}{(\alpha^2 L_p^2 - 1)} \times \left\{ -\alpha L_p e^{-\alpha x_j} + \frac{(s_p + \alpha D_p) - e^{-\alpha x_j}\left[s_p \cosh\left(\frac{x_j}{L_p}\right) + \left(\frac{D_p}{L_p}\right)\sinh\left(\frac{x_j}{L_p}\right)\right]}{s_p \sinh\left(\frac{x_j}{L_p}\right) + \left(\frac{D_p}{L_p}\right)\cosh\left(\frac{x_j}{L_p}\right)} \right\} \end{aligned} \tag{12.11}$$

where $J_p(\lambda)$ denotes the hole current density generated in the n-region by photons of wavelength λ absorbed in this region.

The photocurrent density due to electrons generated in the p-base region can be derived in a similar way to that of the hole current density in the n-region derived above. The continuity equation for electrons in the p-base region is obtained by replacing Δp by Δn, D_p by D_n, and

τ_p by τ_n in Eq. (12.7). However, the boundary conditions for this case are given by

$$\Delta n = 0, \qquad \text{at } x = x_j + W$$

$$-D_n \frac{d\Delta n}{dx} = s_n \Delta n, \qquad \text{at } x = d \tag{12.12}$$

where W is the depletion layer width, d is the thickness of the solar cell, and $\Delta n = (n_p - n_{p0})$ is the excess electron density. The photocurrent density per unit bandwidth due to electrons collected at the depletion edge of the p-region is thus given by

$$\begin{aligned} J_n(\lambda) &= qD_n \frac{d\Delta n}{dx}\bigg|_{x=x_j+W} \\ &= \frac{q\phi_0(1-R)\alpha L_n \exp[-\alpha(x_j+W)]}{(\alpha^2 L_n^2 - 1)} \\ &\times \left\{ \alpha L_n - \frac{s_n\left[\cosh\left(\frac{d}{L_n}\right) - e^{-\alpha d}\right] + \left(\frac{D_n}{L_n}\right)\sinh\left(\frac{d}{L_n}\right) + \alpha D_n e^{-\alpha d}}{s_p \sinh\left(\frac{d}{L_n}\right) + \left(\frac{D_n}{L_n}\right)\cosh\left(\frac{d}{L_n}\right)} \right\} \end{aligned} \tag{12.13}$$

In addition to the diffusion components of the photocurrent collected in both n- and p-quasi-neutral regions given by Eqs. (12.11) and (12.13), the drift component of the photocurrent generated in the depletion region must also be considered. The electron–hole pairs generated in the depletion region are swept out by the built-in electric field in this region. Thus, the drift component of the photocurrent density generated in this region can be expressed by

$$\begin{aligned} J_d(\lambda) &= q \int_{x_j}^{x_j+W} g_E(\lambda)\, dx \\ &= q\alpha\phi_0(1-R)\, e^{-\alpha x_j}(1 - e^{-\alpha W}) \end{aligned} \tag{12.14}$$

Therefore, the total photocurrent (or the short-circuit current) density generated by the incident monochromatic light in a p–n junction solar cell at a given wavelength λ is equal to the sum of Eqs. (12.11), (12.13), and (12.14), i.e.,

$$J_{sc}(\lambda) = J_p(\lambda) + J_n(\lambda) + J_d(\lambda) \tag{12.15}$$

The quantum efficiency, which is defined as the number of electron–hole pairs generated per absorbed photon, for a p–n junction solar cell can be expressed by

$$\eta = \frac{J_{sc}(\lambda)}{q\phi_0(1-R)} \times 100\% \tag{12.16}$$

where $J_{sc}(\lambda)$ is given by Eq. (12.15).

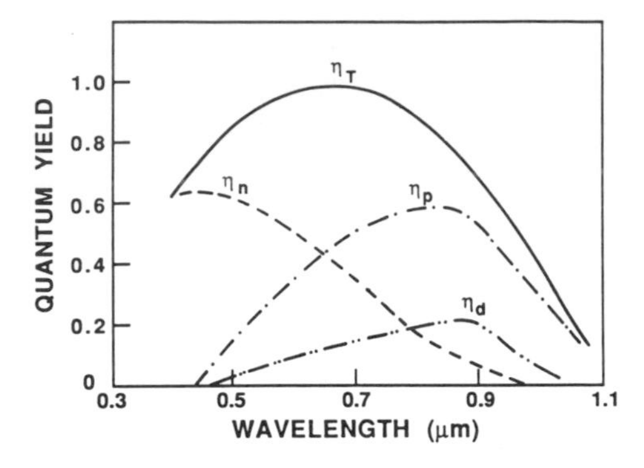

FIGURE 12.4. Normalized quantum yield versus wavelength in the n-emitter, p-base, and space-charge region of a silicon p–n junction solar cell. The solid line denotes the total quantum yield, while the dashed and dotted lines are the quantum yields in different regions of the solar cell, assuming zero reflection loss at the top surface of the cell.

Figure 12.4 shows quantum yield versus wavelength in the n^+-emitter (η_n), p-base (η_p), and junction space-charge (η_d) regions as well as total quantum yield (η_T) of a silicon n^+–p junction solar cell. In this plot the reflection loss at the front surface of the solar cell is assumed equal to zero.

The total photocurrent density generated in the solar cell under the one-sun condition can be obtained by integrating Eq. (12.15) over the entire solar spectrum, which yields

$$J_{ph} = \int_{\lambda_1}^{\lambda_2} J_{sc}(\lambda)\, d\lambda \tag{12.17}$$

where λ_1 and λ_2 denote the cutoff wavelengths at the short- and long-wavelength limits of the solar spectrum, respectively. For a typical p–n junction solar cell, λ_1 can be set at 0.3 μm, and λ_2 is determined by the cutoff wavelength of the semiconductor (i.e., $\lambda_2 = \lambda_g = 1.24\ \text{eV}/E_g$).

Solar Cell Parameters. The equivalent circuit for a p–n junction solar cell is shown in Fig. 12.3d. The photocurrent is represented by a constant current source I_{sc}, which flows in the opposite direction to the dark current I_D. The shunt resistance is represented by R_p, and R_s denotes the series resistance. If we neglect the effects of shunt resistance ($R_{sh} \to \infty$), series resistance ($R_s \sim 0$), and the recombination current in the depletion region, then the I–V characteristics of a p–n junction solar cell under illumination can be expressed by

$$I = I_{ph} - I_{01}[\exp(qV/k_BT) - 1] \tag{12.18}$$

where I_{ph} is given by Eq. (12.17), and I_{01} is the saturation current given by Eq. (12.2). The short-circuit current can be obtained by setting $V = 0$ in Eq. (12.18), so that

$$I_{sc} = I_{ph} \tag{12.19}$$

which shows that the short-circuit current is equal to the photogenerated current. The open-circuit voltage V_{0c} can be obtained by setting $I = 0$ in Eq. (12.18), which yields

$$V_{0c} = V_T \ln\left[\left(\frac{I_{sc}}{I_{01}}\right) + 1\right] \tag{12.20}$$

where $V_T = k_BT/q$ is the thermal voltage. It is seen from Eq. (12.20) that V_{0c} depends on the ratio of the short-circuit current to the dark current. Thus, the open-circuit voltage is increased by keeping the latter ratio as large as possible. This can be achieved by reducing the dark current, either by increasing the substrate doping density or increasing the minority carrier lifetime in the solar cell. Increasing the short-circuit current can also enhance the open-circuit voltage. In practice, the open-circuit voltage can be improved by incorporating a n–n^+ back surface field structure in the p–n junction solar cell. Values of V_{0c} for a silicon p–n junction solar cell may vary between 0.5 and 0.7 V depending on cell structure, dopant densities, and other physical parameters used in the cell fabrication.

The maximum output power of a p–n junction solar cell can be calculated by using the formula

$$P_m = V_m I_m \tag{12.21}$$

where

$$I_m = (I_{sc} + I_{01})\left[\frac{(V_m/V_T)}{(1 + V_m/V_T)}\right] \tag{12.22}$$

is the current corresponding to maximum power output. We note that V_m can be obtained by solving the equation given below using iteration procedures:

$$\exp(V_m/V_T)[1 + V_m/V_T] = \exp(V_{0c}/V_T) \tag{12.23}$$

Another important solar cell parameter known as the *fill factor* (F.F.), which measures the squareness of the photo-I–V curve shown in Fig. 12.2b, is defined by

$$\text{F.F.} = \frac{V_m I_m}{V_{0c} I_{sc}} = \left(\frac{V_m}{V_{0c}}\right)\left[1 - \frac{e^{V_m/V_T} - 1}{e^{V_{0c}/V_T} - 1}\right] \tag{12.24}$$

Depending on the values of the diode ideality factor and the shunt and series resistances, the fill factor for a silicon p–n junction solar cell varies between 0.70 and 0.85, while for a GaAs solar cell it varies between 0.79 and 0.87.

The conversion efficiency of a p–n junction solar cell is defined by

$$\eta_c = \left(\frac{P_{out}}{P_{in}}\right) \times 100\% \tag{12.25}$$

where P_{in} is the input power from the sunlight and P_{out} is the output power from the solar cell. The input powers from sunlight under one-sun AM0, AM1, AM1.5, and AM2 conditions are 135.3, 92.5, 84.4, and 69.1 mW/cm^2, respectively.

Design Considerations. It is clear from the above analysis that the performance of a solar cell is determined by several factors. Thus, in order to obtain optimal cell design, it is important to consider all the key parameters which limit the conversion efficiency of a solar cell. These are discussed now.

Spectral Response. An important consideration in material selection for solar cell fabrication is to select a semiconductor with an energy bandgap which is matched with the peak emission power density of the solar insolation spectrum. This will provide maximum absorption of the incident solar power by the solar cell, and hence will enable the cell to produce the optimum spectral response. Figure 12.5 shows the maximum theoretical conversion efficiency of an ideal p–n junction solar cell versus energy bandgap for different semiconductor materials.[2] We note that a single-junction GaAs solar cell has a maximum theoretical conversion efficiency of around 28% under the AM0 condition, while a silicon p–n junction solar cell has a maximum theoretical conversion efficiency of around 21% under the AM0 condition and at room temperature. The reasons that a GaAs solar cell has a higher conversion efficiency than a silicon cell are the following: (a) the bandgap energy (E_g = 1.43 eV) for GaAs is better matched with the peak solar insolation spectrum than that of silicon, and (b) GaAs is a direct bandgap material while silicon is an indirect bandgap material.

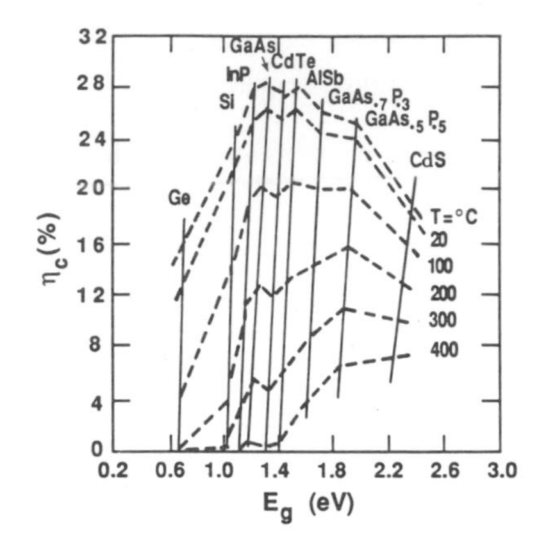

FIGURE 12.5. Maximum theoretical AM0 conversion efficiency versus energy bandgap for different semiconductor materials. Dashed lines denote conversion efficiencies under different operating temperatures (0 to 400 °C). After Wysocki and Rappaport,[2] by permission.

Series Resistance. The series resistance, which is contributed by the contact and bulk resistances of the cell, can affect the shape of the photo-I–V curve, fill factor, and conversion efficiency of a solar cell. For example, a large series resistance will increase the internal power dissipation and reduce the fill factor and output power of the solar cell. To decrease the effect of series resistance, both contact and bulk resistances must be minimized. The contact resistance can be greatly reduced if optimal front contact grids are used. Unfortunately, reducing the sheet resistance of a solar cell is not an easy task. One way to reduce sheet resistance is to employ a heavily doped surface layer in the cell. However, this in turn will reduce the minority carrier lifetime and diffusion length in the surface layer. This is not desirable, since it will also decrease the short-circuit current. Therefore, a compromise between the doping density and junction depth is necessary in order to achieve optimum design. There are a number of ways of making front contact grids. The most common such grids are formed with rectangular-shaped metal grids (fingers), each grid line is equally spaced and set on top of the front surface of the solar cell. This is shown in Fig. 12.2a for a n^+–p–p^+ silicon p–n junction solar cell. It is seen that using the contact grid structure allows the exposure of a major portion of the solar cell surface to sunlight and at the same time keeps the series resistance to a minimum value. The area covered by the front metal contact grids is usually less than 10% of the total solar cell area.

Antireflection (AR) Coatings. Another important factor which must be considered in solar cell design is reflection loss at the solar cell surface. For example, as much as 30 to 35% of the incident sunlight in the visible spectral range is reflected back from the bare surface of a silicon solar cell without AR coating. Therefore, it is important to reduce the surface reflection loss by using proper AR coatings on the solar cell. To illustrate the effect of reflection loss on the quantum efficiency of a solar cell, Fig. 12.6 shows the spectral response of a normal

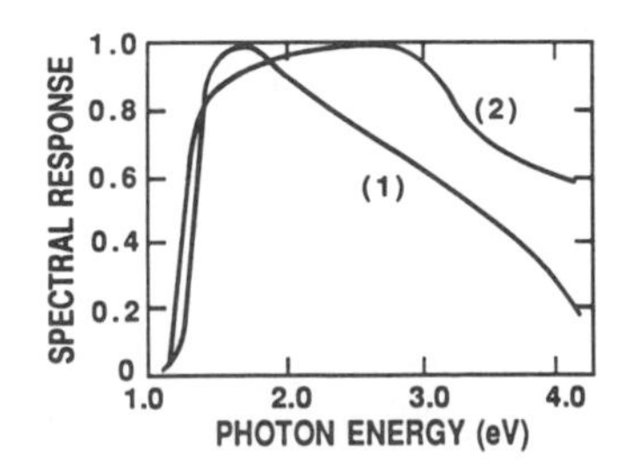

FIGURE 12.6. Relative spectral responses of (1) a normal silicon p–n junction solar cell, and (2) a violet (i.e., nonreflecting) silicon p–n junction cell which utilizes a texturized front surface to reduce reflection loss to near-zero. After Lindmayer and Allison,[3] by permission.

silicon p–n junction solar cell (curve 1) and a violet silicon solar cell (curve 2).[3] The reflection loss in the violet cell is reduced to near-zero from the visible to the UV wavelength regime by using a texturized front surface. The texturized grooves on the front surface of the cell are obtained by using a preferential etching technique. As shown in Fig. 12.6, the short-wavelength response of a violet cell is significantly improved over the normal cell as a result of using the texturized surface.

The most widely used technique for achieving near-zero reflection loss is by applying an antireflection (AR) coating on the front surface of a solar cell. This is usually accomplished by depositing a thin dielectric film of Ta_2O_5, SiO_2, or Si_3N_4 on the front side of the solar cell with thickness equal to one-quarter wavelength of selected incident monochromatic light corresponding to the peak of the solar insolation spectrum (i.e., $\lambda \sim 0.45$–$0.5\ \mu$m). The thickness of the dielectric film for a single AR coating can be calculated by the expression

$$d = \frac{\lambda_0}{4n_1} \tag{12.26}$$

where λ_0 is the wavelength of incident sunlight at a selective wavelength, and n_1 is the refractive index of the dielectric film used for AR coating. For example, using SiO_2 film ($n_1 = 1.5$) for AR coating on a silicon solar cell, the film thickness calculated from Eq. (12.26) is found to equal 800 Å at $\lambda_0 = 4800$ Å. The minimum reflection loss for a quarter-wavelength AR coating may be calculated from the expression

$$R_{min} = \left(\frac{n_1^2 - n_0 n_2}{n_1^2 + n_0 n_2}\right)^2 \tag{12.27}$$

where n_0, n_1, and n_2 are the refractive indices of the air, AR coating film, and the solar cell material, respectively. From Eq. (12.27), it is found that a silicon solar cell coated with a 1100-Å-thick SiO_2 film has a reflection loss of only 7%, which is a drastic improvement over that of a silicon solar cell without using the AR coating ($R = 35\%$). Among the various AR coating materials used today, Ta_2O_5 (with $n_1 = 2.25$) is probably the most widely used dielectric film for AR coating on solar cells. For example, by applying a 700-Å Ta_2O_5 AR coating on a silicon solar cell, the reflection loss can be reduced to about 5%. Therefore, it is evident that by carefully selecting a suitable AR coating film, it is possible to reduce the reflection loss of a solar cell to near-zero. It is noteworthy that aside from the methods cited above for improving solar cell performance, there are other means which may be employed to further improve the performance and conversion efficiency of a solar cell. For example, the short-wavelength spectral response may be improved by using a shallow junction structure with a thin emitter in a p–n junction solar cell, and the open-circuit voltage can be increased by incorporating the back surface-field structure (i.e., n/n^+ or p/p^+) and increasing the doping density in the base region. Theoretical calculations reveal that a maximum open-circuit voltage of around 0.7 V for a silicon n^+–p junction solar cell can be achieved by using a 0.1 ohm · cm silicon material for the p-base.

For a GaAs p–n junction solar cell, the short-wavelength spectral response and contact resistance can be greatly improved with the aid of a wide-bandgap p^+-$Al_{0.95}Ga_{0.05}As$ window layer of 0.3 to 0.5 μm thickness on top of the GaAs p–n junction structure. Figure 12.7 shows the cross-section view of a high-efficiency AlGaAs/GaAs p–n junction solar cell. The reason for using a thin highly doped p^+-AlGaAs wide-bandgap window layer on top of the p^+-GaAs emitter layer is to reduce the surface recombination velocity and series resistance of the GaAs solar cell. Since the $Al_{0.9}Ga_{0.1}As$ window layer is a wide-bandgap ($E_g = 2.1$ eV) material, it is transparent to most of the visible sunlight. Figure 12.8 shows spectral response curves for a GaAs p–n junction solar cell with (curve 2) and without (curve 1) an AlGaAs window layer. It is clearly shown that adding an AlGaAs window layer to the GaAs p–n junction solar cell can indeed produce a significant improvement in the short-wavelength response.

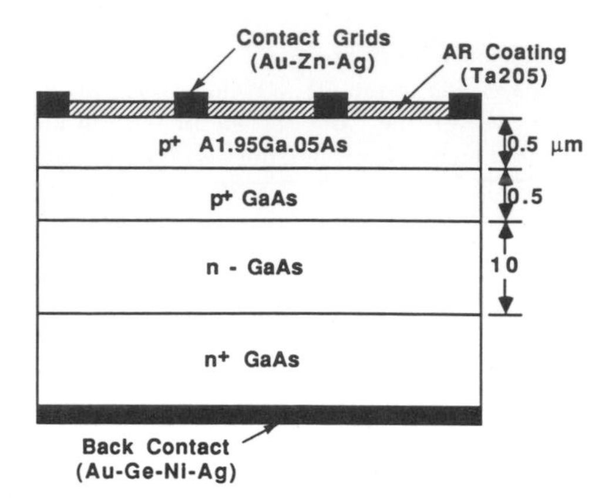

FIGURE 12.7. Cross-section view of a GaAs p–n junction solar cell with a p^+-AlGaAs window layer and a Ta_2O_5 layer for antireflection coating.

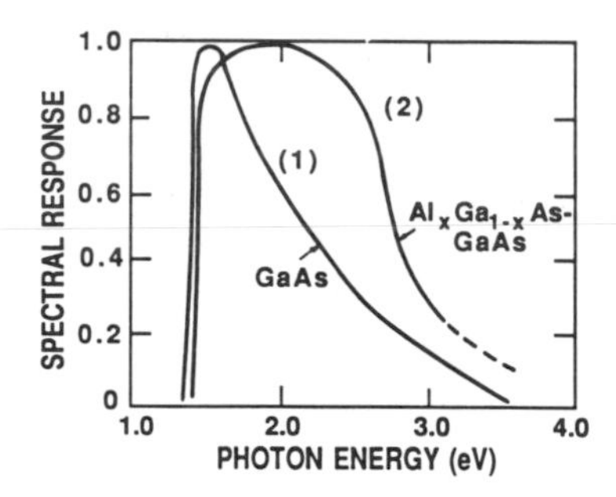

FIGURE 12.8. Spectral response curves (i.e., normalized quantum yield) for a GaAs p–n junction solar cell: (1) without window layer and (2) with an AlGaAs window layer.

12.2.2. Schottky Barrier and MIS Solar Cells

Although most commercial solar cells use a p–n junction structure, other structures such as Schottky barrier, MIS (metal–insulator–semiconductor), heterojunction, and multijunction structures have also been used. The Schottky barrier solar cell is easy to fabricate and possesses the simplest structure. It has a better spectral response at the shorter-wavelength regime, and hence can produce high short-circuit current. However, the conversion efficiency of a Schottky barrier solar cell is usually lower than that of a p–n junction solar cell due to its low open-circuit voltage, which is a result of the high dark current due to the inherent barrier-height limitation (i.e., the thermionic emission current is directly related to the barrier height of a Schottky barrier solar cell).

A Schottky barrier solar cell can be fabricated by using either a thin semitransparent metal film or a grating-type structure deposited on a semiconductor substrate to form Schottky contacts. Figure 12.9a shows the cross-section view of a Schottky barrier solar cell with a 100-Å semi-transparent metal film for Schottky contact, and Fig. 12.9b presents its energy band diagram under illumination conditions.

In a Schottky barrier solar cell, the photocurrents are generated in the depletion and base regions of the solar cell. The collection of electron–hole pairs in the depletion region is similar to that of a p–n junction solar cell discussed in the previous section. The excess carriers generated in the depletion region are swept out by the built-in electric field in this region, leading to a photocurrent density per unit bandwidth given by

$$J_d(\lambda) = q \int_0^W T(\lambda)\alpha\phi_0(\lambda)\, e^{-\alpha x}\, dx$$

$$= qT(\lambda)\phi_0(\lambda)(1 - e^{-\alpha W}) \qquad (12.28)$$

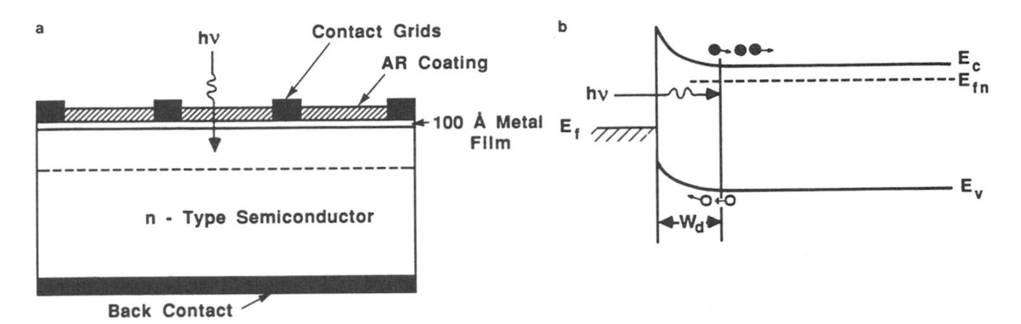

FIGURE 12.9. (a) Cross-section view and (b) energy band diagram of a metal–n-type semiconductor Schottky barrier solar cell under illumination conditions.

where $\phi_0(\lambda)$ is the incident photon flux density at wavelength λ, $T(\lambda)$ is the transmission coefficient of the metal film, and W is the depletion layer width given by

$$W = \sqrt{\frac{2\varepsilon_0\varepsilon_s(V_d - V)}{qN_D}} \tag{12.29}$$

The photocurrent given by Eq. (12.29) is similar to that given by Eq. (12.14) for a p–n junction cell, except that in the latter case the transmission coefficient of light through the metal film [i.e., $T(\lambda)$] for the Schottky contact is replaced by the transmission $(1 - R)$ coefficient of light though a p–n junction solar cell.

The collection of photocurrent in the quasi-neutral base region of a Schottky barrier solar cell is similar to that in the base region of a p–n junction cell. Thus, the photocurrent density due to holes collected in the n-base region can be expressed by

$$J_p(\lambda) = \frac{q\phi_0\alpha L_p}{(1 + \alpha L_p)} T(\lambda)\, e^{-\alpha W} \tag{12.30}$$

This equation is obtained by assuming that the cell thickness is much larger than the hole diffusion length in the base region. The total photocurrent generated in a Schottky barrier solar cell due to monochromatic light of wavelength λ is equal to the sum of Eqs. (12.28) and (12.30). Thus, the total photocurrent generated by sunlight can be obtained by integrating the single-wavelength photocurrent from the UV (λ_1) to the cutoff wavelength (λ_2) of the semiconductor material, namely,

$$J_{ph} = \int_{\lambda_1}^{\lambda_2} [J_d(\lambda) + J_p(\lambda)]\, d\lambda \tag{12.31}$$

where $\lambda_1 \sim 0.3\ \mu$m and $\lambda_2 = \lambda_g$, the cutoff wavelength of the semiconductor.

Under forward-bias conditions, the dark current in a Schottky barrier solar cell is due primarily to thermionic emission of majority carriers in the bulk semiconductor, and is given by

$$J_D = J_s[\exp(V/nV_T) - 1] \tag{12.32}$$

where $J_s = A^*T^2 \exp(-\phi_{Bn}/V_T)$ is the saturation current density, while A^* is the effective Richardson constant equal to 110 A/cm$^2 \cdot$ K^2 for n-type silicon and 4.4 A/cm$^2 \cdot$ K^2 for n-type GaAs. Since the barrier height is generally lower than the bandgap energy of the semiconductor, one expects that the saturation current for a Schottky barrier solar cell will be much higher than that of a p–n junction solar cell. As a result, the open-circuit voltage for a Schottky barrier solar cell is expected to be lower than that of a p–n junction cell. To overcome this problem, barrier height enhancement techniques described in Section 10.8 may be applied to the Schottky barrier solar cell in order to obtain high open-circuit voltage and conversion efficiency.

The open-circuit voltage, fill factor, maximum power output, and conversion efficiency for a Schottky barrier solar cell can be calculated in a similar way to that of a p–n junction solar cell discussed earlier. The short-circuit current and dark current can be calculated using Eqs. (12.28) through (12.33). Typical values of V_{0c} for a silicon Schottky barrier cell may vary between 0.4 and 0.55 V. A fill factor of 0.6 to 0.76 and AM1 conversion efficiency of 12 to 16% are achievable for this type of solar cell.

Another method of enhancing the open-circuit voltage of a Schottky barrier solar cell is by using a MIS structure. In this structure, a thin insulating layer of thickness 10 to 20 Å is formed between the metal Schottky contact and the semiconductor, which results in a MIS solar cell structure. This structure can increase the effective barrier height ($\Delta\phi_B = \delta\chi^{1/2}$), and hence reduce the dark current of the cell. As a result, the open-circuit voltage of a MIS solar cell is usually higher than that of a conventional Schottky barrier cell. Figure 12.10a shows the cross-section view of a MIS Schottky barrier solar cell and Fig. 12.10b presents its energy band diagram under illumination conditions. In a MIS solar cell, current conduction under dark conditions is due to majority carriers tunneling through the thin insulating layer. This tunneling current is given by

$$J_t = A^*T^2 \exp(-\phi_{Bn}/V_T) \exp(-\delta\chi^{1/2}) \exp(V/nV_T) \tag{12.33}$$

where δ is the oxide thickness in Å, χ is the mean incremental barrier height, and ϕ_{Bn} is the barrier height without the thin insulating layer (i.e., $\delta = 0$). Equation (12.33) reduces to Eq. (12.32) for $\delta = 0$. It is seen from Eq. (12.33) that the thin insulating layer in a MIS structure will only limit the majority carrier flow and not the minority carrier flow (or the photocurrent) as long as the thickness of the insulating film remains very thin (e.g., ≤ 30 Å). Thus, the open-circuit voltage of a MIS solar cell will be higher than that of a conventional Schottky barrier solar cell. The open-circuit voltage of a MIS solar cell can be derived from Eqs. (12.20) and (12.33) in the form

$$V_{0c} = nV_T\left[\ln\left(\frac{J_L}{A^*T^2}\right) + \frac{\phi_{Bn}}{V_T} + \delta\chi^{1/2}\right] \tag{12.34}$$

Conversion efficiencies as high as 15% for a Au–Si MIS cell and 17% for a Au–GaAs MIS cell under AM1 conditions have been reported. The main drawback for the MIS solar

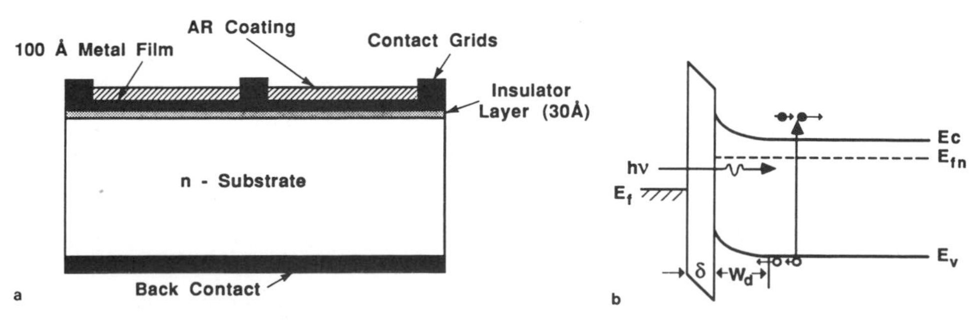

FIGURE 12.10. (a) Cross-section view and (b) energy band diagram of a MIS solar cell under illumination conditions.

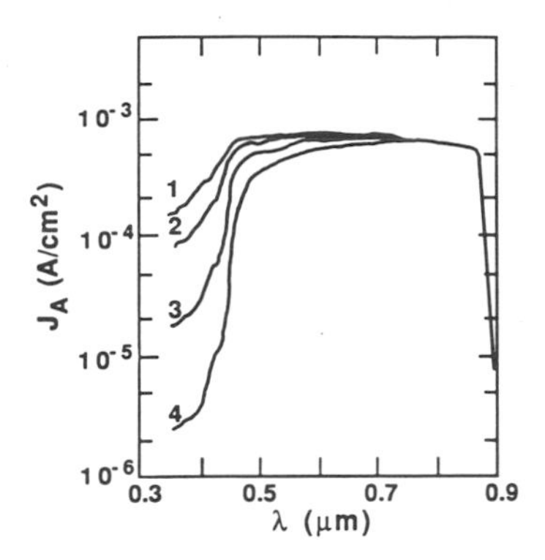

FIGURE 12.11. Calculated photocurrent density versus wavelength for a Au–p$^+$–n GaAs Schottky barrier solar cell under AM0 conditions. The dopant density of the n-GaAs substrate is $N_d = 10^{16}$ cm^{-3}, and the thicknesses and dopant densities for curves 1 through 4 are given by: curve 1, $N_a = 8.2 \times 10^{16}$ cm^{-3} and $W_p = 100$ Å; curve 2, $N_a = 2.2 \times 10^{16}$ cm^{-3} and $W_p = 200$ Å; curve 3, $N_a = 4.4 \times 10^{16}$ cm^{-3} and $W_p = 500$ Å; curve 4, $N_a = 8.2 \times 10^{16}$ cm^{-3} and $W_p = 1000$ Å. After Li,[4] by permission.

cell lies in the precision control of the thin insulating film thickness. If the insulating film thickness exceeds 30 Å, photocurrent suppression results. This, in turn, will lower the open-circuit voltage and the conversion efficiency of the MIS solar cell. An alternative approach for solving problems associated with low barrier height and high dark current in a Schottky barrier solar cell is to introduce a thin semiconductor layer of opposite doping type to the substrate in order to form a metal–p$^+$–n or metal–n$^+$–p Schottky barrier structure, as depicted in Section 10.8. Using this method, enhancement of the effective barrier height for a Au–p$^+$–n and Au–n$^+$–p GaAs Schottky solar cell can be easily achieved, as shown earlier in Figs. 10.21 and 10.22, respectively. Figure 12.11 shows the calculated photocurrent density versus wavelength for an ideal Au–p$^+$–n GaAs Schottky barrier solar cell under AM0 conditions for four different p-layer dopant densities and thicknesses. Theoretical conversion efficiencies as high as 21% for such a solar cell structure can be obtained under AM0 conditions.

Finally, since a Schottky barrier solar cell offers several advantages such as low cost, simple structure, ease of fabrication, and low-temperature processing, it is clear that using a Schottky barrier structure can be an attractive and viable approach to fabricating low-cost photovoltaic systems for terrestrial power generation.

12.2.3. Heterojunction Solar Cells

A p–n heterojunction solar cell can be formed by using two different types of semiconductor material with different bandgaps and opposite dopant impurities. For example, a p–n heterojunction solar cell can be fabricated by using a p$^+$-AlGaAs/n-GaAs, p$^+$-GaAs/n-Ge, or p$^+$-Cu$_x$S/n-CdS material system. Figure 12.12a shows the energy band diagram for a n$^+$-GaAs/p-Ge heterojunction solar cell in equilibrium. In this heterojunction structure, wide-bandgap ($E_{g1} = 1.43$ eV) n-GaAs is used as an emitter layer, while the smaller-bandgap ($E_{g2} = 0.67$ eV) p-Ge is used as a base layer.[5] The distinct feature of a p–n heterojunction solar cell lies in its window effect in which photons possessing energies between E_{g1} and E_{g2} can pass through the wide-bandgap window layer, and are absorbed in the smaller-bandgap base layer. The window layer is usually heavily doped, and has a thickness of a few tenths of a micrometer. Therefore, with the addition of a window layer, the sheet resistance of the heterojunction solar cell can be reduced. This is important for reducing the internal power loss of the solar cell. In general, the output power and conversion efficiency of a heterojunction solar cell is determined primarily by the photocurrent produced in the smaller-bandgap base layer.

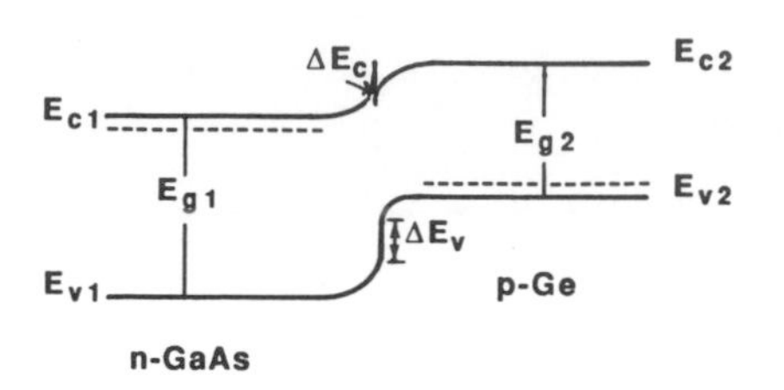

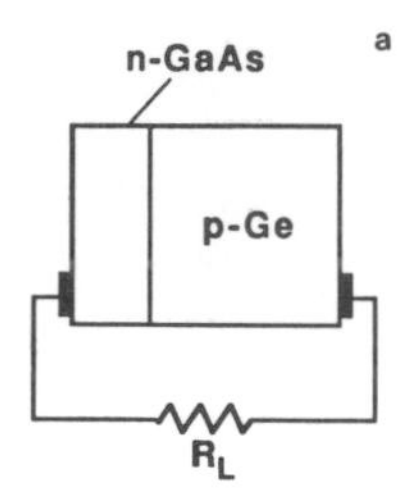

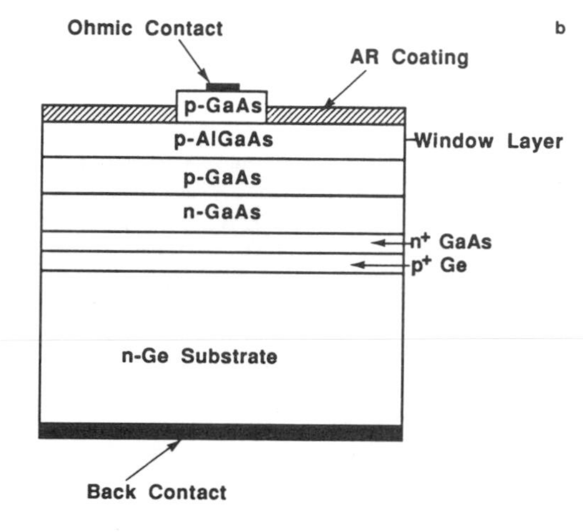

FIGURE 12.12. (a) Energy band diagram of a n-GaAs/p-Ge heterojunction solar cell in equilibrium. (b) Cross-section view of a high-efficiency MOCVD-grown GaAs/Ge monolithic tandem solar cell. After Tobin *et al.*,[5] by permission, © IEEE-1988.

The current collection mechanism for a p–n heterojunction solar cell is similar to that of a p–n homojunction solar cell. The main contribution to the photocurrent comes from the base region, with smaller contributions coming from the top emitter layer and the depletion region. The photocurrent for a p–n heterojunction solar cell can be derived in a similar way to that of a p–n homojunction solar cell discussed earlier.

The p–n heterojunction solar cell usually has better short-wavelength response, lower series resistance, and better radiation tolerance than a conventional p–n homojunction solar cell. In order to obtain the maximum short-circuit, open-circuit voltage, and conversion efficiency, it is essential that materials selected for fabricating the heterojunction cell must have good lattice match and compatible thermal expansion coefficients. Energy band discontinuities at the heterointerface of a p–n heterojunction cell must be minimized to avoid barrier formation at the heterointerface where photocurrent collection can be severely degraded. Several heterojunction pairs with good lattice matches have been reported in the literature. These include AlGaAs/GaAs, AlAs/GaAs, GaP/Si, ZnS/Si, Cu_2S/CdS, and $CuInSe_2$/CdS, among others.

It is interesting to compare the characteristics of a heterojunction solar cell with a Schottky barrier solar cell. The most striking similarity is that short-wavelength photons can be absorbed within or very near the surface region of both cells, leading to an excellent short-wavelength response. However, the open-circuit voltage of a heterojunction solar cell can be much higher than that of a Schottky barrier solar cell due to the use of a larger-bandgap material for the top layer of the solar cell. As a result high conversion efficiency can be expected in a heterojunction solar cell. In fact, an $Al_xGa_{1-x}As$–GaAs p–n heterojunction solar cell with AM1 conversion efficiency as high as 21.5% has been reported. It has been shown that a heterojunction solar cell is more radiation-tolerant to low-energy protons and 1-MeV electron irradiation than that of a conventional p–n junction solar cell, because a thicker window layer can be used in a heterojunction solar cell to cut down radiation damage on the cell without losing short-circuit current and conversion efficiency.

Another type of heterojunction solar cell using wide-bandgap conducting glass such as indium oxide (In_2O_3), tin oxide (SnO_2), or indium tin oxide (ITO) has also been reported in the literature. These highly conducting glasses which have bandgap energies varying from 3.5 to 3.7 eV are n-type semiconductors, and can be deposited on top of a p-type silicon substrate to form a n^+-ITO/p-Si heterojunction solar cell. The ITO film has a typical thickness of around 4000 Å and resistivity of 5×10^{-4} ohm · cm. Conversion efficiencies of 12 to 15% have been reported for the ITO/Si n–p junction solar cells.

In order to fully utilize the solar spectrum and to increase the conversion efficiency of a solar cell, multijunction (or cascade) solar cell structures have also been widely investigated in recent years. For example, a high-efficiency GaAs/Ge monolithic tandem solar cell grown by a metallorganic chemical vapor deposition (MOCVD) technique has been reported recently. This tandem cell structure is shown in Fig. 12.12b, which consists of a 4-μm front AR coating, a 0.46-μm p^+-GaAs front contact layer, a 0.03-μm p^+-AlGaAs window layer, a 0.5-μm p-GaAs emitter and a 2.6-μm n-GaAs base layer for the top cell, a 1.7-μm-thick n^+-GaAs buffer layer, and a p^+ (1 μm)–n (200 μm) Ge bottom cell. AMO conversion efficiency of 21.7% and AM1.5 conversion efficiency of 24.3% have been obtained for this tandom solar cell.

12.2.4. Thin Film Solar Cells

The thin film solar cell is another potential candidate for low-cost terrestrial photovoltaic power-generation applications. A thin film solar cell (film thickness ≤ 10 μm) uses thin active layers to form a p–n junction or a Schottky barrier structure on a foreign substrate. The active layer may be a polycrystalline thin film or an amorphous film. A number of semiconductors including CdS, $CuInSe_2$, polycrystalline silicon, GaAs, InP, and amorphous silicon (a-Si) have been investigated for thin film solar cell fabrication. Substrate materials including a variety of low-cost materials such as ceramic, glass, graphite, aluminum, and metallurgical grade silicon have been studied for depositing thin active layers on solar cells.

Among the various thin film solar cells, the a-Si thin film solar cell has shown great potential for low-cost large-scale photovoltaic power generation. However, the problems associated with long-term stability in the a-Si solar cells need to be resolved before large-scale production of this type of solar cell system can be realized. Most of the a-Si solar cells made today are being used in powering calculators, watches, toys, and cameras. The a-Si solar cells are formed by depositing a 1- to 3-μm-thick a-Si thin film by RF glow-discharge decomposition of silane (which produces 10% hydrogenated a-Si) onto metal or ITO (indium–tin–oxide) coated glass substrates. Figure 12.13a shows a schematic drawing of an a-Si p–i–n solar cell

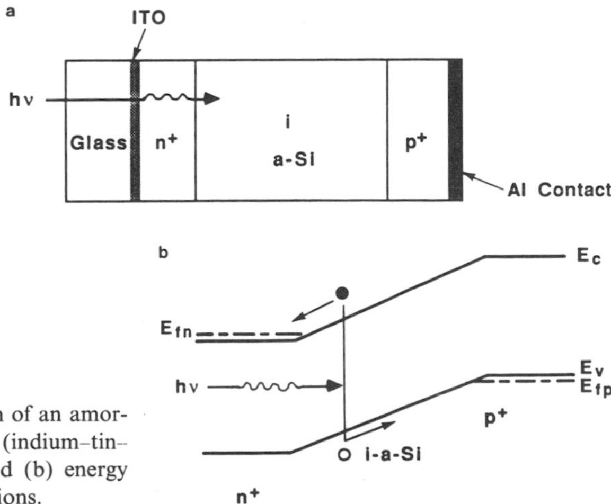

FIGURE 12.13. (a) Schematic diagram of an amorphous-Si p–i–n solar cell grown on ITO (indium–tin–oxide conducting film) coated glass, and (b) energy band diagram under illumination conditions.

fabricated on ITO coated glass substrate, and Fig. 12.13b shows its energy band diagram under illumination conditions. It is of interest that the optical characteristics of a hydrogenated a-Si resembles a direct bandgap material with energy bandgap E_g equal to 1.7 eV. As a result, a-Si has a much higher absorption coefficient at $h\nu = 1.7$ eV than that of single-crystal silicon. Thus, the short-wavelength spectral response (i.e., blue-green) for an a-Si solar cell is much better than that of a single-crystal silicon solar cell. The most commonly used structure for an a-Si solar cell is a p–i–n structure deposited on the conducting ITO coated glass substrate. In this structure, the p^+ and n^+ layer thickness is about 100 Å, and the intrinsic (i) layer thickness varies between 0.5 and 1.0 μm. AM1 conversion efficiency as high as 11% has been reported for such an a-Si solar cell.

12.3. PHOTODETECTORS

Photodetectors and laser diodes are two important active devices for applications in optoelectronics. Since a large number of semiconductor laser diodes and LEDs have been developed for use in a broad wavelength range from the visible to the infrared spectrum, it is equally important to have a variety of photodetectors available for detection in the corresponding wavelengths of laser diodes and LEDs. In millimeter-wave fiber optic communications, the detectors must possess such features as low noise, high responsivity, and large bandwidth. Although high-sensitivity photomultipliers and traveling-wave phototubes are widely used for detecting modulated optical signals at microwave frequencies, however, recent trends are toward the use of various solid state photodetectors including Schottky barrier photodiodes, p–i–n photodiodes, and avalanche photodiodes fabricated from elemental and compound semiconductors. A GaAs Schottky barrier photodetector with cutoff frequency exceeding 100 GHz has been reported recently. High-speed photodetectors are particularly attractive for millimeter-wave fiber optic links. Long-wavelength $In_{0.53}Ga_{0.47}As$ p–i–n photodiodes with bandwidth greater than 30 GHz have also been developed for 1.3- to 1.6-μm optical communications. Avalanche photodiodes fabricated from InGaAs/InP, silicon, and germanium with high internal current gain have also been widely reported.

In this section, various photodetectors including p–n junction, p–i–n diode, avalanche photodiode, Schottky barrier, point contact, and heterojunction photodiodes will be discussed. Since most of these photodiodes are based on depletion-mode operation, they offer high-speed and high-sensitivity detection. A comparison of various solid state photodetectors reveals that the intrinsic photoconductor can provide the highest internal gain ($\simeq 10^4$), while the Schottky barrier photodiode has the shortest response time (10^{-11} sec) and largest bandwidth. On the other hand, the avalanche photodiode has the highest gain–bandwidth product.

A depletion-mode photodiode usually operates under small reverse-bias conditions. Under depletion-mode operation, the reverse saturation current (or dark current) is superimposed by the photocurrent due to electron–hole pairs generated by incident photons in the photodiode. The applied reverse bias is in general not high enough to cause avalanche multiplication, and hence no internal current gain is expected in this operation mode. This is in contrast to the avalanche photodiode in which an internal current gain is expected as a result of avalanche multiplication near the breakdown conditions.

Before presenting the various photodetectors, we shall first discuss some of the key performance parameters such as spectral response, response speed, and noise figure for a photodiode under depletion-mode operation. The spectral response of a photodiode is determined in the wavelength range in which an appreciable photocurrent can be measured in a photodiode. The key physical parameter which affects the spectral response is the optical absorption coefficient of the semiconductor from which the photodiode is fabricated. The cutoff wavelength λ_c of a photodiode is determined by the energy bandgap of the semiconductor. For example, the energy bandgap of silicon is equal to 1.12 eV at room temperature, and hence the cutoff wavelength for a silicon photodiode is around 1.1 μm (i.e., $\lambda_c = 1.24\ \text{eV}/E_g$). Germanium has an energy bandgap of 0.67 eV at 300 K, and hence its cutoff wavelength is around 1.8 μm. The short-wavelength limit is set by the wavelength in which the absorption coefficient of the semiconductor is in excess of $10^5\ \text{cm}^{-1}$. For wavelengths shorter than this value, the absorption of photons takes place mostly near the surface of the photodiode, and hence the electron–hole pairs generated in this region may recombine right near the surface. Therefore, for detectors with large surface recombination velocity, the photocurrent produced by short-wavelength photons can be greatly reduced.

The three most commonly used figures of merit for evaluating the performance of a photodetector are quantum efficiency (η), noise equivalent power (NEP), and detectivity (D^*).

Quantum efficiency, which is widely used in assessing the spectral response of a photodiode, can be defined by

$$\eta = \left(\frac{I_{ph}/q}{P/h\nu}\right) \times 100\% = \left(\frac{I_{ph}}{P\lambda}\right) \times 124\% \qquad (12.35)$$

where I_{ph} (A) is the photocurrent generated when a light beam of input power P (watt) and frequency ν falls onto the active area of the photodiode. The quantum efficiency η is determined at low reverse-bias voltage in which no avalanche multiplication takes place. In Eq. (12.35), h is the Planck constant, q is the electronic charge, and λ is the wavelength of the incident light. As an example, let us consider a silicon p–n junction photodiode. If 0.8-μm monochromatic light with input power of 5 mW is illuminated on this photodiode, and the quantum efficiency of the photodiode is equal to 80%, then the photocurrent produced in this photodiode is equal to 2.58 mA. Figure 12.14 shows relative spectral response curves for some selected Schottky barrier and p–i–n photodiodes. Another important figure of merit related to the quantum efficiency is known as the responsivity, which is defined by the ratio of the photocurrent to the input optical power (i.e., $R = I_{ph}/P$ A/W).

The noise equivalent power (NEP) is another figure of merit widely used in evaluating the performance of an infrared (IR) photodetector. By definition, the NEP of an IR detector and its associated amplifier is the RMS (root mean square) value of the sinusoidally modulated optical power falling on a detector which gives rise to a RMS noise voltage referred to the detector terminal at a reference bandwidth of 1 Hz. For a monochromatic radiant flux (Φ_p) with wavelength λ necessary to produce a RMS signal-to-noise ratio of 1 at frequency f, the

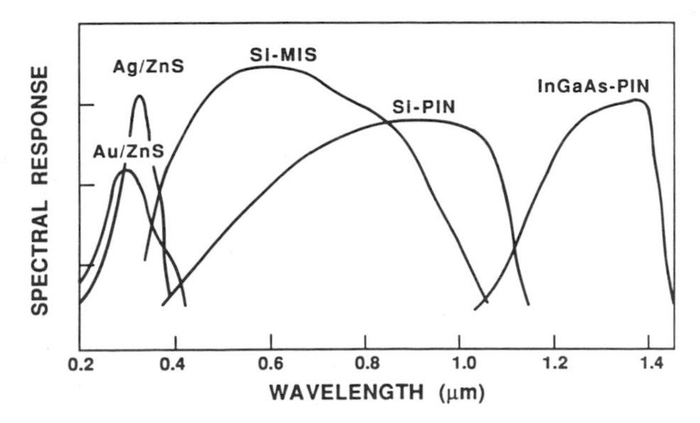

FIGURE 12.14. Relative spectral responses for several selected Schottky barrier, MIS, and p–i–n photodiodes.

NEP is defined by

$$\mathrm{NEP}(\lambda, f) = \frac{hcq\Phi_p}{\lambda} \qquad \text{watt} \tag{12.36}$$

where hc/λ is the incident photon energy in eV, and Φ_p is the RMS photon flux (i.e., photons/sec) required to produce the signal-to-noise ratio (SNR) of 1. Equation (12.36) enables one to estimate the values of NEP due to different noise sources for a detector. It should be noted that the major limitation of the NEP is due to two additional parameters, namely, the noise bandwidth (Δf) and the detector area (A_d), which must be given. Both are related to noise considerations— different Δf gives different noise values and smaller areas collect less power. In addition to the spectral NEP defined above, we can also define the blackbody NEP, which is defined as the blackbody radiant flux necessary to produce a RMS signal-to-noise ratio of 1 at frequency f. As an example, we consider a background-noise-limited detector performance which is common in the IR spectral region. The spectral NEP for a background-limited IR detector (BLIP) can be expressed by

$$\mathrm{NEP}(\lambda, f) = \frac{hcq\Phi_p}{\lambda} = \frac{hc}{\lambda}\left(\frac{2\phi_p^{BG}\Delta f}{\eta}\right)^{1/2} \qquad \text{(watt)} \tag{12.37}$$

where ϕ_p^{BG} is the background photon flux (photons/sec) that falls on the detector, and η is the quantum efficiency.

Another figure of merit commonly used in an IR detector is known as the spectral detectivity $D^*(\lambda, f)$, which is defined as the signal-to-noise ratio normalized per unit area per unit noise bandwidth, and is given by

$$D^*(\lambda, f) = \frac{A_d^{1/2}(\Delta f)^{1/2}}{\mathrm{NEP}} \qquad (\mathrm{cm} \cdot \mathrm{Hz}^{1/2}/\mathrm{W}) \tag{12.38}$$

The spectral detectivity is usually used as a figure of merit for comparing the signal-to-noise performance of photodiodes having different active areas and operating at different noise bandwidths. In general, both D^* and NEP are frequently used in assessing the performance of an IR detector, while the quantum efficiency and responsivity are often used in assessing the spectral response of a photodiode operating in the visible to near-infrared spectral ranges. Equations (12.37) and (12.38) enable the spectral detectivity for a BLIP to be expressed in the form

$$D^*(\lambda, f) = \frac{hc}{\lambda}\left(\frac{\eta A_d}{2\phi_p^{BG}}\right)^{1/2} \tag{12.39}$$

We next discuss the response speed of a photodetector. In general, the response speed of a photodetector depends on three key factors, namely, the carrier diffusion time in the bulk quasi-neutral regions, the carrier drift transit time across the depletion layer, and the RC time constant of the detector system. In a depletion-mode photodiode, excess electron–hole pairs are generated inside the depletion region and the quasi-neutral regions of the photodiode, and are collected as photocurrent across the junction of the photodiode. Since the minority carrier diffusion in the quasi-neutral region is usually slower than the drift of excess carriers in the depletion region, high-speed detection is achieved by generating excess carriers inside the depletion region of the junction or close to the junction so that the diffusion time of the excess carriers is comparable to the transit time across the depletion range. For most semiconductors, the saturation limited velocity of the photogenerated excess carriers generated inside the junction space-charge region of a photodiode is about 1 to 2×10^7 cm/sec. Since the depletion layer width for most p–n junction photodiodes is only a few μm or less, the carrier transit time in the picosecond range can be readily obtained for a depletion-mode photodetector. Since the response speed or bandwidth of a depletion-mode photodetector is determined by the three time constants discussed above, the 3-dB cutoff frequency for a p–n junction or Schottky barrier photodiode can be calculated from the expression

$$f_c = \frac{0.35}{(t_{tr}^2 + t_{dif}^2 + t_{RC}^2)^{1/2}} \tag{12.40}$$

where

$$t_{tr} = \frac{W_d}{2.8 v_s} \tag{12.41}$$

$$t_{dif} \simeq \frac{W_p}{2.43 \tau_n} \tag{12.42}$$

and

$$t_{RC} = \frac{1}{RC} \tag{12.43}$$

Here, t_{tr} is the carrier transit time across the depletion layer region of width W_d, v_s ($\sim 10^7$ cm/sec) is the saturation velocity, t_{dif} and W_p denote the electron diffusion time constant and the width of the p-base region, τ_n is the electron lifetime, and t_{RC} is the RC time constant.

In a practical detector system, however, the cutoff frequency of a photodetector is usually lower than that predicted by Eq. (12.40) due to finite load resistance and stray capacitances from the load resistance and the amplifier circuit. Fast photodiodes may be fabricated by using a planar structure on a semi-insulating substrate with small active area (e.g., diameter less than 10 μm) to keep the diode capacitance and series resistance (or RC time constant) low. The point-contact Schottky barrier photodiode has the highest response speed and bandwidth among all photodetectors discussed in this section.

For any photodetector, the ultimate limitation on its performance is the noise generated in the detector. In general, the noises generated in a photodiode under reverse-bias conditions consist of the shot noise, $1/f$ noise (or flicker noise), and thermal noise (or Johnson noise). The shot noise is created by the reverse leakage current flowing through the photodiode and is given by

$$i_s^2 = 2qI_D\Delta f \tag{12.44}$$

where I_D is the dark current and Δf denotes the noise-equivalent bandwidth. For frequencies below 1 kHz, the noise of a photodiode is usually dominated by $1/f$ noise (e.g., $I_f^2 = BI_{dc}\Delta f/f$), which has a current-dependent power spectrum inversely proportional to the signal frequency. The origin of the flicker noise can be attributed to fluctuation associated with generation–recombination of excess carriers in a photodiode. In the intermediate frequency range (1 kHz $< f <$ 1 MHz), the generation–recombination noise becomes the dominant component. At high frequencies ($f >$ 1 MHz), the photodetector is dominated by white noise (i.e., independent of frequency) which includes shot noise, thermal noise, and generation–recombination noise. It should be noted that the break points of the frequencies for each of these noise sources may vary from material to material.

Thermal noise is usually generated by random motion of carriers through the series resistance of the detector and load resistance. For photodiodes using a guard-ring structure, the channel resistance must also be included. The thermal noise of a photodiode can be calculated using the expression

$$i_{th}^2 = 4k_BTG\Delta f \tag{12.45}$$

which shows that the thermal noise of a photodiode varies with the square root of the product of temperature, noise bandwidth, and diode conductance.

12.3.1. p–n Junction Photodiodes

In this section, the basic principles and general characteristics of a p–n junction photodiode are discussed. Figure 12.15a shows the schematic diagram of a reverse-bias p–n junction photodiode. Electron–hole pairs are generated by the internal photoelectric effect in the photodiode to a depth of the order of $1/\alpha$, where α is the optical absorption coefficient at the wavelength of interest. These photogenerated electron–hole pairs are separated in the depletion region by the built-in electric field and collected as a photocurrent in the external circuit, as shown in Fig. 12.15b under reverse-bias conditions. The small-signal equivalent circuit for a p–n junction photodiode is shown in Fig. 12.15c, which consists of a constant photocurrent source I_{ph}, junction capacitance C_j, series resistance R_s, and shunt resistance R_p. The shunt resistance is usually very high and can be neglected for a typical photodiode operating in the visible spectral range, but is included to account for the possible leakage current path (i.e., low shunt resistance) in a photodiode fabricated from small-bandgap semiconductors. Quantities $\sqrt{I_s^2}$ and $\sqrt{I_R^2}$ shown in Fig. 12.15c are the equivalent noise current sources due to the shot noise and the thermal noise of the photodiode, respectively.

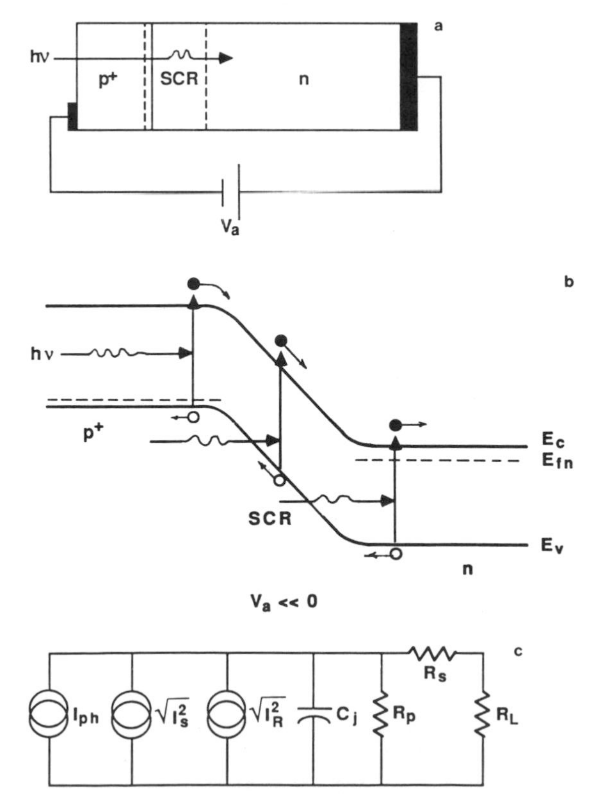

FIGURE 12.15. (a) Schematic representation of a p–n junction photodiode, (b) energy band diagram under illumination and reverse-bias conditions, and (c) equivalent circuit diagram: I_{ph} is the photocurrent, I_s is the shot-noise current source, I_R is the thermal-noise current source, C_j is the junction capacitance, R_s and R_p denote respectively the series and shunt resistances, and R_L is the load resistance.

A high-speed p–n junction photodiode is usually constructed in such a way that most of the photons are absorbed in the p-emitter region. The junction is placed as deep as possible so that efficient separation of photogenerated electron–hole pairs can be obtained. This ensures that most of the photocurrent is carried by electrons whose speed, either by diffusion or drift, is always faster than that of holes. The conditions for achieving excellent low-frequency response in a p–n junction photodiode are that $sW_p/D_n < W_p/(D_n\tau_n)^{1/2} < 1$ and $W_d/(v_s\tau_n) < 1$, where s is the surface recombination velocity, W_p is the width of the p-region, D_n is the electron diffusion constant and τ_n is the electron lifetime in the p-region, W_d is the depletion layer width, and v_s is the saturation velocity of electrons in the depletion region. The diffusion time constant for photogenerated electron–hole pairs in the p-region is given by Eq. (12.40), which is valid for $\alpha W_p < 1$ and for uniform doping in the p-region. If an impurity concentration gradient is present in the p-region, then faster detection can be expected due to the built-in drift field created by the impurity concentration gradient. Since a large impurity concentration gradient is difficult to obtain in the thin diffused p-region, the maximum transit-time reduction introduced by the field-assisted diffusion is about a factor of five to ten. The drift transit time governed by the electric field in the depletion region is given by Eq. (12.41).

The power available from a p–n junction photodiode may be characterized by the power available in a conjugate matched load. Using the equivalent circuit shown in Fig. 12.15c, this can be written as

$$P(\omega_m) = \frac{I_{ph}^2 R_p}{4(1 + R_s/R_p + R_s R_p C^2 \omega_m^2)} \tag{12.46}$$

where I_{ph} is the photocurrent, and ω_m is the frequency at which the photodiode is conjugately matched. For high-frequency operation, a match of the photodiode parameters with load impedance is normally required at frequencies $\omega_m \geq 1/C(R_p R_s)^{1/2}$ so that the maximum power output is given by

$$P(\omega_m) = \frac{I_{ph}^2}{4R_s C_j^2 \omega_m^2} \tag{12.47}$$

where C_j, R_p, and R_s are the junction capacitance, shunt resistance, and series resistance of the photodiode shown in Fig. 12.15c.

12.3.2. p–i–n Photodiodes

The p–i–n photodiode is the most common photodetector structure used in the visible to near-infrared spectral range. Silicon p–i–n photodiodes are widely used in the 0.4- to 1.06-μm spectral range, while InGaAs/InP p–i–n photodiodes are used for detection in the 1.3- to 1.55-μm wavelength range. A p–i–n photodiode consists of a highly doped p-emitter region, a wide undoped intrinsic region (i-region), and a highly doped n-base region. Figure 12.16a shows a schematic drawing of a p–i–n photodiode, and Fig. 12.16b presents photogeneration rates versus distance for two different wavelengths. The reason for the p–i–n photodiode being so popular is because its spectral response can be tailored to a specific need. For example, its long-wavelength spectral response can be controlled by varying the thickness (W) of the undoped i-region, as shown in Fig. 12.17. In addition to the p–i–n structure shown in Fig. 12.16a (i.e., p^+–π–n^+), other structures such as n^+–ν–p^+, n^+–π–p^+, and p^+–ν–n^+ junctions can also be fabricated for photodiode applications (note that ν denotes n^- and π denotes p^-).

A p–i–n photodiode usually operates under the depletion-mode condition in which a sufficiently large reverse bias is applied to the photodiode such that the entire i-region is fully depleted. When photons with energies greater than the bandgap energy of the semiconductor

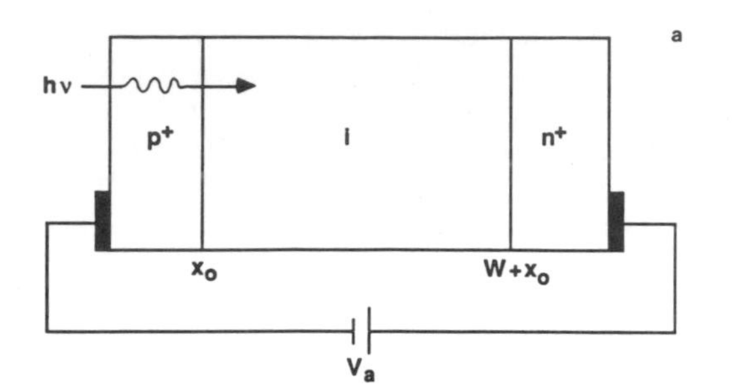

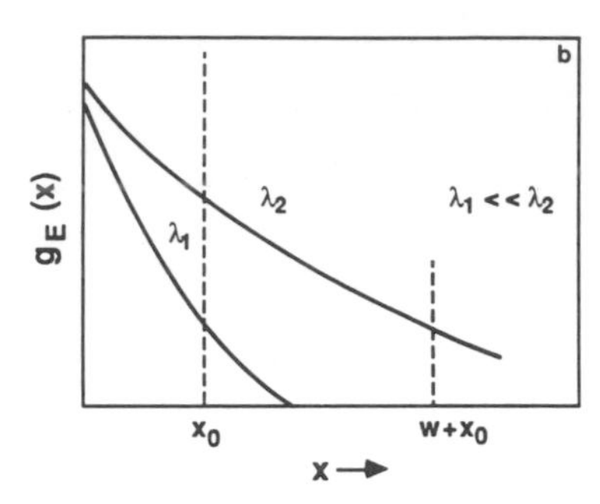

FIGURE 12.16. (a) Schematic diagram of a p–i–n photodiode, and (b) generation rate versus distance for two different wavelengths where λ_2 denotes long-wavelength photons and λ_1 short-wavelength photons.

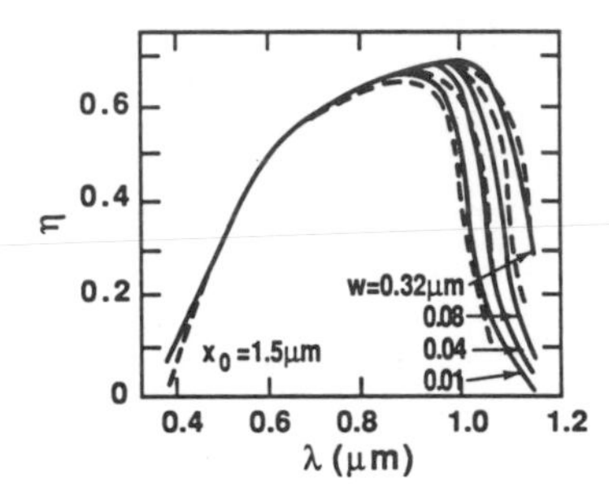

FIGURE 12.17. Quantum yield versus wavelength for a silicon p–i–n photodiode with different depletion layer widths. The solid lines are computed from Eqs. (12.62) and (12.63), and the dashed lines are experimental data. After Li and Lindholm,[6] by permission.

(i.e., $h\nu \geq E_g$) impinge on the photodiode, a small number of photons (primarily short-wavelength photons) will be absorbed in the p^+-region while the majority of photons are absorbed in the i-region. Excess electron–hole pairs generated in the i-region are swept out by the high electric field created by the applied reverse-bias voltage across the photodiode. The photogenerated electron–hole pairs are then collected at the ohmic contacts of the photodiode. The carrier transit time across the i-region can be calculated by the expression

$$t_r = \frac{W}{v_s} \tag{12.48}$$

where W is the thickness of the i-layer, and v_s is the thermal velocity of the excess carriers $[v_s = (3k_BT/m^*)^{1/2}]$ in the i-region.

We next derive expressions for the photocurrents generated in each region of a p–i–n photodiode. The spectral response of such a photodiode can be derived as follows. As in Fig. 12.16a, the thickness of the p-region is denoted by x_0 ($W \gg x_0$). When monochromatic light impinges on a p–i–n photodiode at $x = 0$, the rate of generation of excess carriers is given by

$$g_E(x) = \alpha\phi_0(1 - R)\, e^{-\alpha x} \tag{12.49}$$

where ϕ_0 is the photon flux density, R is the reflection coefficient at the surface of the p-region, and α is the absorption coefficient. Under steady-state conditions, the total photocurrent density J_{ph} is equal to the sum of electron and hole current components produced by incident photons at a given plane along the x-direction of the photodiode. This can be expressed by

$$J_{ph} = J_n(x_0) + J_p(x_0)$$
$$= J_n(x_0) + J_p(W) + J_i \tag{12.50}$$

where

$$J_i = J_p(x_0) - J_p(W) \tag{12.51}$$

is the current density due to hole generation in the i-region.

The spectral dependence of the quantum yield η (i.e., η equals the number of electron–hole pairs produced per absorbed photon) can be obtained by solving $J_n(x_0)$, $J_p(W)$, and J_i as functions of λ, α, W, and x_0. To derive expressions for J_n, J_p, and J_i in the three regions of a p–i–n photodiode, it is assumed that (1) the photogenerated excess carrier densities are small in both n^+- and p^+-regions (i.e., $\Delta p \ll n_0$ and $\Delta n \ll p_0$), (2) the reverse-bias voltage is not large enough to cause avalanche multiplication in the i-region, (3) the surfacc recombination velocity is very high at the illuminated surface such that $\Delta n(0) = 0$, (4) the excess carrier density at the edge of the i-region is small enough so that the boundary condition $\Delta n(x_0) = 0$ holds at $x = x_0$, (5) the effect of the built-in electric field in the emitter region (i.e., the p-region) is neglected, and (6) recombination of excess carriers in the depletion region is negligible. Assumption (3) is valid for most silicon p–n junction photodiodes, because impurity concentration at the surface of the p-emitter region is usually several orders of magnitude higher than that of the i-region. As a result, the carrier lifetime at the surface is also expected to be much shorter than that in the bulk. Therefore, excess carriers generated at the surface will usually recombine before they are able to diffuse to the junction.

The photocurrents produced by the absorbed incident photons in the three regions of a p–i–n photodiode can be derived as follows:[6]

The p-Region ($0 < x \le x_0$). In this region, the contribution of photocurrent is mainly due to electron diffusion current generated in the p-region, and is evaluated at $x = x_0$. This photocurrent component can be derived by solving the continuity equation of excess electrons in the p-region, given by

$$D_n \frac{d^2 \Delta n}{dx^2} - \frac{\Delta n}{\tau_n} = -\alpha \phi_0 (1 - R)\, e^{-\alpha x} \tag{12.52}$$

The general solution of Eq. (12.52) is given by

$$\Delta n(x) = A \sinh\left(\frac{x_0 - x}{L_n}\right) + B \cosh\left(\frac{x_0 - x}{L_n}\right) - \frac{\alpha \phi_0 (1 - R) \tau_n e^{-\alpha x}}{(\alpha^2 L_n^2 - 1)} \tag{12.53}$$

Constants A and B in Eqs. (12.52) and (12.53) can be determined from the boundary conditions: $\Delta n = 0$ at $x = 0$ and $x = x_0$, which yields

$$A = \frac{\alpha \phi_0 (1 - R) \tau_n [1 - \cosh(x_0/L_n)\, e^{-\alpha x_0}]}{(\alpha^2 L_n^2 - 1) \sinh(x_0/L_n)} \tag{12.54}$$

and

$$B = \frac{\alpha \phi_0 (1 - R) \tau_n e^{-\alpha x_0}}{(\alpha^2 L_n^2 - 1)} \tag{12.55}$$

The electron diffusion current density is obtained by solving Eqs. (12.53) through (12.55) and evaluating current density at $x = x_0$ by assuming that $\alpha L_n \gg 1$. The result yields

$$J_n(x_0) = qD_n \left.\frac{d\Delta n(x)}{dx}\right|_{x=x_0}$$

$$= q\phi_0(1 - R)\left\{e^{-\alpha x_0} - \left[\left(\frac{1}{\alpha L_n}\right)\sinh\left(\frac{x_0}{L_n}\right)\right]\left[1 - \cosh\left(\frac{x_0}{L_n}\right)e^{-\alpha x_0}\right]\right\} \quad (12.56)$$

which shows the functional dependence of $J_n(x_0)$ on the absorption coefficient α and thickness x_0.

The i-Region $(x_0 \le x \le W)$. In the intrinsic (i) region, the drift current density contributed by the excess carriers generated in this region is given by

$$J_i = q\int_{x_0}^{x_0+W} g_E(x)\,dx$$

$$= q\phi_0(1 - R)(e^{-\alpha W} - e^{-\alpha x_0}) \quad (12.57)$$

where it is assumed that $W \gg x_0$ and $W + x_0 \simeq W$. Here, $g_E(x) = \alpha\phi_0(1 - R)\,e^{-\alpha x}$ is the photon generation rate shown in Fig. 12.16b for two different wavelengths.

The n-Region $(x \ge W)$. In this region, the photogenerated excess carriers (i.e., holes) contribute to the hole diffusion current. The hole current density can be derived by solving the continuity equation for the excess hole density in the n-region given by

$$D_p\frac{d^2\Delta p}{dx^2} - \frac{\Delta p}{\tau_p} = -\phi_0(1 - R)\,e^{-\alpha x} \quad (12.58)$$

where D_p and τ_p denote the hole diffusion constant and hole lifetime, respectively. The hole diffusion current density is obtained by solving Eq. (12.58) and evaluating at $x = W$. The result is

$$J_p(W) = \frac{-q\phi_0(1 - R)\alpha L_p\,e^{-\alpha W}}{(1 + \alpha L_p)} \quad (12.59)$$

which is obtained by employing the boundary conditions (for $V \gg k_BT/q$)

$$\Delta p = -p_{n0} \qquad \text{at } x = W$$

$$\Delta p = 0 \qquad \text{for } x \to \infty \quad (12.60)$$

The total photocurrent density for a p–i–n photodiode is equal to the sum of Eqs. (12.56), (12.57), and (12.59), and is given by

$$J_{ph} = q\phi_0(1 - R)\left\{\frac{1}{\alpha L_n\sinh(x_0/L_n)}\left[1 - \cosh\left(\frac{x_0}{L_n}\right)e^{-\alpha x_0}\right] - \frac{e^{-\alpha W}}{(1 + \alpha L_p)}\right\} \quad (12.61)$$

The quantum efficiency, defined as the number of electron–hole pairs generated per absorbed photon, is given by

$$\eta = \frac{J_{ph}}{q\phi_0} \times 100\% \quad (12.62)$$

where J_{ph} is given by Eq. (12.61).

For a silicon p–i–n photodiode, the p-region is usually very thin ($\leq 1.5\ \mu$m) while the i-region is much wider (from a few μm to a few tens of μm). As a result, excess carriers generated by short-wavelength photons are confined mainly in the p-region. The quantum efficiency for a silicon p–i–n photodiode can be calculated with the aid of published optical absorption coefficient data and carrier diffusion lengths for silicon. Quantum yield versus photon wavelength for a silicon p–i–n photodiode with i-layer width W as parameter and $x_0 = 1.5\ \mu$m is shown in Fig. 12.17,[6] where solid lines refer to results calculated from Eq. (12.62) for $W = 0.32$, 0.08, 0.04, and 0.01 cm, respectively. The reflection coefficient R for silicon is assumed equal to 0.3 in these calculations.

The sharp decrease of quantum yield in the short-wavelength regime can be explained as follows. It is implied by Eq. (12.53) and Fig. 12.17 that excess carriers generated by short-wavelength photons are confined mainly in the p-region. Therefore, the photocurrent due to photon excitations in the short-wavelength regime must come from excess carriers generated in the p-region. However, only those excess carriers generated near the junction will diffuse toward the i-region. In fact, only a fraction of the excess carriers generated by the short-wavelength photons in the p-region will be collected and contributed to the photocurrent $J_n(x_0)$. Excess carriers generated near the surface region are usually lost via surface recombination at $x = 0$.

An estimate of the maximum cutoff frequency for a p–i–n photodiode operating under the condition limited by the load impedance R_L can be obtained by using the equivalent circuit shown in Fig. 12.15c. We note that excess carriers generated in the i-region and collected by the built-in electric field are represented by a current generator I_{ph}, which is in parallel with the junction capacitance C_j. The series resistance is denoted by R_s, and R_L is the load impedance. The junction capacitance of the photodiode is given by

$$C_j = \frac{A\varepsilon_0\varepsilon_s}{W} \tag{12.63}$$

where A is the cross-section area of the junction, ε_s is the dielectric constant of the semiconductor, and ε_0 is the free-space permittivity. The maximum cutoff frequency for a p–i–n photodiode can be calculated using the expression

$$f_c = \frac{2.4}{2\pi\tau_{tr}} \approx 0.4\alpha v_s \tag{12.64}$$

where v_s is the average thermal velocity of electrons in the i-region. For a germanium p–i–n photodiode with $v_s = 6 \times 10^6$ cm/sec, $\varepsilon_s = 16$, $A = 2 \times 10^{-4}$ cm^2, and $R_L = 10$ ohm, the cutoff frequency f_c is found to equal 41.84×10^9 Hz.

If the light modulation frequency approaches that of the transit time of excess carriers across the entire i-region, a phase shift between the photon flux and photocurrent will appear in the photodiode. This effect is severe for the case where incident photons are absorbed very close to the outer edge of the depletion layer or near the surface of the photodiode. However, for most semiconductors the absorption coefficients vary between 10^1 and 10^5 cm^{-1}. Thus, optical absorption takes place quite deep inside the bulk of the photodetector. If the modulation frequency is around 10^9 Hz, then the depletion layer width must be a few microns. This means that excess carriers are generated throughout the entire volume of the depletion layer, and hence have a distribution of transit times. The phase-shift effect in this case is less severe than in the case of surface generation.

12.3.3. Avalanche Photodiodes

Another important photodiode which has been extensively studied is the avalanche photodiode (APD). An APD can have the dual function of serving as a detector for detecting incident photons or optical signals, and as an amplifier (with internal current gain obtained via avalanche multiplication) for photogenerated carriers produced by the incident photons. An APD is known to produce the highest gain–bandwidth product among all the solid state photodetectors discussed in this chapter. APDs can be fabricated from a wide variety of semiconductor materials with different structures. Besides conventional germanium and silicon APDs, several APDs fabricated from III–V compound semiconductors have also been reported, in particular long-wavelength (1.3 μm) APDs using the InGaAs/InP material system.

Figure 12.18a shows the cross-section view of a planar silicon APD. In this structure, the guard ring is employed to prevent the occurrence of microplasmas (i.e., small regions where the breakdown voltage is lower than that of the p–n junction as a whole) around the edge of the active region of the photodiode. The occurrence of bulk microplasmas can be reduced by using semiconductor material with low defect densities. Figure 12.18b presents the equivalent circuit for the APD, where M denotes the multiplication factor, i_s is the shot-noise current source, while R_s and C_j denote respectively the series resistance and transition capacitance.

The basic principle underlying an APD is discussed next. If the reverse-bias voltage across a photodiode is smaller than the breakdown voltage under dark conditions, a small reverse leakage current will flow through the diode. This reverse leakage current I_D is due to thermally generated carriers in the quasi-neutral regions of the diode, and should be kept as low as possible. If the reverse-bias voltage continues to increase, the electric field in the depletion region will eventually become strong enough to cause both thermally generated electrons and holes to gain sufficient kinetic energy, and impact ionization will occur when electron–electron or hole–hole collisions take place. These electrons and holes will undergo further impact ionizations which produce more electron–hole pairs and more impact ionization, resulting in a runaway condition limited only by the series resistance and circuitry external to the diode. This runaway condition is called avalanche breakdown. Thus, it is understandable that the current flow in an APD during avalanche breakdown, I_{MD}, may be orders of magnitude

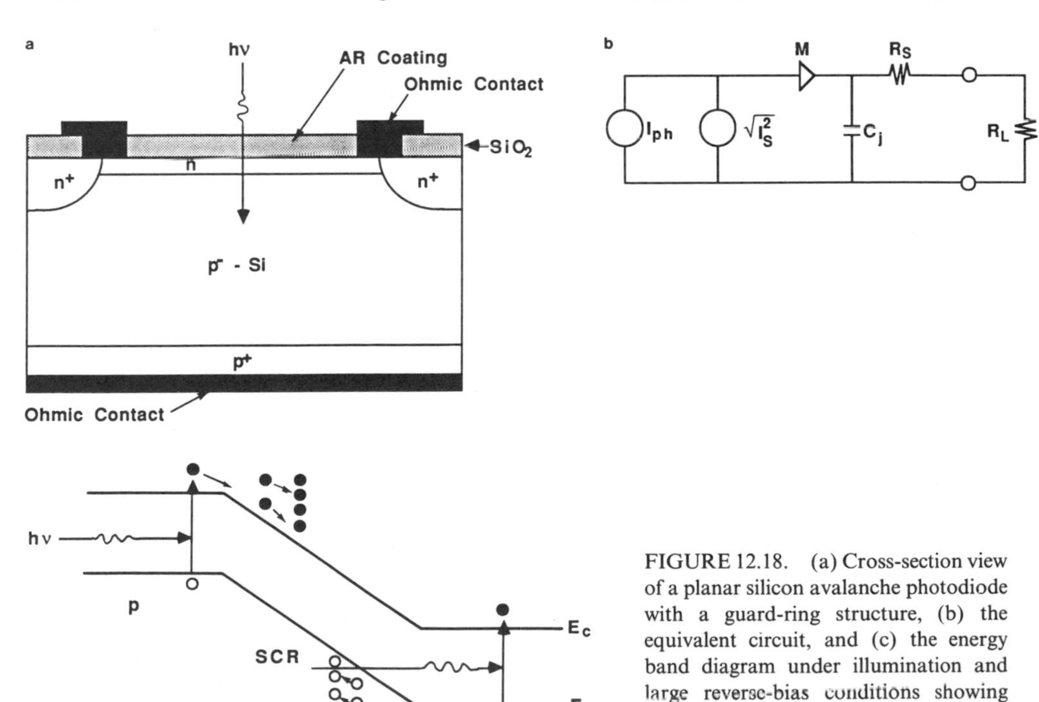

FIGURE 12.18. (a) Cross-section view of a planar silicon avalanche photodiode with a guard-ring structure, (b) the equivalent circuit, and (c) the energy band diagram under illumination and large reverse-bias conditions showing avalanche multiplication in the space-charge region.

larger than I_D, the initial thermally generated dark current. Figure 12.18c shows avalanche multiplication taking place inside the junction space-charge region of a silicon APD. When an APD is illuminated by light under a small reverse-bias condition, the primary current flowing through the APD is denoted by I_p, which consists of the dark current I_D and photocurrent I_{ph} (due to electron–hole pairs generated optically within the depletion region). If the reverse-bias voltage is increased, avalanche multiplication of the primary current I_p occurs, but the onset of avalanche multiplication is usually gradual. This is clearly illustrated by the photo-I–V characteristics of a germanium APD shown in Fig. 12.19, with prime current I_p as parameter.[7]

In general, the I–V characteristics of an APD can be predicted by using an empirical formula

$$I = MI_p \tag{12.65}$$

where$M$ is the multiplication factor given by

$$M = \frac{1}{\left\{1 - \left[\frac{(V - I_p R)}{V_B}\right]^n\right\}} \tag{12.66}$$

We note that I_p $(=I_D + I_{ph})$ is the primary current flowing in the photodiode before the onset of ionizing collisions; $R = R_s + R_c + R_T$ is the total resistance of the APD where R_s is the series resistance of the contacts and the bulk material, R_c is the resistance due to carrier drift through the depletion layer, and R_T is the thermal resistance which heats the junction and increases the diode breakdown voltage V_B. The factor n in the exponent depends on the semiconductor material used, the doping profile of the junction, and the wavelength of incident photons. Low values of n correspond to high photomultiplication at a given voltage, and can be obtained if the multiplication is initiated by carriers with high ionization coefficients.

Despite the fact that large internal current gains can be achieved by the avalanche multiplication process, shot noise also increases rapidly with the multiplication process in an APD. It

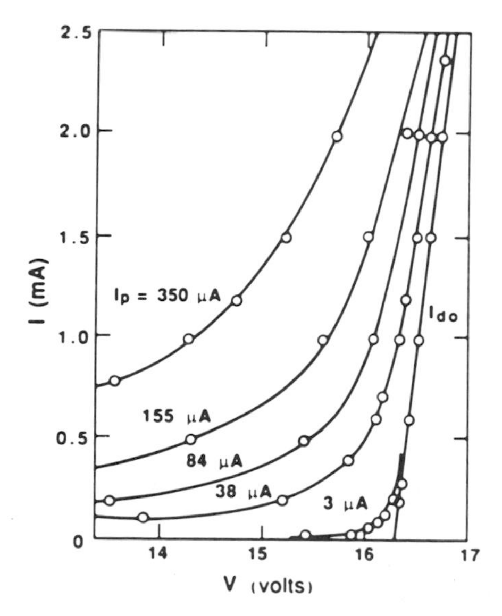

FIGURE 12.19. Current–voltage characteristics of a germanium avalanche photodiode with different primary currents. After Melchior and Lynch,[7] by permission, © IEEE-1966.

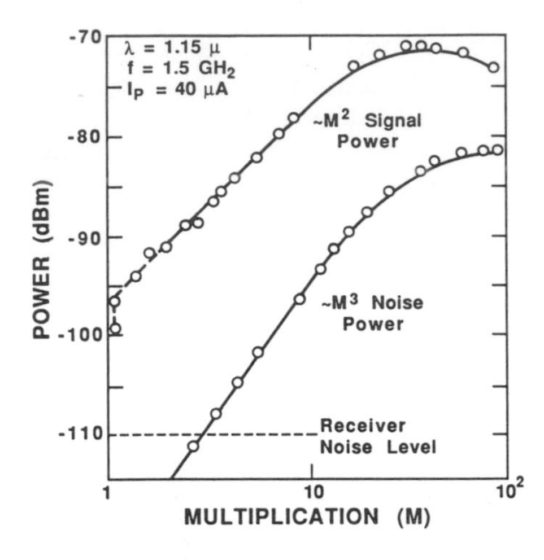

FIGURE 12.20. Signal and noise power outputs of a germanium avalanche photodiode. Both are measured with an input light intensity of 1 mW. After Melchior and Lynch,[7] by permission, © IEEE-1966.

is noteworthy that significant improvement in overall sensitivity has been achieved in both silicon and germanium APDs with wide instantaneous bandwidths. Figure 12.20 shows the signal and noise power outputs of a germanium APD operating at 1.5 GHz and $\lambda = 1.15\ \mu$m.[7]

Recently, a novel planar SAM–APD structure using InGaAs/InP grown by a molecular beam epitaxy (MBE) technique has been reported for long-wavelength detection (e.g., $\lambda \geq 1.3\ \mu$m); this is shown in Fig. 12.21.[8] The SAM (separate absorption and multiplication) APD uses a narrow bandgap (i.e., $E_g = 0.74$ eV) $In_{0.53}Ga_{0.47}As$ active layer to absorb long-wavelength photons ($\lambda = 1.3\ \mu$m) and a separate wide-bandgap (InP) p–n junction grown on top of the $In_{0.53}Ga_{0.47}$ epilayer to achieve avalanche multiplication. Long-wavelength photons impinging on the antireflection coating layer of a SAM–APD will pass through the top layers of the InP p–n junction, and then absorb in the smaller-bandgap InGaAs active layer. Electron–hole pairs generated in the InGaAs region by the absorbed photons will move into the upper InP p–n junction where a large reverse-bias voltage is applied to produce avalanche multiplication by impact ionization. It should be noted that the dopant density and the thickness of each layer of the SAM–APD must be calculated such that the electric field at the heterointerface remains sufficiently small to avoid a significant tunneling current, but is large enough to deplete the entire absorbing region. Precise control of the device parameters for a planar SAM–APD is very important for achieving optimum performance.

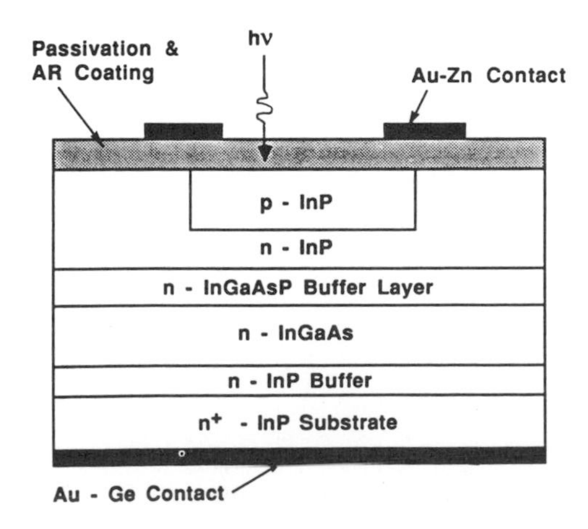

FIGURE 12.21. Structure of an InGaAs/InP SAM–APD (separate absorption and multiplication–avalanche photodiode), showing the n-InGaAs absorber layer and the upper p-InP/n-InP junction for avalanche multiplication. After Nishida *et al.*,[8] by permission.

12.3.4. Schottky Barrier Photodiodes

Schottky barrier photodiodes (SBDs) are particularly attractive for high-speed detection. The basic principles of a SBD for photodetection were described in detail in Chapter 10. Depending on the modes of detection, SBD may be used to detect photons or optical signals with wavelengths extending from ultraviolet (UV) to infrared (IR) spectral ranges. When a SBD is operating in the depletion mode under reverse-bias conditions, electron–hole pairs are generated by incident photons with energy greater than the bandgap energy of the semiconductor. In this case, the cutoff wavelength of the photodiode is determined by the bandgap energy of the semiconductor. Depletion-mode Schottky barrier photodiodes fabricated from larger-bandgap semiconductors such as Si, GaAs, and ZnS are used primarily to detect shorter-wavelength photons (e.g., from UV to near-IR), while Schottky barrier and p–i–n photodiodes fabricated from smaller-bandgap semiconductors such as Ge, InGaAs, InAs, and CdHgTe are used mainly for mid- to long-wavelength IR detection. Schottky barrier photodetectors may be fabricated by depositing a variety of metals on different semiconductors. These detectors cover the wavelengths from UV to IR spectral ranges. For example, a Ag/ZnS Schottky barrier photodetector, which has a peak photoresponse at 0.3 μm, is mainly used for UV light detection. A Au/GaAs Schottky barrier photodiode, which has a peak response at 0.85 μm, may be used for visible to near-IR detection. Infrared (IR) detectors using Schottky barrier structures such as a Au/p-$In_{0.53}Ga_{0.47}As/p^+$-InP Schottky barrier photodiode for 1.3 to 1.5 μm and PtSi on a p-Si Schottky barrier photodiode array operating at 77 K for 3–5 μm photodetection have been reported recently. Figure 12.22a shows top and cross-section views of a high-speed Au/n-GaAs/n^+-GaAlAs planar Schottky barrier photodiode. The detector is capable of detecting modulating optical signals up to 45 GHz at $\lambda = 0.8$ μm. Figure 12.22b presents the spectral response, and Fig. 12.22c is the impulse response for such a detector. A risetime of 8.5 psec and a FWHM of 16 psec are measured for this detector by using the sampling/correlation technique. This corresponds to a 3 dB cutoff frequency of 45 GHz.[9]

Schottky barrier photodiodes can also be operated in the hot electron detection mode. In this case, photons are absorbed inside the metal film, and photogenerated hot electrons in the metal film are then swept across the Schottky barrier and injected into the bulk semiconductor. The cutoff wavelength under this detection mode is determined by the barrier height. Therefore, under this detection mode, a Schottky barrier diode with small barrier height can be used for long-wavelength infrared (IR) detection. A typical example for this type of photodetector is the PtSi/p-type silicon Schottky barrier photodiode, which has a barrier

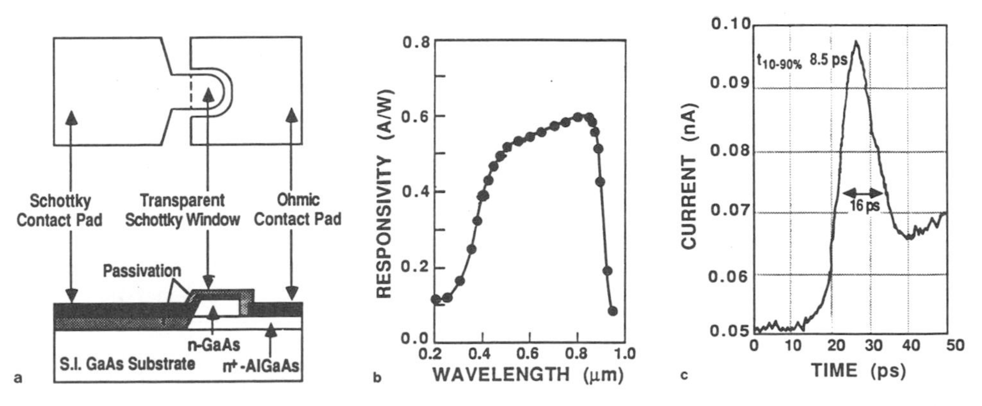

FIGURE 12.22. (a) Top and cross-section views of a Au/GaAs/AlGaAs heterostructure planar Schottky barrier photodiode, (b) external quantum efficiency and responsivity curves, and (c) impulse response of the photodiode showing a risetime of 8.5 psec and a FWHM of 16 psec, which corresponds to a cutoff frequency of 45 GHz. After Lee *et al.*,[9] by permission.

height of around 0.2 eV and a cutoff wavelength of 5.6 μm. PtSi/p-Si Schottky barrier photodiode arrays have been widely used for 3- to 5-μm IR image sensor applications. To reduce the dark current and noise, the PtSi/p-Si Schottky barrier detector is usually cooled down to 77 K.

12.3.5. Point-Contact Photodiodes

A point-contact photodiode may be constructed as a Schottky barrier or as a p–i–n photodiode, depending on the device structures. The active area of a point-contact photodiode is usually very small, and as a result both the junction capacitance and transit time are extremely small. The point-contact photodiode is used mainly to detect optical signals at very high modulation frequencies. As an example, let us consider a germanium point-contact photodiode. The diode is formed by using a p^--epitaxial layer 8-μm thick grown on a p^+-germanium substrate and a 0.6 mil arsenic-doped gold foil alloyed on the p^--region by using a short current pulse. The alloy region is approximately 4 μm in diameter and depth. Light impinges on the surrounding area of the p^--region. Using a 4-μm deletion layer width, the carrier transit time for this photodiode is less than 4×10^{-11} sec. The junction capacitance and series resistance for such a diode are 4.5×10^{-4} pF and 10 ohm, respectively, which yield an RC time constant equal to 4.5×10^{-15} sec. Therefore, the bandwidth (or cutoff frequency) for this photodiode is limited by the transit time, which is of the order of 10^{-11} sec. A cartridge-type point-contact photodiode capable of responding to modulated optical signals with frequencies up to 30 GHz has also been reported in the literature. Point-contact photodiodes are particularly attractive for applications in fiber optic communications in which incident light is confined inside the optic fiber with a light spot of a few microns in diameter.

12.3.6. Heterojunction Photodiodes

A heterojunction photodiode is formed by using two types of semiconductor material with different energy bandgaps and opposite dopant impurities. To reduce the dark current and noise of a heterojunction photodiode, the lattice constants of both semiconductor materials should be chosen as closely matched as possible. There are several semiconductor pairs with good lattice match that can be used to fabricate heterojunction photodiodes. These include AlGaAs/GaAs, GaAs/Ge, and InGaAs/InP, among others. For example, a heterojunction photodiode made from a wide-bandgap n-type GaAs on a narrow-bandgap p-type Ge can be used to detect infrared radiation in the 1.1- to 1.8-μm wavelength regime. The detector is designed in such a way that long-wavelength photons can pass through the top wide-bandgap n-GaAs layer and absorb in the depletion region of the bottom narrow-bandgap p-Ge layer. Carrier generation takes place as a result of absorption of long-wavelength photons in the p-Ge base layer. The optical absorption coefficients for GaAs and Ge are less than 10 cm^{-1} and $2.4 \times 10^4\ \text{cm}^{-1}$ at 1.6 μm, respectively. This implies that less than 1% of the incident photons are absorbed in the n-GaAs layer, while more than 99% of the incident photons are absorbed within 1 μm from the depletion edge of the p-Ge layer. For this n-GaAs/p-Ge heterojunction photodiode, a narrow-peak spectral response will occur at $h\nu =$ 1.38 eV. Since germanium is an indirect bandgap material, the quantum efficiency and responsivity for a n-GaAs/p-Ge heterojunction photodiode are usually low. A superior IR detector using a p-$In_{0.53}Ga_{0.47}As$–ν-$In_{0.53}Ga_{0.47}As$–n-InP heterojunction structure can produce excellent spectral response, high quantum efficiency and responsivity for wavelengths between 1.0 and 1.6 μm. Figure 12.23 shows the cross-section view and spectral response of such a detector.[10] Responsivity greater than 0.5 A/W and quantum efficiency between 55 and 70% are obtained for this detector in the wavelength range from 1.0 to 1.6 μm. In this photodetector, back-illumination (i.e., incident photons impinging from the wide-bandgap InP substrate into the narrow-bandgap InGaAs active layer) is used so that most of the long-wavelength photons

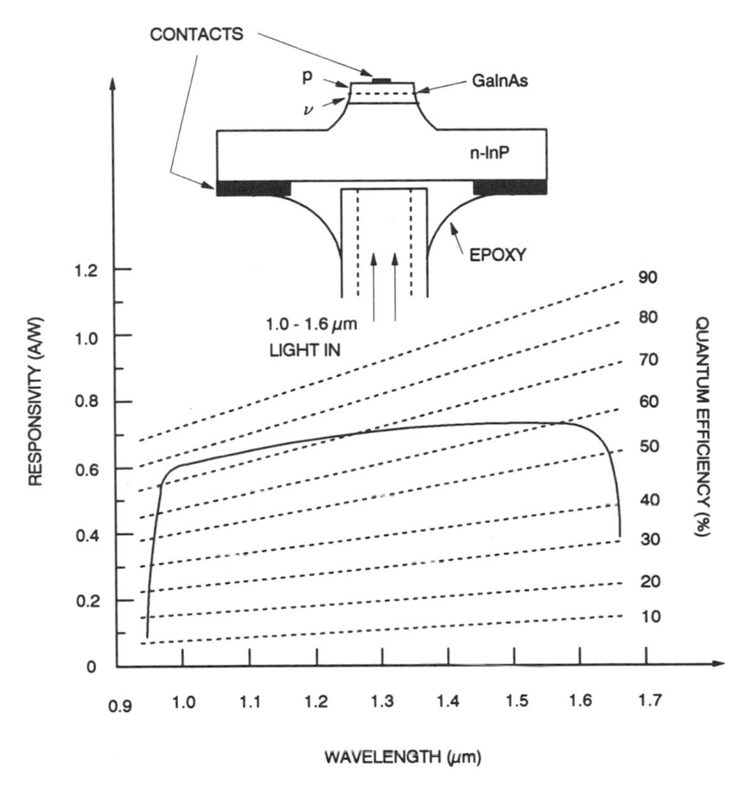

FIGURE 12.23. A p-$In_{0.53}Ga_{0.47}As$–ν-$In_{0.53}Ga_{0.47}As$–n-InP heterojunction p–i–n infrared photodetector for 1.1- to 1.6-μm detection: the cross-section view, and quantum efficiency and responsivity curves. After Lee *et al.*,[10] by permission, © IEEE-1981.

are absorbed in the ν-$In_{0.53}Ga_{0.47}As$ active layer. Since most of the incident photons are absorbed in the i-region, a p–i–n heterojunction photodiode is relatively insensitive to the surface condition. It should be noted that the transit time of the photogenerated carrier across the i-region is usually smaller than the RC time constant of the detector. As a result, the frequency response for such a photodiode is usually limited by the RC time constant rather than the transit time of the detector.

12.3.7. Photomultipliers

The photomultiplier is another type of photodetector and is known as the most sensitive detector available in the visible spectral region. This type of photodetector is particularly useful for photon counting applications. Figure 12.24 shows a partition-type electron multiplier in

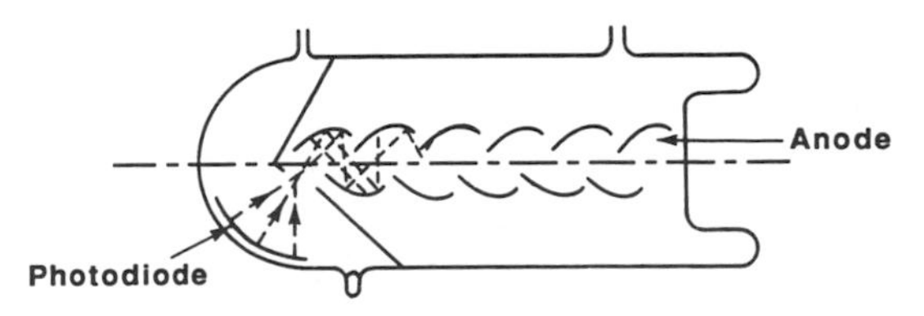

FIGURE 12.24. A typical photomultiplier electrode arrangement showing the linear configuration of the dynode section, with a partitioned transparent photocathode.

which the photocathode is transparent and mounted at the end of the tube. The incident light falls on the front face of the photocathode, and electrons emitted from the cathode surface are multiplied by nine dynode stages. In a conventional P2 phototube, the photocathode is of the S4 class, and the sensitivity is about 40 μA/lm. The overall gain for the nine dynode stages is 2×10^6, corresponding to an average gain of about 5 per stage. Conventional dynode materials for such a photomultiplier include Cs_3Sb, Mg–Ag, and Be–Cu. It is screened from the secondary-emission dynode electrodes by a partition and an aperture that provides convenient separation during activation. Other electron multiplier structures such as Venetian screen, box-type, cross-field, and diode arrangements are used in different photomultiplier applications.

Since the gain per stage is only about 5 for a conventional secondary emitter, the statistical fluctuation in the number of secondary electrons emitted by the first dynode usually limits the tube performance. In order to provide discrimination between signals representing the emission of one or two photoelectrons, it is necessary that the first dynode has a gain of 15 to 20. Even higher gains are needed to distinguish between n and $(n + 1)$ photoelectrons, where n is greater than 1. For a cesium-coated GaP photocathode with the first dynode operating at 600 V, a gain of 20 to 40 has been achieved.

Photomultipliers provide high sensitivity and fast response speed for wavelengths from the UV to the visible spectral range. Photomultipliers have many applications in detection and measurement of very low light-level scintillations. Recent development of photocathode materials using III–V compound semiconductors such as $InAs_xP_{1-x}$ and $In_xGa_{1-x}As$ has extended useful detection wavelengths to the 1–2-μm IR spectral range.

12.3.8. Long-Wavelength Infrared Detectors

Due to the increasing use of infrared (IR) technologies for various image sensor systems and optical communications, a wide variety of IR photodetectors covering wavelengths from 1–2 μm, 3–5 μm, 8–14 μm, and beyond have been extensively investigated and developed in the last two decades using different material systems. For example, InGaAs/InP and InGaAsP/InP p–i–n photodiodes are used mainly for 1.3 and 1.55 μm detection, PtSi/Si Schottky barrier photodetectors have been widely used in the 3–5 μm image sensor arrays, and more recently have been extended to 10 μm applications. Impurity (Cu, In, Hg, etc.) doped germanium and (In, Ga) Si-photoconductive detectors have also been developed for detection in the 8 to 40 μm wavelength range. CdHgTe photoconductive and photovoltaic detectors have been developed for 8–20 μm focal plane arrays and IR sensing applications. In order to reduce the dark current in the long-wavelength IR detectors, most IR detectors are operated in the cryogenic temperatures (i.e., 77 K or below 20 K). In spite of tremendous efforts in the development of various IR detectors, further improvement in the quality of IR material, such as the CdHgTe system, and development of new IR detectors are needed for long-wavelength IR applications. Although CdHgTe material can cover a wide-wavelength range, the quality of the material needs further improvement; in particular, the uniformity of alloy composition across the wafer remains a serious problem. Figure 12.25 shows the energy bandgap versus cutoff wavelength for a wide variety of IR materials.

The development of new and improved long-wavelength (8–12 μm) infrared (IR) detectors for focal plane array (FPA) technology is a very important step toward meeting the challenges and needs of future infrared applications including remote sensing, forward looking infrared (FLIR) image sensors highly sensitive staring IR sensor systems, atmospheric optical communication, environmental studies, and space exploration. Extending detection to longer wavelengths offers several advantages. These include: (1) IR radiation in the 8–12 μm atmospheric window can travel a longer distance through the atmosphere with small attenuation, (2) enabling the use of an existing powerful CO_2 laser ($\lambda \sim 10.6$ μm) and maturing technology, (3) reducing the interference radiation reflected from the background and eliminating susceptibility of false signals triggered by sunlight and other background radiations, and (4) enabling detection and tracking of cooler targets, such as satellites (see Fig. 9.16).

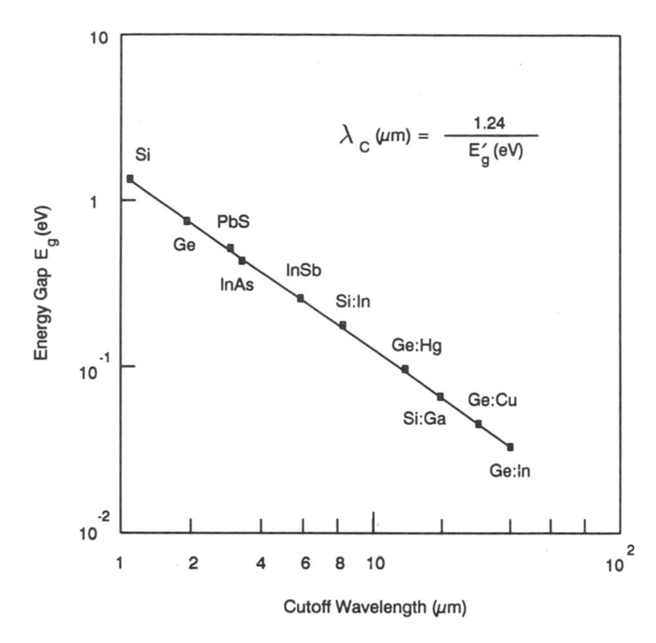

FIGURE 12.25. Energy bandgap versus cutoff wavelength for some important impurity-doped semiconductor materials.

A new class of long-wavelength III–V quantum well infrared photodetectors (QWIPs) based on intersubband transitions have been recently reported. The advances in MBE technology regarding growth of III–V quantum well/superlattice structures makes it possible to design and fabricate various novel quantum effect devices for a wide variety of applications. The existence of a large dipole matrix element between the subbands of the quantum well makes such a structure extremely attractive for long-wavelength IR detection and modulation, especially in the atmospheric window of 8 to 14 μm wavelength range. A great deal of work has been reported on the lattice-matched GaAs/AlGaAs and InGaAs/InAlAs quantum well systems for QWIP fabrications. QWIPs based on bound-to-bound states, bound-to-continuum states, and bound-to-miniband[11] intersubband transitions have been extensively investigated in recent years.

Figure 12.26a shows the energy band diagram of a standard GaAs/AlGaAs QWIP, which uses bound-to-continuum band transition to achieve charge transport and infrared photodetection. Typical device structure for a standard GaAs/AlGaAs QWIP consists of 20 to 50 periods of GaAs (30–40 Å) quantum well doped to 1×10^{18} cm^{-3}, and an undoped $Al_{0.3}Ge_{0.7}As$ (400–500 Å) barrier layer; the heavily-doped (2×10^{18} cm^{-3}) GaAs buffer layer and GaS cap layer are deposited on the bottom and top of the active quantum well layers for ohmic contacts. As shown in Fig. 12.26a, in this QWIP structure only one bound state exists in the quantum well that is filled with electrons, and the next empty band is the continuum band state located slightly above the AlGaAs barrier layer. The conduction band offset (i.e., potential barrier height) for the GaAs/AlGaAs quantum well is about 190 meV, and the energy separation between the bound state and the continuum states is about 120 meV, which corresponds to a peak spectral response of around 10 μm. The standard QWIP operates on photoconductive mode detection using the bound-to-continuum band states intersubband absorption. Since the dark current in this QWIP is controlled by the thermionic emission from the ground states in the quantum well to the continuum states, the detector is required to cool down to 77 K or lower in order to reduce the dark current. Excellent responsivity and detectivity have been obtained in this QWIP. Detectivity greater than 10^{10} cm · Hz$^{1/2}$/W has been reported in this detector at $\lambda = 10$ μm and T = 77 K.

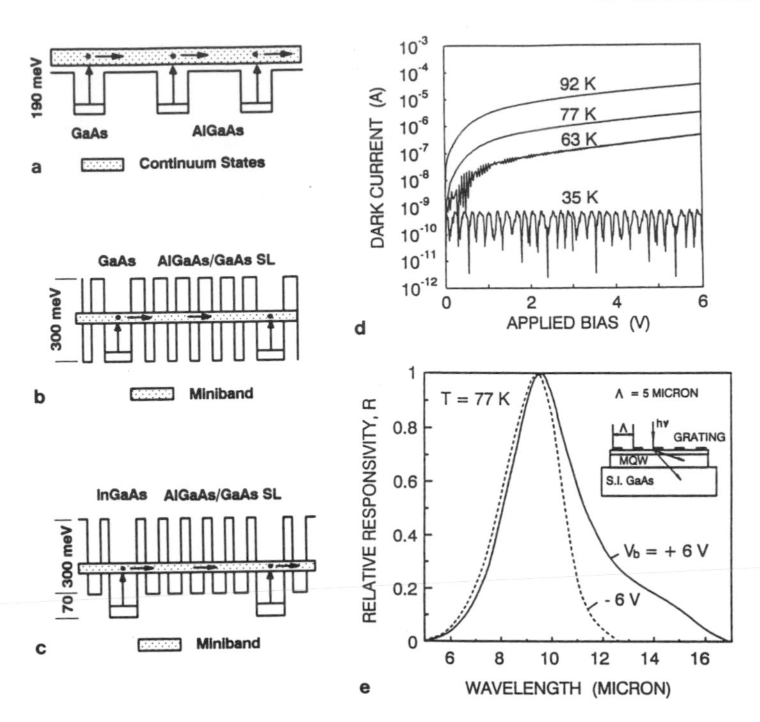

FIGURE 12.26. Energy band diagrams for (a) a standard GaAs/AlGaAs (40/480 Å) quantum well infrared photodetector (QWIP) using bound-to-continuum state transition, (b) a GaAs (88 Å QW)/AlGaAs-GaAs (58–29 Å SL) QWIP using bound-to-miniband transition, (c) an InGaAs (106 Å QW)/AlGaAs-GaAs (58–29 Å SL) QWIP using step-bound-to-miniband transition, (d) dark current versus bias voltage for QWIP shown in (c) with temperature as parameter, and (e) relative responsivity versus wavelength for QWIP shown in (c), measured at $V_B = 6$ V and T = 77 K.

Another new type of QWIP aimed at reducing the dark current in the standard QWIP has recently been reported by Yu and Li.[11] Figure 12.26b shows a bound-to-miniband (BTM) GaAs/AlGaAs QWIP. In this QWIP structure the AlGaAs bulk barrier used in the standard QWIP is replaced by a short period (5-period) AlGaAs (58 Å)/GaAs (29 Å) superlattice barrier layer, and the width of the GaAs quantum well is increased to 88 Å. The BTM QWIP differs from the standard QWIP in that (a) the potential barrier is increased to 300 meV, and (b) a global miniband (coinciding with the first excited state of the quantum well) of the superlattice barrier layer is formed inside the quantum well to facilitate intersubband IR detection. The energy separation between the ground state and the miniband determines the spectral bandwidth of the IR detection (typically 90–120 meV). The current conduction mechanism in this miniband transport is based on thermionic-assisted resonant tunneling from the ground state to the global miniband states as shown in Fig. 12.26b. As a result the dark current in such a QWIP is expected to be lower than a standard QWIP. Good detectivity and responsivity have been obtained for this BTM QWIP. To further reduce the dark current in the BTM QWIP, an InGaAs (with less than 10% of In) enlarged quantum well (106 Å) is introduced to replace the GaAs quantum well, and the resulting structure is shown in Fig. 12.26c. This modified QWIP is also referred to as the step-bound-to-miniband (SBTM) QWIP. The dark current for the SBTM QWIP is generally lower than both the BTM QWIP and standard QWIP. Figure 12.26d shows the dark current versus bias voltage for a SBTM QWIP with temperature as a parameter. The results show that thermionic-assisted tunneling current dominates at 77 K, while resonant tunneling current prevails for temperatures below 50 K. A typical spectral response curve for a SBTM QWIP is presented in Fig. 12.26e, which shows a peak response around 10 μm. It is interesting to note that the spectral response for the miniband transport QWIPs is usually bias-tunable, as is evidenced in Fig. 12.26e.

Although detectivity and responsivity for the QWIPs discussed above are generally not as high as HgCdTe IR detectors, the GaAs/AlGaAs QWIP has the advantages of low noise and high uniformity and number of operating pixel (99%) that result in excellent imaging performance and a noise equivalent temperature difference of $NE\Delta T = 10$ mK. In fact, a low noise 128×128 GaAs/AlGaAs multiple quantum well FPA for staring IR (10 μm) sensor systems has been demonstrated recently. QWIP fabrication utilizes mature GaAs processing technology that exhibits very uniform and controlled growth and processing. In addition, QWIP technology also offers the benefits of wavelength selectivity, multiple-band sensitivity, compatibility for hybridization with GaAs IC electronics, and the possibility of full optical and electronic monolithic integration. QWIP arrays are comparable in complexity to existing GaAs devices and are expected to similarly be producible and low-cost.

12.4. LIGHT-EMITTING DIODES (LEDs)

In this section the basic principles and device characteristics of a light-emitting diode (LED) are discussed. A LED may be considered as an electroluminescent device. The emission of photons in such a device is accomplished by applying a sufficiently large forward-bias

TABLE 12.1. Energy Bandgap, Emission Wavelength, and Methods of Excitation for Some Semiconductor Laser Materials

Material	Photon energy (eV)	Wavelength (microns)	Method of excitation
ZnS	3.82	0.32	electron beam
ZnO	3.30	0.37	electron beam
ZnSe	2.70	0.46	electron beam
CdS	2.50	0.49	electron beam
GaSe	2.09	0.59	electron beam
CdS_xSe_{1-x}	1.80–2.50	0.49–0.69	electron beam
CdSe	1.82	0.68	electron beam
CdTe	1.58	0.78	electron beam
$GaAs_xP_{1-x}$	1.41–1.95	0.63–0.88	p–n junction
GaAs	1.47	0.84	p–n junction, electron beam, optical, avalanche
InP	1.37	0.90	p–n junction
$In_xGa_{1-x}As$	1.5	0.82	p–n junction
GaSb	0.82	1.5	p–n junction, electron beam
InP_xAs_{1-x}	0.77	3.1	p–n junction, electron beam
InSb	0.23	5.2	electron beam
Te	0.34	3.64	electron beam
PbS	0.29	4.26	p–n junction, electron beam
PbTe	0.19	6.5	p–n junction, electron, optical
PbSe	0.145	8.5	p–n junction, electron beam
$Hg_xCd_{1-x}Te$	0.30–0.33	3.7–4.1	optical
$Pb_xSn_{1-x}Te$	0.09–0.19	6.5–13.5	optical

voltage across a p–n junction diode in which minority carrier injection and radiative recombination take place in the quasi-neutral regions of the diode. Materials suitable for LED fabrication include most of the III–V compound semiconductors and some II–VI compound semiconductors, such as those listed in Table 12.1. LEDs have found many practical applications in the areas of optical display, signal processing, and optical communications. These applications are discussed briefly in this section.

12.4.1. Injection Mechanisms

The basic requirement for radiative recombination to take place in a semiconductor is injection of minority carriers into the bulk semiconductor. To explain the injection mechanism in an LED, Fig. 12.27a shows the schematic drawing and Fig. 12.27b the energy band diagram of an AlGaAs/GaAs p–n junction infrared (IR) emitter under thermal equilibrium. The p^+-$Al_{0.3}As$ window layer is used to reduce the surface recombination velocity and to increase the luminescent efficiency of the GaAs IR emitter. Figure 12.27c shows the energy band diagram of the same LED under forward-bias conditions. It is seen that under forward-bias conditions, electrons are injected from the n- into the p-region while holes are injected from the p- into the n-region of the junction. If the band-to-band radiative recombination process is dominant on both sides of the junction, then the emission of optical radiation can be readily achieved.

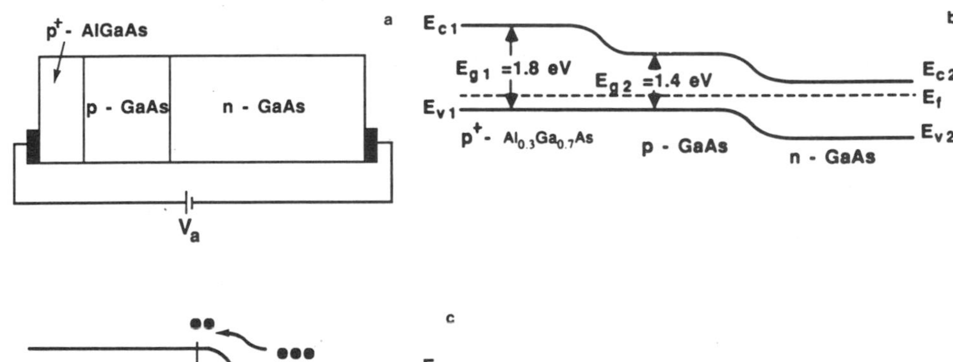

FIGURE 12.27. (a) Schematic drawing of a GaAs IR emitter with AlGaAs window layer, (b) energy band diagram in equilibrium, and (c) energy band diagram under forward-bias condition showing light emission from both p- and n-regions.

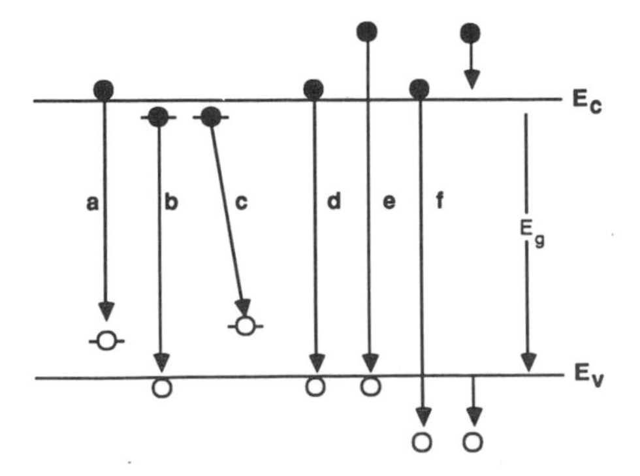

FIGURE 12.28. Possible electronic transitions involving radiative recombination in a semiconductor. (a) Conduction band to acceptor states, (b) donor states to valence band, (c) donor to acceptor states (pair emission), (d) conduction band to valence band (intrinsic emission), (e) hot carrier or avalanche emission, and (f) intraband transition. After Ivey,[12] by permission, © IEEE–1981.

12.4.2. Electronic Transitions

Figure 12.28 shows some possible electronic transitions in a semiconductor due to external excitations.[12] These transitions may lead to either the radiative and nonradiative recombination processes. We note that for an efficient luminescent material, radiative transition usually dominates over the nonradiative process.

In a direct bandgap semiconductor such as GaAs, the emission of optical radiation results mainly from band-to-band radiative recombination. This is shown by process d in Fig. 12.28. The emission spectrum for such a transition can be expressed by

$$I(h\nu) = \nu^2(h\nu - E_g)^{1/2} \exp\left[-\frac{(h\nu - E_g)}{k_B T}\right] \tag{12.67}$$

which shows that peak intensity occurs near the bandgap energy of the semiconductor, and the width of the emission spectrum is proportional to $k_B T$.

If the transition of carriers is from the band edge of one energy band to the impurity level near the opposite band, then the energy of the emitted photons will be slightly smaller than the bandgap energy of the material. The emission spectrum for electronic transition from the conduction band edge to the acceptor level near the valence band is given by

$$I(h\nu) = \nu^2(h\nu - E_g + E_a)^{1/2}\left\{\exp\left[\frac{(h\nu - E - E_g - E_{fn})}{k_B T}\right] + 1\right\}^{-1} \tag{12.68}$$

where E_a is the ionization energy of the acceptor level. The peak intensity occurs near $(E_g - E_a)$, and the width of the emission spectrum is proportional to $k_B T$. Figure 12.29 shows the photoluminescent spectra for a very pure n-type InSb sample ($n_0 = 5 \times 10^{13}\ \text{cm}^{-3}$) at 4.2 K.[13]

The emission wavelength for a LED with band-to-band radiative recombination can be calculated by using the formula

$$\lambda = \frac{hc}{E_g} = \frac{1.24}{E_g} \quad \mu\text{m} \tag{12.69}$$

where E_g is the bandgap energy of the semiconductor in eV. For a GaAs IR emitter, with $E_g = 1.43$ eV, the emission wavelength is equal to 0.873 μm at 300 K.

12.4.3. Luminescent Efficiency and Injection Efficiency

The luminescent efficiency of a LED is defined as the ratio of the total optical radiation output power associated with the radiative process to the total input power. For a given input

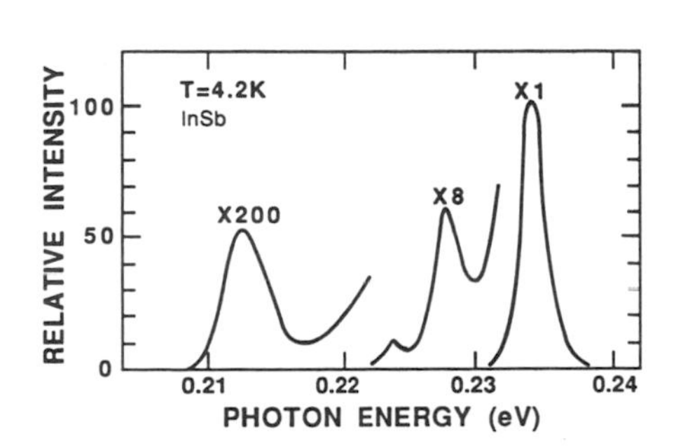

FIGURE 12.29. Photoluminescent spectrum at 4.2 K for a pure n-type InSb specimen with $n_0 = 5 \times 10^{13}\ \text{cm}^{-3}$, due to direct intrinsic interband recombination and neglecting exciton formation. After Mooradian and Fan,[13] by permission.

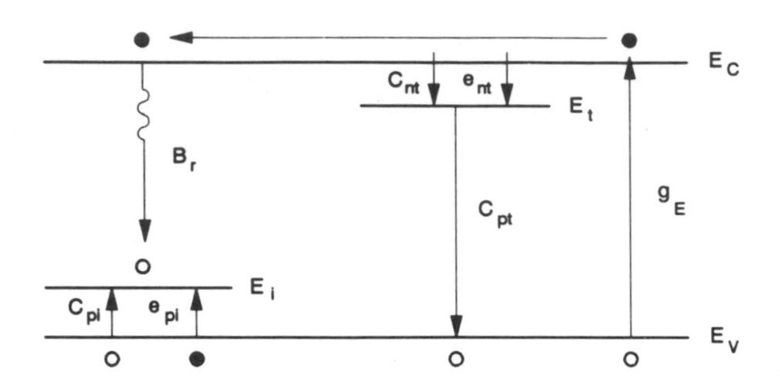

FIGURE 12.30. Energy band diagram showing emission and capture processes of a LED material via a single electron trap and a luminescence center in the forbidden energy bandgap. After Ivey,[14] by permission.

power, the radiative recombination process is in direct competition with nonradiative processes such as the Auger and Shockley–Read–Hall recombination processes in the device. Therefore, in order to increase the luminescent efficiency it is important to increase the radiative recombination process and meanwhile to reduce the nonradiative recombination process in the LED material.

To derive the luminescent efficiency of an LED, let us consider radiative and nonradiative transition processes for a typical LED material as shown in Fig. 12.30.[14] In our present treatment, it is assumed that only one electron trap level with activation energy E_t and density N_t exists below the conduction band edge. Furthermore, it is assumed that there is only one luminescent center with density N_l and activation energy E_l above the valence band edge. It is noteworthy that the recombination of electron–hole pairs via the electron trap E_t is nonradiative, while recombination via the luminescent center is radiative. Under steady-state conditions, the rate equations for electrons in the conduction band and in the trap level are given respectively by

$$\frac{dn}{dt} = g_E - C_{nt}n(N_t - n_t) - B_r np_l + e_{nt}n_t = 0 \tag{12.70}$$

and

$$\frac{dn_t}{dt} = C_{nt}n(N_t - n_t) - C_{pt}n_t p - e_{nt}n_t = 0 \tag{12.71}$$

where g_E is the external generation rate, C_{nt} is the electron capture rate at the E_t trap center, n_t is the electron density in the E_t level, B_r is the radiative capture rate at the E_l luminescent center, e_{nt} is the electron emission rate from E_t, C_{pt} is the hole capture rate of E_t, p_l is the hole density in E_l, and N_l is the density of E_l.

Similar rate equations can also be written for holes in the valence band and in the luminescent center. The solution of Eqs. (12.70) and (12.71) yields the external generation rate in the form

$$g_E = B_r np_l + C_{pt}n_t p \tag{12.72}$$

The first term on the right-hand side of Eq. (12.72) is due to radiative recombination, while the second term is attributed to the nonradiative recombination process. Therefore, the luminescent efficiency derived from the above simple model can be written as

$$\eta_l = \left(\frac{B_r np_l}{g_E}\right) \times 100\% = \frac{1}{(1 + C_{pt}n_t p/B_r np_l)} \times 100\% \tag{12.73}$$

For the low injection case, we can assume that a thermal equilibrium condition prevails between the electron trap level and the conduction band, as well as between the luminescent center and the valence band. Thus,

$$e_{nt}n_t = C_{nt}n(N_t - n_t) \tag{12.74}$$

and

$$e_{pl}p_l = C_{pl}p(N_l - p_l) \tag{12.75}$$

Furthermore, it is assumed that the Fermi level is located between E_t and E_l such that

$$(N_t - n_t) \simeq N_t \tag{12.76}$$

and

$$(N_l - p_l) \simeq N_l \tag{12.77}$$

From Chapter 6, the relationships between e_{nt} and C_{nt}, and e_{pl} and C_{pl} are given, respectively, by

$$e_{nt} = n_1 C_{nt} \tag{12.78}$$

and

$$e_{pl} = p_1 C_{pl} \tag{12.79}$$

where

$$n_1 = n \exp[-(E_t - E_f)/k_B T] \tag{12.80}$$

and

$$p_1 = p \exp[-(E_f - E_l)/k_B T] \tag{12.81}$$

If Eqs. (12.74) through (12.81) are solved, then we obtain

$$\frac{n_t}{p_l} = \left(\frac{N_t}{N_l}\right) \exp[-(E_t - E_l)/k_B T] \tag{12.82}$$

Now, substituting the latter ratio n_t/p_l from Eq. (12.82) into Eq. (12.73) yields the luminescent efficiency

$$\eta_l = \frac{1}{1 + (C_{pt}pN_t/B_r nN_l) \exp[-(E_t - E_l)/k_B T]} \tag{12.83}$$

It is seen from this equation that the luminescent efficiency will increase with increasing density of luminescent centers and decreasing temperature. To increase the luminescent efficiency, the energy level of the luminescent center should be as close to the valence band as possible, and the density of electron traps must be kept as low as possible.

The minority carrier injection efficiency is another important parameter which governs the internal quantum efficiency of a LED. This parameter is directly related to the radiative recombination current, which is the dominant current component in a LED. Depending on the impurity profile and the external applied bias voltage, there are four current components in an LED which should be considered under forward-bias conditions, namely, the electron diffusion current on the p-side, the hole diffusion current on the n-side, the recombination current in the depletion region, and the tunneling current across the junction barrier. The tunneling current is important only for a heavily doped p–n junction under small forward-bias conditions and can be neglected in a LED operating under a moderate forward-bias condition. Since most of the luminescence is usually produced by the electron diffusion current inside the quasi-neutral p-region, we can define the current injection efficiency by

$$\gamma = \frac{I_n}{(I_n + I_p + I_r)} \tag{12.84}$$

where

$$I_n = \left(\frac{qD_n n_i^2}{L_n N_A}\right) A[e^{qV/k_BT} - 1] \tag{12.85}$$

$$I_p = \left(\frac{qD_p n_i^2}{L_p N_D}\right) A[e^{qV/k_BT} - 1] \tag{12.86}$$

and

$$I_r = \left(\frac{qn_i W}{2\tau_0}\right) A e^{qV/2k_BT} \tag{12.87}$$

The hole diffusion current component given in Eq. (12.86) is usually negligible compared to electron current component in a practical LED due to a high electron–hole mobility ratio, and thus Eq. (12.84) can be further simplified. The overall internal quantum efficiency of a LED is equal to the product of Eqs. (12.83) and (12.84) and can be expressed by

$$\eta_I = \eta_i \gamma \tag{12.88}$$

Finally, the single most important physical parameter for assessing the performance of a LED is the external quantum efficiency η_E. Even though the internal quantum efficiency η_I given by Eq. (12.88) for an LED can be quite high (e.g., $\eta \geq 50\%$), the external quantum efficiency is usually very low (i.e., less than 5%). This is due to significant losses of internal absorption and reflection taking place during light emission from the LED. A simple expression relating the external quantum efficiency to the internal quantum efficiency is given by

$$\eta_E = \frac{\eta_I}{(1 + \bar{\alpha} V / A\bar{T})} = \frac{\eta_I}{(1 + \bar{\alpha} x_j / \bar{T})} \tag{12.89}$$

where $\bar{\alpha}$ is the average absorption coefficient, and $\bar{T}$ is the total light emission within the critical angle θ_c. The quantity $\bar{T}$ is related to the transmissivity T by

$$\bar{T} = T \sin^2(\theta_c/2) \tag{12.90}$$

with

$$T = \frac{4n_1n_2}{(n_1+n_2)^2} \tag{12.91}$$

where n_1 and n_2 denote the refractive indices of the semiconductor and ambient, respectively; V is the volume of the LED, A is the active area of the diode, and x_j is the junction depth. For most LEDs, n_1 may vary between 3.3 and 3.8. From Eq. (12.89), it is seen that the external quantum efficiency can be increased by reducing x_j or by increasing $\bar{T}$. However, reducing the junction depth to less than one minority carrier diffusion length will increase the number of minority carriers diffusing toward the surface. This may not be desirable, since a high surface recombination loss will reduce the internal quantum efficiency. For example, to reduce the surface recombination loss in a GaAs IR emitter, it is common practice to incorporate an AlGaAs window layer in the GaAs LED structure. Since the bandgap energy of AlGaAs is larger than that of GaAs, the AlGaAs window layer is transparent to light emitting from the GaAs active region. It is worth noting that the interface state density between the AlGaAs and GaAs layers is usually much lower than at the GaAs surface due to excellent lattice match between these two material systems. Therefore, the junction depth of an AlGaAs/GaAs LED can be greatly reduced. Figure 12.31 shows external quantum efficiency versus junction depth for a GaAs IR emitter at 300 K.[15] The optimal junction depth for a GaAs IR emitter is found to be about 25 μm.

The reduction of absorption in GaAs can be achieved by shifting the luminescence peak beyond the absorption edge with photon energies $h\nu < E_g$. A high external quantum efficiency can be obtained for this case, since the emitted photons fall beyond the absorption edge of the semiconductor where the absorption coefficient in the LED is very small. For a GaP red LED, the layer structure can be fabricated by using a double liquid-phase epitaxial (LPE) technique, and an external quantum efficiency exceeding 5% for such an LED has been reported.[16]

Other loss mechanisms, which may reduce the number of emitted photons and the external quantum efficiency, include: absorption loss within the LED, Fresnel loss, and critical angle loss. For example, the absorption loss for a GaAsP LED grown on a GaAs substrate will be quite large, since GaAs is opaque to visible light and can absorb about 85% of the photons emitted from a GaAsP LED. However, for a GaAsP LED grown on GaP substrate, the absorption loss can be greatly reduced. In fact, only about 25% of the photons emitted from the active region of the GaAsP LED are absorbed by the GaP substrate. Therefore, the external quantum efficiency for such a LED is greatly improved. Fresnel loss arises from the fact that when photons emit from a medium with a high index of refraction (e.g., for GaAs, $n_2 = 3.66$) to a medium with a low index of refraction (e.g., $n_1 = 1$ for air), a portion of the

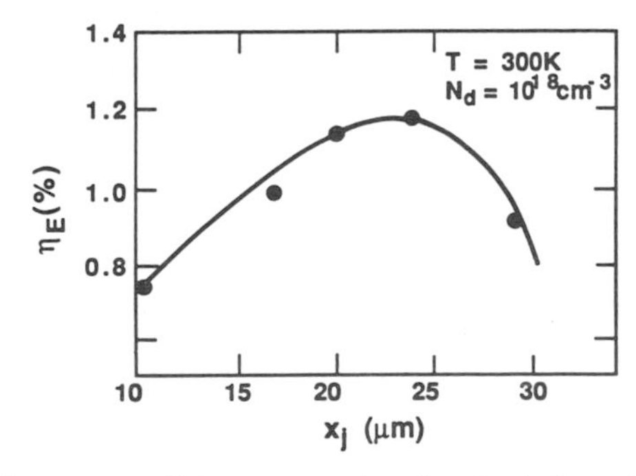

FIGURE 12.31. External quantum efficiency versus junction depth for a GaAs infrared (IR) emitter. After Yang,[15] by permission.

light will be reflected back to the medium interface [i.e., $R = (n_2 - n_1)^2/(n_2 + n_1)^2$]. Finally, critical angle loss is caused by total internal reflection of incident photons impinging on the surface of a LED at an angle greater than the critical angle θ_c, defined by Snell's law, $\theta_c = \sin(n_1/n_2)$.

12.4.4. *Application of LEDs*

In designing a LED, one needs to first characterize the physical parameters discussed above so that the performance of a LED can be optimized for a specific application. The design considerations are similar for LEDs fabricated from both indirect and direct bandgap materials. However, there are distinct differences between direct and indirect bandgap materials which include a larger optical absorption coefficient for the direct bandgap semiconductor in the case of light generation at the junction, and the need to introduce luminescence centers in an indirect bandgap semiconductor to produce radiative recombination.

The most efficient LEDs using indirect bandgap materials are red and green GaP LEDs, while LEDs fabricate from direct bandgap materials including SiC, ZnSe, GaAs, $GaAs_{1-x}P_x$ ($x < 0.45$), and $Ga_{1-x}Al_xAs$ ($x < 0.44$), among others. Applications of LEDs for optical display in the visible spectral range require that wavelengths of the emitted photons from these LEDs fall between 0.45 and 0.68 μm. Therefore, materials useful for this spectral range should have energy bandgaps varying between 1.8 and 2.7 eV. However, for optical communication applications, infrared (IR) driven LEDs such as GaAs and $In_{0.53}Ga_{0.47}As$ IR emitters are the prime candidates. Table 12.2 lists spectral characteristics of some practical LEDs along with their radiation emission peak wavelengths. Figure 12.32 shows different types of lens geometries used in the LEDs, which are aimed at increasing light extraction and optical efficiency.[16] The effectiveness of various geometries used in these LEDs is also summarized in Table 12.2. Figure 12.33a shows emission spectra for a GaAs IR emitter operating at 77 and 295 K, while Fig. 12.33b shows the external efficiency as a function of temperature for a GaAs IR emitter.[17] It should be noted that most commercial LEDs are fabricated from GaP, GaAs, $Ga_xAl_{1-x}As$, and $GaAs_{1-x}P_x$ materials. We shall next discuss some common types of LEDs which are available commercially today.

GaP LEDs. Gallium phosphide (GaP), which is an indirect wide-bandgap material with an energy bandgap of $E_g = 2.26$ eV at 300 K, is widely used for fabricating red and green LEDs. By doping GaP LEDs with certain shallow impurities, green or red light emission can be obtained from these doped GaP LEDs. For example, a red LED can be fabricated from Zn–O-doped GaP, while a green GaP LED may be obtained by using nitrogen-doped GaP. In general, radiative recombination in an indirect bandgap material such as GaP is achieved

TABLE 12.2. Figures of Merit for Various LED Geometries per Unit Internal Photon Flux Generation ($n = 3.6$)

Geometry	Radiant flux p	Maximum radiant intensity $J(\theta) = 0°$	Average radiant intensity $J(\theta) = 26°$
Flat plane diode			
Area emission	0.013	0.0042	0.0039
Hemisphere	0.34	0.054	0.054
Weierstrasse sphere	0.34	1.4	0.52
Truncated ellipsoid	0.25	9.8	0.39
Truncated cone	0.20	0.063	0.059
Paraboloid source			
$R_j/F_p = 0.1$	0.34	0.84	0.52
$R_j/F_p = 0.05$	0.34	3.3	0.52

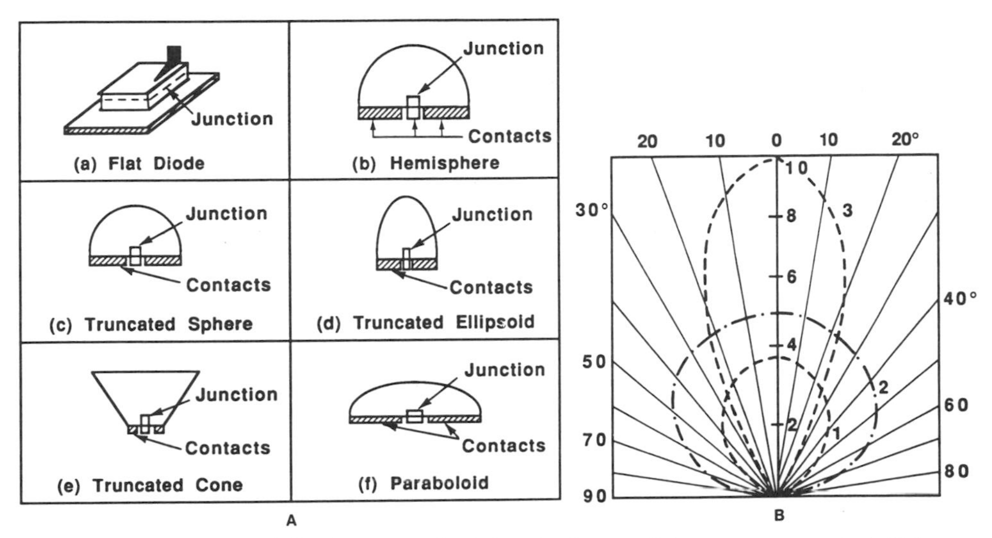

FIGURE 12.32. (a) Some LED lens geometries designed to increase light extraction or optical efficiency; (b) radiation patterns of LEDs with (1) rectangular, (2) hemispherical, and (3) parabolic geometries. After Galginaitis,[16] with permission.

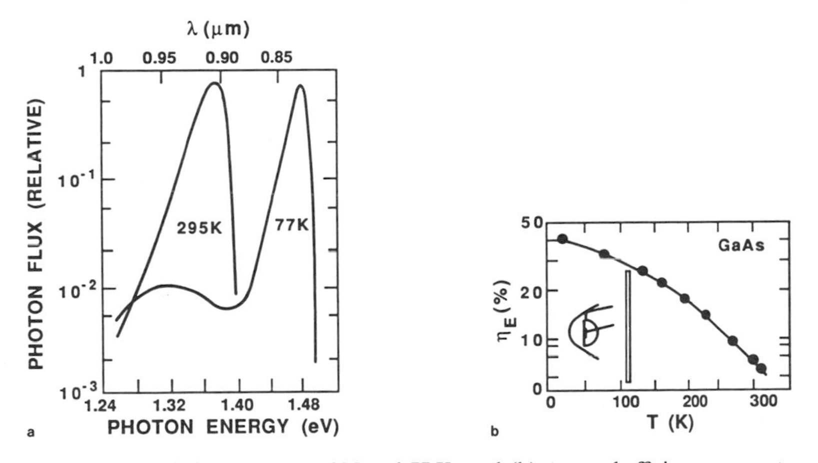

FIGURE 12.33. (a) Emission spectra at 295 and 77 K; and (b) external efficiency versus temperature for a GaAs infrared (IR) emitter. After Carr,[17] by permission, © IEEE–1965.

by the excitonic radiative recombination process via luminescent impurity centers in the forbidden gap. The physical principles and characteristics of a red and green GaP LED are now discussed separately.

Red GaP LED. The zinc- and oxygen- (Zn–O) doped GaP p-n junction diode grown by the LPE technique is the most efficient red LED available commercially. The Zn–O pair impurity is an isoelectronic trap in GaP, which can replace an adjacent Ga–P pair of atoms to form a recombination center with an activation energy of 0.3 eV below the conduction band edge. Radiative recombination of electrons and holes in the Zn–O centers will lead to the emission of red light ($h\nu = 1.95$ eV) from such a LED. The energy of the emitted photons

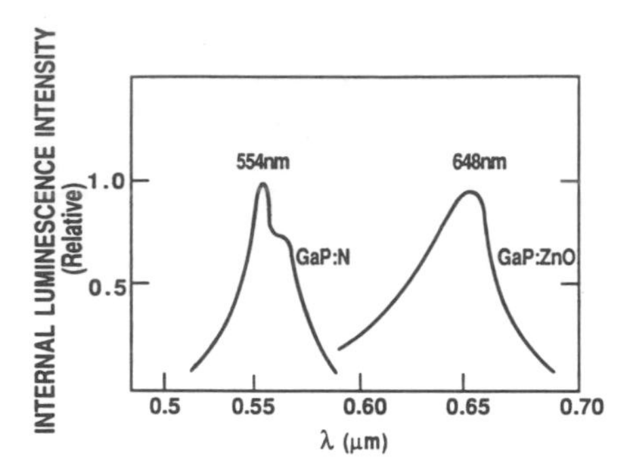

FIGURE 12.34. Normalized internal luminescence intensity spectra of a red GaP:ZnO LED and a green GaP:N LED.

from such a radiative recombination process is given by

$$h\nu = E_g - E_{DA} + \frac{q^2}{4\pi\varepsilon_0\varepsilon_s r} \tag{12.92}$$

where $h\nu$ is the photon energy, E_{DA} is the activation energy of the Zn–O center, and the last term in Eq. (12.92) is due to Coulomb potential energy of the Zn–O pair separated by distance r. The GaP:ZnO LED produces an emission peak at 648 nm and has an emission half-width of 93 nm. This is shown in Fig. 12.34. Typical external quantum efficiencies of 2 to 3% have been obtained at a current level of 10 A/cm^2. The switching speed for a typical red GaP LED is about 100 ns.

Green GaP LED. The emission of green light from a LED can be achieved in nitrogen-doped (5×10^{18} cm^{-3}) GaP LEDs grown by using the vapor-phase epitaxy (VPE) or liquid-phase epitaxy (LPE) technique. The emission mechanism is due to radiative recombination of electron–hole pairs (exciton recombination) at a nitrogen impurity center on a phosphorus site. The nitrogen impurity is an isoelectronic trap in GaP, which can replace the phosphorus impurity to achieve green emission ($h\nu = 1.95$ eV and $\lambda = 554$ nm) in a GaP LED. A nitrogen isoelectronic trap is a highly localized potential well which can trap an electron and become charged. The resulting Coulomb field attracts a hole which pairs with the trapped electron to form an exciton (i.e., a hydrogen-like bound electron–hole pair). The annihilation of this exciton via a radiative recombination process gives rise to a green emission with wavelength equal to 554 nm at 300 K. Due to other nonradiative recombination processes, the external quantum efficiency for a green nitrogen-doped GaP LED is usually less than 1%. However, despite its low external quantum efficiency, the nitrogen-doped GaP LED provides the highest brightness since green emission is near the peak of human-eye sensitivity. The normalized luminesence intensity spectra for a red GaP:ZnO LED and a green GaP:N LED are shown in Fig. 12.34.[18]

Yellow GaP LED. Emission of yellow light can also be accomplished in a nitrogen-doped GaP LED if the nitrogen doping density is greater than 2×10^{19} cm^{-3}. Using a high nitrogen-doping density in a GaP LED will lead to a shift in the emission peak to a longer wavelength (from green to yellow light) due to excitonic recombination at the nitrogen–nitrogen nearest-neighbor complexes.

$GaAs_{1-x}P_x$ LEDs. For a direct bandgap semiconductor, the color of light emission from a LED depends on the energy bandgap of the material. In III–V ternary compound semiconductors such as $Al_xGa_{1-x}As$ and $GaAs_{1-x}P_x$, the energy bandgap of these materials can be altered by varying the AlAs and GaAs alloy composition x. By changing the value of x and hence the energy bandgap, we can change the color of light emitted from such a

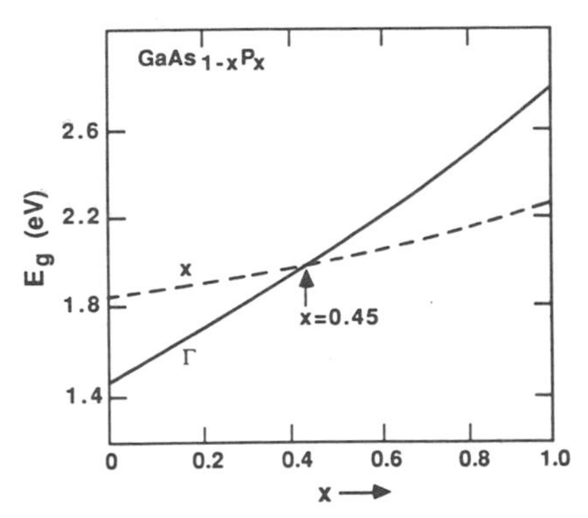

FIGURE 12.35. Energy bandgap versus alloy composition x for $GaAs_{1-x}P_x$ material.

LED. $Al_xGa_{1-x}As$ and $GaAs_{1-x}P_x$ are the two most commonly used ternary compound semiconductors for LED fabrications in the visible spectral range. Figure 12.35 shows energy bandgap versus alloy composition x for the $GaAs_{1-x}P_x$ material. It is seen that $GaAs_{1-x}P_x$ is a direct bandgap semiconductor for $x < 0.45$, and becomes an indirect bandgap material for $x > 0.45$. Therefore, if the phosphorus mole fraction x is chosen equal to 0.3 (i.e., $E_g =$ 1.8 eV), then red emission can be obtained from a $GaAs_{0.7}P_{0.3}$ LED due to band-to-band radiative recombination. A similar bandgap variation versus aluminum composition x as shown in Fig. 12.35 also exists in the $Al_xGa_{1-x}As$ material system. Therefore, a red LED can also be fabricated from $Al_xGa_{1-x}As$ material with x equal to 0.3.

The energy bandgap for a $GaAs_{1-x}P_x$ material varies from 1.43 eV for $x = 0$, to 2.26 eV for $x = 1$. Since band-to-band radiative recombination is the dominant recombination process in a direct bandgap material, it is expected that the external quantum efficiency for a $GaAs_{1-x}P_x$ LED will decrease with increasing alloy composition and energy bandgap. Figure 12.36 shows the external quantum efficiency for a $Ga_xAs_{1-x}P$ LED with and without nitrogen doping.[19] Nitrogen is found to enhance radiative recombination and the external quantum efficiency of a GaAsP LED. Figure 12.37 presents brightness versus photon energy for a GaAsP LED.[19] The brightness of a $GaAs_{1-x}P_x$ LED is seen to peak around $E_g = 1.9$ eV. From Fig. 12.35, this peak corresponds to a phosphorus mole fraction of $x = 0.4$. The peak emission wavelength for this LED is 660 nm, which falls in the red spectral range. External efficiency greater than 1% can be obtained for an encapsulated GaAsP LED with a brightness of 300 fL/A · cm^2. By increasing the phosphorus composition x (and hence the energy bandgap) in a $GaAs_{1-x}P_x$ LED, the emission peak will shift toward the orange color with decreasing external luminescent efficiency and brightness.

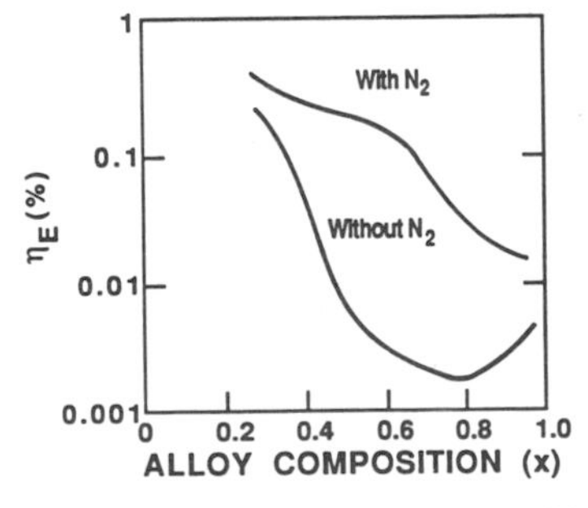

FIGURE 12.36. External quantum efficiency versus alloy composition for $GaAs_{1-x}P_x$ LEDs with and without nitrogen (N_2) doping. After Bhargava,[19] by permission, © IEEE–1975.

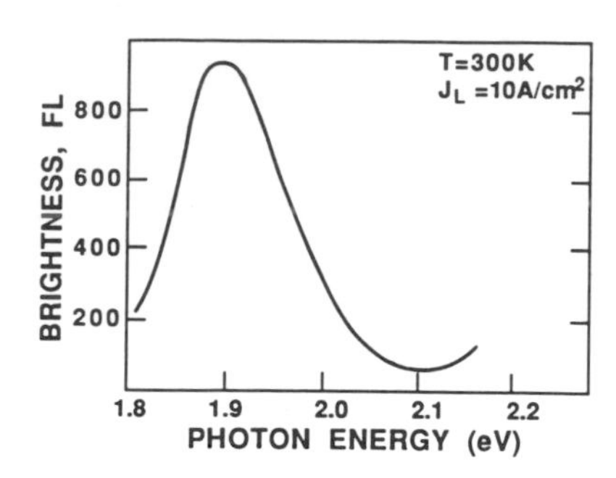

FIGURE 12.37. Brightness versus photon energy for a $GaAs_{1-x}P_x$ LED without nitrogen doping. After Bhargava,[19] by permission, © IEEE–1975.

A planar technology has been employed to fabricate $GaAs_{0.6}P_{0.4}$ LED arrays for numeric and alphanumeric displays. In this technology the GaP substrate is used for fabricating the $GaAs_{1-x}P_x$ LEDs (with $x \geq 0.5$) to avoid absorption of light emitted from the $GaAs_{1-x}P_x$ active layer by the GaP substrate. If GaAs substrate is used for growing the $GaAs_{1-x}P_x$LEDs, then due to lattice mismatch between the GaAs and GaP material system it is necessary to grow a $GaAs_{1-x}P_x$ graded epilayer by gradually increasing the alloy composition x from the surface of GaAs substrate to the top of the $GaAs_{1-x}P_x$ epilayer in order to create lattice match with the GaAs substrate during the epitaxial layer growth. The junction can be formed by changing the dopants during vapor-phase deposition, or by zinc diffusion into a uniformly doped structure of the graded composition. Figure 12.38 shows the cross-section view of a $GaAs_{1-x}P_x$ LED grown on a GaP substrate. A graded n^+-$GaAs_{1-x}P_x$ buffer layer is grown between the n-$GaAs_{1-x}P_x$ and n^+-GaP substrate to provide smooth transition and lattice match. A heavily doped p^+-$GaAs_{1-x}P_x$ is usually grown on top of the p-$GaAs_{1-x}P_x$ active layer to lower the contact resistance of the LED.

GaAs IR Emitter. Gallium arsenide, a direct bandgap material with an energy bandgap of 1.43 eV at 300 K, is widely used for fabricating near-infrared (IR) emitting diodes with emission peak at a wavelength of 890 nm. The GaAs IR emitter is the most efficient and widely used near-IR light source for a variety of applications ranging from optical communications, signal processing, fiber optic links, to optical computing. A GaAs IR emitter can be readily fabricated by using zinc diffusion in an n-type GaAs substrate to form a p–n junction. High external quantum efficiency (e.g., higher than 5%) in a GaAs IR emitter is obtained by using a typical dopant density of around 10^{18} cm^{-3} in both the n- and p-type regions. If a Si-doped GaAs substrate is used for fabricating a GaAs IR emitter, then its emission peak will shift to 1.32 eV, which is below the absorption edge of the GaAs material. As a result, the self-absorption effect in such a GaAs IR emitter is greatly reduced and the external quantum efficiency can be improved significantly ($\eta_{ext} \sim 20\%$). Other features of a GaAs IR emitter include high switching speed and fast recovery time (i.e., 2 to 10 ns), which make GaAs IR emitters ideal for data transmission applications.

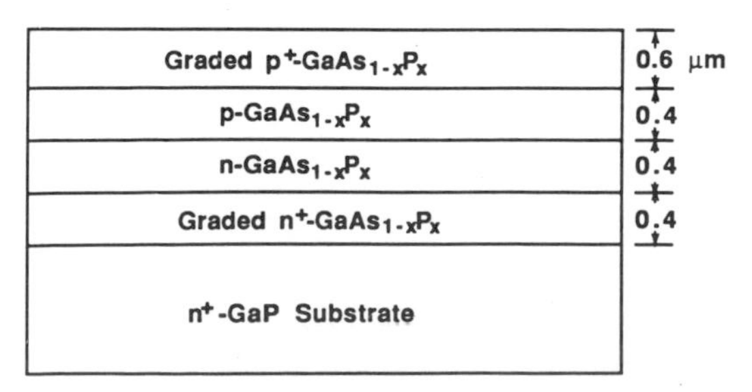

FIGURE 12.38. Cross-section view of a diffused $GaAs_{1-x}P_x$ LED with graded composition grown on an n^+-GaP substrate. Graded layer $x > 0.7$; p-layer $x = 0.7$ and n-layer $x = 0.6$.

It is well known that a direct bandgap GaAsP LED has the fastest switching speed, the best luminescence efficiency and color coverage among all the solid state LED family. Phosphor-coated GaAs IR emitters have the edge in color coverage, but fall well behind in both switching speed and light conversion efficiency.

In addition to high performance, low cost, and reliability, another factor in favor of LEDs is their compatibility with modern electron devices as well as the increasingly important applications in visual displays. Low power, low operating voltage, small size, fast switching speed, and long life are some attractive features of LEDs. It is noteworthy that the manufacturing technology for LEDs is compatible with silicon integrated circuit technology. With the widespread use of computers, the need to display symbolic information (i.e., letters, numbers, and special symbols) is rapidly increasing. Depending on the complexity of visual tasks, LEDs are being used as solid state lamps, symbolic and picture displays, data transmission, and optical communications.

LEDs coupled with silicon photodiodes can be used as optically isolated switches and sensing elements (e.g., to read holes in punched data cards). Flat panel picture display using LEDs is a long-range application goal, and is presently under development. The main obstacle here is the lack of blue LEDs available for optical display. Successful development of blue LEDs using ZnSe or other large-bandgap semiconductors is crucial for future flat panel picture display applications.

12.5. SEMICONDUCTOR LASER DIODES

The first coherent light source became available when Maiman introduced a pulsed solid-state ruby laser in 1960. Since then a wide variety of lasers including gas, solid state, semiconductor, and dye lasers has been subsequently developed. Today, lasers offer coherent light sources with wavelengths extending from extremely short-wavelength (i.e., X-ray) to long-wavelength infrared (LWIR) spectral regimes.

A semiconductor laser diode differs from conventional solid state lasers (e.g., ruby) or gas lasers (e.g., He–Ne, CO_2, or a N-laser) in several aspects: (1) a semiconductor laser diode is extremely small (typically 5 by 5 by 20 mils or smaller), (2) it exhibits high quantum efficiency, and (3) the intensity of its output light signals can be easily modulated by varying current (or bias voltage) in the laser diode. Furthermore, a semiconductor laser diode is usually operating at a much lower power level than that of a solid state laser or gas laser. Therefore, a semiconductor laser diode can be used as a portable and easily controlled coherent radiation source. Semiconductor laser diodes play an important role in many new areas of application including optical communications, optical computing, optical displays, and optoelectronic integrated circuits (OEICs). In this section, the basic device structure, operation principles, and characteristics of a semiconductor laser diode are depicted.

12.5.1. Population Inversion

A semiconductor laser diode is formed by using a heavily doped p–n junction structure on a direct bandgap semiconductor. The dopant density in both regions is usually greater than $10^{19}\ cm^{-3}$. When a large forward-bias voltage is applied to the laser diode, a state of population inversion (i.e., the conduction band states are filled and the valence band states are empty) is established near a narrow region of the p–n junction. Under this condition, radiative recombination takes place between electrons and holes in a narrow population inversion region near the junction, and laser action is followed if a resonant cavity is provided and the oscillation condition is satisfied. We note that the radiative recombination is not confined merely to the conduction–valence band transition. In fact, transitions from impurity band states have also been used in many laser diodes.

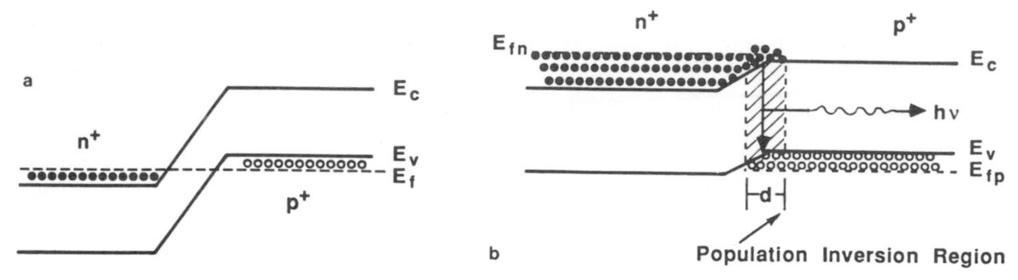

FIGURE 12.39. Energy band diagram for a heavily doped p–n junction laser diode: (a) in thermal equilibrium ($V_a = 0$) and (b) under large forward-bias conditions ($V_a \gg 0$).

Figure 12.39 shows energy band diagrams of a laser diode in (a) thermal equilibrium and (b) population inversion conditions. Since the total rate of radiative recombination is directly proportional to the product of electron and hole densities available in the conduction and valence bands, the radiative recombination is most intense in a narrow region near the metallurgical junction of the laser diode, as shown in Fig. 12.39b.

Population inversion in a p–n junction laser diode may be obtained by injection (or pumping) of minority carriers under a sufficiently large forward-bias voltage. To understand how the population inversion condition is accomplished, let us consider a GaAs laser diode. The applied forward-bias voltage V_a is chosen such that

$$V_a > \frac{h\nu}{q} \tag{12.93}$$

where $h\nu = E_1 - E_2$ is the emitted photon energy, E_1 is the electron energy in the conduction band, and E_2 is the hole energy in the valence band. As shown in Fig. 12.39b, the distribution functions for electrons in both the conduction and valence bands under a forward-bias condition are given respectively by

$$f_c(E_1) = \frac{1}{1 + \exp(E_1 - E_{fn})/k_B T} \tag{12.94}$$

and

$$f_v(E_2) = \frac{1}{1 + \exp(E_2 - E_{fp})/k_B T} \tag{12.95}$$

where $E_2 = E_1 - h\nu$; E_{fn} and E_{fp} are the quasi-Fermi levels for electrons and holes, respectively. Now consider the rate of stimulated emission at frequency ν due to transition from energy state E_1 in the conduction band to energy state E_2 in the valence band. The rate of the stimulated emission is proportional to the product of the density of occupied states in the conduction band, $g_c(E)f_c(E)$, and the density of unoccupied states in the valence band, $g_v(E)[1 - f_v(E)]$. Therefore, the total stimulated emission rate is obtained by integrating over all energy states in the population inversion region. This can be expressed by

$$W_{emission} = \int P_{cv} g_c g_v f_c (1 - f_v)\, dE \tag{12.96}$$

and

$$W_{absorption} = \int P_{vc} g_c g_v (1 - f_c) f_v \, dE \tag{12.97}$$

In these equations it is assumed that the rate of transition probability from the valence band to the conduction band and its inverse are equal (i.e., $P_{cv} = P_{vc}$), where P_{cv} denotes the rate of transition probability from the conduction band to the valence band, and P_{vc} is the rate of transition probability from the valence band to the conduction band. The condition for laser action to occur is that $W_{emission}$ must be greater than $W_{absorption}$. Now by solving Eqs. (12.94) through (12.97) we obtain

$$f_c(1 - f_v) > f_v(1 - f_c) \qquad \text{or} \qquad f_c > f_v \tag{12.98}$$

which implies that more states are occupied in the conduction band than in the valence band, and this is a condition for population inversion. The solution of Eqs. (12.94), (12.95), and (12.98) yields

$$E_{fn} - E_{fp} > (E_1 - E_2) = h\nu \tag{12.99}$$

We note that Eq. (12.99) is identical to Eq. (12.93) since $(E_{fn} - E_{fp})$ is equal to qV_a. Thus, the condition for population inversion in a laser diode is given by either Eq. (12.93) or Eq. (12.99).

12.5.2. Oscillation Conditions

Two conditions must be met to achieve sustained oscillation in a laser diode. First, a resonant cavity must be provided so that photons generated via radiative recombination may make several passages within the cavity to be further amplified in the active medium before leaving the cavity. A typical resonant cavity consists of two parallel reflecting surfaces perpendicular to the junction plane of a laser diode. This type of resonant cavity is known as a Fabry–Perot cavity or an interferometer. The air–semiconductor interface boundary may serve adequately as a reflecting surface if the refractive index of the semiconductor is large enough (e.g., for GaAs, $n = 3.46$). The second requirement is that the overall amplification constant per round trip through the cavity must be positive. If R is the reflection coefficient at the two reflecting surface boundaries and L is the distance between the two reflecting boundaries of the cavity, then for each trip between the boundaries the radiation power density is reduced by a factor of $2R$ at the interface boundaries. If the gain through stimulated emission is designated by $g(\nu)$, and the loss in the laser medium due to free-carrier absorption and defect-center scattering is denoted by l, then the condition for sustained oscillation is given by

$$\Gamma g(\nu) = l + \left(\frac{1}{L}\right) \ln(1/R) \tag{12.100}$$

where Γ is the carrier confinement factor, L is the length of the laser cavity, and $g(\nu) = \alpha_d(f_c - f_v)$ is the gain factor due to stimulated emission, α_d being the absorption coefficient for direct transition. The distribution functions f_c and f_v in the conduction and valence bands are given by Eqs. (12.94) and (12.95), respectively.

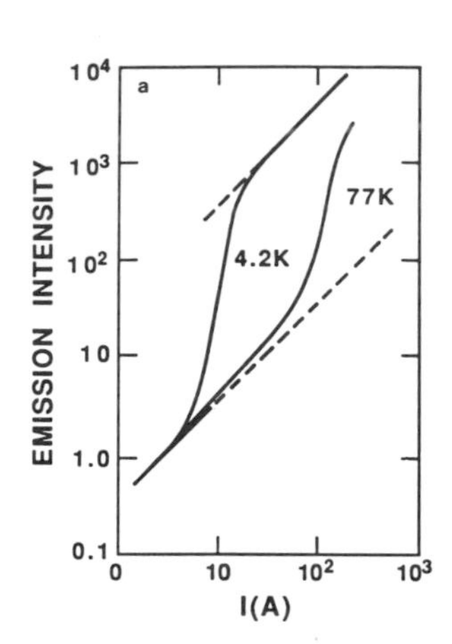

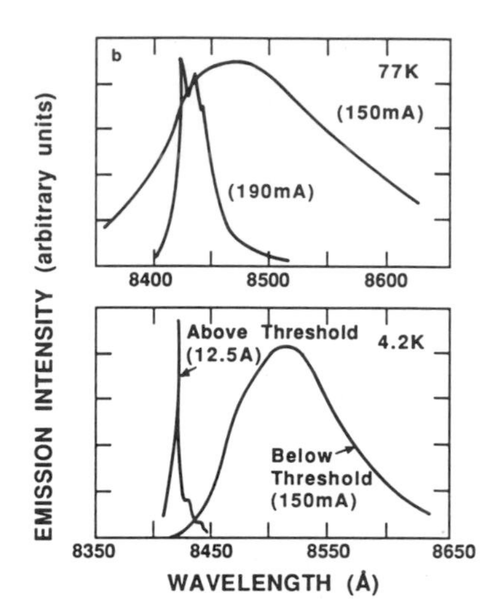

FIGURE 12.40. (a) Relative emission intensity versus diode current for a GaAs laser diode at 4.2 and 77 K. (b) Emission peak intensity versus wavelength before and after reaching threshold oscillation conditions. After Kressel and Butler,[18] by permission.

The physical interpretation of Eq. (12.100) may be described as follows. In laser operation, the condition for a complete population inversion is that $f_c = 1$ and $f_v = 0$. In general, there exist three distinct regimes in a laser diode under different forward-bias conditions. When the nonequilibrium condition is established by current injection (i.e., $V_a > 0$), spontaneous emission occurs. As the injection current increases, a condition for population inversion (i.e., for $f_c > f_v$) is reached near the physical junction (shown in Fig. 12.39b) which marks the beginning of stimulated emission [i.e., $g(\nu)$ becomes positive]. Upon further increase of the injection current, the difference between f_c and f_v widens and hence $g(\nu)$ increases. When the threshold condition is met, the laser diode undergoes an oscillating mode of operation, resulting in the emission of coherent light from the laser diode. As the operating temperature is lowered, the difference in f_c and f_v will increase for the same amount of injection current. Therefore, the threshold current decreases with decreasing diode temperature, as is evident from the experimental results of GaAs and other laser diodes. Figure 12.40a shows experimental curves of emission intensity versus diode current for a GaAs laser diode operating at 4.2 K and 77 K. Figure 12.40b presents emission intensity versus wavelength at 77 K and 4.2 K for conditions below and above the threshold current. The results show that a significant improvement in performance (such as lowering the threshold current and reducing the emission bandwidth) can be obtained at 4.2 K.[18]

12.5.3. *Threshold Current Density*

Figure 12.41 shows the device structure of a GaAs p–n junction laser diode for analyzing the lasing threshold current density. The laser diode is constructed by using a Fabry–Perot cavity with a pair of parallel planes (left and right sides) cleaved or polished to serve as end mirrors. The two remaining sides (front and back) of the laser diode are roughened to reduce emission of photons from both faces.

To facilitate analysis, it is assumed that the junction area is equal to A, the thickness of the population inversion region is d, and the distance between the two end mirrors is equal to L. For simplicity, it is further assumed that the reflection coefficients R_1 and R_2 at the two parallel plane mirrors are equal. At $T = 0$ K, the population inversion condition requires that, in the population inversion region, N_2 is equal to n (where n is the electron density in the

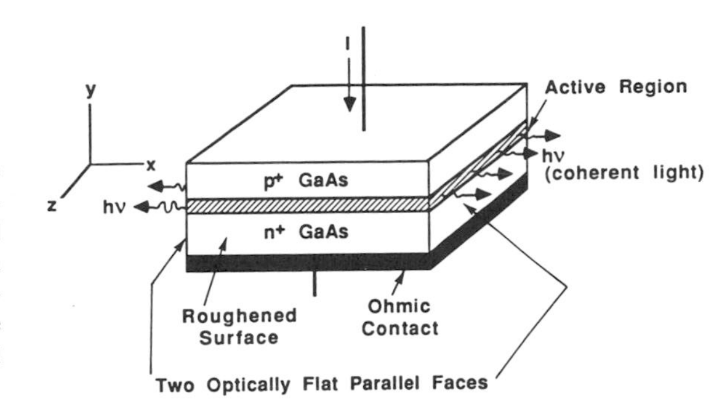

FIGURE 12.41. Structure of a GaAs p–n junction laser diode with a Fabry–Perot cavity. The optically flat parallel faces serve as the optical cavity (with length equal to L) along the x-direction, while the other two faces are roughened to prevent positive optical feedback along the z-direction.

conduction band) and N_1 is zero (where N_1 is the electron density in the valence band). The total number of conduction electrons in the population inversion region is therefore equal to N_2Ad. If the conduction electron making a downward transition from the conduction band to the valence has a lifetime of τ_2 and with each such decay requiring the injection of one electron into the junction region, then the total current flow in the laser diode needed to maintain the conduction band population density of N_2 is given by

$$I = \frac{qN_2Ad}{\tau_2} \quad \text{or} \quad J = \frac{qnd}{\tau_2} \tag{12.101}$$

The inverted population density N_2 is related to the lasing frequency ν_0, linewidth $\Delta\nu_0$, and cavity decay time constant τ_c by

$$N_2 = \left(\frac{4\pi^2}{3}\right)\left(\frac{\Delta\nu_0}{\nu_0}\right)\left(\frac{\tau_r}{\tau_c}\right)\left(\frac{n_0}{\lambda}\right)^3 \tag{12.102}$$

where τ_r is the total lifetime due to radiative and nonradiative recombination processes, and the cavity decay time τ_c is given by

$$\tau_c = \frac{(n_0/2cL)}{(2lL + \ln l/R)} \tag{12.103}$$

where c is the velocity of light, n_0 is the index of refraction for the laser material, and l is the loss associated with free-carrier absorption and scattering events in the cavity media. Now by solving Eqs. (12.101), (12.102), and (12.103) we obtain the threshold current density J_T, which reads

$$J_T = \frac{I_T}{A} = (4\pi^2/3)(2qcdL/\eta)(\Delta\nu_0/\nu_0)(n_0^2/\lambda_0^3)[2lL + \ln(R^{-1})] \tag{12.104}$$

where $\eta = \tau_2/\tau_r$ is the quantum efficiency of the laser diode, and measures the ratio of the radiative lifetime to the total decay lifetime by both radiative and nonradiative recombination processes.

As an example, calculations of threshold current density for a p–n junction laser diode using Eq. (12.104) are discussed as follows. The physical parameters for a GaAs laser diode operating at very low temperatures are given by: quantum yield, $\eta = 1$ (i.e., $\tau_r = \tau_2$); index of refraction for GaAs, $n_0 = 3.46$; linewidth, $\Delta\lambda = 200$ Å; laser wavelength, $\lambda = 8400$ Å; cavity length, $L = 0.3$ mm; junction depth, $d = 10^{-4}$ cm; reflection coefficient, $R = 0.32$; junction area, $A = 3 \times 10^{-4}\ \text{cm}^2$, and losses, $l = 0$. On substituting the values of these physical parameters into Eq. (12.104), we obtain the threshold current density $J_{th} = 120\ \text{A/cm}^2$ or $I_{th} =$ 36 mA. Experimental data for a GaAs laser diode taken at 4.2 K agree well with this calculation. It is noteworthy, however, that for $T > 60$ K the threshold current density for lasing is found to increase with T^3 due to the increase of absorption in the bulk GaAs near the junction region. This in turn increases the density of electrons in the valence band states. The measured threshold current at room temperature is about two orders of magnitude higher than the above predicted value.

12.5.4. GaAs Laser Diodes

The structure of a typical GaAs p–n junction laser diode with a Fabry–Perot resonant cavity is shown in Fig. 12.41. To obtain the population inversion condition in the active region of the p–n junction, both p- and n-regions are heavily doped (i.e., the doping concentrations in both regions are greater than $10^{19}\ \text{cm}^{-3}$). A pair of parallel planes are cleaved and polished perpendicular to the junction to act as an optical resonance cavity, while the two remaining sides of the laser diode are roughened to eliminate possible lasing in directions other than the two main parallel mirror planes.

Several types of conventional GaAs p–n junction laser diode structures have been reported in the literature. These include: (a) epitaxial homostructure, (b) single heterostructure (SH), (c) double heterostructure (DH), and (d) large optical cavity (LOC) DH laser diode structure. The schematic diagrams for these structures and their corresponding energy band diagrams are shown in Fig. 12.42a through d, respectively. The GaAs p–n homojunction laser diode structure is gradually becoming obsolete due to its high lasing threshold current density and poor confinement of both charge carriers and photons in the junction region. These shortcomings can be partially corrected in the SH laser diode structure in which a GaAlAs layer, which has a significantly different refractive index from that of GaAs, is incorporated into the GaAs laser diode structure. Furthermore, the inherent energy bandgap difference in both materials confines the carriers to one side (i.e., the p-GaAs side) of the junction. Significant improvement can be made if the injected carriers and emitted light can be kept near the active region of the p–n junction. These techniques are known as carrier and optical confinement, and they can be achieved by using the double heterostructure (DH) laser diode shown in Fig. 12.42c. The basic device structure consists of a thin p-GaAs active layer sandwiched between p-AlGaAs and n-AlGaAs cladding layers. Electrons injected into p-GaAs under forward-bias conditions are prevented from reaching the p-AlGaAs layer by the conduction band discontinuity. For example, the potential barrier (ΔE_c) for $Al_{0.3}Ga_{0.7}As$ is equal to 0.28 eV, which povides excellent carrier confinement. The optical confinement of the DH laser is due to the difference in the refractive index of GaAs and AlGaAs. Although the refractive index of GaAs is only 5% larger than that of AlGaAs, the optical confinement is excellent for the GaAs/AlGaAs DH laser diode. Population inversion in a GaAs/AlGaAs DH laser diode can be readily reached, and radiative recombination is limited to the p-GaAs active layer. Figure 12.43 shows the CW output power as a function of current and heat-sink temperature for an oxide-defined GaAs/AlGaAs DH laser diode, while Fig. 12.44 shows the typical CW laser spectrum for a GaAs/AlGaAs DH laser diode with a cavity length of 250 μm.[18]

The double heterostructure (DH) laser diode has by far shown the best performance in the laser diode operation. Some DH laser diodes mounted on diamond-II heat sinks can operate continuously at room temperature and above. A comparison of laser diode structures shown in Fig. 12.42 reveals that the LOC laser diode has the advantage of reducing the diffraction of the laser beam at the face of its p–n junction active region. This is achieved by allowing the laser beam to emerge from an opening which is much larger than the 1.2-μm size openings of the other types of laser diodes shown in Fig. 12.42.

Figure 12.45 shows a GaAs–AlGaAs double heterojunction (DH) laser diode with a stripe geometry contact. The substrate material is (111) or (100) oriented GaAs with a dopant

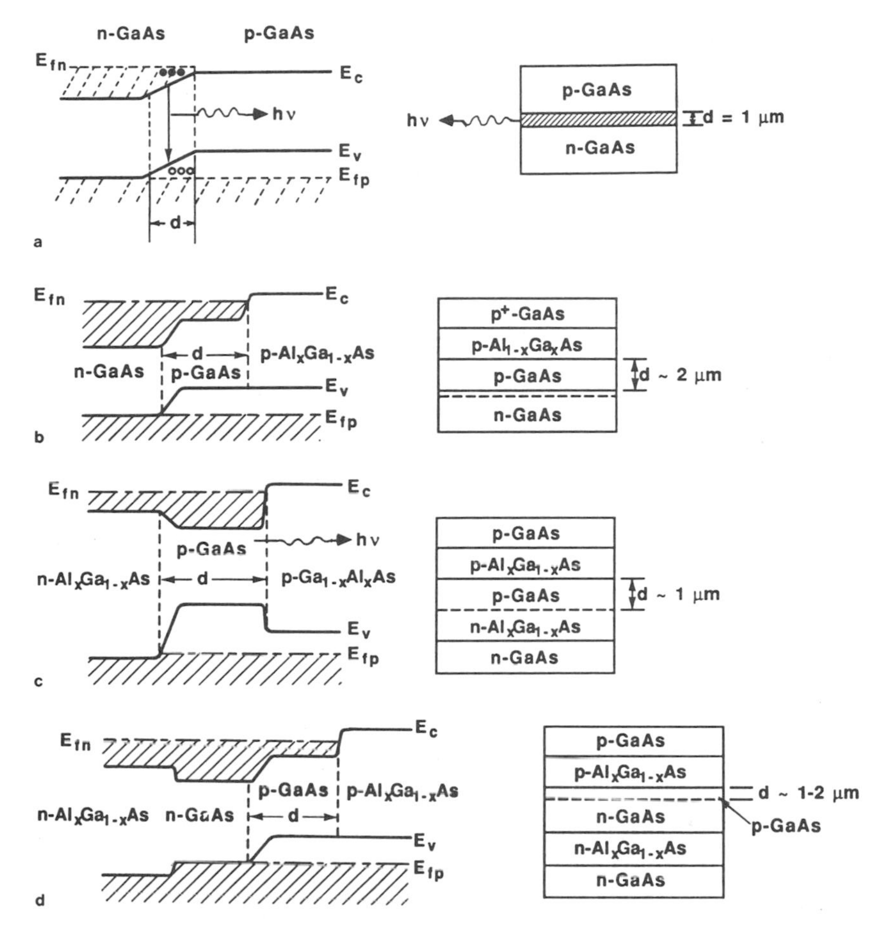

FIGURE 12.42. Multilayer heterostructures for carrier confinement in an injection laser diode: (a) homojunction GaAs laser diode, (b) single heterostructure (SH) laser diode, (c) double heterostructure (DH) laser diode, and (d) large optical cavity DH laser diode (LOC).

density of around 1 to 4×10^{18} cm^{-3}.[21] The first layer grown on the GaAs substrate is n-$Al_xGa_{1-x}As$, 2–5 μm thick, typically Sn-doped, with x varying between 0.2 and 0.4. The second layer is the p-GaAs active region (0.4 to 2 μm thick) doped with Si, and usually contains a small amount of Al, either deliberately provided or carried over from the first layer. The third layer is p-$Al_xGa_{1-x}As$ with a dopant density of around 3.8×10^{18} cm^{-3} and thickness 1 to 2 μm; the composition x for this layer varies between 0.2 and 0.4. The fourth and final layer is a p$^+$-GaAs layer with dopant density 3.5×10^{18} cm^{-3}. The main function of this layer is to provide better ohmic contact. The third and fourth layers are kept quite thin, since the main heat sink for a laser diode is normally provided by the fourth layer. A type-II diamond heat sink is used to improve the thermal performance of the laser diode and possesses thermal conductivity up to five times better than that of copper. The stripe geometry contact provides a convenient way of attaining a small device area with some lateral heat flow which allows a high continuous wavelength (CW) operating temperature. The optimum stripe width is usually between 10 and 15 μm for a typical GaAs laser diode. Figure 12.46 shows the cross-section views of several stripe-geometry InGaAsP infrared (IR) laser diode structures for 1.3 to 1.55 μm wavelength.[22] Figure 12.46a is an inverted-rib laser structure. The wide-bandgap (e.g., AlGaAs) waveguide layer underneath the active layer (GaAs) has a rib-like structure

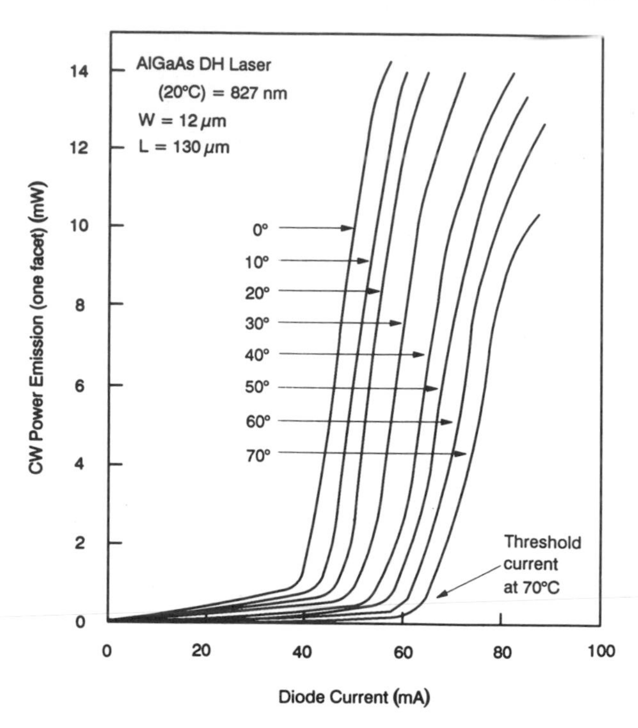

FIGURE 12.43. CW power output versus diode current for a GaAs/AlGaAs DH laser diode for different diode temperatures. After Kressel,[20] by permission.

inside the optical cavity. The change in thickness produces a larger effective index of refraction in the rib region than on both sides of the rib stripe, and results in waveguiding along the rib region. In this structure, the active layer is a planar structure and the injection of current through the active layer is limited to the narrow stripe of the rib region. The ridge-waveguide structure shown in Fig. 12.46b employs the same principle as the rib structure for waveguiding, which occurs underneath the ridge region. Figure 12.46c shows the etched-mesa buried heterostructure laser diode, which is quite different from those shown in Fig. 12.46a and b. In this structure the active layer (InGaAsP) is first etched into a narrow stripe, and InP is then regrown over the active stripe. Since the InP layer has a lower refractive index and larger energy bandgap than the InGaAsP active stripe, both optical and carrier confinement can be achieved using such a structure. It is seen that current injection for structure (c) is confined by the SiO_2 on the top of the device and the reverse-bias p–n junction on both sides of the active stripe, which results in more efficient use of injected carriers and hence lowers the lasing threshold. Figure 12.46d presents another InGaAsP laser diode structure. In this laser structure a V-groove is first etched into the substrate, and then a crescent stripe of InGaAsP active layer and InP cladding layers are grown on top of the V-groove in the wide-bandgap InP by a liquid-phase epitaxy (LPE) technique to achieve optical and carrier confinement in such a structure.

In the double heterostructure (DH) laser diodes discussed above, the typical thickness of the active layer is about 0.1 μm. If the thickness of the active layer is reduced to less than 200 Å, quantum size effects will occur. A new class of quantum well (QW) lasers based on such effects has been investigated extensively in recent years. These new QW lasers display characteristics which are quite different from conventional DH lasers. In a QW laser structure, the confinement of carriers in one dimension causes quantization in the allowed energy levels along the direction perpendicular to the quantum well. The density-of-states function changes

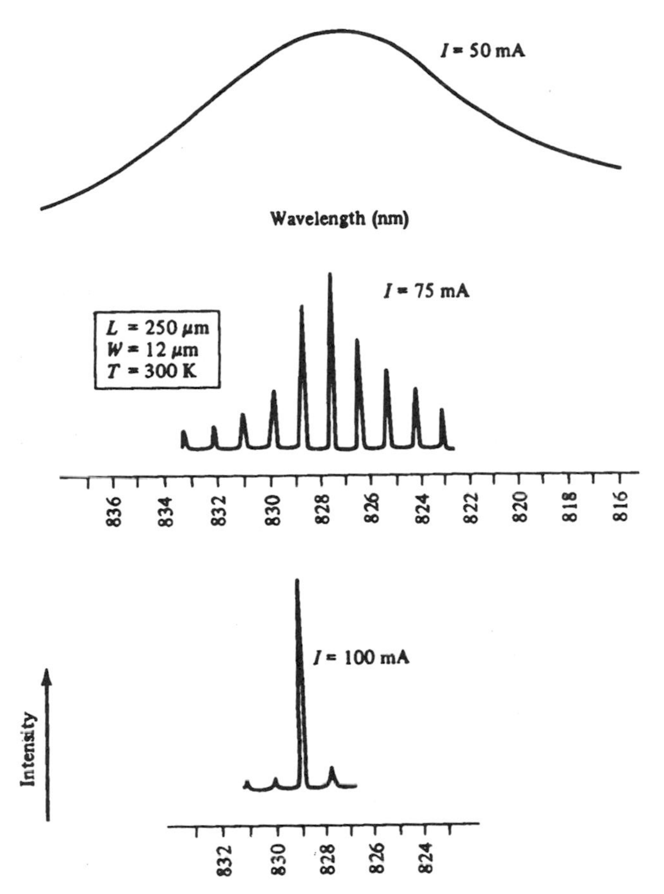

FIGURE 12.44. Lasing spectra of an oxide-defined GaAs–AlGaAs DH laser with cavity length 250 μm and different driving currents. After Kressel,[20] by permission.

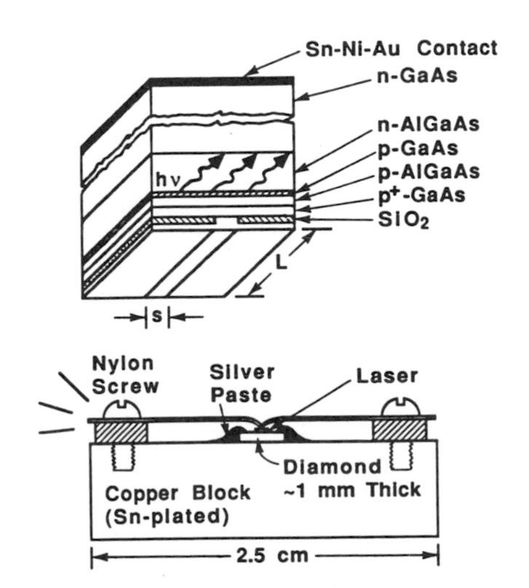

FIGURE 12.45. A GaAs/AlGaAs double heterojunction laser diode with a stripe geometry contact. The laser diode is mounted with the stripe down on a metallized diamond heat sink having five times the thermal conductivity of copper. Light emits from the p-GaAs active layer. After D'Asaro,[21] by permission.

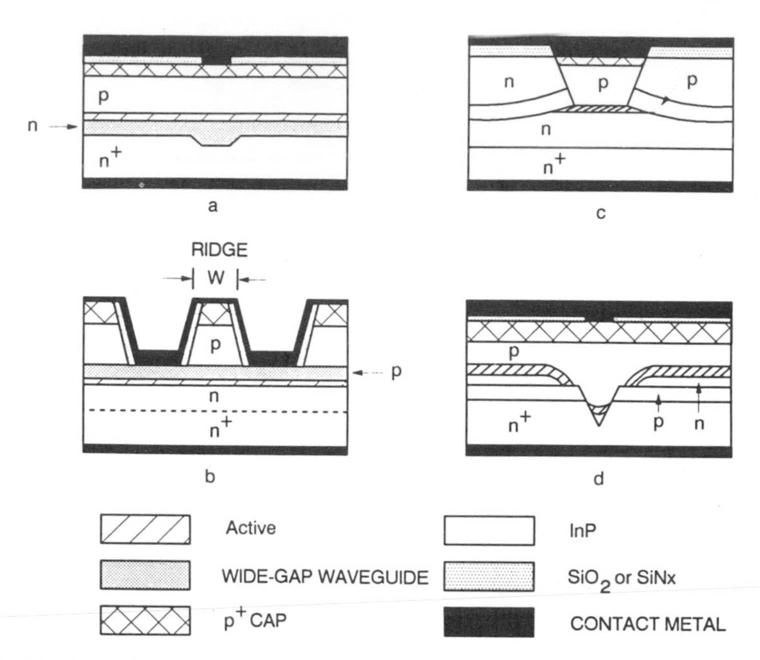

FIGURE 12.46. Some important stripe-geometry GaInAsP/InP laser structures: (a) inverted rib, (b) ridge waveguide, (c) etched-mesa buried heterostructure, and (d) channeled substrate buried structure. After Bowers and Pollack,[22] by permission.

from a square-root dependence on energy in a DH laser to a steplike dependence in a QW laser. If carriers are confined in two or three dimensions (i.e., a quantum wire or quantum box), the peak density of states of the QW lasers becomes even larger at each discrete level. Such modifications in the density-of-states function of the QW lasers can greatly reduce the threshold temperature dependence, lower the threshold current, and narrow the laser linewidth. Of the two most well-developed QW lasers, AlGaAs QW lasers are found to have superior performance characteristics than the InGaAsP QW lasers and conventional DH lasers. As an example, a very efficient AlGaAs/GaAs QW laser can be formed by using the graded-index waveguide, separate-confinement heterostructure (GRIN–SCH) with one or multiple quantum wells in the active layer, as shown in Fig. 12.47,[23] which illustrates that both optical and carrier confinement efficiencies can be greatly improved in a GRIN–SCH AlGaAs/GaAs QW laser structure. The optical confinement efficiency is improved by using the parabolic refractive index profile, which can focus more optical energy to the active quantum well. The improvement in carrier collection efficiency may be attributed to the fact that change in the density of states in the graded layers is reduced. It should be noted that the GRIN–SCH laser has the lowest threshold current ever reported for any semiconductor laser.

12.5.5. Semiconductor Laser Materials

In the previous section, we focused our attention mainly on laser diodes fabricated from GaAs material. The reasons that GaAs is chosen for laser diode fabrication are due to the fact that it is a direct bandgap material and high-purity GaAs epilayers with relatively low defect density can be grown routinely by using the MBE and MOCVD techniques. This is important from the standpoint of fabricating a reliable laser diode. Besides GaAs, infrared laser diodes have been fabricated using direct bandgap compound semiconductors such as InGaAs, InAsP, InGaAsP, $PbSn_xTe_{1-x}$, and $CdHg_xTe_{1-x}$, with emitting wavelengths varying

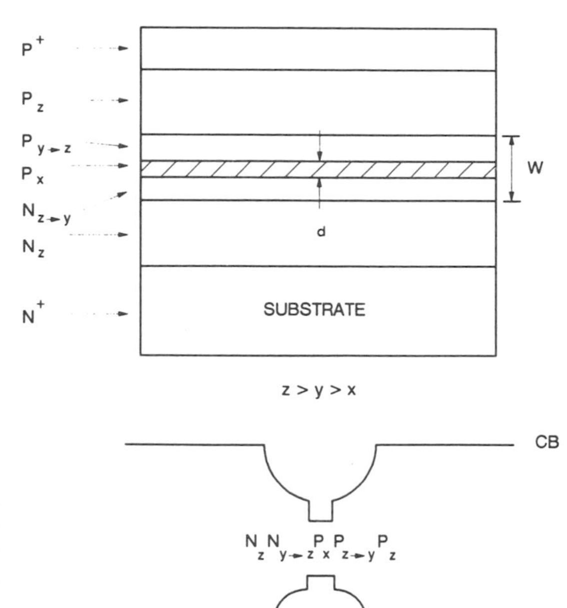

FIGURE 12.47. (a) Schematic diagram of a graded-index waveguide, separate-confinement heterostructure (GRIN-SCH) laser; (b) energy band diagram of a GRIN-SCH laser. After Tsang *et al.*,[23] by permission.

from 1.1 μm to more than 30 μm, depending on the selected alloy compositions of these materials. For example, the energy bandgap for $Cd_{1-x}Hg_xTe$ may vary from −0.142 eV to about 1.49 eV at 300 K as x changes from 1 to 0.

A GaAs laser diode, which emits coherent infrared radiation at 0.84 μm, is an important IR light source for many applications including data transmission, signal processing, optical computing, and optical communications. However, in the visible spectrum where excellent detectors are available, visible coherent light sources can be obtained from laser diodes fabricated on large-bandgap ternary and quaternary compound semiconductor materials such as $Al_xGa_{1-x}As$, $Ga_xIn_{1-x}As_yP_{1-y}$, and $Al_xGa_{1-x}As_yP_{1-y}$. These materials can be grown on either InP or GaAs substrates. Laser diodes fabricated from these materials can extend the useful wavelength of coherent radiation from 0.6 to about 3 μm.

Although a change of energy bandgap can be obtained from a list of compound semiconductors for laser diode applications, the choice is limited by two requirements, namely, the material must be a direct bandgap semiconductor and is capable of forming a p–n junction on such a material system. III–V compound semiconductors such as GaAs, GaSb, InP, InSb, and II–VI compounds such as ZnS, ZnSe are potential candidates for laser diode fabrication. III–V compound semiconductors can be grown in both n-type and p-type, and hence p–n junction laser diodes can be readily fabricated from these materials. Furthermore, since the bandgap energy of III–V ternary compounds such as $Al_xGa_{1-x}As$, $GaAs_xP_{1-x}$, and $In_xGa_{1-x}P$ can be varied by changing their alloy compositions, laser diodes fabricated from these materials will cover the emission wavelengths from the visible to the infrared regime. For example, the energy bandgap of a GaP material is 2.26 eV, while the energy bandgap for GaAs is 1.43 eV at 300 K. Thus, the mixed compounds of $GaAs_xP_{1-x}$ can produce coherent light emissions from the red to the near-infrared spectrum.

Unfortunately, fabrication of II–VI compound semiconductor laser diodes is not as simple and straightforward as III–V compound semiconductor laser diodes. The self-compensation problem in II–VI compound semiconductors prevents possible fabrication of p–n junction laser diodes from several of these materials. For example, CdS and ZnSe can be produced only in n-type, while materials such as ZnS and ZnTe are only available in p-type. In fact, CdTe is the only II–VI compound semiconductor which can be produced in both n- and p-type. Therefore, it is extremely difficult to form a p–n homojunction laser diode directly from

II–VI compound semiconductors. Although it is possible to fabricate a p–n heterojunction laser structure from these materials, lattice mismatch at the interface of two different II–VI semiconductors creates further complication for efficient laser operation. In the absence of an adequate heterostructure, laser diodes made from II–VI materials must utilize an optical or electron-beam pumping technique. Recently, successful fabrication of ZnSe blue LED operating at 77 K and 300 K has been reported. Successful conversion of n-ZnSe into p-type ZnSe has been reported by using nitrogen or oxygen implantation followed by thermal annealing with an IR lamp, while n-type conduction is obtained in Ga-doped ZnSe. Most ZnSe LEDs reported are grown on GaAs substrate using the MOCVD or MBE technique.

It should be noted that the number of semiconductor materials exhibiting laser action has been continually growing in recent years. Table 12.1 is a partial list of semiconductor materials which exhibit lasing action. The photon energy of oscillation, the corresponding wavelength, and methods of excitations are also included in this table.

12.5.6. Applications of Lasers

Applications of lasers to engineering, medicine, and scientific problems are a fast-changing subject. In this section, a brief description of a wide variety of laser applications is given. These include laboratory experiments using coherent laser sources, optical communication systems using laser diodes to transmit video or audio signals, optical computing or optical interconnects between chip and chip, and board and board, and industrial and medical applications using short and powerful pulsed or CW infrared (IR) lasers.

Lasers were first introduced into laboratories for performing experiments involving diffraction, interference, and for studies of material properties. In holography, lasers can be used to photograph three-dimensional objects by storing both amplitude and phase information on a film known as a hologram. The hologram can be illuminated by a laser to reproduce the object in three dimensions.

Another important application of laser diodes is in optical communication systems in which millions of telephone conversations or television programs could be transmitted simultaneously through a single laser beam. Since the carrier frequency in the laser beam is very high (e.g., from 10^{13} to 10^{16} Hz), it is possible to transmit a large amount of information in a single channel using optical communication. In fact, the optical technique for transmitting digital data in local distribution has now become available commercially by using GaAs injection lasers as well as LEDs. In fact, many commercial optical communication systems using InGaAsP LEDs or lasers to transmit modulated electrical or optical signals at 1.3 μm via low-loss optical fibers have been installed in various locations around the world. Furthermore, efforts have also been made to develop optoelectronic integrated circuits (OEICs) for applications in optical computing, signal processing, and optical communication systems. III–V compound semiconductors such as InGaAs/InP and AlGaAs/GaAs epitaxial layers grown by the MBE or MOCVD technique are most widely used for fabricating laser diodes, detectors, and modulators as well as other optical components in OEIC chips.

A pulsed laser diode can also be used as a range finder. If a pulsed laser beam from a Q-switched laser is aimed at a certain target, then the time delay between sending a laser pulse and the detection of a reflected laser pulse from the target gives the time it takes for the laser pulse (traveling at the speed of light) to make a round trip. The time constant can be measured very accurately by using a counting circuit. With this technique, accurate distances can be measured by a pulsed laser range finder. Such a device is important for both space and military applications.

Finally, high-power lasers (such as pulsed ruby lasers and CO_2 lasers) have also been used in cutting and welding small parts and in various small-scale melting vaporizing work. For example, a typical ruby laser pulse delivers its energy in microseconds and can release a tremendous amount of energy to a very small region. Applications of this include drilling holes through a diamond and other hard materials, balancing turbines, and watch movements. In the semiconductor industry, lasers are used in scribing metal interconnects between integrated circuits on a silicon wafer before separating them into individual chips. Lasers have also found many applications in the medical fields, particularly in microsurgery and other medical treatments.

PROBLEMS

12.1. Using Eqs. (12.59) and (12.60) calculate the quantum yield versus photon wavelength for $x_0 =$ 0.4, 0.8, and 1.2 μm, and for $W = 0.01$ cm for a silicon p–i–n photodiode. Assume that $R =$ 0.3, $L_n = 6 \times 10^{-2}$ cm, and $L_p = 4 \times 10^{-3}$ cm.

12.2. (a) Explain the key factors which need to be considered in the design of a photodetector.
(b) Suppose you have the following detectors available for a specific wavelength detection: a Au-n-type Si Schottky photodiode, a Ge–GaAs heterojunction photodiode, and a Cu-doped germanium photoconductor. Which of these detectors would you pick (and explain why) for:
(1) 10.6 μm detection?
(2) Maximum sensitivity needed at 1.06 μm detection?
(3) Low noise, high speed, and high sensitivity in the visible spectra?

12.3. Design a silicon p–n junction solar cell which will produce a conversion efficiency of 19% under AM1 conditions. Choose your own design parameters. What is the main difference in the performance characteristics between a p-on-n and an n-on-p junction solar cell?

12.4. The conversion efficiency for a Schottky barrier solar cell is given by

$$\eta_c = V_{mp}^2 I_0 (q/k_B T) \exp(qV_{mp}/k_B T)(P_{in} A_j)^{-1} \tag{1}$$

where V_{mp} is the voltage at maximum power output, $P_{in} = 93$ mW/cm^2 for AM1 sunlight, A_j is the cell area,

$$I_0 = A_j A^{**} T^2 \exp(-q\Phi_{Bn}/k_B T) \tag{2}$$

is the reverse saturation current; V_{mp} is related to I_0 and I_{ph} through the following relation:

$$\left(1 + \frac{qV_{mp}}{k_B T}\right) e^{qV_{mp}/k_b T} = 1 + \frac{I_{ph}}{I_0} \tag{3}$$

Equation (3) can be solved iteratively for V_{mp}. By substituting Eq. (2) into Eq. (1), the conversion efficiency can be calculated as a function of Φ_{Bn}. Calculate the conversion efficiency η_c for an Al–n–Si Schottky barrier solar cell. Given $\Phi_{Bn} = 0.71$ V, $A^{**} = 110$ A/cm$^2 \cdot$ K^2, $A_j =$ 4 cm^2, and $I_{ph} = 140$ mA, repeat for a Au–n-type Si Schottky barrier solar cell if $\Phi_{Bn} = 0.81$ V.

12.5. (a) Draw the energy band diagram for an AlGaAs/GaAs p–n junction solar cell shown in Fig. 12.7. If x for the $Al_xGa_{1-x}As$ window layer (top layer) is changed from $x = 0$, 0.3, 0.5, 0.7, to 0.9, plot the relative spectral response for this solar cell. *Given*: $E_g = 1.43$ eV for $x = 0$ and $E_g = 2.1$ eV for $x = 0.9$. It is assumed that a linear relation exists between E_g and x.

12.6. Calculate the dark current of an Al–n–Si Schottky barrier solar cell and a Si p–n junction solar cell. *Given*: $\Phi_{Bn} = 0.71$ V, $N_D = 10^{16}$ cm^{-3}, $N_A = 5 \times 10^{18}$ cm^{-3}, $L_n = 100$ μm, $L_p = 20$ μm, $n_i = 1.4 \times 10^{10}$ cm^{-3}, $A_j = 4$ cm^2, $\mu_n = 1000$ cm^2/V $\cdot$ s, and $\mu_p = 100$ cm^2/V $\cdot$ s. If the photocurrents generated in both cells are the same (e.g., $I_{ph} = 140$ mA), what are the open-circuit voltages for both cells?

12.7. If band-to-band radiative recombination is responsible for the emission of photons, what color of light may be expected from an LED made of the following materials and explain why: $GaAs_{0.5}P_{0.5}$, $Ga_{0.5}Al_{0.5}As$, $In_{0.5}Al_{0.5}As$, $InAs_{0.5}P_{0.5}$, $In_{0.5}Al_{0.5}P$, $In_{0.5}Ga_{0.5}As$, $In_{0.5}Ga_{0.5}P$, and $AlSb_{0.5}P_{0.5}$?

12.8. What is the optical power density (per unit volume) generated in a typical GaAs injection laser diode at threshold for $T = 4$, 77, and 300 K? If 1% of this power density is absorbed and converted into heat in the junction volume, what would be the rate of the temperature rise in this GaAs laser diode?

12.9. (a) Show that the short-circuit current produced in the base region of a metal-n-type semiconductor Schottky barrier cell is given by Eq. (12.32), assuming that $d \gg L_p$ and $s_n = \infty$.
(b) Calculate the quantum efficiency versus wavelength for a Au–n-type Si Schottky barrier cell for $L_p = 0.05$ cm, $R = 0.3$, and $W = 0.01$, 0.05, and 0.1 cm.

12.10. Design a Au–GaAs Schottky barrier photodiode for demodulating a 20-GHz modulation optical signal at 0.84 μm (select your own design parameters: diode area, dopant density, AR coating, etc.). If you are asked to design this Au–GaAs Schottky diode having a bandwidth of 100 GHz, what device parameters should be modified in this photodiode in order to meet this specification?

12.11. An $In_{0.53}Ga_{0.47}As$ PIN photodiode is used to detect 1.3 μm infrared radiation. If the dopant density is 1×10^{16} cm^{-3} in the n-region and 2×10^{18} cm^{-3} in the p$^+$-region, the diode area is 50 μm^2, and the n-layer is 1.5 μm thick, calculate R_s, C_j, and the RC time constant of this photodiode. What is the maximum cutoff frequency for this photodetector?

REFERENCES

1. M. P. Thekaekara, "Data on Incident Solar Energy," Suppl. Proc. 20th Annual Meeting Inst. Environ. Sci., p. 21 (1974).
2. J. J. Wysocki and P. Rappaport, "Effect of Temperature on Photovoltaic Energy Conversion," *J. Appl. Phys.* **31**, 571 (1961).
3. J. Lindmayer and J. F. Allison, "The Violet Cell: An Improved Silicon Solar Cell," Conf. Record, 9th IEEE Photovoltaic Spec. Conf., New York, p. 83 (1972).
4. S. S. Li, "Theoretical Analysis of a Novel MPN GaAs Schottky Barrier Solar Cell," *Solid-State Electron.* **21**, 435 (1978).
5. S. P. Tobin, S. M. Vernon, C. Bajgar, V. E. Haven, and L. M. Geoffroy, "High-Efficiency GaAs/Ge Monolithic Tandem Solar Cells," *IEEE Electron. Dev. Lett.* **9**, 256 (1988).
6. S. S. Li and F. A. Lindholm, "Quantum Yield of a Silicon p–i–n Photodiode," *Phys. Status Solidi A* **15**, 237 (1973).
7. H. Melchior and W. T. Lynch, "Signal and Noise Response of High Speed Germanium Avalanche Photodiode," *IEEE Trans. Electron. Dev.* **ED-13**, 829 (1966).
8. K. Nishida, K. Taguchi, and Y. Matsumoto, *Appl. Phys. Lett.* **53**, 251 (1979).
9. D. H. Lee, S. S. Li, and N. Paulter, "A 45 GHz High-Speed GaAs/AlGaAs Heterostructure Schottky Barrier Photodiode," Proc. Int. Conf. on Solid State Devices and Materials, Tokyo Japan, Aug. 23–26 (1988).
10. T. P. Lee, C. A. Burrus, and A. G. Dentai, "InGaAs/InP p–i–n Photodiodes for Lightwave Communications at 0.95–1.65 μm Wavelength," *IEEE J. Quantum Electron.* **17**, 232 (1981).
11. L. S. Yu, and S. S. Li, "A Novel High Sensitivity, Low Dark Current Grating Coupled Multiquantum Well/Superlattice IR Detector Using Bound-to-miniband Transition," *Appl. Phys. Lett.* **59**(11), 1332 (1991).
12. H. F. Ivey, "Electroluminescence and Semiconductor Lasers," *IEEE J. Quantum Electron.* **QE-2**, 713 (1966).
13. A. Mooradian and H. Y. Fan, "Recombination Emission in InSb," *Phys. Rev.* **148**, 873 (1966).
14. H. F. Ivey, "Electroluminescence and Related Effects," Suppl. 1, in: *Advances in Electronics and Electron Physics* (L. Marton, ed.), Academic Press, New York (1963).
15. E. S. Yang, *Microelectronic Devices*, p. 393, McGraw-Hill, New York (1988).
16. S. V. Galginaitis, "Improving the External Efficiency of Electroluminescent Diodes," *J. Appl. Phys.* **36**, 460 (1965).

17. W. N. Carr, "Characteristics of a GaAs Spontaneous p–n Junction Infrared Source with 40% Efficiency," *IEEE Trans. Electron Dev.* **ED-12**, 531 (1965).
18. H. Kressel and J. K. Butler, *Semiconductor Lasers and Heterojunction LEDs*, Academic Press, New York (1977).
19. R. N. Bhargava, "Recent Advances in Visible LEDs," *IEEE Trans. Electron Dev.* **ED-22**, 691 (1975).
20. H. Kressel, in: *Fundamentals of Optical Fiber Communications* (M. K. Barnoski, ed.), 2nd ed., Chap. 4, Academic Press, New York (1981).
21. L. A. D'Asaro, "Advances in GaAs Junction Lasers with Stripe Geometry," *J. Lumin.* **7**, 310 (1973).
22. J. E. Bowers and M. A. Pollack, in: *Optical Fiber Telecommunications II* (S. E. Miller and I. P. Kaminow, eds.), Academic Press, New York (1988).
23. W. T. Tsang, R. A. Logan, and J. P. Van der Ziel, "Low Current Threshold Stripe Buried Heterostructure Lasers with Self-Aligned Current Injection Stripes," *Appl. Phys. Lett.* **34**, 644 (1979).

BIBLIOGRAPHY

K. Aiki, M. Nakamura, T. Kuroda, and J. Umeda, "Lasing Characteristics of Distributed-Feedback GaAs–GaAlAs Diode Laser with Separate Optical and Carrier Confinement," *IEEE J. Quantum Electron.* **QE-12**, 597 (1976).

M. C. Amenn, "New Stripe-Geometry Laser with Simplified Fabrication Process," *Electron Lett.* **15**, 441 (1979).

A. A. Bergh and P. J. Dean, "Light-Emitting Diodes," *Proc. IEEE* **60**, 156 (1972).

A. A. Bergh and P. J. Dean, *Light-Emitting Diodes*, Clarendon Press, Oxford (1976).

D. Boetz, "Laser Diodes are Power-Packed," *IEEE Spectrum* **43**, June (1985).

C. O. Bozler, J. C. C. Fan, and R. W. McClelland, "Efficient GaAs Shallow Homojunction Solar Cells on Ge Substrates," in: *Proceedings of 7th International Symposium on GaAs and Related Compounds* (Inst. Phys. Conf. Ser., 45), pp. 429–436 (1979).

F. Capasso, "Multilayer Avalanche Photodiodes and Solid State Photomultipliers," Laser Focus/Electro-Optics, July (1984).

H. C. Casey, Jr. and M. B. Panish, *Heterojunction Lasers*, Academic Press, New York (1978).

V. Diadiuk and S. H. Groves, "Double-Heterostructure InGaAs/InP PIN Photodiodes," *Solid-State Electron.* **29**, 229 (1986).

N. K. Dutta, R. B. Wilson, D. P. Wilt, P. Besomi, R. L. Brown, R. J. Nelson, and R. W. Dixon, "Performance Comparison of InGaAsP Lasers Emitting at 1.3 and 1.55 μm," *AT & T Tech. J.* **64**, 1857 (1985).

M. Ettenberg, H. Kressel, and J. P. Eittke, "Very High Radiance Edge-Emitting LED," *IEEE J. Quantum Electron.* **QE-12**, 360 (1976).

L. Figueroa and C. W. Slayman, "A Novel Heterostructure Interdigital Photodetector (HIP) with Picosecond Optical Response," *IEEE Electron Dev. Lett.* **EDL-2**, No. 8, Aug. (1981).

K. Gillessen and W. Shairer, *Light Emitting Diodes*, Prentice-Hall, New York (1987).

I. Holonyak, Jr. and S. F. Bevacqua, "Coherent Light Emission from $GaAs_{1-x}P_x$," *Appl. Phys. Lett.* **1**, 91 (1962).

N. Holonyak, Jr., R. M. Kolbas, R. D. Dupuis, and P. D. Dapkus, "Quantum-Well Heterostructure Lasers," *IEEE J. Quantum Electron.* **QE-16**, 170 (1980).

H. J. Hovel, in: *Solar Cells, Semiconductors and Semimetals*, Vol. 11 (R. K. Willardson and A. C. Beer, eds.), Academic Press, New York (1975).

J. H. Kim and S. S. Li, "A High-Speed Au/InGaAs/InP Schottky Barrier Photodiode for 1.3–1.55 μm Photodetection," in: *High-Speed Electronics* (B. Kallback and H. Beneking, eds.), pp. 214–217, Springer-Verlag, Berlin (1986).

H. Kressel, in: *Fundamentals of Optical Fiber Communications* (M. K. Barnoski, ed.), 2nd ed., Chap. 4, Academic Press, New York (1981).

H. Kroemer, "A Proposed Class of Heterojunction Injection Lasers," *Proc. IEEE* **51**, 1782 (1963).

C. H. Lee, *Picosecond Optoelectronic Devices*, Academic Press, New York (1984).

B. F. Levine, G. Hasnair, C. G. Bethea, and N. Chand, *Appl. Phys. Lett.* **54**, 2704 (1989).

S. S. Li, "Process-Induced Defects in AlGaAs p–n Junction Solar Cells Fabricated by LPE, MOCVD, and MBE Techniques," Proc. 15th Photovoltaic Spec. Conf., pp. 1283–1288 (1981).

S. S. Li, C. T. Wang, and F. A. Lindholm, "Quantum Yield of Metal-Semiconductor Photodiodes," *J. Appl. Phys.* **43**, 4123–4129 (1972).

S. S. Li, W. L. Wang, P. W. Lai, and R. Y. Loo, "Deep-level Defects, Recombination Mechanisms, and Performance Characteristics of Low Energy Proton Irradiated AlGaAs Solar Cells," *IEEE Trans. Electron Dev.* **ED-27**, 857–864 (1980).

S. S. Li, C. G. Choi, and R. Y. Loo, "Studies of Radiation-Induced Defects in Proton Irradiated AlGaAs and Germanium Solar Cells," Proc. 19th IEEE Photovoltaic Specialists Conf., pp. 640–645 (1985).

F. A. Lindholm, S. S. Li, and C. T. Sah, "Fundamental Limitations Imposed by High Doping on the Performance of p–n Junction Silicon Solar Cells," Proc. 11th IEEE Photovoltaic Spec. Conf., pp. 3–11 (1975).

R. J. McIntyre, "The Distribution of Gains in Uniformly Avalanche Photodiodes: Theory," *IEEE Trans. Electron Dev.* **ED-19**, 703 (1972).

H. Melchior, M. P. Lepselter, and S. M. Sze, "Metal–Semiconductor Avalanche Photodiode," IEEE Device Research Conf., Boulder, Colo., June 17–19 (1968).

H. Melchior, A. R. Hartman, D. P. Schinke, and T. E. Seidel, "Planar Epitaxial Silicon Avalanche Photodiode," *Bell Syst. Tech. J.* **57**, 1791 (1978).

D. L. Miller and J. S. Harris, Jr., "Molecular Beam Epitaxial GaAs Heteroface Solar Cell Grown on Ge," *Appl. Phys. Lett.* **37**, 1104–1106 (1980).

J. Muller, "Photodiodes for Optical Communication," Adv. Electron. Electron. Phys. **55**, 189 (1981).

H. Namizaki, "Transverse-Junction Stripe Lasers with a GaAs p–n Homojunction," *IEEE J. Quantum Electron.* **QE-11**, 427 (1975).

M. I. Nathan, "Semiconductor Lasers," *Proc. IEEE* **54**, 1276 (1966).

M. I. Nathan, W. P. Dumke, G. Burns, F. J. Dill, Jr., and G. J. Lasher, "Stimulated Emission of Radiation from GaAs p–n Junction," *Appl. Phys. Lett.* **1**, 62 (1962).

L. D. Partain, M. S. Kuryla, R. E. Weiss, R. A. Ransom, P. S. McLeod, L. M. Fraas, and J. A. Cape, "26.1% Solar Cell Efficiency for Ge Mechanically Stacked under GaAs," *J. Appl. Phys.* **62**, 3010 (1987).

J. S. Smith, L. C. Chiu, S. Margalit, A. Yariv, and A. Y. Cho, *J. Vac. Sci. Technol.* **B1**, 376 (1983).

R. J. Stirn, K. L. Wang, and Y. C. M. Yeh, "Epitaxial Thin Film GaAs Solar Cells Using OM–CVD Techniques," Conf. Rec. 15th IEEE Photovoltaic Specialists Conf., pp. 1045–1050 (1981).

G. E. Stillman, L. W. Cook, N. Tabatabaie, G. E. Bulman, and V. M. Robbins, "InGaAsP Photodiodes," *IEEE Electron Dev.* **30**, 364 (1983).

S. M. Sze, *Physics of Semiconductor Devices*, 2nd ed., Wiley, New York (1981).

G. H. B. Thompson, *Physics of Semiconductor Laser Devices*, Wiley, New York (1980).

W. T. Tsang, "Lightwave Communication Technology: Photodetectors," in: *Semiconductors and Semimetals*, Vol. 22-D, Academic Press, New York (1985).

T. Tsukada, "GaAs–AlGaAs Buried Heterostructure Injection Lasers," *J. Appl. Phys.* **45**, 4899 (1974).

D. Wake, L. C. Blank, R. H. Walling, and I. D. Henning, "Top Illuminated InGaAs/InP p–i–n Photodiodes with a 3-dB Bandwidth in Excess of 26 GHz," *IEEE Electron Dev. Lett.* **9**, 226 (1988).

C. T. Wang, and S. S. Li, "A New Grating Type Au-n Silicon Schottky Barrier Photodiode," *IEEE Trans. Electron Dev.* **ED-20**, 522–527 (1973).

D. R. Wight, "Green Luminescence Efficiency in GaP," *J. Phys. D* **10**, 431 (1977).

R. K. Willardson and A. C. Beer, *Infrared Detectors, Semiconductors and Semimetals*, Vol. 12, Academic Press, New York (1976).

E. S. Yang, *Microelectronic Devices*, McGraw-Hill, New York (1988); Prentice-Hall, Englewood Cliffs, NJ (1976).

13

Bipolar Junction Transistor

13.1. INTRODUCTION

The first germanium alloy bipolar junction transistor (BJT) was invented by Bardeen, Brattain, and Shockley in 1948. The bipolar junction transistor is considered to be one of the most important electronic components used in integrated circuits (ICs) for computers, communications and power systems, and in many other digital and analog electronic circuit applications. The subsequent developments of silicon BJTs, the metal–oxide–semiconductor field-effect transistors (MOSFETs), and the ICs based on BJTs and MOSFETS have revolutionized the entire electronics industry. As a result, silicon BJTs and FETs have replaced bulky vacuum tubes for various electronic circuits, computers, microwave and power systems applications. Furthermore, advances in silicon processing technologies such as the development of optical and electron-beam (E-beam) lithographies, new metallization and etching techniques, as well as ion-implantation enable the fabrication of high-performance silicon BJTs with submicron geometries for very-large-scale integrated circuit (VLSIC) applications. Recent development of new Si–Si/Ge heterojunction bipolar transistors (HBTs) grown by molecular beam epitaxy (MBE) and organic-chemical vapor deposition (MOCVD) techniques on silicon substrates promises to offer even higher speed performance for next generation supercomputer applications.

Conventional n^+–p–n or p^+–n–p BJTs may be fabricated by using either alloying, thermal-diffusion, or ion-implantation techniques. Various semiconductor processing technologies such as epitaxy, planar, beam-lead, optical and E-beam lithographies, oxidation, passivations, and dry etching techniques have been developed in the past to facilitate fabrication of silicon devices and ICs. Recent advances in III–V compound semiconductor materials and device processing technologies have made it possible to develop new high-speed and high-frequency devices using GaAs- and InP-based III–V compound semiconductors. For example, high-speed HBTs have been developed by using the AlGaAs/GaAs and InAlGa/InGaAs material systems. In general, semiconductor devices fabricated from GaAs and other III–V compound semiconductors can be operated at a much higher frequency and speed than silicon devices, due to the inherent high electron mobilities for the III–V semiconductors. This will be discussed further in Chapter 15.

A BJT may be operated as an amplifier or as an electronic switch, depending on its bias condition. Unlike an unipolar field-effect transistor (FET), which is a majority carrier device, the BJT is a bipolar device since its current conduction is contributed to by both the minority and majority carriers (i.e., electrons and holes) across the p–n junctions. Therefore, the p–n junction theories described in Chapter 11 can be used to derive the minority and majority carrier distribution, current conduction, and the static and dynamic characteristics for a BJT.

If we add another p–n junction to the n–p–n BJT structure, a four-layer p–n–p–n switching device is formed. A p–n–p–n structure is a bistable device whose operation depends on internal feedback mechanisms, which can produce high- and low-impedance stable states under bias conditions. This enables the p–n–p–n device to operate as a switching device. The p–n–p–n devices are available for a wide range of voltage and current ratings. The low-power p–n–p–n device is designed mainly for use in switching and logic circuitry, while the high-power p–n–p–n device finds wide application in AC switching, DC choppers, phase-control devices, and power inverters.

In this chapter, the basic principles that govern the operation of a BJT and a four-layer p–n–p–n device are depicted. A general description of the basic BJT structure and modes of operation is presented in Section 13.2. The distribution of excess carrier densities, current components, and current–voltage (I–V) characteristics for a BJT under bias conditions are discussed in Section 13.3. The current gain, base transport factor, and emitter injection efficiency of a BJT are examined in Section 13.4. In Section 13.5, we present the Ebers–Moll and Gummel–Poon models, which provide a powerful means for elucidating the physical insights of the transistor action under different operation modes and biasing conditions. The frequency response and switching properties of a BJT as well as the effect of heavy doping on current gain and limitations due to base resistance and junction breakdown are also discussed. Section 13.6 describes basic device theory and performance characteristics of a silicon BJT switching device. Finally, the device structure, operational principles, and performance characteristics of a p–n–p–n four-layer switching device are depicted in Section 13.7.

13.2. BASIC STRUCTURES AND MODES OF OPERATION

The physical make-up of a typical n^+–p–n (or p^+–n–p) BJT consists of three distinct regions, namely, a heavily doped n^+ (or p^+) emitter region, a thin (0.5 μm or less) p (or n) base region, and a lightly doped n (or p) collector region. Figure 13.1a and b show a vertical n^+–p–n and a vertical p^+–n–p BJT, respectively. Figure 13.1c is a schematic representation of

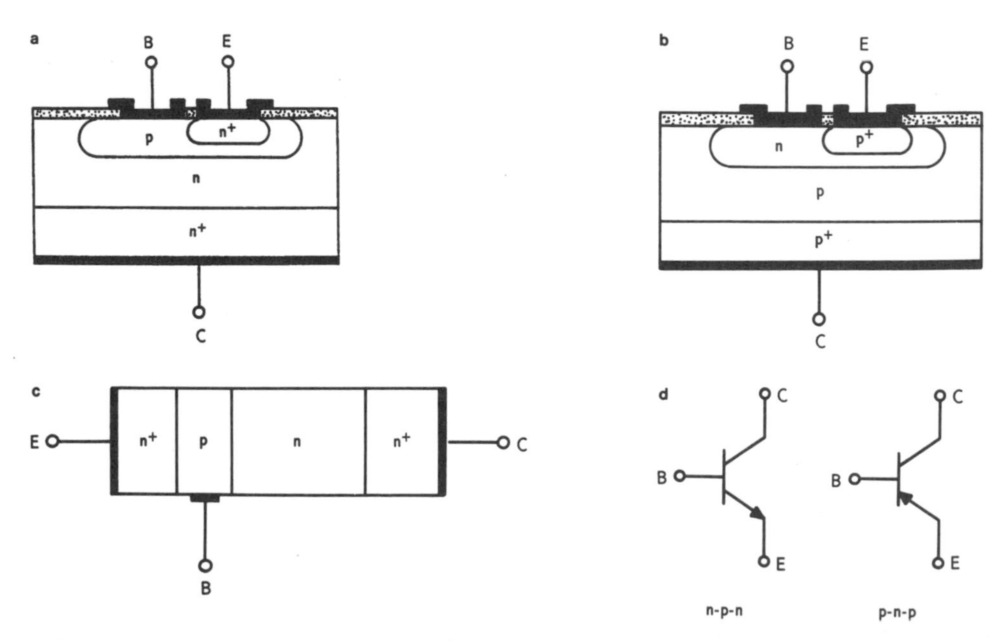

FIGURE 13.1. (a) Cross-section view of a n^+–p–n bipolar junction transistor (BJT), (b) a p^+–n–p BJT, (c) schematic diagram of a n^+–p–n BJT, and (d) circuit symbols for a n–p–n and p–n–p transistor.

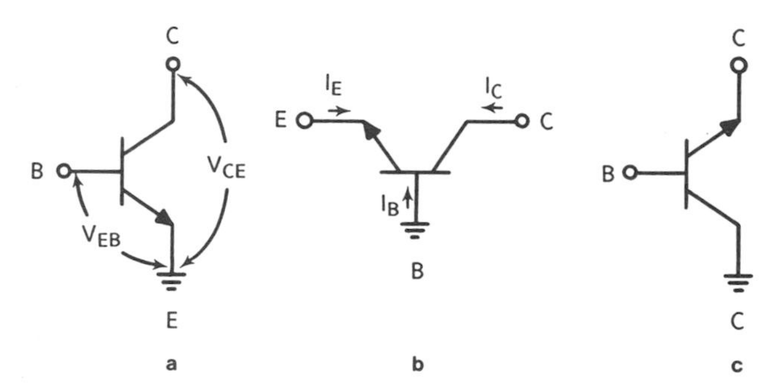

FIGURE 13.2. Three different configurations of a n^+–p–n BJT: (a) common-base, (b) common-emitter, and (c) common-collector connections.

an n^+–p–n transistor and Fig. 13.1d shows the circuit symbols of an n^+–p–n and a p^+–n–p transistor. In general, a BJT may be operated in three different configurations: the common-emitter mode, the common-base mode, and the common-collector mode, as shown in Fig. 13.2. In normal active-mode operation, the emitter-base junction is forward-biased and the collector–base junction is reverse-biased. Under this condition, the transistor is operated as an amplifier. In saturation- mode operation, both the emitter–base and collector–base junctions are forward-biased, while in cutoff-mode operation both the emitter–base and collector–base junctions are reverse-biased. Figures 13.3a and b show the energy band diagram and current components of a p^+–n–p BJT operating as an amplifier under normal active mode-conditions, respectively.

13.3. CURRENT-VOLTAGE CHARACTERISTICS

The current-voltage (I–V) relations for a BJT under bias conditions can be derived from the p–n junction theories described in Chapter 11. To analyze the dc characteristics of a BJT, it is assumed that the I–V relation in the emitter–base (E–B) junction and the collector–base

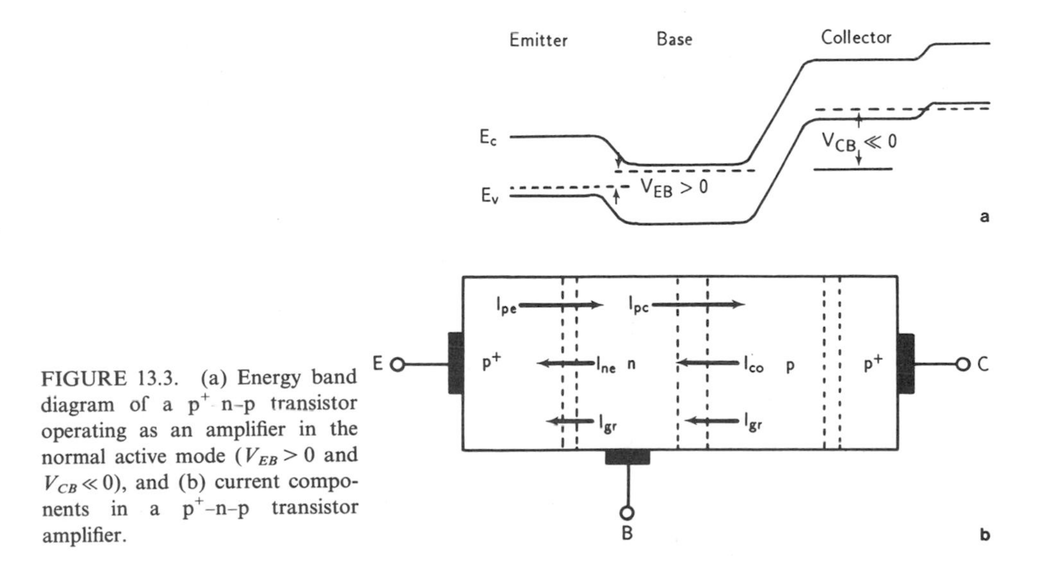

FIGURE 13.3. (a) Energy band diagram of a p^+–n–p transistor operating as an amplifier in the normal active mode ($V_{EB} > 0$ and $V_{CB} \ll 0$), and (b) current components in a p^+–n–p transistor amplifier.

(C–B) junction follows the ideal diode equation. Under this assumption, effects due to surface recombination–generation, series resistance, and high-level injection may be neglected. Figure 13.4 shows spatial distributions of excess carrier densities in the emitter, base, and collector regions. Here, the excess electron and hole densities at the edge of the emitter–base junction of a p$^+$–n–p transistor are given respectively by

$$p'_{nb}(0) = p_{nb}(0) - p_{0b} = p_{0b}(e^{qV_{BE}/k_BT} - 1) \qquad \text{at } x = 0 \tag{13.1}$$

and

$$n'_{pe}(-x_E) = n_{pe}(-x_E) - n_{0e} = n_{0e}(e^{qV_{BE}/k_BT} - 1) \qquad \text{at } x = -x_E \tag{13.2}$$

where p_{0b} and n_{0e} denote the equilibrium hole and electron densities at the edge of the emitter–base junction, respectively. Similarly, the excess hole and electron densities at the edge of the base–collector junction are given respectively by

$$p'_{nb}(W_b) = p_{nb}(W_b) - p_{0b} = p_{0b}(e^{qV_{CB}/k_BT} - 1) \qquad \text{at } x = W_b \tag{13.3}$$

and

$$n'_{pc}(x_C) = n_{pc}(x_C) - n_{0c} = n_{0c}(e^{qV_{CB}/k_BT} - 1) \qquad \text{at } x = x_C \tag{13.4}$$

where n_{0c} is the equilibrium density of electrons in the collector region. Since the potential drop occurs mainly across the depletion region, the continuity equation for holes in the quasi-neutral n-base region can be written as

$$D_{pb}\frac{\partial^2 p'_{nb}}{\partial x^2} - \frac{p'_{nb}}{\tau_b} = 0 \tag{13.5}$$

The general solution of Eq. (13.5) for the excess hole density in the quasi-neutral n-base region (i.e., $0 < x < W_b$) is given by

$$p'_{nb}(x) = C_1\, e^{-x/L_{pb}} + C_2\, e^{x/L_{pb}} \tag{13.6}$$

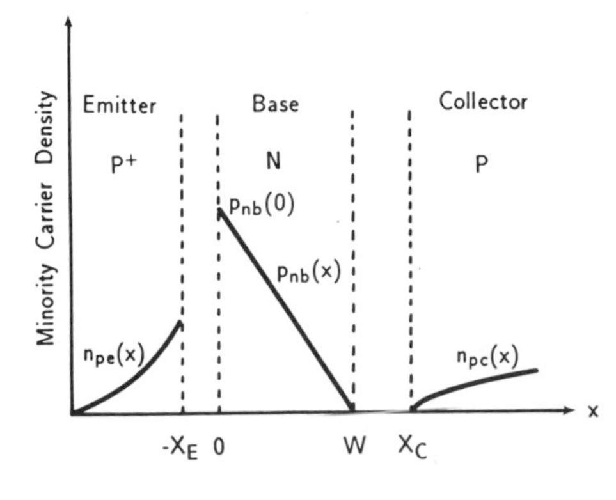

FIGURE 13.4. Minority carrier density distributions in a p$^+$ n–p transistor with E–B junction forward-biased and C–B junction reverse-biased (assuming $W_E \gg L_{pe}$ and $W \ll L_{pb}$).

where C_1 and C_2 are constants to be determined. If the boundary conditions given by Eqs. (13.1) and (13.3) are now substituted into Eq. (13.6), we obtain

$$p'_{nb}(x) = p_{n0}(e^{qV_{EB}/k_BT} - 1)\frac{\sinh\left(\frac{W_b - x}{L_{pb}}\right)}{\sin\left(\frac{W_b}{L_{pb}}\right)} + p_{n0}(e^{qV_{CB}/k_BT} - 1)\frac{\sinh\left(\frac{x}{L_{pb}}\right)}{\sinh\left(\frac{W_b}{L_{pb}}\right)} \tag{13.7}$$

where $L_{pb} = (D_{pb}\tau_b)^{1/2}$ is the hole diffusion length in the n-base region. Equation (13.7) is important, because it relates the minority hole density in the base region to the base width W_b. For example, if W_b approaches infinity (i.e., $W_b/L_{pb} \gg 1$), then Eq. (13.7) reduces to the case of a p–n junction diode, and the transistor action is ceased. If the emitter–base (E–B) junction is forward-biased and the collector–base (C–B) junction is reverse-biased, then the second term in Eq. (13.7) is negligible, and the hole density profile in the base region becomes

$$p'_{nb}(x) \approx p_{n0}(e^{qV_{EB}/k_BT} - 1)\frac{\sinh\left(\frac{W_b - x}{L_{pb}}\right)}{\sinh\left(\frac{W_b}{L_{pb}}\right)} \tag{13.8}$$

Equation (13.8) can be further simplified for most practical transistors since the base width is much smaller than the hole diffusion length L_{pb}, and hence for $V_{EB} \gg k_BT/q$ Eq. (13.8) reduces to

$$p'_{nb} \approx p_{n0}\, e^{qV_{EB}/k_BT}\left(1 - \frac{x}{W_b}\right) \tag{13.9}$$

which shows that the minority hole density in the base region decreases linearly with distance x from the edge of the emitter–base junction to the edge of the base–collector junction, as shown in Fig. 13.4. Deviation from this linear dependence with distance can be attributed to recombination loss in the base region. The hole current in the base region may be derived from the diffusion equation for the excess hole density, and is given by

$$I_{pb} = -\,qAD_{pb}\frac{dp'_{nb}(x)}{dx} \tag{13.10}$$

where $p'_{nb}(x)$ is given by Eq. (13.7). Thus, the injected hole current entering the base can be evaluated at $x = 0$ using Eq. (13.10), and the hole current flowing out of the base is evaluated at $x = W_b$ using Eq. (13.10). The results yield

$$I_{pb}(0) = \frac{qD_{pb}p_{n0}}{L_{pb}}\coth\left(\frac{W_b}{L_{pb}}\right)(e^{qV_{EB}/k_BT} - 1) - \frac{qD_{pb}p_{n0}}{L_{pb}\sinh\left(\frac{W_b}{L_{pb}}\right)}(e^{qV_{CB}/k_BT} - 1) \tag{13.11}$$

and

$$I_{pb}(W_b) = \frac{qD_{pb}p_{n0}}{L_{pb}\sinh\left(\frac{W_b}{L_{pb}}\right)}(e^{qV_{EB}/k_BT} - 1) - \frac{qD_{pb}p_{n0}}{L_{pb}}\coth\left(\frac{W_b}{L_{pb}}\right)(e^{qV_{CB}/k_BT} - 1) \tag{13.12}$$

It is seen that the hole current in the base region is, in general, a function of the applied bias voltages at both the E–B and C–B junctions. The polarity of the bias voltages at both junctions can be changed depending on the operation modes of the transistor.

Similarly, the excess electron densities in the p$^+$-emitter and p-collector regions of the transistor can be determined by solving the continuity equations for the excess electron densities in both regions; they are given respectively by

$$n'_{pe}(x) = n'_{pe}(-x_E)\, e^{(x + x_E)/L_{ne}} \qquad \text{for } x < -x_E \tag{13.13}$$

and

$$n'_{pc}(x) = n'_{pc}(x_C)\, e^{-(x - x_C)/L_{nc}} \qquad \text{for } x > x_C \tag{13.14}$$

where $n'_{pe}(-x_E)$ and $n'_{pc}(x_C)$ are the excess electron densities at the edges of the E–B and C–B junctions defined by Eqs. (13.2) and (13.4), respectively. Equations (13.13) and (13.14) show that the excess electron densities decrease exponentially with distance from the edges of the depletion regions of both junctions, as illustrated in Fig. 13.4. The electron current in the emitter and collector regions can be derived from the diffusion equation given by

$$I_n = qAD_n \frac{dn'_p}{dx} \tag{13.15}$$

The total emitter current, which consists of the electron injection current from the emitter to the base and the hole injection current from the base to the emitter, is given by

$$\begin{aligned} I_E &= A'(J_{pE} + J_{nE}) \\ &= -A'qD_{pb} \left.\frac{dp'_{nb}}{dx}\right|_{x=0} + A'qD_{ne} \left.\frac{dn'_{pe}}{dx}\right|_{x=-x_E} \\ &= I_{BO} \coth\left(\frac{W_b}{L_{pb}}\right)\left[(e^{qV_{EB}/k_BT} - 1) - \frac{1}{\cosh\left(\frac{W_b}{L_{pb}}\right)}(e^{qV_{CB}/k_BT} - 1)\right] \\ &\quad + I_{EO}(e^{qV_{EB}/k_BT} - 1) \end{aligned} \tag{13.16}$$

where

$$I_{BO} = A'qn_i^2 \left(\frac{D_{pb}}{N_{db}L_{pb}}\right) \tag{13.17}$$

and

$$I_{EO} = A'qn_i^2 \left(\frac{D_{ne}}{N_{ae}L_{ne}}\right) \tag{13.18}$$

are the saturation currents in the base and emitter regions of the transistor, respectively. Similarly, the collector current can be expressed by

$$\begin{aligned} I_C &= A(J_{pC} + J_{nC}) \\ &= -AqD_{pb} \left.\frac{dp'_{nb}}{dx}\right|_{x=W_B} + AqD_{nc} \left.\frac{dn'_{pc}}{dx}\right|_{x=x_C} \\ &= \frac{I_{BO}}{\sinh\left(\frac{W_b}{L_{pb}}\right)}\left[(e^{qV_{EB}/k_BT} - 1) - \coth\left(\frac{W_b}{L_{pb}}\right)(e^{qV_{CB}/k_BT} - 1)\right] \\ &\quad + I_{CO}(e^{qV_{CB}/k_BT} - 1) \end{aligned} \tag{13.19}$$

where

$$I_{CO} = Aqn_i^2\left(\frac{D_{nc}}{N_{ac}L_{nc}}\right) \tag{13.20}$$

is the collector saturation current, A is the C–B junction area, and A' is the E–B junction area. If the direction of current flow into the emitter, base, and collector junctions shown in Figs. 13.3b is defined as positive, then the base current I_B is related to the emitter and collector currents by

$$I_B = -I_E - I_C \tag{13.21}$$

where I_E and I_C are the emitter and collector currents given by Eqs. (13.16) and (13.19), respectively. Since the emitter current is nearly equal to the collector current, the base current I_B is usually very small. We note that Eq. (13.21) does not include the recombination current in the base region. As shown by Eq. (13.9), for a uniformly doped base with negligible base recombination (i.e., $W_b \ll L_{pb}$) the injected excess hole density is a linear function of x in the base region. In general, the recombination current in the base region can be expressed by

$$I_r = \left(\frac{qA'}{\tau_p}\right)\int_0^{W_b} p'_{nb}(x)\,dx \approx \left(\frac{qA'n_i W}{2\tau_0}\right)e^{qV_{EB}/2k_BT} \tag{13.22}$$

If the recombination current component given by Eq. (13.22) is not negligible, then Eq. (13.22) should be added to Eq. (13.21) to obtain the total base current.

It is seen that the equations for the current components derived above are valid only for a BJT with a uniformly doped base. However, for a practical BJT in which the emitter–base and base–collector junctions are formed by double diffusion or ion implantation, the base impurity doping profile is no longer uniform and a built-in electric field exists in the base region. This is illustrated in Fig. 13.5 for a double-diffused planar p$^+$–n–p–p$^+$ BJT. The

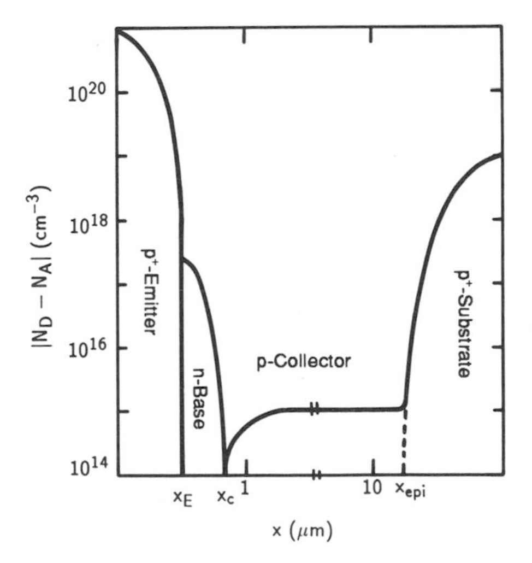

FIGURE 13.5. Impurity density profile for a double-diffused planar epitaxial structure of a p$^+$–n–p–p$^+$ transistor.

built-in electric field due to the nonuniform doping profile in the base region of a BJT can be expressed by

$$\mathscr{E} = \frac{k_B T}{qN_d(x)} \frac{dN_d(x)}{dx} \tag{13.23}$$

The polarity of the built-in electric field given by Eq. (13.23) is such that it assists the transport of injected holes in the base region. Thus, the hole current density in this case is given by

$$\begin{aligned} J'_{pb} &= q\mu_p p_{nb}\mathscr{E} - qD_{pb}\frac{dp_{nb}}{dx} \\ &= qD_{pb}\left[\left(\frac{p_{nb}}{N_{db}}\right)\frac{dN_{db}(x)}{dx} + \frac{dp_{nb}}{dx}\right] \end{aligned} \tag{13.24}$$

This equation is obtained by substituting Eq. (13.23) for the electric field and using the Einstein relation for μ_p (i.e., $D_p/\mu_p = k_B T/q$) in the first term of Eq. (13.24). Multiplying both sides of Eq. (13.24) by N_{db} and integrating the equation yields

$$p'_{nb}(x) = \frac{J'_{pb}}{qD_{pb}N_{db}} \int_x^{W_b} N_{db}\, dx \tag{13.25}$$

We note that Eq. (13.25) is obtained by using the boundary condition $p'_{nb}(W_b) = 0$ and $x = W_b$ and by assuming that J'_{pb} is constant (i.e., the recombination current is negligible in the base region). From Eq. (13.25), the hole density at $x = 0$ is given by

$$\begin{aligned} p'_{nb}(0) &= \frac{n_i^2}{N_{db}(0)} e^{qV_{BE}/k_BT} \\ &= \frac{J'_{pb}}{qD_{pb}N_{db}(0)} \int_0^{W_b} N_{db}(x)\, dx \end{aligned} \tag{13.26}$$

Therefore, the hole current in the base region can be deduced from Eq. (13.26), which yields

$$I'_{pb} = J'_{pb}A' = \frac{(qA'D_{pb}n_i^2)\, e^{qV_{BE}/k_BT}}{\int_0^{W_b} N_{db}(x)\, dx} \tag{13.27}$$

The integral in the denominator of Eq. (13.27) represents the total number of impurity atoms in the base and is known as the Gummel number. For silicon BJTs the Gummel number may vary between 10^{12} and 10^{13} cm^{-2}. Therefore, a larger electron current flow can be realized with a smaller Gummel number which corresponds to a narrower base width. Figure 13.6 shows the base and collector currents versus the emitter–base bias voltage for a silicon BJT.[1] Four regions are observed in this plot: (i) the low V_{BE} nonideal region in which the base current is dominated by the recombination current and I_B varies with $e^{qV_{EB}/2k_BT}$; (ii) the ideal region in which both the base and collector currents are dominated by the diffusion current (i.e., I_B and $I_C \sim e^{qV_{EB}/k_BT}$); (iii) the moderate injection region in which a significant voltage drop occurs across the base resistance (i.e., $r_b I_B$ drop), and (iv) the high injection region in which I_C and I_B vary with $e^{qV_{EB}/2k_BT}$. In general, the base recombination current can be controlled by reducing the process-induced defects in the base region, while the high-injection and base resistance effects can be minimized by modifying the base doping profile and the transistor structure.

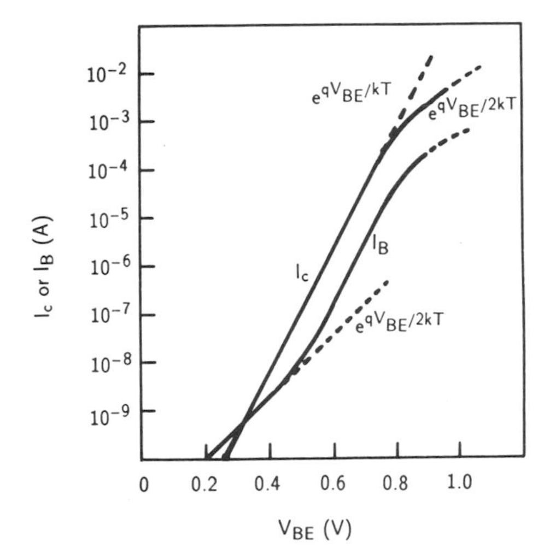

FIGURE 13.6. Base and collector currents as a function of the emitter–base bias voltage for a silicon BJT under forward-bias conditions. After Jespers,[1] by permission.

The output I–V (i.e., I_C versus V_{CE}) characteristics for a silicon p^+–n–p BJT with a common-emitter (C–E) configuration is shown in Fig. 13.7.[2] Also shown in this figure are the Early voltage V_A, collector saturation current I'_{CO} (also known as I_{CEO}), and breakdown voltage V_{CEO}. The breakdown voltages and saturation currents for a p^+–n–p silicon BJT with a common-base (C–B) and common-emitter (C–E) configuration are shown in Fig. 13.8. It is seen that the saturation current I_{CO} (also known as I_{CBO}) for the C–B configuration is substantially smaller than I_{CEO} for the C–E configuration (i.e., by a factor of β_0). In the C–B configuration, the current gain α_0 is close to unity and I_C is nearly independent of V_{CB}. From the current equations derived earlier, we see that both the emitter current I_E and collector current I_C are functions of the applied voltage across the emitter–base and collector–base junctions. In the C–E configuration the current gain can be quite large, and I_C usually increases as V_{CE} increases. The collector saturation current I'_{CO} for the C–E configuration (i.e., base opened and $I_B = 0$) is related to the saturation current I_{CBO} for the C–B configuration by

$$I'_{CO} = I_{CEO} = \frac{I_{CO}}{(1-\alpha_0)} = \beta_0 I_{CBO} \tag{13.28}$$

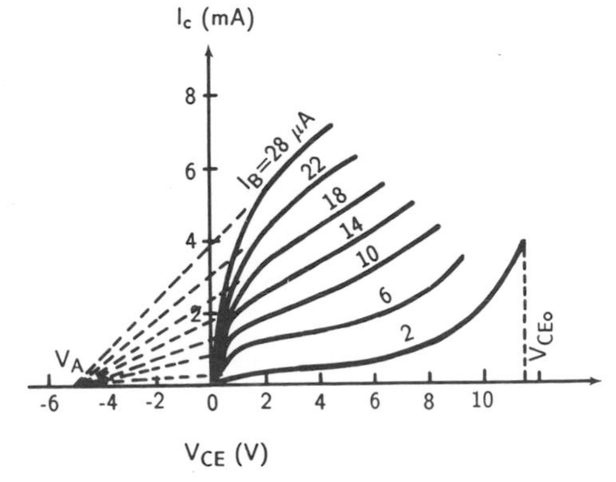

FIGURE 13.7. I_c versus V_{CE} for a p^+–n–p silicon BJT with common-emitter configuration. Also shown in the figure are the Early voltage V_A, saturation current I'_{CO}, and breakdown voltage V_{CEO} for the common-emitter configuration. After Gummel and Poon,[2] by permission.

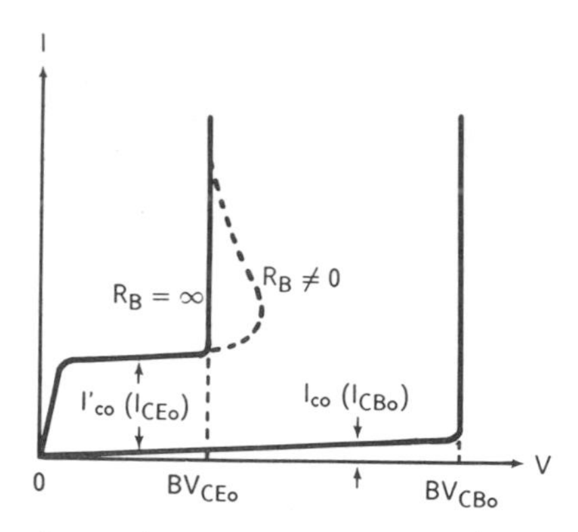

FIGURE 13.8. Breakdown voltage and saturation current for a p$^+$–n–p silicon BJT with common-base and common-emitter configurations.

The base width W_b will decrease with increasing V_{CE}, which in turn will cause an increase in the current gain. The continued increase of I_C with increasing V_{CE} is attributed to a large increase of β_0 with V_{CE} in the C–E mode of operation. This phenomenon is called the Early effect, which is a direct result of the base-width modulation by the collector–base junction bias voltage variation. We note that in a BJT, a change of collector–base junction bias voltage will result in a change of space-charge layer width at the collector–base junction and consequently will modify the width of the quasi-neutral base region. This variation will result in several effects which further complicate the performance of a BJT as a linear amplifier. The voltage V_A at which the extrapolated I_C versus V_{CE} curves (see Fig. 13.7) meet the negative V_{CE} axis is called the Early voltage, which is given by

$$V_A = \frac{qN_B W_b^2}{\varepsilon_0 \varepsilon_s} \tag{13.29}$$

This equation is valid for a BJT whose base width W_b is much larger than its depletion layer width in the base. To reduce the influence of the collector–base voltage on the collector current, the value of V_A must be increased. From Eq. (13.29), it is seen that this can be accomplished by increasing the base doping density which, in turn, will reduce the collector–base depletion layer width and hence the Early effect. This reduces the movement of the collector–base boundary to the base region of a BJT.

The breakdown voltage BV_{CEO} for the C–E configuration can be related to the breakdown voltage BV_{CEO} for the C–B configuration by

$$BV_{CEO} = BV_{CBO}(1 - \alpha_0)^{1/m} \tag{13.30}$$

where m is an integer. Since α_0 is very close to unity for most BJTs, BV_{CBO} is usually much larger than BV_{CEO}. We note that BV_{CBO} under open-base conditions can be related to the multiplication factor M by

$$M = \frac{1}{1 - (V/BV_{CBO})^m} \tag{13.31}$$

where V is the applied bias voltage. When the base is opened, the emitter and collector currents are equal (i.e., $I_E = I_C = I$). Both I_{CO} and $\alpha_0 I_E$ are multiplied by M as they flow across the collector–base junction. From Eq. (13.30) it is seen that for $\alpha_0 \sim 1$, BV_{CBO} becomes much larger than BV_{CEO}. This is clearly illustrated in Fig. 13.8, which shows the breakdown voltage BV_{CBO} and saturation current I_{CO} for the C–B configuration, and the corresponding BV_{CEO} and I'_{CO} for the C–E configuration.

The current–voltage equations derived in this section for a BJT will be used in the Ebers–Moll model for large-signal and transient analysis, which is discussed in detail in Section 13.5.

13.4. CURRENT GAIN, BASE TRANSPORT FACTOR, AND EMITTER INJECTION EFFICIENCY

When a BJT is biased in the normal active mode, it operates as an amplifier and hence a current gain results. For a p^+–n–p transistor, the emitter current consists of two components: a hole current I_{pE}, which is due to hole injection from the p^+-emitter into the n-base region, and an electron current I_{nE}, which is due to electron injection from the n-base into the p-emitter region. The collector current also consists of two components, namely, a hole current I_{pC} injecting from the n-base to the p-collector region, and an electron current I_{nC} injecting from the p-collector to the n-base region. Expressions for the emitter and collector current components are given respectively by Eqs. (13.16) and (13.19). For a common base BJT amplifier, the key parameters affecting its performance include the emitter injection efficiency, the base transport factor, and the current gain. If the base recombination current component is included, then the emitter injection efficiency γ can be expressed by

$$\gamma = \frac{I_{pE}}{I_E} = \frac{I_{pE}}{(I_{nE} + I_{pE} + I_r)} \tag{13.32}$$

and the base transport factor β_T is given by

$$\beta_T = \frac{I_{pC}}{I_{pE}} \tag{13.33}$$

The common-base current gain α_0 is defined by

$$\begin{aligned} \alpha_0 = h_{FB} &= \frac{dI_C}{dI_E} \\ &= \frac{-(I_C - I_{CO})}{I_E} = \frac{I_{pC}}{(I_{nE} + I_{pE} + I_r)} = \gamma\beta_T \end{aligned} \tag{13.34}$$

which shows that for a common-base BJT the current gain is equal to the product of emitter injection efficiency and base transport factor. Since I_{pC} is smaller than I_{pE}, the common-base current gain α_0 is always smaller than unity. However, for a well-designed BJT, this gain factor can be very close to unity (e.g., $\alpha_0 = 0.9999$), and from Eq. (13.34) we obtain

$$I_C = -\alpha_0 I_E + I_{CO} \tag{13.35}$$

which relates the collector current to the emitter current with the base as a common terminal. We note that I_{CO} is the collector reverse saturation current for the C–B configuration.

To obtain current amplification in a BJT, the transistor is usually operated in the common-emitter (C–E) configuration. In this configuration, the emitter terminal is used as a common ground; the base–emitter terminal is used as an input port, and the collector–emitter terminal serves as an output port. The common-emitter current gain β_0 (or h_{FE}) is defined by

$$\beta_0 = h_{FE} = \frac{dI_C}{dI_B} = \frac{\alpha_0}{(1 - \alpha_0)} \tag{13.36}$$

which is obtained by solving Eqs. (13.21) and (13.35). Since value of α_0 for a well-designed BJT is very close to unity, β_0 for the common-emitter operation is usually much larger than unity (e.g., if $\alpha_0 = 0.99$, then $\beta_0 = 99$).

For a p^+–n–p transistor operating under normal active-mode conditions (i.e., $V_{BE} > 0$ and $V_{CB} \ll 0$), the emitter injection efficiency γ can be derived by using Eq. (13.16), and the result is

$$\begin{aligned}\gamma &\approx \frac{I_{pE}}{(I_{pE} + I_{nE})}\\ &= \frac{1}{1 + \left(\frac{N_{db} D_{ne} L_{pb}}{N_{ae} D_{pb} L_{ne}}\right) \tanh\left(\frac{W_b}{L_{pb}}\right)}\end{aligned} \tag{13.37}$$

This equation neglects the base recombination current. Similarly, the base transport factor can be obtained by solving Eqs. (13.16) and (13.19), which yield

$$\beta_T = \frac{I_{pC}}{I_{pE}} = \frac{1}{\cosh\left(\frac{W_b}{L_{pb}}\right)} \approx 1 - \frac{W_b^2}{2L_{pb}^2} \tag{13.38}$$

Here we have assumed that the base width is much smaller than the hole diffusion length in the base region. An interesting physical insight of β_T can be obtained if Eq. (13.38) is expressed in terms of the transit time τ_B and minority carrier lifetime τ_p in the base region. It can be shown that in order to have a base transport factor close to unity, the base transit time must become so short (i.e., $\tau_B \ll \tau_p$) that the injected holes have little chance to recombine with electrons in the base region. For a practical silicon BJT, β_T is very close to unity, and hence the current gain β_0 can be obtained by solving Eqs. (13.36) and (13.37), which yield

$$\begin{aligned}\beta_0 = h_{FE} &= \frac{\alpha_0}{1 - \alpha_0} = \frac{\gamma \beta_T}{1 - \gamma \beta_T}\\ &\approx \frac{\gamma}{(1 - \gamma)}\\ &= \left(\frac{N_{ae} D_{pb} L_{ne}}{N_{db} D_{ne} L_{pb}}\right) \coth\left(\frac{W_b}{L_{pb}}\right) \approx \frac{N_{ae}}{Q_B}\end{aligned} \tag{13.39}$$

where Q_B is the Gummel number defined by the denominator of Eq. (13.27). Thus, for a given emitter doping density N_{ae}, the static common-emitter current gain β_0 is inversely proportional to the base charge density Q_B. Figure 13.9 shows the current gain h_{FE} versus collector current I_C for the silicon BJT shown in Fig. 13.6. As shown in Fig. 13.9, β_0 is small when I_C is small. This can be attributed to bulk and surface recombination losses that occurred in the base

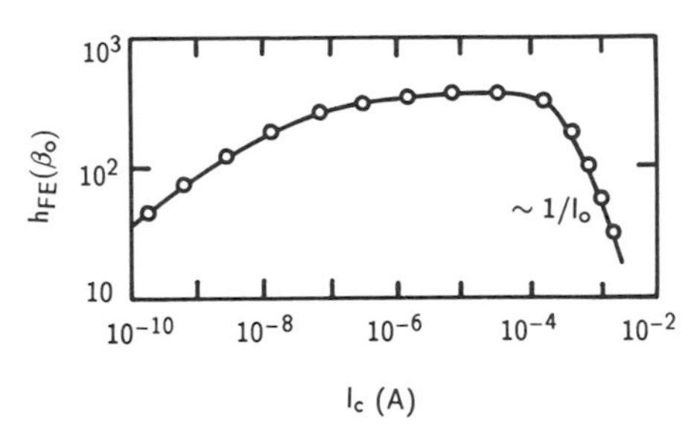

FIGURE 13.9. Common-emitter current gain versus collector current for the silicon transistor shown in Fig. 13.6.

region at low bias voltage. The recombination current in the base region may be larger than the diffusion current component at low current level. Thus, by reducing the bulk trap density and surface recombination loss, the current gain can be increased substantially at low collector current. As the collector current continues to increase, h_{FE} will also increase and eventually reach a saturation value. At a still higher collector current, the minority carrier density injected into the base approaches the majority carrier density, and the injected carriers effectively increase the base doping density which, in turn, will cause the emitter injection efficiency to decrease. This is the so-called high-injection condition. Under a high-injection level, the current gain varies inversely with collector current (i.e., $h_{FE} \sim e^{-qV_{EB}/2k_BT} \sim I_C^{-1}$), as is shown in Fig. 13.9.

Equation (13.39) shows that h_{FE} is directly proportional to the emitter doping density. Therefore, in order to increase h_{FE} it is necessary to increase the doping density in the emitter region. However, two adverse effects associated with heavy doping in the emitter region may result, namely, bandgap narrowing and Auger recombination. Both of these effects can severely affect the current gain of the BJT. The effect of bandgap narrowing on current gain can be evaluated by examining the bandgap narrowing effect on the effective intrinsic carrier density in the heavily doped emitter region. The square of the effective intrinsic carrier density under heavy doping conditions is given by

$$n_{ie}^2 = N_c N_v \exp[-(E_g - \Delta E_g)/k_B T] = n_i^2 \exp(\Delta E_g/k_B T) \tag{13.40}$$

where N_c and N_v are the effective densities of the conduction and valence band states, respectively, while n_i is the intrinsic carrier density for the nondegenerate case. The quantity ΔE_g is the bandgap shrinkage due to the heavy doping effect in the emitter region. The minority carrier densities in the base and emitter regions are given respectively by

$$p_{nb} = \frac{n_i^2}{N_{db}} \tag{13.41}$$

and

$$n_{pe} = \frac{n_{ie}^2}{N_{ae}} = \frac{n_i^2}{N_{ae}} e^{\Delta E_g/k_B T} \tag{13.42}$$

Therefore, the effect of bandgap narrowing on current gain can be estimated qualitatively from Eqs. (13.39) through (13.42), which yield

$$h_{FE} \sim \frac{p_{nb}}{n_{pe}} \sim e^{-\Delta E_g/k_B T} \tag{13.43}$$

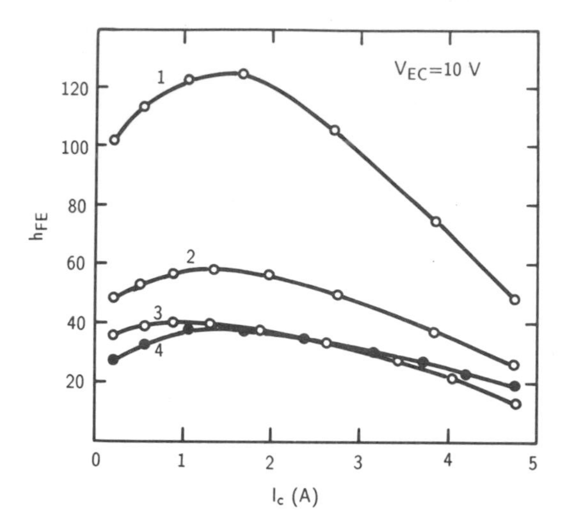

FIGURE 13.10. Calculated and measured common-emitter current gain h_{FE} versus collector current for a silicon power transistor by considering: (1) the SRH process only, (2) SRH and bandgap narrowing, (3) measured values, and (4) Auger, SRH, and bandgap narrowing. After McGrath and Navon,[3] by permission, © IEEE–1977.

This equation reveals that h_{FE} will decrease exponentially with increasing bandgap narrowing, ΔE_g.

Another heavy doping effect which can greatly degrade the transistor performance is associated with the reduction of minority carrier lifetime with increasing doping density in the emitter region of the BJT. As the doping density in the emitter region increases, Auger recombination becomes the dominant recombination process for the minority carriers. In this case, the minority carrier lifetime in the emitter region of a BJT is controlled by Auger recombination instead of the Shockley–Read–Hall (SRH) process. As a result, the minority carrier lifetime in the emitter region will decrease with the square of the majority carrier density. This, in turn, will reduce the emitter minority carrier diffusion length and degrade the emitter injection efficiency. Figure 13.10 shows the effects of bandgap narrowing and Auger recombination on the current gain of a silicon power transistor.[3] The results clearly show that in order to accurately predict the measured current gain data, the effects of bandgap narrowing and Auger recombination in the heavily doped emitter region should be taken into account. The relative importance of these effects as regards the SRH recombination process depends on the emitter junction depth and the dopant density as well as the injection level.

The base spreading resistance is another important parameter which will affect the performance of a BJT at very high frequencies. In general, it will increase the base–emitter voltage drop at high base current for power and switching transistors. A close examination of the cross section of the BJT shown in Fig. 13.1 reveals that the base current must flow some distance from the base terminal to the bulk base region between the emitter–base and collector–base junctions. Since the base region is very thin (i.e., $\leq 0.5\ \mu$m) and not highly doped, a parasitic resistance known as the base spreading resistance, $r_{b'b}$, exists in this region. This spreading resistance must be included in the BJT device modeling to account for its adverse effect on transistor performance at high frequencies and at high injection level.

13.5. MODELING OF A BIPOLAR JUNCTION TRANSISTOR

In this section, we present the Ebers–Moll model and Gummel–Poon model for the BJT. In order to predict the performance of a BJT for large signals or transient behavior under any biasing conditions, it is necessary to develop a simple and accurate device model so that its electrical output characteristics can be correlated to the physical parameters of the transistor. This is particularly important for IC designs, since an accurate device model is needed in the design of any integrated circuit. The first BJT device model for large-signal circuit simulation was introduced by Ebers and Moll in 1954, and later modified by Gummel and Poon to account for various physical effects which were not included in the Ebers–Moll model. The Ebers–Moll model is the simplest device model for the BJT, which can be used to predict carrier injection and extraction phenomena in a BJT.

The BJT device model developed by Gummel and Poon is based on the integral charge equation which relates terminal electrical characteristics to base charge. By taking into account many physical effects in the device modeling parameters, the Gummel–Poon model can much more accurately predict the transistor behavior than the Ebers–Moll model. To implement the Gummel–Poon model for computer circuit simulation, many physical parameters must first be determined. It can be shown that a simplified version of the Gummel–Poon model can be reduced to the basic Ebers–Moll model.

Figure 13.11a shows the equivalent circuit of the simplest Ebers–Moll model for a BJT.[4] This large-signal transistor model consists of two diodes connected back to back, and each diode is connected in parallel with a current source. The current sources are driven by the diode currents which are assumed to have ideal diode characteristics. Using the results derived in Section 13.2 for a p^+–n–p BJT, the terminal current equations for the Ebers–Moll model can be written as

$$I_E = I_F - \alpha_R I_R \tag{13.44}$$

$$I_C = I_R - \alpha_F I_F \tag{13.45}$$

$$I_B = -(I_E + I_C) = -(1 - \alpha_F)I_F - (1 - \alpha_R)I_R \tag{13.46}$$

where

$$I_F = I_{ES}(e^{qV_{BE}/k_BT} - 1) \tag{13.47}$$

and

$$I_R = I_{CS}(e^{qV_{BC}/k_BT} - 1) \tag{13.48}$$

I_F being the forward current flowing through the E–B junction and I_R the reverse current flowing through the C–B junction; α_F and α_R denote the forward and reverse common-base current gains, respectively, while I_{ES} and I_{CS} denote the emitter and collector saturation currents, respectively. Expressions for α_F, α_R, I_{ES}, and I_{CS} can be derived from Eqs. (13.16) through (13.20) for a p^+–n–p BJT, and are given by

$$I_{ES} = \frac{qA'D_{pb}n_i^2}{N_{db}L_{pb}}\coth\left(\frac{W_b}{L_{pb}}\right) + \frac{qA'D_{ne}n_i^2}{N_{ae}L_{ne}} \tag{13.49}$$

$$I_{CS} = \frac{qAD_{pb}n_i^2}{N_{db}L_{pb}}\coth\left(\frac{W_b}{L_{pb}}\right) + \frac{qAD_{nc}n_i^2}{N_{ac}L_{nc}} \tag{13.50}$$

$$\alpha_F = \frac{1}{I_{ES}}\frac{qA'D_{pb}n_i^2}{N_{db}L_{pb}}\frac{1}{\sinh\left(\dfrac{W_b}{L_{pb}}\right)} \tag{13.51}$$

and

$$\alpha_R = \frac{1}{I_{CS}}\frac{qAD_{pb}n_i^2}{N_{db}L_{pb}}\frac{1}{\sinh\left(\dfrac{W_b}{L_{pb}}\right)} \tag{13.52}$$

Equations (13.44) and (13.45) relate currents I_E and I_C to the terminal voltages V_{EB} and V_{CB}, and the four transistor parameters I_{ES}, I_{CS}, α_F, and α_R. The current equations for the emitter

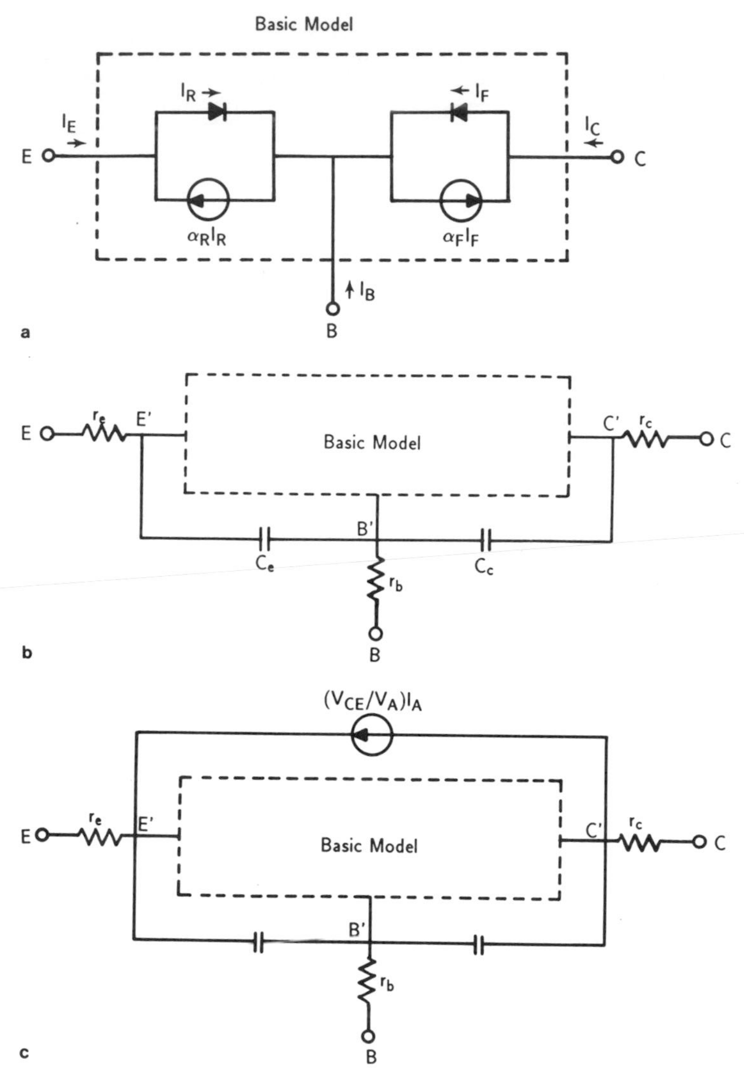

FIGURE 13.11. Equivalent circuit diagrams of a n–p–n transistor based on the Ebers–Moll model: (a) basic model, (b) modified model including series resistance and depletion capacitances, and (c) additional current source for the Early effect. After Ebers and Moll,[4] by permission, © IEEE–1961.

and collector junctions given above enable the general expressions for the emitter and collector currents to be rewritten as

$$I_E = a_{11}(e^{qV_{EB}/k_BT} - 1) + a_{12}(e^{qV_{CB}/k_BT} - 1) \tag{13.53}$$

and

$$I_C = a_{21}(e^{qV_{EB}/k_BT} - 1) + a_{22}(e^{qV_{CB}/k_BT} - 1) \tag{13.54}$$

where

$$a_{11} = -I_{ES}, \qquad a_{12} = \alpha_R I_{CS}, \qquad a_{21} = \alpha_F I_{ES}, \qquad a_{22} = -I_{CS} \tag{13.55}$$

Based on the reciprocity property of a two-port device we obtain $a_{12} = a_{21}$, and hence $\alpha_R I_{CS} = \alpha_F I_{ES}$. Therefore, only three unknowns are involved in the basic Ebers–Moll model shown in Fig. 13.11a. The accuracy of this basic model can be increased by adding the emitter and collector series resistances (r_e and r_c) and the emitter and collector depletion capacitances (C_e and C_c) to the equivalent circuit shown in Fig. 13.11a; the result is shown in Fig. 13.11b. In this case, the diode is controlled by the internal junction voltages $V_{E'B'}$ and $V_{C'B'}$ but not by the external voltages. If we add the Early effect (i.e., the base-width modulation) to the model, then an extra current source must be included between the internal emitter and collector terminals, as is shown in Fig. 13.11c. A comparison of Fig. 13.11a and b shows that, in order to improve the model accuracy from Fig. 13.11a to b, the unknown physical parameters must increase threefold. This makes the model more complicated to handle and more difficult to solve. Furthermore, the model shown in Fig. 13.11b can be improved by adding a diode to the base lead to account for the two-dimensional current crowding effect along the emitter–base junction. Therefore, it is evident that the basic Ebers–Moll model can provide a first-order solution for relating the device physical parameters to the large-signal dc and transient characteristics of a BJT. The accuracy and complexity of this model depend on the number of physical effects being considered in the model. This can be best illustrated by the Gummel–Poon model in which more than twenty physical parameters are incorporated in the equivalent circuit of this model.

It should be noted that the Ebers–Moll equations given above can also be applied to a n^+–p–n transistor provided that the polarities defined for I_E, I_C, I_B, V_{EB}, and V_{CB} are reversed. An examination of the Ebers–Moll model given by Eqs. (13.53) and (13.54) reveals that there are three operating regions for the common-base (C–B) or common-emitter (C–E) configuration. As shown in Fig. 13.7, the three regions of operation for a common-emitter configuration are: (i) the cutoff region with $V_{EB} < 0$ and $V_{CB} \ll 0$, (ii) the active region with $V_{EB} > 0$ and $V_{CB} \ll 0$, and (iii) the saturation region with $V_{EB} > 0$ and $V_{CB} \gg 0$. In region (i) both diodes are reverse-biased and only leakage currents flow through the transistor. This region corresponds to the "off" state in the switching transistor operation. In region (ii), the transistor operates as an amplifier. In this region of operation (i.e., normal active mode), a change in base current due to a small change in input voltage V_{EB} across the emitter–base junction at the input terminal will result in a large change in the collector current, and hence a voltage drop across the load resistance in the collector output terminal with consequent voltage and power amplification. In region (iii), both junctions are forward-biased, and V_{CE} is nearly equal to zero but with a large collector current. This region corresponds to the "on" state in the switching transistor operation.

The Gummel–Poon model is widely used in modeling BJTs for various IC designs.(2) It is based on the integral charge model which relates the terminal electrical characteristics to the base charge. This device model is very accurate since it takes many physical effects into consideration. For example, over two dozen physical parameters are needed to cover a wide range of transistor operation. In the Gummel–Poon model, the current that flows from the emitter to the collector terminals with unit current gain is given by

$$I_{CC} = \frac{(qn_i A)^2}{Q_b}\left(e^{qV_{EB}/k_BT} - e^{qV_{CB}/k_BT}\right) \tag{13.56}$$

where

$$Q_b = qA \int_0^{W_b} p_b(x)\,dx \tag{13.57}$$

Q_b being the base charge and A the junction area. The Gummel–Poon model is based on the control of base charge given by Eq. (13.57) which links junction voltages, collector current, and base charge. The base charge consists of five components and is given by

$$\begin{aligned} Q_b &= Q_{b0} + Q_{je} + Q_{jc} + Q_{de} + Q_{dc} \\ &= Q_{b0} + Q_{je} + Q_{jc} + \tau_F I_F + \tau_R I_R \end{aligned} \tag{13.58}$$

where Q_{b0} is the zero-bias charge in the base region, Q_{je} and Q_{jc} are charges associated with the emitter and collector junction depletion capacitances, respectively, Q_{de} $(=\tau_F I_F)$ and Q_{dc} $(=\tau_R I_R)$ represent minority carrier charges associated with the emitter and collector diffusion capacitances, respectively. As the injection level increases, the diffusion capacitance also increases, which results in high-injection gain degradation. The current flow from the emitter region to the collector region may be written as

$$I_{CC} = I_F - I_R \tag{13.59}$$

where

$$I_F = \frac{I_s Q_{b0}}{Q_b}(e^{qV_{BE}/k_BT} - 1) \tag{13.60}$$

and

$$I_R = \frac{I_s Q_{b0}}{Q_b}(e^{qV_{BC}/k_BT} - 1) \tag{13.61}$$

Equations (13.60) and (13.61) resemble Eqs. (13.47) and (13.48) of the Ebers–Moll model.

The base current I_B, which is related to the base charge and base recombination current, can be expressed by

$$I_B = \frac{dQ_b}{dt} + I_{rB} \tag{13.62}$$

where I_{rB} denotes the base recombination current and consists of two components given by

$$I_{rB} = I_{EB} + I_{CB} \tag{13.63}$$

where

$$I_{EB} = I_1(e^{qV_{EB}/k_BT} - 1) + I_2(e^{qV_{EB}/n_ek_BT} - 1) \tag{13.64}$$

and

$$I_{CB} = I_3(e^{qV_{CB}/n_ck_BT} - 1) \tag{13.65}$$

Here I_{EB} is the emitter part of the base current and I_{CB} is the collector part of the base current; n_e and n_c denote the diode ideality factors for the emitter–base and collector–base junctions, respectively. Values of n_e and n_c may vary between 1 and 2, depending on whether the diffusion or recombination current is the predominant component in the base region. Thus, referring to Fig. 13.12, the total emitter and collector currents can be written respectively as

$$I_E = I_{CC} + I_{EB} + \tau_F \frac{dI_F}{dt} + C_{jE}\frac{dV_{EB}}{dt} \tag{13.66}$$

and

$$I_C = I_{CC} - I_{CB} - \tau_R \frac{dI_R}{dt} + C_{jC}\frac{dV_{CB}}{dt} \tag{13.67}$$

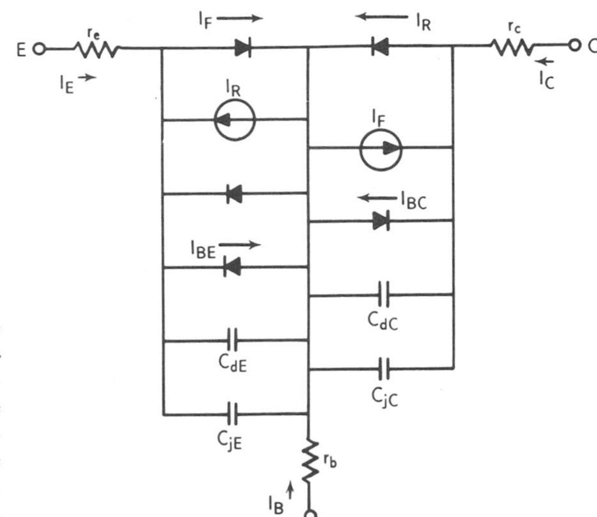

FIGURE 13.12. Equivalent circuit diagram of a p–n–p BJT based on the Gummel–Poon model: C_{jE} and C_{jC} denote the emitter and collector depletion capacitances, C_{dE} and C_{dC} are the emitter and collector diffusion capacitances, while r_e, r_b, and r_c are the emitter, base, and collector resistances, respectively. After Gummel and Poon,[2] by permission.

Figure 13.12 presents the equivalent circuit diagram for the Gummel–Poon model which includes the junction depletion and diffusion capacitances of the emitter–base and base–collector junctions as well as series resistances r_e, r_b, and r_c.[2] Since Q_b is voltage dependent, the effect of high injection in the base (i.e., $\tau_F I_F \ll Q_{b0}$) is included. The Early effect is also included in the model by using the voltage dependence of charge Q_{jC} $(=C_{jC}V_{CB})$. The emitter part of the base current I_{EB} is represented by the two diodes connected in parallel, one ideal and one with a diode ideality factor greater than one (i.e., to account for the bulk or surface recombination current) which makes the current gain bias-dependent at low current levels. Other effects, such as current-induced base push-out (i.e., the Kirk effect), can be incorporated into the model by adding a multiplication factor B to the $\tau_F I_F$ term given by Eq. (13.58). Therefore, the Gummel–Poon model is indeed a very accurate device model for predicting large-signal dc or transient behavior in a BJT. It allows one to predict the device terminal characteristics with good physical insight over a wide range of operation. For a complete description of this device model, readers are referred to the original paper published by Gummel and Poon.[2]

13.6. SWITCHING TRANSISTOR

As pointed out earlier, depending on the bias conditions and modes of operation, a BJT can be operated either as an amplifier or as a switching device. In this section, we discuss the switching properties of a BJT.

When a BJT is operated as a switching device, the transistor has to change its bias condition from the low-current, high-voltage state (off) to the high-current, low-voltage (on) state within a very short period of time (e.g., tens of nanoseconds or shorter). Figure 13.13 shows the operation regions and switching modes of a BJT.[5] The switching behavior of a BJT is seen to be a large-signal transient phenomenon. Since switching speed is a key parameter in the operation of a switchng transistor, one must include both junction depletion and diffusion capacitances, as well as the base spreading resistance in the Ebers–Moll model shown in Fig. 13.11. The junction depletion capacitance is important under reverse-bias conditions, while the diffusion capacitance becomes dominant under forward-bias conditions. The diffusion capacitance is related to the excess carrier stored charge in the transistor. In the active

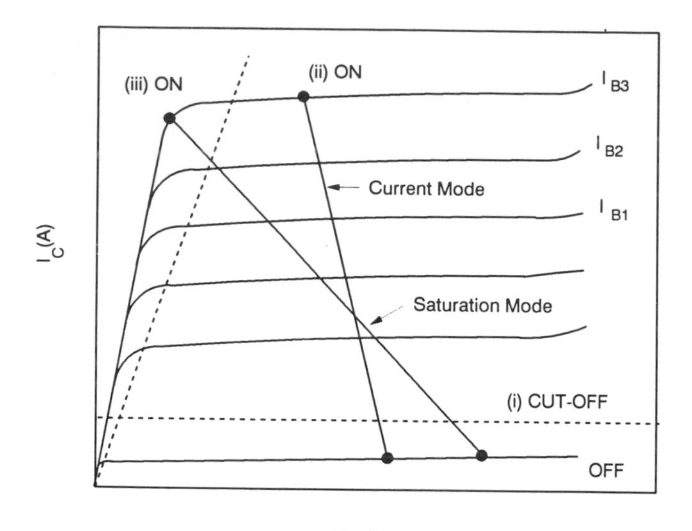

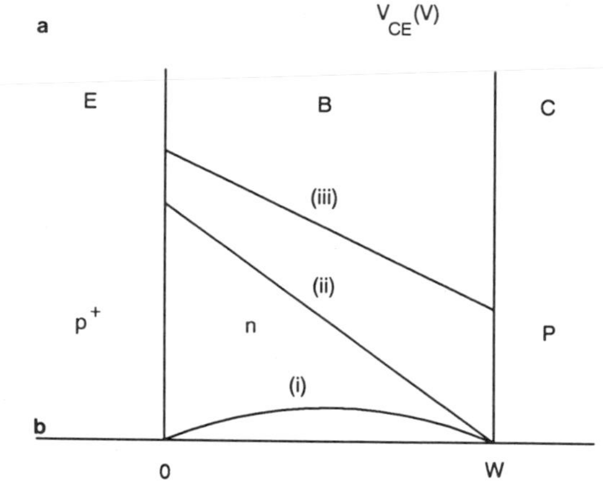

FIGURE 13.13. (a) Operation regions and switching modes of a silicon switching transistor, (b) distribution of minority carrier densities in the base for cutoff (i), active (ii), and saturation (iii) modes. After Moll,[5] by permission, © IEEE–1954.

mode this charge is stored in the base, but in the saturation mode a large part of this charge is stored in the collector region. Nonlinear computer programs like SPICE are available for computer-aided simulation of digital bipolar transistors for all regions (i.e., cutoff, active, and saturation) of operation.

A switching transistor can be operated in several different modes. The saturated mode and current mode are the two most commonly used modes of operation for switching applications. Figure 13.13 shows these two basic modes of operation and their corresponding load lines. If the transistor is to be used as a current switch in digital circuits, it is always operated in the common-emitter (C–E) configuration. In the C–E configuration, current amplification (h_{FE}) is obtained. As shown in Fig. 13.13a, for the current switch mode, the large collector current I_C flowing through the load resistance is switched by controlling the smaller base current I_B at the input. The static "on" and "off" states can be analyzed by using the modified Ebers–Moll model shown in Fig. 13.14.

The operation of a switching transistor is determined by its output characteristic curve, as illustrated in Fig. 13.13a. In the cutoff region, the collector current is off and both the emitter and collector junctions are reverse-biased. In the active region, the emitter junction is forward-biased and the collector junction is reverse-biased. In the saturation region, the emitter and collector regions are both forward-biased. The minority carrier density distributions in

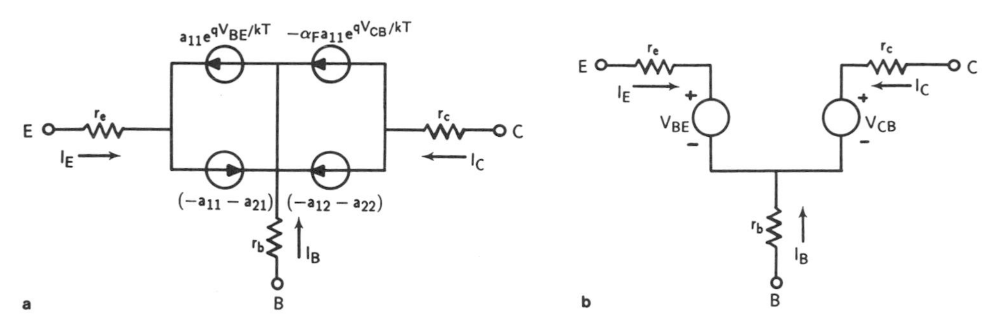

FIGURE 13.14. Equivalent circuit diagrams of a switching transistor (a) in regions (i) and (ii), and (b) in region (iii). After Ebers and Moll,[4] by permission, © IEEE–1961.

the base region corresponding to (i) cutoff, (ii) active, and (iii) saturation regions are shown in Fig. 13.13b.

The switch-off condition of a switching transistor for all switching modes is obtained by extending the load line into the cutoff regime of the transistor. Therefore, the operating mode of a switching transistor is determined mainly by the dc current at the switch-on condition and the location of the operating point. The most common mode of operation for a switching transistor is the saturation mode. The transistor is nearly open-circuited between the emitter and the collector terminals during the switch-off condition, and is short-circuited during the switch-on condition. The current-mode operation is suitable for high-speed switching applications, since the delay associated with the excursion of the transistor into the saturation regime is eliminated.

In designing a switching transistor, two important factors must be considered: the switching time and the current gain. The switching time is normally controlled by the minority carrier lifetime which controls the charge storage time in the case and the collector. For example, gold is commonly used in silicon switching transistors to shorten its switching time, because gold impurity introduces a mid-gap acceptor level ($E_{Au}^{-} = E_c - 0.55$ eV) in silicon. The gold acceptor center is known as the most effective recombination center in silicon. Thus, by doping silicon transistors with a high concentration of gold impurities, the minority carrier lifetime can be drastically reduced, and hence the switching speed of a silicon BJT can be greatly increased. Finally, it should be noted that the current gain of a switching transistor can be improved by lowering the doping density in the base region of the transistor.

The switching behavior of a BJT may be analyzed by using the Ebers–Moll model discussed in the previous section. From Eqs. (13.53) and (13.54), the four coefficients a_{11}, a_{12}, a_{21}, and a_{22} can be related to the measurable parameters I_{EO}, I_{CO}, α_F, and α_R and are given by

$$a_{11} = \frac{-I_{EO}}{(1-\alpha_F\alpha_R)}, \qquad a_{12} = \frac{\alpha_R I_{CO}}{(1-\alpha_F\alpha_R)}, \qquad a_{21} = \frac{\alpha_F I_{EO}}{(1-\alpha_F\alpha_R)}, \qquad a_{22} = \frac{-I_{CO}}{(1-\alpha_F\alpha_R)} \tag{13.68}$$

where I_{EO} is the reverse saturation current of the emitter junction with collector opened, and I_{CO} is the reverse saturation current of the collector junction with emitter opened; α_F and α_R denote the forward and reverse common-base current gains, respectively. For switching operation, the collector junction is reverse-biased in the cutoff and active regimes, and Eqs. (13.53) and (13.54) reduce to

$$I_E = \frac{-I_{EO}\, e^{qV_{EB}/k_BT}}{(1-\alpha_F\alpha_R)} + \frac{(1-\alpha_F)I_{EO}}{(1-\alpha_F\alpha_R)} \tag{13.69}$$

and

$$I_C = \frac{\alpha_F I_{EO}\, e^{qV_{EB}/k_BT}}{(1-\alpha_F\alpha_R)} + \frac{(1-\alpha_R)I_{CO}}{(1-\alpha_F\alpha_R)} \tag{13.70}$$

The equivalent circuit of a switching transistor corresponding to Eqs. (13.69) and (13.70) is shown in Fig. 13.14a.[4] We note that the emitter resistance r_e, base resistance r_b, and collector resistance r_c are included in the equivalent circuit shown in Fig. 13.14a to account for the finite resistances in each region of the transistor. As for the saturation regime, both the emitter and collector junctions are under forward-bias conditions, and the collector–base and emitter–base junction voltages can be derived from Eqs. (13.53) and (13.54) in terms of the emitter and collector currents. The results are

$$V_{EB} = \left(\frac{k_BT}{q}\right) \ln[-(I_E + \alpha_R I_C)/I_{EO} + 1] \tag{13.71}$$

and

$$V_{CB} = \left(\frac{k_BT}{q}\right) \ln[-(I_C + \alpha_F I_E)/I_{CO} + 1] \tag{13.72}$$

Figure 13.14b shows the equivalent circuit of a switching transistor operating in the saturation regime. Equations (13.69) through (13.72) may be used to analyze the nonlinear large-signal switching characteristics of a switching transistor.

In order to characterize a switching transistor, several key parameters such as the current carrying capability, maximum open-circuit voltage, on- and off-impedances as well as the switching time must be considered. The current carrying capability is determined by the maximum power dissipation allowed in the transistor. The maximum open-circuit voltage is determined by the breakdown or punch-through voltage. The impedance during on and off conditions can be determined from Eqs. (13.69) through (13.72) using proper boundary conditions. For example, for a common-base configuration, the on- and off-impedances of the transistor are given respectively by

$$Z_c(\text{on}) = \frac{V_C}{I_C} = \left(\frac{k_BT}{qI_C}\right) \ln[-(I_C + \alpha_F I_E)/I_{CO}] \tag{13.73}$$

and

$$Z_c(\text{off}) = \frac{V_C}{I_C} = \frac{V_C(1-\alpha_F\alpha_R)}{I_{CO} - \alpha_F I_{EO}} \tag{13.74}$$

Equation (13.73) shows that the "on" state impedance varies inversely with the collector current. The "on" impedance is very small when the collector current is large. On the other hand, the "off" impedance is very low when the reverse saturation currents, I_{EO} and I_{CO}, are small.

Let us analyze the switching behavior of a transistor switch. Figure 13.15a shows the circuit diagram of a n^+–p–n BJT operating in the common-base configuration.[5] The transistor is assumed to be driven by a square current pulse from the emitter terminal whose wave form is shown in Fig. 13.15b. The corresponding output collector current response is shown in Fig. 13.15c. In the time interval from $t = 0$ to $t = t_1$, the transistor is turned on and the transient is determined by the transistor parameters in the active regime. At time t_1, the operating point of the transistor is in the saturation regime. The time required for the current to reach 90% of its saturation current (i.e., $I_{C1} = V_{CC}/R_L$) is called the turn-on time, τ_0. At time t_2, the

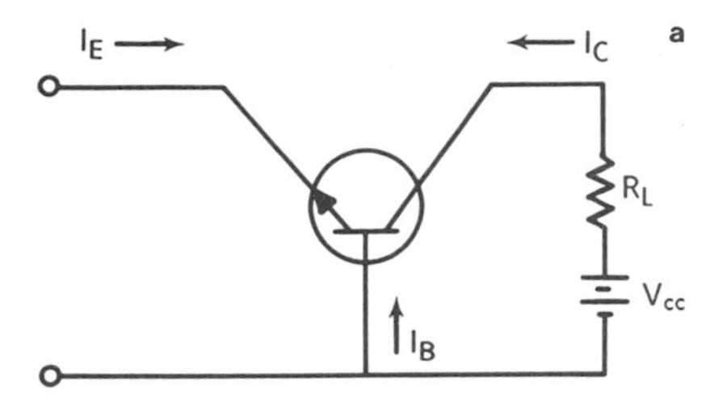

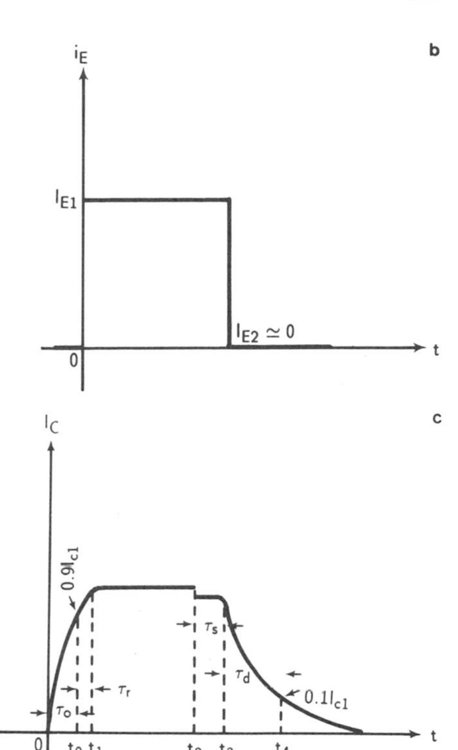

FIGURE 13.15. (a) Circuit diagram of a n–p–n switching transistor, (b) input emitter current pulse, and (c) collector output current response. τ_0 is the turn-on delay time, τ_r is the rise time, τ_s is the storage time, and τ_d is the decay time. After Moll,[5] by permission, © IEEE–1954.

emitter current is reduced to zero (i.e., $I_{E2} \sim 0$) and the turnoff transient begins. From time t_2 to t_3 the minority carrier density in the base region is large. This corresponds to operation in regime III, except that the minority carrier density decays toward zero. During time τ_1, the collector is in the low-impedance state, and the collector current is determined by the external circuit. At time t_3, the carrier density near the collector junction is close to zero. At this point, the collector junction impedance increases rapidly and the transistor begins to operate in the active regime (II). The time interval τ_1 is called the carrier storage time. After time t_3, the transient behavior is calculated from the active regime parameters. At time t_4, the collector current has decayed to 10% of its peak value. The time between t_3 and t_4 is called the decay time, τ_d.

The turn-on time τ_0 can be determined from the transient response in the active regime. From a step input current pulse I_{E1}, the Laplace transform is given by I_{E1}/s. If the common-base current gain is expressed in terms of $\alpha_F/(1 + j\omega/\omega_N)$, where ω_N is the alpha cutoff frequency at which $\omega/\omega_N = 0.707$, then the Laplace transform of the current gain is equal to $\alpha_F/(1 + s/\omega_N)$. Thus, the Laplace transform of the collector current is given by

$$I_C(s) = \frac{\alpha_F I_{EI}}{(1 + s/\omega_N)} \tag{13.75}$$

and the inverse transform of Eq. (13.75) can be written as

$$I_C(t) = \alpha_F I_{E1}(1 - e^{-\omega_N t}) \tag{13.76}$$

If one sets $I_{C1} = V_{CC}/R_L$ as the saturation value of the collector current, then τ_0 is obtained by setting $I_C = 0.9I_{C1}$ in Eq. (13.76), which yields

$$\tau_0 = \left(\frac{1}{\omega_N}\right) \ln\left[\frac{I_{E1}}{(I_{E1} - 0.9I_{C1}/\alpha_F)}\right] \tag{13.77}$$

where τ_0 is the time constant for the collector current to reach 90% of its peak value. Similarly, the storage time τ_1 and decay time τ_d for the common-base configuration can be written, respectively, by

$$\tau_1 = \frac{(\omega_N + \omega_I)}{\omega_N \omega_I (1 - \alpha_F \alpha_R)} \ln\left[\frac{(I_{E1} - I_{E2})}{(I_{C1}/\alpha_N - I_{E2})}\right] \tag{13.78}$$

and

$$\tau_d = \left(\frac{1}{\omega_N}\right) \ln\left[\frac{(I_{C1} - \alpha_F I_{E2})}{(0.1I_{C1} - \alpha_F I_{E2})}\right] \tag{13.79}$$

where ω_I is the inverted alpha cutoff frequency, while I_{E1} and I_{E2} (≈ 0) are the peak and bottom of the emitter input current pulse. It is seen that the turnoff time is equal to the sum of τ_1 and τ_d. From Eqs. (13.78) and (13.79) it is clear that both switching times (i.e., turn-on time τ_0 and turnoff time $\tau_1 + \tau_d$) are inversely proportional to the cutoff frequency of the transistor. Therefore, in order to increase the switching speed, one must increase the cutoff frequency of the transistor. Since the cutoff frequency for most switching transistors is limited by the collector storage capacitance, it is important that this capacitance be maintained at its minimum value.

13.7. ADVANCED BIPOLAR TRANSISTOR

If we wish to increase the current gain of a BJT, then the emitter region is usually doped very heavily and the base region is made very thin. Recently, silicon BJTs have been fabricated by using the polysilicon-emitter structure heavily doped *in situ* with phosphorous impurities on the lightly doped ion-implanted base at a temperature low enough (630°C) to prevent dopant diffusion. Common-emitter current gains in excess of 10^4 and emitter Gummel number greater than 10^{14} cm^{-4} have been achieved in a silicon BJT with polysilicon emitter. Other emitter structures such as the MIS tunnel junction emitter transistor with high current gain, have also been reported in the literature. Polysilicon is widely used in bipolar technology for the emitter and base contacts of advanced silicon BJTs. Vertical scaling of silicon BJTs can be greatly simplified by using polysilicon emitter contacts, since the base saturation current and emitter junction depth can be effectively reduced. Self-aligned polysilicon-emitter BJTs have rapidly become the predominant bipolar structure used in very large scale integrated circuits (VLSI). The advantages of using the polysilicon-emitter structure over the conventional metal-emitter contact structure includes superior process yields, higher packing densities, and better device performance.

Fabrication of BJTs using the polysilicon-emitter structure is quite different from that of the conventional metal-contacted emitter BJTs. For example, after the emitter window is opened, a polysilicon layer is deposited and doped by ion implantation or, alternatively, an *in situ* doped polysilicon layer is deposited onto the underlying emitter region. This polysilicon layer serves as the emitter contact and, at the same time, as a dopant source for the underlying emitters during post-implant activation annealing. Process yields are enhanced because the polysilicon layer prevents implantation damage from the underlying emitter. Figure 13.16

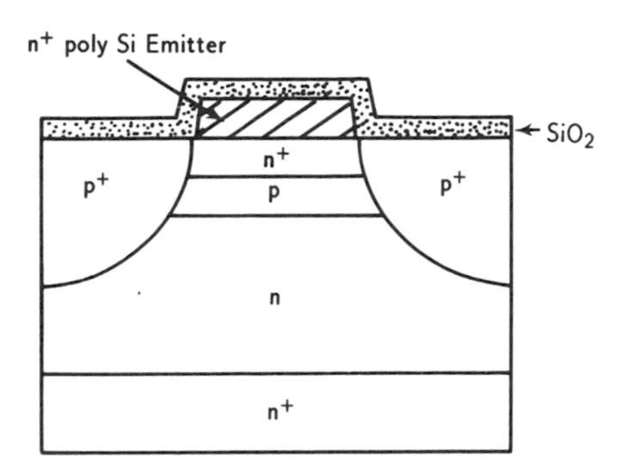

FIGURE 13.16. Cross-section view of a self-aligned silicon BJT with polysilicon emitter contact. After Cuthbertson and Ashburn,[6] by permission, © IEEE–1985.

shows a self-aligned silicon BJT with polysilicon-contacted emitter.[6] The polysilicon emitter is formed by arsenic implantation, following which the polysilicon is selectively etched to form the emitter contact. The structure is then oxidized, resulting in a thicker oxide over the polysilicon and a thinner oxide layer over the silicon. The p^+ base contact region is formed by using high-energy boron implantation. A high-temperature diffusion step is used to produce the emitter region and the extrinsic base region. Packing densities are increased substantially by realizing the self-aligned structures. Device performance is significantly improved by the self-aligned structure and reduction of base current, because the former reduces the device parasitics while the latter is traded for low base resistance. As a result, the speed–power performance of a polysilicon-emitter BJT is greatly improved. Using polysilicon-emitter BJTs, emitter-coupled logic (ECL) circuits with propagation delay times in the sub-100 ps have been reported recently.

13.8. THYRISTORS

When an extra p–n junction is added to a p–n–p or n–p–n bipolar transistor, a n^+–p–n–p or p^+–n–p–n four-layer thyristor is formed. A thyristor is a semiconductor device which exhibits bistable characteristics and can be switched between a low-impedance, high-current on-state to a high-impedance, low-current off-state. The operation of a thyristor is very similar to the operation of a BJT in that both electrons and holes participate in the transport process. Typical doping densities in a p^+–n–p–n^+ four-layer structure are 10^{19} cm^{-3} for the p_1^+ region, 5×10^{14} for the n_1 region, 10^{16} to 10^{17} for the p_2 region, and 10^{19} for the n_2 region.

The schematic diagrams of a p–n–p–n device with two, three, and four terminals are shown in Fig. 13.17a, b, and c, respectively. The device consists of three junctions, J_1, J_2, and J_3 (i.e., p_1^+–n, n_1–p_2, and p_2–n_3), in series. The contact electrode connected to the outer p_1-layer is called the anode, and the contact electrode connected to the outer n_2-layer is called the cathode. Figure 13.17a shows the two-terminal p–n–p–n diode with the gate terminal opened. If a gate electrode is connected to the inner p_2-layer to form a three-terminal p^+–n–p–n^+ device, then the device is called a semiconductor-controlled rectifier (SCR) or a thyristor. This is shown in Fig. 13.17b. An additional gate electrode may be connected to the inner n_1-layer of a p^+–n–p–n diode with two gate electrodes, as shown in Fig. 13.17c. If no gate electrode is provided, then the device is operated as a two-terminal p^+–n–p–n Shockley diode as shown in Fig. 13.17a.

The current–voltage characteristics of a typical p–n–p–n thyristor are shown in Fig. 13.18a. We note that there are four distinct regions shown in this plot. In region I ($0 \Rightarrow 1$) at

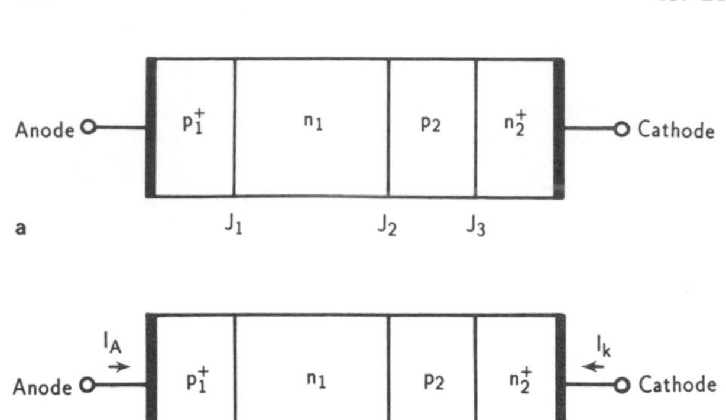

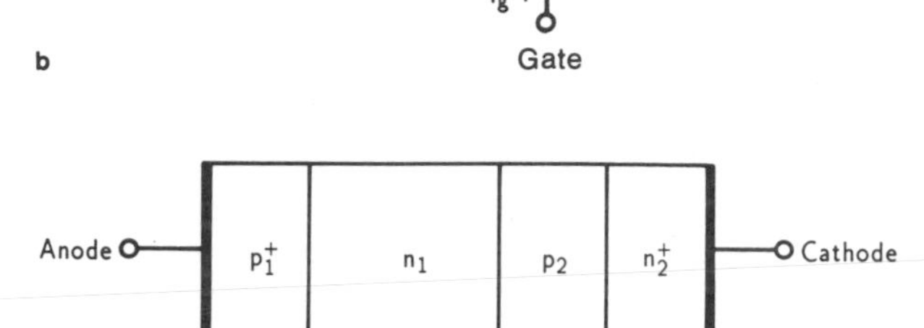

FIGURE 13.17. (a) Schematic diagram of a two-terminal p^+–n–p–n diode, (b) three-terminal thyristor (SCR) with a controlled gate, and (c) four-terminal p–n–p–n device with two controlled gates. The device has three junctions, J_1, J_2, and J_3, in series. The current gain α_1 is for the p–n–p transistor, and α_2 is for the n–p–n transistor. Under the forward blocking condition, the center junction J_2 is reverse-biased and serves as a common collector for the p–n–p and n–p–n transistors.

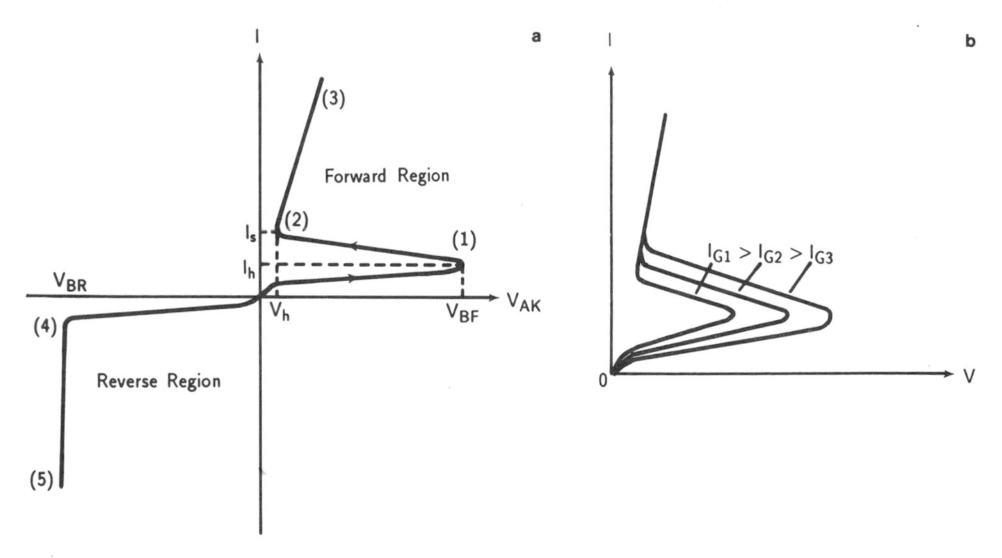

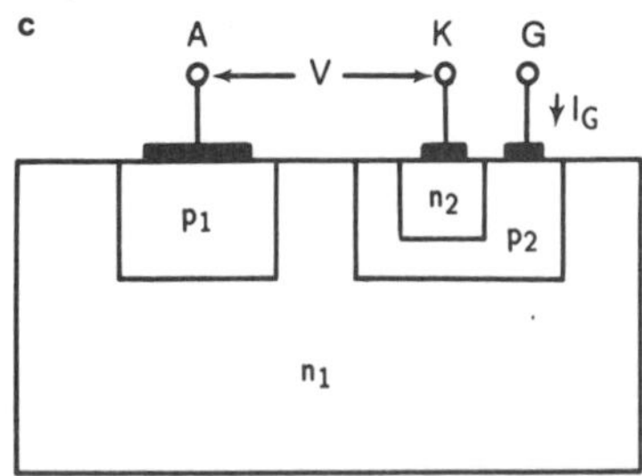

FIGURE 13.18. (a) Current–voltage (*I–V*) characteristics of a thyristor showing the forward and reverse regions. Region (1): forward blocking or "off" state (high impedance, low current); (2) negative resistance regions; (3) forward conducting or "on" state (low impedance, high current); (4) reverse blocking state; (5) reverse breakdown region. (b) The effect of the gate current on the current–voltage characteristics of a thyristor (SCR). (c) A low-power SCR device structure.

low bias voltages, junctions J_1 and J_3 are forward-biased and junction J_2 is reverse-biased. Therefore, the external voltage drop is almost entirely across the J_2 junction, and the device behaves like a reverse-biased p–n junction diode. In this region, the device is in the forward blocking or high-impedance, low-current "off" state. In region I, the forward breakover occurs when $dV/dI = 0$, and a breakover voltage V_{bo} and a switching current I_s can be defined in this region. These parameters are shown in Fig. 13.18a. Region II ($1 \Rightarrow 2$) is the negative differential resistance region in which the current increases with decreasing applied voltage. In region III ($2 \Rightarrow 3$), the current increases rapidly as the applied voltage increases slowly. In this region, junction J_2 is forward-biased, and the voltage drop across the device is that of a single p–n junction diode. The device is in the low-impedance, high-current "on" state. When the current flow in the diode is reduced, the device will remain in the "on" state until it reaches a current level, I_h. The current I_h and its corresponding voltage V_h are called the holding current and holding voltage, respectively. When the current drops below I_h, the diode switches back to its high-impedance state and the cycle repeats. If a negative bias voltage (in region $0 \Rightarrow 4$) is applied to the p_1-terminal and a positive voltage to the n_2-terminal, then both J_1 and J_3 become reverse-biased. Zener or avalanche breakdown may occur when the applied reverse-bias voltage causes the breakdown of junctions J_1 and J_3. This region is usually avoided in thyristor operation. A thyristor operating in the foward-bias region is thus a bistable device which can switch from a high-impedance, low-current "off" state to a low-impedance, high-current "on" state.

Thyristors are the most widely used four-layer p–n–p–n devices with applications ranging from speed control in home appliances to switching and power inversion in high-voltage transmission lines. In p–n–p–n diode operation, it is necessary to increase the external applied voltage so that junction J_2 is in the avalanche multiplication region. The breakover voltage of a p–n–p–n diode is fixed during fabrication. However, the shape of the I–V characteristic curve can be controlled by using a third terminal or a gate in the p_2 region of the thyristor as shown in Fig. 13.17c. A thyristor can be fabricated by using standard silicon planar technology; the p_1 and p_2 regions are formed by using thermal diffusion (or implantation) of boron dopant followed by diffusion of phosphorus impurity to form the n_2 region to complete the four-layer p–n–p–n structure as shown in Fig. 13.17. The p_1–n_1–p_2 structure is known as the lateral transistor and the n_2–p_2–n_1 structure as the vertical transistor. The gate electrode is connected to the p_2 region to control the I–V characteristics of the thyristor. Figure 13.18b shows the I–V characteristics of a typical silicon-controlled rectifier (SCR) under different gate currents I_g.

The predominant effects of increasing I_g in a SCR device are the increase of off-current and the decrease of both breakover voltage and holding current. These effects can be explained qualitatively in terms of the two-transistor equivalent circuit shown in Fig. 13.19a and b. In the off-state, the device behaves essentially like a normal n–p–n transistor with a p–n–p transistor acting as an emitter follower having a very small forward current gain. Increasing the gate current I_g increases the collector current (and hence the anode current) of the n–p–n transistor. The larger anode current will result in an increase of the transistor current gain. When $\alpha_1 + \alpha_2 = 1$, the avalanche multiplication factor decreases and the breakover voltage decreases. In the on-state, the flow of gate current will again increase the value of α. Thus, the holding current can reach a lower value before it switches back to the off-state.

In general, the basic I–V characteristics of a thyristor can be best explained by using a two-transistor analog developed by Ebers in which the n-base of a p–n–p transistor is connected to the emitter of a n–p–n transistor to form a four-layer p–n–p–n device. Figure 13.19a shows a three-terminal thyristor, and Fig. 13.19b is its equivalent circuit representation. It is seen from Fig. 13.19b that the collector current of the n–p–n transistor provides the base drive for the p–n–p transistor, while the collector current and gate current of the p–n–p transistor supply the base drive for the n–p–n transistor. Thus, a regeneration condition occurs when the total loop gain is greater than one. The base current I_{B1} of the p–n–p transistor is equal to the

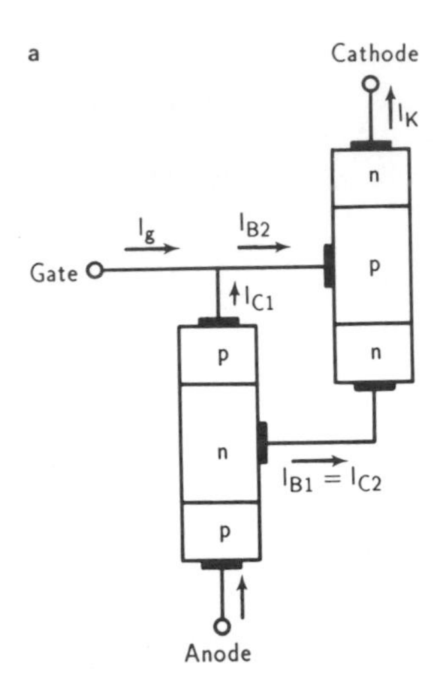

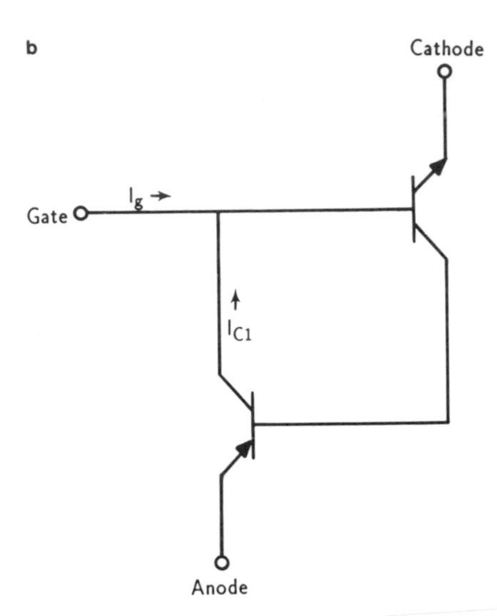

FIGURE 13.19. (a) Two-transistor approximation of a three-terminal thyristor. (b) Same as (a) using transistor symbols.

collector current I_{C2} of the n–p–n transistor, and is given by

$$I_{B1} = I_{C2} = (1 - \alpha_1)I_A - I_{CO1} \tag{13.80}$$

where I_A is the anode current of the p–n–p transistor and α_1 is the dc common-base current gain. The collector current of the n–p–n transistor is given by

$$I_{C2} = \alpha_2 I_K + I_{CO2} \tag{13.81}$$

where α_2 is the dc common-base current gain, and $I_K = I_A + I_g$ is the cathode current of the n–p–n transistor. By solving Eqs. (13.80) and (13.81) we obtain

$$I_A = \frac{(\alpha_2 I_g + I_{CO1} + I_{CO2})}{(1 - \alpha_1 - \alpha_2)} \tag{13.82}$$

which predicts the dc characteristics of a thyristor up to the breakover voltage, and the device behaves as a p–i–n diode beyond the breakover voltage. It is seen that all the current components in the numerator of Eq. (13.82) are very small except when $\alpha_1 + \alpha_2$ approaches unity. At this point, the denominator of Eq. (13.82) becomes zero, and the current I_A increases without limit. As a result, the forward breakover or switching takes place (i.e., when $dV_{AK}/dI_A = 0$). The transistor current gain is seen to increase with collector voltage and collector current at low current levels. The effect of collector voltage on α is particularly pronounced as it approaches the avalanche voltage. Therefore, as the voltage across the thyristor is increased, the collector current and the α values in the two equivalent transistors will also increase. When $\alpha_1 + \alpha_2$ approaches unity, I_A increases sharply which, in turn, increases the values of α. When the sum of avalanche-enhanced α values is equal to one (i.e., $\alpha_1 + \alpha_2 = 1$), breakover will occur. Because of the regenerative nature of these processes, the device is eventually switched to its on-state. Upon switching, the current through the thyristor must be limited by the external load resistance, or the device will be destroyed if the supplied high voltage is large enough. In the on-state, all three junctions in the device are forward-biased, and normal transistor action is no longer effective. The voltage across the device is nearly equal to the sum of the three saturation junction voltage drops, on the order of 1 V for a silicon thyristor. In order to keep the device in its low-impedance on-state, the condition that $\alpha_1 + \alpha_2 = 1$ must be satisfied. In this case, the holding current corresponds to the minimum current at which $\alpha_1 + \alpha_2 = 1$ is satisfied. Further reduction of current will result in the device being switched back to the high-impedance off-state.

Figure 13.20 shows the energy band diagrams of a p–n–p–n thyristor under different bias conditions: Fig. 13.20a is for equilibrium conditions, Fig. 13.20b for the forward off-state in which most of the voltage drop is across junction J_2, and Fig. 13.20c for the forward on-state in which all three junctions are forward-biased. In practice, when a positive voltage is applied to the anode to turn the thyristor from the off-state to the on-state, the junction capacitance across J_2 is charged. This charging current flows through the emitter junctions of the two transistors. If the rate of change of applied voltage with time is large, the charging current may be large enough to increase the α values of the two transistors sufficiently to turn on the device. This rate effect may reduce the breakover voltage to half or less than half of its static value. The voltage at which a SCR device goes from the "on" to the "off" state is usually controlled by a small gate signal. In a low-power SCR, the gate electrode can be used to turn the device to the on- and off-states. However, for high-power SCRs, once the device is in the "on" state, the gate circuit has little effect on the device operation.

The SCR is usually a large-area device since it needs to handle a large amount of current. As a result, lateral gate current flow can give rise to a substantial voltage drop across the device, and the current-crowding effect tends to turn on the periphery of the device first. This turn-on condition may propagate through the entire device. During the turn-on transient, all

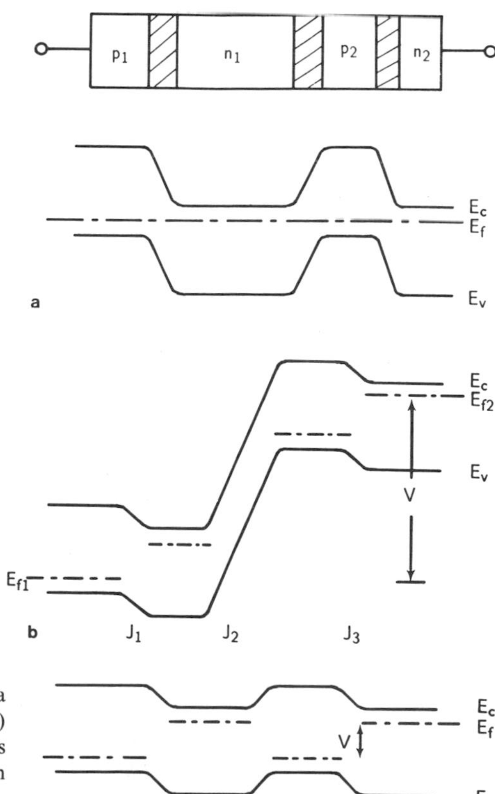

FIGURE 13.20. Energy band diagrams of a p–n–p–n diode in (a) equilibrium conditions, (b) forward off-state where most of the voltage drop is across the J2 junction, (c) forward on-state in which all three junctions are forward-biased.

the anode current passes through the small peripheral area momentarily, and the high current density could cause the device to burn out. To prevent this problem, the interdigitated structure is often used to reduce the lateral effect.

PROBLEMS

13.1. (a) Draw the energy band diagram for a n–p–n transistor in thermal equilibrium and in the normal active mode of operation.
(b) Sketch a schematic diagram to represent this transistor, and show all the current components in the three regions of the transistor.

13.2. Plot the minority carrier density profiles for a n^+–p–n BJT for the following cases:
(a) the E–B junction is forward-biased and the C–B junction is reverse-biased,
(b) both the E–B and C–B junctions are reverse-biased,
(c) both junctions are forward-biased.
Plot minority carrier distributions in the base region for the cases $W \ll L_{nb}$ and $W > L_{nb}$, assuming that $W_E > L_{pe}$ and $W_C > L_{pc}$.

13.3. Consider a double-diffused silicon p–n–p planar transistor, where the impurity profile after base diffusion is given by

$$N_D(x) = \frac{Q_0}{\sqrt{\pi D t}} \exp\left(-\frac{x^2}{4Dt}\right)$$

where $Q_0 = 10^{16}\ \text{cm}^{-2}$, $t = 5$ h, $D_{As} = 5 \times 10^{-14}\ \text{cm}^2/\text{sec}$, and N_A (substrate) $= 10^{15}\ \text{cm}^{-3}$.
(a) Calculate the collector junction depth for the above transistor.
(b) If the emitter junction is obtained by an additional short period of boron diffusion which yields an emitter junction depth of 1 μm, what is the base dopant density $N_B(0)$ near the emitter junction, assuming that the collector junction depth remains unchanged after the emitter diffusion?

13.4. Using the results given by Eqs. (13.10) through (13.16), show that the base current I_B for a n^+–p–n transistor can be expressed approximately by

$$I_B = qA'n_i^2\left[\left(\frac{D_{pe}}{N_{de}W_e} + \frac{D_{nb}W_b}{2N_aL_{nb}^2}\right)(e^{qV_{EB}/k_BT} - 1) + \left(\frac{x_E}{2\tau_0 n_i}\right)e^{qV_{EB}/2k_BT}\right]$$

and the inverse common-emitter current gain is given by

$$\frac{1}{h_{FE}} = \frac{N_{ab}W_bD_{pe}}{N_{de}W_bD_{nb}} + \frac{W_b^2}{2L_{nb}^2} + \left(\frac{N_{ab}W_bx_E}{2D_{nb}n_i\tau_0}\right)e^{-V_{EB}/2k_BT}$$

where x_E is the emitter depletion layer width. Note that the second term in the square bracket of the above equation represents the recombination current in the depletion region of the forward-biased emitter junction. We assume that the base width W_b is much smaller than the electron diffusion length L_{nb} in the base region.

13.5. An interesting insight into the base transport factor β_T can be obtained if Eq. (13.33) is expressed in terms of the transit time τ_B of the minority carriers through the base.
(a) Show that the base transport factor for a p–n–p transistor can be written as

$$\beta_T = \frac{1}{1 + \tau_B/\tau_{pb}}$$

where $\tau_B = W_b^2/2D_{pb}$ and τ_{pb} denote the base transit time and hole lifetime in the n-base region, respectively, where W_b is the base width.
(b) Explain the physical significance of the result given in (a).

13.6. When a BJT is operated in the normal active mode, the C–B junction is reverse-biased. The collector voltage, which determines the depletion layer width of the C–B junction, can thereby affect the actual base width.

(a) Plot the minority carrier charge (i.e., holes) in the base region of a p–n–p transistor for two values of V_{CB}, assuming that I_E is kept constant. How is the base transport factor affected by this base width modulation (i.e., by the change of V_{CB})?

(b) Derive an expression governing the output resistance for the transistor given in (a).

(c) For a planar Si p^+–n–p transistor with $L_{pb} = 12\ \mu m$, $W_b = 1\ \mu m$, $V_{CB} = 10$ V, $I_C = 1$ mA, $N_D = 10^{16}\ cm^{-3}$, and $N_A = 5 \times 10^{15}\ cm^{-3}$, calculate the output resistance for this transistor using the result derived in (b).

13.7. The Gummel number can be calculated from the denominator of Eq. (13.27) if the impurity profile in the base region is known. Calculate the Gummel number for a silicon p–n–p transistor with

(a) uniformly doped base with $N_d = 5 \times 10^{-16}\ cm^{-3}$ and a base width of 1 μm,

(b) $N_d(x) = N_0\, e^{-x/W}$ where $N_0 = 10^{18}\ cm^{-3}$ with a 1 μm base width.

13.8. (a) Show that the general expression for the base transport factor of a p^+–n–p BJT with an arbitrary base impurity doping profile is given by

$$\beta_T = 1 - \left(\frac{1}{L_{pb}^2}\right)\int_0^W \left[\frac{1}{N_d}\int_x^W N_d(x)\right] dx \qquad (13.83)$$

Note that the above equation will reduce to Eq. (13.33) if the base doping profile is uniform.

(b) Using the expression for the base transport factor defined by (a), find values of the base transport factor for the base impurity dopant profiles given by (a) and (b) of Problem 13.7.

13.9. If the space-charge recombination current is negligible, show that the exact expression for the common-emitter output characteristics of a BJT is given by

$$-V_{CE} = \left(\frac{k_BT}{q}\right)\ln\frac{-I_{CO} + \alpha_F I_B - I_C(1-\alpha_F)}{-I_{EO} + I_B + I_C(1-\alpha_R)} + \left(\frac{k_BT}{q}\right)\ln\left(\frac{\alpha_R}{\alpha_F}\right)$$

13.10. (a) There are three possible ways of keeping a BJT switch in the "off" state. These include: (i) the open base ($I_B = 0$), (ii) the base–emitter junction short ($V_{BE} = 0$), and (iii) the base–emitter junction reverse-biased ($V_{BE} < 0$). Plot the equivalent circuit diagrams of a p–n–p BJT for these three cases showing the polarity of V_{CC}, V_{CB}, and V_{EB}, the current flow, and the load resistance R_L.

(b) Find an expression for I_{CEO} (the open-base, collector–emitter leakage current) in terms of I_{CBO} (the open emitter, collector–base leakage current) and the forward current gain α_F for case (i).

(c) Find an expression for I_{CES} (the shorted base, collector–emitter leakage current) in terms of I_{CBO}, α_F, and α_R for case (ii).

(d) Find an expression for I_{CER} (the emitter–base junction reverse-biased) in terms of I_{CBO}, α_F, and α_R for case (iii).

(e) If $\alpha_F = 0.99$ and $\alpha_R = 0.1$, calculate the ratio of the leakage current to I_{CBO} for cases (i), (ii) and (iii).

REFERENCES

1. P. G. Jespers, "Measurements for Bipolar Devices," in: *Process and Device Modeling for Integrated Circuit Design* (F. Van de Wiele, W. L. Engl, and P. G. Jespers, eds.), Noordhoff, Leyden (1977).
2. H. K. Gummel and H. C. Poon, "An Integral Charge Control Model of Bipolar Transistors," *Bell Syst. Tech. J.* **49**, 827 (1970).
3. E. J. McGrath and D. H. Navon, "Factors Limiting Current Gain in Power Transistors," *IEEE Trans. Electron Devices* **ED-24**, 1255 (1977).
4. J. J. Ebers and J. L. Moll, "Large Signal Behavior of Junction Transistors," *Proc. IRE* **49**, 834 (1961).

5. J. L. Moll, "Large-Signal Transient Response of Junction Transistors," *Proc. IRE* **42**, 1773 (1954).
6. A. Cuthbertson and P. Ashburn, "Self-Aligned Transistors with Polysilicon Emitters for Bipolar VLSI," *IEEE Trans. Electron Devices* **ED-32**, 242 (1985).

BIBLIOGRAPHY

A. Bar-Lev, *Semiconductors and Electronic Devices*, 2nd ed., Prentice-Hall, Englewood Cliffs (1984).

J. Bardeen and W. H. Brattain, "The Transistor, A Semiconductor Triode," *Phys. Rev.* **74**, 230 (1948).

J. Early, "Effects of Space-Charge Layer Widening in Junction Transistors," *Proc. IRE*, **40**, 1401 (1952).

J. J. Ebers and J. L. Moll, "Large Signal Behavior of Junction Transistors," *Proc. IRE*, **49**, 834 (1961).

P. E. Gray and C. L. Searle, *Electronic Principles: Physics, Models and Circuits*, Wiley, New York (1969).

P. E. Gray, D. DeWitt, A. R. Boothroyd, and J. F. Gibbons, *Physical Electronics and Circuit Models of Transistors*, p. 145, SEEC Vol. II, Wiley, New York (1964).

A. S. Grove, *Physics and Technology of Semiconductor Devices*, Wiley, New York (1967).

C. T. Kirk, "A Theory of Transistor Cutoff Frequency Falloff at High Current Density," *IEEEE Trans. Electron Devices* **ED-9**, 164 (1962).

S. Konaka, Y. Yamamoto, and T. Sakai, "A 30 ps Bipolar IC Using Super Self-Aligned Process Technology," *IEEE Trans. Electron Devices* **ED-33**, 526 (1986).

R. S. Muller and T. I. Kamins, *Device Electronics and for Integrated Circuits*, Wiley, New York (1977).

T. H. Ning and R. D. Isaac, "Effect of Emitter Contact on Current Gain of Silicon Bipolar Devices," *IEEE Trans. Electron Devices* **ED-27**, 2051 (1980).

H. K. Park, K. Boyer, C. Clawson, G. Eiden, A. Tang, Y. Yamaguchi, and J. Sachitano, "High-Speed Polysilicon Emitter–Base Bipolar Transistor," *IEEE Electron Devices Lett.* **ED1-7**, 658 (1986).

W. Shockley, "The Theory of p–n Junctions in Semiconductors and p–n Junction Transistors," *Bell Syst. Tech. J.* **28**, 435 (1949).

S. M. Sze, *Physics of Semiconductor Devices*, 2nd ed., Wiley, New York (1981).

M. Vora, Y. L. Ho, S. Bhanre, F. Chien, G. Bakker, H. Hingarh, and C. Schmitz, *A Sub-100 Picosecond Bipolar ECL Technology*, IEDM Tech. Digest, p. 34, 1985.

E. S. Yang, *Microelectronic Devices*, McGraw-Hill, New York (1988).

14

Metal–Oxide–Semiconductor Field-Effect Transistors

14.1. INTRODUCTION

The metal–oxide–semiconductor (MOS) system is by far the most important device structure used in both large scale integration (LSI) and very large scale integration (VLSI) technologies. The present LSI and VLSI digital circuits are based almost entirely on n-channel MOSFETs and complementary MOSFETs (CMOSFETs). The MOS structure is a basic building block for several key integrated-circuit active components, namely, MOS field-effect transistors (MOSFETs), insulated-gate field-effect transistors (IGFETs), and charge-coupled devices (CCDs). Most commerically available MOSFETs and CCDs are fabricated from the Si–SiO_2 system. Therefore, it is pertinent to devote this chapter to discussing silicon-based MOS capacitors, MOSFETs and CCDs. The FETs fabricated from III–V compound semiconductors will be depicted in Chapter 15.

As discussed in Chapter 11, the operation of a junction field-effect transistor (JFET) is based on control of the channel current by a reverse-bias p–n junction gate. In contrast to a JFET, the channel current of a MOSFET is controlled by the voltage applied across the gate electrode through a thin gate oxide grown on top of the channel. The current–voltage characteristics of a MOSFET are very similar to those of JFET. However, there are several advantages of a MOSFET over a JFET including lower power consumption, simpler structure, smaller size, higher packing density, and higher compatiblity with VLSI technologies.

In this chapter, we present the basic device theories and general characteristics of silicon-based MOS capacitors, MOSFETs, and CCDs. Section 14.2 describes the physical properties of the surface space-charge region and the capacitance–voltage (C–V) behavior of an ideal MOS capacitor. The oxide charges and interface traps associated with the SI–SiO_2 interface of a nonideal silicon MOS capacitor are discussed in Section 14.3. Section 14.4 is concerned with basic device physics, current–voltage characteristics, small-signal device parameters, and the equivalent circuit of a MOSFET. Some of the problems associated with a scaled-down MOSFET used in VLSI circuits are also discussed in this section. Finally, the operation principles and basic characteristics of a charge-coupled device (CCD) are described in Section 14.5.

14.2. AN IDEAL METAL–OXIDE–SEMICONDUCTOR SYSTEM

In this section, the formation of a surface space-charge region and energy band diagrams for an ideal MOS capacitor under different bias conditions are discussed. The MOS structure

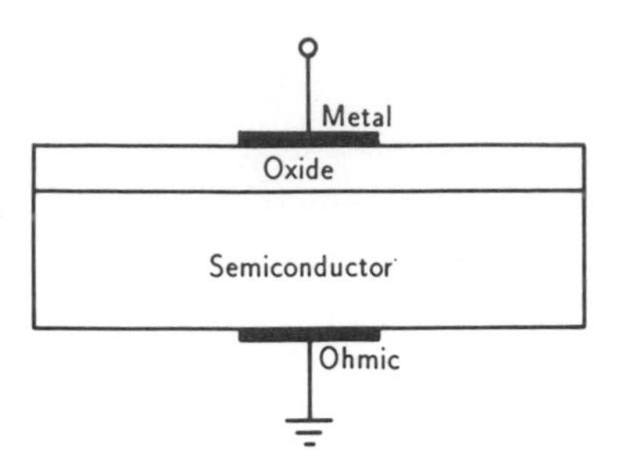

FIGURE 14.1. Cross-section view of a MOS capacitor.

has been used extensively for investigating the physical and electrical properties of the semiconductor surface as well as for various integrated-circuit applications. Since the reliability and stability of a MOSFET and CCD are closely related to the conditions of the semiconductor surface, understanding the physical and electrical properties of the semiconductor surface is essential for improving the performance of MOS devices. Although extensive studies of the Si–SiO_2 interface have been reported in the literature, new physical phenomena associated with the use of an ultrathin SiO_2 layer in the scaled down MOSFETs need to be studied, and investigation of the top and bottom interface properties of the SOI (Silicon-On-Insulator) MOSFETs has also been of considerable interest recently.

Figure 14.1 shows the cross-section view of a simple MOS capacitor. The energy band diagrams for an ideal MOS structure with n-type and p-type semiconductor substrates under equilibrium conditions ($V = 0$) are shown in Fig. 14.2a and b, respectively. An ideal MOS system is defined by the conditions that (i) the work function difference between a metal and a semiconductor is assumed equal to zero at zero-bias conditions (i.e., $\phi_{ms} = 0$), (ii) the flat-band condition prevails, (iii) at any given bias conditions, an equal amount of charge with opposite sign can only exist in the bulk semiconductor and at the metal–insulator interface, and (iv) no dc current can flow through an insulator (i.e., infinite oxide resistance). Condition (i) can be described by

$$\phi_{ms} = \phi_m - \left(\chi_s + \frac{E_g}{2q} - \Psi_B\right) = 0 \qquad \text{for n-type} \tag{14.1}$$

$$\phi_{ms} = \phi_m - \left(\chi_s + \frac{E_g}{2q} - \Psi_B\right) = 0 \qquad \text{for p-type} \tag{14.2}$$

where ϕ_m is the metal work function, χ_s is the electron affinity of the semiconductor, E_g is the energy bandgap, Ψ_B is the bulk potential, and q is the electronic charge. As shown in Fig. 14.2a, χ_0 denotes the electron affinity of the oxide, ϕ_B is the potential barrier between the

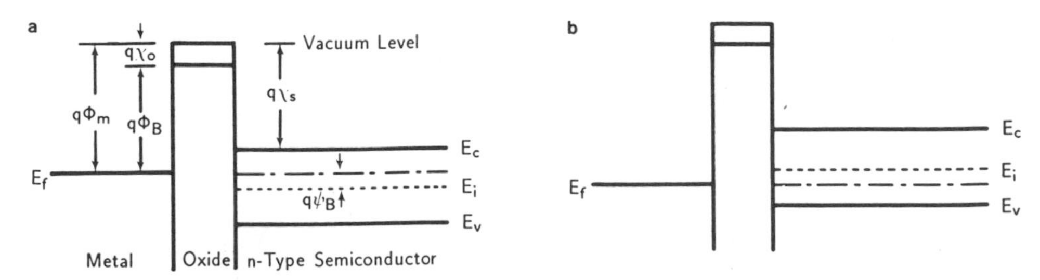

FIGURE 14.2. Energy band diagrams for an ideal MOS capacitor with (a) n-type and (b) p-type semiconductor substrates under equilibrium conditions ($V = 0$).

metal and oxide, E_f is the Fermi level, and E_i is the intrinsic Fermi level. When a bias voltage is applied to an ideal MOS capacitor, three different surface charge conditions (i.e., accumulation, depletion, and inversion) can be created in the semiconductor surface; these are illustrated in Fig. 14.3a, b, and c for a metal–oxide–p-type semiconductor structure. When a negative voltage is applied to the metal gate, the valence band bends upward and moves closer to the Fermi level. This results in an exponential increase in the majority carrier density (holes) at the semiconductor-oxide interface and the semiconductor surface is in *accumulation*, as shown in Fig. 14.3a. When a small positive voltage is applied to the metal gate electrode, the valence band bends downward and the semiconductor surface becomes depleted; this is shown in Fig. 14.3b. Finally, if a large positive voltage is applied to the metal gate, the valence band bends downward even more and the Fermi level moves above the intrinsic Fermi level. In this case, an inversion layer is formed at the semiconductor surface, as is shown in Fig. 14.3c. Therefore, depending on the polarity and the applied bias voltage, an accumulation, depletion, or inversion region can be created at the semiconductor surface of a MOS system. If the MOS structure is formed on an n-type semiconductor substrate, similar surface conditions to those of a p-type substrate discussed above can be obtained, provided that the polarity of the applied bias voltage is changed. The charge distributions under different bias conditions are also shown on the right-hand side of Fig. 14.3a, b, and c. We shall next discuss the physical

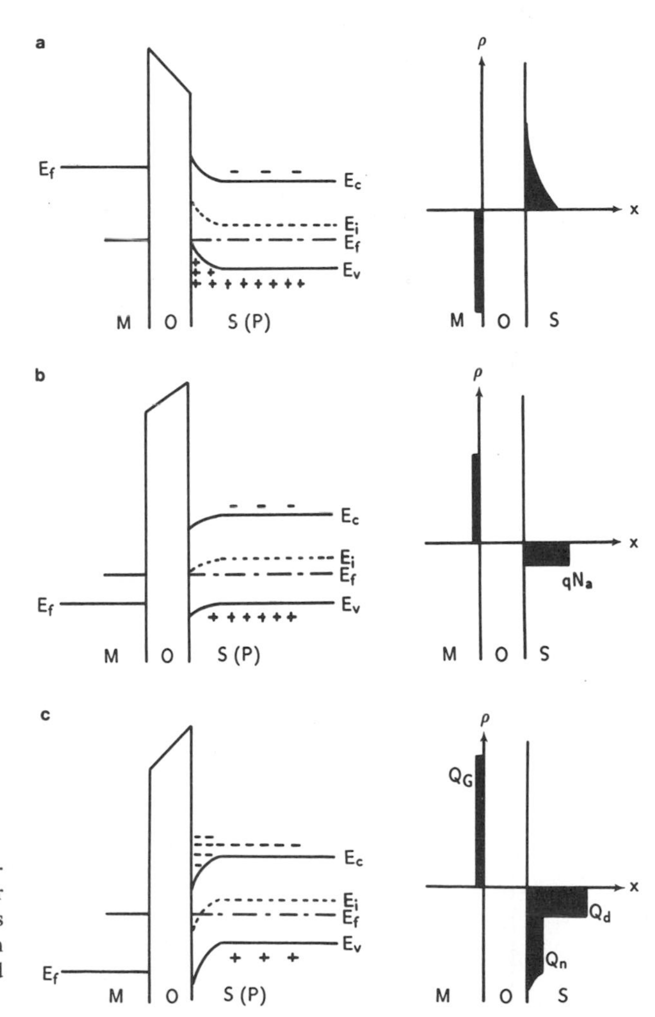

FIGURE 14.3. Energy band diagrams and charge distribution for a p-type MOS capacitor under bias conditions: (a) accumulation ($V < 0$), (b) depletion ($V > 0$), and (c) inversion ($V \gg 0$).

properties of the surface space-charge region and the high- and low-frequency capacitance–voltage (C–V) behavior for an ideal MOS capacitor.

14.2.1. Surface Space-Charge Region

In order to predict the capacitance versus applied voltage (C–V) characteristics of an ideal MOS capacitor, we first derive the expressions for the space-charge density and electric field, which depend on the surface potential of the semiconductor. Figure 14.4 shows the energy band diagram for a p-type semiconductor surface. The potential Ψ is measured with respect to the intrinsic Fermi level (i.e., $q\Psi_B = E_f - E_i$) in the bulk, which reduces to zero inside the bulk semicoductor. At the semiconductor surface, $\Psi = \Psi_s$, where Ψ_s is called the surface potential.

In a bulk semiconductor, electron and hole densities as a function of potential Ψ can be expressed by

$$p_p = p_{p0} \exp\left(-\frac{q\Psi}{k_B T}\right) \tag{14.3}$$

and

$$n_p = n_{p0} \exp\left(\frac{q\Psi}{k_B T}\right) \tag{14.4}$$

where p_{p0} and n_{p0} denote the equilibrium densities of holes and electrons in a p-type semiconductor, respectively. We note that Ψ is positive when the band bends downward. At the semiconductor surface, the densities of electrons and holes are given respectively by

$$p_s = p_{p0} \exp\left(-\frac{q\Psi_s}{k_B T}\right) \tag{14.5}$$

and

$$n_s = n_{p0} \exp\left(\frac{q\Psi_s}{k_B T}\right) \tag{14.6}$$

Equations (14.5) and (14.6) relate the carrier density at the semiconductor surface to the surface potential Ψ_s. Depending on the polarity and magnitude of the surface potential, different surface conditions can be established. These include: (i) for $\Psi_s < 0$, accumulation of holes results, with the band bending upward; (ii) for $\Psi_s = 0$, the flat-band condition is

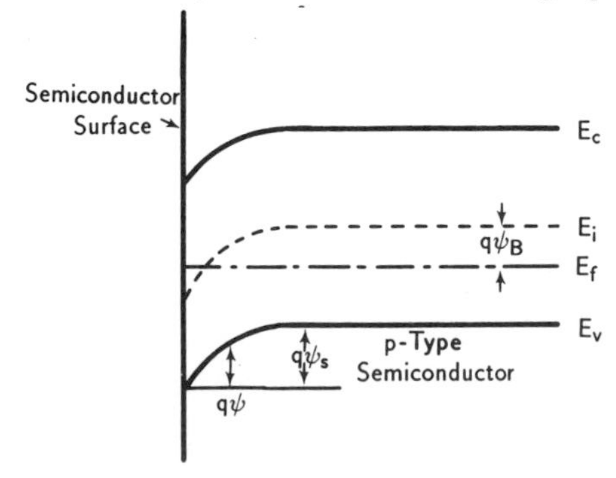

FIGURE 14.4. Energy band diagram at the surface of a p-type semiconductor. The potential Ψ is measured with respect to the intrinsic Fermi level and is equal to zero in the bulk semiconductor; $\Psi_B = (E_f - E_i)/q$ denotes the bulk potential. At the semiconductor surface, $\Psi = \Psi_s$, where Ψ_s is the surface potential. Accumulation occurs when $\Psi_s < 0$, depletion occurs when $\Psi_B > \Psi_s > 0$, and inversion occurs when $\Psi_s > \Psi_B$.

obtained; (iii) for $\Psi_B > \Psi_s > 0$, the depletion of holes results, with the band bending downward; and (iv) for $\Psi_s > \Psi_B$, an inversion region is created near the surface, with the band bending downward. In general, the potential and electric fields as a function of distance from the interface to the bulk semiconductor can be obtained by solving Poisson's equation

$$\frac{d^2\Psi}{dx^2} = q(N_D^+ - N_A^- + p_p - n_p) \tag{14.7}$$

where N_D^+ and N_A^- denote the ionized donor and acceptor densities, respectively. Since the potential Ψ is zero in the bulk we obtain $N_D^+ - N_A^- = n_{p0} - p_{p0}$. By substituting Eqs. (14.3) and (14.4) into Eq. (14.7) and using the condition that $(N_D^+ - N_A^-) = (n_{p0} - p_{p0})$, the electric field as a function of distance from the surface into the bulk of the semiconductor can be obtained by solving Eqs. (14.3), (14.4), and (14.7), and the result is given by

$$\mathscr{E} = \pm \frac{\sqrt{2}k_BT}{qL_D}\left[(e^{-q\Psi/k_BT} + q\Psi/k_BT - 1) + \frac{n_{p0}}{p_{p0}}(e^{-q\Psi/k_BT} + q\Psi/k_BT - 1)\right]^{1/2} \tag{14.8}$$

where the plus sign is for $\Psi > 0$ and the minus sign for $\Psi < 0$; L_D is the extrinsic Debye length for holes. The space charge per unit area required to produce this electric field can be obtained by using Gauss's law, which is

$$Q_s = -\varepsilon_0\varepsilon_s\mathscr{E}_s = \pm \frac{\sqrt{2}k_BT}{qL_D}\left[(e^{-q\Psi_s/k_BT} + q\Psi_s/k_BT - 1) + \frac{n_{p0}}{p_{p0}}(e^{-q\Psi_s/k_BT} + q\Psi_s/k_BT - 1)\right]^{1/2} \tag{14.9}$$

where $\mathscr{E}$ is the electric field at the surface, and Ψ_s is the surface potential. Detailed derivation of the above equations as well as the variation of the space-charge density with the surface potential for p-type silicon can be found in the classic paper by Garrett and Brattain.(1)

It is interesting to note that the onset of strong inversion in a MOS device occurs at a surface potential given approximately by

$$\Psi_{si} \simeq 2\Psi_B = \left(\frac{2k_BT}{q}\right)\ln\left(\frac{N_A}{n_i}\right) \tag{14.10}$$

where Ψ_B is the bulk potential.

14.2.2. Capacitance–Voltage Characteristics

In an ideal MOS capacitor, the effects due to interface traps, oxide charges, and work function difference are negligible. The energy band diagram for an ideal MOS device formed on a p-type silicon substrate is shown in Fig. 14.3b for $V > 0$. The charge distributions in the

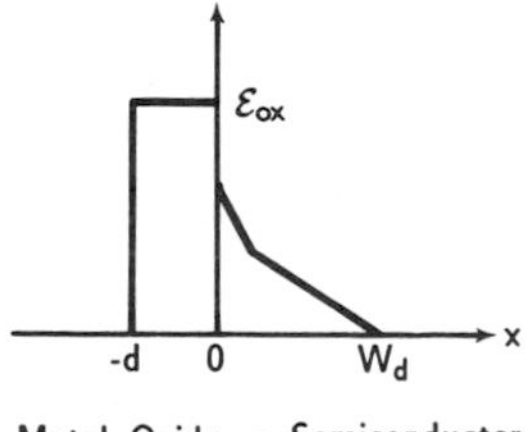

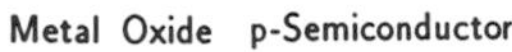

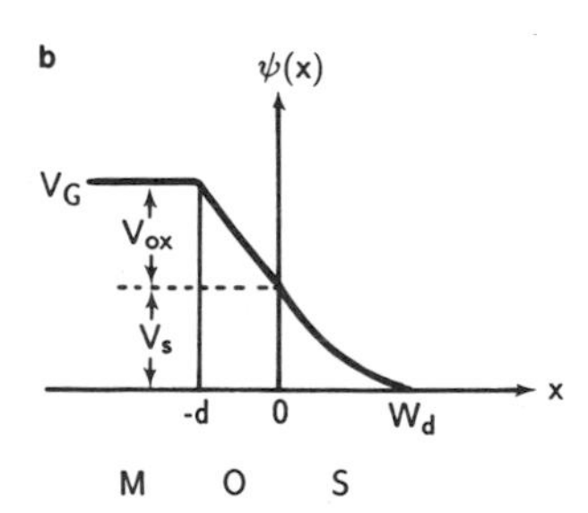

FIGURE 14.5. (a) Electric field distribution and (b) potential distribution of an ideal MOS capacitor under inversion conditions. The energy band diagram and charge distribution are shown in Fig. 14.4.

bulk semiconductor and across the metal–oxide and oxide–semiconductor interfaces are shown in Fig. 14.3b. From the charge neutralty condition we obtain

$$Q_M = Q_n + qN_A W_d = Q_s \tag{14.11}$$

where Q_M is the charge per unit area in the metal, Q_n is the electron charge per unit area in the inversion region, $qN_A W_d$ is the ionized acceptors per unit area in the space-charge region of width W_d, and Q_s is the total charge per unit area in the bulk semiconductor. The electric field and potential distributon for an ideal MOS capacitor are shown in Fig. 14.5a and b, respectively.

If the work function difference between the metal and semiconductor is neglected, then the applied voltage across the MOS capacitor is equal to the sum of the voltage drops across the oxide and semiconductor. This can be expressed by

$$V = V_{0x} + \Psi_s \tag{14.12}$$

where V_{0x} is the potential drop across the oxide and is given by

$$V_{0x} = \mathscr{E}_{0x} d_{0x} = \frac{Q_s}{C_{0x}} \tag{14.13}$$

We note that $C_{0x} = \varepsilon_{0x}\varepsilon_0/d_{0x}$ is the oxide capacitance per unit area. The total capacitance per unit area, C, is equal to the series combination of the oxide capacitance C_{0x} and the depletion layer capacitance C_d $(=\varepsilon_s\varepsilon_0/W_d)$, namely,

$$C = \frac{C_{0x}C_d}{C_{0x} + C_d} \tag{14.14}$$

Since C_d depends on the applied voltage, it is evident that the total capacitance of the MOS capacitor is a function of the applied bias voltage. Figure 14.6 shows the low- and high-frequency small-signal capacitance versus voltage (C–V) plot for an ideal MOS capacitor formed on a p-type substrate. At high frequencies (typically 1 MHz), an accumulation of holes occurs near the semiconductor surface when a large negative bias voltage is applied to the metal electrode, and a strong inversion region is formed near the semiconductor surface when a large positive bias voltage is applied to the metal gate. A depletion region is created below the semiconductor surface when a small positive bias voltage is applied to the MOS capacitor. It is seen that in the strong accumulation region (i.e., $V \ll 0$), C_d becomes very large and the

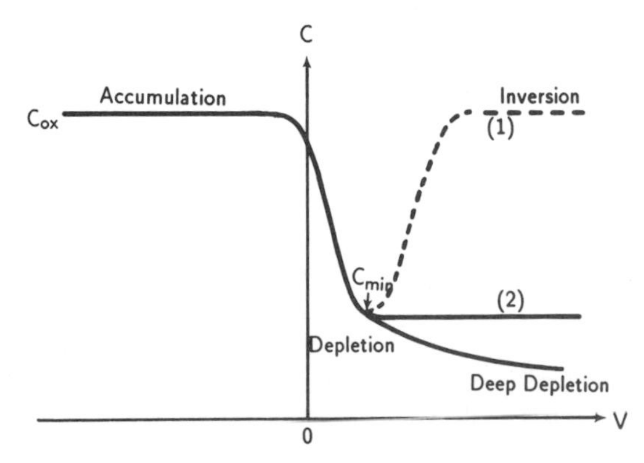

FIGURE 14.6. High- and low-frequency capacitance versus voltage (C–V) curves for a p-type MOS capacitor: (1) low-frequency C–V curve, (2) high-frequency C–V curve, and (3) high-frequency C–V curve in deep depletion.

total capacitance is equal to the oxide capacitance C_{0x}. This corresponds to the maximum capacitance of the MOS capacitor. In the strong inversion region (i.e., $V \gg 0$), the depletion layer reaches a maximum width and remains constant for further increase in the applied bias voltage. Thus, the total capacitance in the strong inversion region is also constant. If the applied voltage becomes more positive than the flat-band voltage, then holes are pushed away from the semiconductor surface and the surface becomes depleted. In this region, the depletion layer width varies with the applied voltage, and the total capacitance is also a function of the applied voltage. Of particular interest in the depletion region is the total capacitance per unit area under the flat-band condition (i.e., $\Psi_s = 0$); it is given by

$$C_{FB} = \frac{1}{d_{0x}/\varepsilon_{0x}\varepsilon_0 + L_D/\varepsilon_s\varepsilon_0} \tag{14.15}$$

where $L_D = \sqrt{2k_BT\varepsilon_s\varepsilon_0/q^2N_A}$ is the extrinsic Debye length. For an ideal MOS capacitor, by neglecting interface traps, oxide charges, and the work function diffeience, the flat-bad capacitance occurs at $V = \Psi_s = 0$. It is noteworthy that a depletion rcgion is formed in the device when the surface potential Ψ_s is greater than zero but less than Ψ_B, where $\Psi_B = (k_BT/q)\ln(N_A/n_i)$ is the bulk potential. Thc weak inversion region begins at $\Psi_s = \Psi_B$, and the onset of strong inversion occurs at $\Psi_s \simeq 2\Psi_B$.

The high-frequency C–V behavior (typically at 1 MHz) for a silicon MOS capacitor shown in Fig. 14.6 can be explained by using the one-sided abrupt junction approach. When the silicon surface is depleted, the number of ionized acceptors in the depletion region is equal to $-qN_AW_d$, where W_d is the depletion layer width. In this case, the potential distribution in the depletion region is a quadratic function of distance, and by solving Poisson's equation we obtain

$$\Psi = \Psi_s\left(1 - \frac{x}{W_d}\right)^2 \tag{14.16}$$

where Ψ_s is the surface potential which is equal to

$$\Psi_s = \frac{qN_AW_d^2}{2\varepsilon_s\varepsilon_0} \tag{14.17}$$

From Eqs. (14.16) and (14.17), it is seen that both Ψ_s and W_d will increase with increasing applied voltage. Eventually, the strong inversion condition is reached at $\Psi_{si} \simeq 2\Psi_B$. When strong inversion occurs, the depletion layer width reaches a maximum. This maximum depletion layer width can be deduced from Eqs. (14.10) and (14.17), and the result yields

$$W_{d\,max} \simeq \sqrt{\frac{2\varepsilon_0\varepsilon_s\Psi_{si}}{qN_A}} = \sqrt{\frac{4k_BT\varepsilon_0\varepsilon_s\ln(N_A/n_i)}{q^2N_A}} \tag{14.18}$$

The threshold (or turn-on) voltage V_{TH} at the onset of strong inversion is given by

$$V_{TH} = \frac{Q_s}{C_i} + 2\Psi_B \tag{14.19}$$

where $Q_s = qN_A W_{d\,max}$ is the total charge in the depletion region at strong inversion. The corresponding total capacitance at the onset of strong inversion is

$$C'_{min} = \frac{1}{1/C_{0x} + W_{d\,max}/\varepsilon_s\varepsilon_0} \tag{14.20}$$

where C'_{min} is the minimum capacitance at the onset of strong inversion. Therefore, the high-frequency C–V behavior can be predicted by using the above equations for different bias voltages. Values of the capacitance measured in the strong inversion region depend on the ability of minority carriers (e.g., electrons in a p-type substrate) to follow up the applied ac signals in the small-signal capacitance measurements. This is usually accomplished by the low-frequency C–V measurements in which the generaton–recombination rates of minority carriers can keep up with the small ac signals. The simplest case arises when both the dc gate bias voltage and the small-signal measuring voltage are changed very slowly such that the semiconductor is near equilibrium. In this case, the signal frequency is low enough that the inversion layer population can follow it. The measured capacitance is just that of the stored charge on either side of the oxide. Its value is thus equal to the oxide capacitance C_{0x}. Under this condition, the C–V curve follows the low-frequency behavior as shown in Fig. 14.6. The capacitance, which is equal to the oxide capacitance C_{0x} in the accumulation region, decreases while the surface is depleted, and moves back up to C_{0x} when the surface becomes inverted. For silicon MOS devices, the onset of low-frequency C–V behavior occurs when $f \leq 100$ Hz. In general, the capacitance in the inversion region increases from C_{min} to C_{0x} as the signal frequency decreases from the high-frequency regime to the quasi-static regime. Another interesting capacitance behavior shown in Fig. 14.6 is the deep-depletion region. This corresponds to the experimental situation in which both the gate voltage and the small ac signals vary at a faster rate than the minority carrier generation–recombination rates in the surface depletion region, since in this case the inversion layer cannot form and the depletion region becomes wider than $W_{d\,max}$. Consequently, the deep depletion is used to describe this region. The deep-depletion phenomenon can be relaxed by using higher bias voltages or by illuminating the MOS device during the C–V scan. A final note on the high-frequency C–V curve of the MOS capacitor is that the series resistance of the semiconductor substrate can also affect the capacitance value in the accumulation region under high-frequency conditions. It is generally observed that, in the accumulation region, values of the capacitance will decrease with increasing frequencies when the series resistance effect becomes important.

14.3. OXIDE CHARGES AND INTERFACE TRAPS

The oxide charges and interface traps play an important role in affecting the physical and electrical properties of a MOS device. To illustrate the importance of these charges, consider an Al–SiO_2-Si MOS capacitor structure shown in Fig. 14.7. For a thermally grown SiO_2 layer on silicon substrate, the transition from silicon to the stoichiometric SiO_2 is sharp. The transition region consists of SiO_x, where x may vary between 1 and 2. From X-ray photospectroscopy (XPS) measurements, this region has been found to be approximately 10 Å thick. A tail of silicon atoms bonding to only three oxygen atoms extends about 30 Å into the SiO_2 layer.

There are four major types of charges in SiO_2 and at the Si–SiO_2 interface which need to be considered. These are the mobile ionic charges (Q_m), the oxide trapped charges (Q_{0t}), the fixed oxide charges (Q_f), and the interface trapped charges (Q_{it}). The mobile ionic charges are usually caused by sodium or potassium ions which become mobile at high temperatures under an applied electric field. These positively charged ions can migrate from the bulk of the oxide layer to the Si–SiO_2 interface over a period of time, slowly increasing the oxide charge there. The oxide trapped charges arise from defects in the oxide. These defects can be structural, chemical, or impurity related. The defects, initially neutral, capture electrons or holes

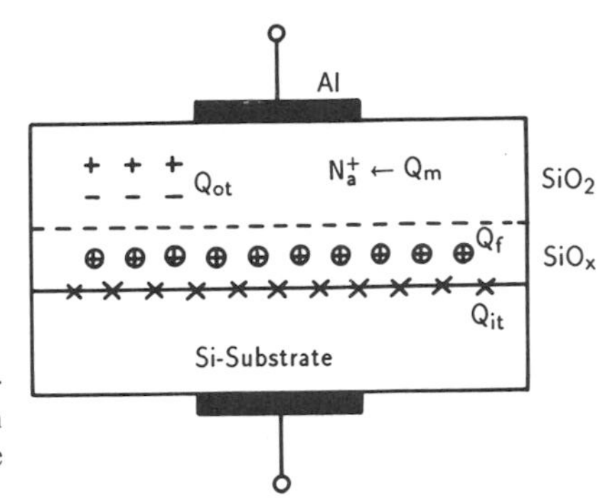

FIGURE 14.7. The thermally grown silicon MOS structure shown in the figure comprises a fixed oxide charge, an oxide trap charge, a mobile ion charge, and an interface trap charge at the Si–SiO_2 interface.

and become charged. Since very little current flows through the oxide layer during normal device operation, the traps usually remain neutral. However, if carriers are injected into the oxide, or ionizing radiation travels through the oxide, these traps can become charged. Both Q_m and Q_{0t} are distributed randomly throughout the oxide. The exact nature of the fixed oxide charges is not known. However, more than one type of defect may cause fixed charges. Although 90% of the charge is located within 30 to 40 Å of the Si–SiO_2 interface, fixed charges are not mobile and are independent of the applied voltage. The density of Q_f is highly dependent on the process used to create the oxide layer and on the orientation of silicon substrate. Finally, interface traps are generally caused by trivalent silicon, which occurs when silicon atoms bond to only three oxygen atoms instead of four. This defect is amphoteric. As the Fermi level rises from the valence band toward the mid-gap, the interface traps capture electrons and become neutral. As the Fermi level rises toward the conduction band, the traps accept additional electrons and become negatively charged. It is also possible for positive charges near the interface to induce interface traps. The energies of the traps vary continuously throughout the silicon bandgap. Therefore, the probability of trapping an electron is dependent on the applied voltage. The positions of the oxide charges and the interface traps in a Si–SiO_2 system are shown in Fig.14.7.

14.3.1. *Interface Trap Charges*

Interface states in a Si–SiO_2 system can be regarded as fast states which can exchange charges with silicon in a very short period of time. They may be created by trivalent silicon (i.e., the so-called P_b center), excess oxygen, or impurities. The density profile of these interface traps across the forbidden gap of silicon is generally found to be of a "U" shape, relatively flat near the mid-gap and increasing very rapidly toward the band edges. Since the interface trap states are distributed across the silicon forbidden gap, it is important to find out the distribution of the interface trap density across the bandgap. The density of the interface trap states can be expressed by

$$D_{it} = \left(\frac{1}{q}\right)\left(\frac{dQ_{it}}{dE}\right) \tag{14.21}$$

where D_{it} has units of $cm^{-2} \cdot eV^{-1}$ and Q_{it} is the total number of interface charges per unit area ($C \cdot cm^{-2}$). Integration of Eq. (14.21) with respect to energy E from the valence band edge E_v to the conduction band edge E_c should yield the total interface state density per unit area in the forbidden gap.

When a voltage is applied to a MOS capacitor, the interface trap levels will move up or down with the valence and conduction bands while the Fermi level remains constant. A change of interface trap charge density will cause a change in capacitance, and hence will alter the C–V curve of an ideal MOS device. Figure 14.8a shows the equivalent circuit of a MOS capacitor when the interface state traps are included, where C_{0x} denotes the oxide capacitance, C_d is the depletion layer capacitance, C_{it} is the capacitance associated with the interface traps, and R_{it} is the resistance associated with the interface traps. Quantities C_{it} and R_{it} are functions of the surface potential. The product $R_{it}C_{it} = \tau$ is defined as the interface state lifetime which determines the frequency behavior of the interface traps. The parallel branch of the equivalent circuit shown in Fig. 14.8a can be converted into a parallel frequency-dependent capacitance C_p and a parallel conductance G_p equivalent circuit as shown in Fig. 14.8b, with both components given by

$$C_p = C_d + \frac{C_{it}}{1 + \omega^2\tau^2} \tag{14.22}$$

and

$$G_p = \frac{C_{it}\omega^2\tau}{1 + \omega^2\tau^2} \tag{14.23}$$

Thus, the input admittance of an ideal MOS capacitor can be written as

$$Y_{in} = G_{in} + j\omega C_{in} \tag{14.24}$$

where G_{in} and C_{in} are given by

$$G_{in} = \frac{\omega^2 C_{it}\tau C_{0x}^2}{(C_{0x} + C_d + C_{it})^2 + \omega^2\tau^2(C_{0x} + C_d)^2} \tag{14.25}$$

and

$$C_{in} = \frac{C_{0x}}{(C_{0x} + C_d + C_{it})}\left[C_d + C_{it}\frac{(C_{it} + C_d + C_{0x})^2 + \omega^2\tau^2 C_d(C_{0x} + C_d)}{(C_{it} + C_d + C_{0x})^2 + \omega^2\tau^2(C_{0x} + C_d)^2}\right] \tag{14.26}$$

Since both the input conductance G_{in} and input capacitance C_{in} given by Eqs. (14.25) and (14.26) contain similar information with regard to the interface state traps, we can determine

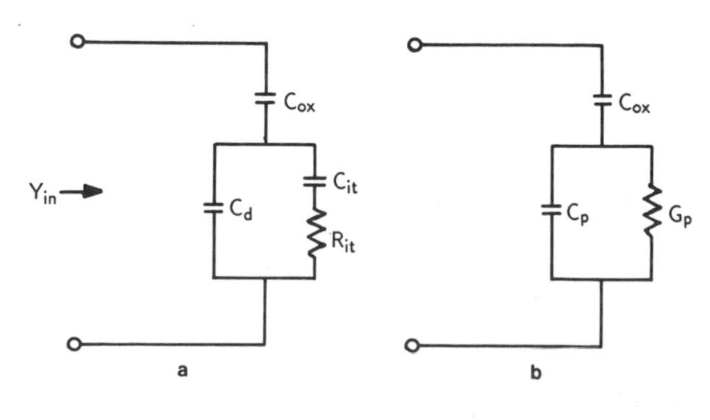

FIGURE 14.8. Equivalent circuit of a MOS capacitor taking into account the interface states effect: (a) C_{it} and G_{it} are the capacitance and conductance associated with the interface traps, respectively; C_d is the depletion capacitance and C_{0x} is the oxide capacitance. Quantities C_p and G_p shown in (b) are given respectively by $C_p = C_d + C_{it}/(1 + \omega^2\tau^2)$ and $G_p = C_{it}\omega^2\tau/(1 + \omega^2\tau^2)$.

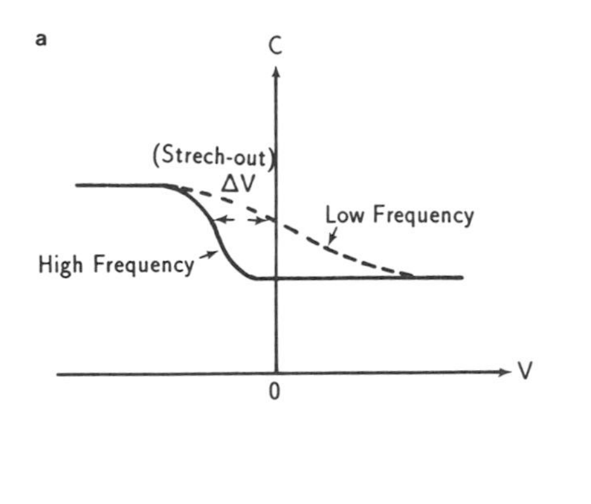

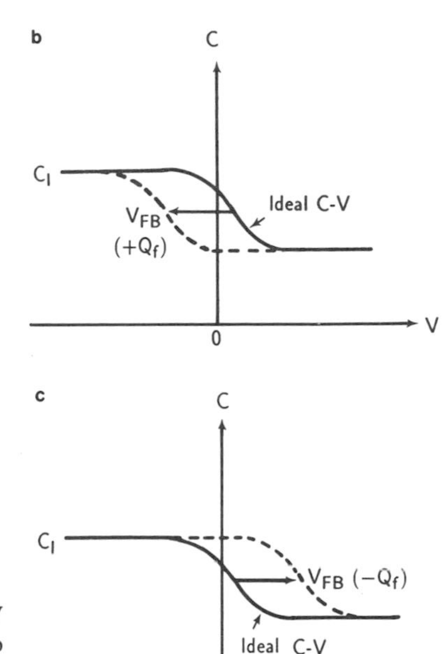

FIGURE 14.9. (a) The stretch-out of high-frequency C–V curve for a p-type MOS capacitor due to the interface trap charges, and the shift of C–V curves due to (b) the positive fixed charge and (c) the negative fixed charge in the oxide.

the interface state traps by using either the capacitance or conductance measurement technique. It can be shown that, for MOS devices, when the interface state density is low, the conductance technique is more accurate than the capacitance method. On the other hand, the capacitance technique can give a rapid evaluation of the flat-band shift and the total interface trap charge, Q_{it}. Figure 14.9 shows the stretch-out of the C–V curve of a MOS capacitor due to the increase in the interface trap charges. At high frequencies (i.e., $\omega\tau \gg 1$), the interface traps cannot follow the ac signal. As a result, the expression for the capacitance given by Eq. (14.26) reduces to Eq. (14.14). In this case, the effect of interface traps on the high-frequency capacitance curve is negligible. As shown in Fig. 14.9, in the presence of interface traps the ideal MOS C–V curve will stretch out along the voltage axis, because in the presence of interface trap charges, more charges on the metal gate are needed for a given surface potential.

Measurements of the interface trap density using capacitance and conductance techniques in a Si–SO_2 system have been reported extensively in the literature.[(2)] Figure 14.10 shows the distribution of interface trap densities in the forbidden gap of silicon for thermally oxidized silicon along the (111) and (100) orientations. The results clearly show that the interface trap density at mid-gap is about one order of magnitude higher for the (111) orientation than for the (100) orientation. This result has been correlated to the difference in the available dangling bonds per unit area on the (111) and (100) silicon surfaces [e.g., the density of dangling bonds is $11.8 \times 10^{14}\ \text{cm}^{-2}$ for the (111) silicon surface and equals $6.8 \times 10^{14}\ \text{cm}^{-2}$ for the (100) surface].

14.3.2. Oxide Charges

We shall next consider the oxide charges in silicon dioxide. There are three different types of oxide charges which can be presented in the SiO_2. These include the oxide trap charge Q_0t, the mobile ionic charge Q_m, and the fixed oxide charge Q_f. As discussed earlier, the fixed oxide charge is located within 30 Å of the Si–SiO_2 interface. It cannot be charged or discharged over a wide range of the surface potential. The density of the fixed oxide charge is virtually independent of the oxide thickness and the type or density of impurities in the bulk silicon.

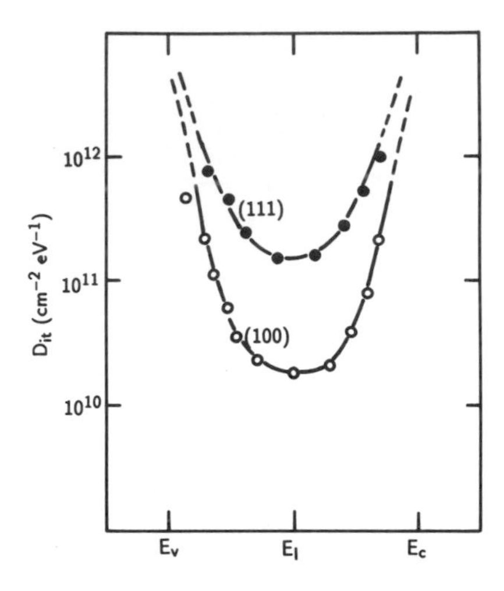

FIGURE 14.10. Distribution of interface trap densities in the forbidden gap of silicon for a thermally grown silicon dioxide along the (111) and (100) orientations. After White and Cricchi,[3] by permission, © IEEE-1972.

Its charge state is generally positive and depends on the oxidation annealing conditions and on the orientation of silicon crystal. The physical origins of the fixed oxide charge have been attributed to the trivalent silicon and nonbridging oxygen (excess oxygen) near the Si–SiO_2 interface. Figure 14.9b shows the shift of a high-frequency C–V curve along the voltage axis when the positive or negative fixed charges are present near the Si–SiO_2 interface of a p-type silicon substrate. The voltage shift is measured with respect to an ideal C–V curve when the fixed charge is equal to zero. For a negative fixed charge, the C–V curve shifts to more positive bias voltages, and a positive fixed charge shifts the C–V curve toward more negative bias voltages with respect to the ideal C–V curve. The reason for the shift of the C–V curve in the presence of fixed oxide charges is due to the fact that charge neutrality in a practical MOS capacitor requires that every negative charge on the metal gate must be compensated by an opposite charge in the oxide and in the bulk silicon substrate. This implies that in the presence of a positive fixed oxide charge in the oxide, the net shallow ionized donor density in silicon must be reduced which, in turn, will decrease the depletion layer width. Thus, the capacitance will be higher than that of the ideal case for all values of the applied gate voltages in the depletion and weak inversion regions. The result is a shift of the C–V curve toward a more negative bias region for the positive fixed charges and toward a more positive gate bias for the negative fixed charges. The magnitude of the C–V shift due to fixed oxide charges with respect to the ideal C–V curve can be estimated by using the expression

$$\Delta V_f = \frac{Q_f}{C_{0x}} \tag{14.27}$$

where Q_f is the fixed oxide charge density and C_{0x} is the oxide capacitance per unit area.

Mobile ionic charges such as sodium ions (which are present in the thermally grown SiO_2) can cause surface instability of passivated silicon devices. Reliability problems of silicon devices operating at high temperatures and high bias conditions may be related to the trace contamination of sodium ions in these devices. Mobile ionic charges due to sodium can move in and out of the oxide with changes in biasing and temperature conditions. These problems can be reduced or eliminated if a cap layer of Si_3N_4, Al_2O_3, or phosphosilicate glass is used as a sodium barrier layer. The shift of the C–V curve due to the mobile ionic charge can be calculated by using the expression

$$\Delta V_m = \frac{Q_m}{C_{0x}} \tag{14.28}$$

where Q_m is the mobile ionic charge per unit area at the Si–SiO_2 interface.

Oxide trap charges can also cause a voltage shift in the ideal C–V curve. The oxide traps are associated with defects created either by impurities or by radiation damage in the oxide layer. They are usually neutral and become charged when electrons or holes are captured by these traps. The voltage shift due to the oxide trap charges can be calculated from

$$\Delta V_{0t} = \frac{Q_{0t}}{C_{0x}} \tag{14.29}$$

where Q_{0t} is the net oxide trap charge per unit area in the SiO_2. Therefore, the total voltage shift of the ideal C–V curve can be expressed by

$$\Delta V_t = \Delta V_f + \Delta V_m + \Delta V_{0t} = \frac{Q_0}{C_{0x}} \tag{14.30}$$

where $Q_0 = Q_f + Q_m + Q_{0t}$ is sum of the effective net oxide charges per unit area in the SiO_2 layer.

From the above analysis, it is clear that oxide charges play an important role in affecting the stability and reliability of a MOS capacitor. Therefore, effective control of the oxide charges in silicon MOS devices is essential for stable device operation.

Finally, the metal–semiconductor work function difference affects the flat-band voltage shift. The work function difference between a metal and an n-type semiconductor is obtained from Eq. (14.1), which reads

$$\phi_{ms} = \phi_m - \left(\chi_s - \frac{E_g}{2q} - \Psi_B\right) \tag{14.31}$$

For an ideal MOS device, ϕ_{ms} is equal to zero. If ϕ_{ms} and Q_0 are not equal to zero, then the measured C–V curve of a practical MOS capacitor will be shifted from the ideal C–V curve by an amount equal to

$$V_{FB} = \phi_{ms} - \frac{Q_0}{C_{0x}} \tag{14.32}$$

where V_{FB} is known as the flat-band voltage shift, and $Q_0 = Q_f + Q_m + Q_{0t}$ is the total oxide charge. If the mobile ionic charge Q_m and the oxide trap charge Q_{0t} are negligible, then the flat-band voltage is reduced to

$$V_{FB} = \phi_{ms} - \frac{Q_f}{C_{0x}} \tag{14.33}$$

It is seen from this equation that the work function difference ϕ_{ms} can have a significant influence on both surface potential and voltage shift. For example, for an Al–SiO_2–Si MOS capacitor with $\phi_m = 4.1$ eV and $\phi_s = 4.35$ eV, the work function difference ϕ_{ms} is found equal to $-0{\cdot}25$ eV. This work function difference must be included in the calculation of voltage shift in the measured C–V curve.

14.4. THE MOS FIELD-EFFECT TRANSISTORS

In this section we present the basic principles, device structure, and characteristics of a long channel metal–oxide–semiconductor field-effect transistor (MOSFET). The MOSFET is a unipolar device in which the current conducton is due to the majority carriers (i.e., electrons for an n-channel MOSFET and holes for a p-channel MOSFET). The silicon MOSFET is probably the most important active component used in a wide variety of silicon integrated-circuit (ICs) applications such as microprocessors, logic and memory chips, power devices, and many other digital ICs. The MOSFET is usually referred to the field-effect transistor (FET) formed on the Si/SiO_2 system. Other acronyms such as IGFET (Insulated-Gate FET) or MISFET (Metal-Insulator-Semiconductor FET) have also been used for FETs fabricated on different insulators or semiconductors. Although semiconductors such as Si, GaAs, and InP, and insulators such as SiO_2, Al_2O_3, and Si_3N_4 have been used in the fabrication of IGFETs, the majority of IGFETs used in present-day VLSI technologies are almost entirely based on the Si–SiO_2 system. Therefore, in this section, only silicon-based MOSFETS will be depicted, while MESFETs (MEtal–Semiconductor FETs) using III–V semiconductors (e.g., GaAs) will be depicted in Chapter 15. In addition to conventional MOSFETs, scaled-down MOSFETs will also be discussed in this section.

14.4.1. General Characteristics of a MOSFET

Figure 14.11 shows the structure of a n-channel silicon MOSFET. The basic structure of a n-channel MOSFET consists of a p-type silicon substrate, the heavily doped n^+ source and drain regions formed by ion implantation or by thermal diffusion on the p-substrate, and a thin gate oxide (MOS structure) deposited on the p-substrate which serves as the gate electrode to control the current flow through the channel region underneath the gate oxide and between the source and drain regions. The gate electrode deposited on top of the SiO_2 gate oxide is formed by using either heavily doped polysilicon or a combination of polysilicon and silicide metal. In addition to the gate oxide, a much thicker field oxide surrounding the MOSFET is also deposited on the outer edge of the device to isolate it from other devices on the same IC chip. A conducting channel between the source and drain can be formed by using a buried implanted layer or a channel can be induced by applying a gate voltage. The distance between the metallurgical junctions of the source and drain regions is defined as the channel length L (i.e., along the y-direction), and the channel width along the Z-direction is designated as Z. The gate oxide thickness is denoted by d_{ox} (i.e., ≤ 1000 Å), and N_A is the substrate dopant density. The n^+-p junctions formed in the source and drain regions are electrically isolated from one another when the gate voltage is equal to zero. If a positive voltage is applied to the gate electrode, a n-type inversion layer is induced at the surface of the semiconductor, which

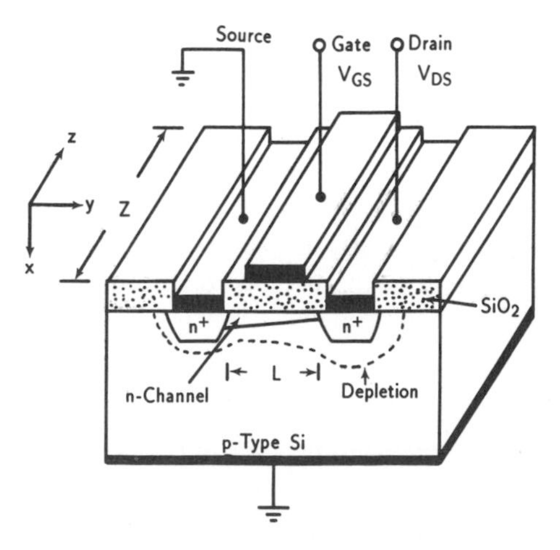

FIGURE 14.11. Device structure of a n-channel silicon MOSFET.

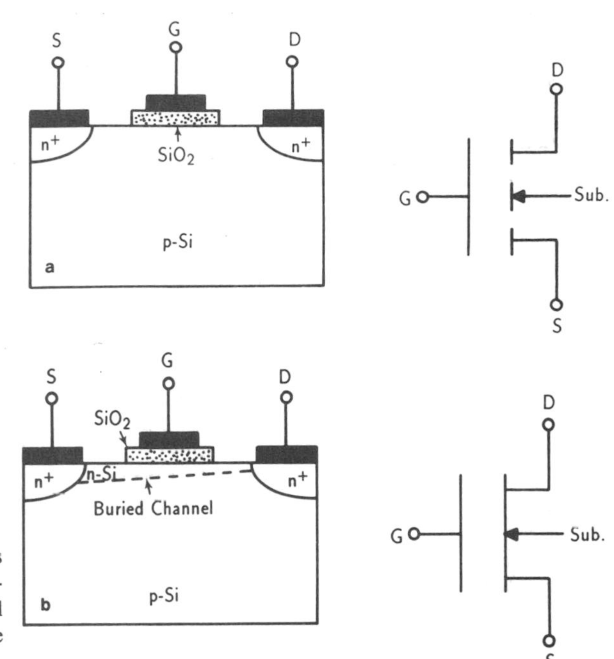

FIGURE 14.12. Cross-section views and circuit symbols of (a) an enhancement-mode (normally-off) n-channel MOSFET and (b) a depletion-mode (normally-on) n-channel MOSFET.

in turn creates a conducting channel between the source and drain regions. When the semiconductor surface under the gate oxide is inverted, and a voltage is applied between the source and drain junctions, electrons can enter the inverted channel from the source junction and leave at the drain junction. Simiarly, a p-channel MOSFET can be fabricated by using a n-type substrate and implantation of boron to form heavily doped p^+-source and drain regions. In a p-channel MOSFET, holes are the majority carriers flowing through the channel between the source and drain regions.

Under thermal equilibrium conditions (i.e., zero bias), if a work function difference and oxide charges exist in the Si–SiO_2, then an inverted surface or channel between the source and drain regions is formed in the MOSFET. In this case, the device is called the *depletion-mode* MOSFET because a negative bias voltage must be applied to the gate in order to deplete the carriers from the channel region to reduce the channel conductance. This type of MOSFET is also known as the normally-on depletion-mode MOSFET. However, in most MOSFETs, a positive-bias voltage must be applied to the gate to induce a channel under the gate oxide of the advice. This type of device is usually called the *enhancement-mode* MOSFET or the normally-off MOSFET. Figure 14.12 shows the cross-section views of (a) an enhancement-mode n-channel MOSFET and (b) a depletion-mode n-channel MOSFET. Enhancement-mode MOSFETs are more widely used in IC applications than depletion-mode MOSFETs. The depletion-mode MOSFET is also called the buried-channel MOSFET because the channel conduction occurs inside the bulk silicon.

The MOSFET is a four-terminal device with electrical contacts to the source, drain, gate, and substrate. Under normal operating conditions, the source and substrate terminals are connected to a common ground. However, when a bias voltage is applied to the substrate it can also change the channel conductance of the MOSFET.

14.4.2. Channel Conductance

The channel conductance is a very important parameter for MOSFET operation. As shown in Fig. 14.11, when a positive voltage is applied to the gate electrode of a MOS transistor, an inversion layer is formed in the semiconductor surface under the gate oxide. The inversion layer provides a conducting path between the source and drain, and is known as the channel. The channel conductance can be calculated by using the expression

$$g_I = \frac{Z}{L}\int_0^{x_I} q\mu_n n_I(x)\,dx \tag{14.34}$$

where $n_I(x)$ is the density of electrons in the inversion channel, Z/L is the gate width to gate length ratio, μ_n is the electron mobility in the channel, q is the electronic charge, and x_I is the channel depth along the x-direction. The total charge density per unit area in the inversion channel, Q_I, is obtained by integrating $n_I(x)$ over the channel depth. This yields

$$Q_I = -\int_0^{x_I} qn_I(x)\,dx \tag{14.35}$$

Now by solving Eqs. (14.34) and (14.35), the channel conductance becomes

$$g_I = -\left(\frac{Z}{L}\right)\mu_n Q_I \tag{14.36}$$

It is noteworthy that Q_I is a function of the applied gate voltage and that the induced mobile charges in the channel become the current carriers in the MOSFET. The threshold voltage V_{TH} is defined as the gate voltage required to bring about strong inversion in the MOSFET and is given by

$$V_{TH} = -\frac{Q_B}{C_{0x}} + \Psi_{si} \tag{14.37}$$

where Q_B is the bulk charge, and $\Psi_{si} \simeq 2\Psi_B$ is the surface potential under strong inversion. If a voltage is established in the channel by the applied voltages from the source and drain electrodes, then Q_B under strong inversion can be written as

$$Q_B = -\sqrt{2q\varepsilon_0\varepsilon_s N_A(V_c + \Psi_{si})} \tag{14.38}$$

Equation (14.38) shows that Q_B depends on the applied voltage V_c in the channel. If the effects of the work function difference and oxide charges are included, then the threshold voltage given above must be modified, and Eq. (14.37) becomes

$$\begin{aligned} V'_{TH} &= \phi_{ms} + \Psi_{si} - \frac{Q_0}{C_{0x}} - \frac{Q_B}{C_{0x}} \\ &= V_{FB} + \Psi_{si} + \sqrt{2q\varepsilon_s\varepsilon_0\Psi_{si}}/C_{0x} \end{aligned} \tag{14.39}$$

where ϕ_{ms} is the metal–semiconductor work function difference, and Q_0/C_{0x} is the voltage shift due to the oxide charges. Quantity V_{FB} is the flat-band voltage defined by Eq. (14.32). Furthermore, if a bias voltage is applied to the substrate (body) of the MOSFET, the threshold voltage becomes

$$V''_{TH} = V_{FB} + \Psi_{si} + \frac{\sqrt{2q\varepsilon_s\varepsilon_0 N_A(\Psi_{si} + V_{sb})}}{C_{0x}} \tag{14.40}$$

where V_{sb} denotes the substrate bias voltage. Beyond strong inversion, the charge condition in the surface region becomes

$$Q_s = Q_I + Q_B = Q_I - qN_A W_{d\,max} \tag{14.41}$$

where Q_I and Q_B denote the channel and bulk charges, respectively; $W_{d\,max}$ is the maximum depletion layer width at the onset of strong inversion and is given by Eq. (14.18). The applied gate voltage V_{GS}, which corresponds to strong inversion, is given by

$$V_{GS} = -\frac{Q_s}{C_{0x}} + \Psi_{si} \tag{14.42}$$

As an approximation, the channel charge Q_I may be related to the threshold voltage V_T by

$$Q_I = -C_{0x}(V_{GS} - V_{TH}) \tag{14.43}$$

Therefore, the channel conductance g_I can be expressed in terms of the gate voltage and threshold voltage. From Eq. (14.43), the channel conductance is given by

$$g_I = -\frac{Z\mu_n Q_I}{L} = \frac{Z}{L}\mu_n C_{0x}(V_{GS} - V_{TH}) \tag{14.44}$$

Equation (14.44) accurately describes the channel conductance in the strong inversion region (i.e., $V_G > V_{TH}$). In this region, the channel conductance is a linear function of the applied gate voltage. In fact, most MOSFETs operate in this region.

14.4.3. Current–Voltage Characteristics

To analyze the current–voltage relation of a MOSFET, let us consider a n-channel silicon MOSFET shown in Fig. 14.11. For a conventional silicon MOSFET, typical channel lengths may vary between 3 to 10 μm. With the advances in electron-beam lithography technology, silicon MOSFETs with submicron channel length have been developed in recent years. In fact, for VLSI design submicron device geometries are routinely employed in present-day IC layouts.

A qualitative description of the current–voltage characteristics for n-channel silicon MOSFET is given first. To facilitate our analysis, it is assumed that the source and substrate terminals are gounded and a gate voltage V_{GS} greater than the threshold voltage V_{TH} is applied to the gate electrode in order to induce an inversion surface channel. When a small drain voltage V_{DS} is applied to the drain electrode, current flows from the source to the drain electrodes via the inversion channel. For small V_{DS}, the channel acts as a variable resistor and the drain current I_{DS} varies linearly with V_{DS}. This corresponds to *linear region* operation (i.e., $V_{GS} > V_{TH}$ and $V_{DS} \simeq 0$). As the drain voltage continues to increase, the space-charge region under the channel and near the drain region widens and eventually reaches a pinch-off condition in which the channel depth x_I at $y = L$ becomes zero. Beyond the pinch-off point, the drain current remains essentially constant as the drain voltage continues to increase. This region is called the *saturation region* since the drain current becomes saturated. Figure 14.13 shows schematic diagrams of a MOSFET operating under different drain voltage conditions. The effect of drain voltage on width of the depletion region across the source, channel, and drain junctions is clearly demonstrated in this figure. Figure 14.13a shows the MOSFET operating in the linear region, and Fig. 14.13b shows the onset of saturation operation. When the drain voltage is increased beyond saturation (i.e., $V_{DS} > V_{D\,sat}$), the effective channel length L' is reduced (i.e., $L' < L$); however, as shown in Fig. 14.13c, the saturation drain current flow from source to drain remains unchanged.

For a long-channel MOSFET (i.e., $L \gg W_d$), the drain current versus drain voltage relation can be derived from the gradual channel approximation. In this approximation, it is assumed that: (i) the transverse electric field along the x-direction inside the channel is much larger than the longitudinal electric field in the y-direction (see Fig. 14.13a), (ii) both electric fields are independent of each other, (iii) the effects of fixed oxide charges, interface traps,

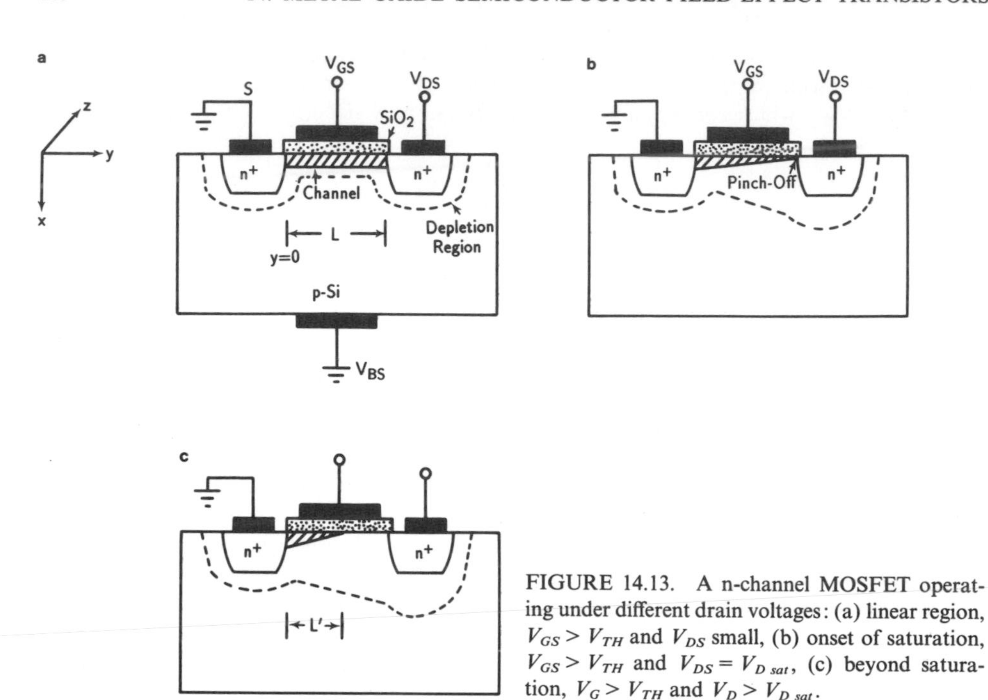

FIGURE 14.13. A n-channel MOSFET operating under different drain voltages: (a) linear region, $V_{GS} > V_{TH}$ and V_{DS} small, (b) onset of saturation, $V_{GS} > V_{TH}$ and $V_{DS} = V_{D\,sat}$, (c) beyond saturation, $V_G > V_{TH}$ and $V_D > V_{D\,sat}$.

and metal–semiconductor work function difference are negligible, (iv) carrier mobility in the inversion layer is constant, and (v) current conduction in the channel consists of only the drift component. It should be noted that the transverse electric field in the x-direction is to induce an inversion channel, while the longitudinal electric field in the y-direction is to produce a drain current which flows through the surface inversion channel.

Linear Region Operation $(V_{DS} \simeq 0)$. The drain current versus drain voltage for a MOSFET operating in the linear region under strong inversion and small drain voltage conditions can be derived as follows. Let us consider a small incremental section along the channel in the y-direction as shown in Fig. 14.13a for $V_{GS} > V_{TH}$. Under this bias condition, mobile carriers are induced in the inversion layer. If the channel voltage is equal to zero, the relationship between the mobile charge Q_I in the inversion layer and the gate voltage is given by Eq. (14.43). When a drain voltage V_{DS} is applied to the drain electrode (source electrode grounded), a channel potential V_c is established along the y-direction in the inversion channel. Thus, the new expression for the induced charge Q_I in the channel is given by

$$Q_I(y) = -C_{ox}[V_G - V_{TH} - V_c(y)] \tag{14.45}$$

and the drain current I_{DS}, which is due to the majority carrier (electrons) flow, can be written as

$$I_{DS} = Z\mu_n Q_I \mathscr{E}_y \tag{14.46}$$

Now, substituting $\mathscr{E}_y = -dV_c/dy$ and Eq. (14.45) into Eq. (14.46) and integrating over distance from $y = 0$ to $y = L$ are potential from $V_c = 0$ to $V_c = V_{DS}$, we obtain

$$I_{DS} = \frac{C_{0x}\mu_n Z}{L}\left(V_{GS} - V_T\text{H} - \frac{V_{DS}}{2}\right)V_{DS} \tag{14.47}$$

We note that in deriving Eq. (14.47) it is assumed that V_{TH} is independent of V_c [see Eqs. (14.37) and (14.38)]. This approximation could lead to a substantial error, since V_{TH} generally increases as one moves closer to the drain region due to the increase of bulk charge Q_B with applied drain voltage. If Eq. (14.38) is used for Q_B, then a more accurate drain current versus drain voltage relation can be derived, and the result is given by

$$I_{DS} = C\frac{C_{0x}\mu_n Z}{L}\left\{\left(V_{GS} - \phi'_{ms} - \Psi_{si} + \frac{Q_{0x}}{C_{0x}} - \frac{V_{DS}}{2}\right)V_{DS} - \frac{2}{3}\frac{\sqrt{2q\varepsilon_s\varepsilon_0 N_A}}{C_{0x}}\left[(V_{DS} + \Psi_{si})^{3/2} - \Psi_{si}^{3/2}\right]\right\} \tag{14.48}$$

This equation shows that for a given gate voltage V_{GS}, the drain current initially increases linearly with drain voltage (i.e., linear region), and then gradually levels off, reaching a saturated value (i.e., saturation region). It is seen that the simplified expression given by Eq. (14.47) usually predicts a higher value of I_{DS} at large V_{DS} than predicted by Eq. (14.48). However, the simple expression given by Eq. (14.47) for I_{DS} offers better physical insight for the device operation, and hence it is easier to obtain a first-order prediction of the MOSFET performance using this equation in a digital circuit design.

Saturation Region Operation ($V_{DS} > V_{D\,sat}$). In the linear region, it is seen that the inversion layer is formed throughout the semiconductor surface between the source and drain. As the drain voltage increases, the inversion layer at the drain side of the channel will gradually diminish. When the drain voltage is increased to the point such that the charge in the inversion layer at $y = L$ becomes zero, the so-called pinch-off condition is reached. The saturation drain voltage and drain current at the pinch-off point are designated by $V_{D\,sat}$ and $I_{D\,sat}$, respectively. Beyond the pinch-off point, further increase of drain voltage will not increase the drain current significantly, and the saturation region is reached. This is shown in Fig. 14.13b. Under the saturation condition, the channel charge at the drain side of the channel is reduced to zero (i.e., $Q_I = 0$ at $y = L$). The saturation drain voltage $V_{D\,sat}$ is obtained from Eqs. (14.37) and (14.38) by setting $V = 0$, and the result yields

$$V_{D\,sat} = V_{GS} - V_{TH} = V_{GS} - \Psi_{si} + \frac{\sqrt{2q\varepsilon_0\varepsilon_s N_A \Psi_{si}}}{C_{0x}} \tag{14.49}$$

The saturation drain current is obtained by substituting Eq. (14.49) (i.e., $V_{D\,sat} = V_{GS} - V_{TH}$) into Eq. (14.47). Hence

$$I_{D\,sat} = \frac{\mu_n Z C_{0x}}{2L}(V_{GS} - V_{TH})^2 \tag{14.50}$$

Equation (14.50) predicts that the drain current in the saturation region is a quadratic function of the gate voltage. We note that Eq. (14.50) is valid at the onset of saturation. However, beyond this point the drain current can be considered a constant. Thus, this equation

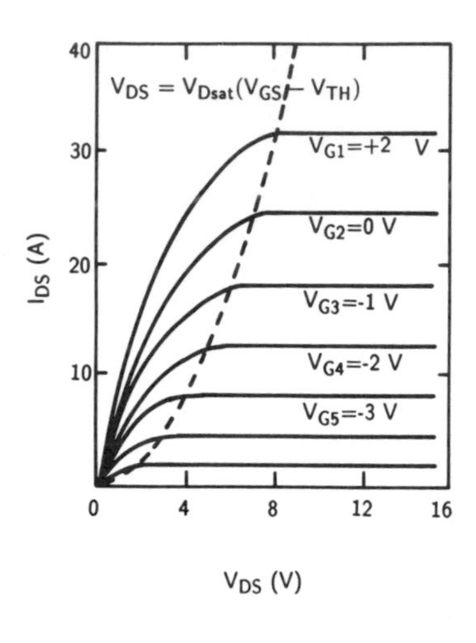

FIGURE 14.14. Drain current versus drain voltage curves for a n-channel silicon MOSFET. The dashed line denotes the locus of the onset of current saturation under different gate bias voltages.

is still valid for $V_{DS} > V_{D\,sat}$. Figure 14.14 shows typical drain current versus drain voltage curves for a n-channel silicon MOSFET under different gate bias conditions. The dashed line denotes the locus of the $I_{D\,sat}$ versus $V_{D\,sat}$ plot which represents the onset of current saturation in the drain region.

Cutoff Region Operation ($V_{GS} \ll V_{TH}$). If the gate voltage is much smaller than the threshold voltage, then there would be no inversion layer formed in the channel region. Under this condition, the MOSFET acts like two p–n junction diodes connected back to back with no current flow in either direction between the source and drain electrodes. Thus, in the cutoff region, the MOSFET is open-circuited.

Subthreshold Region Operation. Another important region of operation for a MOSFET is known as the subthreshold region. In this region, the gate voltage is smaller than the threshold voltage, and the semiconductor surface is in weak inversion (i.e., $\Psi_s < 2\Psi_B$). The drain current in the weak inversion region is called the subthreshold current. The subthreshold region operation is particularly important for low-voltage and low-power applications such as switching devices used in digital logic and memory applications.

In the weak inversion region, the drain current is dominated by diffusion, and hence the drain current can be derived in a similar way as the collector current is derived in a BJT with a uniformly doped base (i.e., the MOSFET can be treated as a n^+–p–n BJT). It can be shown that in the subthreshold region, the drain current varies exponentially with gate voltage (i.e., $I_{DS} \sim e^{gV_{GS}/k_BT}$). For drain voltages greater than $3k_BT/q$, the drain current becomes independent of drain voltage. A detailed derivation of the drain current versus drain voltage relation for a silicon MOSFET operating in the subthreshold region can be found in Ref. 5

14.4.4. Small-Signal Equivalent Circuit

The small-signal equivalent circuit of a MOSFET operating in a common-source configuration is shown in Fig. 14.15. An ideal MOSFET has an infinite input resistance (R_i) and a current generator ($g_m V_{GS}$) at the output terminal of the device. However, in a practical MOSFET, several physical parameters must be included to reflect the nonideality and parasitic effects of the device.

In small-signal analysis two important device parameters must be considered, namely, the channel conductance g_d and mutual transconductance g_m. In linear region operation, both

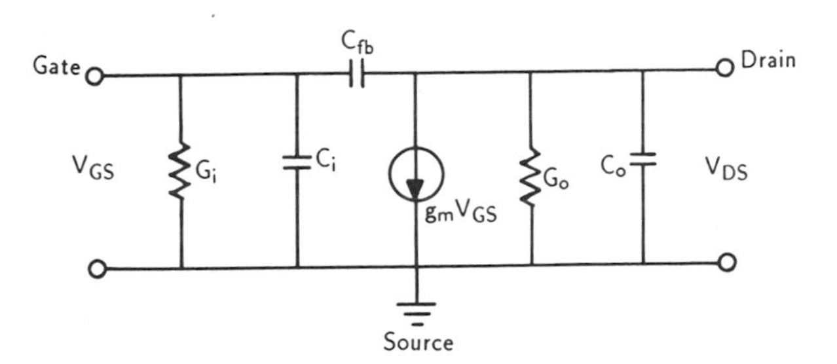

FIGURE 14.15. The small-signal equivalent circuit for a MOSFET in a common-source configuration.

parameters can be derived directly from Eq. (14.47) and the results are given by

$$g_d = \left.\frac{\partial I_{DS}}{\partial V_{DS}}\right|_{V_{GS}=\text{const.}} = \frac{\mu_n Z C_{0x}}{L}(V_{GS} - V_{TH}) \tag{14.51}$$

and

$$g_m = \left.\frac{\partial I_{DS}}{\partial V_{GS}}\right|_{V_{DS}=\text{const.}} = \frac{\mu_n Z C_{0x}}{L} V_{DS} \tag{14.52}$$

It is seen that in the linear region, the drain conductance g_d is ohmic and depends linearly on the gate voltage (except at high gate voltages for which mobility decreases with increasing gate voltage). The inverse of the drain conductance is usually referred to as the "on" resistance. Thus, in linear region operation, the MOSFET is operated essentially as a voltage-controlled resistor in circuit applications.

In the saturation region, the mutual transconductance can be derived by differentiating the saturation drain current given by Eq. (14.50) with respect to the gate voltage, and the result yields

$$g_m = \left.\frac{\partial I_{DS}}{\partial V_{GS}}\right|_{V_{DS}=\text{const.}} = \frac{Z\mu_n C_{0x}}{L}(V_{GS} - V_{TH}) \tag{14.53}$$

which shows that the transconductance varies linearly with gate voltage in the saturation region.

In the saturation region, the drain current remains constant for any drain voltage beyond the pinch-off point. This is true only for the ideal case. In practice, the drain resistance r_d has a finite value for $V_{DS} > V_{D\,sat}$, and hence we can define a drain resistance in the saturation region as

$$r_{d\,sat} = \left.\frac{\partial V_{DS}}{\partial I_{DS}}\right|_{V_{GS}=\text{const.}} \tag{14.54}$$

Thus, from the slope of an I_{DS} versus V_{DS} plot, $r_{d\,sat}$ can be determined. We note that the small-signal equivalent circuit shown in Fig. 14.15 includes all the device parameters discussed above. However, it is worth noting that the gate-to-drain capacitance C_{gd} is the key parameter which controls the high-frequency characteristics of a MOSFET. This is known as the Miller effect.

Another important parameter relating to the small-signal operation of a MOSFET is the maximum frequency for which its short-circuit current gain drops to unity. This frequency is usually referred to as the unity-gain cutoff frequency, f_0. Under the unit gain condition, the input current through the input capacitance ($=2\pi f_0 C_g V_{GS}$) is equal to the output drain current ($\approx g_m V_{GS}$), as shown in Fig. 14.15. Thus, in the linear region of operation (i.e., $V_{DS} \leq V_{D\,sat}$), the unity current gain cutoff frequency can be expressed by

$$f_0 = \frac{g_m}{2\pi C_g} \simeq \frac{\mu_n V_{DS}}{2\pi L^2} \tag{14.55}$$

where C_g is the total gate capacitance. Equation (14.55) is obtained by using Eq. (14.52) for g_m and $C_g = ZLC_{0x}$. Thus, for small-signal high-frequency operation, short gate length and high electron mobility in the channel are highly desirable for a MOSFET.

In the saturation region, the unity current gain cutoff frequency is given by

$$f_0 = \frac{g_m}{2\pi C_g} \simeq \frac{v_s}{2\pi L} \tag{14.56}$$

where $v_s = \mu_n V_{D\,sat}/L$ is the saturation velocity in the channel. Equation (14.56) is obtained by using Eq. (14.53) for g_m and $C_g = ZLC_{0x}$.

In the saturation region operation, the MOSFET device may be used as an amplifier or a closed switch. A closed switch operates in the region where the gate voltage is smaller than the threshold voltage. Under this operation condition, no inversion layer is formed. As a result, the MOSFET behaves like two p–n junction diodes connected back to back, and no current is expected to flow in either direction. The device acts as an open-circuit switch in this region of operation.

The theoretical expressions presented in this section are valid for the long-channel MOSFET in which the channel length is much larger than the depletion layer width of the source and drain junctions. However, with recent advances in silicon VLSI technologies, reduction of channel lengths to less than 1 μm has become a reality, and gate lengths in the submicron region have also been achieved. As the size of MOSFETs is reduced, the channel length becomes equal to or less than the depletion layer width of the source and drain junctions, and hence departure from long-channel behavior occurs in short-channel devices. The short-channel effects are the results of the two-dimensional potential distribution and the high electric field in the channel region. The gradual channel approximation used in analyzing the long-channel MOSFET presented in this section is no longer valid, and must be modified to take into account short-channel effects. For example, the two-dimensional potential distribution will cause degradation of the subthreshold behavior, the dependence of threshold voltage on channel length and biasing voltages, as well as the failure of current saturation due to punch-through. As the electric field increases, the channel mobility becomes field-dependent, and eventually velocity saturation takes place. At high electric fields, carrier multiplication occurs near the drain region, leading to substrate current and parasitic bipolar transistor action. High fields can also cause hot-carrier injection into the oxide. This can lead to oxide charging, transconductance degradation, and a shift in the threshold voltage. In short, in order to further the advancement of short-channel MOSFETs with submicron geometries, it is essential that these short-channel effects be eliminated or minimized.

14.4.5. *Scaled-Down MOSFETs*

In addition to the basic enhancement-mode and depletion-mode MOSFET structures discussed above, a variety of new MOSFET structures such as high-performance MOSFET (HMOS), double-diffused MOSFET (DMOS), vertical or V-shaped grooved MOSFET (VMOS), U-shaped grooved MOSFET (UMOS), Schottky-barrier source and drain MOSFET, lightly-doped-drain (LDD) FETs, and thin film transistors (TFTs) have been widely investigated. Furthermore, recent development of several new silicon-on-insulator

(SOI) technologies [e.g., Separation by IMplantation of OXygen (SIMOX), Zone Melting Recrystalizaton (ZMR)] enables the fabrication of MOSFETs and BJTs on these SOI substrates for various radiation-hard digital IC applications. Using these new structures, high-performance FETs have recently been obtained. Performance improvements include higher speed, lower power consumption, higher packing density, higher radiation resistance, and higher power handling capability.

For VLSI applications, the dimensions and voltages used in conventional MOSFETs must be reduced drastically. One way to avoid the undesirable short-channel effects discussed in the foregoing section while maintaining the long-channel behavior of the MOSFET is to simply scale down all dimensions and voltages of the long-channel MOSFET. The basic idea underlying the scaling-down theory is to keep the electric field strength invariant while reducing the device sizes and voltages. Smaller device geometries will then be translated to shorter transit times (higher device speed) and lower voltages. Both vertical and horizontal dimensions must be scaled down accordingly. The source and drain junction depths must be reduced proportionally to prevent sidewall diffusion from encroaching upon the effective channel diffusion length and the substrate dopant density must be increased so that the depletion region can be scaled down accordingly, otherwise punch-through between the source and drain may occur. In addition, the oxide thickness must also be scaled down to maintain the gate field at a reduced gate voltage and the height of the oxide steps on the surface must be reduced, otherwise it may cause breaks in the thinner interconnects.

The scaling down of dimensions and device parameters in a MOSFET can be obtained as follows. If the channel length (L), channel width (Z), gate oxide thickness (d_{0x}), and voltages (V_{GS}, V_{DS}, V_B) (substrate voltage) are all scaled down by a factor K ($K > 1$), and the substrate dopant density (N_A) is increased by the same factor, then the device parameters for the scaled-down MOSFET are modified from the long-channel MOSFET according to the following formulas:

$$C'_{0x} = \frac{\varepsilon_{0x}\varepsilon_0}{d_{0x}/K} = KC_{0x} \tag{14.57}$$

$$W'_d = \sqrt{\frac{2\varepsilon_s\varepsilon_0(V_{bi}/K)}{qKN_A}} = \frac{W_d}{K} \tag{14.58}$$

$$\Psi'_B = E_F - E_i = -k_BT\ln\left(\frac{KN_A}{n_i}\right) \tag{14.59}$$

$$Q'_B = -q(KN_A)(W_d/K) \simeq Q_B \tag{14.60}$$

$$V'_{TH} = \phi_{Ms} + \frac{2\Psi_B}{q} - \frac{(Q_{0x} + Q_B)}{KC_{0x}} \simeq \frac{V_{TH}}{K} \tag{14.61}$$

$$I'_{DS} = \frac{\mu_n(KC_{0x})(Z/K)}{L/K}[(V_GS/K - V_TH/K)(V_{DS}/K) - \tfrac{1}{4}(V_{DS}/K)^2] = \frac{I_{DS}}{K} \tag{14.62}$$

$$I'_{Dsat} = \frac{\mu_n(KC_{0x})(Z/K)}{2L/K}(V_{GS}/K - V_{TH}/K)^2 = \frac{I_{D\,sat}}{K} \tag{14.63}$$

$$g'_m = \frac{\partial(I_{DS}/K)}{\partial(V_{GS}/K)} = g_m \tag{14.64}$$

$$C'_{gs} = \tfrac{2}{3}(L/K)(Z/K)(KC_{0x}) = \frac{C_{gs}}{K} \tag{14.65}$$

$$t'_d = \frac{C_{gs}/K}{g_m} = \frac{t_d}{K} \tag{14.66}$$

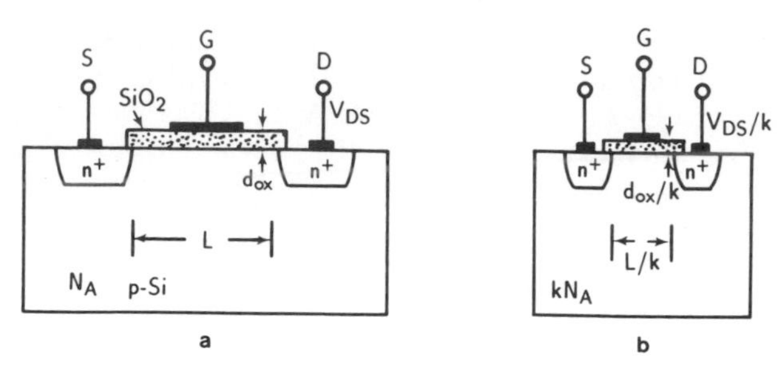

FIGURE 14.16. (a) A conventional long-channel MOSFET and (b) a scaled-down (by a factor K) short-channel MOSFET.

Based on these equations, it can be shown that the device area (A) is reduced by a factor of K^2, power dissipation (IV) is reduced by a factor of K^2, and the power-delay product (IVt'_d) is reduced by a factor of K^3. The power dissipation per unit area (IV/A), however, remains unchanged, although considerable gain in the area and power-delay product (an important figure of merit) is expected by scaling down the MOSFET dimensions. However, a drastic reduction in device size will also increase the importance of other secondary effects, such as narrow-channel and short-channel effects on the threshold voltage, etc. These effects must also be considered in the device modeling. Finally, weak inversion occurs at $V_{GS} < V_{TH}$, and hence subthreshold conduction must also be considered.

Figure 14.16a and b show a conventional long-channel MOSFET and a scaled-down MOSFET, respectively. In the scaled-down MOSFET, both drain voltage V_{DS} and threshold voltage V_T are scaled down by a factor of K, and the number of devices per unit area and power dissipation per unit cell are increased by a factor of K^2. Also, the delay time due to transit across the channel is decreased by a factor of K. It is interesting to note that the subthreshold current remains essentially the same for both devices, since the subthreshold voltage swing remains the same. The junction built-in potential and surface potential at the onset of weak inversion do not scale down with the dimensions, and change by only about 10% for a tenfold increase in substrate dopant density. Additionally, the range of gate voltages between depletion and strong inversion is about 0.5 V. However, the parasitic capacitance may not scale down, and the interconnect resistance will usually increase in a scaled-down MOSFET.

14.5 CHARGE-COUPLED DEVICES

In Fig. 14.17, a charge-coupled device (CCD) is referred to an array of closely spaced (<2.5 μm) MOS capacitors formed on a continuous oxide layer (1000 to 2000 Å thick) grown on a semiconductor substrate. An input gate and an output gate are added to both sides of the MOS capacitor array for the purpose of injecting and detecting the signal charges in the CCD array. The MOS capacitors are built closely enough to one another so that minority carriers stored in the inversion layer associated with one MOS capacitor can be transferred to the surface channel region of an adjacent capacitor. The operation of a CCD is based on the storage and transfer of the minority carriers (known as the charge packet) between the potential wells created by the voltage pulses applied to the gate elecrode of the MOS capacitors. When a controlled sequence of clock voltage pulses is applied to the CCD array, the MOS capacitor is biased into deep depletion, and the charge packet from input signals can be stored and transferred from one potential well to another in a controlled manner across the semiconductor substrate. The basic types of CCDs include the surface-channel CCD (SCCD) and buried-channel CCD (BCCD). In a SCCD the charge packet is transferred along the

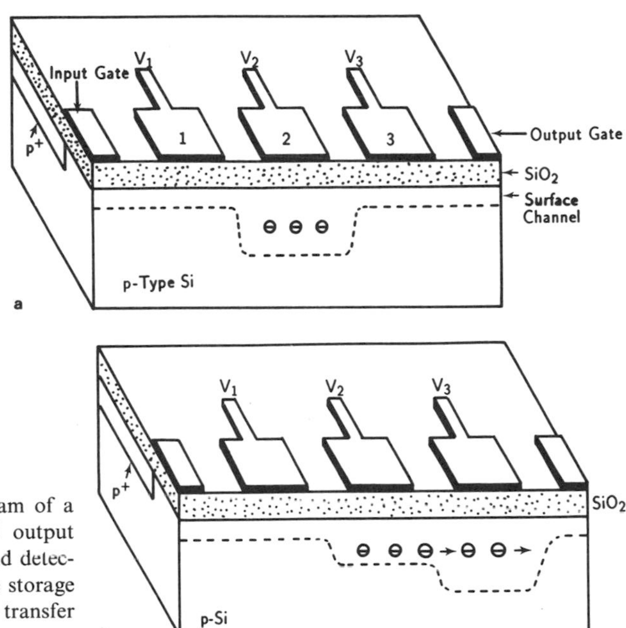

FIGURE 14.17. Schematic diagram of a three-phase CCD with input and output gates fo- signal charge injection and detection. Operating under a (a) charge storage mode ($v_1 = v_3 > v_2$) and (b) charge transfer mode ($v_3 > v_2 > v_1$).

surface channel, while in a BCCD doping of the semiconductor substrate is modified so that storage and transfer of the charge packet can take place in the buried channel of the semiconductor substrate.

In this section, we present the basic structure, operational principles, and characteristics of a surface-channel CCD. The operation of a SCCD requires that the MOS capacitors in the CCD array be biased into deep depletion. Since the storage and transfer of charge packets in a CCD are achieved mainly by controlling a sequence of voltage pulses on the closely spaced MOS capacitors, it is important to understand the transient behavior of a MOS capacitor operating in deep depletion under pulsed bias conditions. In a practical CCD, there are several different types of electrode configurations and clocking techniques which can be used to control its operation. Depending on the electrical performance, fabrication difficulty, and cell size, two-, three-, and four-phase CCDs have been made for digital and analog circuit applications. For example, a two-phase CCD has two MOS gates per cell, while a three-phase CCD has three MOS gates per cell. Figure 14.18 shows a three-phase SCCD with potential wells and their phase timing diagrams at $t = 0$.

14.5.1. Charge Storage and Transfer

Figure 14.17a and b show the cross-section views of a three-phase CCD operating in storage and transfer modes, respectively. The basic storage and transfer mechanisms for a CCD can be explained as follows. Figure 14.17a presents the storage mode for a three-phase SCCD. By applying gate voltage pulses with $v_2 > v_1 = v_3$, the charge packet is stored under the middle gate which has a channel beneath it. If the applied gate pulses are such that $v_3 > v_2 > v_1$, then the right-hand gate causes the transfer of the charge packet from the channel region of the middle gate to the right. If we reduce the voltage pulse on the middle gate to v_1, then the vias voltage on the right-hand gate will be reduced to v_2. The net result is a shift of the charge packet one stage to the right.

Figure 14.19a shows the schematic drawing of a surface-channel MOS capacitor built on a p-type silicon substrate. If a large positive pulsed bias voltage is applied to the gate, then

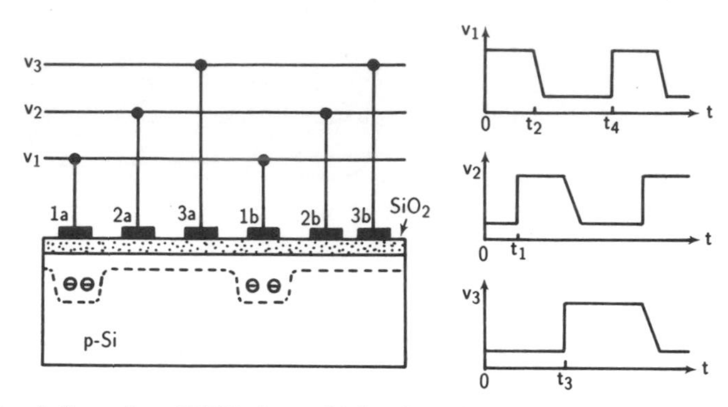

FIGURE 14.18. A three-phase SCCD along with its phase timing diagrams; the potential wells shown are for $t = 0$.

the condition of deep depletion is established under the gate. We note that no inversion layer is formed initially since no minority carriers are available in the depletion layer. Under deep-depletion conditions, a major portion of the applied voltage is across the depletion layer, and hence the surface potential Ψ_s is large. Figure 14.19b shows the energy band diagram for the MOS capacitor in deep depletion under zero signal charge conditions (i.e., $Q_{si} = 0$). In this case, the potential well is empty with a height equal to Ψ_{s0}. The surface potential as a function of the gate voltage V_G for a MOS capacitor under deep depletion is given by

$$V_G - V_{FB} = -\frac{Q_s}{C_{0x}} + \Psi_s \tag{14.67}$$

where V_G is the applied gate voltage and V_{FB} is flat-band voltage shift. The total surface charge Q_s is equal to the sum of the depletion layer charge and the signal charge Q_{si}. This can be written as

$$Q_s = -qN_A W_d - Q_{si} \tag{14.68}$$

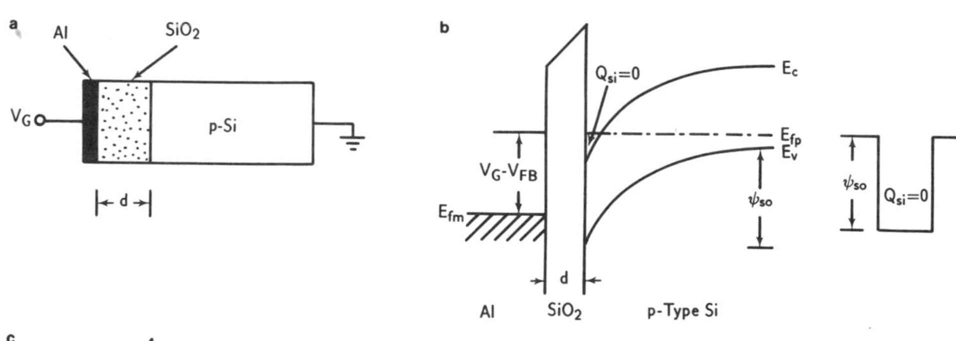

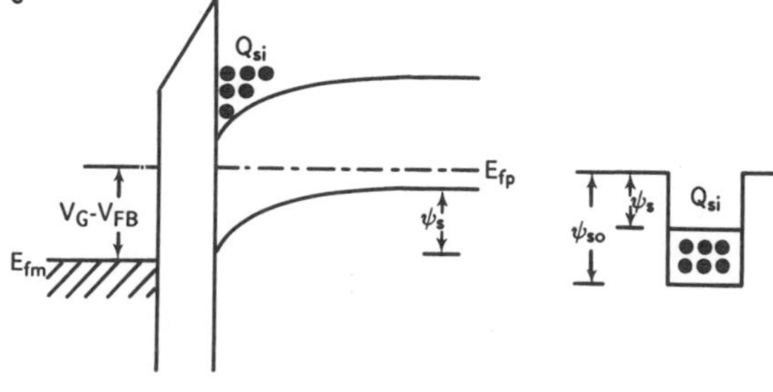

FIGURE 14.19. (a) Schematic drawing of a SCCD using an Al–SiO_2-p-type Si MOS capacitor structure. (b) Band bending into the deep depletion and the empty potential well under a large gate bias voltage. (c) Band bending at the Si–SiO_2 interface and the potential well partially filled with signal charges. After Barbe,[4], by permission, © IEEE–1975.

where

$$W_d = \sqrt{2\varepsilon_s \varepsilon_0 \Psi_s / qN_A} \tag{14.69}$$

is the depletion layer width. Now on substituting Eqs. (14.68) and (14.69) into Eq. (14.67), we obtain

$$V_G - V_{FB} - \frac{Q_{si}}{C_{0x}} = \frac{\sqrt{2q\varepsilon_s \varepsilon_0 N_A \Psi_s}}{C_{0x}} \times \Psi_s \tag{14.70}$$

The solution of the above equation for Ψ_s yields

$$\Psi_s = V'_G - B\left[\left(1 + \frac{2V'_G}{B}\right)^{1/2} - 1\right] \tag{14.71}$$

where

$$V'_G = V_G - V_{FB} - \frac{Q_{si}}{C_{0x}} \tag{14.72}$$

and

$$B = \frac{q\varepsilon_s \varepsilon_0 N_A}{C_{0x}^2} \tag{14.73}$$

which shows that the surface potential is a function of the stored signal charge, the gate voltage, the substrate dopant density, and the oxide thickness. For a given gate voltage, Ψ_s decreases linearly with increasing stored signal charge. The linear relationship between Ψ_s and Q_{si} provides a simple explanation of the charge storage mechanism in the potential well of a CCD. The magnitude of Ψ_s specifies the depth of the potential well (W_d) as defined by Eq. (14.69). As shown in Fig. 14.19c, filling the potential well with signal charges will result in a linear reduction of the surface potential. Equation (14.71) is very important for CCD design, since the gradient of Ψ_s controls the movement of minority carriers in the potential well.

There are three basic charge transfer mechanisms in a CCD: thermal diffusion, self-induced drift, and the fringing field effect. When the signal charge packet is small, the predominant charge transfer mechanism is usually due to thermal diffusion. In this case, the total charge under the storage electrode decreases exponentially with time, and the decay time constant is given by

$$\tau_{th} = \frac{4L^2}{\pi^2 D_n} \tag{14.74}$$

where D_n is the minority carrier diffusion constant and L is the length of the gate electrode. For example, if we assume $D_n = 10\ \text{cm}^2/\text{sec}$ and $L = 10\ \mu\text{m}$, then the decay time constant τ_{th} is 4 μsec.

When the signal charge packet is very large (typically $> 10^{10}\ \text{cm}^{-2}$), the charge transfer is dominated by the self-induced drift produced by electrostatic repulsion of the minority carriers in the potential well. In most cases the transfer of the first 99% of the charge is due to this mechanism. In some CCD operation, to improve transfer efficiency, the entire channel is filled with a large background charge known as fat zero. Self-induced drift is most important under this operation mode. The magnitude of the self-induced longitudinal electric field can be estimated by taking the gradient of the surface potential given by Eq. (14.71), which governs the transfer of charge carries in a CCD. The decay of the initial charge packet Q_i due to self-induced drift can be calculated from

$$Q(t) = \left(\frac{t_0}{t + t_0}\right) Q_i \tag{14.75}$$

where

$$t_0 = \frac{\pi L^3 W C_{0x}}{2\mu_n Q} \tag{14.76}$$

Here L and W denote the gate electrode length and width, respectively, and μ_n is the electron mobility.

The surface potential under the storage electrode is influenced by the voltage applied to the adjacent electrodes due to the two-dimensional coupling of the electrostatic potential. The applied gate voltage results in a surface fringing field, which is present even when the signal charge is zero. The charge transfer process can be speeded up by the fringing field established between the gate electrodes. This fringing field has maxima at the boundaries between adjacent electrodes and minima at the centers of transferring gate electrodes. The magnitude of this fringing field increases with gate voltage and oxide thickness, and decreases with gate length and substrate doping density. For a surface channel CCD, with clock frequencies of several tens of megabits per second, charge-transfer efficiencies greater than 99.99% (or transfer inefficiencies of less than 10^{-4}) can be obtained in the presence of a fringing field.

14.5.2. Charge Injection and Detection

In this section, we discuss charge injection and detection methods in a CCD. Charge packets can be injected by electrical or optical means. If a CCD is used as an image sensor, then the injection is carried out by optical means. On the other hand, in the shift register or delay line application, electrical injection is used instead. Figure 14.20a shows a p–n junction diode used to inject minority carries into the potential well of a CCD. The n-type source is

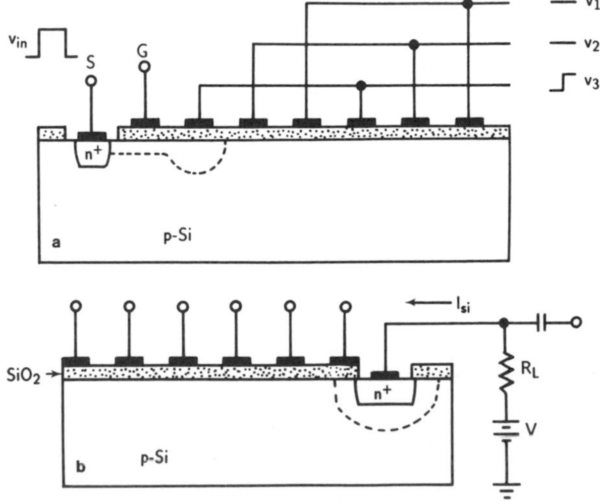

FIGURE 14.20. (a) Injection of minority carrier charges into the potential well of a CCD by using a p–n junction diode. The n-type source is short-circuited to the substrate. (b) Detection of signal charge by the current-sensing technique.

short-circuited to the substrate. When a positive pulse is applied to the input gate (V_{in}), the electrons injected from the source will flow into the potential well under the ϕ_1 gate electrode, and the current source keeps filling the potential well for the duration, Δt, of the input signal. Efficient injection can be achieved by biasing the source and input gates. Charge injection by optical means can be achieved by impinging light from the back side of the p-substrate. It should also be noted that optically generated minority carriers are attracted by the gate electrodes and accumulate in the potential wells.

The detection of signal charge packets in a CCD can be achieved by using either current-sensing or charge-sensing methods. Figure 14.20b presents the current-sensing method. In this technique, a reverse-bias p–n junction diode is used as a drain electrode at the end of the CCD array to collect the signal charge packet. When the signal charge packet reaches the drain junction, a current spike is detected at the output gate as a capacitive charging current. The charge-sensing detection method employs a floating diffusion region to periodically reset the voltage to a reference potential, V_D. When the signal charge packet arrives in this region, the voltage there becomes a function of the signal charge, and the change of voltage at the floating diffusion region is detected by a MOSFET amplifier.

14.5.3. Buried-Channel CCDs

In the previous section we discussed the operation principles and general property of a surface-channel CCD (SCCD), in which charge storage and charge transfer occur in the potential wells created in the surface inversion layer. In a SCCD, the interface traps play an important role in controling the charge transfer loss (i.e., transfer inefficiency) and noise in the CCD, particularly when the signal charge is small. To overcome problems associated with

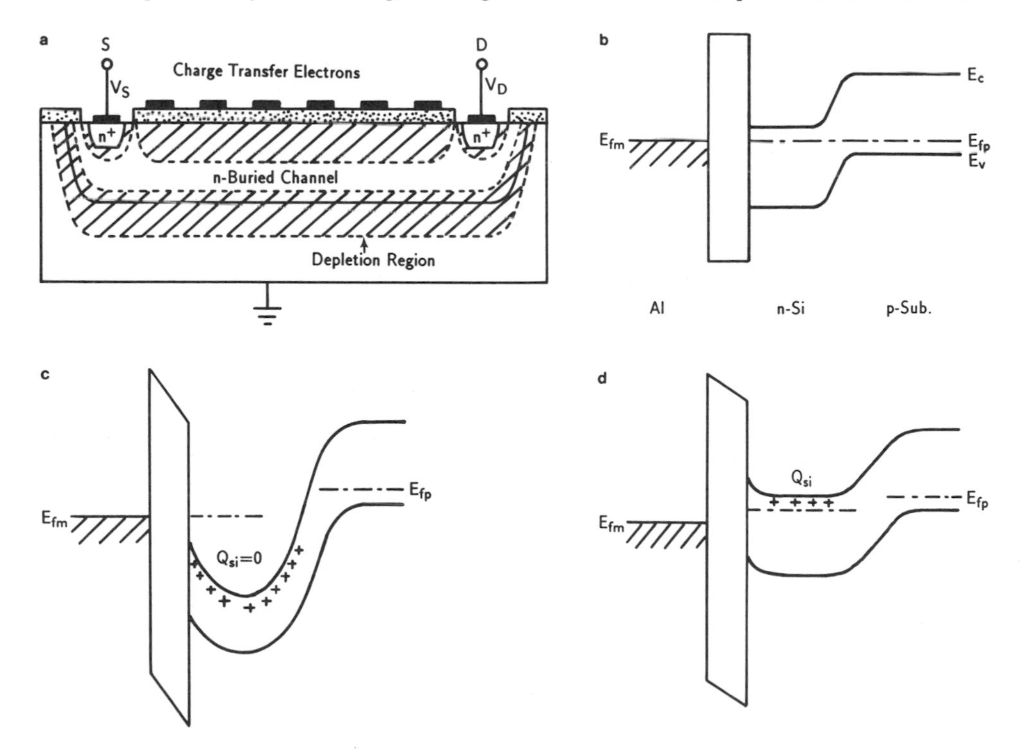

FIGURE 14.21. (a) Cross-section view of a buried-channel CD (BCCD). (b) Energy band diagram of a BCCD under equilibrium, (c) under reverse bias with an empty potential well and zero signal charge (i.e., $Q_{sig} = 0$), and (d) under reverse bias and in the presence of a signal charge packet (i.e., $Q_{sig} \neq 0$).

interface traps, it is common practice to move the channel away from the Si–SiO_2 interface. This results in the bulk- or buried-channel charge-coupled device (BCCD) as shown in Fig. 14.21a. The channel is formed by depositing a thin n-type epitaxial or diffused layer on a p-type substrate. Figure 14.21b shows energy band diagrams of a BCCD under equilibrium ($V_S = V_D = 0$), Fig. 14.21c under reverse-bias ($V_S = V_D = V$) and zero-signal charge conditions, and Fig. 14.21d under reverse-bias conditions with signal charge. It is shown in Fig. 14.21c that when a large positive bias voltage (reverse-bias) is applied to both the source and drain gates, the majority carriers in the channel become completely depleted and a potential well is formed in the n-buried channel. When a signal charge packet is injected from the source into the channel, a flat region in the mid-portion of the potential well results, as is shown in Fig. 14.21d. This potential well can store and transfer signal charge packets by using the controlled clock pulses applied to the MOS gate electrodes. This action is similar to that of a SCCD discussed earlier. The advantages of a BCCD over a SCCD include the elimination of problems associated with interface traps and the improvement of carrier mobility in the channel. Thus, transfer efficiency and channel mobility in a BCCD are expected to be higher than those of a SCCD. However, the drawbacks of a BCCD are the additional process complications and a small capacitance which reduces the signal handling capability.

When CCDs are used in digital applications, ones and zeros are usually represented by the presence or absence of a charge packet in the channel under the closely spaced MOS gates. The signals are clocked through the CCDs at a rate set by the timing of the gate voltages. In analog applications, CCDs are frequently used as image sensors. For this application, the whole CCD is biased into deep depletion and exposed to a focused image for a time interval. The generation of charge carriers under the CCD gate is enhanced during the exposure time according to the brightness of the image. As a result, each channel will become charged to a level that represents the brightness at its location. The analog information can then be sent to the output gate and amplifier built on the edges of the CCD image array. This scheme has been demonstrated successfully in a portable TV camera. Finally, analog signals can also be stored in the CCD arrays, since the amount of charge packet in the channel can be varied continuously. Therefore, CCDs are also used in analog delay line applications.

PROBLEMS

14.1. Draw energy band diagrams and charge distributions for an Al–SiO_2-n-type Si MOS capacitor under (a) accumulation, (b) depletion, and (c) inversion conditions. Assume that the interface traps and work function difference are negligible.

14.2. Consider an Al-gate MOSFET fabricated on a n-type silicon substrate with dopant density $N_d = 2 \times 10^{15}\ cm^{-3}$. If the thickness of the gate oxide is 1000 Å and the interface trap density at the Si–SiO_2 interface is $2 \times 10^{11}\ cm^{-2}$, calculate:

(a) the work function difference between aluminum and silicon (the modified work function $\phi'_m = 3.2$ eV for Al, and the modified electron affinity $\chi'_s = 3.5$ eV for Si);

(b) the threshold voltage, V_{TH}.

14.3. Using Eqs. (14.38), (14.39), (14.45), and (14.46), show that the general expression for I_D is given by Eq. (14.48).

14.4. Consider a p-channel Al-gate MOSFET with dimensions and physical parameters given by $x_0 = 1000$ Å, $L = 10\ \mu m$, $W = 4\ \mu m$, $N_d = 10^{15}\ cm^{-3}$, $Q_{ss} = 2 \times 10^{11}\ cm^{-2}$, and $\mu_p = 250\ cm^2/V \cdot sec$.

(a) Calculate I_D for $V_G = -2, -4, -6$, and -8 V and plot I_D versus V_D for the four different gate voltages.

(b) If $V_G - V_{TH} = 1$ V, calculate the oxide capacitance and cutoff frequency for this transistor.

(c) What are the transconductances of this MOSFET for $V_G = -2, -4$, and -6 V?

14.5. Calculate the drain saturation current $I_{D\,sat}$ and drain saturation voltage for a n-channel MOSFET with $x_0 = 1000$ Å, $W/L = 15$, $\mu_n = 1100\ cm^2/V \cdot sec$ $V_{TH} = 0.5$ V, and for $V_G = 3$ V and 5 V.

14.6. Consider a MOSFET built on a p-type silicon substrate with $N_A = 1 \times 10^{15}\ \text{cm}^{-3}$ with an aluminum metal gate with $\phi_{ms} = -0.27$ V and an oxide thickness of 1000 Å. Calculate:
(a) Ψ_{si} and Q_B at the onset of strong inversion,
(b) threshold voltages V_{TH} and V'_{TH} using Eqs. (14.37) and (14.39),
(c) V''_{TH} using Eq. (14.40) for $V_{sb} = -1$ V and -3 V.

14.7. The oxide of a p-channel MOSFET contains mobile sodium ions which can move slowly toward the Si–SiO_2 interface under the influence of an applied electric field. Discuss the effect of these mobile sodium ions on the characteristics of such a MOSFET if a positive gate voltage is applied to the device.

14.8. The mobile charge in the channel per unit area, $Q_I(y)$, in a n-channel MOSFET is given by Eq. (14.45), and the potential $V(y)$ in the channel is given by Eq. (14.47) [with V_D replaced by $V(y)$]. Show that the small-signal capacitance measured between the gate and source in the saturation region is given by

$$C_{GS} = \frac{dQ_I}{dV_g} = \tfrac{2}{3} WLC_{0x}$$

where Q_I is the total charge in the channel and $V(y) = V_G - V_{TH}$ for $y = L$.

14.9. The potential distribution for the BCCD shown in Fig. 14.21 can be obtained analytically by using the depletion approximation for the case where the impurity concentrations are constant in the n- and p-regions of the BCCD. Poisson's equations for the potential are given by:

$$\frac{d^2\Psi}{dx^2} = 0 \qquad -d_{0x} < x < 0$$

$$\frac{d^2\Psi}{dx^2} = \frac{-qN_d}{\varepsilon_0 \varepsilon_s} \qquad 0 < x < W_n$$

$$\frac{d^2\Psi}{dx^2} = \frac{qN_A}{\varepsilon_0 \varepsilon_s} \qquad W_n < x < W_n + W_p$$

The boundary conditions are given by;
(1) $\Psi = (V_G - V_{FB})$ at $x = -d_{0x}$,
(2) $\Psi = 0$ at $x = W_n + W_p$,
(3) the potential and electric displacement are continuous at $x = 0$ and $x = W_n$.
Show that the maximum potential well displayed in Fig. 14.21b is given by

$$\Psi_{max} = \left(\frac{qN_A}{2\varepsilon_s \varepsilon_0}\right) x_p^2 \left(1 + \frac{N_A}{N_D}\right)$$

REFERENCES

1. C. G. Garrett and W. H. Brattain, "Physical Theory of Semiconductor Surfaces." *Phys. Rev.* **99**, 376 (1955).
2. D. K. Schroder, *Semiconductor Material and Device Characterization*, Chap. 6, Wiley Interscience, New York (1990).
3. M. H. White and J. R. Cricci, "Characterization of Thin-Oxide MNOS Memory Transistors," *IEEE Trans. Electron Devices* **ED-19**, 1280 (1972).
4. D. F. Barbe, "Imaging Devices Using the Charge-Coupled Concept," *Proc. IEEE* **63**, 38 (1975).
5. J. Brews, "Physics of the MOS Transistor," in *Silicon Integrated Circuits*, Part A (D. Kahng, ed.), Academic Press, New York (1981).

BIBLIOGRAPHY

G. F. Amelio, W. J. Bertram, Jr., and M. F. Tompsett, "Charge-coupled Image Devices: Design Considerations," *IEEE Trans. Electron Devices* **ED-18**, 986 (1971).

A. Bar-Lev, *Semiconductors and Electronic Devices*, 2nd ed., Prentice-Hall, Englewood Cliffs (1984).

W. J. Betram, M. Mohsen, F. J. Morris, D. A. Sealer, C. H. Sequin, and M. F. Tompsett, "A Three-Level Metallization Three-Phase CCD," *IEEE Trans. Electron Devices* **ED-21**, 758 (1974).

J. Y. Chen, "CMOS- The Emerging VLSI Technology," *IEEE Circuits & Device Magazine* **2**, 16 (1986).

P. E. Gray and C. L. Searle, *Electronic Principles: Physics, Models and Circuits,* Wiley, New York (1969),

A. S. Grove, *Physics and Technology of Semiconductor Devices*, Wiley, New York (1967).

C. K. Kim and M. Lenzlinger, "Charge Transfer in Charge Coupled Devices," *J. Appl. Phys.*, **42**, 3856 (1971).

R. S. Muller and T. I. Kamins, *Device Electronics for Integrated Circuits*, 2nd ed., Wiley, New York (1986).

L. C. Parrilo, "VLSI Process Integration," in: *VLSI Technology* (S. M. Sze, ed.), McGraw-Hill, New York (1983).

C. H. Sequin and M. F. Tomsett, *Charge Transfer Devices*, Chap. 5, Academic Press, New York (1975).

S. M. Sze, *Physics of Semiconductor Devices*, 2nd ed., Wiley, New York (1981).

M. F. Tomsett, "A Simple Regenerator for Use with Charge-Transfer Devices and the Design of Functional Logic-Arrays," *IEEE J. Solid-State Circuits* **SC-7**, 237 (1972).

R. R. Troutman, *Latchup in CMOS Technology*, Kluwer, Boston (1986).

Y. Tsividis, *Operation and Modeling of the MOS Transistor*, McGraw-Hill, New York (1987).

E. S. Yang, Microelectronic Devices, McGraw-Hill, New York (1988).

15

High-Speed III–V Semiconductor Devices

15.1. INTRODUCTION

In this chapter we present a new class of high-speed III–V compound semiconductor devices, and discuss the basic device physics, operation principles, and general characteristics of these devices. The devices to be discussed here include the GaAs-based metal–semiconductor field-effect transistors (MESFETs), modulation doped field-effect transistors (MODFETs), heterojunction bipolar transistors (HBTs), hot-electron transistors (HETs), resonant tunneling devices (RTDs), and Gunn effect devices. In addition to GaAs based devices, the InP-based MODFETs and HBTs will also be discussed. The GaAs-based high-speed devices are usually fabricated by using the GaAs/AlGaAs material system grown on a semiinsulating GaAs substrate, while the InP-based devices use the InAlAs/InGaAs or InGaAs/InP material grown on a semiinsulating InP substrate. Although the GaAs/AlGaAs material technology is more mature than that of the InP/InGaAs material system, the InP-based devices can be operated at a higher frequency and speed than that of the GaAs-based devices. This is so because the InGaAs/InP material system has a higher electron mobility and smaller electron effective mass than that of the AlGaAs/GaAs material system.

In silicon IC technologies, in order to increase the operating frequency and speed of a silicon BJT or MOSFET, it is common practice to use approaches such as scaling down the device dimensions and changing the dopant densities and device geometry to achieve these goals. However, increasing device speed by scaling down the device dimensions has its physical limitations. For example, the parasitic capacitance and resistance of a device do not scale down linearly with the device dimensions. Interconnects can also pose another problem for scaled-down devices. These problems are particularly severe in the very large scale integrated circuits (VLSIs). In addition, there are other limitations on scaling down dimensions due to lithographic (optical or E-beam) techniques. In view of these limitations, the GaAs and other III–V compound semiconductors such as InP and InGaAs have been developed for high-speed device applications. The III–V materials generally have higher electron mobilities and higher peak saturation velocities than silicon, as well as the availability of semiinsulating substrates. With the advent of both MBE and MOCVD growth techniques, high quality AlGaAs/GaAs or InGaAs/InAlAs heterostructure epitaxial layers and quantum-well/superlattice structures can be readily grown on either the semiinsulating GaAs or InP substrates for fabricating the high-speed devices. FETs fabricated on the semiinsulating substrates can greatly reduce the leakage current and parasitic capacitances, and hence enabling integration of both lumped and distributed microwave components on the same substrate.

The successful development of GaAs monolithic microwave integrated circuits (MMICs) is a good example of using GaAs in high-speed digital integrated-circuits applications. In fact, impressive results have been obtained in both the GaAs- and InP-based FETs and HBTs, this is due to the availability of high-quality epitaxial layers prepared by using MBE or MOCVD growth techniques as well as new device processing techniques. For example, high current (600 mA/mm) and high cutoff frequency ($f_T > 80$ GHz) MODFETs using delta doped-heterostructures have been fabricated in AlGaAs/GaAs material systems. Thin channel and highly doped (2×10^{18} cm^{-3}) refractory gate self-aligned GaAs MESFETs fabricated by rapid thermal annealing (RTA) have produced transconductance values exceeding 550 mS/mm. A self-aligned AlGaAs/GaAs HBT with an InGaAs emitter cap layer has yielded a transconductance of 16 mS/μm^2 and a cut-off frequency (f_T) of 80 GHz. Propagation delay of 5.5 psec/gate for a 21-stage ring oscillator has also been demonstrated using this structure. InGaAs/InP MODFETs with f_{max} greater than 100 GHz and Si/GeSi HBTs formed by using the chemical vapor deposition (CVD) technique with f_T equal to 80 GHz have been recently demonstrated.

Section 15.2 describes device structure, *I–V* characteristics, small-signal device parameters, and some second-order effects in a GaAs MESFET. Section 15.3 presents the equilibrium properties of the two-dimensional electron gas (2-DEG) in the triangular potential well of an AlGaAs/GaAs heterostructure. The basic device theory, operation principle, and *I–V* characteristics of a MODFET are also depicted. In Section 15.4, we present the device physics and structure, dc characteristics, and small-signal device parameters of a HBT. The hot-electron transistor (HET) and its physical limitations are discussed in Section 15.5. In Section 15.6, we describe the device physics, general characteristics, and performance limitations of a resonant tunneling device (RTD). Finally, the basic physical principles and general properties of a GaAs Gunn effect device are discussed in Section 15.7.

15.2. METAL–SEMICONDUCTOR FIELD-EFFECT TRANSISTORS

The maturity of the GaAs metal–semiconductor field-effect transistor (MESFET) technology has created a major impact on the advancement of solid state microwave technologies. The GaAs MESFET has made great strides in both microwave amplifier and digital applications for commercial and military systems. GaAs MESFET amplifiers operating up to 40 GHz are now commercially available. It is the most important microwave device in existence today with one of the highest unity gain cutoff frequencies (f_T) available in a transistor. For example, a GaAs MESFET with a 0.25 μm gate length exhibits a cutoff frequency of 80 GHz at 300 K. In fact, GaAs MESFETs, which form the basis of microwave monolithic integrated circuit (MMIC) technologies, are now emerging from research to a pilot line production. 16 K static random access memory (SRAM) and ring oscillators have been built for different digital circuit applications. The GaAs MESFET applications include both the discrete and integrated circuit technologies. In this section, the device structures, *I–V* characteristics, and small-signal device parameters, as well as some second-order effects of a GaAs MESFET are depicted.

15.2.1. Basic Device Structure and Characteristics

A GaAs MESFET is a rather simple field-effect transistor (FET). It is a three-terminal majority carrier device which consists of two ohmic contacts (i.e., source and drain) separated by a Schottky barrier gate contact electrode. Figure 15.1a and b show the cross-section views of two different GaAs MESFET device structures formed (a) by epitaxial layer growth and (b) by the ion-implantation technique. The basic structure for a GaAs MESFET consists of a n-type GaAs active layer of 0.2 to 0.3 μm with a dopant density of 2×10^{17} cm^{-3} deposited on a high resistivity buffer layer which is grown on top of an undoped- or Cr-doped semi-insulating (i.e., $\rho \geq 10^7 \Omega \cdot$ cm), (100) oriented GaAs substrate. An undoped high-resistivity

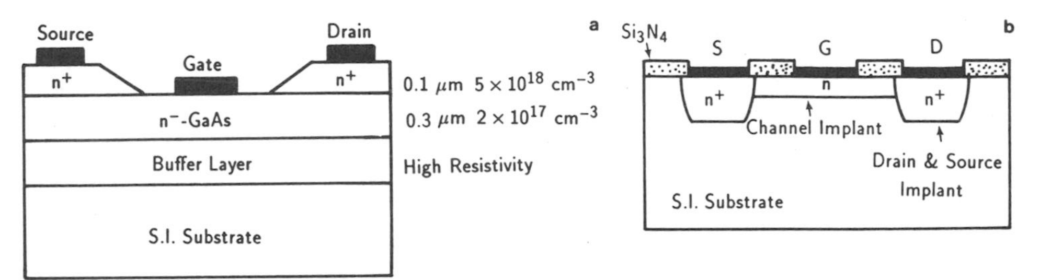

FIGURE 15.1. Cross-section views of a GaAs MESFET: (a) with a high-resistivity buffer layer, a n^- channel epilayer and a n^+ cap layer; (b) with a n^+ source, drain, and n^- channel regions created by direct implantation on a semi-insulating GaAs substrate.

buffer layer of 3 to 5 μm thick is grown between the active layer and substrate to prevent out-diffusion of the residual impurities from the substrate into the active layer. Two ohmic contact regions, separated by 5 μm or less, form the source and drain electrodes, and a Schottky barrier contact between the source and drain regions (i.e., the channel region) forms the control-gate electrode. To reduce the source and drain contact resistance, a heavily doped n^+ GaAs contact layer typically 0.1 μm thick and doping density of $2 \times 10^{18}\,\text{cm}^{-3}$ is grown on top of the active layer to reduce parasitic resistances. However, as shown in Fig. 15.1b, most GaAs MESFETs available today are being fabricated by direct-ion implantation onto a semi-insulating GaAs substrate. Selective implantation enables an active channel region of dopant density about $10^{17}\,\text{cm}^{-3}$ to be realized. Heavily doped source and drain regions can be implanted immediately adjacent to the channel in order to minimize source resistance by using the self-aligned process. In this case, implantation is restricted to those areas necessary for MESFET fabrication, leaving the rest of the wafer in its semi-insulating condition.

In general, a GaAs MESFET differs from a silicon JFET in several respects. The channel region consists of a very thin ($\sim$0.2 μm) n-type GaAs layer grown on a Cr-doped or undoped semi-insulating GaAs substrate. For a n-channel GaAs MESFET, the controlled gate is formed by a Schottky barrier metal contact (e.g., Ti/Pt/Au, or WSi_x for a self-aligned process) with gate lengths varying from a few microns to one tenth of a micron. At a channel doping density of around $2 \times 10^{17}\,\text{cm}^{-3}$, the very thin channel under the gate is only partially depleted by the Schottky barrier built-in voltage ($\sim$0.8 V). When a positive drain voltage V_{DS} is applied to the FET, the drain current I_{DS} will flow through the channel producing current–voltage characteristics similar to those of a JFET. A negative gate voltage (V_{GS}) will further deplete the channel, reducing I_{DS} until the device cuts off when V_{GS} reaches the pinch-off voltage. On the other hand, a small positive gate voltage V_{GS} ($<$0.8 V), at which the Schottky gate starts forward conduction, will cause I_{DS} to increase dramatically. The carrier transit time τ through a channel of length L is approximately equal to L/v_s, where v_s is the saturation velocity of electrons in the channel. For a GaAs MESFET with $v_s = 2 \times 10^7$ cm/sec and a channel length of 0.5 μm, the device unity current gain cutoff frequency f_T is about 70 GHz.

The GaAs MESFET structure shown in Fig. 15.1a has been widely used as a low-noise, small-signal microwave amplifier. In this structure, the active layer is etched briefly before the Schottky gate contact is made. The recessed gate structure along with a n^+ ion implantation into the source and drain regions will reduce the source and drain parasitic series resistances and greatly improve the ohmic contacts in both regions. Reducing the parasitic resistances will also decrease the noise figure and increase the transconductance and cutoff frequency of the MESFET. For power amplification at microwave frequencies, high current is needed which, in turn, requires a very wide gate. Therefore, a high-power MESFET usually consists of many small MESFETs connected in parallel to increase the total current and power output.

Although overlays are frequently used to reduce contact resistances and thicken up the bonding pads, the metallizations required to complete the basic device structure of a GaAs

MESFET are the source and drain ohmic contacts and the Schottky gate contact. The gate length for a microwave FET usually varies between 0.25 and 1.5 μm. Such small dimensions can be controlled by a lift-off process using either the optical or electron beam lithographic technique. Typical Schottky barrier gate metallizations include using Al or a multilayer structure such as Ti/Pt/Au or WSi_4 contact. Au/Ge/Ni is commonly used for ohmic contacts on n-type GaAs, while Au/Zn is widely used for ohmic contacts on p-type GaAs.

Depending on the channel doping density and channel height, there are two types of MESFETs which are commonly used in GaAs digital IC applications. These are the normally-on or depletion mode (D-) MESFET and the normally-off or enhancement mode (E-) MESFET. The former requires a negative gate-source voltage (i.e., V_{GS}) to cutoff the drain current under all conditions, while the later requires a positive gate voltage to open up the channel. By controlling the channel doping density N_D and channel height a, both E/D MESFETs can be fabricated on a semi-insulating GaAs substrate. The solution of Poisson's equation in the channel depletion region under the Schottky gate can be employed to show that a normally-on MESFET can be obtained if

$$qN_D a^2/2\varepsilon_0\varepsilon_s > V_{bi} \tag{15.1}$$

and a normally-off MESFET is obtained if

$$qN_D a^2/2\varepsilon_0\varepsilon_s < V_{bi} \tag{15.2}$$

where V_{bi} is the built-in potential of the MESFET and other parameters have their usual meanings. It is seen that Eqs. (15.1) and (15.2) are the basic equations which may be used to guide the design of E/D MESFETs on a GaAs substrate.

We next discuss the velocity-field relation and the general characteristics for a GaAs MESFET. Figure 15.2a and b present the velocity-field relation for undoped GaAs at 300 K and 77 K, respectively, and shows a negative differential mobility regime under steady-state conditions and at moderate field strength. This velocity-field relation may be represented by using a two-piecewise linear approximation as shown in curve 2 or a more accurate relation as shown in curve 3 of Fig. 15.2a. The two-piecewise linear approximation is often used in the analysis of MESFET characteristics due to its simplicity. This is understandable by noting that at low fields the electron drift velocity v_d is linearly related to the electric field with a proportionality constant equal to the low-field mobility μ which satisfies

$$v_d = \mu\mathscr{E} \qquad \text{for } \mathscr{E} < \mathscr{E}_c \tag{15.3}$$

and in the saturation regime

$$v_d = v_s \qquad \text{for } \mathscr{E} \geq \mathscr{E}_c \tag{15.4}$$

where $\mathscr{E}_c$ is the critical field which divides the linear and saturation regions of drift velocity versus the electric field curve. A more precise equation describing the velocity-field relation shown in curve 3 of Fig. 15.2a is given by

$$v_d = \frac{\mu\mathscr{E}}{[1 + (\mu\mathscr{E}/v_s)^2]^{1/2}} \tag{15.5}$$

Since the transit time of electrons in the high-field regime is comparable to the time constant characterizing relaxation to steady-state velocity-field characteristics, the electron dynamics in a microwave FET are much more complex than the dynamics derived from steady-state velocity-field characteristics. Figure 15.3a shows the device structure of a GaAs

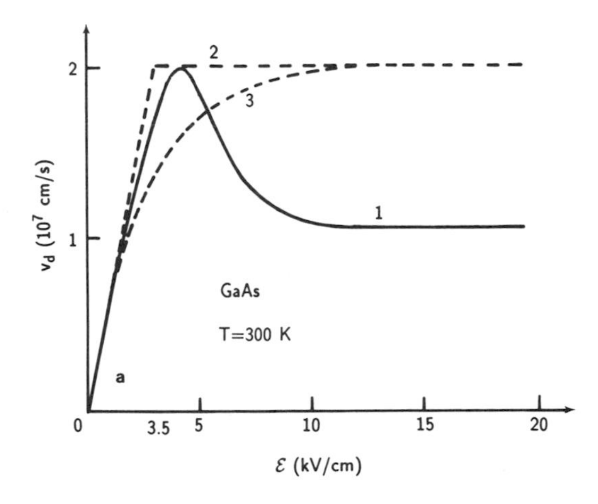

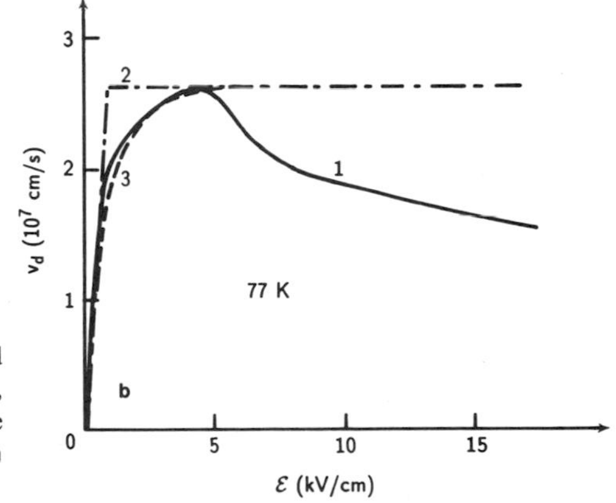

FIGURE 15.2. Drift velocity versus electric field (curve 1) in bulk GaAs at (a) 300 K and (b) 77 K, respectively. Curve 2 is calculated from a two-piece linear approximation. Curve 3 is calculated from Eq. (15.5).

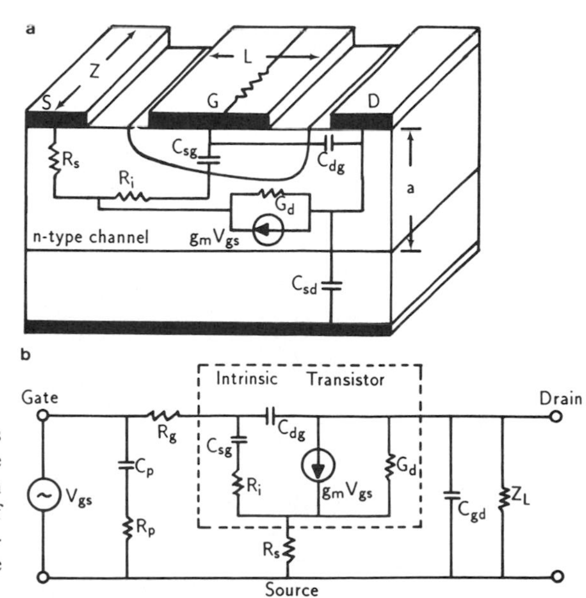

FIGURE 15.3. (a) Sketch of a GaAs MESFET device structure along with the origins of each circuit element shown in (b). (b) Small-signal equivalent circuit of the GaAs MESFET operating in the saturation region in a common-source configuration.

MESFET along with the origins of each circuit element and Fig. 15.3b presents its small-signal equivalent circuit. The dc device parameters are derived by examining the GaAs MESFET structure shown in Fig. 15.3a, which has a n-type active layer (channel) of thickness a and doping density N_D, gate width Z, and gate length L. The saturation current I_{Dsat} which can be carried by the channel is given by

$$I_{sat} = qN_D Z a v_s \tag{15.6}$$

where q is the electronic charge and v_s is the saturation velocity. If the metal gate forms a Schottky contact to this active layer and $V_{DS} = 0$, then by using a one-sided abrupt junction approximation the depletion layer width W_d of the Schottky gate to the active channel is given by

$$W_d = \sqrt{\frac{2\varepsilon_0 \varepsilon_s(-V_{GS} + V_{bi})}{qN_D}} \tag{15.7}$$

where V_{bi} is the built-in potential of the Schottky barrier contact gate and V_{GS} is the gate to source voltage. Thus, the channel height $b(x)$ $[=a - W_d(x)]$ can be controlled by the gate voltage V_{GS}. When the channel is pinched off, the depletion layer depth extends from the surface ($y = 0$) completely through the n-type active layer ($y = a$), and V_{GS} equals the pinch-off (threshold) voltage V_{p0}. Equations (15.1) and (15.2) yield the pinch-off voltage required to deplete the channel at the drain side:

$$V_{p0} = \frac{qN_D a^2}{2\varepsilon_0 \varepsilon_s} \tag{15.8}$$

It is seen that the relative importance of the roles of velocity saturation and pinch-off can be measured by using the saturation index ($\mathscr{E}_s L/V_{p0}$), which is equal to the ratio of potential drop along the gate region at the saturation field to the pinch-off potential V_{p0} required to totally deplete the channel. The smaller the index, the greater the importance of velocity saturation in limiting the source–drain current. We shall next derive the dc current–voltage relation and the small-signal device parameters for a GaAs MESFET.

15.2.2. Current–Voltage Characteristics

A MESFET operation can be described analytically if some simplified assumptions concerning the electric field distribution within the device, the carrier dynamics, and the channel profile are made. For example, if the drain voltage V_{DS} is below the knee voltage of the I_{DS} versus V_{DS} characteristic curves, then electrons travel with constant mobility in the channel. In this case, the current–voltage characteristics can be derived by using the gradual channel approximation as in the case of a JFET (see Chapter 11). As shown in Fig. 15.4a, the operation of a MESFET above the knee voltage may be considered by dividing the channel into three regions. In region I, carriers travel with constant mobility between the source end of the gate and the velocity saturation point $x = x_1$, where the electric field rises to the value of the saturation field $\mathscr{E}_s$. In regions II and III, the carriers travel at their saturation velocity. These two regions meet at $x = L$, i.e., the drain end of the gate. Region III terminates when the electric field in the channel falls to the saturation field. The longitudinal electric field $\mathscr{E}_x$ as a function of position in the three regions of the channel is shown in Fig. 15.4b. The electric field in region I initially increases as the carriers enter the channel beneath the gate. The electric field continues to increase until it reaches the saturation field at the boundary between regions I and II. At the knee voltage, the velocity saturation point is located at the drain end of the gate. At higher drain–source potentials, the velocity saturation point moves toward the source end of the gate, and the electric field in regions II and III increases to accommodate

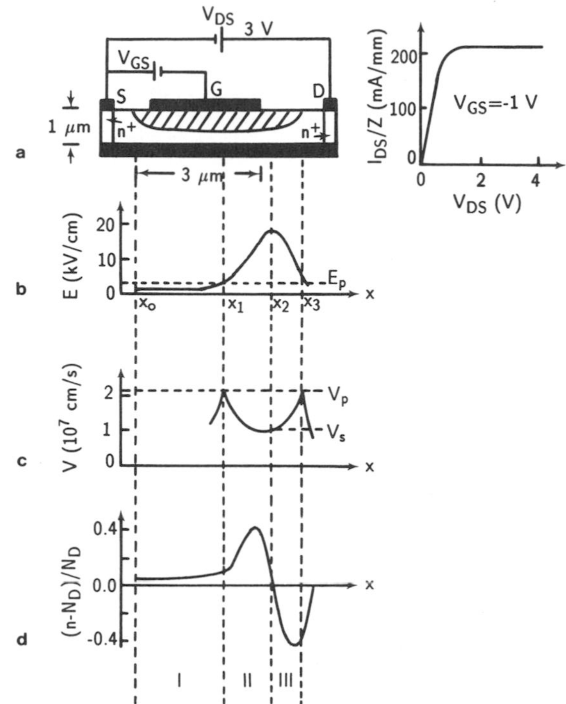

FIGURE 15.4. Distribution of electric field, drift velocity, and space charge in the channel of a GaAs MESFET. (a) Cross-section view showing the three regions of the channel: (I) the depletion region, (II) the electron accumulation region, and (III) formation of the Gunn domain. (b) Electric field versus distance in the channel. (c) Drift velocity versus distance in the channel. (d) Normalized space-charge density versus distance in the channel. Also shown in (a) is the normalized I_{DS}/Z versus V_{DS} plot. After Liechti,[1] by permission, © IEEE–1976.

the drain–source potential. In region II, carriers travel at their saturation velocity v_s. To satisfy Gauss's law and current continuity in the channel, the increase in $\mathscr{E}_x$ must be accompanied by a reduction in channel height and by carrier accumulation beyond the charge-neutral value of region I. Therefore, the electric field in the channel should reach its peak value near the drain end of the gate. In region III, the carrier density falls as the channel height increases beyond the drain end of the gate; and as the channel height rises above that at the velocity saturation point, the mobile carrier density drops below the charge-neutral value. This transition from carrier accumulation to carrier depletion is accomplished by a sharp drop in the electric field in the channel. Charge neutrality is restored when the electric field falls below the saturation field and electrons re-enter the constant mobility regime. The strong saturation in the drain–source current above the knee voltage is a direct result of the fact that most of the V_{DS} voltage drop is across the accumulation–depletion regions of the channel. Increasing V_{DS} will emphasize these regions and also increases the length of the velocity saturation region in the channel. Figure 15.4c shows the drift velocity versus the electric field in the three regions, where v_p is the peak velocity. Figure 15.4d presents the space-charge distribution in the channel.

As mentioned above, the current–voltage characteristics of a MESFET below the knee voltage, at which the electric field is low and the mobility is constant, can be analyzed by using the gradual channel approximation. Figure 15.5 shows the symmetrical structure for the gradual channel analysis of a GaAs MESFET. It is shown in this figure that the channel current I_{ch} per half device is related to the channel potential $V(x)$ by

$$I_{ch} = G_0 L\left(1 - \frac{W_d(x)}{a}\right)\frac{dV(x)}{dx} \tag{15.9}$$

where G_0 is the conductance of a fully opened channel and can be expressed by

$$G_0 = \frac{qN_D\mu Za}{L} \tag{15.10}$$

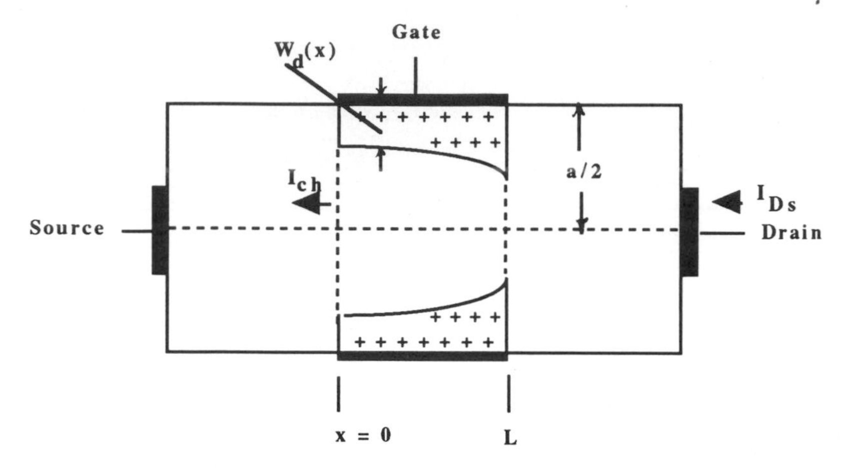

FIGURE 15.5. The symmetrical structure used in the gradual channel analysis of a GaAs MESFET.

where Z, a, and L are the device dimensions defined in Fig. 15.3a and Fig. 15.5; $W_d(x)$ is the depletion layer width in the channel under the gate, while μ and N_D denote the low-field mobility and dopant density, respectively. The general expression for the depletion layer width $W_d(x)$ as a function of position in the channel can be obtained by including $V(x)$ in Eq. (15.7), and the result yields

$$W_d(x) = \sqrt{\frac{2\varepsilon_0\varepsilon_s(V(x) - V_{GS} + V_{bi})}{qN_D}} \tag{15.11}$$

If the source and drain series resistances are neglected in the analysis, then the channel current can be obtained by substituting Eq. (15.11) into (15.9), and integrating x from 0 to L and $V(x)$ from 0 to V_i. Hence

$$I_{ch} = G_0\left\{V_i - \frac{2}{3}\frac{[(V_i - V_{GS} + V_{bi})^{3/2} + (-V_{GS} + V_{bi})^{3/2}]}{V_{p0}^{1/2}}\right\} \tag{15.12}$$

where I_{ch} is the channel current, V_i is the voltage drop across the gate region, V_{bi} is the built-in voltage, V_{GS} is the gate voltage, G_0 is defined by Eq. (15.10), and V_{p0} is the pinch-off voltage given by Eq. (15.8). We note that Eq. (15.12) is valid when $V_i \ll V_s = \mathscr{E}_s L$, where $\mathscr{E}_s = v_s/\mu$ is the average field under the gate to reach the sustaining domain in the drain region. For a typical GaAs MESFET with $L \sim 1\ \mu$m and $V_s \ll (V_{bi} - V_{GS})$, the channel current varies almost linearly with channel voltage up to the saturation point and is given by

$$I_{ch} \simeq G_0\left[1 - \left(\frac{V_{bi} - V_{GS}}{V_{p0}}\right)^{1/2}\right]V_{ch} = G_d V_{ch} \tag{15.13}$$

where

$$G_d \simeq G_0\left[1 - \left(\frac{V_{bi} - V_{GS}}{V_{p0}}\right)^{1/2}\right] \tag{15.14}$$

is the drain conductance. From Eq. (15.13), the channel saturation current I_{sat} can be written as

$$I_{sat} = G_d V_s \tag{15.15}$$

For a GaAs MESFET with low pinch-off voltage (e.g., $V_{p0} \leq 2V$), the current–voltage relation operating in the saturation region can be described accurately by using the "square law" model, which is given by

$$I_{Dsat} = K(V_{GS} - V_T)^2 \tag{15.16}$$

where I_{Dsat} denotes the drain–source saturation current, V_T is the threshold voltage, and

$$K = \frac{2\varepsilon_0\varepsilon_s\mu v_s Z}{a(\mu V_{p0} + 3v_s L)} \tag{15.17}$$

The source series resistance effects are neglected in Eq. (15.16).

Figure 15.6 shows plots of I_{DS} versus V_{DS} for (a) low pinch-off voltage GaAs MESFET ($V_{p0} = 1.8$ V, $N_D = 1.81 \times 10^{17}$ cm^{-3}, $L = 1.3$ μm, and $Z = 20$ μm), and (b) high pinch-off voltage GaAs MESFET ($V_{p0} = 5.3$ V, $N_D = 6.5 \times 10^{16}$ cm^{-3}, $L = 1.0$ μm, and $Z = 500$ μm).

15.2.3. Small-Signal Device Parameters

The small-signal ac device parameters for the MESFET will be derived in this section. The general expression for the transconductance g_m in the saturation region can be derived from Eq. (15.12) by taking the derivative of the channel current with respect to the gate voltage at $V_i = V_s$, and the result yields

$$g_m = \left.\frac{\partial I_{ch}}{\partial V_{GS}}\right|_{Vi = Vs} = G_0 \frac{(V_s + V_{bi} - V_{GS})^{1/2} - (V_{bi} - V_{GS})^{1/2}}{V_{p0}^{1/2}} \tag{15.18}$$

If $V_s \ll (V_{bi} - V_{GS})$, then Eq. (15.18) is reduced to

$$g_m \simeq \frac{G_0 V_s}{2[V_{p0}(V_{bi} - V_{GS})]^{1/2}} = v_s Z \left[\frac{qN_D\varepsilon_0\varepsilon_s}{2(V_{bi} - V_{GS})}\right]^{1/2} \tag{15.19}$$

It is of interest that Eq. (15.19) is completely different from that predicted by Shockley theory according to which the transconductance in the saturation region should equal the drain conductance in the linear region as given by Eq. (15.14). However, it has been shown that the theoretical prediction given by Eq. (15.19) agrees well with experimental data for a GaAs MESFET. If the effects due to the gate-to-source series resistance R_s and the source contact

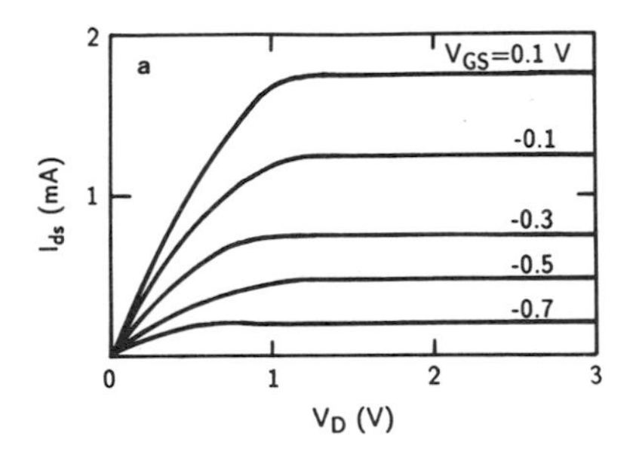

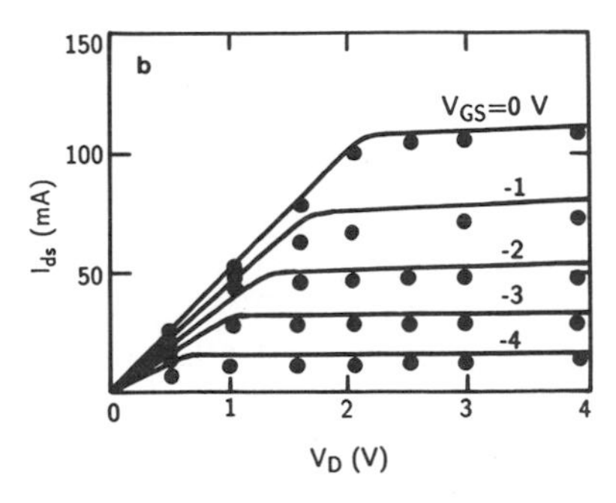

FIGURE 15.6. I_{DS} versus V_{DS} for (a) a low pinch-off voltage GaAs MESFET ($V_{p0} = 1.8$ V, $N_D = 1.81 \times 10^{17}$ cm^{-3}, $L = 1.3$ μm, and $Z = 20$ μm), and (b) a high pinch-off voltage GaAs MESFET ($V_{p0} = 5.3$ V, $N_D = 6.5 \times 10^{16}$ cm^{-3}, $L = 1.0$ μm, and $Z = 500$ μm). After Pucel *et al.*,[(2)] by permission.

resistance R_{sc} are taken into account, then the intrinsic transconductance given by Eq. (15.19) should be modified to

$$g_m' = \frac{g_m}{1 + (R_s + R_{sc})g_m} \tag{15.20}$$

which is smaller than the value of g_m predicted by Eq. (15.19).

To derive the unity current gain cutoff frequency for a MESFET, we should first derive expressions for the total charge under the gate and the gate-source capacitance, using the depletion layer width under the gate given by Eq. (15.11). If we assume that $V(x)$ is a linear function of position in the channel under the gate [i.e., $V(x) = V_i(x/L)$], then Eq. (15.11) can be rewritten as

$$W_d(x) = \frac{a(V_i x/L + V_{bi} - V_{GS})^{1/2}}{2V_{p0}^{1/2}} \tag{15.21}$$

Now using Eq. (15.21), the total charge under the gate in the linear region when $V_i \leq V_s$ can be obtained with the aid of Eq. (15.21), in which case

$$Q_d = qN_D Z \int_0^L W_d(x)\, dx = \frac{2ZL(2\varepsilon_0\varepsilon_s N_D)^{1/2}}{3V_{ch}}[(V_{ch} + V_{bi} - V_{GS})^{3/2} - (V_{bi} - V_{GS})^{3/2}] \tag{15.22}$$

For $V_i \ll (V_{bi} - V_{GS})$, the total charge under the gate can be approximated to

$$Q_d \simeq (qN_D ZLa/2)\left(\frac{V_{bi} - V_{GS}}{V_{p0}}\right)^{1/2} \tag{15.23}$$

Equation (15.22) allows one to derive expressions for the gate-to-source and drain-to-gate capacitances of a GaAs MESFET shown in Fig. 15.3a, and the results yield

$$\begin{aligned} C_{gs} &= \left|\frac{\partial Q_d}{\partial V_{GS}}\right|_{(V_{ch} - V_{GS} = \text{const.})} \\ &= \frac{2ZL(2\varepsilon_0\varepsilon_s N_D)^{1/2}}{3V_{ch}^2}\left[(V_{ch} + V_{bi} - V_{GS})^{3/2} - (V_{ch} + V_{bi} - V_{GS})^{3/2}\right. \\ &\qquad \left. - \frac{3}{2}(V_{bi} - V_{GS})^{1/2}V_{ch}\right] \end{aligned} \tag{15.24}$$

Similarly,

$$\begin{aligned} C_{dg} &= \left|\frac{\partial Q_d}{\partial V_{ch}}\right|_{(V_{GS} = \text{const.})} \\ &= \frac{2ZL(2\varepsilon_0\varepsilon_s N_D)^{1/2}}{3V_{ch}^2}\left[\frac{3}{2}V_{ch}(V_{ch} + V_{bi} - V_{GS})^{3/2} - (V_{bi} - V_{GS})^{3/2}\right. \\ &\qquad \left. - (V_{bi} - V_{GS})^{3/2}V_{ch}\right] \end{aligned} \tag{15.25}$$

If $V_i \ll (V_{bi} - V_{GS})$, then Eqs. (15.24) and (15.25) can be simplified to

$$C_{gs} = C_{dg} = \frac{ZL}{2}\sqrt{\frac{\varepsilon_0 \varepsilon_s q N_D}{2(V_{bi} - V_{GS})}} \tag{15.26}$$

The current gain β in the common-source configuration is given by

$$\beta = \frac{i_{DS}}{i_{GS}} = \frac{g_m}{\omega C_{gs}} \tag{15.27}$$

where i_{DS} and i_{GS} denote the small-signal ac drain-to-source current and gate-to-source current, respectively. The unity current gain cutoff frequency is obtained from Eq. (15.27) by setting β equal to one. Hence

$$f_T \simeq \frac{g_m}{2\pi C_{gs}} \tag{15.28}$$

For $V_i \ll (V_{bi} - V_{GS})$, Eq. (15.28) reduces to

$$f_T \simeq \frac{v_s}{\pi L} = \frac{1}{\pi \tau} \tag{15.29}$$

where v_s is the saturation velocity of electrons, and $\tau = L/v_s$ is the transit time for electrons to travel the length of the gate at saturation velocity. For a GaAs MESFET with 1 μm gate length, $f_T \sim 25.5$ GHz, which is in good agreement with experimental data.

Equation (15.29) shows that in order to maintain high-frequency operation in a GaAs MESFET, it is imperative that the transit time be kept as short as possible. In practice, the unity current gain cutoff frequency is usually lower than that predicted by Eq. (15.29) due to the gate fringing capacitance, interelectrode capacitances, and other parasitic effects. Figure 15.7 shows the calculated f_T as a function of gate length for silicon, GaAs, and InP FETs. We note that the InP FET has a higher f_T than that of a GaAs FET because of its higher peak velocity. For GaAs MESFETs with gate lengths less than 0.5 μm, one can expect f_T to be greater than 30 GHz.

Experimental results show that the ratio g_m/C_{gs} depends on both V_{GS} and V_{DS} even above the knee voltage. In practice, velocity saturation is not attained at the source end of the gate, and the properties of the constant mobility region of the channel must be taken into account.

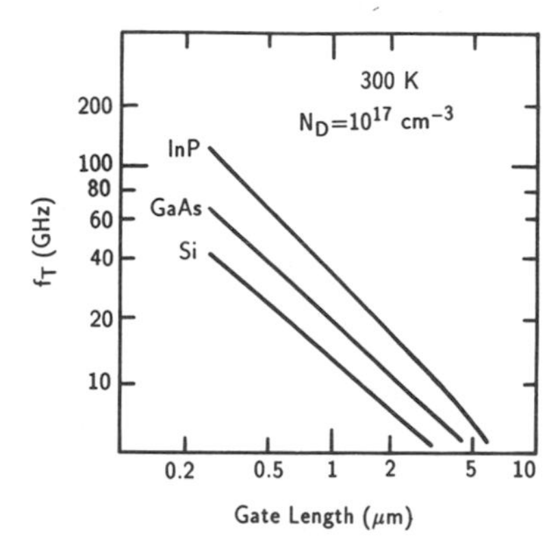

FIGURE 15.7. Calculated unit-gain cutoff frequency f_T as a function of gate length for silicon, GaAs, and InP FETs at 300 K. After Sze,[3] with permission by John Wiley & Sons Co.

This causes g_m to decrease more rapidly with V_{GS} than predicted by Eq. (15.16). The reduction of carrier density and carrier mobility near the active layer/substrate interface as well as the conduction in the substrate or buffer layer can also cause a reduction of g_m. The combined effect of these deviations from the simple model is to cause the current gain to diminish as the gate potential approaches pinch-off and the channel current approaches zero.

In order to maintain high transconductance, it is important to reduce the effective source resistance of the MESFET. Further degradation of g_m may be caused by the inductance of a bond wire between the source and the ground. The reduction of such an inductance is an important feature of GaAs power MESFETs where many small FET elements are connected in parallel.

The small-signal equivalent circuit of a MESFET operating in the saturation region in a common-source configuration is shown in Fig. 15.3b. It is seen that in an intrinsic FET, the total gate-to-channel capacitance is equal to the sum of C_{gd} and C_{gs}, and the input resistances R_i and R_{ds} under the gate show the effects of channel resistance. The extrinsic (or parasitic) elements include the source resistance R_s, drain resistance R_d, and substrate capacitance C_{sd}.

The gate current flowing through the Schottky barrier gate-to-channel junction of the MESFET is given by

$$I_G = I_s[e^{qV_G/nk_BT} - 1] \tag{15.30}$$

where $I_s = A^*T^2A\exp(-q\phi_{Bn}/k_BT)$ is the saturation current of the Schottky barrier contact at the gate; n is the diode ideality factor. The input resistance R_i for the MESFET can be expressed by

$$R_i = \left(\frac{\partial I_G}{\partial V_G}\right)^{-1} = \left(\frac{nk_BT}{q}\right)(I_G + I_s)^{-1} \tag{15.31}$$

As I_G approaches zero, the input resistance becomes very high (e.g., $R_i = 250\ \text{M}\Omega$ for $I_s = 10^{-10}$ A). The source and drain series resistances, which cannot be modulated by the gate voltage, will introduce a voltage drop between the gate-source and gate-drain electrodes. As a result, they will reduce the drain conductance as well as the transconductance in the linear region (i.e., the low-field constant-velocity regime) of operation. In the saturation region, however, the transconductance is only affected by the source resistance because, for $V_D > V_{DS}$, increasing V_D will have little or no effect on the drain current.

The characteristic switching time of a GaAs MESFET operating in the saturation region can be derived from Eqs. (15.14), (15.15), and (15.23), and the result is given by

$$\tau_s = \frac{Q_d(V_s)}{I_{sat}} \simeq \frac{L}{v_s}\left(\frac{V_{bi} - V_{GS}}{V_{p0} + V_{bi} - V_{GS}}\right)^{1/2} \tag{15.32}$$

Equation (15.32) shows that the switching time is proportional to the transit time under the gate and that the saturation velocity rather than the peak velocity determines the switching time. It is also shown that decreasing the gate length can reduce the switching time and increase the cutoff frequency of the MESFET. However, there are some physical limitations which can limit the switching speed of the MESFET, particularly the parasitic capacitances between the gate, drain, and source contacts. For a typical GaAs MESFET, the switching time in the picosecond range can be easily obtained.

Finally, we consider the power required by a MESFET at the saturation point. This is given by

$$P = (qN_Dv_sLZaE_s)\left[1 - \left(\frac{V_{bi} - V_{GS}}{V_{p0}}\right)^{1/2}\right] = \left(\frac{qN_DZL^2aE_s}{\tau}\right)\left(\frac{V_{bi} - V_{GS}}{V_{p0}}\right)^{1/2} \tag{15.33}$$

from which the power–delay product is equal to

$$P\tau = qN_D Z a L_2 E_s \left(\frac{V_{bi} - V_{GS}}{V_{p0}}\right)^{1/2} = ZL^2 E_s \sqrt{2\varepsilon_0 \varepsilon_s q N_D (V_{bi} - V_{GS})} \tag{15.34}$$

where $E_s = v_s/\mu$ is the average electric field under the gate which reaches domain sustaining field in the drain region. Using Eq. (15.34), the value of the power-delay product for a typical GaAs MESFET is estimated to be in the femtojoule range and is in good agreement with experimental data.

15.2.4. Second-Order Effects in a GaAs MESFET

It is noteworthy that the behavior of a practical GaAs MESFET does not always follow the theoretical predictions given in the previous section. There are several second-order effects observed in MESFETs which are attributed to the nonideality (e.g., high concentration of carbon acceptors and EL2 deep donor centers) of the semi-insulating GaAs substrate. The second-order effects include backgating (or sidegating), light sensitivity, low output resistance, low source–drain breakdown voltage, low output power gain at RF frequencies, drain current transient lag effects, temperature dependence, and the subthreshold current effect. Among these problems, backgating is the most significant for both digital and analog circuit applications.

The backgating (or sidegating) effect refers to the reduction of drain current in a MESFET as a result of the presence of other nearby neighboring MESFETs which happen to be negatively biased with respect to the source of the device under consideration. In response to changes in voltage on the substrate or adjacent devices, the substrate conducts enough current to modulate the interface space-charge region. When this interfacial depletion region widens into the active channel, the drain–source current I_{DS} is reduced.

The degree of backgating in a MESFET can vary significantly from substrate to substrate, making the prediction of backgating threshold unreliable. One approach, which has often been used to alleviate this problem, is the use of proton (or oxygen) implantation between MESFETs devices. The unannealed implantation produces a high concentration of defects which acts as electron traps at the surface down to a depth of 300–400 Å. The backgating threshold voltage is significantly increased through this process step, and the effect of backgating is reduced considerably. The proton bombardment is usually carried out after the ohmic contact alloying step.

The effect of backgating can also be reduced by growing a high-resistivity buffer layer on a semi-insulating substrate, as shown in Fig. 15.1a. A number of possible buffer layers which have been suggested for this purpose include undoped GaAs, AlGaAs, and superlattices (GaAs/GaAlAs). Recently, a new buffer layer grown by the MBE technique at a low substrate temperature (LT; $T = 150$ to 300 °C) using Ga and As_4 beam fluxes has been developed. It is highly resistive, optically inactive, and crystalline. High-quality GaAs active layers have been grown on top of this LT buffer layer. GaAs MESFETs fabricated in the active layer grown on top of such a LT GaAs buffer layer have shown total elimination of backgating and sidegating effects with a significant improvement in output resistance and breakdown voltages, while other characteristics of MESFET performance remain about the same as the other MESFETs reported in the literature, using alternative approaches.

The drain lag effect refers to the drain current to overshoot and recover slowly when a positive step voltage V_{DS} is applied to the drain electrode under saturation operation. The effect is attributable to the presence of deep-level defects (e.g., EL2 centers) in the semi-insulating substrate below the channel of the MESFET. In saturation, there is an accumulation of electrons beyond the velocity saturation point where the channel becomes very narrow and the electric field is very high at the drain end of the gate. Therefore, the drain current becomes very sensitive to the small variation in channel height. If a positive voltage step is applied to the drain electrode, the capacitance through the substrate between the drain electrode and the channel will cause a sudden widening of the channel, leading to an abrupt increase in drain current. Another manifestation of this effect can be observed in the frequency domain, which shows a considerable increase of small-signal output conductance g_{ds} with frequency in the saturation region for frequencies between 100 Hz and 1 MHz at room temperature. The effect can be explained in terms of trapping and capture mechanisms taking place at the channel/substrate interface. At high frequencies, the traps are too slow to capture and release electrons during one cycle of the ac signal, and hence they do not counteract the effect of drain capacitance on the channel–substrate interface, and thereby the drain conductance is large in the saturation region. On the other hand, the traps can follow the ac signal at low frequencies and effectively shield the channel from the drain capacitance through the substrate, and thus the drain conductance is decreased.

The subthreshold current flow in the channel from source to drain electrode beyond the pinch-off voltage is a well-known phenomenon in a MESFET device. The pinch-off is a transition between a region of normal conduction in which the current conduction in the channel is due to the drift of electrons and a region of subthreshold conduction in which the currents are due to both drift and diffusion. For small V_{DS}, electrons can be transported by diffusion (via a concentration gradient between the source and drain electrodes) and the current flow is characterized by an exponential dependence of I_{DS} on V_{DS} and V_{GS}.

Finally, the temperature effect should also be considered in a MESFET device operation. The temperature dependence of the drain current of a MESFET is influenced by two related mechanisms, namely, the variation of the built-in voltage (V_{bi} of the channel–substrate interface) and the variation in the channel transconductance factor K. In fact, there are two built-in voltages of interests which exhibit temperature dependence. Both built-in potentials for the Schottky barrier gate and the channel–substrate interface are affected by the temperature dependence of V_n $[=(k_BT/q)\ln(n_0/N_c)]$. Any change in these built-in voltages will affect the threshold voltage of the MESFET. The channel transconductance parameter K, as defined in the square-law relationship $I_D = K(V_{GS} - V_T)^2$, can also vary with temperature. The channel transconductance factor K $(=Z\varepsilon\mu/2La)$ is found to decrease with increasing temperature, since the mobility decreases with increasing temperature and the effective channel thickness increases with temperature due to the temperature dependence of V_{bi} discussed above.

15.3. MODULATION-DOPED FIELD-EFFECT TRANSISTORS (MODFETS)

Besides GaAs MESFETs discussed in the previous section, several newly developed high-speed devices such as MODFETs, HBTs, RTDs, and HETs using lattice-matched AlGaAs/GaAs and InGaAs/InAlAs material systems have been developed for high-speed and high-frequency microwave applications. Furthermore, the non-lattice-matched pseudomorphic quantum well structure such as AlGaAs/InGaAs/GaAs has also been successfully grown in conventional AlGaAs/GaAs HEMTs without extensive crystal defects if the InGaAs layer is thin enough (i.e., less than 200 Å) so that lattice mismatch can be accommodated by the elastic strain rather than by the formation of dislocations. In fact, significant improvement in low-noise microwave performance has been accomplished in the pseudomorphic HEMT relative to conventional HEMTs. In this section we shall consider the basic device principles, structure, and characteristics of an AlGaAs/GaAs MODFET.

The AlGaAs/GaAs modulation-doped (or high electron mobility) field-effect transistor (MODFET or HEMT) introduced in 1981 has offered both high speed and excellent gain, noise, and power performance at microwave and millimeter-wave (30 to 300 GHz) frequencies. This device, using novel properties of the two-dimensional electron gas (2-DEG) at the interface between the GaAs and $Al_xGa_{1-x}As$ epitaxial layer and an evolutionary improvement over the GaAs MESFET, has been used extensively in both hybrid and monolithic circuits. The concept evolves from the fact that high electron mobilities in the undoped 2-DEG GaAs layer can be achieved if electrons are transferred across the heterointerface from the heavily doped, wider-bandgap $Al_xGa_{1-x}As$ layer ($x \sim 0.3$) to the nearby undoped GaAs buffer layer. This process is now known as *modulation doping*. In addition to the name MODFET, other acronyms such as HEMT (high electron mobility transistor), TEGFET (two-dimensional electron gas field-effect transistor), and SDFET (selectively-doped field-effect transistor) have also been used in the literature. These acronyms are all descriptive of various aspects of the same device. The most commonly used name for this device, however, is MODFET or HEMT. It is worth noting that the MODFET or HEMT is comparable to Josephson junction devices for high-speed applications with very short switching times and low power dissipation. The conventional AlGaAs/GaAs HEMT is very similar to a GaAs MESFET. To reduce confusion, we shall use here either MODFET or HEMT to represent this class of devices.

In this section we describe theoretical aspects of the 2-DEG in a modulation-doped AlGaAs/GaAs heterostructure. The basic characteristics and factors affecting the performance of a MODFET as well as the model for predicting current–voltage (I–V) and capacitance–voltage (C–V) behavior will be discussed in this section.

The switching speed of a FET is increased by minimizing the carrier transit time and maximizing the current level I_{DS} as well as the transconductance g_m. In order to design a high-speed FET, we need to optimize the device parameters. This procedure includes using small gate length L, high carrier concentration n_0, high saturation velocity v_s, and large gate width-to-length ratio (i.e., aspect ratio). Increasing the dopant density beyond $5 \times 10^{18}\ \text{cm}^{-3}$ while maintaining a high saturation velocity v_s, however, cannot be achieved simultaneously in a MESFET. Increasing the dopant density will reduce the electron velocity due to the increase of ionized impurity scattering. Therefore, to meet the requirements of large n_0 and v_s, the MODFET structure is employed. Figure 15.8 shows the cross-section view of an AlGaAs/GaAs MODFET. The structure consists of an undoped GaAs buffer layer, undoped $Al_xGa_{1-x}As$ spacer layer, and Si-doped n^+-$Al_xGa_{1-x}As$ ($x = 0.33$) layer grown sequentially on a semi-insulating (S.I.) GaAs substrate by using the molecular beam epitaxy (MBE) technique. The source and drain regions are formed by using ion implantation of a n^+-GaAs layer on the doped n^+-$Al_xGa_{1-x}As$ layer. A Schottky barrier contact is formed on the doped $Al_xGa_{1-x}As$ between source and drain contacts to serve as the gate electrode. Since the bandgap energy for the $Al_xGa_{1-x}As$ is larger than the bandgap energy of GaAs (e.g., $E_g =$ 1.8 eV for $x = 0.3$) and the energy level of the GaAs conduction band is lower than the conduction band edge of $Al_xGa_{1-x}As$, electrons will diffuse from the doped $Al_xGa_{1-x}As$ top layer into the undoped GaAs buffer layer. These electrons cannot drift away from the AlGaAs–GaAs interface in the GaAs layer due to the Coulombic attractive force of the ionized donor impurities in the AlGaAs layer. Nor can they go back to the doped AlGaAs layer due to the potential barrier at the heterointerface of the AlGaAs/GaAs layer. A triangle potential well (quantum well) is formed in the undoped GaAs layer with well width less than the de Broglie

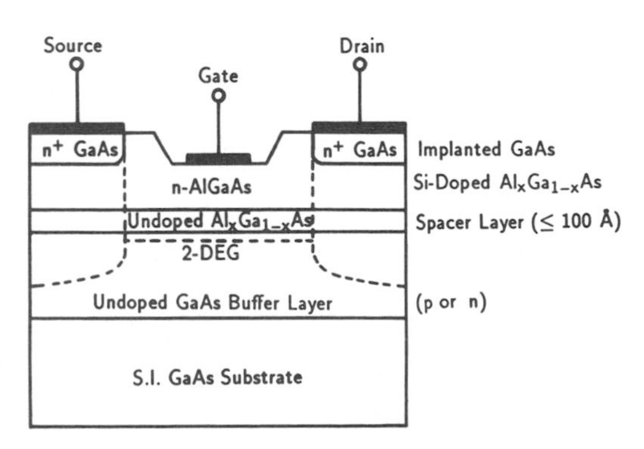

FIGURE 15.8. Cross-section view of a GaAs MODFET, showing the heavily doped n^+-GaAs implanted regions under source and drain ohmic contacts, Si-doped AlGaAs layer, undoped AlGaAs spacer layer, 2-DEG region formed between the undoped AlGaAs/GaAs interface and undoped GaAs buffer layer grown on a S.I. GaAs substrate.

wavelength (i.e., ~250 Å at 300 K) and quantization of energy levels results in the direction perpendicular to the heterointerface. As a result, a 2-DEG sheet charge is accumulated in the triangle potential well at the undoped GaAs layer near the GaAs/AlGaAs interface, and forms a conducting channel for the MODFET. The concentration of the 2-DEG sheet charge can be modulated by the applied gate voltage. The dopant densities and dimensions are chosen such that the AlGaAs layer is fully depleted of free electrons and, as a result, conduction is mainly attributed to the 2-DEG sheet charge in the undoped GaAs buffer layer. In general, the density of the 2-DEG sheet charge will depend on the doping density and aluminum mole fraction x of the silicon-doped $Al_xGa_{1-x}As$ layer, as well as the thickness of the undoped $Al_xGa_{1-x}As$ spacer layer.

Since the 2-DEG sheet charge in the channel of the undoped GaAs buffer layer is spatially separated from the ionized donor impurities by a thin undoped AlGaAs spacer layer (≤ 60 Å), they will experience no ionized impurity scattering inside the triangle potential well in the undoped GaAs buffer layer. Consequently, both the electron mobility and electron velocity in the channel are expected to be very high. For example, the 2-DEG electron mobility of around 8500 $cm^2/V \cdot sec$ at 300 K and 50,000 $cm^2/V \cdot sec$ at 77 K have been obtained in the channel region of the GaAs buffer layer. This, along with the small separation (~300 Å) between the gate and conducting channel, leads to extremely high transconductance (e.g., >500 mS/mm at 77 K), large current-carrying capabilities, small source resistances, and very low noise figures. For example, a MODFET amplifier operating at 77 K has a noise figure of 0.25 dB at 10 GHz and 0.35 dB at 18 GHz. Since the transconductance g_m is large, one expects that the unity current gain cutoff frequency f_T will be very high for a MODFET. Since the maximum oscillation (for unity power gain) frequency f_{max} of the MODFET is strongly influenced by the parasitics (i.e., gate and source resistances, rf drain conductance, feedback and input capacitances), it is essential that these parasitic components be minimized in order to further improve the high-frequency performance. We shall next derive theoretical expressions for the Fermi level, density of 2-DEG sheet charge, electric field at the interface, and current–voltage characteristics of a MODFET.

15.3.1. Equilibrium Properties of the 2-DEG in GaAs

In this section, the Fermi level, density of 2-dimensional electron gas (2-DEG) charge sheet, and electric field at the interface of a modulation-doped AlGaAs/GaAs heterostructure are described. Only the lowest (ground state) and first excited subbands (i.e., E_0 and E_1) in the triangular potential well of the undoped GaAs buffer layer will be considered. If the electric field in the triangular potential well is assumed quasi-constant, then the solution for the longitudinal quantized energy levels can be expressed by

$$E_n \simeq \left(\frac{\hbar^2}{2m_l^*}\right)^{1/3}\left(\frac{3}{2}\pi q\mathscr{E}\right)^{2/3}\left(n+\frac{3}{4}\right)^{2/3} \tag{15.35}$$

where m_l^* is the longitudinal effective mass, n is the quantum number, and $\mathscr{E}$ is the electric field. For GaAs, the two subbands E_0 and E_1 in which the 2-DEG sheet charge resides are given by

$$E_0 \simeq 1.83 \times 10^{-6}\mathscr{E}^{2/3} \quad \text{eV} \qquad \text{and} \qquad E_1 \simeq 3.23 \times 10^{-6}\mathscr{E}^{2/3} \quad \text{eV} \tag{15.36}$$

where $\mathscr{E}$ is the electric field in V/m.

In order to derive a current–voltage relation for the MODFET, it is important to first establish a relationship between the interface electric field $\mathscr{E}_{i1}$ and the 2-DEG sheet charge concentration. For an AlGaAs/GaAs modulation-doped heterostructure, the electric field in the undoped GaAs buffer layer obeys Poisson's equation, which is given by

$$\frac{d\mathscr{E}_1}{dx} = -\frac{q}{\varepsilon_1}[n(x) + N_{a1}] \tag{15.37}$$

where $n(x)$ is the bulk electron concentration and N_{a1} is the ionized acceptor density in the undoped p-GaAs buffer layer. Integration between the limit of the depletion region ($\mathscr{E}_1 = 0$) and the interface ($\mathscr{E}_1 = \mathscr{E}_{i1}$) yields

$$\varepsilon_1 \mathscr{E}_{i1} = qn_s + qN_{a1}W_1 \tag{15.38}$$

where ε_1 is the dielectric permittivity of the undoped p-GaAs, n_s is the 2-DEG sheet charge density, and W_1 is the depletion layer width. In a MODFET, N_{a1} in the undoped p-GaAs layer is usually very small, and hence the second term in Eq. (15.38) is negligible. Thus, Eq. (15.38) becomes

$$\varepsilon_1 \mathscr{E}_{i1} \simeq qn_s \tag{15.39}$$

Equation (15.39) can also be applied to the undoped n-type GaAs layer. The subband positions can also be expressed in terms of the 2-DEG sheet charge concentration, and is given by

$$E_0 = \gamma_0 n_s^{2/3} \quad \text{and} \quad E_1 = \gamma_1 n_s^{2/3} \tag{15.40}$$

where γ_0 and γ_1 are parameters which can be adjusted to obtain the best fit with experimental data. To deal with the 2-DEG sheet charge in the undoped GaAs buffer layer, the relation between n_s and the Fermi level position must be derived first. The density of states associated with a single quantized energy level for a 2-DEG gas system is a constant, and can be expressed by

$$D = \frac{qm^*}{\pi\hbar^2} \tag{15.41}$$

where a spin degeneracy of 2 and a valley degeneracy of 1 have been used. Between the two subbands E_0 and E_1, the 2-dimensional density of states is given by D, and for energies between E_1 and E_2 it is equal to $2D$. Now, using Fermi–Dirac statistics, the 2-DEG sheet charge concentration as a function of the Fermi level position and temperature can be written as

$$\begin{aligned} n_s &= D\int_{E_0}^{E_1} \frac{dE}{1 + e^{(E - E_F)/k_BT}} + 2D\int_{E_1}^{\infty} \frac{dE}{1 + e^{(E - E_F)/k_BT}} \\ &= D\left(\frac{k_BT}{q}\right)\ln[(1 + e^{(E_F - E_0)/k_BT})(1 + e^{(E_F - E_1)/k_BT})] \end{aligned} \tag{15.42}$$

which at low temperatures reduces to

$$n_s = D(E_F - E_0) \tag{15.43}$$

when the second subband is empty, and

$$n_s = D(E_1 - E_0) + 2D(E_F - E_1) \tag{15.44}$$

when the second subband is occupied. From published data taken at low temperatures using Shubnikov de Hass or cyclotron resonance experiments, values of γ_0 and γ_1 for p-type GaAs are found equal to 2.5×10^{-12} and 3.2×10^{-12}, respectively. From the measured cyclotron mass, D is found to be $3.24 \times 10^{21}\ \text{cm}^{-2} \cdot \text{V}^{-1}$. It is seen from Eq. (15.43) that the 2-DEG sheet charge density n_s is equal to the product of the density of states D, and the energy difference between the Fermi level and ground state when the second subband is empty.

Figure 15.9a and b show energy band diagrams of a single-period modulation-doped n^+-$Al_xGa_{1-x}As$/p-GaAs heterostructure: (a) in equilibrium and isolated from the influence of any external contact, and (b) under an applied gate bias voltage. We note that the position for the two presumed subbands in the quasi-triangular potential well shown is only for illustrative purpose. The structure consists of a Si-doped $Al_xGa_{1-x}As$ ($x \leq 0.3$) layer of thickness d_d and an undoped $Al_xGa_{1-x}As$ spacer layer of thickness d_i (i.e., 20 to 60 Å) which serves as a buffer layer to further reduce the scattering of 2-DEG in the undoped GaAs layer (thickness d_1) by the ionized impurities in the space-charge region of the Si-doped $Al_xGa_{1-x}As$ layer. The electric displacement vector at the interface of $Al_xGa_{1-x}As$/GaAs can be calculated by using the depletion approximation in the space-charge layer. In this case, the potential $V_2(x)$ in the space-charge region of the $Al_xGa_{1-x}As$ can be derived from Poisson's equation, which is given by

$$\frac{d^2V_2(x)}{dx^2} = -\frac{q}{\varepsilon_2} N_{d2}(x) \tag{15.45}$$

If the heterojunction interface is chosen as origin, then the following boundary conditions prevail:

$$V_2(0) = 0, \qquad \left(\frac{dV_2}{dx}\right)_{x=-W_2} = 0, \qquad \left(\frac{dV_2}{dx}\right)_{x=0} = -\mathscr{E}_{i2} \tag{15.46}$$

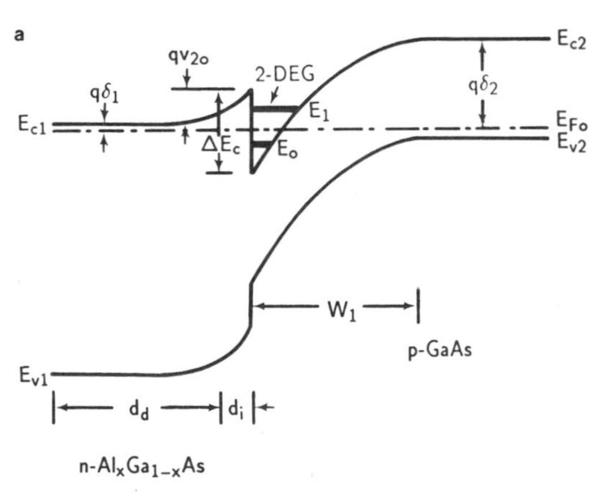

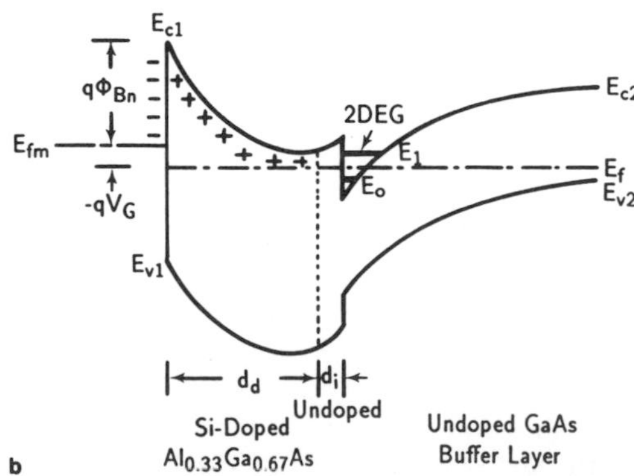

FIGURE 15.9. Energy band diagrams for an $Al_xGa_{1-x}As$/GaAs MODFET shown in Fig. 15.8: (a) in equilibrium and (b) under applied bias conditions ($V = -V_G$).

where W_2 is the space-charge layer width in the Si-doped $Al_xGa_{1-x}As$ layer, and $\mathscr{E}_{i2}$ and $V_2(-W_2)$ are given respectively by

$$\mathscr{E}_{i2} = -\left(\frac{q}{\varepsilon_2}\right)\int_0^{-W_2} N_{d2}(x)\,dx \tag{15.47}$$

and

$$V_2(-W_2) = v_{20} = \mathscr{E}_{i2}W_2 - \frac{q}{\varepsilon_2}\int_0^{-W_2} dx \int_0^x N_{dd}(x')\,dx' \tag{15.48}$$

For the MODFET structure shown in Fig. 15.9a, we can write

$$\begin{aligned} N_{d2}(x) &= 0 \qquad \text{for } -d_i < x < 0 \\ N_{d2}(x) &= N_{d2} \qquad \text{for } x < -d_i \end{aligned} \tag{15.49}$$

where N_{d2} is the dopant density in the Si-doped AlGaAs layer. Now, solving Eqs. (15.47) through (15.49) yields

$$\varepsilon_2\mathscr{E}_{i2} = qN_{d2}(W_2 - d_i) \tag{15.50}$$

and

$$v_{20} = \frac{qN_{d2}}{2\varepsilon_2}(W_2^2 - d_i^2) \tag{15.51}$$

Since the band bending of the AlGaAs layer at the heterointerface is denoted by v_{20}, we can easily show from Eqs. (15.50) and (15.51) that

$$\varepsilon_2\mathscr{E}_{i2} = \sqrt{2q\varepsilon_2 N_{d2}v_{20} + q^2N_{d2}^2d_i^2} - qN_{d2}d_i \tag{15.52}$$

From Fig. 15.9a, we obtain

$$v_{20} = \Delta E_c - \delta_2 - E_{F0} \tag{15.53}$$

Using Gauss's law and neglecting interface traps, the 2-DEG sheet charge density, which is related to the dielectric constant and the electric field, can be expressed by

$$qn_s = \varepsilon_1\mathscr{E}_{i1} = \varepsilon_2\mathscr{E}_{i2} \tag{15.54}$$

Now, the solution of Eqs. (15.50) through (15.54) yields a general expression for the 2-DEG sheet concentration, which reads

$$\begin{aligned} n_s &= \sqrt{2\varepsilon_2 N_{d2}v_{20}/q + N_{d2}^2d_i^2} - N_{d2}d_i \\ &= \frac{Dk_BT}{q}\ln(1 + e^{(E_{F0}-E_0)/k_BT})(1 + e^{(E_{F0}-E_1)/k_BT}) \end{aligned} \tag{15.55}$$

The Fermi level E_{F0} can be solved numerically from Eq. (15.55) by iteration procedures. It is seen that the value of n_s calculated from Eqs. (15.55) and (15.42) should be the same.

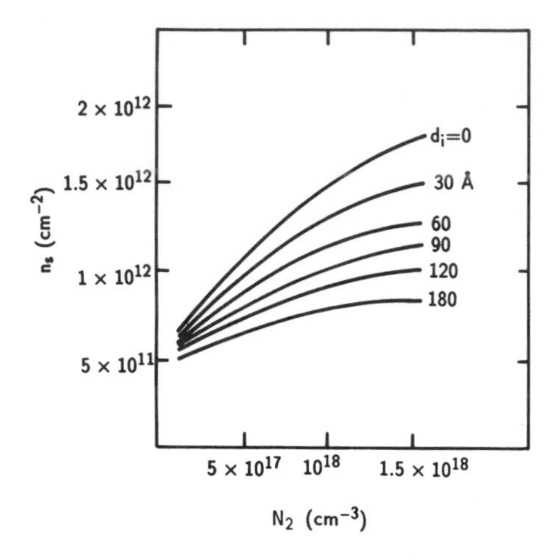

FIGURE 15.10. 2-DEG sheet concentration in the p^--GaAs layer (with $N_a = 10^{14}\ \text{cm}^{-3}$) as a function of donor impurity density (N_2) in the Si-doped $Al_{0.3}Ga_{0.7}As$ layer for different undoped AlGaAs spacer layer thicknesses calculated at 300 K. After Delagebeaudeuf and Ling,[4] by permission, © IEEE–1982.

Otherwise, values of the Fermi energy E_{F0} must increase until this condition is met. Figure 15.10 shows the 2-DEG sheet concentration in the p^--GaAs layer (with $N_{a1} = 10^{14}\ \text{cm}^{-3}$) as a function of donor impurity density (N_{d2}) in the Si-doped $Al_{0.3}Ga_{0.7}As$ layer for different undoped AlGaAs spacer layer thicknesses d_i at 300 K. The 2-DEG sheet concentration is seen to increase as the thickness of the undoped AlGaAs spacer layer is decreased, and to increase as the dopant concentration of the Si-doped AlGaAs layer is increased.

15.3.2. 2-DEG Charge Control Regime

The charge control regime, i.e., the region between the Schottky contact on the Si-doped AlGaAs layer and the AlGaAs/GaAs heterointerface, is totally depleted, and the electrostatic potential obeys Poisson's equation subject to the conditions

$$\begin{aligned} N_{d2}(x) &= 0 \qquad \text{for } -d_i < x < 0 \\ N_{d2}(x) &= N_{d2} \qquad \text{for } -d_d < x < -d_i \end{aligned} \tag{15.56}$$

If we choose the origin at the AlGaAs/GaAs interface and let $V_2(0) = 0$ at $x = 0$, then $V_2(-d_d)$ can be written as

$$V_2(-d_d) = -v_2 = \mathscr{E}_{i2} d_d - \frac{q}{\varepsilon_2} \int_0^{-d_d} dx \int_0^x N_{d2}(x')\, dx' \tag{15.57}$$

Equation (15.57) can be easily solved using the boundary conditions given by Eq. (15.56). The result yields

$$v_2 = \frac{qN_{d2}}{2\varepsilon_2}(d_d - d_i)^2 - \mathscr{E}_{i2} d_d \tag{15.58}$$

From Fig. 15.9b and Eq. (15.58) we can establish the relation

$$\varepsilon_2 \mathscr{E}_{i2} = \frac{\varepsilon_2}{d_d}(V_{p2} - v_2) \tag{15.59}$$

where

$$V_{p2} = \frac{qN_{d2}}{2\varepsilon_2}(d_d - d_i)^2 \tag{15.60}$$

By examining Fig. 15.9b, the potential v_2 is given by

$$v_2 = \phi_M - V_G + E_F - \Delta E_c \tag{15.61}$$

Now substituting Eq. (15.61) into Eq. (15.59) yields

$$\varepsilon_2 \mathscr{E}_{i2} = \left(\frac{\varepsilon_2}{d_d}\right)(V_{p2} - \phi_M - E_F + \Delta E_c + V_G) \tag{15.62}$$

Therefore, in the absence of interface states, the total charge in the 2-DEG GaAs buffer layer can be obtained from Eqs. (15.62) and (15.54) in the form

$$Q_s = qn_s = \frac{\varepsilon_2}{d_d}(V_{p2} - \phi_M - E_F + \Delta E_c + V_G) \tag{15.63}$$

Since E_F, which is a function of V_G, is usually much smaller than the other terms given in Eq. (15.63), we can approximate Eq. (15.63) by

$$Q_s \simeq \frac{\varepsilon_2}{d_d}(V_G - V_{off}) \tag{15.64}$$

where

$$V_{off} = \phi_M - \Delta E_c - V_{p2} \tag{15.65}$$

is the "off voltage" which eliminates the 2-DEG. Both Eqs. (15.64) and (15.65) are obtained by neglecting the Fermi energy E_f, so it is evident that they are insensitive to the exact positions of the two subbands. Therefore, Eqs. (15.64) and (15.65) are valid for both p^-- and n^--type GaAs buffer layers. If the interface states charge Q_i is included, then Eq. (15.65) becomes

$$V_{off} = \phi_M - \Delta E_c - V_{p2} - \frac{d_d}{\varepsilon_2}Q_i \tag{15.66}$$

For a given AlGaAs layer width, there exists a threshold voltage V_{Gth}, which separates the charge control regime from the equilibrium state. This can be obtained by equating the two expressions for $\varepsilon_2 \mathscr{E}_{i2}$ given by Eqs. (15.52) and (15.62); the result yields

$$V_{Gth} = \phi_M - \delta_2 - \left(\sqrt{\frac{qN_{d2}d_d^2}{2\varepsilon_2}} - \sqrt{(\Delta E_c - \delta_2 - E_{F0}) + \frac{qN_{d2}d_i^2}{2\varepsilon_2}}\right)^2 \tag{15.67}$$

15.3.3. *Current–Voltage Characteristics*

The current–voltage (I–V) characteristics of a MODFET can be derived by using the charge control model and the gradual channel approximation. If $V_c(x)$ is the channel potential under the gate at position x, and V_{GS} is the applied gate voltage, then the effective potential V_{eff} for charge control at x is given by

$$V_{eff}(x) = V_{GS} - V_c(x) \tag{15.68}$$

From Eqs. (15.64) and (15.68), the 2-DEG sheet charge in the channel can be rewritten as

$$Q_s(x) = qn_s = \frac{\varepsilon_2}{d}[V'_{GS} - V_c(x)] \tag{15.69}$$

where $V'_{GS} = (V_{GS} - V_{off})$ and $d = d_d + d_i + \Delta d$; $\Delta d = \varepsilon_2 a/q \sim 80$ Å for $a = 1.25 \times 10^{-21}$ V · cm^2. The channel current at position x is given by

$$I_{DS} = Q_s(x)Zv(x) \tag{15.70}$$

where Z is the gate width and $v(x)$ is the electron velocity at position x. It is seen that $v(x)$ $(=\mu\mathscr{E})$ for a 2-DEG sheet charge in the channel is generally field-dependent and the electron mobility (μ) as a function of the electric field can be expressed by

$$\mu = \frac{\mu_0}{1 + \dfrac{1}{\mathscr{E}_c}\dfrac{dV}{dx}} \tag{15.71}$$

where $\mathscr{E}_c$ is the critical field strength above which velocity saturation occurs, and μ_0 is the low-field electron mobility. For field strengths less than $\mathscr{E}_c$, $v(x)$ varies linearly with $\mathscr{E}$ $(= -dV/dx)$ with a proportionality constant equal to μ_0, and $v(x)$ becomes saturated [i.e., $v(x) = v_s$] if $\mathscr{E}$ is equal to or greater than $\mathscr{E}_c$. The current–voltage (I–V) relation of a MODFET can be derived by employing a simple two-piecewise linear approximation for the v_d versus $\mathscr{E}$ relation, which may be written as

$$v_d = \mu_0\mathscr{E} \qquad \text{for } \mathscr{E} < \mathscr{E}_c \tag{15.72}$$

$$v_d = v_s \qquad \text{for } \mathscr{E} \geq \mathscr{E}_c \tag{15.73}$$

This is the so-called gradual channel approximation, which is used widely in modeling the I–V relation of a FET. Therefore, using the two-piecewise linear approximation and assuming that the electric field in the channel is parallel to the heterointerface, we can derive an analytical expression for the I–V relation of a MODFET in the linear region ($V_{DS}/L < \mathscr{E}_c$) and in the saturation region ($V_{DS}/L > \mathscr{E}_c$). We note that this approximation does not include the diffusion current component in the channel.

Linear Region ($\mathscr{E} < \mathscr{E}_c$). In the ohmic regime where the electric field is much smaller than the critical field strength $\mathscr{E}_c$, the drain current given by Eq. (15.70) can be expressed in the form

$$I_{DS} = Q_s(x)Zv(x) = Q_sZ\mu\mathscr{E}_x = Q_sZ\mu\left(-\frac{dV_c(x)}{dx}\right) \tag{15.74}$$

Now if $Q_s(x)$, given by Eq. (15.69), is substituted into (15.74), the expression for I_{DS} can be obtained by integrating both sides of Eq. (15.74) from $x = 0$ to $x = L$:

$$\int_0^L I_{DS}\,dx = \int_{V_c(0)}^{V_c(L)} \mu Z\frac{\varepsilon_2}{d}[V'_G - V_c(x)]\left(-\frac{dV_c(x)}{dx}\right) \tag{15.75}$$

If the source and drain resistances are neglected, then Eq. (15.75) reduces to

$$I_{DS} = \frac{\varepsilon_2\mu Z}{dL}(V'_{GS}V_{DS} - V_{DS}^2/2) \tag{15.76}$$

Equation (15.76) is obtained by using the condition that $V_c = 0$ at $x = 0$ and $V_c = V_{DS}$ at $x = L$. We see that I_{DS} is constant in the channel and $V_c(x)$ increases with distance from source to drain. The electric field reaches a maximum near the drain side of the channel, and velocity saturation will occur first at the drain side of the gate region. In the linear region, where the drain voltage V_{DS} is very small, Eq. (15.76) can be simplified to

$$I_{DS} = \frac{\varepsilon_2 \mu Z}{dL}(V'_{GS} V_{DS}) \tag{15.77}$$

which shows that the drain current varies linearly with drain voltage, and the MODFET acts like a pure voltage-controlled resistor (by V'_{GS}). If the source and drain resistances are not negligible, then the channel voltages at the source and drain sides of the gate region, $V_c(0)$ and $V_c(L)$, are given respectively by

$$V_c(0) = R_s I_{DS} \tag{15.78}$$

and

$$V_c(L) = V_{DS} - R_D I_{DS} \tag{15.79}$$

where R_s and R_D denote the source and drain series resistances, respectively. Now by solving Eqs. (15.75), (15.78), and (15.79) we obtain

$$I_{DS} = \frac{\varepsilon_2 \mu Z}{dL}\{V'_{GS}[(R_s + R_D)I_{DS} - V_{DS}] - \tfrac{1}{2}[(R_s + R_D)I_{DS} - V_{DS}]^2\} \tag{15.80}$$

For small V_{DS} and I_{DS}, the first-order approximation enables Eq. (15.80) to be reduced to

$$I_{DS} = V_{DS}\left[-(R_s + R_D) + \frac{Ld}{Z\mu\,\varepsilon_2 V'_{GS}}\right]^{-1} \tag{15.81}$$

which again shows that a linear relation exists between the drain current and drain voltage.

Saturation Regime ($\mathscr{E} \geq \mathscr{E}_c$). In the saturation regime, the velocity saturation occurs first at the drain side of the gate region with $\mathscr{E}(L) = \mathscr{E}_c$. The drain current under the saturation condition is given by

$$I_{Dsat} = \frac{Z\varepsilon_2 v_s}{d}\left(\sqrt{(V'_{GS} - R_s I_{Dsat})^2 + \mathscr{E}_c^2 L^2} - \mathscr{E}_c L\right) \tag{15.82}$$

For a long gate MODFET, Eq. (15.82) can be approximated by

$$I_{Dsat} \simeq \frac{Z\varepsilon_2 \mu}{2dL}(V'_{GS} - R_s I_{Dsat})^2 \tag{15.83}$$

and for a short gate MODFET, Eq. (15.82) becomes

$$I_{Dsat} \simeq \frac{Z\varepsilon_2 v_s}{d}(V'_{GS} - R_s I_{Dsat} - \mathscr{E}_c L) \tag{15.84}$$

which is valid if $(V'_{GS} - R_s I_{Dsat}) \gg \mathscr{E}_c L$. In fact, the experimental results for a short gate GaAs MODFET confirm the linear relation between I_{DS} and V_{GS} in the saturation region.

The two-piecewise linear model for the channel charge and current as functions of the gate and drain voltages presented in this section gives a first-order description of the *I–V* characteristics of a MODFET in the linear and saturation regions of operation. A more realistic three-piecewise linear model for predicting the *I–V* characteristics of a MODFET has also been developed. Such a model uses a three-piecewise linear approximation for the v_d versus $\mathscr{E}$ relation to derive the current–voltage characteristics of the MODFET. A comparison of two- and three-piecewise linear approximations with the actual velocity versus electric field relation shows that the three-piecewise linear approximation yields roughly a 10 to 20% improvement in accuracy over the two-piecewise linear approximation. Figure 15.11 shows the *I–V* characteristics of a normally-on MODFET. The solid lines are calculated from the three-piecewise linear model and the dot and dashed lines are experimental data. A more rigorous model using numerical simulation of the current–voltage characteristics of a GaAs MODFET has also been reported in the literature.

From the above description, we see that some advantages associated with a MODFET include the large 2-DEG sheet charge density ($\sim 10^{12}\ \text{cm}^{-2}$), high electron mobility, and high saturation velocity. These unique features have resulted in significant improvement in device performance (i.e., high speed, low noise) compared to conventional GaAs MESFETs. In the past few years, successful development of 0.25 μm gate-length AlGaAs/GaAs HEMTs have offered new promise for low-noise applications at microwave frequencies. Further improvement in current gain and noise performance of the MODFET can be achieved by further reducing the gate length to 0.1 μm or less. However, a 0.1 μm gate length generates undesirable short-channel effects. The effects are largely due to the space-charge injection of carriers (which is inversely proportional to the square of the effective gate length) into the buffer layer under the channel. This increases the HEMT output conductance and results in a shift of the pinch-off voltage and transconductance reduction near the pinch-off region. To overcome this problem, an AlGaAs/InGaAs/GaAs pseudomorphic HEMT structure has been proposed recently using a 0.1 μm gate length. Using the MBE growth technique, the AlGaAs/InGaAs/GaAs pseudomorphic HEMT is grown at a lower temperature than that of the AlGaAs/GaAs conventional HEMT structure. In this structure a thin strained superlattice (TSSL) of an undoped $In_{0.35}Ga_{0.65}As$ (50 Å)/GaAs (15 Å)/$In_{0.35}Ga_{0.65}As$ (50 Å) pseudomorphic active layer structure is grown on an undoped GaAs buffer layer. Due to the InGaAs quantum-well channel structure, this pseudomorphic HEMT structure has greatly improved the carrier confinement and hence reduced short-channel effects. In addition, due to the superior transport properties in the InGaAs channel and the large conduction band discontinuity with the AlGaAs spacer layer (50 Å), the pseudomorphic HEMT also provides a very high electron velocity and 2-DEG sheet charge density. A maximum extrinsic transconductance g_m

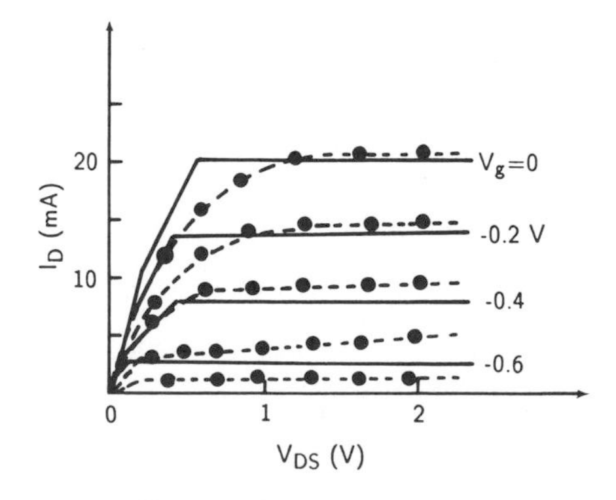

FIGURE 15.11. Comparison of calculated and measured I_D versus V_{DS} curves in a normally-on GaAs MODFET using a three-piece linear model. After Lee *et al.*,[(5)] by permission, © IEEE–1983.

of 930 mS/mm with excellent pinch-off characteristics at room temperature has been reported recently for a 0.1-μm gate-length planar-doped pseudomorphic HEMT. The device has a maximum stable gain of 19.3 dB measured at 18 GHz. At 60 GHz the device has demonstrated a minimum noise figure of 2.5 dB with an associated gain of 8 dB. The unity current gain cutoff frequency f_T for this device is around 100 GHz. These are the best gain and noise results reported to date for HEMTs.

Figure 15.12 presents a comparison of a conventional AlGaAs/GaAs HEMT with a GaAs-based pseudomorphic HEMT. The difference between these two structures is that, in the latter, a thin (typically 50–200 Å) layer of $In_xGa_{1-x}As$ ($x = 0.15$–0.35) is inserted between the doped AlGaAs barrier layer and the GaAs buffer layer. It is clear that there will be a lattice constant mismatch between AlGaAs/InGaAs/GaAs layers introduced by the thin InGaAs channel layer. The strain from this lattice mismatch wll be taken up totally by the InGaAs quantum well layer. If the thickness of the InGaAs layer is less than the critical thickness, then the lattice mismatch strain between the InGaAs and GaAs can be accomodated elastically, and the InGaAs layer is compressed to mirror the structure of the GaAs and AlGaAs layers (hence the name "pseudomorphic"). This critical thickness is dependent on the InAs mole fraction. For example, at 35% InAs mole fraction, the critical thickness is about 50 Å. Since ΔE_c increases and carrier confinement and transport properties improve with higher InAs mole fraction, it is desirable to increase the InAs mole fraction to the highest possible level in the pseudomorphic HEMT. However, the lattice mismatch strain between InGaAs and GaAs layers has precluded the use of a very high InAs mole fraction in the pseudomorphic channel for device applications.

In addition to the AlGaAs/GaAs HEMTs discussed above, HEMTs have also been fabricated from InAlAs/InGaAs heterostructure on InP substrate. The advantages of using a $In_{0.52}Al_{0.48}As/In_{0.53}Ga_{0.47}As$ material system for HEMT fabrication include: (1) a large Γ–L

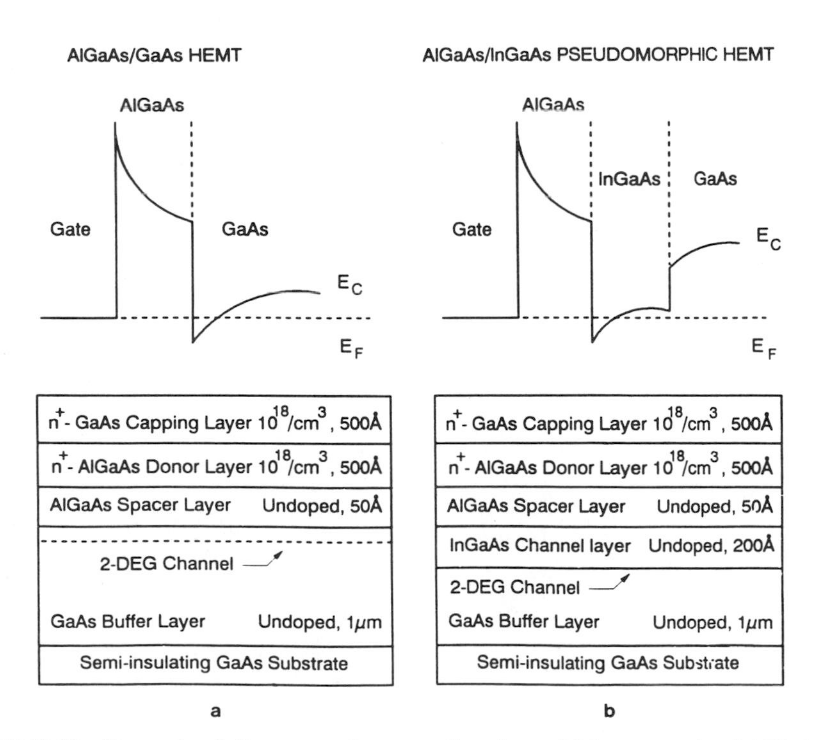

FIGURE 15.12. Energy band diagrams and cross-section views of (a) a conventional AlGaAs/GaAs HEMT and (b) a AlGaAs/InGaAs pseudomorphic HEMT.

valley separation ($\Delta E_{\Gamma-L} = 0.55$ eV) in $In_{0.53}Ga_{0.47}As$, which leads to low intervalley noise; (2) high density of 2-DEG with $n_s > 3 \times 10^{12}\ cm^{-2}$; (3) low sheet resistance of 2-DEG ($\rho \cong 150\ \Omega/\square$) which reduces thermal noise; and (4) high electron velocity in short-channel GaInAs ($v_p \sim 6$–7×10^7 cm/sec) which leads to high f_T and g_m. A transconductance g_m equal to 800 mS/mm and a cutoff frequency f_T of 130 GHz have been demonstrated for a AlInAs/InGaAs on InP HEMT with a gate length of $L = 0.2\ \mu$m and $W = 25\ \mu$m, at a pinchoff voltage $V_p = -1.3$ V. Figure 15.13 presents a comparison of the unity current gain cutoff frequency f_T versus gate length L_g for several III–V MODFETs, GaAs MESFET, and Si NMOS at 300 K. The results show that a value of f_T greater than 300 GHz can be achieved from AlInAs/GaInAs MODFETs with a gate length of 0.1 μm.

Power capabilities of HEMTs at microwave and millimeter-wave frequencies have also made steady progress over the past few years. High-gain and high-efficiency performance HEMTs operating from 10 to 60 GHz have been reported in the literature. The advantages that a HEMT device has over a GaAs MESFET for power applications are higher power gain and higher efficiency. These advantages are due to higher current gain cutoff frequencies resulting from higher electron velocities, which lead to greater output power. For example, a double heterojunction type HEMT with a 1 μm gate length could generate 1.05 W output power at 20 GHz and 0.58 W at 30 GHz. Recently, millimeter-wave power HEMTs employing a single-chip multiple-channel AlGaAs/GaAs structure have generated output power of 1.0 W with 3.1 dB gain and 15.6% efficiency at 30 GHz using a 0.5 μm gate length and 2.4 mm gate periphery.

The conventional AlGaAs/GaAs HEMT has demonstrated significantly better low-noise performance than the GaAs MESFET, at frequencies up to 60 GHz. It has replaced the GaAs MESFET for microwave low-noise applications in many cases at both room temperature and low temperatures. The conventional HEMT has also established technology for high-speed digital applications. It is expected that the strong trend toward the integration of HEMTs into both hybrid and monolithic low-noise circuits will continue to blossom. As HEMT fabrication technology matures further, it is anticipated that GaAs MESFETs may be replaced entirely by HEMTs for low noise applications. As for the pseudomorphic HEMT, this device offers even better noise performance than the conventional AlGaAs/GaAs HEMT, and its power performance is also superior to that of either the conventional HEMT or the GaAs MESFET. The GaAs-based pseudomorphic HEMTs, with higher carrier density, are very attractive for microwave power applications. Finally, the InP-based InAlAs/InGaAs HEMTs will most

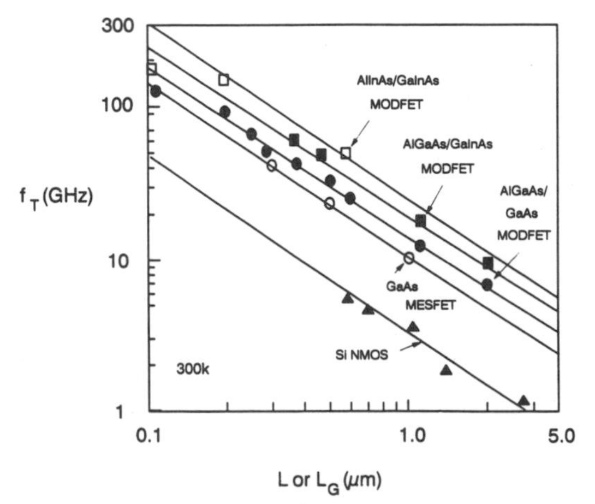

FIGURE 15.13. Comparison of f_T versus gate length for several III–V MODFETs, GaAs MESFET, and Si NMOS devices. After Sze,[6] with permission by John Wiley & Sons Co.

likely make a great impact on ultrahigh-speed digital and millimeter-wave low-noise applications.

15.4. HETEROJUNCTION BIPOLAR TRANSISTOR

In this section, we present the device structure, fabrication technology, operation principles, current–voltage behavior, performance characteristics such as current gain and cutoff frequency, and applications of an AlGaAs/GaAs HBT.

15.4.1. Device Structure and Fabrication Technology

The concept of a heterojunction bipolar transistor (HBT) was first proposed by Shockley, and the basic device theory describing the operation principles of a HBT was subsequently developed by Kroemer. In spite of the great potential for using HBTs in high-speed digital and microwave circuit applications, the technology for fabricating HBTs did not exist until the 1970s. With the advances of MBE and MOCVD growth techniques of III–V epitaxial layers, significant progress in HBT device fabrication technology has been made in recent years although it is still not as mature as the FETs. For example, frequency divider circuits using AlGaAs/GaAs HBTs with clock frequencies exceeding 20 GHz and a maximum oscillation frequency f_{max} of 105 GHz have been reported recently. The main motivation for using the HBT structure is to overcome some of the limitations found in homojunction bipolar transistors (BJTs). The advantages of a HBT over a BJT include using (i) a wide-bandgap emitter to suppress minority carrier back injection, (ii) a lightly doped emitter to reduce the emitter–base junction capacitance, and (iii) a heavily doped base to lower the base resistance. As a result both the speed and frequency performance of the HBT can be significantly improved over the conventional BJT. In this section, we will present the device structure, operation principles, and dc characteristics of an AlGaAs/GaAs HBT.

In a MESFET or HEMT the current flow is parallel to the surface, and hence the critical control dimension (i.e., the gate length) is established by the fine-line lithography. In a HBT, the current path (i.e., speed-limiting factor) is perpendicular to the surface and the epilayers. Therefore, to first order, the speed of a HBT is governed mainly by the thickness of the epilayer. Since epilayer thickness can be easily made much smaller by the MBE or MOCVD technique than the horizontal lithography dimensions, for a given horizontal dimension there is a higher speed potential for the HBT structure than for MESFETs. The HBT using the AlGaAs/GaAs material system has shown great promise for high-speed device applications. The use of a wide-bandgap $Al_xGa_{1-x}As$ emitter for the HBT results in an injection efficiency of close to unity even if the doping density in the GaAs base region is much higher than the emitter. This provides an extra degree of freedom in transistor design, which helps to achieve high-speed operation in such a device. Its major limitations include technological problems related to reproducible and stable processing and the device physics related to gain degradation mechanisms. Historically, the AlGaAs/GaAs emitter-up HBTs have been fabricated on emitter, base, and collector epilayers grown sequentially by the MBE or MOCVD technique, with ohmic contacts being made on the emitter, base, and collector regions by sequential etching. Etching through the emitter to the base, and the emitter–base–n^- collector to the n^+ collector usually leads to steps in the GaAs surface ranging from 0.4 to 1.0 μm in depth. Although high-quality HBTs can be readily fabricated in this manner, the resulting mesa structure is a severe topographical obstacle to integrating these HBTs with a multilevel metal system into a densely packed integrated circuit. High levels of integration have been achieved with HBTs using a planar HI^2L technology, which relies on an emitter-down AlGaAs/GaAs structure with implanted base and extrinsic p^+ base regions. Other advantages of III–V HBTs over silicon BJTs include possible transient electron velocity overshoot, radiation hard, and compatibility with optoelectronic integrated circuits (OEICs).

Figure 15.14a shows a cross-section view of an AlGaAs/GaAs HBT fabricated with a self-aligned base process, and Fig. 15.14b shows the AlGaAs/GaAs HBT fabricated by using an ion-implanted process. The advantages of an ion-implanted process include low base contact resistance, flexibility of layer structure, and low collector-base capacitance, and the problems associated with this process are dopant diffusion, anneal uniformity and parasitic base resistance. As for the self-aligned base process, the advantages include a simpler, faster, and low-temperature process, while etch control, higher base contact resistance, and lower current gain are some of the problems associated with this process. Figure 15.15a shows the energy band diagram of an AlGaAs/GaAs HBT with an abrupt emitter–base junction, and Fig. 15.15b is the energy band diagram for an AlGaAs/GaAs HBT with a graded emitter–base junction. The effects of using the graded E–B junction shown in Fig. 15.15b include (i) reducing space-charge recombination in the E–B junction, (ii) increasing injection electron velocity, (iii) being less effective in suppressing hole injection from the base to the emitter, and (iv) being more susceptable to base dopant diffusion.

The typical dopant densities and layer thicknesses for an AlGaAs/GaAs HBT structure shown in Fig. 15.14a and b are as follows: The device structure consists of a 0.2 μm heavily doped (3×10^{18} cm^{-3})n^+-GaAs cap layer grown on top of the wide-bandgap $Al_{0.3}Ga_{0.7}As$ emitter layer to reduce the emitter contact resistance, a 0.1 μm $Al_{0.3}Ga_{0.7}As$ emitter layer of dopant density around 5×10^{17} cm^{-3}, a 0.1 μm p^+-type GaAs base layer with dopant density 1×10^{19} cm^{-3}, and a 0.3 μm n-type GaAs collector layer with dopant density 10^{17} cm^{-3} grown on top of a n^+ GaAs buffer layer with dopant density 3×10^{18} cm^{-3}. These GaAs/AlGaAs active layers were grown on a semi-insulating GaAs substrate by using the MBE or MOCVD technique.

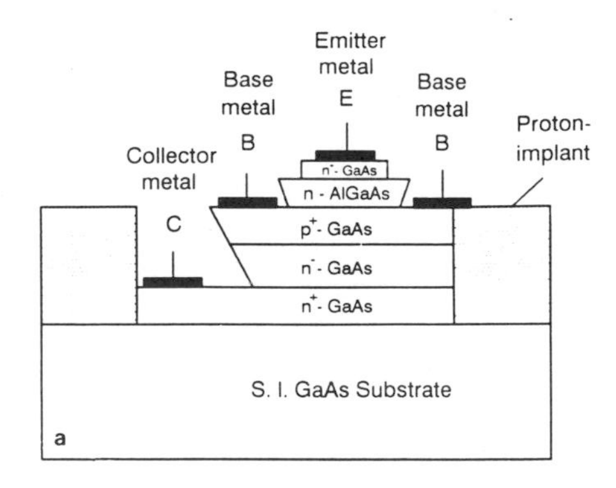

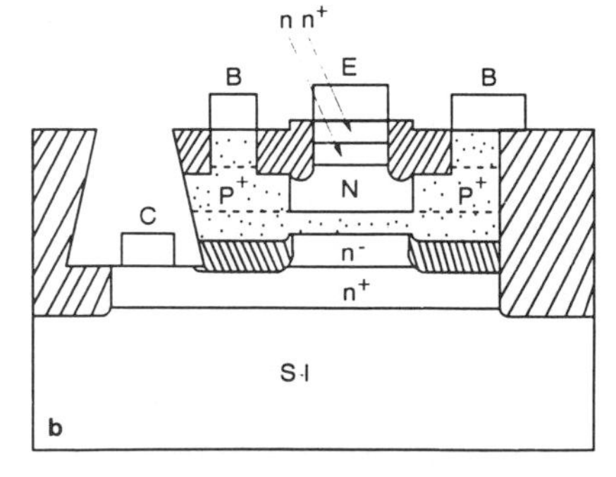

FIGURE 15.14. Schematic cross-section view of an AlGaAs/GaAs n–p$^+$–n–n$^+$ HBT (a) with a self-aligned process and (b) fabricated by an ion-implanted process. After Asbeck,[7] by permission, © IEEE–1988.

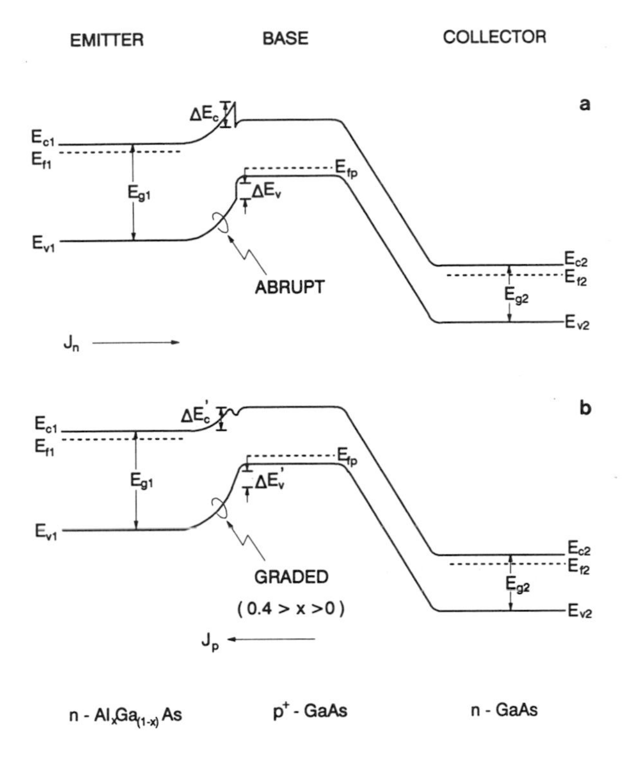

FIGURE 15.15. Energy band diagram for an AlGaAs/GaAs HBT (a) with an abrupt E–B junction and (b) a graded E–B structure. The electron injection is from the wide-bandgap AlGaAs emitter region ($n \sim 5 \times 10^{17}\ \mathrm{cm}^{-3}$) into the narrow-bandgap GaAs base region ($p \sim 10^{19}\ \mathrm{cm}^{-3}$).

15.4.2. Current Gain and Device Parameters

The current gain expression for a HBT can be derived by using the theory developed by Kroemer for a wide-bandgap emitter HBT. For example, the electron current injected from emitter to base (i.e., J_n) and the hole current (J_p) injected from base to emitter for a n–p–n AlGaAs/GaAs graded E–B junction HBT shown in Fig. 15.15b can be written as follows:

$$J_n = q\left(\frac{D_n}{W_B}\right)[N_E \exp(-\Delta E_c'/kT)], \qquad J_p = q\left(\frac{D_p}{W_E}\right)[N_B \exp(-\Delta E_v'/kT)] \tag{15.85}$$

From Eq. (15.85) we can estimate the maximum current gain of the HBT shown in Fig. 15.14b from the ratio of the electron and hole current density, which is given by

$$\beta_{\max} \sim \frac{J_n}{J_p} \simeq \frac{N_E v_{nB}}{N_B v_{pE}} \exp[-(\Delta E_c' - \Delta E_v')/k_B T] = \frac{N_E v_{nB}}{N_B v_{pE}} \exp(\Delta E_g/k_B T) \tag{15.86}$$

From Eq. (15.85) it is seen that a very high value of $\beta_{\max}$ can be achieved even when N_E is smaller than N_B. To obtain a current gain of $\beta > 100$ in the AlGaAs/GaAs HBT structure shown in Fig. 15.15b with a base-to-emitter dopant density ratio (N_B/N_E) of 50 to 100, the value of $\Delta E_v' - \Delta E_c' = \Delta E_g$ should be equal to or greater than 0.24 eV, which corresponds to a wide-bandgap $Al_{0.22}Ga_{0.78}As$ (i.e., with 22% of AlAs) emitter. A typical AlAs mole fraction used in an AlGaAs/GaAs HBT is about 25%. We note that $\Delta E_v'$ ($=\Delta E_g' + \Delta E_c'$) is the bandgap discontinuity in the valence band edge of the wide-bandgap AlGaAs emitter. It is clear that a substantial increase in current gain may be achieved in a HBT due to the exponential increase of β with $\Delta E_v'$. A decrease in $\beta_{\max}$ due to a decrease in the potential barrier for holes may be partially compensated by the increase of electron velocity in the base caused by ballistic injection of electrons from the spike-notch structure at the conduction band edge of the wide-bandgap emitter near the emitter–base junction (see Fig. 15.15a). Smoothing out the conduction band spike can be achieved by grading the composition of the wide-bandgap emitter near the heterointerface of the emitter–base junction, as shown in Fig. 15.15b.

Since the emitter injection efficiency of a HBT can be made very high, its current gain is essentially equal to the base transport factor. For a n–p$^+$–n HBT, this is given by

$$\beta \approx \frac{\tau_n}{\tau_B} \tag{15.87}$$

where τ_n and τ_B denote the electron lifetime in the base and the transit time across the base, respectively. For a uniformly doped base the electron transport across the base is by diffusion and $\tau_B \simeq W_B^2/2D_n$, while for a graded composition base the transport of electrons in the base is by drift and $\tau_B \simeq W_B/\mu_n\mathscr{E}$. Thus, in order to obtain a high current gain, τ_B should be as small as possible. For example, for a sufficiently short base HBT with $W_B = 0.1\ \mu$m, $\beta > 10^3$ can be obtained even if τ_n in the base is of the order of 1 nsec. From Eq. (15.87) it is interesting to note that the current gain in a HBT does not depend on the emitter doping level, and is sensitive to the base doping density only through the variation of τ_n with the base doping density N_B. Therefore, it is possible to shape the doping profiles of a HBT such that the emitter doping density N_E is smaller than the base doping density N_B. As a result, the base spreading resistance $r_{b'b}$ and the emitter depletion capacitance C_{TE} can be greatly reduced. The base spreading resistance $r_{b'b}$ for a circular and a rectangular geometry is given respectively by

$$r_{b'b} = \frac{1}{8\pi\mu_p Q_B} \quad \text{(circular)} \tag{15.88}$$

and

$$r_{b'b} = \frac{1}{12(h/l)\mu_p Q_B} \quad \text{(rectangular)} \tag{15.89}$$

where

$$Q_B = q\int_0^{W_b} N_B(x)\,dx \tag{15.90}$$

is the Gummel number. The emitter junction transition capacitance is given by

$$C_{TE} = A_E\sqrt{\frac{q\varepsilon N_E}{2(V_{bi} - V_{BE})}} \tag{15.91}$$

The above equations can be used to improve the high-frequency performance of a HBT, such as increasing the cutoff frequency f_T and power gain G. This will be discussed further in a later section.

15.4.3. Current–Voltage Characteristics

In this section, we discuss the behavior of the collector current (I_C) and base current (I_B) as well as the current gain of a single heterojunction AlGaAs/GaAs HBT. In general, the current–voltage (I–V) behavior for an AlGaAs/GaAs HBT is similar to that of a silicon BJT but with a few distinct differences. Figure 15.16 shows the collector current versus collector–emitter bias voltage with base current as parameter for a single heterojunction AlGaAs/GaAs HBT used in digital and analog-to-digital (A/D) converter circuits. The HBT has an emitter area of $2 \times 3.5\ \mu\text{m}^2$. Several distinct features which are absent in a Si BJT are displayed in this figure. First, the nonzero offset voltage V_{CE} to produce positive I_C for the HBT is due to the difference in the turn-on voltage of the emitter–base junction and collector–base junction. Second, there exists a negative differential output conductance at a higher I_B and V_{CE}, which is attributed to the heating effect at a higher current level or higher temperature. In general, the current gain of a HBT decreases with increasing temperature. Third, the dc current gain increases with increasing collector current ($\sim I_C^{1/2}$) and becomes saturated at high collector current. To understand the basic mechanisms governing the current conduction in a HBT, we next analyze the collector current and base current separately.

Figure 15.17 shows the Gummel plot (I_C, I_B versus V_{BE}) for the HBT shown in Fig. 15.16. As shown in this figure, the ideality factor for the collector current is equal to unity for low to medium values of V_{BE}, implying that the diffusion current is the dominant component, while the diode ideality factor for I_B at low V_{BE} is equal to 2, indicating that the recombination current is dominant in the base. At high V_{BE} the series resistance effect becomes dominant for both I_C and I_B.

The collector current for a HBT can be explained by using the Moll–Ross–Kroemer relation. If the collector current is base transport limited, then the collector current density J_C can be expressed by

$$J_C = \frac{qD_n n_{ie}^2 \exp(qV_{BE}/k_BT)}{\int_0^{W_B} p(x)\,dx} \tag{15.92}$$

The integral in the denominator represents the number of impurity atoms per unit area (cm^{-2}) in the base, and is known as the Gummel number. This, a large collector current can be realized with a smaller Gummel number, which corresponds to a narrow base width.

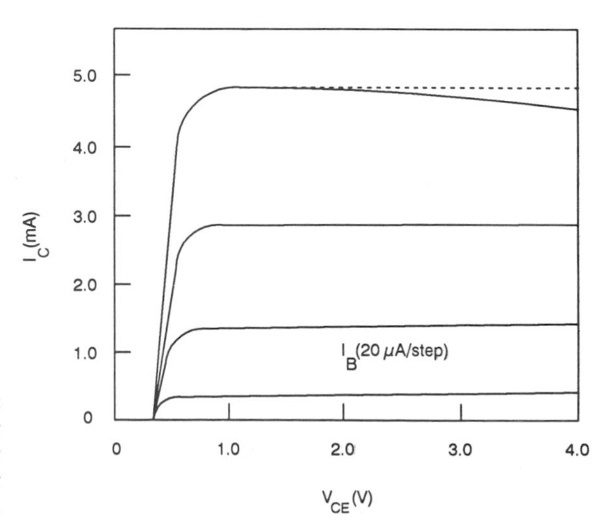

FIGURE 15.16. Collector current versus collector–emitter bias voltage for an AlGaAs/GaAs HBT with emitter area $2 \times 3.5\ \mu\text{m}^2$; I_B steps: 20 μA. After Asbeck,[7] by permission, © IEEE–1988.

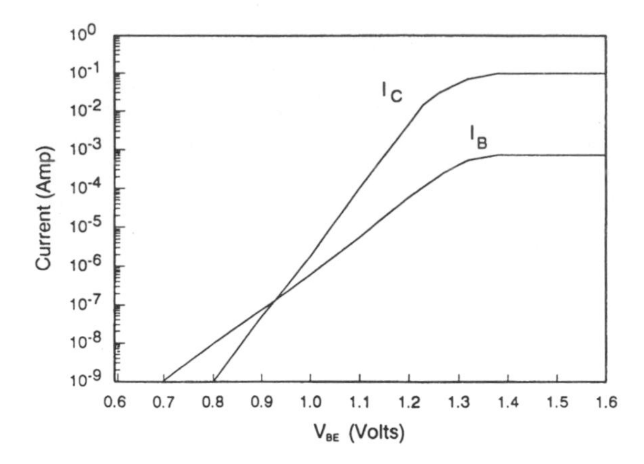

FIGURE 15.17. Collector and base currents versus emitter–base bias voltage (Gummel plot) for the HBT shown in Fig. 15.16.

The base current I_B in an AlGaAs/GaAs HBT is more complex than that of a silicon BJT due to the use of the wide-bandgap AlGaAs emitter and the narrow-gap GaAs base. In general, deep-level defects such as DX centers in AlGaAs play an important role in controlling the recombination current in the emitter–base junction of the HBT. For example, the base current of a HBT may consist of four components: (i) the recombination current in the base, (ii) the recombination current in the emitter–base junction space-charge region, (iii) the recombination current in the emitter, and (iv) the periphery current. A general expression for these current components is given by

$$I_B \sim \exp(qV_{BE}/nk_BT) \tag{15.93}$$

where n is the diode ideality factor and may vary between 1 and 2. When the recombination current is dominant in the base due to the short minority carrier lifetimes, the value of n is equal to unity and the current gain β $(=I_C/I_B)$ is constant. If recombination in the emitter–base junction space-charge region dominates due to the high density of deep level centers, such as in the case of the graded E–B junction HBT shown in Fig. 15.15b, then the value of n is equal to 2, and β increases with I_C and decreases with increasing temperature. If recombination in the emitter is dominant, then the value of n is equal to unity, and β decreases with increasing temperature. Finally, the periphery current is attributed to the high surface recombination velocity around the emitter edge if the AlGaAs emitter surface is not properly passivated. In general, the current gain of the AlGaAs/GaAs HBT scales with the length of the emitter but not the area. Figure 15.18 shows the current gain versus collector current for the HBT shown in Fig. 15.16. The results show that β greater than 100 can be obtained at higher collector current (i.e., $I_C \geq 10$ mA) for this device.

15.4.4. High-Frequency Performance

The cut-off frequency f_T is an important figure of merit for assessing the performance of a HBT in high-speed applications. The values of f_T for a HBT can be calculated by using the expression

$$\frac{1}{2\pi f_T} = \tau_E + \tau_C + \tau_B + \tau_{TC} \tag{15.94}$$

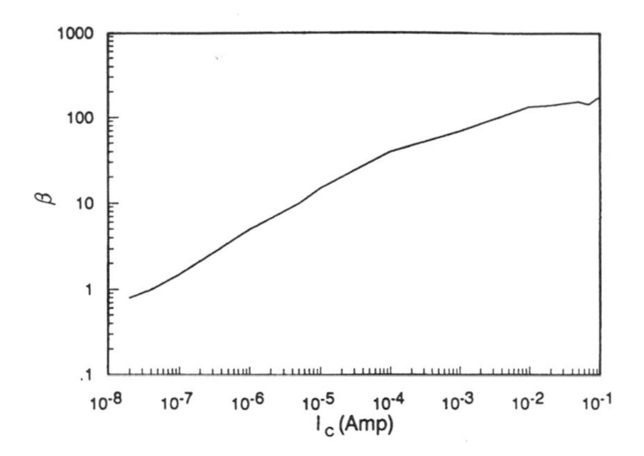

FIGURE 15.18. The dc current gain versus collector current for the HBT shown in Fig. 15.16.

where

$$\tau_E = r_e(C_{TE} + C_{DE}) \approx \frac{4k_BT}{qI_E}C_{TE}(0) \tag{15.95}$$

is the emitter capacitance charging time; r_e is the emitter–base junction resistance and C_{DE} is the emitter diffusion capacitance. The collector charging time τ_C is given by

$$\tau_C = r_{cc'}C_{TC} \tag{15.96}$$

where

$$C_{TC} = A_C\sqrt{\frac{q\varepsilon N_C}{2(V_{bi} + V_{CB})}} \tag{15.97}$$

is the collector–base junction depletion capacitance and $r_{cc'}$ is the collector series resistance. Thus, to reduce τ_C, the doping density between the collector region and the collector contact should be as large as possible so that $r_{cc'}$ can be minimized. The effective base transit time τ_B is related to the effective electron velocity v_n and base width W_B by

$$\tau_B = \frac{W_B}{v_n} \approx \frac{W_B^2}{2D_n} \tag{15.98}$$

where D_n $(=k_BT\mu_n/q)$ is the electron diffusion constant in the base. For an AlGaAs/GaAs HBT with a p^+ GaAs base of $W_B = 500$ Å and $v_n = 1 \times 10^7$ cm/sec, the value of τ_B is found to be 0.5 psec. The transit time of carriers across the collector–base junction τ_{CT} is given by

$$\tau_{CT} = \frac{x_C}{v_s} \tag{15.99}$$

where x_c is the depletion layer width of the collector–base junction and v_s is the saturation velocity of carriers in the collector–base junction. Finally, the power gain of a HBT can be written as

$$G = \frac{f_T}{8\pi f^2 r_{bb'}C_{TC}} \tag{15.100}$$

Equation (15.100) shows that the power gain of a HBT is directly proportional to the cutoff frequency f_T, and varies inversely with the parasitic base resistance and collector–base junction capacitance. It is evident that high electron mobility in GaAs is essential for high-frequency performance of the HBT, because an increase in μ_n will lower the values of both τ_B and $r_{cc'}$, which in turn will increase f_T and hence the power gain G. A value of f_T equal to 75 GHz can be achieved for an AlGaAs/GaAs HBT with a 1.2 μm emitter width. In addition to high electron mobility, the lower doping density in the wide-bandgap AlGaAs emitter region and the higher doping density of the GaAs base region will result in a smaller emitter–base junction capacitance and a smaller base spreading resistance. These two factors are essential for high-speed and high-frequency operation of the HBT. An AlGaAs/GaAs HBT with a very short base width (≤ 0.1 μm) can have a current gain of several thousands or higher, provided that the electron lifetime τ_n in the base is of the order of a nanosecond.

The high electron mobility and high base doping density in the GaAs base region will have additional beneficial effects on the performance of an AlGaAs/GaAs HBT. For example, parasitics mechanisms such as emitter current crowding and base widening in the collector region can be greatly reduced with the HBT structure shown in Fig. 15.15a. The additional advantage is the increase of critical current density when the base widening becomes important. This critical current density is related to the electron mobility by

$$J_{bwc} = q\mu_n N_{dc} \frac{V_{CB}}{W_c} \tag{15.101}$$

where W_c is the width of the collector region. Another important consideration in the design of a HBT is the emitter current crowding effect, which becomes important when the emitter current density exceeds J_{ec}, where J_{ec} is given by

$$J_{ec} = \frac{8}{l^2} D_{pb} Q_b h_{FE} \tag{15.102}$$

This equation shows that a higher base doping density (i.e., a higher Q_b) will reduce the emitter current crowding in the HBT.

In many circuit applications in which a large load capacitance is required, bipolar junction transistors (BJTs) are preferred over field-effect transistors (FETs) because of their large current carrying capability, high transconductance, and excellent threshold voltage control. The main advantage for developing a HBT is to reduce the base resistance $r_{bb'}$, which severely limits the high-speed performance of a BJT in the bipolar digital and microwave circuits. For example, the maximum oscillation frequency f_{max} for a HBT is given by

$$f_{max} = \frac{1}{4\pi r_{bb'} C_{TC} \tau_{EC}} \tag{15.103}$$

which clearly shows that f_{max} is controlled by the base resistance $r_{bb'}$. We note that the collector junction capacitance C_{TC} can be reduced by reducing the collector junction area; τ_{EC} is the total emitter-to-collector delay time, which is given by

$$\tau_{EC} = \tau_E + \tau_B + \tau_{TC} + \tau_C \tag{15.104}$$

Since τ_E is inversely proportional to the emitter current density, a large emitter current will improve the frequency response. The base resistance $r_{bb'}$ can also have a profound effect on the noise and performance of the HBT. Values of $r_{bb'}$ can be reduced by increasing the base width W_B and base doping density N_B. Increasing base width W_B is not desirable since it will increase the base transit time τ_B, which in turn will reduce the base transport factor γ

and current gain β. It is well known that increasing N_{ab} in a BJT will increase the unwanted carrier injection from the base into the emitter, which in turn will reduce the emitter injection efficiency. In a n–p–n HBT, however, due to the presence of two different bandgap materials, the energy barriers for injection of electrons and holes are quite different. The barrier is larger for holes, and the injection efficiency is nearly independent of the dopant density in the base. Furthermore, the gain is only limited by the base transport factor. As a result, the base region of a HBT can be heavily doped without significantly affecting the current gain. In fact, the dopant densities in the emitter and collector regions can be adjusted to minimize the junction capacitance and series resistance of a HBT. This is a very attractive feature of the HBT.

The conduction band spike at the emitter–base junction shown in Fig. 15.15a is due to the abrupt transition at the AlGaAs/GaAs interface. This spike can be smoothed out by using compositional grading across the interface (i.e., by changing the aluminum mole fraction x gradually in the $Al_xGa_{1-x}As$ layer). The spike can be used for near-ballistic injection of electrons into the base region to reduce the base transit time τ_B (e.g., from 1 psec to 0.2 psec for a base width of 0.1 μm). Without ballistic injection, the best reported unit current gain cutoff frequency f_T $(=1/2\pi\tau_{EC})$ for a HBT is about 40 GHz with an emitter width of 1.6 μm. Therefore, with ballistic injection, further improvement in f_T is possible. To further reduce τ_{EC}, the device area and parasitics of HBTs will have to be reduced and the current level increased. An important factor for increasing the speed of a HBT is the reduction of base resistance $r_{bb'}$. Efforts to reduce the base resistance of a HBT have led to the development of the metal base hot electron transistors (HETs), which will be discussed in the next section.

Finally, additional advantages of HBTs over BJTs include: (1) the suppression of hole injection into the collector in saturating logic, (2) emitter–collector interchangeability leading to an improvement in VLSI circuit design, packing density, and interconnects, and (3) better control of the emitter–collector offset voltage. The collector of the HBT can also be a wide-gap AlGaAs material resulting in a double heterojunction transistor (DHBT). Besides III–V semiconductor HBTs, several new types of silicon HBTs using materials such as hydrogenated-amorphous silicon (a-Si:H and a-SiC:H) and hydrogenated microcrystalline silicon (μc-Si:H) as wide-bandgap emitters have been reported recently. Among these devices, the μc-Si:H n–p–n silicon HBT has the best overall performance characteristics. The device shows a much higher common emitter current gain than the conventional homojunction BJT. Recently, using molecular beam epitaxy (MBE) and chemical vapor deposition (CVD) techniques, a high-performance Si/Ge_xSi_{1-x} HBT with f_T exceeding 80 GHz has been demonstrated. This latest development has offered new promise for silicon-based HBTs to compete directly with III–V compound HBTs and HEMTs for high-speed and high-frequency circuit applications. Figure 15.19 presents a comparison of the unit current gain cutoff frequency f_T as a function of

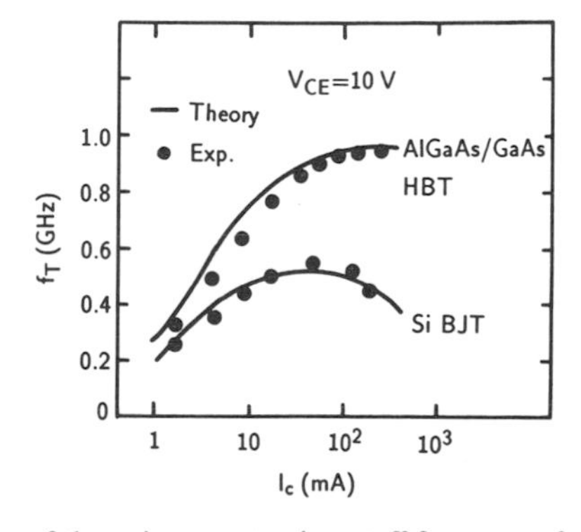

FIGURE 15.19. Comparison of the unit current gain cutoff frequency f_T versus collector current for an AlGaAs/GaAs HBT and a Si BJT with similar geometries. After Beilbe *et al.*,[8] by permission, © IEEE–1980.

collector current for an AlGaAs/GaAs HBT and a silicon BJT with similar geometries. The results clearly show that the former has a much higher f_T than the latter.

15.5. HOT ELECTRON TRANSISTORS

Hot electron transistors (HETs) are based on an old concept first proposed by C. A. Mead in 1960. The main objective for the HET is to reduce both the base resistance and transit time and to increase the current density of the BJTs for high-frequency performance. To achieve these goals various metal/insulator/semiconductor structures have been proposed, but due to difficulty in fabricating these structures success has so far been limited.

The capability of growing very thin epitaxial layers on semi-insulating substrates has been greatly enhanced by the availability of the MBE growth technique. For example, using GaAs substrate and lattice-matched large-bandgap $Al_xGa_{1-x}As$, it is possible to grow a high-resistivity undoped $Al_xGa_{1-x}As$ barrier layer on a GaAs base layer with barrier height adjusted by the Al mole fraction x, or using modulation doping to form a metal-like 2-DEG sheet charge on the GaAs base layer. As a result, several different HET structures have been reported recently. The main difference in these structures is the method by which the hot electrons (i.e., electrons with energies a few k_BT above the conduction band edge) are injected into the thin base region. The hot electron injection can be achieved by injection over the barrier or by tunneling. For base thickness smaller than the mean free path of hot electrons, the majority of the hot electrons injected into the thin base wll be collected by the collector where they will thermalize to the lattice temperature. The electrons lost in the base constitute the base current, which can usually be removed very rapidly from the base in a HET since the base resistance $r_{bb'}$ is very low, and the transit time of the majority electrons across the thin base is extremely small.

Figure 15.20a shows the cross-section view and conduction band diagram for an AlGaAs/GaAs tunneling HET in equilibrium, and Fig. 15.20b presents the conduction band diagram under bias conditions. Figure 15.21 shows the conduction band diagrams of a modified AlGaAs/GaAs HET with two-dimensional electron gas (2-DEG) charge sheet in the GaAs base formed by (a) modulation doping and (b) applied collector-base voltage. The device structures shown in Figs. 15.20 and 15.21 have a GaAs base layer with thickness 0.1 μm or less. In both cases, electrons in the base, being hot, travel at such a high velocity ($\sim 5 \times 10^7$ cm/sec) that the base transit time τ_B is negligible. One expects the speed of the HET to be limited by the emitter capacitance charging time through the emitter resistance. Its value depends on the current injection mechanism, and is larger for tunneling HETs. Although the base conductivity in both structures can be made very high by either heavy doping or using 2-DEG, it is usually difficult to make good ohmic contacts to the base. HETs are usually unsuitable for operation at room temperature due to high leakage current caused by the thin injection barrier with low barrier height (i.e., $\phi_B \sim 0.25$ eV for HETs versus 1.3 eV for HBTs). Thus, the predicted subpicosecond performance has yet to be realized in a practical HET device.

The tunneling HET structure shown in Fig. 15.20a consists of a n^+-GaAs emitter, an undoped thin (50 Å) $Al_xGa_{1-x}As$ barrier layer, a n^--GaAs base layer (1000 Å), an undoped thick (3000 Å) $Al_xGa_{1-x}As$ barrier layer, and a n^+-GaAs collector. As shown in this figure, the carrier injection from the emitter to the base relies on tunneling through the AlGaAs barrier layer, which occurs when the base is biased positively with respect to the emitter. If the collector barrier height is smaller than the energy of hot electrons, then the electrons are collected by the collector. The effective barrier width for electron tunneling can be varied by the applied bias voltage. It is seen that the emitter barrier must be thin enough to allow tunneling, and the collector barrier thick enough to minimize the leakage current. The common base current gain α can be equal to one if losses due to the spread of energy, scattering in the base, and reflection from the collector barrier are prevented. The first HET based on this

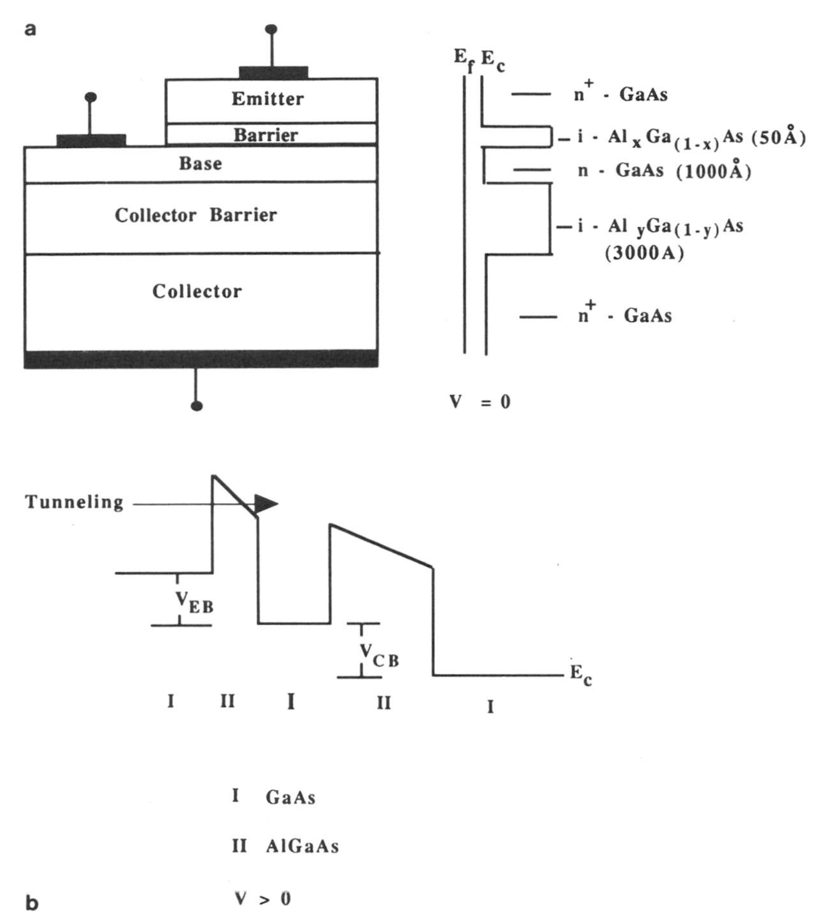

FIGURE 15.20. Conduction band diagrams of a tunneling hot-electron transistor (HET) (a) in equilibrium and (b) under bias conditions. After Capasso,[9] by permission, © IEEE–1988.

structure with a common emitter current gain of 1.3 at 40 K was demonstrated recently. The low current gain may bc attributed to significant loss of carriers in the base, which can be minimized by reducing the base width or collector barrier height. Reducing the base width will, however, increase the base resistance. A new approach to solve the problems associated with low current gain in such a tunneling HET has been reported recently. This involves the use of a resonant tunneling double-barrier quantum-well structure in such a HET. For example, a resonant tunneling HET can be obtained if the AlGaAs injection barrier shown in Fig. 15.20a is replaced by an AlGaAs (50 Å)/GaAs (56 Å)/AlGaAs (50 Å) double-barrier quantum-well structure between the emitter and the thin base of this HET.

The second HET structure shown in Fig. 15.21a has the potential to produce a very small base resistance. In this structure, the GaAs base is undoped or lightly doped; the AlGaAs barrier between the emitter and base is triangular in shape and doped with donor impurities in Fig. 15.21a and undoped in Fig. 15.21b. A 2-DEG charge sheet is formed in the base by modulation doping in Fig. 15.21a and by base collector voltage in Fig. 15.21b, as in a MODFET. Since the base region is undoped, the carrier mobility of the 2-DEG is very high (~8000 to 9000 $cm^2/V \cdot sec$ at 300 K), and is much higher at 77 K due to the reduction in ionized impurity scattering. With such a high electron mobility, the base will behave like a metal and the device will operate like a metal-base transistor [i.e., a permeable base transistor (PBT)].

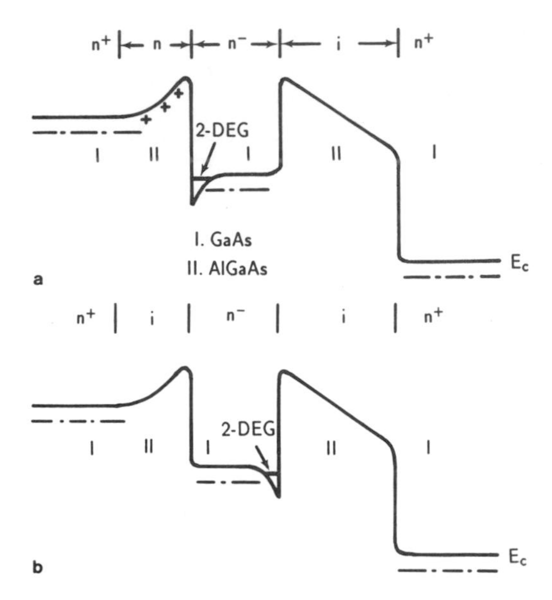

FIGURE 15.21. Conduction band diagrams of a modified HET with a 2-DEG in the base region formed by (a) modulation doping and (b) base collector voltage. The undoped GaAs base thickness is less than 0.1 μm.

However, since the base is undoped or lightly doped (to form 2-DEG), one expects the base resistance to be very high which makes ohmic contact to the base very difficult. To overcome this problem, the base region has to be doped heavily so that the base resistance can be sharply reduced.

Another type of HET, which has been reported recently for high-speed and high-frequency applications, is called the permeable base transistor (PBT). Figure 15.22 shows a cross-section view of a GaAs PBT. The PBT is basically a vertical MESFET similar to a vacuum triode. The emitter and collector regions of the device are separated by a parallel array of metal stripes, which are connected to an external terminal of the base. The voltage applied to this terminal controls the current flow from the collector to the emitter terminal. In a GaAs PBT, the metal stripes are imbedded in the GaAs by an epitaxial overgrowth process. The device structure consists of a n^+-GaAs substrate, a n^--GaAs emitter, a thin tungsten grating Schottky contact on GaAs which forms the base, and a n^--GaAs collector. The doping densities in the emitter and in the collector layers are adjusted so that the depletion region due to the tungsten–GaAs Schottky barrier extends across the openings in the grating. As an example, a tungsten grid consisting of a linewidth and spacing of 1600 Å in a 300-Å-thick layer of tungsten have been used in such a structure. The flow of electrons from the emitter to the collector is only through the tungsten grating and is controlled by varying the tungsten base potential.

The advantages of a PBT for high-frequency and high-speed operation can be explained in terms of its transconductance g_m, output resistance R_0, base resistance R_B, base-emitter capacitance C_{BE}, and base–collector junction capacitance C_{BC}, which are due to the capacitive

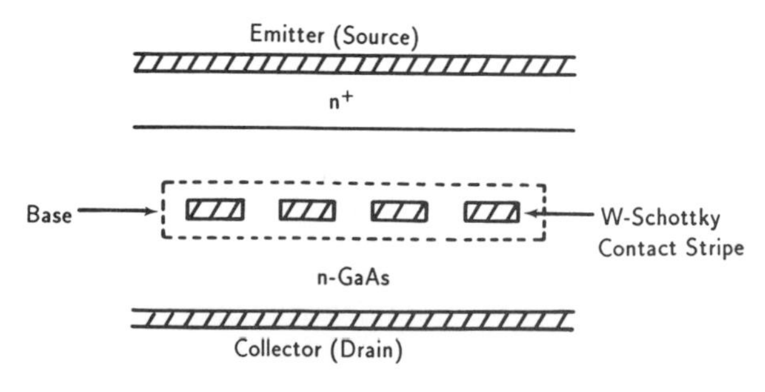

FIGURE 15.22. Cross-section view of a GaAs permeable base transistor (PBT). The grating tungsten/GaAs Schottky barrier contacts form the base region. After Bozler and Alley,[10,11] by permission.

coupling across the depletion width surrounding the Schottky barrier base electrode. The unity current gain cutoff frequency f_T of the PBT is given by

$$f_T = \frac{g_m}{2\pi(C_{BE} + C_{BC})} \tag{15.105}$$

The predicted value of f_T for a GaAs PBT is greater than 200 GHz. The maximum oscillation frequency f_{max} can be expressed by

$$f_{max} = \frac{f_T}{2\left(\frac{R_B + R_E}{R_0} + \frac{R_B g_m C_{BC}}{C_{BE} + C_{BC}}\right)^{1/2}} \tag{15.106}$$

Based on the above equation, a value of f_{max} near 1000 GHz and a power–delay product of less than 1 fJ are predicted. A value of f_{max} around 100 GHz and a gain of 16 dB at 18 GHz have already been reported in the literature for a GaAs PBT.[10] Although HETs show promise for high-speed and high-frequency performance, many obstacles remain to be solved before practical HETs can be built for high-frequency and high-speed applications.

15.6. RESONANT TUNNELING DEVICES

Resonant tunneling through a double-barrier quantum-well structure (e.g., AlGaAs/GaAs/AlGaAs) was first reported by Chang *et al.*[12] in 1974. Subsequently, a variety of two- and three-terminal resonant tunneling devices (RTDs) has been reported. In general, RTDs can be implemented with less devices per function, and hence they have potential for high-speed applications with reduced circuit complexity because the intrinsic speed of a tunneling device is much faster than devices operating on a drift or diffusion process. Since carrier transport in a FET or HBT is limited either by drift or by the diffusion process, devices such as RTDs operating on a tunneling process offer an attractive advantage for high-speed applications. An RTD operates on the principle of quantum mechanical tunneling through a multibarrier structure consisting of alternating layers of potential barriers (e.g., wide-bandgap AlGaAs) and quantum wells (e.g., GaAs). In a RTD, the maximum tunneling current occurs when the injected carriers have certain resonant energies which are determined by the Fermi energy in the doped cap region and the electron energy levels in the quantum wells. Energy band diagrams for a two-terminal double-barrier AlGaAs/GaAsAlGaAs RTD structure under different bias conditions are shown in Fig. 15.23a, b, and c, along with the resonance tunneling process. The current–voltage characteristic of a HET is shown in Fig. 15.23d. Electrons originated from the conduction band of the doped GaAs are on the left-hand side of the RTD, which is at the ground potential. These injected electrons tunnel through the AlGaAs barrier into the GaAs quantum well, and finally tunnel through the second barrier into the unoccupied states of the doped GaAs at the positive potential. Resonance occurs (i.e., the tunneling current reaches a maximum) when the energy of the electrons injected into the well becomes equal to the discrete quantum states in the well, as shown in Fig. 15.23b. The tunneling current decreases rapidly when the discrete energy level in the well drops below the conduction band edge of the left-hand-side GaAs layer due to an increase in applied bias voltage, as shown in Fig. 15.23c. This leads to a negative differential resistance (NDR) in the I–V characteristics, as shown in Fig. 15.23d. The solid circles shown in this figure correspond to different biases applied to the RTD. The NDR effect becomes more prominent at low temperatures, and hence it can be used for microwave generation and amplification. RTDs with a negative differential resistance and a peak-to-valley ratio exceeding 15 have been reported recently. Figure 15.24a, b, and c show the energy band diagrams of a three-terminal resonant tunneling

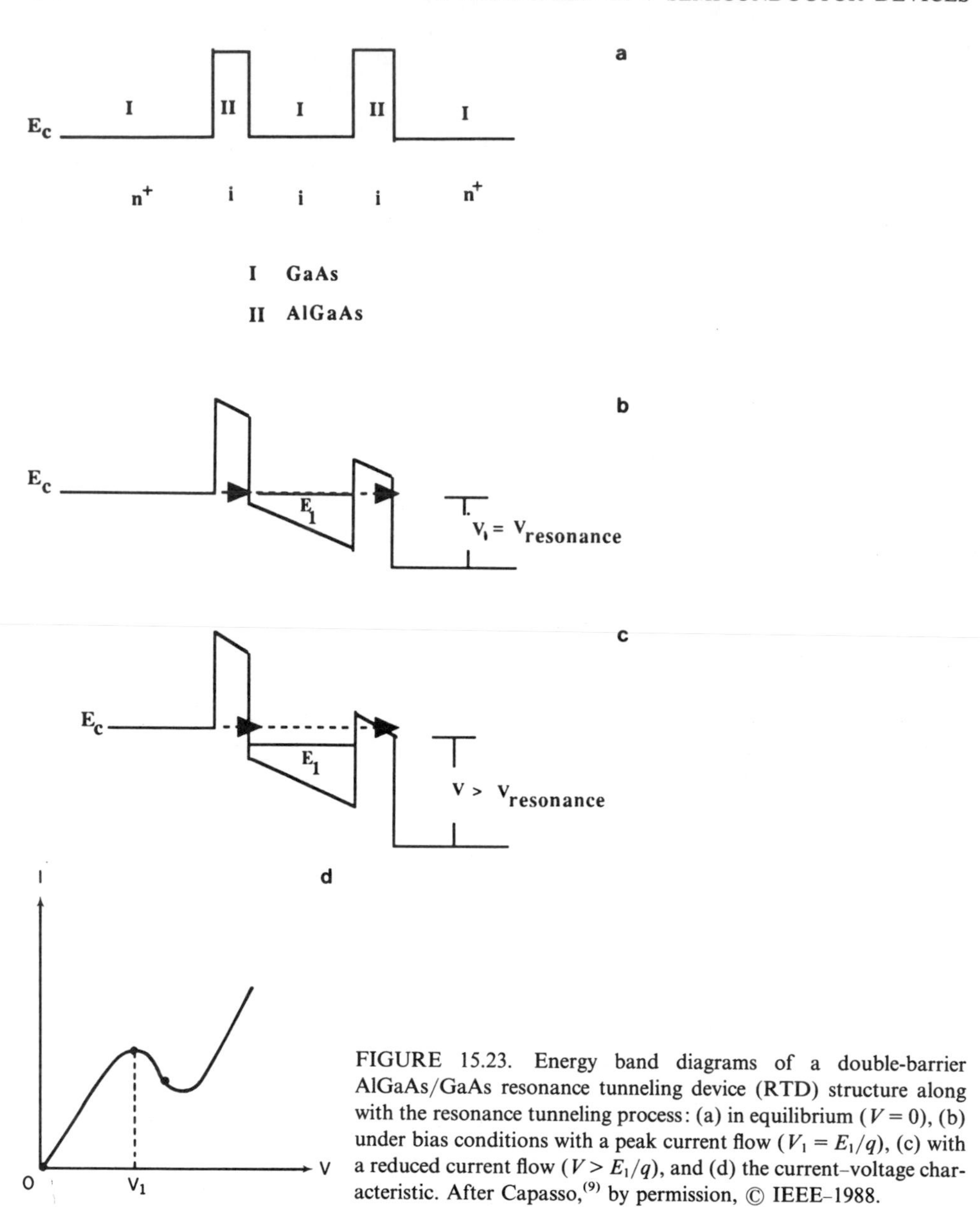

FIGURE 15.23. Energy band diagrams of a double-barrier AlGaAs/GaAs resonance tunneling device (RTD) structure along with the resonance tunneling process: (a) in equilibrium ($V = 0$), (b) under bias conditions with a peak current flow ($V_1 = E_1/q$), (c) with a reduced current flow ($V > E_1/q$), and (d) the current–voltage characteristic. After Capasso,[9] by permission, © IEEE–1988.

transistor (RTT) with an emitter tunneling injection barrier and double-barrier quantum-well base under different bias conditions. Resonant tunneling occurs when the applied bias voltage is equal to the energy of the ground state in the quantum-well base layer, as shown in Fig. 15.24b. The I–V characteristic curve similar to that of Fig. 15.23d is expected for the RTT. A room-temperature current gain of 7 (i.e., $\beta = \Delta I_C/\Delta I_B$) has been obtained for the AlGaAs/GaAs RTT shown in Fig. 15.24. Other types of resonant tunneling transistors, including a graded emitter RTT with electrons ballistically launched into the base, a RTT with a parabolic quantum-well in the base, and a RTT with a superlattice base, have also been reported with improved performance over the RTD shown in Fig. 15.24.

Resonant tunneling devices (RTDs) are capable of achieving an intrinsic speed as high as 10^{-1} fsec and an oscillation frequency of a few hundred GHz against a 100-GHz limit set

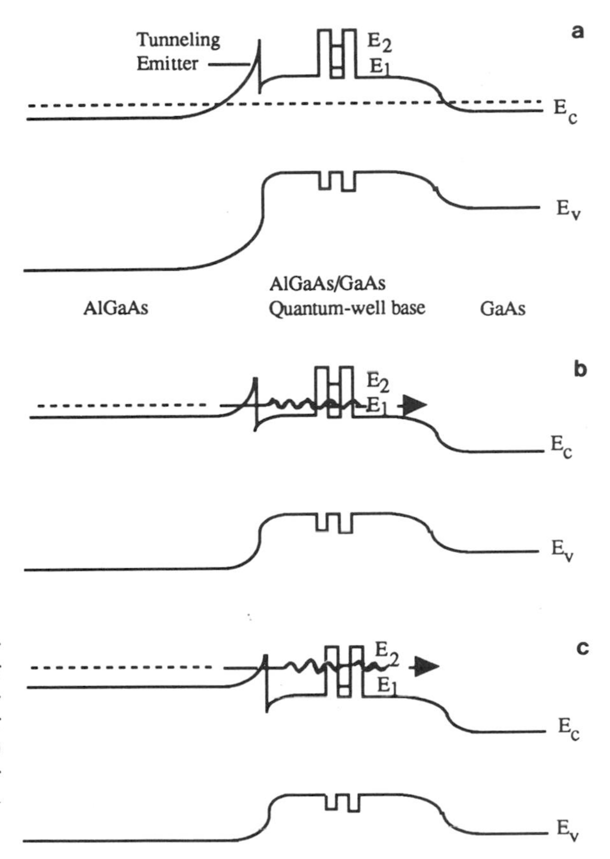

FIGURE 15.24. Energy band diagrams of a resonant tunneling transistor (RTT) with a tunneling emitter and double-barrier quantum-well base under different bias conditions: (a) $V = 0$, (b) $V = E_1/q$, and (c) $V = E_2/q$. After Capasso,[9] by permission, © IEEE-1988.

by a Gunn device, Impatt diode, and Esaki diode. Recently, an AlGaAs/GaAs two-terminal negative differential resistance (NDR) RTD was successfully used to generate oscillation frequency at 18 GHz with an output power of 5 μW at 200 K. Detecting and mixing studies at 2.5 THz demonstrated that the charge transport was faster than 0.1 psec. The RTD has also been used as an oscillator and in frequency multiplier circuits.

15.7. TRANSFERRED-ELECTRON DEVICES

The transferred-electron device (TED) has been extensively used as a local oscillator and power amplifier in the frequency range from 1 to 100 GHz. The TED, also known as the Gunn effect diode, was first discovered by J. B. Gunn in 1963.[13] Gunn found that coherent microwave output was generated when a dc electric field was applied across a short n-type GaAs or InP sample with a critical field strength of a few thousand volts per cm. The oscillation frequency is approximately equal to the reciprocal of the carrier transit time across the length of the sample. The mechanism responsible for the negative differential resistivity (or mobility) is due to a field-induced transfer of electrons from the low-energy, high-mobility conduction valley (i.e., the Γ-valley) to the higher-energy, lower-mobility satellite conduction valleys (L-conduction valleys), as first proposed by Ridley, Watkins, and Hilsum (RWH model).[14,15] Therefore, the transferred-electron effect has also been referred to as the Ridley–Watkins–Hilsum (RWH) effect. Several experiments performed on GaAs and $GaAs_{1-x}P_x$ samples revealed that the threshold field decreases with decreasing energy separation between the valley minima. The results provide convincing evidence that the transferred-electron effect is indeed responsible for the Gunn oscillation observed in GaAs and other III–V compound semiconductors.

To understand the physical mechanisms of the transferred-electron effect, which produces the negative differential resistance in a bulk semiconductor, let us consider the energy–momentum diagrams for the GaAs and InP crystals shown in Fig. 15.25. The band structures consists of a low-energy, high-mobility central conduction band valley located at the Γ-point and several satellite valleys of higher energy and lower mobility located at L-points along the [111] axes. The energy separation ($\Delta E_{\Gamma-L}$) between the upper satellite valleys (L-bands) and the lower conduction valley (Γ-band) is 0.31 eV for GaAs and 0.53 eV for InP. If the densities of electrons in the upper and lower valleys are designated by n_2 and n_1, respectively, and the total carrier density is given by $n = n_1 + n_2$, then the steady-state current density of the bulk semiconductor is given by

$$J = q(\mu_1 n_1 + \mu_2 n_2)\mathscr{E} = qnv_d \tag{15.107}$$

where μ_1 and μ_2 denote the electron mobilities in the lower and upper conduction valleys, respectively, and v_d is the average drift velocity defined by

$$v_d = \left(\frac{\mu_1 n_1 + \mu_2 n_2}{n_1 + n_2}\right)\mathscr{E} \approx \frac{\mu_1 \mathscr{E}}{1 + (n_2/n_1)} \tag{15.108}$$

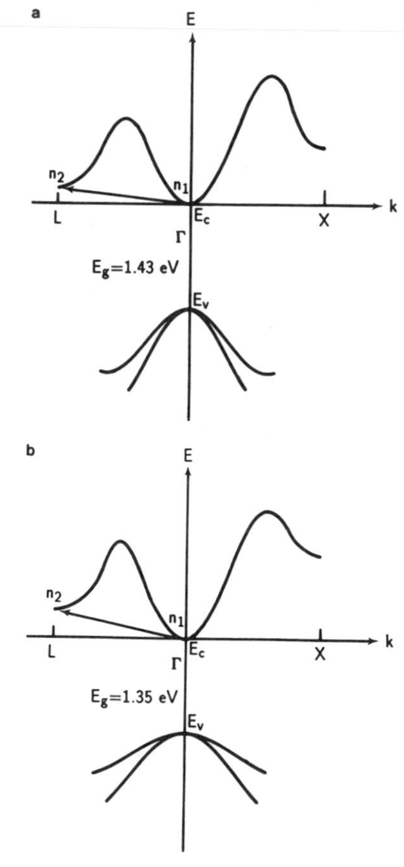

FIGURE 15.25. Energy band diagrams of (a) GaAs and (b) InP showing the transferred-electron (Gunn) effect from the central Γ-valley to the L-satellite valleys along {111} axes. The energy separation $\Delta_{\Gamma-L}$ is equal to 0.31 eV for GaAs and 0.53 eV for InP. The population in the Γ- and L-valleys are given by n_1 and n_2, respectively.

We have made use of the fact that $\mu_1 \gg \mu_2$. The population ratio n_2/n_1 between the upper and lower conduction valleys separated by an energy of ΔE_{21} is given by

$$\frac{n_2}{n_1} = R \exp(-\Delta E_{21}/k_B T_e) \tag{15.109}$$

where

$$R = \left(\frac{\nu_2}{\nu_1}\right)\left(\frac{m_2^*}{m_1^*}\right)^{3/2}$$

is the density-of-states ratio for the upper (L-band) and lower (Γ-band) conduction band valleys, ν_1 and ν_2 denote the number of lower and upper valleys, and m_1^* and m_2^* are the effective masses of electrons for the lower and upper valleys, respectively. For GaAs, $\nu_1 = 1$ and $\nu_2 = 4$, $m_1^* = 0.067m_0$ and $m_2^* = 0.55m_0$, and $R = 94$.

The concept of energy relaxation time allows the electron temperature to be expressed in the form

$$T_e = T + \frac{2q\mathscr{E} v_d \tau_e}{3k_B} \tag{15.110}$$

where τ_e is the energy relaxation time, which is of the order of 10^{-12} sec. We note that the electron temperature T_e given by Eq. (15.108) is larger than the lattice temperature T, since the kinetic energy of an electron is increased by the accelerated electric field. Now substituting v_d from Eq. (15.108) and n_2/n_1 from Eq. (15.109) into Eq. (15.110) yields

$$T_e = T + \left(\frac{2q\tau_e \mu_1}{3k_B}\right)\mathscr{E}^2\left[1 + R\exp\left(-\frac{\Delta E_{12}}{k_B T}\right)\right]^{-1} \tag{15.111}$$

which shows that the electron temperature will increase with the square of the electric field above the critical field. For a given temperature T, we can calculate T_e as a function of the electric field $\mathscr{E}$ using Eq. (15.111). The drift velocity versus electric field relation can also be derived from Eqs. (15.108) and (15.109), and the result is given by

$$v_d = \frac{\mu_1 \mathscr{E}}{1 + R\exp(-\Delta E_{12}/k_B T)} \tag{15.112}$$

Figure 15.26 shows drift velocity electric field curves calculated from Eq. (15.112) for GaAs at three different lattice temperatures; also shown is the population ratio n_2/n_1 versus

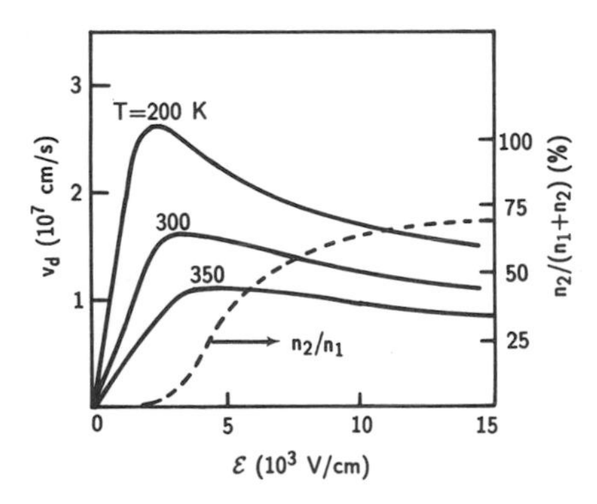

FIGURE 15.26. Drift velocity versus electric field curves for GaAs when $T = 200$, 300, and 350 K; also shown is the population ratio n_2/n_1 versus electric field at 300 K. After Sze,[3] p. 647, with permission by John Wiley & Sons Co.

electric field at 300 K. It is of interest that at $\mathscr{E} = 15$ kV/cm, approximately 70% of the total electron population is contributed by the upper-satellite valleys. As shown in Fig. 15.26, in the low-field regime, the velocity varies linearly with electric field (ohmic regime) and attains a peak value at critical field strength ($\mathscr{E}_c$). It then decreases with further increases in the electric field strength (i.e., the NDR regime).

It is seen from the simple model presented above that (1) there is a well-defined threshold field $\mathscr{E}_c$ for the onset of negative differential resistivity or mobility, (2) the threshold field increases with increasing lattice temperature (see Fig. 15.26), and (3) the NDR disappears if the lattice temperature is too high or the energy separation ΔE_{12} between the satellite and central valleys is too small. Therefore, in order to create a NDR effect via the electron transfer mechanism, the following conditions must be met: (1) the lattice temperature must be low enough such that, at thermal equilibrium, most of the electrons reside in the lower conduction valley (i.e., the Γ-band), or $k_B T < \Delta E_{12}$; (2) the electron mobility μ_1 is much larger than μ_2 (i.e., $m_1^* \ll m_2^*$), and the density of states for upper valleys is much higher than for the lower valleys; and (3) $\Delta E_{12} \ll E_g$ so that avalanche breakdown does not occur before electrons are transferred into the upper satellite valleys by the applied field. Among semiconductors satisfying these conditions, n-type GaAs and InP crystals are the most widely studied materials for the NDR devices. However, the transferred-electron effect has also been observed in many other compound semiconductors. Of particular interest are $Ga_xIn_{1-x}Sb$ ternary compounds, which have very low threshold fields and high electron velocities. For example, the critical field $\mathscr{E}_c$ for $Ga_{0.5}In_{0.5}Sb$ is only 600 V/cm and the peak velocity v_p is 2.5×10^7 cm/sec.

Room-temperature experimental results show that the critical electric field which defines the onset of NDR is approximately 3.2 kV/cm for GaAs and 10.5 kV/cm for InP. The peak velocities are about 2.2×10^7 and 2.5×10^7 cm/sec for high-purity GaAs and InP, respectively, while the maximum negative differential mobilities are equal to -2400 cm^2/V · sec for GaAs and -2000 cm^2/V · sec for InP.

Fabrication of TEDs requires extremely pure and uniform materials with very low defect densities. Early TEDs were fabricated from bulk GaAs and InP materials using alloyed ohmic contacts. Modern TEDs are usually fabricated on epitaxial films grown on n^+ substrates by the VPE, MOCVD, or MBE technique. Typical donor densities are in the 10^{14} to 10^{16} cm^{-3} range, and device lengths are in the range of a few microns to several hundred microns. Some high-power TEDs are made by selective metallization and mesa etching. To improve device performance, injection-limited cathode contacts have been used instead of the n^+ ohmic contacts. By using injection-limited contacts (e.g., Schottky barrier contact with low barrier height), the threshold field for the cathode current can be adjusted to a value approximately equal to the threshold field at the onset of NDR, resulting in uniform electric fields. For ohmic contacts, the accumulation or dipole layer grows some distance from the cathode due to finite heating of the lower-valley (Γ) electrons. This dead zone can be as large as 1 μm, which may limit the minimum device length and hence the maximum operating frequency. In an injection-limited contact, hot electrons are injected from the cathode and hence the dead zone is reduced. Since transit time effects can be minimized, the device can exhibit a frequency-independent negative conductance shunted by its parallel-plate capacitance. If an inductance and a sufficiently large conductance are connected to the device, it can be expected to oscillate in a uniform-field mode at the resonance frequency.

Figure 15.27a, b, and c show the cross section, dopant density profile, energy band diagram, and electric field distributions of three different cathode contacts of a Gunn device: (a) ohmic, (b) Schottky barrier, and (c) two-zone Schottky-barrier contacts. For the ohmic contact, there is always a low-field region near the cathode, and the field is nonuniform across the length of the device, as is clearly shown in Fig. 15.27a. The Schottky barrier contact shown in Fig. 15.27b consists of a low barrier-height (0.15 to 0.3 eV) contact, which is generally very difficult to make in GaAs. The device in this case can only be operated in a very narrow temperature range due to its exponential temperature-dependent injection current. The two-zone cathode contact shown in Fig. 15.27c consists of a high-field zone and a n^+ zone. In this

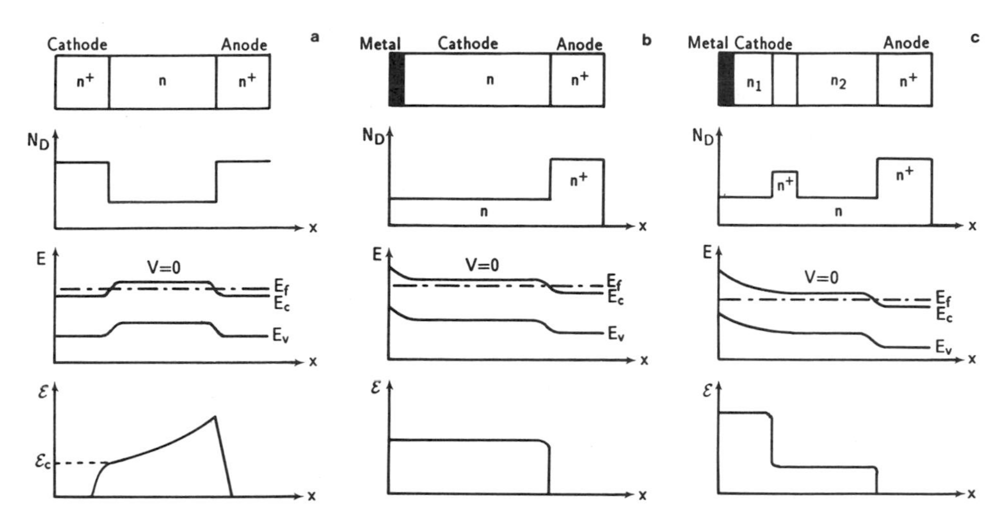

FIGURE 15.27. Doping profiles (N), energy band diagrams (E), and electric field distributions ($\mathscr{E}$) of a Gunn effect device for three different cathode contacts: (a) ohmic, (b) Schottky barrier (low barrier), and (c) two-zone Schottky-barrier contact. After Sze,[3] p. 669, with permission by John Wiley & Sons Co.

case, electrons are heated up in the high-field zone and subsequently injected into the active region with a uniform field. The structure in Fig. 15.27c has been successfully used over a wide range of temperatures. The maximum efficiency obtained from an InP TED with a two-zone cathode contact is 24%. However, a GaAs TED with an injection-limited cathode contact has yet to be realized because of the Fermi-level pinning effect.

PROBLEMS

15.1. Calculate the necessary thickness of an active n-channel layer with dopant density of 2×10^{17} cm^{-3} grown on a semi-insulating GaAs substrate for a GaAs MESFET with a threshold voltage of -2 V. Assume that the surface states pin the Fermi level at $\frac{2}{3}E_g$ below the conduction band edge at the metal–GaAs interface under the Schottky contact gate.

15.2. The lateral electric field in the channel of a GaAs MESFET under saturation conditions is usually high enough for electrons to drift with their scattering limited velocity v_s.
(a) Show that the drain saturation current in this case is given by

$$I_{Dsat} = \tfrac{1}{3}G_0 V_p = \tfrac{2}{3}ZqbN_D v_s$$

where b is the undepleted active layer thickness at $V_{GS} = 0$.
(b) Calculate b for $N_D = 1.5 \times 10^{17}$ and $I_{Dsat} = 250$ mA/mm gate width (Z), assuming $v_s = 1.1 \times 10^7$ cm/sec.

15.3. (a) Construct the energy band diagram of a n$^+$-$In_{0.51}Ga_{0.49}P$/p-GaAs heterojunction diode, assuming that the conduction band offset $\Delta E_c = 0.21$ eV and the valence band offset $\Delta E_v = 0.25$ eV.
(b) Plot the energy band diagram for a n$^+$-$Al_{0.3}Ga_{0.7}As$/p-GaAs/n-GaAs HBT with an abrupt emitter–base interface by including the effect of the energy band offsets in the diagram.

15.4. (a) Derive an expression for the threshold voltage V_T in terms of channel dopant density N_D, channel height a, and built-in potential V_{bi}, for a GaAs MESFET under pinch-off conditions.

(b) If the channel of a GaAs MESFET is uniformly doped to $2 \times 10^{17}\ \text{cm}^{-3}$, and the built-in potential of the Ti–Pt–Au Schottky barrier gate contact on n-GaAs is 0.8 V, find the channel thickness required to obtain a threshold voltage of -1 V.

15.5. Using a three-piecewise linear relation between drift velocity and electric field, derive an expression for the current–voltage characteristics of a MODFET device [see the paper by K. Lee *et al.*, *IEEE Trans. Electron Devices* **ED-30** (3), 207–212 (1983)].

15.6. For a short-channel FET (i.e., gate length $L \leq 1\ \mu\text{m}$), a semiempirical formula may be used to define the effective saturation velocity v_s in the channel, which is given by

$$v_s = 59L^{-0.56}\ \text{m/sec}$$

where L is in meters.
(a) Calculate v_s for $L = 0.25$, 0.50, and 1.0 μm.
(b) Calculate the unit current gain cutoff frequency f_T $(= v_s/2\pi L)$ for the gate lengths and saturation velocities given in (a).

15.7. (a) Show that the transconductance g_m of a uniformly doped n-channel MESFET is given by

$$g_m = \frac{\varepsilon_0 \varepsilon_s v_{sat} Z}{W_d}$$

(b) Show that the expression for g_m for a nonuniform doping profile given by $N_D(y) = Ky$ for $y < W_d$ is the same as for the uniform doping profile given by (a).
(c) Derive an expression for the threshold voltage V_T.
(d) Calculate g_m from (a), if $\varepsilon_0 \varepsilon_s = 1 \times 10^{-12}$ F/cm, $v_{sat} = 1.2 \times 10^7$ cm/sec, $Z = 50\ \mu$m, and $W_d = 0.1\ \mu$m.

15.8. (a) Draw a small-signal equivalent circuit diagram for a MESFET showing both the extrinsic and intrinsic circuit elements.
(b) Show that f_T (cutoff frequency for unit current gain with output of FET shorted) of a MESFET can be expressed by

$$f_T = \frac{v_{sat}}{2\pi L}$$

where L is the gate length and v_{sat} is the saturation velocity of electrons.
(c) Calculate f_T for $L = 1.0$, 0.5, 0.1 μm and $v_{sat} = 1.2 \times 10^7$ cm/sec for a GaAs MESFET.
(d) Derive an expression for the extrinsic transconductance g_{em} and drain conductance g_{ds} of a MESFET showing the effect of the source resistance R_s.

15.9. (a) Discuss the second-order effects on the performance of a MESFET (e.g., backgating, drain current lag, temperature, subthreshold current, etc.).
(b) Draw a typical self-aligned process sequence of the GaAs MESFET fabrication steps.
(c) Explain why p-channel MESFETs cannot be achieved in GaAs.

15.10. (a) Construct the energy band diagram for a depletion-mode (D-) AlGaAs/GaAs modulation doped FET (HEMT) including the spacer layer, and explain how the two-dimensional electron gas (2-DEG) is formed in the undoped GaAs layer.
(b) Explain how the high electron mobility is achieved in the 2-DEG GaAs layer.
(c) Explain why the transconductance g_m and cutoff frequency f_T for a MODFET can be higher than those of a MESFET.
(d) Note that both depletion- (D-) and enhancement- (E-) mode GaAs HEMTs can be made, but D-HEMTs are more common. Plot the energy band diagram for an AlGaAs/GaAs E-HEMT when $V_{GS} = 0$ and $V_{GS} > V_T$.

15.11. Consider a GaAs MESFET with device parameters given by: channel length $L = 3\ \mu$m, channel height $a = 1\ \mu$m, $N_D = 2.6 \times 10^{15}\ \text{cm}^{-3}$, and $V_{GS} = -1$ V and $V_{DS} = 3$ V are applied to the device.

(a) Draw the cross-section view of this MESFET showing the depletion region, the channel, and the Gunn domain region.
(b) Plot (i) electric field versus x, (ii) drift velocity versus x, and (iii) space charge versus x in the channel.

REFERENCES

1. C. A. Liechti, *IEEE Trans. Microwave Theory Tech.* **MTT-24**, 286 (1976).
2. R. Pucel, H. Haus, and H. Statz, *Advances in Electronics and Electron Physics*, Vol. 38, p. 195, Academic Press, New York (1975).
3. S. M. Sze, *Physics of Semiconductor Devices*, 2nd ed., Wiley, New York (1981).
4. D. Delagebeaudeuf and N. T. Ling, "Metal-n-AlGaAs/GaAs Two-Dimensional Electron Gas FET," *IEEE Trans. Electron Devices* **ED-29**, 955 (1982).
5. K. Lee, M. S. Shur, T. J. Drummond, and H. Morkoc, "Current–Voltage and Capacitance–Voltage Characteristics of Modulation-doped Field Effect Transistors," *IEEE Trans. Electron Devices* **ED-30**, 207 (1983).
6. S. M. Sze, High Speed Devices, Wiley, New York (1991).
7. P. M. Asbeck, *IEDM Short Course: Heterostructure Transistors*, Institute of Electrical and Electronics Engineers, New York (1988).
8. J. P. Beilbe, A. Marty, P. H. Hicp, and G. E. Rey, "Design and Fabrication of High-speed GaAlAs/GaAs Heterojunction Transistors," *IEEE Trans. Electron Devices* **ED-27**, 1160 (1980).
9. F. Capasso, in: *High-Speed Electronics* (B. Kallback and H. Beneking, eds.), Vol. 22, pp. 50–61, Springer-Verlag, Berlin (1986).
10. C. O. Bozler and G. D. Alley, "Fabrication and Numerical Simulation of the Permeable Base Transistor," *IEEE Trans. Electron Devices* **ED-27**, 1128 (1980).
11. C. O. Bozler and G. D. Alley, "The Permeable Base Tansistor and Its Application to Logic Circuits," *Proc. IEEE* **70**, 46 (1982).
12. L. L. Chang, L. Esaki, and R. Tsu, *J. Appl. Phys.* **24**, 593 (1974).
13. J. B. Gunn, "Microwave Oscillations of Current in III–V Semiconductors," *Solid State Commun.* **1**, 88 (1963).
14. C. Hilsum, "Transferred Electron Amplifiers and Oscillators," *Proc. IRE* **50**, 185 (1962).
15. B. K. Ridley and T. B. Watkins, "The Possibility of Negative Resistance," *Proc. Phys. Soc.* **78**, 291 (1961).

BIBLIOGRAPHY

J. S. Blackmore, "Electron and Hole Traps in GaAs," *J. Appl. Phys.* **53**, R123 (1982).

I. B. Bott and W. Fawcett, "The Gunn Effect in GaAs," in *Advances in Microwaves* (L. Yung, ed.), Vol. 3, pp. 223–300, Academic Press, New York (1968).

F. Capasso, in: *Semiconductors and Semimetals* (R. K. Willardson and A. C. Beer, eds.) Vol. 22, Part D, p. 2, Academic Press, New York (1985).

F. Capasso, J. Allam, A. Y. Cho, K. Mohammed, R. J. Malik, A. L. Hutchinson, and D. Sivco, *Appl. Phys. Lett.* **48**, 1294 (1986).

T. H. Chen and M. S. Shur, "Capacitance Model of GaAs MESFETs," *IEEE Trans. Electron Devices* **ED-32**, 883 (1985).

H. F. Cooke, *Proc. IEEE* **59**, 1163 (1971).

R. Dingle, H. L. Stormer, A. C. Gossard, and W. Wiegmann, *Appl. Phys. Lett.* **37**, 805 (1978).

K. Drangeid, R. Sommerhalder, and W. Walter, *Electron Lett.* **6**, 228 (1970).

T. J. Drummond, H. Morkoc, K. Lee, and M. S. Shur, "Model for Modulation Doped Field Effect Transistor," *IEEE Electron Device Lett.* **EDL-3**, 338 (1981).

T. J. Drummond, W. Kopp, M. Keever, H. Morkoc, and A. Y. Cho, "Electron Mobility in Single and Multiple Period Modulation-Doped AlGaAs/GaAs Heterostructures," *J. Appl. Phys.* **23**, 230 (1984).

W. P. Dumke, J. M. Woodall, and V. L. Rideout, "GaAs–GaAlAs Heterojunction Transistor for High Frequency Operation," *Solid-State Electron.* **15**, 1339 (1972).

L. F. Eastman, "Very High Electron Velocity in Short GaAs Structures," in: *Advances in Solid State Physics* 12 (J. Treush, ed.), p. 173, Vieweg, Braunschweig (1982).
A. A. Grinberg and M. S. Shur, "Density of Two-dimensional Electron Gas in Modulation-doped Structure with Graded Interface," *Appl. Phys. Lett.* **45**, 573 (1984).
M. Heiblum, *Solid-State Electron.* **24**, 343 (1961).
M. Hirano, K. Oe, and F. Yanagawa, "High-Transconductance p-Channel Modulation-Doped AlGaAs/GaAs Heterostructure FETs," *IEEE Trans. Electron Devices* **ED-33**, 620 (1986).
M. A. Hollis, S. C. Palmateer, L. F. Eastman, N. V. Dandekar, and P. M. Smith, *IEEE Electron Device Lett.* **EDL-4**, 440 (1983).
H. Kroemer, "Theory of Wide-gap Emitter Transistors," *Proc. IRE* **45**, 1535 (1957).
H. Kroemer, "Theory of the Gunn Effect," *Proc. IEEE* **52**, 1736 (1964).
H. Kroemer, "The Gunn Effect Under Imperfect Cathode Boundary Conditions," *IEEE Trans. Electron Devices* **ED-15**, 819 (1968).
H. Kroemer, "Heterostructure Bipolar Transistors and Integrated Circuits," *Proc. IEEE* **70**, 13 (1982).
K. Lehovec and R. Zuleeg, "Voltage–Current Characteristics of GaAs-JFETs in the Hot Electron Range," *Solid-State Electron.* **13**, 1415 (1970).
S. Luryi, *Appl. Phys. Lett.* **47**, 490 (1985).
C. A. Mead, *Proc. IRE* **48**, 359 (1980).
A. G. Milnes and D. L. Feucht, *Heterojunctions and Metal–Semiconductor Junctions*, Academic Press, New York (1972).
T. Mimura, S. Hiyamizu, T. Fijii, and K. Nanbu, "A New Field Effect Transistor with Selectively Doped GaAs/n-AlGaAs Heterojunctions," *Jpn. J. Appl. Phys.* **19**, L225 (1980).
H. Morkoc, J. Chen, U. K. Reddy, T. Henderson, P. D. Coleman, and S. Luryi, *Appl. Phys. Lett.* **42**, 70 (1986).
K. Park and K. D. Kwack, "A Model for the Current–Voltage Characteristics of MODFETs," *IEEE Trans. Electron Devices* **ED-33**, 673 (1986).
B. K. Ridley, "Specific Negative Resistance in Solids," *Proc. Phys. Soc.* **82**, 954 (1963).
L. P. Sadwick and K. L. Wang, "A Treatise on the Capacitance–Voltage Relation of High Electron Mobility Transistors," *IEEE Trans. Electron Devices* **ED-33**, 651 (1986).
E. F. Schubert and A. Fischer, "The Delta-Doped Field-Effect Tansistor (δFET)," *IEEE Trans. Electron Devices* **ED-33**, 625 (1986).
B. L. Sharma and R. K. Purohit, "Semiconductor Heterojunctions, Pergamon, London (1974).
W. Shockley, *Bell Syst. Tech. J.* **30**, 990 (1951).
W. Shockley, "A Unipolar Field-effect Transistor," *Proc. IRE* **40**, 1365 (1952).
M. S. Shur, "Analytical Model of GaAs MESFETs," *IEEE Trans. Electron Devices* **ED-25**, 612 (1978).
M. S. Shur, "Analytical Models of GaAs MESFETs," *IEEE Trans. Electron Devices* **ED-32**, 18 (1985).
M. S. Shur, *GaAs Devices and Circuits*, Plenum Press, New York (1987).
M. S. Shur and L. F. Eastman, "Current–Voltage Characteristics, Small-Signal Parameters and Switching Times of GaAs FETs," *IEEE Trans. Electron Devices* **ED-25**, 606 (1978).
M. S. Shur and L. F. Eastman, "A Near Ballistic Electron Transport in GaAs Devices at 77 K," *Solid-State Electron.* **24**, 11 (1981).
T. C. Sollner, W. D. Goodhue, P. E. Tannenwald, C. D. Parker, and D. D. Peck, *Appl. Phys. Lett.* **43**, 588 (1983).
P. M. Solomon and H. Morkoc, *IEEE Trans. Electron Devices* **ED-31**, 1015 (1984).
S. M. Sze, *Physics of Semiconductor Devices*, 2nd ed., Wiley, New York (1981).
G. W. Taylor, H. M. Darley, R. C. Frye, and P. K. Chatterjee, "A Device Model for an Ion Implanted MESFET," *IEEE Trans. Electron Devices* **ED-26**, 172 (1979).
R. Tsu and L. Esaki, *Appl. Phys. Lett.* **22**, 562 (1973).
H. Unlu and A. Nussbaum, "Band Discontinuities at Heterojunction Device Design Parameters," *IEEE Trans. Electron Devices* **ED-33**, 616 (1986).
T. Wada and S. Frey, "Physical Basis of Short-Channel MESFET Operator," *IEEE Trans. Electron Devices* **ED-26**, 476 (1979).
G. W. Wang and W. H. Ku, "An Analytical and Computer-aided Model of the AlGaAs/GaAs High Electron Mobility Transistor," *IEEE Trans. Electron Devices* **ED-33**, 657 (1986).

Index